PRENTICE HALL

WORLD GEOGRAPHY

PRENTICE HALL

WORLD GEOGRAPHY

THOMAS J. BAERWALD

CELESTE FRASER

PRENTICE HALL

Needham, Massachusetts   Englewood Cliffs, New Jersey

Authors

Thomas J. Baerwald received a B.A. degree in geography and history from Valparaiso University in Indiana and earned his M.A. and Ph.D. degrees in geography at the University of Minnesota. He has served on the boards of the Association of American Geographers and the National Council for Geographic Education and has lectured at many universities across the country. Currently Dr. Baerwald is Program Director for Geography and Regional Science at the National Science Foundation in Washington, D.C.

Celeste Fraser received her B.A. and M.A. degrees at the University of Colorado. She serves on the board of the National Council for Geographic Education and is a member of the National Council for Social Studies and the Illinois Geographic Society. Ms. Fraser has taught at the middle and high school levels and has developed a variety of materials for geography education.

Senior Program Consultant

Muncel Chang is a member of the Executive Board of the National Council for Geographic Education. He is a founding member of the California Geographic Alliance and a Geographic Education Teacher Consultant for the National Geographic Society. Formerly a high school social studies teacher, Mr. Chang is an occasional lecturer in the Department of Geography at California State University, Chico.

Area Specialists

Mexico and Other Countries of South America
Dr. Louis B. Casagrande
Senior Vice President
Science Museum of Minnesota
St. Paul, Minnesota

Central America and the Caribbean
Dr. Sam Sheldon
Associate Professor of Geography
Department of Sociology and Social Sciences
St. Bonaventure University
St. Bonaventure, New York

The Middle East and North Africa
John Voll
Professor of History
University of New Hampshire
Durham, New Hampshire

South Asia
Leonard A. Gordon
Professor of History
Brooklyn College
City University of New York
New York, New York

Teacher Reviewers

Dr. Marty Bock
MacArthur High School
San Antonio, Texas

Penny Todd Claudis
Curriculum Supervisor
Caddo Public Schools
Shreveport, Louisiana

Bud Dorholt
South High School
Sheboygan, Wisconsin

Vernon J. Hixon
East Junior High School
Boise, Idaho

Beth Kirk
Florida Geographic Alliance and
McMillan Middle School
Miami, Florida

Credits

Editorial: Mary Ann Gundersen, David Lippman, Nancy Rogier
Design: Betty Fiora, Sue Gerould/Perspectives, L. Christopher Valente
Cover Design: Martucci Studio and L. Christopher Valente
Map Research: Elizabeth Hovinen
Cartography: Richard and John Sanderson
Photo Research: Susan Van Etten
Production: Virginia Shine, Pauline Wright
Marketing: Jeff Ikler, Anne Riccio
Pre-press Production: Martha E. Ballentine
Manufacturing: Bill Wood

PRENTICE HALL
A Division of Simon & Schuster
Englewood Cliffs, New Jersey 07632

ISBN 0-13-969254-1

Printed in the United States of America

4   5   6   7   8   99   98   97   96   95   94   93

TABLE OF CONTENTS

UNIT 1

Physical and Human Geography   xx

Latin America

162

Western Europe 252

UNIT
9

South Asia   572

xiii

Maps

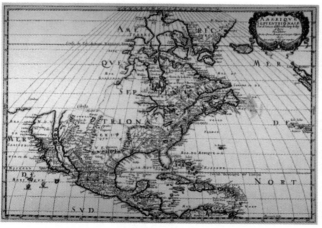

Charts, Graphs, and Diagrams

Daily Life

Skills Check

Map and Globe

Critical Thinking

Social Studies

Reading and Writing

A Geographic View of History

Case Study on Current Issues

Place Location: Where on Earth?

Making Connections: Where Regions Meet

UNIT
1
Physical and Human Geography

CHAPTERS

Vast, parched deserts . . . endless blue oceans . . . lofty, snow–capped mountains . . . golden prairies dotted with farms . . . noisy cities with glaring neon signs. These are the varied landscapes of our world. It is our world to discover.

Geography is a journey of discovery that tantalizes the senses and expands the mind. Our world is full of a wonderful variety of sights, sounds, and textures. As you read this unit, you will learn how geography helps you discover our world.

River rafting,
British Colombia ▶

◀ The Canadian Rockies, Alberta

Geologist in Antarctica ▶

1

1

The Study of Geography

Chapter Preview

Both sections in this chapter provide an introduction to the study of geography.

Sections	Did You Know?
1 THE FIVE THEMES OF GEOGRAPHY	Columbus based his 1492 voyage on the geographer Ptolemy's inaccurate estimates of the distance between Spain and China.
2 THE GEOGRAPHER'S CRAFT	The earliest known map was found on a clay tablet from Babylonia in Southwest Asia. It dates from around 2300 B.C.

A hang glider's view of the earth

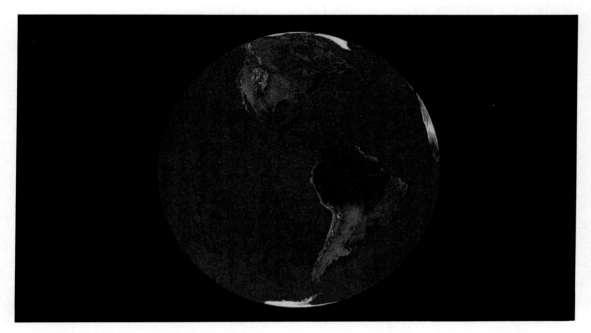

An Astronaut's View of the Earth

The hang glider on the facing page is able to see the earth from a different perspective than he can from the ground. Like hang gliding, geography lets you view the earth and its people from many different perspectives.

1 The Five Themes of Geography

Section Preview

Key Ideas

- Each place on earth has a unique combination of physical and cultural characteristics.
- The relationship of people to their environment may produce both good and bad consequences.
- People depend on the movement of goods, services, and ideas.
- A region is a group of places sharing similar characteristics.

Key Terms

geography, absolute location, Equator, hemisphere, latitude, longitude, Prime Meridian, relative location, culture

For centuries before travel was common, Europeans knew that other peoples and civilizations occupied parts of the world, but information about faraway places was hard to get or was nonexistent. Very few people were willing to risk the hardships and dangers of travel. Even close to home, travelers in the European countryside might be attacked by bandits or wolves. Who knew what greater dangers awaited travelers farther from home? The seas held even worse terrors. When mapmakers reached the limits of their knowledge, they wrote the simple and alarming warning: "There Be Dragons Here."

What information there was about other places, though appealing, was considered unreliable. In the 1200s, for example, an Italian merchant named Marco Polo braved the long and dangerous overland route to Asia, bringing back fantastic stories of a great civilization in China. His book about his travels quickly became a best seller in Europe. But even as they eagerly read the book, most Europeans were convinced that it was fiction.

The Study of Geography

Today we know much, much more about the world's wonders than was known in the 1200s. We have all this information because people are, and always have been, curious creatures. Being curious means asking questions —about the world and how it works. In this sense people have always been geographers.

What is geography? The word *geography* comes from a Greek word that means writing about, or describing, the earth. The study of geography begins with the understanding that each place on earth is different in significant ways and that the people in each place are unique. Geography is the study of the earth's surface and the processes that shape it, the connections between places, and the complex relationships between people and their environments.

Is the earth as it approaches the twenty-first century bigger than it was almost eight hundred years ago because we know that so much more of it exists? Or is it smaller today than it was because it is no longer so frighteningly huge and mysterious? As our perspective changes, so does the scope of geography. Unlike the days when unknown territory was described as lands "where dragons be," today there are few parts of the earth that have not been explored. We can see the whole planet in satellite images sent from space, looking like a small and fragile crystal ball.

In a sense, the world *is* a fragile place. The challenges of growing populations and shrinking resources, global warming, and nuclear waste all stand to threaten our fragile environment—for people and places all over the globe. Even events that are not so critical—the razing of forestland to build a state highway, a town's new incinerator, the decision of one person to recycle paper—have far-reaching effects. One of the most important lessons that the study of geography can give us is the understanding that we are each a piece of a greater whole and that each place on earth is connected to other places.

Geography's Five Themes

Each generation of geographers and students of geography continue to search for new explanations for why the world is so amazingly diverse. In studying places on earth, geographers try to answer five important questions:

- Where is this place?
- What is it like?
- How have people changed it by their interaction with the environment?
- How has the place been affected by the movement of people, goods, and ideas?
- How is this place similar to and different from other places?

Each of these questions is related to one of five themes that form the core of all geographic inquiry into the world and its people. The five geographic themes are location, place, human–environment interaction, movement, and regions. Studying these themes takes geographers into the study of nearly everything on earth. For as one geographer wrote:

> There is no problem in this world . . . that is exclusively geographical, . . . but there are few problems that . . . are not in some way geographical.

Location

The first step in finding out about a place involves finding out where on earth it is located. A place's location can be described in two ways.

Absolute Location The first way of answering the question "Where?" is by finding the absolute location of a place. Every place on earth has an **absolute location**—its position on the globe. Geographers have developed several ways of describing a place's absolute location. One way is by drawing a grid of imaginary lines around the earth. The **Equator** is one such line that circles the globe at its widest point, halfway between the North and South Poles. The Equator divides the earth into two halves, or **hemispheres**. The Northern Hemisphere includes all of the land and water between the Equator and the North Pole. The Southern Hemisphere includes all of the land and water between the Equator and the South Pole.

Latitude Other imaginary lines run around the earth parallel to the Equator. These lines are called lines of **latitude,** or parallels. Parallels measure the distance north and south of the Equator in degrees. The Equator, for example, is designated 0°, and the Poles are at 90° north (N) and south (S).

Two lines of latitude are particularly important to geographers—the Tropic of Cancer, located at 23.5°N, and the Tropic of Capricorn, at 23.5°S. Because the earth is tilted 23.5 degrees as it revolves around the sun, these two lines of latitude mark the boundaries of the places on earth that receive the greatest heat energy from the sun. Look at the diagram on page 6 to locate the Equator and the tropics.

Longitude The second set of imaginary lines are lines of **longitude**, or meridians. These lines run north and south from one Pole to the other. The beginning line, the **Prime Meridian**, is the line of longitude that runs through the Royal Observatory in Greenwich, England. It is designated 0° longitude. To the east and west of the Prime Meridian are other lines of longitude.

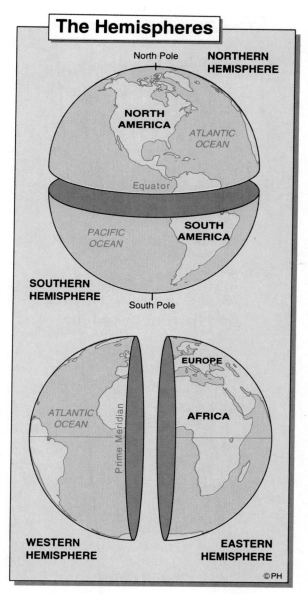

The Hemispheres

Diagram Study

The Equator and the Prime Meridian each divide the earth into halves. In which of the hemispheres is North America located?

Unlike lines of latitude, meridians are not parallel to one another, but they are regularly spaced. As you can see in the diagram on page 6, meridians join at the Poles and are farthest apart at the Equator. All lines of longitude are

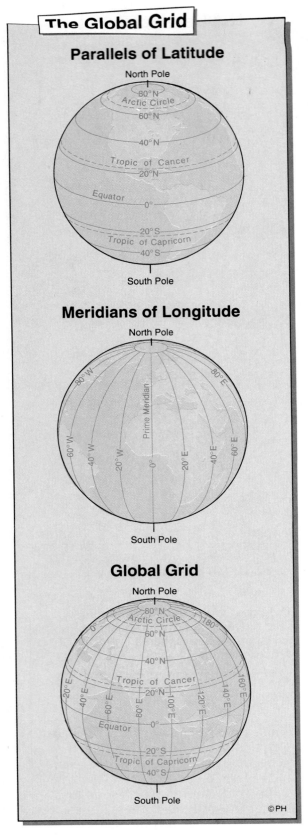

The Global Grid

Parallels of Latitude

North Pole
- 80°N
- Arctic Circle
- 60°N
- 40°N
- Tropic of Cancer
- 20°N
- Equator — 0°
- 20°S
- Tropic of Capricorn
- 40°S

South Pole

Meridians of Longitude

North Pole

80°W, 60°W, 40°W, 20°W, Prime Meridian, 0°, 20°E, 40°E, 60°E, 80°E

South Pole

Global Grid

North Pole
- 80°N
- Arctic Circle — 180
- 60°N
- 40°N
- Tropic of Cancer
- 20°N
- 0°, 20°E, 40°E, 60°E, 80°E, 100°E, 120°E, 140°E, 160°E
- Equator — 0°
- 20°S
- Tropic of Capricorn
- 40°S

South Pole

©PH

numbered in degrees from 0 to 180, either east (E) or west (W) of the Prime Meridian. Look at the map showing the grid formed by latitude and longitude lines. By use of this grid, each place on earth can be located according to its coordinates of latitude and longitude. Paris, France, for example, is located at 48°N latitude and 2°E longitude. Chicago, Illinois, is located at 42°N and 88°W.

Relative Location The second way of answering the question "Where is this place?" is to tell where the place is located in relation to another place on earth—to describe its relative location. Chicago, Illinois, for example, might be said to be located on the southwest bank of Lake Michigan, south of Milwaukee, Wisconsin. Another way to describe Chicago's relative location would be to say it is in the northeast corner of Illinois, northwest of Gary, Indiana. Of course, as the world and its technology change, so do the relative locations of places. Another way to describe Chicago's relative location today would be to say that it is three hours from Houston, Texas, by air.

Place

Geographers know that every place on earth has features that distinguish it from every other place. They usually describe places either by their physical or their cultural characteristics.

Physical Characteristics Physical geographers study the natural features of the earth, such as landforms, bodies

---

Diagram Study

This diagram illustrates the imaginary lines of latitude and longitude. At what coordinates do the Prime Meridian and the Tropic of Cancer intersect?

of water, wind, and temperature. The study of the earth's natural features takes physical geographers into a vast number of different topics. Hurricanes, earthquakes, volcanoes, plant and animal life, deserts, and oceans are only a few of the areas explored by physical geographers.

Cultural Characteristics The second way that geographers describe a place is by its cultural characteristics. Culture includes every part of the way of life of people, including their government, language, religions, customs, and beliefs. Geographers study all aspects of human activity, including such things as the growth of cities and populations, farming methods, architecture, and political and religious beliefs.

The human characteristics of a place are those that make it unique because of the people who live there. Chicago is unique because of its economic role as an industrial city close to the nation's agricultural heartland.

In the same way, every city or town in the world has its own human "flavor." Travelers returning home do not tell friends the latitude and longitude of their vacation spot. They talk about the interesting people, unusual customs, tempting dishes, snow-capped mountains, or sparkling beaches that made their time away special. This is what geographers mean when they talk about *place.*

Human-Environment Interaction

The third geographical theme can best be described as a series of questions: "How do people use their environment?" "How have they changed it?" "Why have they made these changes?" and "What are the consequences of these changes?"

Human beings have the ability to make enormous changes in their environment. Sometimes the changes they

Place: Chicago, Illinois
Describe some physical and human characteristics of Chicago, using this view of the city's skyline.

make are deliberate; sometimes the changes are unplanned. Either type of change can bring good and bad results. For example, geographers study the technological advances that have made the American Southwest an attractive place to live in the second half of the twentieth century. Before air conditioning, swimming pools, massive irrigation, and fast, efficient transportation, the hot, dry lands of the southwestern part of the United States were sparsely populated. Today it is one of the fastest-growing regions in the country. Geographers also ask how the rapid growth in population— which brings new roads, airports, housing developments, and cities—is affecting the delicate natural environment of the Southwest.

Movement

Movement is a characteristic of nature. When geographers talk about movement, they are most often talking about the movement of people, goods, and ideas.

Movement at Chicago's O'Hare Airport
Throughout its history, Chicago has been a hub for the movement of people. Today more than 150,000 people pass through O'Hare each day.

Chicago's history illustrates movement. During the 1800s and early 1900s, thousands of southern blacks and European immigrants came to work in Chicago's industries. Today many new Chicago residents are immigrants from Asia or Latin America.

Today Chicago is connected to every part of the world by various types of movement. O'Hare Airport in Chicago is one of the busiest airports in the world. Trucks and railroad cars carry more goods in and out of Chicago than through any other city in the United States. These are what geographers mean by movement.

Regions

The last of the five geographic themes is used to study how places are alike and unalike. A region is a group of places bound together by one or more similar characteristics. The way a geographer divides the world into re-

gions depends on what he or she is interested in studying.

One way to divide the earth into regions is by political regions—areas ruled by the same government. Political regions include towns, counties, states, and countries. Many world maps show the world divided into countries.

Another way of dividing the earth into regions is according to the physical characteristics of places. For example, places with similar climates, vegetation, or landforms can be grouped together in regions.

Another way of dividing the world into regions is according to cultural characteristics. Mexico is categorized differently depending on whether it is being considered culturally or physically. Physically, Mexico is part of the North American continent. Culturally, it has more in common with the Spanish-speaking countries of Central and South America than it does with the United States and Canada.

■ SECTION **1** REVIEW ■

Developing Vocabulary
1. Define: **a.** geography **b.** absolute location **c.** Equator
   d. hemisphere **e.** latitude
   f. longitude **g.** Prime Meridian
   h. relative location **i.** culture

Place Location
2. Describe both the absolute and relative location of Chicago, Illinois.

Reviewing Main Ideas
3. What are some of the physical and cultural characteristics of the place where you live?

Critical Thinking
4. **Making Comparisons** How does the study of a place's cultural characteristics differ from the study of its physical characteristics?

2 The Geographer's Craft

Section Preview

Key Ideas

- Pictures and the printed word are two important tools that are used by geographers.
- Maps and globes are the geographer's major tools.
- Modern geographers work in business, government, and education.

Key Terms

hypothesis, scientific method, remote sensing, census, great circle, topography, cartographer, distortion, map projection, cardinal directions, intermediate directions

When geographers go about the task of studying a place, they make use of specialized tools. These tools give the geographer a unique perspective on the earth and its inhabitants.

The Geographer as Scientist

Geographers are scientists. They begin, as all scientists do, with a question about the world. Next, they form a hypothesis (hy PAHTH uh suhs), or educated guess, about the answer to that question. For example, geographers might ask why people settled in a particular place. They might hypothesize that one reason was because of the resources and relative location of the place.

The next step in the scientific method—the systematic approach to knowledge followed by scientists—is to collect information that can prove or disprove that hypothesis. Geographers use many methods of collecting data. These methods range from simple observation to the most advanced technological approaches.

Collecting Information

Today, geographers rely upon a variety of information to aid them in their studies. Geographers choose their information according to what questions they are trying to answer.

Pictures from Air and Space Advances in technology have made an extremely valuable tool available to geographers and other scientists. Pictures taken from planes or satellites can be used for an amazing variety of purposes.

Remote sensing is a method that employs airplanes and satellites to produce photographs or computer-generated images of sections of the earth's surface. One of the best-known examples of remotely sensed images are those created by the Landsat satellites. The Landsat image below is of the Nile Delta in Egypt. When the satellite passed over the area it transmitted, or sent, electronic information

The Nile Delta from a Satellite

This picture was taken from an altitude of 570 miles (917 km). Plants appear in red while barren lands are gray. Water appears black.

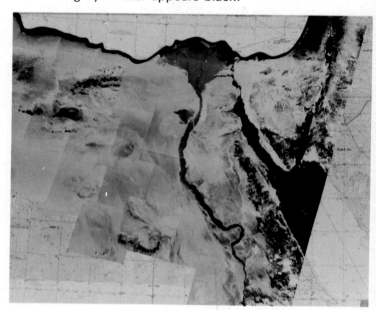

to a ground station, where a computer translated the information and produced this image of the delta.

Pictures taken from air or space provide information that may not be obtainable at ground level. A satellite can provide pictures of large areas detailing rivers, streams, and mountain ranges. It can scan the earth for signs of undiscovered natural resources, such as oil, gas, or mineral deposits. Or it can zero in on smaller areas, providing a view of city streets. Pictures taken by remote sensing devices can be used to detect unhealthy crops, forest fires, and polluted air.

The Printed Word A very different kind of written information that is valuable to geographers is census data. A **census** is a detailed counting of the population. Most governments make some effort to count and gather information about their populations.

United States census reports, which have been made every ten years since 1790, provide geographers with information about the nation's population, such as family size, ethnic origins, language, and religion. Geographers use census data to study the growth and decline of the population in different regions of the United States. Census information also provides geographers with information about changes in things such as birth and death rates, household size, and average family income.

Displaying Data

Once data has been gathered, it is important for the geographer to be able to analyze and display it in a meaningful way so that people can understand and learn from it. Diagrams, charts, graphs, and tables are all valuable methods of displaying information, and all are used by geographers. But without a doubt, the most often-used tools of the geographer are globes and maps.

Globes A globe is the most accurate method of showing the entire surface of the earth. A globe is an exact scale-model of the earth. It shows the actual shapes, sizes, and locations of all the earth's landmasses and bodies of water. Because a globe is shaped like the earth, it also provides accurate information about the distance between two points and about direction from one point to another.

The spherical shape of a globe also allows it to show great circles more clearly than other representations of the earth's surface. A **great circle** is any imaginary line that circles the earth and divides it into two equal halves. The shortest distance between any two points on a globe can be found by stretching a piece of string between them. If you kept going and stretched that string all around the globe, it would make a great circle. Airplanes flying over great distances use great circle routes to save fuel and cut down on travel time.

Applying Geographic Themes

Movement This map compares a great circle route with a straight route. How much shorter is the circle route?

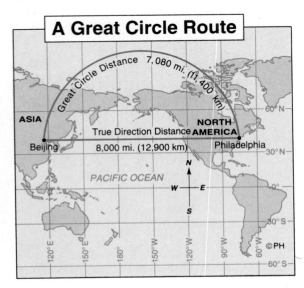

A Great Circle Route

Great Circle Distance 7,080 mi. (11,400 km)

ASIA

True Direction Distance **NORTH AMERICA**

Beijing 8,000 mi. (12,900 km) Philadelphia

60°N
30°N

PACIFIC OCEAN

N
W — E
S

120°E 150°E 180° 150°W 120°W 90°W 60°W

0°
30°S
©PH
60°S

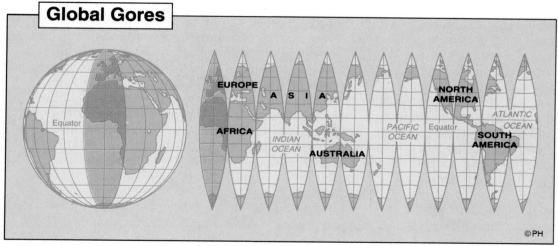

Global Gores

EUROPE
ASIA
NORTH AMERICA
ATLANTIC OCEAN
Equator
AFRICA
INDIAN OCEAN
PACIFIC OCEAN
Equator
SOUTH AMERICA
AUSTRALIA

©PH

Applying Geographic Themes

Location With the help of mathematics, cartographers flatten the surface of the earth. Size, shape, and distance are distorted when curves become straight lines. How is distance across water distorted on this map?

But globes do have some disadvantages. For one thing, when you look at a globe you can see only about half of the earth at a time. For another, because globes show the entire earth, they cannot offer detailed information. Globes are bulky and hard to transport. You certainly can't fold one up and carry it in your pocket!

Maps Because of the drawbacks of globes, geographers use maps for many purposes. A map is a flat drawing of all or part of the earth's surface. The most primitive maps were temporary—sketched out in the sand or soil with a stick. Later, more permanent maps, drawn on stones, bark, animal skins, or paper, used writing to show detailed labels and place names.

Maps are the geographer's most important tool because of their ability to show many different types of information. Many commonly used maps show **topography** (tuh PAHG ruh fee) —features of the earth's surface—and political boundaries. Special-purpose maps are used to provide information about nearly every topic imaginable. A look at a land use map of the world can

show you at a glance which areas of the world are devoted to commercial farming, manufacturing, or forestry. A look at a population density map shows which areas of the earth are most heavily populated and which are completely uninhabited.

Maps often are used in combination with one another. Transparent map overlays can be used to show data from two or more maps of the same area simultaneously. A transparent climate map laid over a population map of the United States provides more information than either map does alone. Combining information in this way helps geographers to identify relationships between different types of information.

Map Projections

Because maps are flat pictures of the earth's curved surface, it is impossible for them to capture the correct size and shape of every landmass or body of water. Imagine trying to flatten out an orange peel—assuming that you could get the peel off in one piece. The sides would split, the shape would change, and you would find it necessary to cut pieces of the peel away to get it to lie flat.

These are the problems that **cartographers**, or mapmakers, face when they try to map the earth. It cannot be done without stretching some places and shrinking others. This shrinking or stretching is called **distortion**—a change made to the original shape. Cartographers draw the earth's surface using different **map projections**—ways of showing the round earth on flat paper. Each type of projection produces some distortion. No single projection can perfectly depict the correct area, shape, distance, and direction for the earth's surface. Cartographers deal with this problem by choosing the projection that least distorts the data they are interested in studying.

Applying Geographic Themes

Location   Each map projection has advantages and disadvantages. The Robinson projection, used throughout this book, shows equal areas with a high degree of accuracy. It is conformal except near the edges where distortion increases. What is another advantage of the Robinson projection?

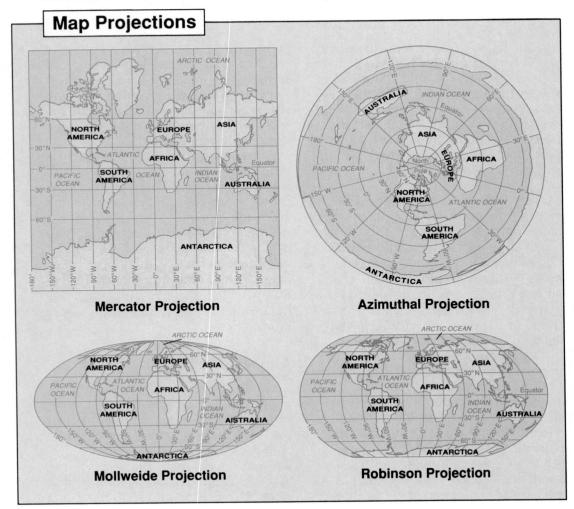

Map Projections

Mercator Projection

Azimuthal Projection

Mollweide Projection

Robinson Projection

DAILY LIFE

Mental Maps

How do you get from place to place every day? Bicycle? Bus? Car? Subway? Your own two feet? No matter the method, you rely on mental maps to get you where you're going.

Mental maps are invisible, detailed mind sketches that show the routes to and around the places where we live. Each of us has a mental file of these maps: At Home, To School, To Job, To Mall. Each map has short and long variations, depending on our moods or on the things we need to do along the way.

Mental maps are so familiar that we are not aware of using them. As we need them during the day, our brains call them up, connect the dots, and our feet follow. Without mental maps, we could not unlock—or go through—the many doors of our daily lives.

1. When are you most aware of using mental maps in your own life?
2. What are the chief differences between a mental map and a regular map printed on paper?

Conformal Maps Four types of projections are shown on page 12. Maps that use a projection that maintains the true shapes of landmasses are called conformal maps. This is because the shapes conform to, or look like, the shapes that are shown on a globe. One way to know when a map is conformal is to check the lines of latitude and longitude. If they cross at right angles, as they do on a globe, the map is a conformal map.

One example of a conformal map is a Mercator projection, named for the cartographer who invented it. The Mercator projection accurately shows direction and shape, but it distorts distance and size. The distortion is least near the Equator and greatest near the Poles. You can see this as you compare the size of Greenland on the Mercator projection with Greenland as it appears (more accurately) on the Mollweide projection.

WORD
ORIGIN

legend
The term *legend* has its root in the Latin word *legere*, which means "to gather." For example, a map legend "gathers" or points out information.

Equal-Area Maps Maps using projections that show the correct size of landmasses in relation to other landmasses are called equal-area maps. Equal-area maps, in order to depict correct size, need to distort shapes. The distortion is usually greater at the edges of equal-area maps than at the center.

The Mollweide projection is an example of an equal-area map. It shows land and water areas in their accurate proportions. To see the distortion in shape compare the shape of North America on the Mollweide projection with that on the Mercator projection.

Equidistant Maps Maps that use projections that show the correct distance between places are called equidistant maps. Maps of the whole world can never show all distances accurately because it is not possible to show the correct lengths of all lines of latitude and longitude. Small areas, however, can be mapped with little distortion of distance.

Azimuthal Maps The fourth type of projection is one that shows true compass direction. This is called an azimuthal map. Azimuthal maps, like the one on page 12, are circular and are often used to map the areas of the North and South Poles. Because north and south follow the curved lines of latitude, directions cannot be shown on a directional indicator as they can with other maps.

Map Components

Most maps used today have four basic components that help the user interpret the information they present. These fundamental components are a title, a legend or key, a scale, and a directional indicator.

Obviously, the title is important because it tells what the map is about. Frequently the title will also contain a date, which is helpful when the map represents information that changes over time, such as political boundaries or shifts in population density.

The legend or key explains what the map's symbols, lines, and colors represent. The map scale is used to show the distance between places shown on the map. You will read more about using map scales in the Skills Check following this section.

The final component, the directional indicator, may be shown simply as an arrow indicating north. Or it may be a full compass rose that shows the four cardinal directions of north, south, east, and west and the intermediate directions of northeast, northwest, southeast, and southwest. In some instances, parallels and meridians may be used to indicate direction. A compass rose showing cardinal directions is used on most of the maps in this book.

SECTION 2 REVIEW

Developing Vocabulary
1. Define: **a.** hypothesis **b.** scientific method **c.** remote sensing **d.** census **e.** great circle **f.** topography **g.** cartographer **h.** distortion **i.** map projection **j.** cardinal directions **k.** intermediate directions

Place Location
2. Identify the locations of the Tropic of Cancer and the Tropic of Capricorn.

Reviewing Main Ideas
3. Explain why flat maps can be deceiving in their depiction of the earth's landforms and bodies of water.

Critical Thinking
4. **Analyzing Information** Based on what you have learned about different map projections, give an example of a purpose for which one type of projection would be unsuitable.

✔ Skills Check

- ☐ Social Studies
- ☑ Map and Globe
- ☐ Reading and Writing
- ☐ Critical Thinking

Comparing Maps of Different Scale

Understanding scale is vital to understanding the information that a map contains. The two maps below depict the Chicago area in two different scales. Map 1 shows the northern half of the state of Illinois, with Chicago claiming a relatively small area. Map 2 focuses on the city and its immediate surroundings, giving much more detail about the city. Use the following steps to study and analyze the two maps.

1. **Locate the area of Map 2 on Map 1.** Look for familiar geographic reference points, such as a lake, an ocean, a city, or a river. Answer the following questions: (a) Name two reference points that appear on both maps. (b) What reference points lie just outside the area shown on Map 2?

2. **Determine the scale of the map.** Check the scale of each map. The scale translates distances on the map (an inch, for example) into distances over land (usually miles). Answer the following questions: (a) What is the scale of Map 1? (b) Of Map 2?

3. **Compare the information you have gathered from the two maps.** On Map 1, determine the distance in miles between Highland Park and Elgin. Now use Map 2 to determine the distance between the same two points. Answer the following questions: (a) What distance does Map 1 show between these two towns? (b) How does the distance shown on Map 2 compare? (c) On which map is it easier to determine the exact distance? Why?

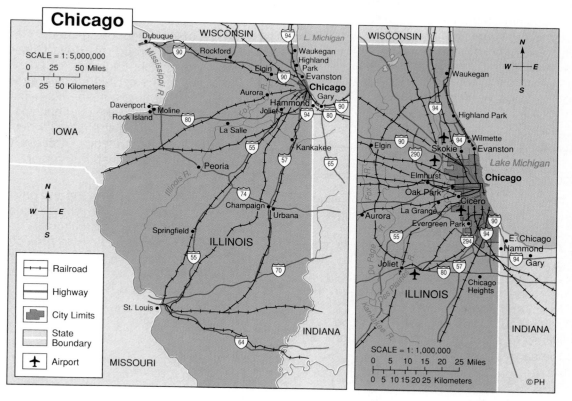

15

1

REVIEW

Section Summaries

SECTION 1 The Five Themes of Geography The five themes of geography are (1) location—absolute or relative; (2) place—the unique physical or human characteristics of a place; (3) human-environment interaction—how people use and change their environment; (4) movement—of people, goods, and ideas; and (5) regions—how the world can be divided according to places with shared characteristics.

SECTION 2 The Geographer's Craft Geographers have many tools with which to collect, analyze, and display information. Among them are pictures, the printed word, graphs, charts, tables, maps, and globes. There are several types of map projections. Most maps have a title, a legend, a scale, and a directional indicator. Today advances in technology have made new and more sophisticated tools available to geographers. These tools include pictures taken from planes or satellites, remote sensing, and census data.

Vocabulary Development

Match the definitions with the terms below.

1. Every part of a group's way of life, including its government, language, religions, customs, and beliefs
2. An educated guess, or answer, to any given question
3. An imaginary line that circles the globe at its widest point, halfway between the North and South Poles
4. A method that employs airplanes and satellites to produce images of sections of the earth's surface
5. An imaginary line that circles the earth and divides it into two equal halves
6. Imaginary lines that run north and south from one Pole to the other
7. A place's specific position on the globe
8. A person who makes or designs maps
9. Where one place is located in relation to another place
10. Imaginary lines that run around the earth parallel to the Equator
11. A way of showing the round earth on flat paper
12. Physical features of the earth's surface
13. The systematic approach to knowledge followed by all scientists
14. North, south, east, and west
15. A detailed counting of the population
16. One of the two halves of the earth, divided by the Equator
17. Northeast, northwest, southeast, and southwest
18. The line of longitude, designated 0° longitude, that runs through the Royal Observatory in Greenwich, England
19. The study of the earth's surface and the processes that shape it, the connections between places, and the relationships between people and their environments
20. The shrinking or stretching process that changes an original shape on a map

a. geography
b. hypothesis
c. Prime Meridian
d. relative location
e. intermediate directions
f. scientific method
g. hemisphere
h. absolute location
i. longitude
j. great circle
k. topography
l. remote sensing
m. culture
n. map projection
o. census
p. cartographer
q. latitude
r. distortion
s. Equator
t. cardinal directions

Main Ideas

1. How can the relationship of people to their environment bring good and bad results?
2. Name three kinds of geographic regions.
3. What four properties of the physical world do maps represent?

Critical Thinking

1. **Distinguishing False from Accurate Images** When most people look at a world map, they assume that every bit of information, including area, shape, and distance, shown is accurate. Explain why this is not true.
2. **Recognizing Bias** Why may paintings and diaries not always be an accurate source of information for geographers?

Practicing Skills

Comparing Maps of Different Scale Draw a map of your classroom. Include the doors, windows, and desks. Make a scale for the map, converting classroom measurements from feet to inches on the map.

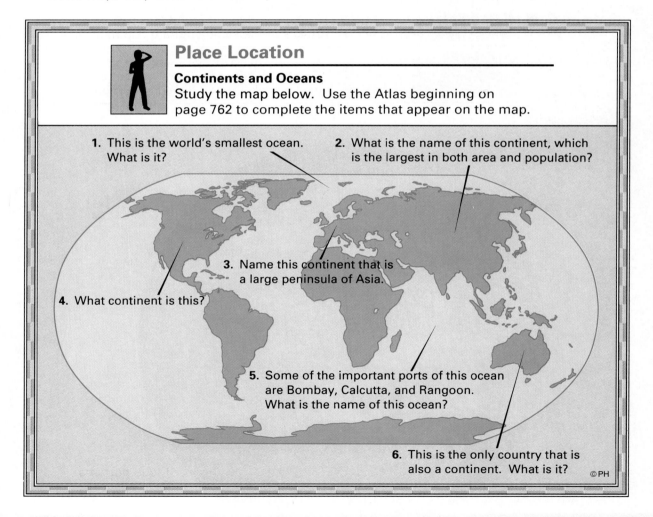

Place Location

Continents and Oceans
Study the map below. Use the Atlas beginning on page 762 to complete the items that appear on the map.

1. This is the world's smallest ocean. What is it?
2. What is the name of this continent, which is the largest in both area and population?
3. Name this continent that is a large peninsula of Asia.
4. What continent is this?
5. Some of the important ports of this ocean are Bombay, Calcutta, and Rangoon. What is the name of this ocean?
6. This is the only country that is also a continent. What is it?

©PH

Land, Climate, and Vegetation

Chapter Preview

Each of these sections provides an overview of physical geography.

Sections	Did You Know?
1 THE CHANGING EARTH	According to an East African myth, an earthquake resulted when a cow shifted the earth from one horn to the other.
2 EXTERNAL FORCES THAT CHANGE THE EARTH'S SURFACE	Erosion of a grain of beach sand is extremely slow because each grain is cushioned by a film of water.
3 WEATHER AND CLIMATE	Scientists in Colorado observed 120,000 lightning bolts and found that two or more strikes in the same spot are common.
4 VEGETATION REGIONS	A variety of palm tree found in the Seychelles Islands has a seed that is bigger than a basketball and weighs 50 pounds.

Cloud patterns from space

The Laws of Nature
Volcanoes have been objects of fear and marvel since ancient times. During a volcanic eruption, red-hot lava, or molten rock, flows freely. Volcanic ash enhances agriculture by adding minerals to the soil.

1 The Changing Earth

Section Preview

Key Ideas

- The earth is a changing planet, affected by geologic processes.
- Forces inside the earth create and change landforms on the surface.
- The theory of plate tectonics answers many questions about the earth's landforms.

Key Terms

geology, core, mantle, magma, crust, continents, relief, plateau, plain, volcanism, lava, fault, plate tectonics, continental drift theory, fossils, rift valley, convection, subduction zone

The earth is not a quiet planet. Earthquakes topple buildings and open up great cracks in the ground. Volcanoes erupt with red-hot lava and dangerous gases. While these are some of the most spectacular ways in which the earth is changing, they are not the only ways. Many processes—some slow, some dramatic—are always at work shaping the earth on which we live.

The Structure of the Earth

Geology—the study of the earth's physical structure and history—is a relatively new science. It deals, however, with very ancient history—that of the earth itself. This history, scientists now think, goes back about 4.6 billion years. Since it began, the earth has been changing. Geologists try to learn what those changes were and to understand why they occurred.

The Earth's Inner Structure Scientists have developed an idea of what the interior of the earth is like. The diagram on page 20 shows the earth's layers as geologists see them.

The Earth's Layers

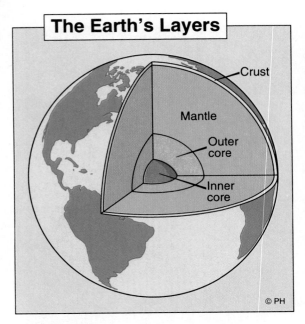

Crust
Mantle
Outer core
Inner core

© PH

Diagram Skill

This diagram shows what geologists think is the internal structure of the earth. What is the outermost layer called?

The center of the earth is a dense core of very hot metal, mainly iron mixed with some nickel. The inner core is thought to be dense and solid, while the metal of the outer core is molten, or liquid. Around the core is the mantle, a thick layer of rock. Scientists believe that the mantle is about 1,800 miles (2,900 km) thick. Mantle rock is mostly solid, but some upper layers may be pliable, like putty. The mantle also contains pockets of magma, or melted rock.

The earth's crust, the rocky surface layer, is surprisingly thin, like frosting on a cake. The thinner parts of the crust, which are only about 5 miles (8 km) thick, are below the oceans. The crust beneath the continents is thicker and very uneven. It averages about 22 miles (35 km) in thickness. It is on the crust that natural forces have changed the earth's original rock to make landforms, or natural features of the earth's surface.

The Earth's Land and Water The first photographs of the earth taken from space show clearly that it is a "watery planet." More than 70 percent of the earth's surface is covered by water, mainly the salt water of oceans and seas. The large landmasses in the oceans are the continents. Although some of these landmasses are not separated by ocean waters but are joined, geographers define seven separate continents. Asia is the largest, Australia the smallest. All the continents have a variety of landforms, although those in Antarctica are hidden by ice.

Landforms help give each place a characteristic shape and landscape. They are commonly classified according to differences in relief—the height or elevation above the surrounding landscape. Another important characteristic is whether they rise gradually or steeply.

The major types of landforms are mountains, hills, plateaus, and plains. Mountains have high relief—that is, they rise more than 2,000 feet (610 m) above sea level. Most mountains also have steep slopes, rising to a peak or summit. Hills are lower, though they always rise at least 500 feet (152 m) above the surrounding land. They tend to be more rounded and not as steep. A plateau is also a raised area, but its surface is generally flat. Many plateaus, however, have deep gulleys or canyons, making the surface seem rough rather than flat. At least one side of a plateau rises steeply above the surrounding land.

Plains are landforms, too, though much less dramatic than mountains. A plain is a flat or gently rolling area where there are few changes in elevation. Many plains are along coasts.

Other landforms include valleys, canyons, and basins. Various geographical features of landscapes include rivers, peninsulas, and islands. Many of the earth's landforms are shown on the diagram on page 21.

Internal Forces That Shape Landforms

One basic question for geologists and geographers is: What forces shaped the rock to make these mountains or that plain? Major landforms are shaped first by internal forces—processes that start in the earth's interior. One of these processes is volcanic activity, or volcanism, which involves the flow of magma, or molten rock. The other major internal force consists of different movements that fold, lift, bend, and break the rock of the earth's crust.

Volcanoes The ancient Romans believed in a god named Vulcan, a skillful craftsman who worked with hot iron or gold at his forge. Not surprisingly, their myths placed the forge below the smoking, fiery mountain on the island of Volcano off the coast of Italy. Both the island and the mountain are named after Vulcan.

A volcano forms when molten rock, or magma, breaks through the surface of the earth as lava. Lava and ash eventually build up in successive layers to form a distinctive cone-shaped mountain. One of the most famous is Mt. Fuji in Japan. Lava may also flow out slowly, forming a shield volcano or a thick plateau.

Changes in the Earth's Crust The movements that bend and fold the earth's crust are varied and complex. One result of such movements and stresses is a fault, a break in the earth's crust. Sometimes the rock on either

---

Diagram Skill

Many of the earth's most common landforms are shown in this diagram. How does a plateau differ from a plain? Use the glossary beginning on page 785 to find the definition of a delta.

Landforms and Water Bodies

side of a fault slips or moves suddenly. Large movement along a fault can send out shock waves through the earth, causing an earthquake.

The Earth's Geologic History

Most changes in the earth's surface take place so slowly that they are not immediately noticeable to humans. Nonetheless, geologists have reconstructed much of the earth's history from the record they read in the rocks. For many years scientists assumed that the basic arrangement of oceans and continents was stable and permanent. Today, however, most accept the idea that the earth's landmasses have broken apart, rejoined, and moved to other parts of the globe. This concept forms part of the plate tectonic theory. While scientists are still working to prove some parts of it, this theory answers many puzzling questions about the earth.

Plate Tectonics According to the theory of plate tectonics, the earth's outer shell is not one solid piece of rock. Instead it is composed of a number of large, moving slabs of rock, called plates. These plates are not anchored, but slide very slowly over a pliable layer of the mantle.

The earth's oceans and continents ride atop the plates as they move in different directions. The map below

Applying Geographic Themes

Movement The world's continents and oceans ride atop moving tectonic plates. In which direction are the Nazca and South American plates moving? What are some of the results of this movement?

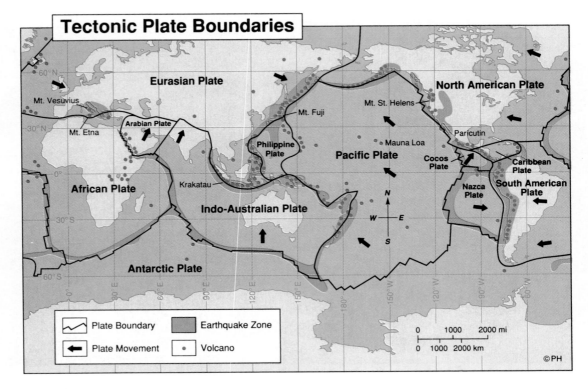

Tectonic Plate Boundaries

Pangaea and the Drifting Continents

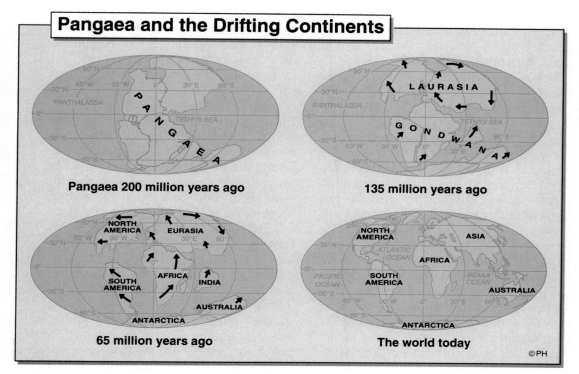

Pangaea 200 million years ago

135 million years ago

65 million years ago

The world today

©PH

Applying Geographic Themes

Place Alfred Wegener was one of several theorists who thought that the earth was once composed of a single "supercontinent" called Pangaea. Between which years did South America break away from Africa? What two giant landmasses existed 135 million years ago?

shows the boundaries of the different plates. It also shows the direction in which the plates are moving. The African Plate and the South American Plate, for example, are moving apart. The Nazca Plate and the South American Plate, however, are moving toward each other. It is along the boundaries where plates meet that most earthquakes, volcanoes, and other geologic events occur.

The plate tectonic theory began to be widely accepted in the 1960s. It was based on earlier ideas and research, however, and includes two important, separate theories.

The Continental Drift Theory As early as the 1600s, people looking at maps noticed how the eastern coast of South America closely matched the western coast of Africa. Other continental coasts also seemed to fit together like jigsaw puzzle pieces. Could they once have been joined as one landmass?

In the early 1900s Alfred Wegener, a German explorer and scientist, suggested the continental drift theory. Wegener proposed that there was once a single "supercontinent" on the earth. He called it Pangaea (pan JEE uh), from the Greek words *pan*, meaning "all," and *gaia*, personifying the earth. Wegener theorized that about 200 million years ago, Pangaea began to break up into separate continents. Today the earth's continents continue to move as the plates move.

Wegener gathered evidence to support his theory. He showed that fossils—the preserved remains or traces of ancient animals and plants—from South America, Africa, India, and Australia were almost identical. The rocks where the fossils were found were also much alike. Despite his evidence, many scientists were skeptical.

Sea-Floor Spreading   The other theory supporting plate tectonics developed from World War II technology. Sonar, which was invented to locate enemy submarines underwater, sends out sounds that bounce back when they strike an object. Following the war, scientists began to use sonar to map the floor of the Atlantic Ocean. They also began to investigate the age of the rocks on the sea floor as well as other features of the ocean landscape.

One unique ocean feature is the underwater ridge system. One of the many ridges in this system, the Mid–Atlantic Ridge, shown on the map on pages 776 and 777, is part of the mid-oceanic ridge, a mountain system that extends around the world. The Pacific Rim is another example of an underwater ridge system. In a few places, the mountaintops of the ridge emerge as islands, for example, Iceland, the Azores, and Easter Island.

A common feature of all underwater ridge systems is the **rift valley,** a large split along the crest of the mountains. Small earthquakes and volcanic eruptions occur here frequently.

When scientists measured the age of the rocks they had taken from the sea floor, they were surprised to find that these rocks were much younger than those of the continents. The youngest rocks of all were those nearest the Mid-Atlantic Ridge.

The explanation suggested in the 1960s by geologists is the theory of sea-floor spreading. According to this theory, molten rock from the mantle rises under the mid-oceanic ridge and breaks through the rift. The rock then spreads out in both directions from the ridge as if it were on two huge conveyor belts. As the sea floor moves away from the ridge, it carries older rocks away. This idea, along with Wegener's older theory of continental drift, became part of the theory of plate tectonics.

Plate Movement   One reason that people in the 1920s doubted the continental drift theory was Wegener's explanation of just *how* the continents moved. What force was powerful enough to send gigantic plates sliding around the globe?

Today, most scientists believe this force is a process called convection. Convection is a circular movement caused when a material is heated, expands and rises, then cools and falls. This process is thought to be occurring in the mantle rock under the plates. The heat energy that drives convection probably comes from the slow decay of radioactive materials under the earth's crust.

Plate Boundaries   As mentioned earlier, the places where plates meet are some of the most restless parts of the earth. Plates meet at their boundaries and react in one of three different ways. They pull away from each other, crash head-on, or slide past each other.

Where plates pull away from one another, they form a diverging plate boundary, or spreading zone. As the Mid-Atlantic Ridge shows, such areas are likely to have a rift valley, earthquakes, and volcanic action.

Several different things can happen when plates meet, or converge. Because continental crust is lighter than oceanic crust, continental plates "float" higher. Therefore, when an oceanic plate meets a continental plate, it slides under the lighter plate and down into the mantle. Any place where this occurs is known as a **subduction zone**

When subduction occurs, volcanic mountain building and earthquakes may also occur on the continental plate. The Andes Mountains, for

Major Types of Plate Movement

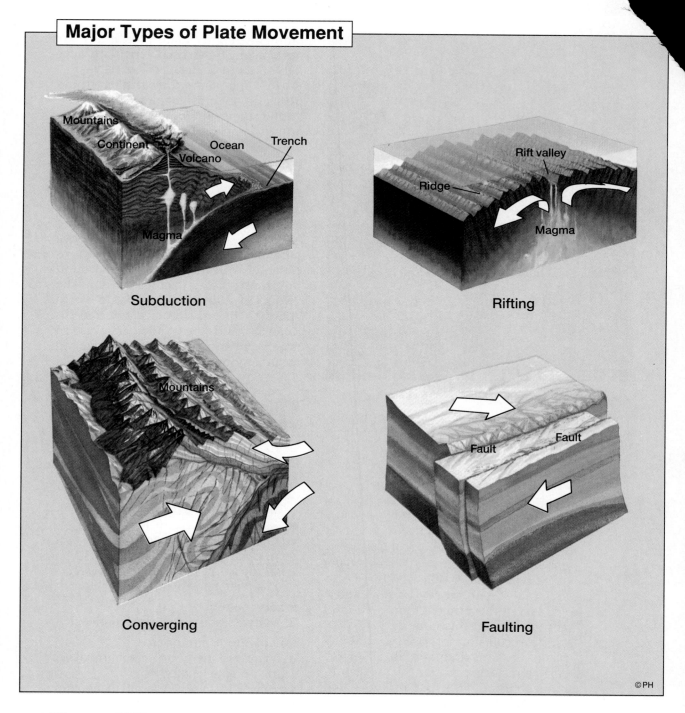

Subduction

Rifting

Converging

Faulting

©PH

Diagram Skill

In a *subduction zone,* one plate slides or dives under another. In a *spreading zone,* two plates move apart from each other creating a rift, or crack, in the earth's crust. In a *collision or convergence zone,* two plates collide and push slowly against each other. At a *transform fault,* plates grind or slide past each other rather than colliding. Which type of plate boundary occurs along the west coast of South America?

The San Andreas Fault
This fault lies along the west coast of North America. Use the map on page 22 to name the two plates that form this fault.

Indo-Australian Plate continues to push against the Eurasian Plate at a rate of about 2 inches (5 cm) a year.

In some places the earth's tectonic plates do not crash against each other. Rather they slip or grind past each other along faults. The San Andreas Fault in California is an example of this type of plate boundary.

Explanations The plate tectonic theory helps to explain so many events and processes that have changed the earth. For example, scientists have long observed the "Ring of Fire," a circle of volcanic mountains around the rim of the Pacific Ocean. The "Ring" includes the Cascades in North America, the islands of Japan and Indonesia, and the Andes in western South America. These areas are all located along plate boundaries.

example, formed over millions of years as the Nazca Plate slid under the South American Plate.

Different kinds of converging zones occur where two plates of the same type collide. When both are oceanic plates, one moves, or is subducted, under the other. Often an island group forms at this boundary.

The collision of two continental plates is more dramatic. In all of the earth's history, perhaps the most stunning collision was that of India and Asia. As the map on page 22 shows, India today is clearly part of Asia. But it is not part of the Eurasian Plate. It is the northern tip of the Indo-Australian Plate, which crashed into the Eurasian Plate millions of years ago. Earth's highest mountain range, the Himalayas, was formed by the collision of these two plates. Even today, the

■ SECTION 1 REVIEW ■

Developing Vocabulary
1. Define: **a.** geology **b.** core
   c. mantle **d.** magma **e.** crust
   f. continents **g.** relief **h.** plateau
   i. plain **j.** volcanism **k.** lava
   l. fault **m.** plate tectonics
   n. continental drift theory
   o. fossils **p.** rift valley **q.** convection
   r. subduction zone

Place Location
2. Where are the Andes Mountains?

Reviewing Main Ideas
3. What are the two internal processes that create landforms?
4. How did World War II technology add to the plate tectonic theory?

Critical Thinking
5. **Making Comparisons** Use the ocean floor map on pages 776 and 777 to identify at least three types of landforms that are found on the ocean floor as well as on land.

2 External Forces That Change the Earth's Surface

Section Preview

Key Ideas
- Mechanical and chemical weathering are forces that change landforms.
- Erosion is another external force that alters the surface of the earth.

Key Terms
weathering, mechanical weathering, frost wedging, chemical weathering, acid rain, erosion, sediment, floodplain, delta, mouth, loess, glacier, moraine

In the soil of the Hawaiian Islands, there is crumbly gray clay that is older than the volcanic rock of the islands themselves. For years, scientists wondered how this had happened. Now they think that the clay comes from a desert in far-off China. Blown across the ocean by the wind, it has been deposited on the islands by centuries of rainstorms. By providing a mineral needed for plant growth, the clay enriches Hawaii's land. This process is still going on. As one conservationist explained:

> So much soil from the Asian mainland blows over the Pacific Ocean that scientists taking air samples at the Mauna Loa observatory in Hawaii can now tell when spring plowing starts in North China.

Wind is only one of several external forces that change the earth's surface, build and destroy landforms, and affect the soil in which plants grow. The actions of all these forces are usually grouped into two categories: weathering and erosion.

Weathering

Changes in the earth's surface usually take place very slowly, over thousands or millions of years. The process of **weathering**, for example, breaks down rock at or near the earth's surface into smaller and smaller pieces. While it may take millions of years, weathering can reduce a mountain to gravel. Depending on the forces involved, weathering is either mechanical or chemical.

Mechanical Weathering When rock is actually broken or weakened physically, the process is termed **mechanical weathering**. The most common type of mechanical weathering takes place when water freezes to ice in a crack in the rock. Because water expands by nearly 10 percent when it freezes, the ice widens the crack and eventually splits the rock. This process is known as **frost wedging**. Frost wedging over time can even cause huge parts of a mountainside to break and fall away.

Another kind of mechanical weathering occurs when seeds take root in cracks in rocks. In the same way as sidewalks crack and rise over tree roots, a rock will split as plants or trees grow within a fracture.

Chemical Weathering While mechanical weathering can destroy rock, it changes only the physical structure, not the original crystals or minerals that make up the rock. It leaves the rock's chemical structure unchanged. **Chemical weathering** alters the rock's chemical makeup by changing the minerals that form the rock or combining them with new chemical elements. Both kinds of weathering cause the breakdown of rock, but chemical weathering can also change one kind of rock into another.

The most important forces in chemical weathering are water and carbon dioxide. Commonly, carbon dioxide

erosion
In the Middle Ages the word *erosion* was used to describe the way acid cut through metal. The term is derived from the Latin *rodere,* which means "to gnaw." The word *rodent* has the same Latin origin.

from the air or soil combines with water to make carbonic acid. When the acidic water seeps into cracks in certain types of rock, such as limestone, it can completely dissolve away the rock. Many caves were formed in this way.

Moisture is an important element in chemical weathering. In a dry region where water is scarce, there is little chemical weathering. But in a damp or wet area, chemical weathering occurs quickly and is widespread.

Another type of chemical weathering is acid rain. Chemicals in the polluted air combine with water vapor and eventually fall back to earth as acid rain. Acid rain not only destroys forests, pollutes water, and kills wildlife, it also eats away the surfaces of stone buildings, statues, and natural rock formations. Industrial pollution, acid–producing agents from the ocean, and volcanic activity are among the known causes of acid rain. But continuing study of the problem periodically reveals new information.

Observing Weathering   Weathering is an extremely slow process. Its effects can be seen, however, on almost any old stone structure—whether natural or made by humans. Weathering blurs the lettering on old tombstones and softens the sharp features on carved stone statues.

Weathering changes natural landforms, too. Over millions of years, entire mountains can be worn from jagged, sharp peaks to long, rounded hills. For example, in an area where temperature changes cause frost wedging, the south side of a mountain in the Northern Hemisphere is likely to be more rugged than the north slope. Because the south side receives more sunlight, water in the cracks of rocks thaws and freezes more often than on the cold north side. As a result, rocks on the southern slope are more likely to split and fall away, making the mountainside uneven and rugged.

Erosion

While weathering changes the rock on the earth's surface, **erosion** is the movement of weathered materials including gravel, soil, and sand. The three most common agents of erosion are water, wind, and glaciers.

Erosion is an important part of the cycle that has made and kept the earth a place where living things can survive. Without this process, the earth's surface would be barren rock, with no soil where plants can grow. Erosion is actually a significant agent in mechanical weathering, described above. The erosive forces, or "wearing away," that created Niagara Falls and the Grand Canyon, for example, are all part of mechanical weathering.

Water   The largest canyons and the deepest valleys on the earth were made by running water. Moving water —rain, rivers, streams, and oceans—is the greatest agent of erosion. Over time, it can cut into even the hardest rock and wear it away.

It is not water alone that carves out valleys and canyons. Water moving swiftly down a streambed carries sediment—small particles of soil, sand, and gravel. Like sandpaper, the sediment helps grind away rocks along the stream's path.

The rocks and soil carried away by water are eventually deposited somewhere else. When the stream or river slows down, sediment settles on the banks or streambed as alluvium. New kinds of landforms are built up by alluvium. A broad floodplain, or alluvial plain, for example, may form on either side of the river, or a delta may form. A delta is a flat, low-lying plain that is sometimes formed at the mouth of a river—the place where the river enters a lake, a larger river, or the ocean.

The Mississippi River, for example, carries an estimated 500 million tons (454 million metric tons) of sediment a year. The river deposits some of this rich sand, silt, gravel, and clay along its floodplain, which is as much as 80 miles (130 km) wide in some places. The rest builds up the delta where the river empties into the Gulf of Mexico.

Rivers and streams play the largest role in water erosion. But crashing ocean surf or the gentler waves along a lakeshore can also erode beach cliffs, carve steep bluffs, and pile up sand dunes. As bluffs are undercut by the force of the water, rocks tumble into the water. Continuing erosion wears rocks into sandy beaches, then carries the sand farther down the shoreline.

Wind Wind is a second major cause of erosion, especially in areas where there is little water and few plants to hold the soil in place. In the 1930s, for example, wind erosion devastated farms on the Great Plains in the central United States. As farmers plowed more farmland, more and more land surface was stripped of its plant life and was exposed to the wind. The upper layers of soil that are usually rich in minerals and nutrients were dry from a long drought. As a result, the wind picked up and carried away the soil in great dust storms. As their farms' fertile soil blew away, several states became part of a "dust bowl." One farmer described his experience:

Soon everything was moving—the land is blowing, both farm and pasture alike. The fine dirt is sweeping along at express-train speed, and when the very sun is blotted out, visibility is reduced to some fifty feet; or perhaps you cannot see at all. . . .

On the other hand, farmers in China, the American Midwest, and other parts of the world have benefited from windblown deposits of mineral-rich dust and silt called loess (LO ess). Because its particles are so fine, loess may be blown thousands of miles.

Sandstorms, or windblown sand, are major causes of erosion, especially near deserts. Just as sandblasting cleans stone buildings, windblown sand carves or smoothes the surfaces of both rock formations and objects made by humans.

Glaciers Another major agent of erosion is glaciers. These are huge, slow-moving sheets or "rivers" of ice. They form over many years as layer after layer of unmelted snow is pressed together, thawing slightly, then turning to ice. As glaciers move, they carry dirt, rocks, and boulders.

In what geologists think of as fairly recent times, the earth was cooler than it is today. Much of the planet's water became locked up in immense glaciers that covered up to a third of the earth's

Arches National Monument, Utah
Sometimes nicknamed "desert windows," this arch was probably caused by both wind and water erosion.

Mendenhall Glacier, Juneau, Alaska
Alaska has thousands of valley glaciers that carve spectacular U-shaped valleys as they slide forward because of their own great weight. Glaciers are the main source of the earth's fresh water supply.

The extensive glaciers of the Ice Ages were mainly continental glaciers, or ice sheets. Today such broad, flat glaciers exist in only a few places. They cover about 80 percent of Greenland and most of Antarctica. The Greenland glacier is estimated to be 9,900 feet (3,020 m) thick. The front of the glacier usually moves a few feet each winter and then recedes during the summer.

Valley or alpine glaciers, on the other hand, are found throughout the world in high mountain valleys where the climate is not warm enough for the ice to melt. In North America, valley glaciers snake through the Rocky and Cascade mountains, the Sierra Nevada, and the Alaskan ranges.

Although glaciers are sometimes described as "rivers of ice," they do not move like water but slide forward because of their great weight. Large valley glaciers in Europe may move nearly 600 feet (180 m) in a year. Glacial landscapes are distinctly different from landscapes formed by water. For example, while rivers cut sharp-sided V-shaped valleys, glaciers carve out valleys that are U-shaped.

surface. Over thousands of years they melted back, then formed again when the earth grew colder. Long periods of these colder temperatures are known as Ice Ages. Geologists believe that there have been at least four Ice Ages in the past 500 million years, the last of which peaked about 18,000 years ago.

If you live in the northern part of the United States you might see the effects of Ice Age glaciers. Over time huge glaciers, like giant bulldozers, scooped out the basins of the Great Lakes, as well as thousands of smaller lakes elsewhere in the United States and Canada. When glaciers melted and receded in some places, they left behind ridgelike piles of rocks and debris called moraines (muh RAYNS). Long Island, New York, has significant moraine deposits.

SECTION 2 REVIEW

Developing Vocabulary
1. Define: **a.** weathering **b.** mechanical weathering **c.** frost wedging **d.** chemical weathering **e.** acid rain **f.** erosion **g.** sediment **h.** floodplain **i.** delta **j.** mouth **k.** loess **l.** glacier **m.** moraine

Place Location
2. Describe where the Mississippi Delta is located.

Critical Thinking
3. **Determining Relevance** How does weathering play a role in the survival of life on the earth?

3 Weather and Climate

Section Preview

Key Ideas

- Climate is the weather that prevails in an area over a long period of time.
- Location, latitude, elevation, and landforms are among the factors that influence the climate of a place.
- Convection is the process by which the sun's heat is distributed throughout the world.
- Climate regions are classified mainly by temperature and precipitation.

Key Terms

weather, atmosphere, climate, rotation, revolution, solstice, equinox, tropical zones, temperate zones, polar zones, Coriolis effect, precipitation, humidity, windward, leeward, rain shadow, continental climate

Everybody talks about the weather. No matter where you live or what language you speak, you probably know some folk beliefs for predicting weather. In India, people sometimes say, "When the frog croaks in the meadow, there will be rain in three hours' time." In Britain and America, the advice is different: "Rain before seven, sun by eleven." Weather seems so important because it affects everyday life—planting, harvests, and sometimes survival.

Weather and Climate

But what is "weather"? Weather is the condition of the bottom layer of the earth's atmosphere in one place over a short period of time. The atmosphere is a multilayered band of gases, water vapor, and dust above the earth. A description of the weather usually mentions temperature, moisture or precipitation, and wind. That is, a day might be "warm, dry, and calm" or "cold, snowy, and windy."

If weather includes all those factors, then what is climate? Climate is the term for the weather patterns that an area or region typically experiences over a long period of time. The climate of a place depends on a number of factors, including its elevation, latitude, and location in relation to nearby landforms and bodies of water.

The Sun and the Earth

The ultimate source of the earth's climates—and of life on earth—lies some 93 million miles (150 million km) away. The sun, an intensely hot star, gives off energy and light that is essential for the survival of plants and animals.

The Greenhouse Effect Only a small amount of solar radiation reaches the earth's atmosphere. Some of the radiation is reflected back into space by the atmosphere and by the earth's surface, but enough remains to warm the earth's land and water. The atmosphere also prevents heat from escaping back into space too quickly.

In this sense, the earth's atmosphere has been compared to the glass walls and roof of a greenhouse, which trap the sun's warmth for growing plants. Without this so-called "greenhouse effect," the earth would be too cold for most living things.

Even inside the greenhouse of the atmosphere, not all places on earth get the same amount of heat and light from the sun. Day and night, seasonal change, and differing climates all depend somewhat on the relative positions of the sun and the earth.

Rotation and Revolution As the earth moves through space it spins like a top on its axis. This movement is

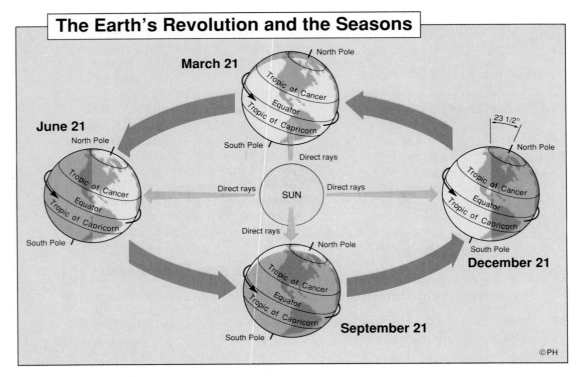

The Earth's Revolution and the Seasons

March 21

North Pole
Tropic of Cancer
Equator
Tropic of Capricorn
South Pole

Direct rays

June 21

North Pole
Tropic of Cancer
Equator
Tropic of Capricorn
South Pole

Direct rays

SUN

Direct rays

Direct rays

23 1/2°

North Pole
Tropic of Cancer
Equator
Tropic of Capricorn
South Pole

December 21

North Pole
Tropic of Cancer
Equator
Tropic of Capricorn
South Pole

September 21

©PH

Diagram Skill

It takes 365¼ days for the earth to make one revolution around the sun. Because of the earth's tilt, some places receive more direct sun rays than others. Where are the sun's rays the most direct?

known as **rotation**. The axis is an invisible line through the center of the earth from pole to pole. The earth completes one rotation in approximately twenty-four hours. On the side that faces the sun, it is daytime, with warmth coming from sunlight. On the side turned away, it is night.

The earth also revolves, or moves, around the sun in a nearly circular path called an orbit. A **revolution** is one complete orbit around the sun. The earth completes one revolution every 365¼ days—the length of a year.

As the earth revolves, its position relative to the sun is not straight up-and-down. Rather, the earth is tilted 23½° on its axis. Because of the earth's tilt, sunlight strikes different parts of the planet more directly at certain times of the year. As the diagram on this page shows, when the North Pole is tilted toward the sun, the sun's rays

fall more directly on the Northern Hemisphere, bringing longer, warmer days. The Northern Hemisphere experiences the season of summer while it is winter in the Southern Hemisphere. As the earth moves halfway around the sun, the Southern Hemisphere tilts closer to the sun, and the situation is reversed. These changes in season are marked by the summer or winter **solstices**—the days when the sun appears directly overhead to observers at the Tropics of Capricorn and Cancer.

The other markers for seasonal change occur on or about March 21 and September 21. These dates are known as the spring and fall **equinoxes**. On those days, the sun appears directly overhead to observers at the Equator. Around these dates, the duration of day and night are nearly equal everywhere on the earth.

Latitude and Climate The angle of the sun's rays affects weather and climate in other ways. Because the earth is round, the sun's rays always fall most directly at or near the Equator. As the diagram on page 32 shows, the rays grow less and less direct as they fall closer to the North and South poles. As a result, most places near the Equator have warm climates, while the areas farthest from the Equator are cold.

Geographers use latitude, or distance from the Equator, to divide the world into zones: tropical, temperate, and polar. The tropical zones are low-latitude zones reaching 23½° north and south of the Equator. Most places in the tropics are hot year-round. The earth's two temperate zones are in the middle latitudes. They extend from 23½° N to 66½° N and from 23½° S to 66½° S. The temperate zones are generally cooler than the tropics and have a wide range of temperatures. The polar zones are in the high latitudes, from 66½° N and 66½° S to the poles. Because sunlight strikes here very indirectly, the sun's rays are spread out over a wide area. Polar climates are always cool or bitterly cold.

Distributing the Sun's Heat

Heat from the sun falls unevenly on different parts of the earth, but it does not all stay where it falls. If it did, the tropics would grow hotter each year and the polar regions colder. Instead, heat is distributed by a process called convection—the transfer of heat from one place to another.

Convection occurs because warm gases or liquids are lighter or less dense than cool gases and liquids. Therefore, warm gases or liquids tend to rise. Cooler, heavier gases and liquids sink and displace the lighter materials.

This process takes place in both air and water as well as in the earth's interior, as described in the previous section. Movements of the air are called winds; in the water movements are called currents. Warm air and warm water both flow from the

---

Applying Geographic Themes

Regions Which zone of latitude lies between the Tropic of Cancer and the Arctic Circle? Which zones lie between the Tropics of Cancer and Capricorn? Which prevailing winds arise in the polar zones?

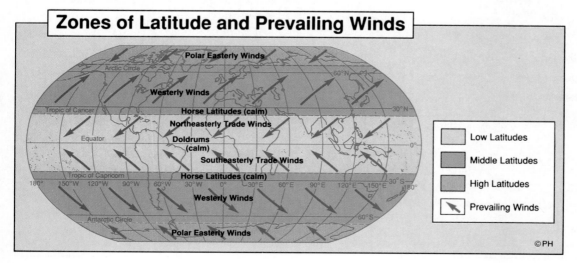

Zones of Latitude and Prevailing Winds

Polar Easterly Winds
Arctic Circle
60°N
Westerly Winds
Tropic of Cancer
30°N
Horse Latitudes (calm)
Northeasterly Trade Winds
Doldrums (calm)
Equator
0°
Southeasterly Trade Winds
Tropic of Capricorn
Horse Latitudes (calm)
180° 150°W 120°W 90°W 60°W 30°W 0° 30°E 60°E 90°E 120°E 150°E 180°
30°S
Westerly Winds
60°S
Antarctic Circle
Polar Easterly Winds

	Low Latitudes
	Middle Latitudes
	High Latitudes
	Prevailing Winds

©PH

Equator toward the poles. Cold air and cold water tend to move from the poles toward the tropics and the Equator. Within these broad movements are complex smaller patterns.

Wind Atmospheric pressure is the weight of the atmosphere overhead. Rising warm air creates areas of low pressure; falling cool air causes areas of high pressure. Winds move from areas of high pressure into areas of low pressure. The movement of winds worldwide redistributes the sun's heat over the earth's surface.

The pattern of winds begins when light, warm air rises from the Equator and flows northward and southward toward the Poles. At the same time, cold air from the polar regions sinks down and moves toward the Equator.

If the world were standing still, these winds would blow in a straight line. But the earth is rotating, and its movement deflects, or bends, them. This deflection is called the Coriolis effect. In the Northern Hemisphere the path of the winds curves to the right, while in the Southern Hemisphere it curves left. The diagram on page 33 shows these patterns.

Wind Patterns In each latitude zone, temperature and pressure combine to create a pattern of dominant, or prevailing, winds. At the Equator, the rising warm air causes calm weather or very light, variable breezes. The doldrums at the Equator are a region of light winds. Two other regions of light and unpredictable winds are at about 30° North and South latitudes, where

WORD ORIGIN

doldrums
When wind and sails were the only source of power for ships, sailors were often stranded near the Equator without enough wind to move. The area became known as the *doldrums* from an old English word meaning "slow and dull."

Applying Geographic Themes

Movement Warm and cold ocean currents, like the winds, help redistribute heat from the regions near the Equator to the polar regions. Which ocean current moves north along South America's west coast?

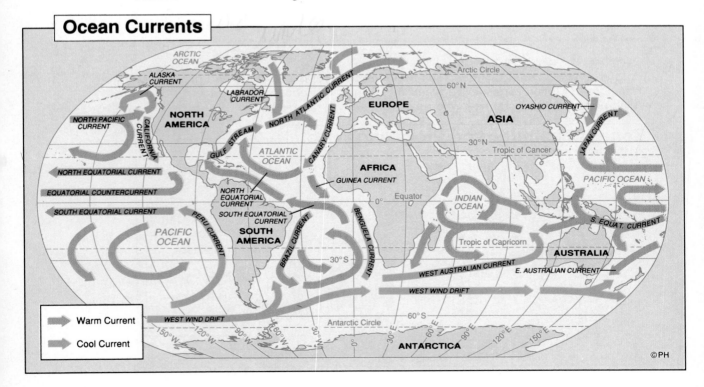

Ocean Currents

DAILY LIFE

Living in a Harsh Climate

People who inhabit difficult environments often must take extraordinary steps to adapt to their harsh living conditions. A small town in South Australia offers a dramatic example of this process of adaptation.

Coober Pedy, a mining community, provides about 90 percent of the world's opals. For four months of the year the average temperature in Coober Pedy is 97°F (36.1°C). In order to escape the heat, residents have created a unique solution: underground homes. They have dug caves in the same hillsides where they work mining for opals.

On the inside, the underground homes at Coober Pedy, like the one shown at right, resemble those you might see anywhere. But they have one great advantage: they are cool.

1. What geographic features characterize Coober Pedy?
2. Why are cave homes cooler than homes above the ground?

---

cool air sinks toward earth. Sailing ships had trouble getting enough wind to travel in these "horse latitudes." Supposedly, this nickname arose when Spaniards sailing to the Americas threw their horses overboard in order to lighten their ships and move faster.

Between the "horse latitudes" and the Equator, the trade winds blow steadily toward the Equator from the northeast and southeast. The "northeast trades" were the winds that carried the first European explorers to the Americas. These winds got their name because they were dependable and useful to merchant trading ships.

On their return to Europe from the Americas, explorers and merchants sailed with the help of the prevailing westerlies. They usually blow from the west, because the Coriolis effect deflects them from their straight path toward the poles.

Currents The waters of the oceans also help to distribute heat. Following similar convection patterns to those of winds, ocean currents carry warm water away from the tropics toward the poles and bring cold water back toward the Equator. The circular patterns of currents in the oceans are influenced by both wind and the Coriolis effect. The map on page 34 shows the major warm and cold ocean currents.

Precipitation

Precipitation refers to all of the forms of water that fall to the earth from the atmosphere, including rain and snow. The amount of precipitation that falls in an area is important in shaping its climate. So is the amount of water vapor normally in the air— the humidity.

Air is a mixture of gases, particles, and moisture in the form of water vapor. Water vapor enters the air when water evaporates from lakes, oceans, and other bodies of water. This process is part of the water cycle, or hydrologic cycle.

Precipitation depends largely on the temperature of the air. Because warm air is less dense, it can carry more moisture than cool air. But when moist, warm air begins to cool, it can no longer hold as much water vapor. The water vapor in the air condenses —it turns back into liquid water. Tiny droplets of water gather together to form clouds. If more water collects than can gather in the clouds, the excess falls as precipitation. Whether the precipitation is rain, snow, sleet, or hail depends on the air temperature and winds.

Convectional Precipitation Geographers divide precipitation into three types, depending on what causes it. One type occurs when hot, humid air rises from the earth's surface and cools, losing its ability to hold as much water. As part of the convection process, this kind of precipitation is known as convectional precipitation. Convectional precipitation is common near the Equator and in the tropics, where the surface air is generally hot and humid. As the air rises and cools, it produces the heavy rainfalls that nourish lush tropical forests.

Orographic Precipitation Sometimes warm, moist air is forced upward when it crosses over higher landforms. This effect—called orographic precipitation—is common on seacoasts, where moist winds blow from the ocean toward coastal mountains. As the warm winds are forced upward and over the mountains, they cool. Clouds form and rain or snow falls. By the time the air reaches the other side of the mountains, it is cool and dry. For example, in the Cascade Range on the Pacific coast of the United States, winds from the Pacific Ocean deposit abundant moisture on the windward slopes, which face into the wind. These areas are thickly forested, and the weather is often foggy and rainy.

But the land on the leeward side of the mountains—away from the wind —lies in a rain shadow. Not only does the air lose its moisture crossing the mountains, but it now drops to a lower elevation and warms up again. This dry, hot air often causes land behind coastal mountains to be very dry. California's Mohave Desert, for example, lies inland behind the Sierra Nevada.

Frontal Precipitation This type of precipitation usually occurs when two fronts, or air masses, of different temperatures meet. The warmer air is forced upward by the heavier, cool air. As the warm air rises and cools, frontal precipitation forms. The diagrams on page 37 illustrate the processes causing each of the three types of precipitation.

Other Influences on Climate

While temperature and precipitation are the major elements in weather and climate, other factors influence specific areas. These include distance from large bodies of water, elevation, and location in relation to nearby landforms.

Distance from Large Bodies of Water
Land and water absorb and store heat at very different rates. Land temperatures can go up and down by many degrees even within a few hours. From season to season the change is even more dramatic. In parts of Siberia on the Asian mainland, for example, land temperatures can vary by as much as 100°F (56°C) from summer to winter. By contrast, water temperatures change much more slowly. The average temperature at the ocean surface varies less than 10°F (5°C) throughout the year.

Because of this difference, large bodies of water—oceans or large lakes —affect the climates of their surrounding areas. Winds that blow over water take on the temperature of the water beneath them. As these winds blow onshore they moderate the temperature on land. For this reason, such areas often have milder climates than areas at the same latitude far from a large body of water.

Several specific types of climate are found in coastal areas. For example, some middle-latitude areas on the west coasts of continents have a mild, humid, marine climate. The prevailing westerlies bring in a steady supply of warm, moist ocean air. Marine climates are found on the Pacific coast of North America, in southern Chile, and in southeastern Australia.

The British Isles and the countries of Western Europe also have a marine climate. Although these countries are

Diagram Skill

The three major types of precipitation are shown on the right. Use the diagram to describe convectional precipitation in your own words.

Convectional Precipitation

Warm air

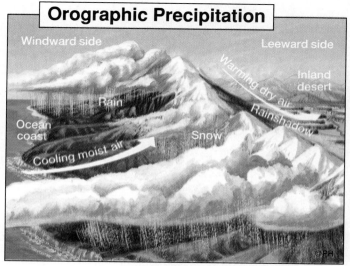

Orographic Precipitation

Windward side
Leeward side
Inland desert
Warming dry air
Rain
Rainshadow
Ocean coast
Snow
Cooling moist air

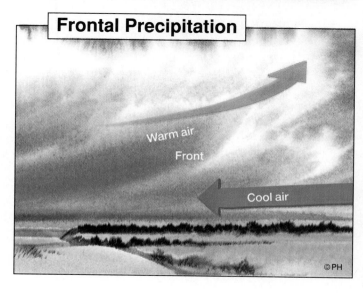

Frontal Precipitation

Warm air
Front
Cool air

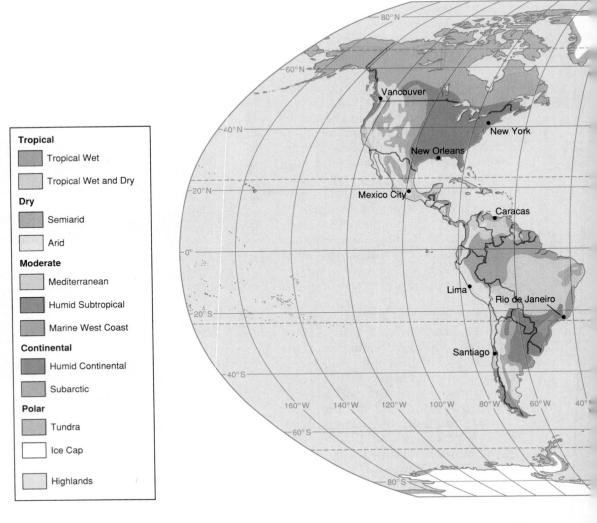

Tropical

- Tropical Wet
- Tropical Wet and Dry

Dry

- Semiarid
- Arid

Moderate

- Mediterranean
- Humid Subtropical
- Marine West Coast

Continental

- Humid Continental
- Subarctic

Polar

- Tundra
- Ice Cap
- Highlands

located relatively far north, the winds that blow onshore from the Atlantic have been warmed by a branch of the warm-water current known as the Gulf Stream.

Away from the moderating influence of the oceans, the great central areas of continents in the Northern Hemisphere have what are known as **continental climates.** Most areas with continental climates have cold, snowy winters and warm or hot summers. Humidity and precipitation vary, and temperatures often reach extremes of hot and cold. Central Europe, Northern Eurasia, parts of China, and much of North America have continental climates.

Elevation Although it is located almost on the Equator in Tanzania, Africa, Mount Kilimanjaro is capped with snow all year-round. The peak's elevation of 19,340 feet (5,895 m) above sea level affects its climate much more than does its location in the tropics. Elevation has a dramatic effect on climate in highland areas throughout the world, regardless of the latitude of the particular area.

Air temperature decreases at a rate of about 3.5°F (2°C) for every 1,000 feet (305 m) in height or elevation. For this reason hikers must use caution. They can leave the base of a mountain in hot weather and face snow at the peak.

The World: Climate Regions

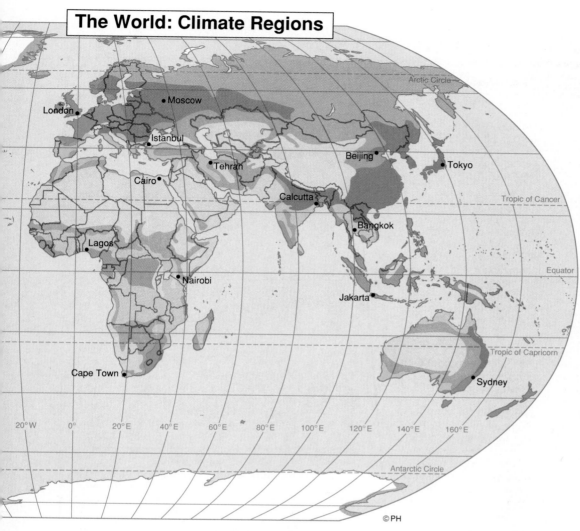

Applying Geographic Themes

1. Regions   The borders between climate regions are usually gradual, not abrupt as shown. Which climate occurs frequently along the Equator?
2. Place   Which of the continents have large areas of arid and semiarid climates?

Landforms   Coastal mountains are not the only landforms that can affect climate. Inland mountains, large desert areas, lakes, forests, and other natural features can influence climate nearby. Even a concentration of tall buildings can influence climate in the surrounding area. Such small variations within a region are called microclimates.

World Climate Regions

Over the years, geographers have not always agreed on how to categorize or define the world's many climate regions. In most systems of classification, temperature and precipitation are the main factors. However, classifying climates is not always easy because climatic conditions change. The

World Climate Regions

Climate Type	Temperature	Precipitation
TROPICAL		
Tropical Wet	Hot all year (avg.): 79°F (26°C)	Yearly: 100 in. (254 cm) Monthly (avg.): 8.3 in. (21 cm)
Tropical Wet and Dry	Hot all year 79°F (26°C)	Yearly: 50 in. (127 cm) Summer: 10.2 in. (26 cm) Winter: 0.5 in. (1 cm)
DRY		
Semiarid	Hot summers, mild to cold winters Summer (avg.): 78°F (26°C) Winter (avg.): 51°F (11°C)	Yearly: 18 in. (46 cm) Monthly (avg.): Summer: 3.4 in. (7 cm) Winter: 0.2 in.(0.5 cm)
Arid	Hot days, cold nights Summer (avg.): 81°F (27°C) Winter (avg.): 55°F (13°C)	Yearly: 5 in. (13 cm) Monthly (avg.): Summer: 0.6 in. (1.5 cm) Winter: 0.1 in. (0.2 cm)
MODERATE		
Mediterranean	Hot summers, cool winters Summer (avg.): 72°F (22°C) Winter (avg.): 52°F (11°C)	Yearly: 23 in. (58 cm) Monthly (avg.): Summer: 0.4 in. (1 cm) Winter: 3.8 in. (10 cm)
Humid Subtropical	Hot summers, cool winters Summer (avg.): 77°F (25°C) Winter (avg.): 47°F (8°C)	Yearly: 50 in. (127 cm) Monthly (avg.): Summer: 6.2 in. (16 cm) Winter: 2.8 in. (7 cm)
Marine West Coast	Warm summers, cool winters Summer (avg.) 60°F (16°C) Winter (avg.): 42°F (6°C)	Yearly: 45 in. (114 cm) Monthly (avg.): Summer: 2.5 in. (6 cm) Winter: 5.5 in. (14 cm)
CONTINENTAL		
Humid Continental	Warm summers, cold winters Summer (avg.): 66°F (19°C) Winter (avg.): 21°F (−6°C)	Yearly: 27 in. (69 cm) Monthly (avg.): Summer: 3.2 in. (8 cm) Winter: 1.6 in. (4 cm)
Subarctic	Cool summers, very cold winters Summer (avg.): 56°F (13°C) Winter (avg.): −8°F (−22°C)	Yearly: 17 in. (43 cm) Monthly (avg.): Summer: 1.8 in. (5 cm) Winter: 1.2 in. (3 cm)
POLAR		
Tundra	Cold summers, very cold winters Summer (avg.): 40°F (4°C) Winter (avg.): 0°F (−18°C)	Yearly: 16 in. (41 cm) Monthly (avg.): Summer: 1.9 in. (5 cm) Winter: 1.2 in. (3 cm)
ICE CAP	Cold all year Summer (avg.): 32°F (0°C) Winter (avg.): −14°F (−25°C)	Yearly: 8 in. (20 cm) Monthly (avg.): Summer: 1.0 in. (2 cm) Winter: 0.4 in. (1 cm)
HIGHLAND	Varies depending on elevation	Yearly: Ranges from 3 in. (8 cm) to 123 in. (312 cm)

absence of accurate data on weather in many regions of the world also makes classification difficult.

Nevertheless, most geographers now generally follow a system devised in the early 1900s by Wladimir Koppen (VLAD uh meer KEPP pen), a German biologist. This system set up six broad types of climate regions: tropical, dry, moderate, continental, polar, and highland. Most of these general climate types also have specific subdivisions.

The map on pages 38–39 shows the world's climate regions. Regional divisions on a map do not, of course, mean that the climate changes abruptly at that place. The lines on maps usually mark areas of transition where one climate region merges with another.

Changing Climates

While some climate changes occur through changes in nature, more and more are caused by human actions.

Many historians and scientists think that changes in climate may have ended some civilizations or brought drastic changes in others. For example, the Vikings first settled on the shores of Greenland and Iceland around the tenth century, when those regions were somewhat warmer. But in the mid-1500s the earth moved into what is now called the Little Ice Age. The cooler climate hurt agriculture and brought a decline in population. The Greenland settlements were gradually abandoned. Today most of Greenland is covered by an ice sheet.

Over hundreds of thousands of years world climates have changed—sometimes in small areas, sometimes over much of the earth.

Chart Skill

On average, how many inches of rain fall annually in a tropical wet climate zone?

The Influence of Climate in Niger
The thick, windowless walls of the buildings in this village keep people cool in a hot climate.

SECTION **3** REVIEW

Developing Vocabulary
1. Define: **a.** weather **b.** atmosphere **c.** climate **d.** rotation **e.** revolution **f.** solstice **g.** equinox **h.** tropical zones **i.** temperate zones **j.** polar zones **k.** Coriolis effect **l.** precipitation **m.** humidity **n.** windward **o.** leeward **p.** rain shadow **q.** continental climate

Place Location
2. Locate one area on the world climate map that has a marine west coast climate.

Reviewing Main Ideas
3. How do winds help distribute the heat that comes from the sun?
4. What are ocean currents?

Critical Thinking
5. **Perceiving Cause-Effect Relationships** Why do coastal mountains often have foggy, rainy climates?

✓ Skills Check

- ☑ Social Studies
- ☐ Map and Globe
- ☐ Reading and Writing
- ☐ Critical Thinking

Using Climate Graphs

In most climates, temperatures slide up or down, and precipitation rises and falls depending on the season. Climate graphs usually show two kinds of basic information: average temperatures along one vertical scale and average precipitation along a second vertical scale. A time scale across the bottom of the graph indicates the months of the year by their initial letter.

The climate graphs below are for the cities of São Paulo, and Beijing. Use the following steps to analyze the information in the graphs and compare climates in those cities.

1. **Determine the annual variation in temperature for each city.** Temperature is shown by the curved line at the top of the graphs. To find the average temperature for any month, locate the point on the temperature line directly above the name of the month. Then look at the markings for "degrees" on the left-hand scale. Answer the following questions: (a) Which city has the largest annual range of temperatures—that is, the greatest extremes of hot and cold? (b) Which has the most even temperatures year-round?

2. **Determine the annual variation in precipitation for each city.** Precipitation is shown by the bar graph at the bottom of each graph. It is measured in inches on the right-hand vertical scale. To read the graph, measure the height of the bar for any given month against that scale. (a) What is the rainiest month in Beijing? (b) About how many inches of rain fall in that month? (c) Which city has the most even precipitation throughout the year?

3. **Compare the values for temperature and precipitation to describe each climate.** Climate graphs show temperature and precipitation side by side because these are the two major factors that determine a climate. Answer the following questions: (a) Which city has the most variable climate—with both wet and dry seasons and great changes in temperature? (b) How would you describe São Paulo's climate?

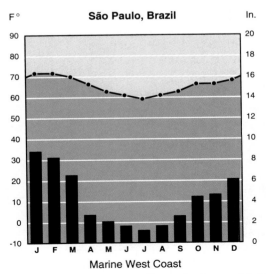

São Paulo, Brazil — Marine West Coast

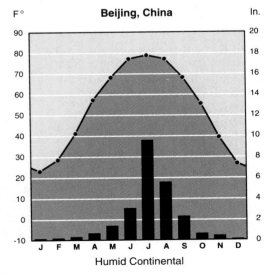

Beijing, China — Humid Continental

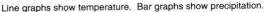

Line graphs show temperature. Bar graphs show precipitation.

4 Vegetation Regions

Section Preview

Key Ideas

- In similar environments, similar groups of plants grow together to form a plant community.
- Geographers classify vegetation regions based on different types of forests, grasslands, desert plants, and tundra.
- Plants in every region adapt to the conditions in their environment.

Key Terms

plant community, environment, biome, natural vegetation, deciduous, coniferous, chaparral, savanna, tundra, permafrost

Except for the polar ice caps and the most barren spots of the deserts, plants grow everywhere on earth. They range from microscopic one-celled plants to gigantic redwood trees. Plants are the oldest and most basic form of life on earth. Green plants in particular are important, because they supply food for humans and animals and help in recycling the earth's water supply.

Plant Communities

Plants seldom live alone. Instead, groups of plants tend to be interdependent—that is, they depend on one another for such things as shade, support, and even nourishment. In the wild, for example, trees provide certain vines with both food and a place to grow.

The mix of interdependent plants that naturally grow in one place is called a plant community. Such a natural grouping consists of plants that can survive successfully in a particular environment—the physical conditions of the natural surroundings. Climate, sunlight, temperature, precipitation, elevation, soil, and landforms are all part of the plant environment. Scientists have found that wherever similar environments exist, the same types of plants ultimately group together to form a plant community.

Biomes and Vegetation Regions The term biome (BY om) is used to describe a region in which the environment, plants, and animal life are suited to one another. For example, in any temperate forest biome, you are likely to find moderate temperatures and rainfall, oak or maple trees, and animals such as deer, squirrels, raccoons, and owls.

Geographers also classify regions by their natural vegetation, or the typical plant life that abounds in areas where humans have not altered the landscape significantly. The map on pages 44–45 shows major vegetation regions of the world. Because these regions have similar plant life, they are often similar in climate, soil, and other characteristics on which plants depend. Vegetation regions are of four general types: forest, grassland, desert, and tundra.

Forest Regions

The word *forest* probably makes you think of whatever kind of forest is most familiar to you—giant groves of redwoods, thick clumps of oaks and maples, hillsides covered with fragrant pines. All these forests exist in North America. In different parts of the world there are other types of forest regions.

Tropical Rain Forest In areas near the Equator, where the temperature is warm and great amounts of rain fall, thick tropical rain forests grow. The largest are in the Amazon River basin in South America and the Congo River basin in central Africa. Within the rain

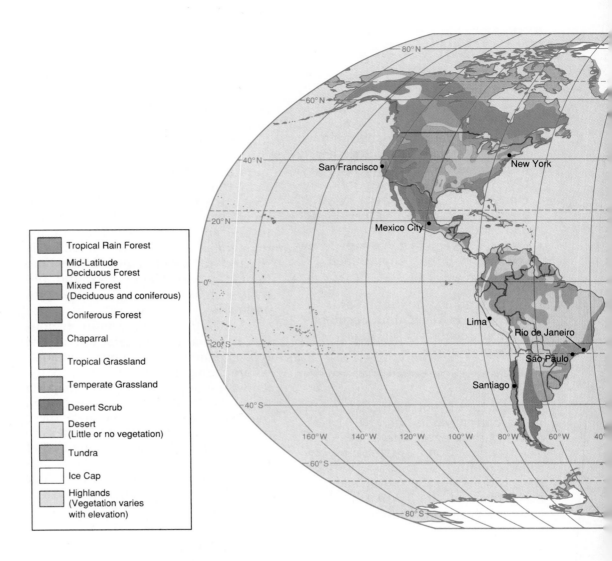

Tropical Rain Forest

Mid-Latitude
Deciduous Forest

Mixed Forest
(Deciduous and coniferous)

Coniferous Forest

Chaparral

Tropical Grassland

Temperate Grassland

Desert Scrub

Desert
(Little or no vegetation)

Tundra

Ice Cap

Highlands
(Vegetation varies
with elevation)

forest, tall trees stretch skyward as they seek sunlight. Their foliage forms a dense layer of leaves that almost entirely blocks out the sun's rays.

Mid-Latitude Forest The trees in the tropical rain forest are broadleaf evergreens, which keep their leaves year-round. By contrast, the dominant trees in the forests of the middle latitudes are deciduous. That is, they shed their leaves when winter approaches. In some parts of the world, the broad leaves of these trees—such as oaks, birch, and maple—turn brilliant colors before they fall.

Broadleaf deciduous forests once covered much of Europe, eastern North America, and eastern Asia. These lands in the middle latitudes have a temperate climate with adequate rainfall, warm summers, and cool or cold winters.

Coniferous Forest In the colder parts of the middle latitudes grow several kinds of trees that can survive long, cold winters. Pines, spruce, fir, and their relatives have long, thin "needles" rather than broad, flat leaves. Needle leaves expose only a small surface to the cold and so can remain

The World: Natural Vegetation

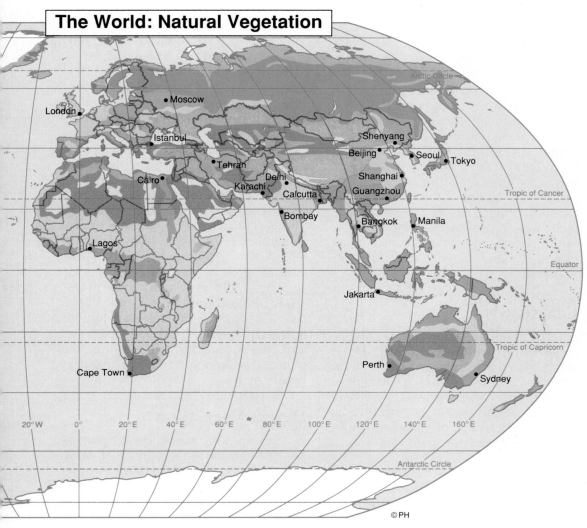

Applying Geographic Themes

Regions People have always adapted to their environment, but they have also learned to change it. This map shows the world's vegetation as it would occur if it were left untouched by people. Which two continents have large areas of tundra?

on the tree in winter without freezing. As a result, these needleleaf trees are "evergreen," using whatever sunlight is available throughout the year.

The northern forests are sometimes called needleleaf forests, but they are better known as **coniferous**, for the cones that carry and protect their seeds. Coniferous forests cover huge areas stretching across northern North America, Europe, and Asia.

Other Forest Types In many places, forest regions overlap. In addition, small areas of the world have unique forest vegetation. A mixed region has coniferous and broadleaf deciduous trees growing together. Clusters of such forests are common in many places, including the northern United States. They grow in cool parts of the middle latitudes or at high elevations, where winters are mild or very cold.

Another distinctive forest type is chaparral, which includes small evergreen trees and low bushes, or scrub. Chaparral is a Spanish word for "an area of small evergreen oak trees." This vegetation is uniquely adapted to a Mediterranean climate, where most of the precipitation falls during the winter and summers are hot and dry. Many chaparral plants have leathery leaves to hold moisture over the dry summer. Regions of chaparral are found on the coasts of the Mediterranean Sea, southern California, Chile, South Africa, and Australia, as well as in a few inland areas of the American Southwest.

Grasslands

The central regions of several continents are covered by grasslands. At the edges, grasslands and forest often mix. In addition, scattered clumps of trees often grow on grasslands where there is enough water. As with forests, the characteristics of grasslands depend a great deal on their latitude.

Tropical Grasslands   Huge tropical grasslands, or savannas, grow in the warm lands nearest the Equator. Savanna grasses thrive in tropical climates. Scattered trees and other plants that can survive the dry periods of tropical wet and dry climates dot the savanna. Although fire is usually considered destructive, it is necessary to the savanna. Fires that break out in the dry season encourage new grasses to grow and maintain the savanna.

Temperate Grasslands   The grasslands in cooler parts of the world are known by several names. They differ in the length and kinds of grasses, which in turn vary with the amount of rainfall and soil types.

The temperate grasslands of North America are called prairies. Because rainfall decreases toward the west, the prairie grasses also are different. In the

east, which gets as much as 40 inches (100 cm) of rain a year, "tallgrass" prairie once grew. This true prairie has tall grasses dotted with wildflowers. Grasses are shorter in the central prairies and in the dry Great Plains. Before settlers learned how to farm the Great Plains, mapmakers labeled this area "The Great American Desert."

In most parts of the prairies, trees and shrubs grow along the banks of rivers and streams. Though prairie grasses once covered the American Midwest, little of the natural prairie vegetation is now left. The grassland region, though, provided fertile farmland for growing grain.

The cool, dry, temperate grasslands of Northern Eurasia and central Asia—called the steppes—are similar to the Great Plains. Other well-known and productive grasslands are the pampas of Argentina and the veld of South Africa.

Desert Vegetation

Despite their name, desert regions are not just barren expanses of sand. Many plants have adapted in different ways to life with almost no water. Cactus plants, for example, store water in thick stems. Their leaves are prickly needles, which protect them and their water supply from animals. Other desert plants have small leaves, which lose little moisture into the air through evaporation. Still others have seeds that can survive for many years until there is enough water for them to sprout.

Roots also help plants find and store water in desert regions. On valley floors and in dry streambeds, long roots can reach deep into the ground to seek water. Other species of plants have long, shallow roots that stretch out over a wide area to gather the water that falls during the brief rainfalls. Desert plants grow widely scattered, for the water in one area cannot support much vegetation.

WORD ORIGIN

prairie
The word *prairie* is used to describe the temperate grasslands of North America. It is derived from the French word meaning "meadow."

Tundra

In **tundra** regions, where temperatures are always cool or cold, only specialized plants can grow. One type of tundra exists in high mountains. Elevation and winds keep the temperature low even when it is sunny. No trees grow at the high elevations of alpine tundra. Small plants and wildflowers grow in sheltered spots. Tiny, brightly colored plants called lichens (LYE kenz) make patterns on the rocks.

In the arctic tundra, plants must also be able to live in cold temperatures and short growing seasons. In addition, they must go without sunlight for most of the winter. Arctic landscapes are treeless, covered with grasses, mosses, lichens, and some flowering plants. In parts of the tundra, a layer of soil just below the surface, known as **permafrost**, stays permanently frozen. The soil above this layer of permafrost is often soggy and waterlogged.

Although the tundra sounds like a bleak place, many people find it beautiful. One naturalist wrote this description of the Alaskan tundra in late June:

> *When we climbed the higher, drier [river] banks we looked over an eternal expanse of green and brown. It was a glorious garden of arctic plants, this summer tundra-delta, and stiff with northern birds, so that never for a moment were we out of sight or hearing of crane, goose, duck, or wader.*

In some extreme climate regions of the world, vegetation is rare. Some highland areas, for instance, are almost bare of plants. The polar ice caps and the great ice sheets of Greenland and Antarctica are considered regions without vegetation. Even there, however, small and simple plants survive in some areas. As scientists explore more and more of the earth's extreme

Richardson Mountains, Yukon, Canada
Which special characteristic of tundra vegetation is apparent in this picture?

climate regions, they are finding that there are few places on earth with no plants at all.

SECTION 4 REVIEW

Developing Vocabulary
1. Define: **a.** plant community **b.** environment **c.** biome **d.** natural vegetation **e.** deciduous **f.** coniferous **g.** chaparral **h.** savanna **i.** tundra **j.** permafrost

Place Location
2. Find two areas where tropical rain forests are located.

Reviewing Main Ideas
3. What are the four major categories of vegetation?

Critical Thinking
4. **Formulating Questions** What questions would you have to ask in order to define a natural vegetation region?

2

REVIEW

Section Summaries

SECTION 1 The Changing Earth The earth is a changing planet. Its interior includes a core of hot metal, a mantle of solid rock, and a relatively thin rock crust. Internal forces alter the rock to create landforms on the earth's surface. These forces are volcanism, the movement of molten rock, and forces that fold, bend, and break the rock. Major landforms shaped by these forces are mountains, hills, plateaus, and plains. According to the theory of plate tectonics, the crust is made of huge, shifting plates. Now widely accepted, this theory explains many natural events, including volcanoes and earthquakes. It includes the theories of continental drift and sea-floor spreading.

SECTION 2   External Forces That Change the Earth's Surface Once they have been formed by interior forces, landforms and other surface features are changed and sometimes destroyed by external forces. Mechanical weathering breaks down rock physically into smaller and smaller pieces. Chemical weathering, caused mainly by water and carbon dioxide, alters the chemical composition of the rock. Both help in forming soil, which is necessary for plant life to grow. Erosion is another external force that alters the earth's surface. The major agents of erosion are water, wind, and glaciers.

SECTION 3   Weather and Climate Climate is the weather that prevails in an area over a long period of time. Climates are influenced by latitude, elevation, landforms, and other factors. The climates at different latitudes are greatly influenced by the amount and intensity of sunlight the areas receive and by the way wind and water redistribute the sun's heat throughout the world by the process of convection. Climate regions are classified mainly on the basis of seasonal temperatures and precipitation.

SECTION 4   Vegetation Regions In similar environments, similar groups of plants form a plant community. Both natural events and human actions, however, can change the environment. Geographers classify natural vegetation regions based on different types of forests, grasslands, desert plants, and tundra. Plants in every region adapt to the conditions in their environment.

Vocabulary Development

Use each key term in a sentence that shows the meaning of the term.

1. plate tectonics
2. convection
3. erosion
4. climate
5. atmosphere
6. biome

Main Ideas

1. How does the theory of plate tectonics explain the location of certain mountain ranges?
2. What role does convection play in influencing climate?
3. Why is weathering important to life on the earth?
4. How do wind and water alter the earth's surface?
5. What activity takes place at the Mid-Atlantic Ridge?
6. How does a coastal mountain range affect the surrounding climate?
7. Why is latitude an important factor in a region's climate?
8. What regions have the least plant life?

Critical Thinking

1. **Expressing Problems Clearly** Living near the boundaries of moving plates can be dangerous. Explain why this statement is true.
2. **Analyzing Information** Why are the earth's winds and ocean currents important to life on the planet?
3. **Predicting Consequences** How could an increase in the greenhouse effect change the way of life on the earth?

Practicing Skills

Using Climate Graphs Study the two climate graphs on page 42 and then answer the following questions.

1. Which city, if either, has a climate that most closely resembles the climate in your community?
2. Which of the two climates has a relatively equal amount of rainfall all year?
3. Which of the two climates clearly has a brief rainy season?

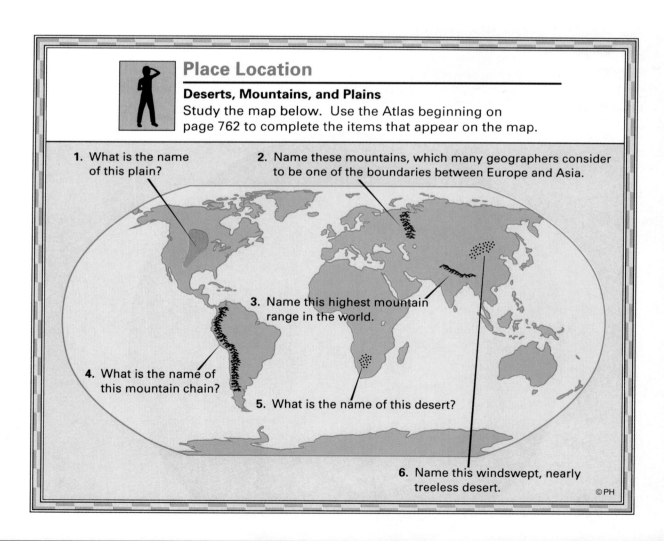

Place Location

Deserts, Mountains, and Plains
Study the map below. Use the Atlas beginning on page 762 to complete the items that appear on the map.

1. What is the name of this plain?
2. Name these mountains, which many geographers consider to be one of the boundaries between Europe and Asia.
3. Name this highest mountain range in the world.
4. What is the name of this mountain chain?
5. What is the name of this desert?
6. Name this windswept, nearly treeless desert.

©PH

3

Population and Culture

Chapter Preview

Both sections in this chapter describe how peoples' ways of life differ throughout the world.

Sections	Did You Know?
1 THE STUDY OF HUMAN GEOGRAPHY	The world's population has quintupled since about 1750, passing five billion in 1988.
2 WORLD POLITICAL AND ECONOMIC SYSTEMS	The Constitution of the United States is one of the oldest plans of government still in use today.

New York, New York

Adapting to the Environment
Extremely high population densities and lack of inexpensive or available land have encouraged many of Hong Kong's residents to look to the water as a place of residence.

1 The Study of Human Geography

Section Preview

Key Ideas

- Population is unevenly distributed throughout the world.
- World population has increased more rapidly during the twentieth century than at any other time.
- Every group of people develops a set of learned customs, beliefs, and actions that make up its culture.

Key Terms

demography, population density, birthrate, death rate, metropolitan area, urbanization, cultural trait, cultural landscape, diffusion, acculturation, culture hearth

For millions of years, humans have been changing the natural environment. At first people cut trees for weapons and firewood. Later they built dams on rivers, grazed sheep on wild grasses, and planted crops. All these human actions changed the earth.

Therefore, although geographers study the physical characteristics of the earth—its landforms, climate, and vegetation—they also are interested in *human,* or cultural, geography. Human geography covers a wide variety of topics such as the study of language, religion, and people's economic and political systems. A special field within cultural geography, **demography,** focuses on the study of populations, including their rates of birth, marriage, and death. Other social scientists study culture, the beliefs and actions that define the way of life of a particular group.

Population Distribution

Sometime in 1988, the world population passed five billion. Those five billion people, however, are very unevenly distributed throughout the world. Some areas have a high population density—the average number of people in a square mile or square kilometer. Other areas have only a few people.

Why do people live where they do? To begin with, less than a third of the earth's surface is land. And deserts, mountains, and bitterly cold climates make about half of that land area nearly uninhabitable. Consequently, almost everyone on earth lives on what amounts to a fairly small percentage of the earth's surface. Most people are concentrated where you would expect them to be—in places where the soil is reasonably fertile, water is reasonably plentiful, and the climate is mild enough to grow crops or raise animals.

People and Their Environments
The physical landscape has always influenced where humans live. People in many parts of the world live in what appear to Americans to be hostile or unappealing environments, such as Arctic or desert regions. Nevertheless, over the years people in each place have developed cultures, or ways of life, that are suited to their environments. At the same time they have changed these environments, sometimes gently, sometimes drastically.

Population Density Figures for average population density are usually calculated simply by dividing the total population by the land area of a country (or other region). The resulting numbers, however, can present a misleading picture. In Egypt, for instance, more than 90 percent of the land is desert. Consequently, nearly all Egyptians live in a narrow strip along the Nile River.

Chart Skills

By the year 2010, more than half the world's population will live in urban centers. Which three European cities were no longer among the ten most populous by 1985?

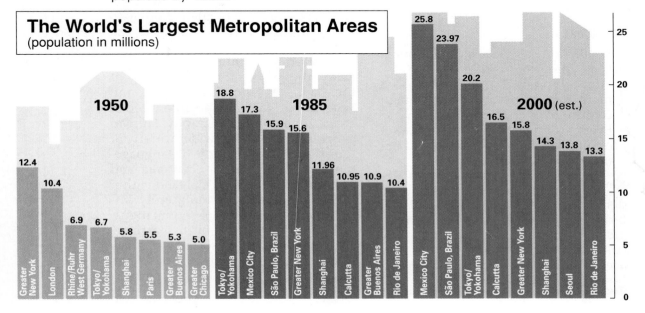

The World's Largest Metropolitan Areas (population in millions)

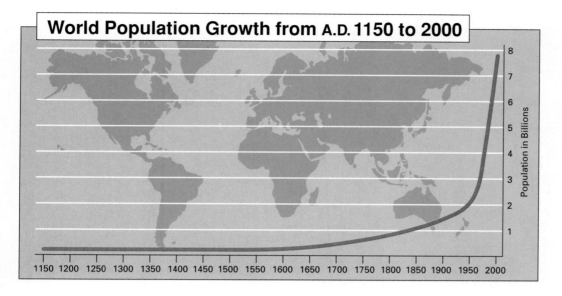

World Population Growth from A.D. 1150 to 2000

Population in Billions

1150 1200 1250 1300 1350 1400 1450 1500 1550 1600 1650 1700 1750 1800 1850 1900 1950 2000

Graph Skills
This graph show how the world's population has grown since A.D. 1150.
Around which date did world population begin to increase dramatically?

Some geographers, then, prefer to figure a country's population density in terms of its arable land—land that can be farmed—rather than its total land area. Egypt had an average population density in 1990 of about 142 people per square mile (54 per sq km). Measured in terms of arable land, the density was about 3,600 people per square mile (1,400 per sq km)!

Population Growth

In the last half of the twentieth century, the world's population has increased dramatically. The current yearly increase alone is equal to the population of Mexico—some 86 million people. This growth is a special problem for poorer countries and a challenge for the world as a whole.

A Historical Perspective The graph above depicts the growth in world population. While the most rapid growth has taken place since 1950, population began to increase rapidly midway through the eighteenth century. At that time there were fewer than one billion people on earth. In only two hundred years, world population *quintupled*, passing five billion in 1988.

Growth Rates Demographers think that the population boom will continue until the year 2100. Some trace a pattern of stages in population growth experienced by different countries. At each stage there is a different balance between the birthrate—the number of live births each year per 1,000 people—and the death rate—the number of deaths each year per 1,000. When the two rates are more or less equal, a country has reached what is called "zero population growth." This situation exists in some highly industrialized countries today.

In many countries, however, birthrates are still high, while death rates have fallen because of improved health. Population in these countries is growing by more than 4 percent a year, compared with the overall world growth rate of about 1.7 percent.

WORD ORIGIN

demography
The word *demography* comes from two Greek roots. *Demos* means "the people" in Greek, and *graphein* means "to write." Thus demography means "writing about the people," or "the study of population."

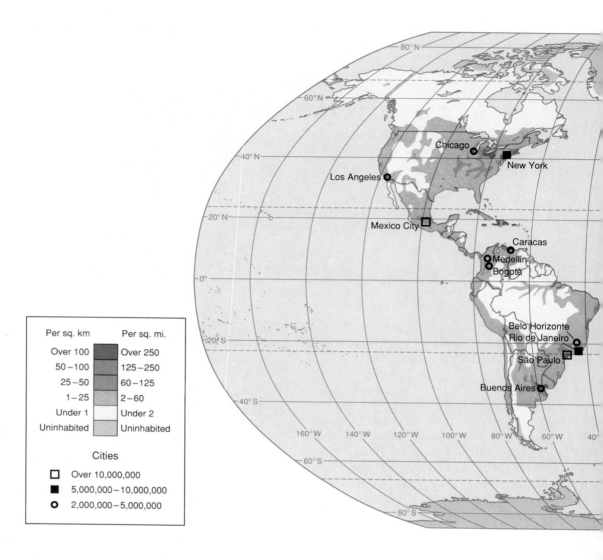

Per sq. km	Per sq. mi.
Over 100	Over 250
50–100	125–250
25–50	60–125
1–25	2–60
Under 1	Under 2
Uninhabited	Uninhabited

Cities

☐ Over 10,000,000
■ 5,000,000–10,000,000
○ 2,000,000–5,000,000

Patterns of Population

Differences in population growth as well as in environment have influenced where people live today. The map above dramatically illustrates the uneven distribution of world populations. The densest clusters, or concentrations of people, are in four regions: East Asia, South Asia, Europe, and eastern North America. Many of the people of these population clusters live in **metropolitan areas**—central cities surrounded by suburbs. Populations in Europe and North America today are already mostly urban. The trend toward **urbanization**—the growth of city populations—is significant throughout the world. In many countries, urban populations are growing twice as fast as rural populations. This rapid growth is a serious problem especially in many of the less industrialized countries, which are home to about three quarters of the world's population.

The Nature of Culture

Differences in population patterns are in part reflections of differences in culture. As children grow up in a given culture, they learn skills, language,

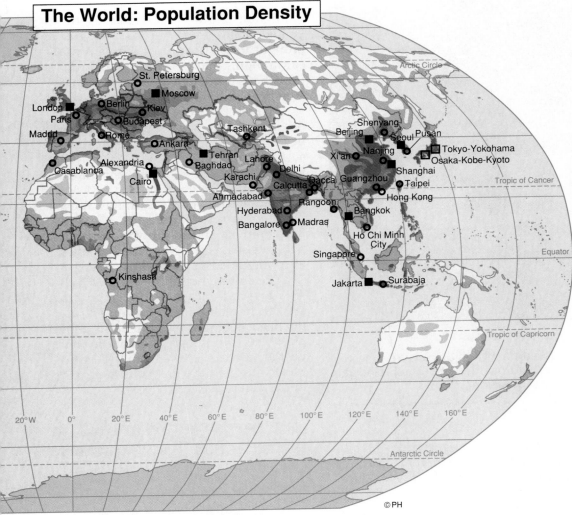

The World: Population Density

St. Petersburg
Moscow
London
Berlin
Kiev
Paris
Budapest
Madrid
Rome
Ankara
Tashkent
Shenyang
Beijing
Seoul
Pusan
Tehran
Lahore
Xi'an
Nanjing
Tokyo-Yokohama
Alexandria
Baghdad
Osaka-Kobe-Kyoto
Casablanca
Karachi
Delhi
Shanghai
Cairo
Calcutta
Dacca
Guangzhou
Taipei
Ahmadabad
Rangoon
Hong Kong
Hyderabad
Bangkok
Bangalore
Madras
Ho Chi Minh City
Singapore
Kinshasa
Jakarta
Surabaja

Arctic Circle
Tropic of Cancer
Equator
Tropic of Capricorn
Antarctic Circle

20°W   0°   20°E   40°E   60°E   80°E   100°E   120°E   140°E   160°E

©PH

Applying Geographic Themes

1. Regions   Where are some of the most densely populated regions of the world? Where are some of the least densely populated regions?
2. Place   Name three cities with populations of over three million.

eating customs, and thousands of other cultural traits. Later they pass these on to their own children. Over time, cultures change, but usually very slowly.

A culture is reflected in both ideas and objects, that is, *nonmaterial* and *material* culture. Nonmaterial culture includes religion, language, spiritual beliefs, superstitions, and ways of viewing or behaving toward the world. Some cultures, for example, value cooperation and group activities; others value individual achievement more. Material culture refers to all the things that people make or build with natural or human-made resources—from teacups to television sets. It includes styles of food, clothing, and architecture, as well as arts, crafts, and technology.

Cultural Landscapes   From the point of view of geography, technology is a significant part of culture. As human

The Spread of Culture
A Pacific Island family enjoys leisure time together. What does this photograph indicate about cultural diffusion?

beings use natural resources or alter the surface of the earth, they produce a cultural landscape. Nearly all landscapes on the earth have been changed somewhat by human activities. Cultural landscapes are also a reflection of differences between cultures. Farming areas in Kansas and in China, for example, look quite different.

Cultural Change Cultures often are changed by both internal and external influences. Within a culture, discoveries and inventions can bring change. Change from outside—from another culture—comes through diffusion, or the spread of cultural traits from one person or society to another. The process of accepting, borrowing, and exchanging traits between cultures is called acculturation

Diffusion often occurs through travel or migration, when people from one country settle in another. Each immigrant group that came to the United States from Europe and Asia brought its own culture. Through acculturation, many of these cultural traits have become part of mainstream American culture.

The term culture hearth is used to describe a place where important ideas begin and from which they spread to surrounding cultures. It is usually used for the areas where, in ancient times, major traits of human culture first developed. In the Middle East, for example, people first learned to tame and herd animals and to grow crops from wild grasses. Writing and mathematics also originated in this culture hearth.

The culture hearth for most of East Asia was ancient China. Its language, arts, technology, and government influenced neighboring lands and peoples. The cultures of Teotihuacán (tay uh tee wah KAHN) and the Maya in Mexico formed the first culture hearth in the Americas.

═══ SECTION **1** REVIEW ═══

Developing Vocabulary
1. Define: **a.** demography
   b. population density **c.** birthrate
   d. death rate **e.** metropolitan area
   f. urbanization **g.** cultural trait
   h. cultural landscape **i.** diffusion
   j. acculturation **k.** culture hearth

Place Location
2. Name two cities located in the most densely populated regions of East Asia.

Reviewing Main Ideas
3. What is zero population growth?
4. Where are the world's four areas of greatest population density?

Critical Thinking
5. **Testing Conclusions** Name three things that contribute to today's "global culture."

✔ Skills Check

☐ Social Studies
☑ Map and Globe
☐ Reading and Writing
☐ Critical Thinking

Reading a Population Density Map

Government officials, business leaders, and social scientists often need to know not only how many people live in an area but *where* they live. For example, where are new roads needed? Where will a business find the most customers? They can find some of the answers on a population density map. The map on this page shows the population density of North America. Use the following steps to help you analyze it.

1. **Know what information is on the map.** Make sure you know what area the map covers. Maps can show population density for small areas, such as a neighborhood or city, or for very large areas, such as a continent or the entire world.

   Next study the map key to find out the colors and codes that are used. Maps usually show density with shading. Other symbols may indicate the sizes of cities. (a) How many categories of population density does this map include? (b) What areas average fewer than 2 people per square mile?

2. **Study the data to draw conclusions.** A population density map can make a very dramatic first impression. It is obvious immediately that population is very unevenly distributed. In this case you can also see at a glance the overall patterns of human settlement in North America. With these first impressions in mind, look at the map in more detail. (a) What areas of North America appear to be quite uninhabited? (b) What different physical factors do you think might keep people from settling there? (c) Name four cities in eastern North America that have populations of more than one million people.

3. **Analyze population density patterns.** Move from your general impressions to think about the data available on the map. For instance, think geographically as you consider the areas of highest population density on this map. How would a geographer explain why human settlements are concentrated here? (a) What region of the United States has the greatest concentration of people? (b) What factors contribute to the population density of this region? (c) Where are densities lighter? Why?

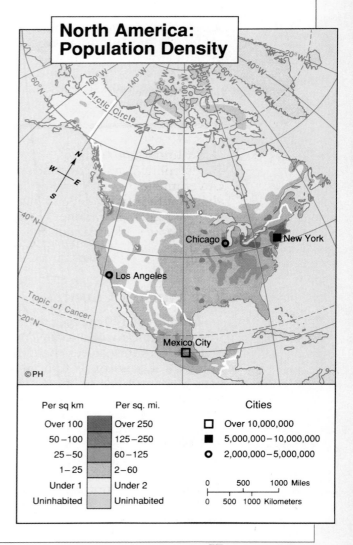

North America: Population Density

Per sq km	Per sq. mi.
Over 100	Over 250
50–100	125–250
25–50	60–125
1–25	2–60
Under 1	Under 2
Uninhabited	Uninhabited

Cities
☐ Over 10,000,000
■ 5,000,000–10,000,000
○ 2,000,000–5,000,000

0   500   1000 Miles
0   500 1000 Kilometers

©PH

2 World Political and Economic Systems

Section Preview

Key Ideas

- The characteristics of a country are territory, population, sovereignty, and government.
- Governments differ in their structure and in the source of their political power.
- Any country's economic system must try to answer three basic economic questions.

Key Terms

sovereignty, unitary, federal, confederation, authoritarian, dictatorship, totalitarianism, monarchy, constitutional monarchy, democracy, capitalist, market economy, communism, socialism

Two important traits of any culture are its political and economic systems. Governments usually reflect beliefs about authority, independence, and human rights. Economic systems reflect people's ideas about the use of resources and the distribution of wealth.

The World's Countries

More than 170 independent countries exist in the world today. Yet each one has certain features that help define it as a country. For social scientists, a country is a political unit with four specific characteristics. These are (1) clearly defined territory; (2) population; (3) sovereignty —the freedom and power to decide on policies and actions; and (4) a government.

Territory A country's territory includes the land and water within its national boundaries as well as all of its natural resources. The territories of modern countries range anywhere in size from Russia, which has more than 6.5 million square miles (16.8 million sq km) of land, to tiny Monaco, with less than 1 square mile (1.9 sq km). While a country's territorial size can contribute to its power and prestige, its resources may be even more important. For example, the territories of the small countries in the Persian Gulf region are mostly barren desert. Yet these countries are wealthy because of their huge reserves of oil. Throughout history, disputes over territory—land and resources—have often been the causes of war.

Population Modern countries also vary widely in both the size and the makeup of their populations. Some, such as India and the United States, contain a wide diversity of people and cultures. In other countries, such as Sweden or Greece, most people share a similar background, language, and culture.

The people of any nation are considered its citizens. This legal status usually assures them of protection by their government. In return, a citizen usually must pay taxes, serve in the military, or carry out other obligations to his or her country.

Sovereignty A sovereign country establishes its own policies and determines its own course of action. A country's sovereignty entitles it to act independently, deal equally with other sovereign countries, and protect its territory and citizens.

Governments and Political Systems

Despite the large number of countries in the world, there are only a few political systems. A country's government can be classified on the basis of its structure, or according to the source of its power and authority.

Managing the Rain

Local governments—whether formal or informal—often make important economic decisions about how to use and protect resources. Such is the case in the Indonesian island of Bali shown at right. There, toward the end of each rainy season, rice farmers give thanks for the fresh water that sustains their terraced fields. The farmers know their lives depend on this precious resource that spills from the skies.

The resource has been carefully managed for the past thousand years. Rice farmers in Bali are organized into hundreds of village-sized farming co-operatives called *subaks*. If a subak wants to tap a new spring, or divert water from a canal into its own fields, it must first consult a local government official. The official weighs the consequences of the move before giving a decision.

Recent studies have shown that the Balinese water management system has succeeded better than any modern approach in conserving water, controlling the spread of pests, and preserving soil quality.

1. Why is water precious in Bali?
2. How is water usage controlled?

Government Structure Nearly every country contains smaller geographic and political units. These may be called states, provinces, prefectures, even republics. One way of classifying governments is based on the relationship between these smaller units and the national government.

If one central government runs the nation, the system is said to be **unitary**. This one central government makes laws for the entire nation and gives local governments only limited power and authority. Great Britain and Japan are among the countries that have unitary governments.

Election Day in Santiago, Chile
Chileans ousted a military ruler in favor of a freely elected president in 1989.

In the United States, on the other hand, the national government shares power with the fifty state governments. This federal system gives the national government certain powers and reserves others for the states.

The third type of government structure is confederation. According to this system, smaller political units keep their sovereignty and give the central government very limited powers.

Government Authority Another way of defining a government is according to where—or from whom—it derives its power. Until fairly modern times, most countries had authoritarian governments. The leaders held all, or nearly all, the political power.

Authoritarian governments take different forms. Today the most common is the dictatorship. Most dictators gain and keep power by military force or political terror. People are not free to express opinions.

The most extreme form of dictatorship is totalitarianism. In countries under totalitarian rule, the leaders try to control every part of society—politics, the economy, and people's personal lives. Nazi Germany under Hitler and the Soviet Union under Stalin are both examples of totalitarian rule.

Throughout much of history, the most common kind of authoritarian government was a monarchy. The monarchs—kings, queens, pharaohs, shahs, sultans—inherit their positions by virtue of being born into the ruling family. In the past, monarchs often ruled with complete and dictatorial power.

Nearly all monarchies today are constitutional monarchies. Their rulers both lead the nation and play an important symbolic role. Real political power, however, rests with an elected lawmaking body—a legislature or parliament. The nation is governed according to its laws and constitution, not the monarch's will.

Any country in which the people choose their leaders and have the power to determine government policy is a democracy. (A constitutional monarchy can also be a democracy.) Today's democratic countries are *representative* democracies. All of the nation's eligible adult citizens have the right to choose the people who will represent them in making the country's laws. In most representative democracies, the elected legislature or parliament not only makes the laws but also sees that they are carried out.

World Economic Systems

Just as a sovereign country can choose its form of government, it also can develop its own economic system. But no matter what type of economic system a government follows, it will have to answer three basic economic questions for the society: What (and how many) goods and services will be produced? How will these products be produced? How will the products and the wealth gained from their sale be distributed?

WORD ORIGIN

totalitarianism
The root of the term *totalitarianism* is the Latin word *totus,* meaning "whole." The term came into widespread use to describe the European dictatorships of the mid-1930s.

The main difference between modern economic systems is the degree of government participation in answering these economic questions.

Capitalism   In a capitalist system, the people answer the three basic economic questions. People, as consumers, help determine what will be produced by buying or not buying certain products. In response, producers try to make more of the products people want. Because all these decisions are made in a free market, a capitalist system is also called a market economy.

According to "pure" capitalism, government takes no part in the economy. In fact, in the United States and in other capitalist countries, governments do play economic roles. For instance, they offer some goods and services such as a postal service, highways, and public education. Government also plays a limited role in regulating businesses to protect the public. Regulations affect areas such as safety, energy conservation, and environmental pollution.

Communism   In contrast to the notion of pure capitalism, in which the government takes no part in the economy, communism requires the state to make *all* the economic decisions. The state owns and operates all the major farms and factories, businesses, utilities, and stores. Government planners decide what products will be made, how much workers will be paid, and how much things will cost. Only a small amount of private enterprise—small farms and stores—is allowed.

Because of their centralized economic planning, communist systems are also called "planned economies." Clearly, communism involves a close relationship between the economy and the political system. By the early 1990s, the weaknesses of many of the world's communist systems were clear. Sweeping changes were taking place in both political and economic life.

Socialism   The basic philosophy of socialism is that, for the good of society as a whole, the state should own and run basic industries such as transportation, communications, banking, coal mining, and the steel industry. Because private enterprise operates most other parts of the economy, most socialist systems could also be described as *mixed economies.* That is, the free market and the government jointly make economic decisions.

Socialists believe that wealth should be distributed more equally and that everyone is entitled to certain goods and services. Socialist countries are sometimes known as "welfare states" because they provide many social services such as housing, health care, child care, and pensions for retired workers. To pay for these services, taxes usually are high.

SECTION 2 REVIEW

Developing Vocabulary
1. Define: **a.** sovereignty **b.** unitary **c.** federal **d.** confederation **e.** authoritarian **f.** dictatorship **g.** totalitarianism **h.** monarchy **i.** constitutional monarchy **j.** democracy **k.** capitalist **l.** market economy **m.** communism **n.** socialism

Place Location
2. Where is Monaco, the world's smallest sovereign country, located?

Reviewing Main Ideas
3. What ideas form the basis of socialism?
4. Why are some economies described as "mixed"?

Critical Thinking
5. **Demonstrating Reasoned Judgment**   Why might a unitary government be more successful in a country like Japan than in the United States?

CHAPTER

3

REVIEW

Section Summaries

SECTION 1   The Study of Human Geography   World population is unevenly distributed over the globe, and there are great differences in population densities. People at first tended to settle where there was adequate water, soil for farming or grazing, and a moderate climate. Nevertheless, people also have settled in many harsh environments, where they have both adapted their ways of living and made changes in their environments. World population has increased more rapidly in the twentieth century than in the past.

Every group develops a set of learned customs, beliefs, and actions that make up its culture. Culture includes both ideas—nonmaterial culture—and objects—material culture. Cultures change slowly, mainly through the exchange of ideas with other cultures.

SECTION 2   World Political and Economic Systems   There are more than 170 independent countries in the world. They share four common characteristics: territory, population, sovereignty, and government. A sovereign country can protect its territory and people and control its dealings with other countries. Governments can be classified according to either their structure or the source of their political power. In a dictatorship, the leader gains and keeps power by force; in a democracy, the people hold political power through their right to elect representatives.

Any country's economic system must try to answer the society's basic economic questions: What and how much shall be produced, by whom, and for whom? Depending mainly on the amount of government participation in the economy, modern economic systems can broadly be described as capitalist, communist, or socialist.

Vocabulary Development

Match the definitions with the terms below.

1. centralized government that makes laws for the entire country
2. the study of population and its growth
3. a type of government in which rulers inherit power through being born into the ruling family
4. an area where cultural innovations begin and from which they spread
5. a central city and its suburbs
6. the spread of cultural ideas from one culture to another
7. a country's right to protect itself and make its own political decisions
8. yearly number of births per 1,000 people

a. birthrate
b. culture hearth
c. demography
d. diffusion
e. metropolitan area
f. monarchy
g. sovereignty
h. unitary

Main Ideas

1. How do population patterns in East Asia differ from those in Europe?
2. What physical factors were most favorable for early human settlements?
3. What is material culture?
4. How does culture change?
5. How do the governments of Japan and the United States differ in structure?
6. What is the relationship between dictatorship and totalitarianism?
7. What basic questions must an economic system answer?
8. Why is capitalism also called a "market economy"?
9. In a socialist economy, what types of industry are government-owned?

Critical Thinking

1. **Identifying Assumptions** What ideas lie behind the concept of a welfare state?
2. **Predicting Consequences** How might continued rapid population growth affect poorer countries of the world?
3. **Recognizing Cause-Effect Relationships** Why would a change in a country's communist economic system have powerful effects on its political system of government as well?

Practicing Skills

Reading a Population Density Map Study the world population density map on page 55. Looking specifically at Mexico and Central America, answer the following questions.

1. What would you conclude about the general landscape and life-style here?
2. What part of the region seems to be an exception to this generalization?
3. What is the largest city in this area?

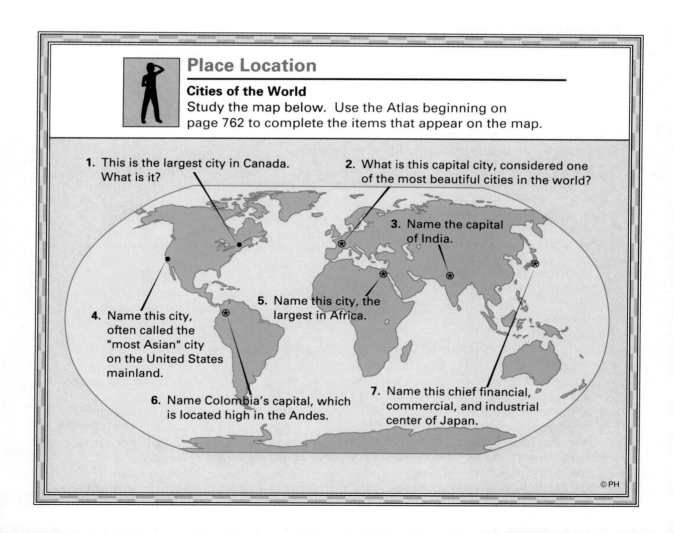

Place Location

Cities of the World
Study the map below. Use the Atlas beginning on page 762 to complete the items that appear on the map.

1. This is the largest city in Canada. What is it?

2. What is this capital city, considered one of the most beautiful cities in the world?

3. Name the capital of India.

4. Name this city, often called the "most Asian" city on the United States mainland.

5. Name this city, the largest in Africa.

6. Name Colombia's capital, which is located high in the Andes.

7. Name this chief financial, commercial, and industrial center of Japan.

©PH

Place Location: WHERE ON EARTH?

Money in the Bank

Congratulations! You have just graduated from high school. Dressed in a long robe, you step across the gleaming stage. The audience applauds as you accept your diploma. Now you're surrounded by family and friends, hugging and congratulating you.

Then your Uncle Jim, one of your favorite relatives, steps up. You can't remember a time when Uncle Jim wasn't fun to be around. With a big smile, he shakes your hand, and everyone suddenly becomes unusually quiet. You sense that something special is about to happen.

"Your family asked me to help them plan something really unusual for your graduation," your uncle tells you. "We thought the best idea would be a graduation fund. Every family member contributed, and so did a lot of old family friends. The contributions added up to a rather sizable amount."

Your uncle holds up a hand as you begin to thank everyone. "Hang on," he says. "There's a catch." And suddenly you remember something else about your uncle. He never misses an opportunity to make you think. There was the Thanksgiving dinner when he taught everyone the names for everything on the table—in German. After the meal, he challenged you to plot the shortest highway route between New York and San Francisco. Your reward was a pair of concert tickets.

Now your uncle tells you that your graduation money has been deposited in a bank in one of ten foreign cities. The cities are:

Tokyo, Japan
Melbourne, Australia
Montreal, Canada
Rio de Janeiro, Brazil
Buenos Aires, Argentina
Singapore
La Paz, Peru
São Paulo, Brazil
Mexico City, Mexico
Cairo, Egypt

Then he hands you a neatly folded piece of paper. As you unfold it, he says, "These are some clues that will help you figure out which city is the one where the money is waiting. I'm asking you to solve the problem by nine o'clock tonight. Do you think you can do it?"

You quickly scan the piece of paper. The heading reads, "YOUR CLUES." Underneath are a few neatly typed lines:

1. The city is on the mainland of a continent.
2. The city is located in the Western Hemisphere.
3. The city is located entirely north of the Tropic of Capricorn.
4. The city is located entirely south of the Tropic of Cancer.
5. The city is located on a coast.

No problem, you think. The puzzle requires a geographic solution. You're good at geography. You look up at

64

your uncle and the rest of your family and smile. You know you can figure out which city has the graduation fund.

As you leave the auditorium with your family, your mind is racing. How should you begin to solve the puzzle? Your geography textbook has an atlas. You might also use an atlas in your school library. You can start by finding out which of the ten cities are located on the mainland of a continent.

Once you begin work, in minutes you learn that all of the cities except Tokyo, Japan, and Singapore are located on the mainland of a continent. Now you must check out the remaining clues and narrow the possibilities until just one city remains. You begin working on the next clue, thinking ahead to other maps you might need. Suddenly you find yourself laughing. You should have the answer figured out in plenty of time. By nine o'clock tonight, your only challenge should be how to use your graduation gift!

Congratulations on your graduation

Your Clues

Montreal, Canada

Cairo, Egypt

Tokyo, Japan

Tropic of Cancer

Mexico

Equator

Singapore, Malaysia

La Paz, Peru

Rio de Janeiro, Brazil

Tropic of Capricorn

São Paulo, Brazil

Buenos Aires, Argentina

Melbourne, Australia

INTER-AMERICAN BANK

SOLVE THE MYSTERY

Which city has the graduation fund? *Hint:* Use political maps of the continents to help you find the answer. You can find such maps throughout this book or in an atlas.

4

Resources and Land Use

Chapter Preview

Both of these sections provide an overview of the world's resources and the ways in which people use them.

Sections	Did You Know?
1 WORLD RESOURCES	About 97 percent of the earth's water is in the oceans and cannot be used for drinking or farming because it is salt water.
2 WORLD ECONOMIC ACTIVITY	According to some scientists, only about 11 percent of the world's land area is suitable for raising crops.

Rice terrace, the Philippines

Open Pit Iron Mining in South Australia
Iron, a nonrenewable resource, enhances the economies of the countries in which it is found. Yet as this picture shows, the mining industries often cause devastating destruction of the landscape.

1 World Resources

Section Preview

Key Ideas

- Natural resources, derived from the environment, are either renewable or nonrenewable.
- Natural resources are unevenly distributed throughout the world.
- Modern civilizations depend on reliable sources of energy.

Key Terms

natural resource, renewable resource, nonrenewable resource, fossil fuels, nuclear energy, geothermal energy, solar energy

Even the earliest people used the resources they found in their environment. They breathed the air, drank the clear water, and caught fish to eat. They gained knowledge and skills and made tools to shape the earth's materials into useful goods. They hammered copper into weapons and ornaments and hollowed out trees to make canoes. People today are just as dependent on materials from the earth. The ways that people use the earth's resources, where the resources are located, how resources are distributed among people, and how the use of resources affects the earth are all subjects that geographers study.

Natural Resources

Different kinds of resources are used to meet people's needs. Capital resources are the money and machines used to produce goods or services. Human resources are the people whose labor and skills carry out production. This discussion is concerned with **natural resources**. These are materials that people take from the natural environment to survive and to satisfy their needs.

Renewable Resources When people use certain natural resources, called renewable resources, the environment continues to supply or replace them. Soil, for example, is always being created by the weathering of rocks and by decomposing plant and animal material. Through the water cycle, new supplies of fresh water return to the land as rain or snow. Our most important energy resource, the sun, although not renewable, will keep the earth warm for five billion years or more.

Being careless with renewable resources, however, is risky. Natural growth takes time, and human activities can get in the way of the process of renewal and change the environment. An oil spill, for example, might affect the wildlife in a bay for years.

Nonrenewable Resources As their name points out, nonrenewable resources are resources that cannot be replaced when they are used. Nonrenewable resources are minerals formed in the earth's crust by geologic forces over millions of years. The earth has only a limited supply of them, and they would take millions of years to be replaced.

Among the most important nonrenewable mineral resources, and the ones that people are using up most quickly, are the fossil fuels. These fuels are coal, oil, and natural gas, which formed from the remains of ancient plants and animals. But the fossil fuels aren't the only mineral resources that are important to people today. Other minerals include iron, copper, aluminum, uranium, and gold.

Supplies of nonrenewable resources vary greatly. Some minerals, such as salt, are in great supply and may last almost forever. On the other hand, the known supply of copper will be used up in less than fifty years. New technology may help people find and use new supplies of minerals from the oceans. But, recycling and using resources wisely is important.

Energy Sources

Modern industrial countries are dependent on various kinds of energy to light cities, power cars and airplanes, and run computers and other machines. The main source is fossil fuels, which are nonrenewable. The search for new energy sources—and the competition for those that already exist—is a strong force in world politics and economics.

Fossil Fuels Nearly all modern industrialized countries, including the United States, depend heavily on oil. Almost none of these countries have large enough supplies of their own and so must import much of what they use.

Oil and natural gas reserves are not spread out throughout the world. More than 60 percent of the world's known oil supply is located in just a few countries in the Middle East. Even with all known reserves, the world's oil is likely to run out in about fifty years at the present rate of use. Reserves of natural gas are also limited. Northern Eurasia has the world's greatest reserves of natural gas, which provides the region with both energy and a valuable export.

Coal deposits are larger and more widespread than oil reserves. The United States, China, and Russia have rich deposits. Industrial areas in Europe also arose near coal supplies. The world's reserves of coal are thought to be enough for at least two hundred years or more. Coal, however, has drawbacks. Coal smoke can create acid rain and air pollution.

Nuclear Energy Many countries now supply part of their energy needs through electricity that is created by

WORD ORIGIN

fossil
Medieval fields were encircled with trenches called *fosses,* from the Latin "to dig." Fossil fuels are found underground and hence are obtained by digging.

nuclear power. Nuclear energy today is produced by fission—the splitting of uranium atoms in a nuclear reactor, releasing their stored energy.

Many questions and concerns surround the use of nuclear power. Safety problems include the danger of leaks or explosions, as well as the disposal of long-lasting radioactive waste and contaminated water. Nuclear power plants are very expensive to build. In addition, nuclear fission uses uranium, which is a limited nonrenewable resource. Scientists hope to find a way to generate energy through fusion, a type of nuclear reaction for which the fuel is plentiful.

Other Energy Sources Many experts think that people must develop other sources of renewable energy and become less dependent on fossil fuels. All these sources are in use today somewhere in the world.

One ancient source of energy, water power, uses the energy of falling water to move machinery or generate electricity. Although new dams must be built from time to time, water power is a renewable energy source. Ocean tides are another source of power.

In areas with volcanic activity, a potential energy source is geothermal energy—energy that comes from the earth's internal heat. The intense heat of magma within the earth heats underground water, producing steam that can be used to heat homes or make electricity. Iceland, Italy, Japan, and New Zealand all make use of geothermal energy.

More research and technology are still needed to make solar energy practical on a large scale. Solar energy is energy produced by the sun. Systems to collect and store the sun's energy have been used for years to heat water and homes. Generating electricity from solar energy has been more difficult. Nevertheless, solar radiation is potentially the greatest renewable energy source available.

Harnessing Nature
Wind is an effective energy source on this wind farm in Palm Springs, California. What are some of the benefits of wind energy?

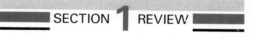

SECTION 1 REVIEW

Developing Vocabulary
1. Define: **a.** natural resource **b.** renewable resource **c.** nonrenewable resource **d.** fossil fuels **e.** nuclear energy **f.** geothermal energy **g.** solar energy

Place Location
2. Use the world map on pages 762–763 to locate the Middle Eastern country of Saudi Arabia.

Reviewing Main Ideas
3. What is the main source of energy for most modern industrial countries?
4. What are the advantages and disadvantages of using coal?

Critical Thinking
5. **Explaining Problems Clearly** Why are alternative energy sources important for the future?

Waste Disposal and Recycling

Imagine a row of ten-ton garbage trucks, one behind the other in a line 145,000 miles (233,000 km) long—more than halfway to the moon! According to the Environmental Protection Agency, that is what it would take to carry what American households throw away in just *one year.*

The amount of waste and the problems involved in disposing of it have increased as many countries of the world have grown wealthier. While waste disposal is a worldwide problem, it is most serious in the United States. Every person in our "throwaway society" discards about 3.5 pounds (1.6 kg) of waste material every day. By contrast, Europeans discard about 2 pounds (0.9 kg), and people in poorer countries throw away even less—a little over a pound (0.5 kg) apiece. Industries and agriculture also produce millions of tons of waste every year.

A Modern Problem

Whenever archaeologists unearth an ancient city, they find its midden, or trash heap. Even very early humans left piles of shells and bones near their campsites. Waste disposal is no longer that simple. As population and industrial production have increased in the last two hundred years, waste has also grown. More people, with more things, are disposing of them at an ever-increasing rate.

In addition, much waste material is now hazardous. Some wastes—paints, pesticides, cleaners—contain poisonous chemicals. Many plastics and other synthetics are not biodegradable. That is, they do not decompose or break down naturally but last almost forever.

Looking Ahead

During the environmental movement of the 1970s, a popular saying was "There is no away." That is, throwing things away is no longer a choice. Most countries are rapidly running out of landfills where solid wastes can be buried. Between 1978 and 1988, the number of landfills in the United States dropped from about 20,000 to about 6,000. Most of those were expected to be full by the mid-1990s; few new sites were available.

Landfills are a far from perfect solution. Not only are they unsightly but they also can pollute water supplies. Simply getting trash to the landfills is also a costly problem. New Jersey's trash, for example, was sent all the way to New Mexico to be buried in a landfill.

Community Responsibility American towns are asking their residents to recycle.

What are the choices that remain? Many environmentalists suggest a multistrand approach: waste reduction, recycling, composting, landfills, and incineration. Each of these methods has certain problems. The incineration process, for example, sends ash and poisonous gases into the air.

One approach to the waste problem is to reduce the amount thrown away. For American manufacturers and consumers, this would mean great changes in the packaging of many products. It would also require a shift in Americans' throwaway habits.

Recycling—another long-term strategy—also depends on motivation and cooperation. Paper, steel and aluminum, glass, and some plastics all can be processed and reused. But the economics of recycling are complex: a market for the material must be found, and revenues must balance against the costs of processing and shipping the waste.

Other countries are far ahead of the United States in recycling efforts. Japan recycles half of its solid waste; the United States, about 11 percent. Gradually, however, American cities, households, and offices seem to be moving toward this approach to a critical problem.

TAKING ANOTHER LOOK

1. Why has the problem of waste disposal increased in modern times?
2. What materials can be recycled?

Critical Thinking

3. **Identifying Alternatives** What new attitudes might reduce America's waste disposal problem?

Landfills and Recycling If future archaeologists judge our society from landfills (left), what might they say? By recycling (right), the public plays an essential role in environmental protection.

✓ Skills Check

☐ Social Studies
☑ Map and Globe
☐ Reading and Writing
☐ Critical Thinking

Interpreting an Economic Activity Map

The economic activity of a given region is an example of the type of data that may be represented on a map. Other thematic maps include those that represent mineral resources, agricultural products, religion, or languages. This map uses a color-coded key to communicate basic information. It is easy to see, for example, that far more land in Africa is used for subsistence farming than in Europe. However, precise measurements in square miles are not shown on this map.

Use the following steps to read and analyze the world map below.

1. **Identify the variety of economic activities shown.** Economic activity can range from gathering nuts and berries to manufacturing cars and trucks on an assembly line. Use the key to answer the following questions: (a) What appears to be the major economic activity in Australia? (b) What appears to be the major economic activity in the northernmost region of Asia? (c) Where are the manufacturing and trade centers in South America?

2. **Look for relationships among the data.** Use the map to look for patterns. Answer the following questions: (a) Does more manufacturing and trade take place in countries in the northern hemisphere or in the southern hemisphere? (b) How does the amount of land used for commercial farming in the United States compare with the amount used in Australia? (c) In South America, is more area used for commercial fishing or for subsistence farming?

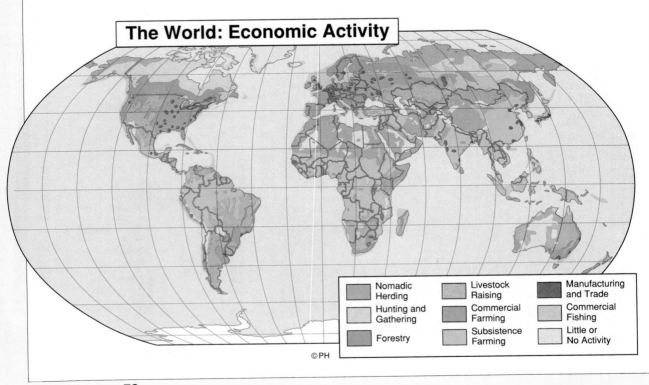

The World: Economic Activity

Key:
- Nomadic Herding
- Hunting and Gathering
- Forestry
- Livestock Raising
- Commercial Farming
- Subsistence Farming
- Manufacturing and Trade
- Commercial Fishing
- Little or No Activity

©PH

2 World Economic Activity

Section Preview

Key Ideas

- Economic activities are the ways in which people use land and resources to earn their living.
- Countries are at different stages of economic development.
- Agriculture is the occupation of about half the workers in the world.

Key Terms

manufacturing, developed country, developing country, gross national product (GNP), per capita GNP, subsistence farming, commercial farming

Every morning, in the cities of Asia, Africa, Europe, and the Americas, millions of people ride trains, cars, buses, or bicycles to work in stores, factories, and offices. Every night they go home. To them this seems a natural way to live and work.

To millions of other people, this way of living would seem strange. For these people, working means farming the land where they live, traveling with herds of animals, or catching fish far out at sea.

Types of Economic Activity

The map on page 72 gives you an idea of the varied ways in which people around the world make their living. Many of these categories of economic activity are related to a region's natural resources and climate.

Farming may be limited by land and climate. Even where the land can be farmed, rainfall and temperature influence what crops farmers will choose to grow. A farmer in the Soviet Union would probably not choose to grow coffee, while a planter in Kenya would probably not choose to grow sugar beets.

Fishing, forestry, and mining also are limited to places where there are enough fish, trees, or minerals to make this activity practical. Because they use natural resources directly, activities like agriculture and forestry are sometimes classified as primary activities.

Industries that use and process natural resources are termed secondary activities. They include processing of raw materials, such as food processing, and manufacturing—the process of turning raw materials into finished products.

Service industries, or tertiary activities, make up the third level of economic activity. Service industries are businesses that are not directly

Rubber Tapping in Brazil

Because of its use in many products, rubber is an important natural resource. Here latex drips from a freshly cut groove in a rubber tree.

DAILY LIFE

Urban Green Spaces

Cities are built on land and, therefore, are a type of land use—urban land use. Land use in cities, however, refers to more than factories, stores, and homes.

At lunch time in most major industrial cities throughout the world, thousands of workers pour out of office buildings and head for the nearest public parks. There they find trees and bushes, open space, and a quiet place to relax and visit with friends.

Many urban dwellers and planners believe that patches of green are as valuable to a city as its banks, offices, and department stores. A survey of seven metropolitan areas in the United States found that the percentage of land set aside for open space and recreation areas was greater in every case than the amount of space devoted to businesses.

1. What are some of the benefits of green space in a city?
2. Why do cities set aside space for recreational purposes?

related to manufacturing or the gathering of raw materials. Service industries include transportation, sales, advertising, education, banking, health care, and government.

Stages of Economic Development

How its people earn a living is one way of measuring a country's economic development. Modern industrial societies with well-developed economies, such as France, the United States, and Japan, are said to be **developed countries.** The poorer countries, which often do not have modern technology and industries and depend on developed countries for many of their manufactured goods, are said to be **developing countries.**

Generally, developed countries are richer than developing countries. One way to compare the wealth of countries is to look at the total value of the goods and services they produce in a year, which is called the **gross national product**, or **GNP**. Another measure of development, **per capita GNP**, is the GNP divided by the country's total population. In the late 1980s the average per capita GNP in developing countries was $640 a year. The average in developed countries was $11,300.

Agriculture

Although populations in world cities have soared, developing countries are still mainly rural. About half of all the world's workers are in agriculture —from nomadic herders to large-scale commercial farmers. The agricultural way of life includes many ways of growing crops and raising animals.

Hunters, Gatherers, and Herders
The Bushmen of the Kalahari and the Mountain Lapps of northern Scandinavia are some of the people who still follow ancient ways of hunting and gathering food and herding animals. Hunters and gatherers live in small family groups in remote areas where population is scarce. They gather berries, roots, nuts, and other wild foods, sometimes catching fish or hunting. Herders follow large herds of animals, such as caribou, over hundreds of miles of grazing land.

Hunters, gatherers, and herders have developed well-organized cultures. Their ways of life are often based on a knowledge of the environment that has been passed down from generation to generation for hundreds of years. They are most often nomadic— they travel from place to place in different seasons.

Subsistence Farming Much of the agriculture in developing countries is **subsistence farming**. People grow only enough for their own family's or village's needs. If they are lucky enough to have a very good crop or an

extra animal, they may sell or trade it. Mainly, though, they grow food to eat, not to sell.

Commercial Farming In developed countries, nearly all farmers raise crops and livestock to sell in the market. This is **commercial farming**. Modern techniques and equipment make these farmers more productive. As a result, only a small part of the labor force in these countries is made up of farm workers. In Japan, for example, about 9 percent of workers are in farming or fishing.

Some commercial farming takes place in the developing countries too. Large plantations in tropical regions produce coffee, sugar, cotton, bananas, and other crops. Until recently, most were owned by foreign investors.

■ SECTION **2** REVIEW ■

Developing Vocabulary
1. Define: **a.** manufacturing **b.** developed country **c.** developing country **d.** gross national product (GNP) **e.** per capita GNP **f.** subsistence farming **g.** commercial farming

Place Location
2. Use the World Economic Activity map on page 72 to locate two areas inhabited by people who live by nomadic herding.

Reviewing Main Ideas
3. What is the difference between subsistence farming and commercial farming?
4. In what ways do climate and resources influence the economic activity in an area?

Critical Thinking
5. **Identifying Central Issues** What reasons might hunters, gatherers, and herders have for moving from place to place with the seasons?

per capita
Per capita comes from the Latin *per*, meaning "by," and *capita*, meaning "head." A country's per capita income is the income each person would have if the country's total income were evenly distributed.

4

REVIEW

Section Summaries

SECTION 1 World Resources To meet their needs, people take natural resources from the environment—from the land, water, and air. Renewable resources are those that grow or can be replenished, such as water, forests, and wildlife. Some renewable resources are being used faster than they are renewed, however. Nonrenewable resources include fossil fuels, metals, and other minerals. Both renewable and nonrenewable resources are unevenly distributed throughout the world. Modern civilizations are dependent on reliable sources of energy, primarily on nonrenewable fossil fuels. Alternative renewable energy sources include nuclear energy and solar energy.

SECTION 2 World Economic Activity Economic activities are the ways in which people use land and resources to earn their living. These activities vary greatly throughout the world. Economic activities depend to some degree on land and resources; those that use resources most directly include farming, forestry, and mining. Countries are at different stages of economic development. The industrial countries of Europe, North America, and Japan are considered to be developed countries; poorer countries are considered to be developing. Agriculture is a way of life for about half the workers in the world; those in developing countries live mainly by subsistence farming. These countries face problems of hunger and economic development.

Vocabulary Development

Match the definitions with the terms below.

1. all the goods and services produced by a country in a year
2. agriculture whose goal is to produce crops or livestock to sell
3. coal, oil, natural gas
4. resources whose supplies are limited and cannot be replaced
5. a country with a low GNP, lack of technology, and little industry
6. making finished products from natural resources

a. commercial farming
b. developing country
c. fossil fuels
d. gross national product
e. manufacturing
f. nonrenewable resources

Main Ideas

1. Why are soil, trees, and fish considered renewable resources?
2. Name at least three types of nonrenewable fossil fuels.
3. Why are fossil fuels so important in modern society?
4. What region has the largest percentage of world oil supplies?
5. What are the advantages and disadvantages of coal as a fuel?
6. What are some energy alternatives to fossil fuels?
7. What types of economic activity depend most on a region's natural resources and climate?
8. What factors distinguish developing and developed countries?
9. In general, how is agriculture in developed countries different from that in developing countries?
10. Why has waste disposal become a problem in many American towns, and what are some possible solutions to this growing problem?

Critical Thinking

1. **Determining Relevance** How do the economic activities of a people influence their culture?
2. **Recognizing Cause and Effect** How might foreign ownership of businesses hurt the economy of a developing country?
3. **Predicting Consequences** Why would some kinds of industrial development be harmful to a developing country?
4. **Identifying Alternatives** What energy sources can be used to replace fossil fuels?

Practicing Skills

Interpreting an Economic Activity Map
Look again at the map on page 72 and then answer the following questions.

1. Where are the manufacturing and trade centers in South America located? Where do you think the major cities in these regions are located?
2. What type of economic activity dominates the interior of Australia?
3. How do you think this economic activity is related to climate in that region?

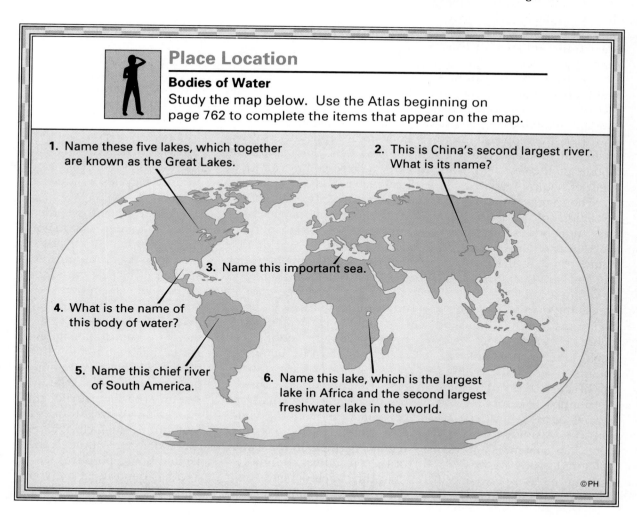

Place Location

Bodies of Water
Study the map below. Use the Atlas beginning on page 762 to complete the items that appear on the map.

1. Name these five lakes, which together are known as the Great Lakes.
2. This is China's second largest river. What is its name?
3. Name this important sea.
4. What is the name of this body of water?
5. Name this chief river of South America.
6. Name this lake, which is the largest lake in Africa and the second largest freshwater lake in the world.

©PH

MAKING
CONNECTIONS
•WHERE REGIONS MEET•

Transportation and Trade

Regions of the world do not exist in isolation. They are connected to each other in countless ways. For example, they are linked by the movement of people, goods, and ideas from one region to another; by the satellite linkages that transport news and information from one region to another; and even by the clothes that are made in one region and worn by people in another. Regions are also connected to one another by the movement of natural resources.

The shapes, colors, and conditions of the earth can vary tremendously from one place to another. Likewise, the many resources that people use and on which they depend are spread unevenly across the planet. Though one region may enjoy a thousand riches, another may possess key resources necessary for the first to enjoy its wealth.

Because of this uneven distribution of resources, the regions of the world have become increasingly interdependent. Raw materials and manufactured goods are in constant motion between different regions of the earth.

Nations around the world trade what they have for what they do not have in a never-ending cycle.

Transportation Linking the Regions

The movement of goods traded between nations is an enormous and impressive undertaking. Most of these goods are shipped by ocean-going vessels. Such ships can carry a large volume of goods for much less than the cost of air transportation. The map on the following page shows the major ocean transportation routes of the world. Note that the ocean route between the Middle East and Europe carries more tonnage than any other route. Oil tankers use this route to carry shipments from the oil-rich nations of the Middle East to the oil-poor nations of Europe. In turn, freighters follow this route carrying European-made consumer goods to the Middle East. These two regions are regular partners in the trade of their resources. But more than that, each depends on the other for resources.

An Example of Transportation and Trade

Consider the bicycle for another example of how regions are connected by the movement of resources. None of the major bicycle-producing nations—the United States, China, France, Great Britain, Italy, Japan, South Korea, and Taiwan—has all the resources necessary to produce bicycles in large numbers.

For example, the United States and Japan lack major deposits of bauxite, the ore used to make aluminum. A number of bicycle parts, such as chainwheels and brakes, are commonly made of aluminum. Both nations must therefore import bauxite in order to make aluminum. In return, the United States and Japan export some of their resources to the bauxite supplier. Many bicycle frames are made from steel, chrome, and molybdenum. Although the United States produces steel and has deposits of molybdenum, it has no chrome. The areas of the world that have major supplies of chromite include Brazil, Eastern Europe, and Turkey.

Many of the bicycles sold in the United States are produced in another country—perhaps more than one country. Some bicycle manufacturers buy components, such as gear shifts, from Japan and ship them to Taiwan for assembly. The finished bicycles are then shipped to the United States for sale. What ocean routes might a freighter use to deliver bicycles assembled in Taiwan?

The View from the U.S.A.

 The manufacture of bicycles is just one example of how trade plays a critical role in United States economy. The United States is the world leader in the value of the goods and services it produces. Its farms and factories produce goods that are in great demand around the world. At the same time, the United States needs resources that are available only from other nations.

TAKING ANOTHER LOOK

1. Why are ocean routes generally used to ship raw materials?
2. How does the example of bicycle production show interdependence? How does it show the importance of transportation?

Critical Thinking

3. **Recognizing Cause and Effect** How might a war or natural disaster in one country affect the manufacture of goods in other nations?

Applying Geographic Themes

Movement   This map shows major railways as well as shipping routes. Which areas of the world have the most complex rail networks?

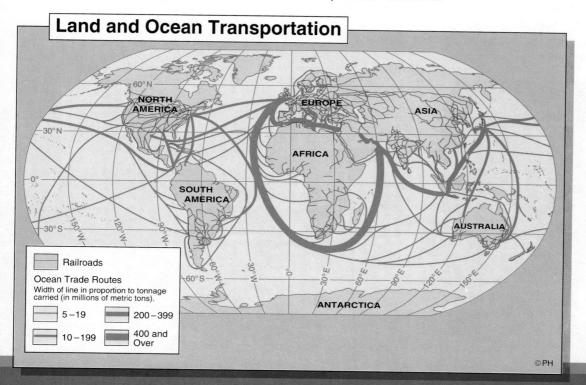

Land and Ocean Transportation

Railroads

Ocean Trade Routes
Width of line in proportion to tonnage carried (in millions of metric tons).

5–19   200–399
10–199   400 and Over

©PH

UNIT
2

The United States and Canada

CHAPTERS

Churning seas, soaring mountains, silent forests, and languid prairies . . . European, Asian, African, Hispanic, and Native American. The United States and Canada are a kaleidoscope of physical and human geography—so much that is different, so much that is the same.

An American writer used those words to describe the geographic realm of the United States and Canada. As you read this unit, you will discover how that description captures both the diversity and similarity that characterize this region.

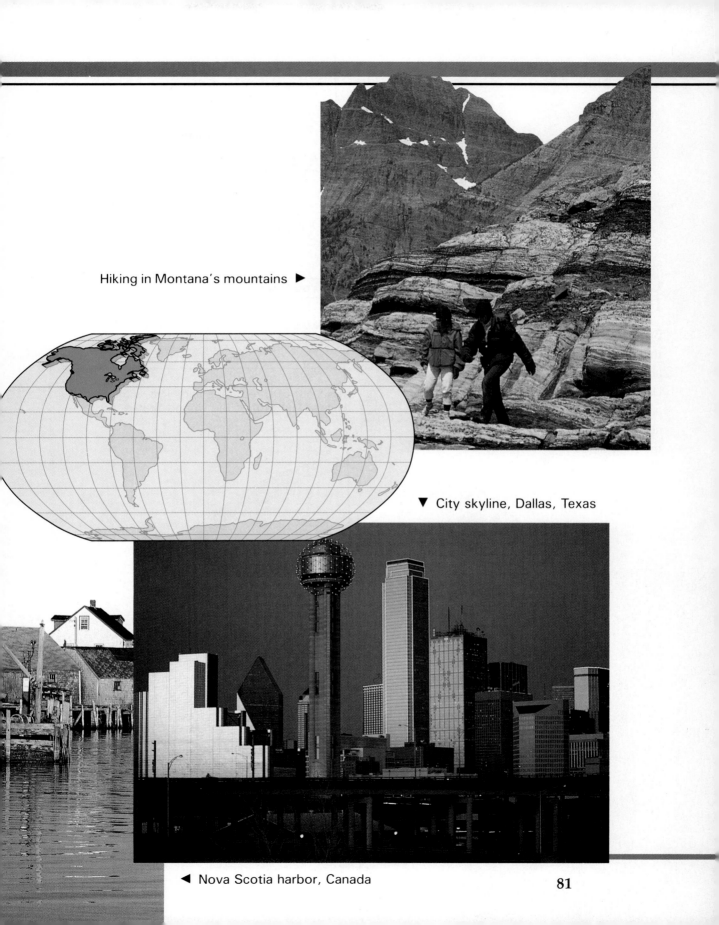

Hiking in Montana's mountains ▶

▼ City skyline, Dallas, Texas

◀ Nova Scotia harbor, Canada

81

5

Regional Atlas:
The United States and Canada

Chapter Preview

Both sections in this chapter are about the United States and Canada, shown in red on the map below.

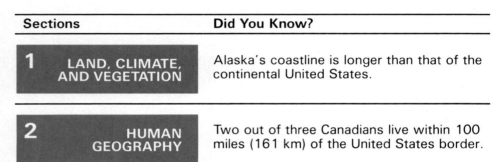

Sections		Did You Know?
1	**LAND, CLIMATE, AND VEGETATION**	Alaska's coastline is longer than that of the continental United States.
2	**HUMAN GEOGRAPHY**	Two out of three Canadians live within 100 miles (161 km) of the United States border.

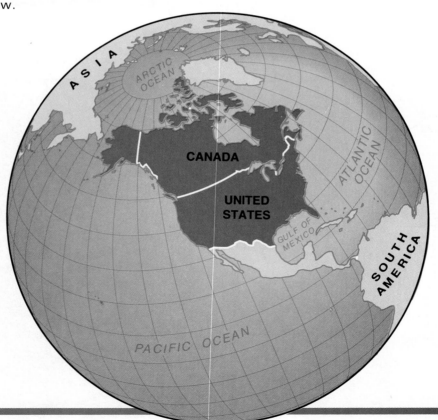

Fall Colors in Virginia

The Blue Ridge Mountains, part of the Appalachian Mountain chain, overlook the Shenandoah River valley of Virginia. How do these eastern mountains compare to the Rocky Mountains shown on the map on page 84?

1 Land, Climate, and Vegetation

Section Preview

Key Ideas

- Canada and the United States share many of the same landforms.
- Canada's climates are generally colder than those of the United States, because Canada is located at more northern latitudes.
- Most of the natural vegetation of Canada and the United States falls into four broad categories: tundra, forest, grassland, and desert scrub.

Key Terms

cordillera, bedrock, province, continental divide, drainage basin, tributaries, prairie

The United States and Canada are both located on the continent of North America. They are huge countries— together covering more than 6 million square miles (15 million sq km). Because these two giants share the same landmass, they also share many of the same landforms: craggy mountains in the west; low, rounded mountains in the east; and golden, rolling plains in the center.

Similar Landforms

Notice on the map on page 84 that much of the United States and Canada border stretches in a straight line from east to west. The major landforms of North America, however, generally extend from north to south. Notice also how the Pacific and Atlantic coastlines angle toward each other toward the south. Located about parallel to these coastlines rise two of the major

The United States and Canada: Physical – Political

ASIA

ARCTIC OCEAN

ICELAND

Bering Strait

Ellesmere I.

Greenland
(Denmark)

BERING
SEA

BEAUFORT
SEA

Aleutian Is.

Alaska
(U.S.)

Mt. McKinley
20,320 ft. (6,194 m)

ALASKA RANGE

Yukon R.

BROOKS RANGE

Victoria I.

Mackenzie R.

Baffin I.

Mt. Logan
19,524 ft. (5,951 m)

GULF OF
ALASKA

Great Bear
Lake

Great Slave
Lake

HUDSON
BAY

GULF OF
ST. LAWRENCE

Newfoundland

PACIFIC
OCEAN

Queen
Charlotte Is.

Vancouver I.

COAST MOUNTAINS

CANADA

Edmonton

Nelson R.

L. Winnipeg

CANADIAN SHIELD

Ottawa R.

LAURENTIAN HIGHLANDS

St. Lawrence R.

INTERIOR
PLAINS

L. Superior

L. Huron

Montreal

Ottawa

Toronto

New York

Fraser R.

ROCKY MOUNTAINS

GREAT PLAINS

Columbia R.

CASCADES

Great
Salt Lake

Missouri R.

L. Michigan

Chicago

L. Erie

L. Ontario

APPALACHIAN MOUNTAINS

Washington, D.C.

SIERRA NEVADA

GREAT
BASIN

Platte R.

CENTRAL
PLAINS

ATLANTIC
OCEAN

COAST RANGES

Mt. Whitney
14,491 ft. (4,417 m)

UNITED
STATES

Ohio R.

Los Angeles

Colorado R.

Arkansas R.

Mississippi R.

GULF-ATLANTIC COASTAL PLAIN

Rio Grande

New Orleans

Straits of Florida

Tropic of Cancer

GULF OF MEXICO

CUBA

MEXICO

160° W   155° W

Hawaii
(U.S.)

20° N

Elevation of Land

Meters	Feet
Over 3,050	Over 10,000
1,525 to 3,050	5,000 to 10,000
610 to 1,525	2,000 to 5,000
305 to 610	1,000 to 2,000
0 to 305	0 to 1,000
Below sea level	Below sea level

Depth of Water

0 to 153	0 to 500
Below 153	Below 500

National
Capital

0   500   1000 Miles

0   500   1000 Kilometers

©PH

mountain chains on the continent—the Rocky Mountains in the west and the Appalachians in the east. Expanses of plains lie between these mountain chains in the central regions of both countries.

High Western Mountains The Rocky Mountains are a cordillera (kawr dill YER uh), a related set of separate mountain ranges. This wall of rock stretches for more than 3,000 miles (4,800 km) from northern Alaska to Mexico, making it the longest mountain chain in North America.

Compared with other mountains of the world, the Rockies are relatively young. They started to rise only about sixty million years ago, when tectonic forces produced their jagged peaks. Because these landforms have had relatively little time to erode, many of the Rocky Mountain peaks still reach over 14,000 feet (4,270 m). As this description by an American writer and adventurer points out, such high elevations have a great effect on the climate.

After two days of riding we reached the foot of Fremont Peak, a monolith lifting 13,730 feet, draped with snow. . . . We dismounted and started our ascent. Soon, my sea-level lungs felt tight as a fist. Halfway up, we began to sink to our thighs in snow; each step demanded a decision. After three hours of exhaustion, we reached the top. I was seeing what I had never seen before: a breathtaking vista of glaciers, frozen lakes, and soaring peaks.

Between the Rockies and the Pacific Coast are several other mountain ranges. Look again at the map on page 84. These ranges include the Alaska Range, the Cascade Range, and the Sierra Nevada. What major landforms lie between the Sierra Nevada and the Rockies?

Central Plains After completing their exploration of the American West in 1806, Meriwether Lewis and William Clark made a difficult crossing over the Rockies. From a summit they looked to the east and noted in their journal, with great relief, that they saw "an immence Plain and level Country." Their ordeal in the Rockies was behind them.

The plains that Lewis and Clark gazed upon were the Great Plains. These are part of a vast plain that stretches across the central part of both Canada and the United States. The Great Plains are a region of grasslands that slope gently eastward from the Rockies to the Central Plains, which lie west of the Appalachian Mountains.

In Canada, part of the plains region is known as the Interior Plains. The Interior Plains have deep rich soils, just like much of the plains in the United States. But the eastern part of the plains region in Canada is a vast area of ancient rock, scattered with boulders and hundreds of small lakes. This region, known as the Canadian Shield, extends into the highlands of eastern Canada.

On the map, the Canadian Shield forms a huge horseshoe around Hudson Bay. Some people describe the bay as the hole that lies in the middle of a

Applying Geographic Themes

1. Place Two sprawling mountain chains dominate the eastern and western landscape of the United States. Name these enormous mountain ranges.

2. Location The Great Plains region is also known as "America's Breadbasket." Where are the Great Plains located?

A Border Town

Every night, Madame Irene Morneau sleeps with her feet in Canada and her head in the United States. How does she do this? Mme. Morneau lives in Estcourt Station, Maine, a town on the United States-Canadian border (shown at right). Some of Estcourt Station's roads and farms are in the United States, some are in Canada. One house—Mme Morneau's—is in both countries.

No one thinks about raising a physical barrier along this friendly border. Says an American official, "The United States-Canadian border is a living organism—with a life and culture. We try not to disturb it."

1. What is special about the United States-Canadian border?

2. How might living in a border town affect your feelings of patriotism?

Appalachian
The Spanish explorer Cabeza de Vaca first used the term *Apalachen* as the name of an Indian province. The English respelling, *Appalachian,* first applied to only the southern part of the mountain system, and later to the whole mountain range.

broken doughnut, or as a hollow in a ring of endless rock. The Canadian Shield is, in fact, an expansive region of exposed bedrock. Bedrock is solid rock that is usually covered by soil, gravel, and sand. However, glaciers moving across the Canadian Shield during the last Ice Age—about 20,000 years ago—scraped the ancient bedrock bare of overlying materials. No other place on the earth has this much exposed bedrock. The land of the shield is rugged, as this geologist describes:

. . . slipping on the rocks, and clambering over cliffs, with a pack on my back or a canoe on my

shoulders, sweat and flies blinding me, I was reminded how hard a country it is.

Low Eastern Mountains and Coastal Plains East of the central plains stand the Appalachian Mountains and the Laurentian (law REN shuhn) Highlands. Find the Laurentian Highlands on the map on page 84. They are located in the Canadian province, a political division, of Quebec, north of the St. Lawrence River. The Appalachians extend from the Canadian province of New Brunswick to Alabama in the United States. These two ancient mountain ranges are much older than the Rockies. Tectonic forces pushed them upward between 650 and 250 million years ago. Over hundreds of millions of years, the once-sharp peaks have been worn down by rain, ice, and wind. As a result, the Appalachian Mountains today are slump-shouldered—rounded and relatively low. Only a few of the Appalachian peaks reach higher than 6,000 feet (1,900 m).

East of the Appalachians in the United States lies a coastal plain. In the southeastern United States, the plain is wide and the soils are rich. In the northeastern United States and southeastern Canada, the coastal plain narrows and soils are poor.

Landforms and Water

Because rivers wind their way downhill, the pattern of landforms determines the direction in which the water systems flow. The high ridge of the Rockies that separates the rivers that flow west from the rivers that flow east is called the Continental Divide. A continental divide is a boundary that separates rivers flowing toward opposite sides of a continent. At one point along the Continental Divide, in Cutbank Pass, Montana, three brooks are so close together that a person can pour water into all of them at the same time. One brook carries water west to the Pacific Ocean, another to Hudson Bay, and the third into the Gulf of Mexico.

A divide also separates one drainage basin from another. A drainage basin is the entire area of land that is drained by a major river and its tributaries —rivers and streams that carry water to a river. The tributaries flow into the main river just as the branches of a tree are joined to the trunk. With the exception of the Mackenzie River system, which drains a large basin in northern Canada and flows into the Arctic Ocean, most of Canada's major rivers flow into Hudson Bay. The Mississippi and Missouri rivers and their

tributaries drain the largest area in the United States. The Mississippi is the longest river in the United States, and some of its tributaries, such as the Illinois and Ohio rivers, are themselves major rivers. Look at the map on page 84 to find the body of water into which the Mississippi River drains.

Differences in Climate

The United States and Canada have similar landform patterns, but the climates of the two countries differ greatly. A journalist writing an article on Canada for *National Geographic* tells a story about fishing on Lake Huron.

Contrasting Climates
While people in Montreal (inset) grapple with snowstorms during the winter, Florida's tropical climate keeps temperatures moderate all year.

Two Toronto businessmen have taken me for a weekend of bass fishing at their cottage on Lake Huron. . . . Gordon is busy changing his lure. 'Let's face it,' he says, 'we made a big mistake. We should have divided North America from north to south, not east to west. Then we'd each . . . have some warm weather.'

The United States and Canada share many characteristics, but they would have had more similar climates if they *had* been divided from north to south.

One important difference between the two lands is that, except for Alaska, Canada lies farther north than the United States, stretching its frigid fingers into the Arctic Circle. Because the average temperature of a place is greatly affected by its distance from the Equator, Canada's climates are generally colder than those of the United States. Nowhere in Canada will you find beaches lined with palm trees. Swimmers looking for warm weather in January head not for Canada but for southern Florida in the United States. Most parts of Canada and the northern United States are frozen in winter. And northern Canada has such a short period of nonfreezing weather that crops cannot grow. Besides latitude, two other factors influence climate in both countries: landforms and nearness to large bodies of water.

Linking Location to Climate Many places in the United States and Canada are far from oceans or other large bodies of water that moderate climate. As a result, these interior locations often have a continental climate. As you read in Chapter 2, places with continental climates are distinguished by their extreme temperatures. They tend to be cold in winter and hot in summer. Places at the same latitude located along the coast have milder climates. This is because temperatures in coastal areas are moderated by winds blowing onto land from the ocean. Water warms and cools more slowly with the changing seasons than nearby land. Thus ocean temperatures are generally cooler than land temperatures in summer and warmer than land temperatures in winter. So, too, are the winds that blow over the oceans onto the land.

The influence of location on climate can be seen by comparing Winnipeg, in the interior of Canada, with San Francisco, on the west coast of the United States. Locate these cities on the climate map, then compare their average temperatures on the climate graphs. Despite being farther north, Winnipeg is actually warmer during the summer. But its frigid January temperatures average 50°F (28°C) less than San Francisco. While people in Winnipeg may not have to wear sweaters in July, as San Franciscans often do, you can be sure that they wear many more sweaters, coats, hats, and mittens in January.

Linking Landforms to Climate As the climate map on page 89 shows, much of the western part of North

---

Applying Geographic Themes
1. Place What type of climate dominates the central regions of the United States and Canada? Which region has a Marine West Coast climate?
2. Location Winnipeg's inland location greatly affects its climate. Use the lines on the climate graphs to compare average monthly temperatures in Winnipeg and San Francisco. Which city has colder winters? Which city has warmer summers? Use the bars on the climate graphs to compare average precipitation. Which city has dry summers and wet winters?

The United States and Canada: Climate Regions

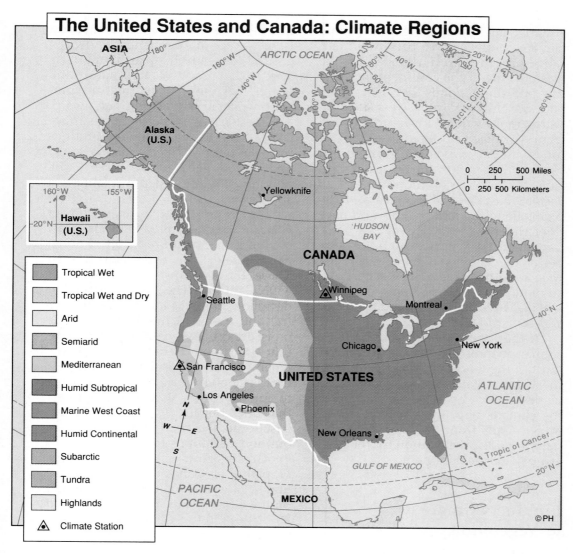

Legend

- Tropical Wet
- Tropical Wet and Dry
- Arid
- Semiarid
- Mediterranean
- Humid Subtropical
- Marine West Coast
- Humid Continental
- Subarctic
- Tundra
- Highlands
- △ Climate Station

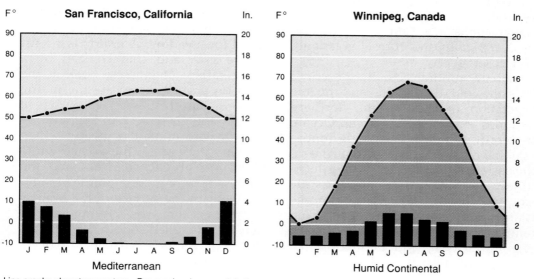

San Francisco, California — Mediterranean

Winnipeg, Canada — Humid Continental

Line graphs show temperature. Bar graphs show precipitation.

Chapter 5, Section 1   **89**

America does not experience the same climates as the eastern part. Mountain ranges greatly affect climates on the western side of the continent. Because the prevailing winds between 30°N and 60°N blow from west to east, warm, moist winds blow onto the land from the Pacific Ocean. As the air moves inland, it rises, cools, and drops heavy precipitation on the windward side of the coastal mountains. The eastern parts of the Rockies and the plains lie in the rain shadow and have drier climates.

Study the climate map for evidence of this effect. Notice that the west coast has moderate Mediterranean and marine west coast climates, whereas areas farther inland have arid and semiarid climates.

If the Pacific Ocean were the only source of moisture for the continent, most of the land east of the Rockies in the United States and Canada would be desert. However, as you can see on the climate map, this is not the case. Moist winds from the Gulf of Mexico help provide precipitation for the eastern part of the continent.

Linking Climate to Vegetation

With a few exceptions, the natural vegetation of the United States and Canada falls into four broad categories: arctic tundra, forest, grassland, and desert scrub.

Arctic tundra actually describes both the cold, dry climate and the low vegetation of the extreme northern latitudes encircling the Arctic Ocean. In North America, the arctic tundra region stretches across most of Alaska, the eastern half of the Northwest Territories, the Ungava Peninsula to the east of Hudson Bay, and all of the islands north of Hudson Bay. Cold temperatures, strong winds, a short growing season, and permafrost severely limit the natural vegetation of

much of the arctic tundra. Permafrost is a layer of permanently frozen ground beneath the earth's surface. The frozen mass of soil, sand, and gravel that make up permafrost prevents water from seeping downward. As a result, when the shallow surface layer of soil thaws during the tundra's brief summer, it often remains soggy and waterlogged. Only lichens (LY kuhnz), mosses, and tiny plants can survive.

Bordering the tundra region of Canada and the United States is a vast area of coniferous forests, also known as needleleaf or evergreen forests. These forests grow mostly north of 45°N latitude. The largest region of deciduous forest is located in the central and eastern United States in the moist areas between 30° and 50°N latitude. Deciduous trees shed their leaves when winter approaches. Use the natural vegetation map on the next page to find a region of deciduous forest that does not fit this general pattern. Also locate the areas of mixed coniferous and deciduous forests.

Grasses grow well in the dry interior plains of Canada and the United States. Blocked by the Rockies, rainfall on the Great Plains averages only 8 to 20 inches (20 to 51 cm) a year. This is just enough moisture for tall, healthy grasses. Early explorers described this region east of the Rocky Mountains as prairie, which comes from the French word for meadow. A prairie is a temperate grassland characterized by a great variety of grasses. Look at the map on page 91 to locate the prairie regions. Short grasses grow where precipitation is scarcer and tall grasses grow where rainfall is greater. Upon viewing the prairie, the French wrote about the absence of trees and the endless flow of golden prairie grasses.

Beneath the surface of the prairies the roots of the grasses form a layer of dense roots called sod. In some places the sod is as much as 6 feet (2 m) deep. Today, much sod has been broken up

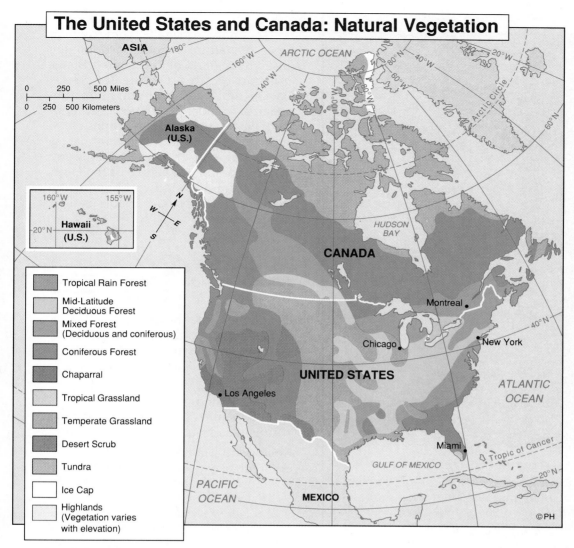

The United States and Canada: Natural Vegetation

ASIA

ARCTIC OCEAN

Alaska
(U.S.)

0   250   500 Miles
0   250   500 Kilometers

Hawaii
(U.S.)

HUDSON
BAY

CANADA

Montreal

Chicago

New York

UNITED STATES

Los Angeles

ATLANTIC
OCEAN

Miami

Tropic of Cancer

GULF OF MEXICO

PACIFIC
OCEAN

MEXICO

©PH

Legend:
- Tropical Rain Forest
- Mid-Latitude Deciduous Forest
- Mixed Forest (Deciduous and coniferous)
- Coniferous Forest
- Chaparral
- Tropical Grassland
- Temperate Grassland
- Desert Scrub
- Tundra
- Ice Cap
- Highlands (Vegetation varies with elevation)

Applying Geographic Themes

Regions   Canada and the United States share many of the same natural vegetation regions shown on the map. Which natural vegetation regions are found in the United States but not in Canada? What type of vegetation is found in the northernmost regions of both countries? In which part of the United States would you find desert scrub vegetation?

by machinery to create farmland. The region is known as America's breadbasket because of the enormous quantity of wheat and other grains that grow in its fertile soil.

Even grasses cannot grow in the hot, dry climate of the southwestern United States, which receives less than 10 inches (25 cm) of rainfall each year. As the natural vegetation map on this page shows, this desert region supports only cactus and leathery shrubs. Relatively few people live in this region, where water is scarce. Those who do live there find the landscape well suited to raising cattle and sheep

Desert Vegetation in the Southwest
The fishhook barrel cactus (foreground) holds water in its stems. The plant's long roots store extra moisture to sustain the plant during long periods of dryness.

on expansive ranches. This region is described below by a geologist who found great beauty in the striking landscape of the desert.

A good way to get acquainted with such a giant landscape is to fly over it at low altitudes—early in the morning, when the air is still, cool, and smooth, and longer shadows emphasize subtle shapes that will be unrecognizable under the high sun of midday. From our altitude we see only a few shrubs and grasses amid the desolation.

The map on page 91 shows which types of natural vegetation dominate different regions of the United States and Canada. In a few areas, such as Hawaii, much of the natural vegetation remains. In most places however, people have changed the environment so dramatically that little natural vegetation remains. In the next section, you will read about the people of the United States and Canada and about the changes that they have made in their environment.

SECTION 1 REVIEW

Developing Vocabulary
1. Define: **a.** cordillera **b.** province **c.** bedrock **d.** continental divide **e.** drainage basin **f.** tributaries **g.** prairie

Place Location
2. Into what major body of water does the Mississippi River flow?

Reviewing Main Ideas
3. In what ways are the landform patterns of Canada like those of the United States?
4. Explain why inland locations experience colder winters than coastal locations at the same latitude.
5. What type of landform dominates the central regions of Canada and the United States?
6. Describe the arctic tundra of North America.
7. Where is the greatest expanse of coniferous forest located in Canada and the United States?

Critical Thinking
8. **Demonstrating Reasoned Judgment** Based on what you have read about the landscapes and climates of North America, which country do you think offered greater opportunities for settlers, Canada or the United States? Explain your answer.

✔ Skills Check

- ☐ Social Studies
- ☑ Map and Globe
- ☐ Reading and Writing
- ☐ Critical Thinking

Using a Cross-Sectional Diagram

A physical map, like the one on page 84, uses color and relief to show the changes in elevation as if the land were being viewed from above. A cross-sectional diagram shows changes in elevation as if the land were cut and you were viewing one slice from the side. The cross-sectional diagram below shows a slice of the United States from east to west. Changes in elevation appear as sharp spikes and deep depressions. Use the following steps to study the diagram.

1. **Study the geographic region shown in the diagram.** Compare the diagram with the political-physical map on page 84. Look for familiar mountain ranges, valleys, basins, and plains. If the diagram were represented as a line on the map, it would be joining Washington, D.C. and San Francisco. Which landforms on the political-physical map would you expect to see on the cross-sectional diagram?

2. **Analyze specific information shown in the diagram.** This cross-section shows the basic pattern of landforms across the United States. (a) What are the highest and lowest elevations shown on the cross-section? (b) What large portion of the country lies below an elevation of 1,000 feet (305 m)? (c) How does the height of the Appalachian Mountains compare with that of the Rocky Mountains?

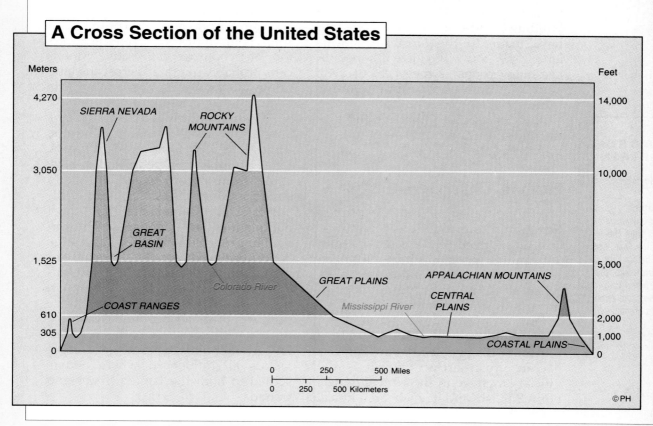

A Cross Section of the United States

© PH

2 Human Geography

Section Preview

Key Ideas

- The populations of the United States and Canada have similar characteristics.
- Population density is related to resources and land use.
- Human interaction provides cultural and economic links between the United States and Canada.

Key Terms

literacy, standard of living, hydroelectricity

The physical geography of a place might be considered a stage on which people work, live, and play. The things people add to their physical environment might be compared to the props used in a play. Think of the roads, buildings, and other additions people have made to the North American landscape. You will realize that people use many props as they act out their daily lives. Just as actors bring unique talents to their roles, the people of the United States and Canada bring distinctive skills and experiences to their activities.

Human Characteristics

The populations of the United States and Canada differ greatly in size. The United States has more than 240 million people, whereas Canada has fewer than 30 million. In many other ways, however, these populations have similar characteristics.

High Standard of Living Three fourths of the people in both countries live in urban areas. English is the official language in the United States; both English and French are official languages in Canada. Look at the table on pages 751–761. Notice that, compared with many of the world's nations, the populations of both countries have long life expectancies, high per capita incomes, and high literacy rates. Literacy is the ability to read and write. Taken together, these and other statistics show that most of the people in these two countries have a high standard of living. Standard of living is a measurement of a person's or group's education, housing, health care, and nutrition.

Statistics like these, however, can be deceiving, for many people in the United States and Canada suffer from poverty, unemployment, lack of education, and inadequate health care.

Cultural Diversity Another similarity between the people of the United States and Canada is the diversity of their cultural backgrounds. In some respects, these countries have no native people. The ancestors of every person who now lives in either country came from another continent at some point in history. Even the ancestors of people we call Native Americans probably migrated across the land bridge of the Bering Strait from Asia about 30,000 years ago.

Most Americans and Canadians have ancestors who came from Europe during the last four hundred years. Large numbers of people also have come from Africa, Asia, and Latin America. Most large cities in the United States and Canada are a mosaic of nationalities and languages. The Canadian province of Quebec has a large concentration of people of French ancestry, while in the southwestern United States there is strong evidence of the region's Spanish heritage. Large numbers of Asian people have settled on the west coasts of both countries. Historical, economic, and geographic factors all influenced where people settled in both the United States and Canada.

WORD ORIGIN

ancestor
The word *ancestor* comes from the Latin for "one who goes before." In modern usage, the word refers to a relative more remote than a grandparent.

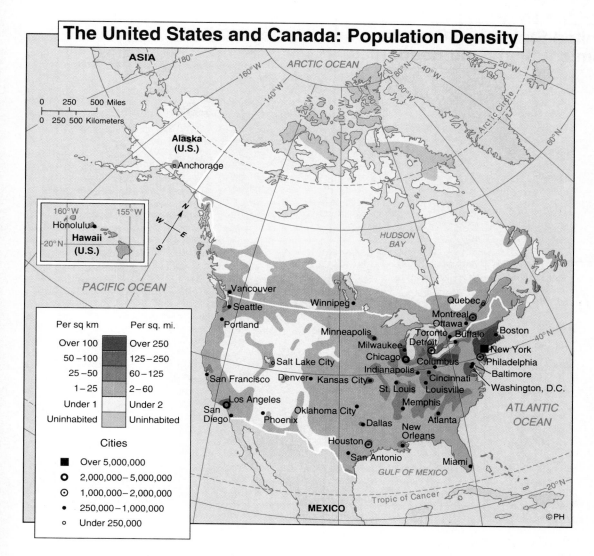

The United States and Canada: Population Density

ASIA

ARCTIC OCEAN

Alaska (U.S.)

Anchorage

Honolulu

Hawaii (U.S.)

PACIFIC OCEAN

HUDSON BAY

Vancouver

Seattle

Portland

Winnipeg

Minneapolis

Milwaukee

Salt Lake City

Chicago

Detroit

Indianapolis

San Francisco

Denver

Kansas City

St. Louis

Los Angeles

Oklahoma City

San Diego

Phoenix

Dallas

Houston

San Antonio

Quebec

Montreal

Ottawa

Toronto

Buffalo

Boston

New York

Philadelphia

Baltimore

Cincinnati

Louisville

Columbus

Washington, D.C.

Memphis

Atlanta

New Orleans

Miami

ATLANTIC OCEAN

GULF OF MEXICO

Tropic of Cancer

MEXICO

©PH

Per sq km	Per sq. mi.
Over 100	Over 250
50–100	125–250
25–50	60–125
1–25	2–60
Under 1	Under 2
Uninhabited	Uninhabited

Cities

■ Over 5,000,000

◉ 2,000,000–5,000,000

⊙ 1,000,000–2,000,000

• 250,000–1,000,000

○ Under 250,000

0 250 500 Miles

0 250 500 Kilometers

Applying Geographic Themes

Place Where are the most densely populated regions of Canada located? Where are the most densely populated regions of the United States? How do the patterns of population density shown on this map compare to climate patterns shown on the map on page 89?

Population Patterns

Look at the population density map above. Nearly two thirds of all Canadians live within one hundred miles of the United States border. Few people live in the northern two thirds of Canada. As you read in the previous section, much of this northern region is cold for most of the year. The desert and mountain regions of the western United States are also thinly populated because of the severe climate and rugged landscape.

Linking Population Density to Resources Regions with low population density most often remain rather unchanged. The tundra region of northern Canada is an example of thinly

The United States and Canada: Economic Activity and Resources

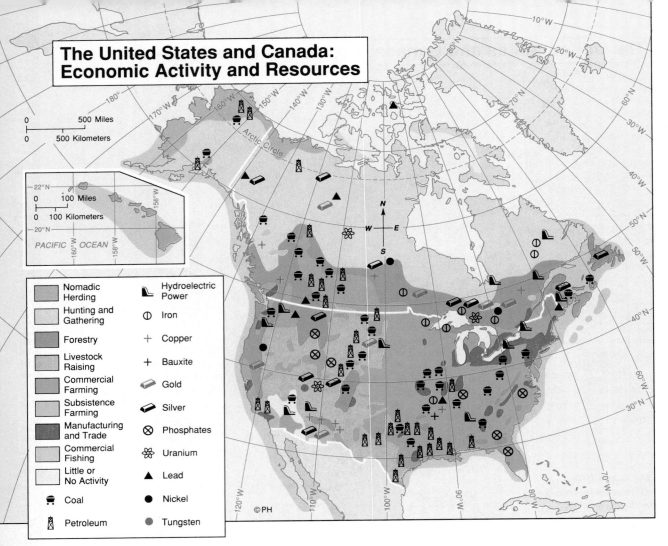

Legend:

- Nomadic Herding
- Hunting and Gathering
- Forestry
- Livestock Raising
- Commercial Farming
- Subsistence Farming
- Manufacturing and Trade
- Commercial Fishing
- Little or No Activity
- Coal
- Petroleum
- Hydroelectric Power
- Iron
- Copper
- Bauxite
- Gold
- Silver
- Phosphates
- Uranium
- Lead
- Nickel
- Tungsten

0 500 Miles
0 500 Kilometers

0 100 Miles
0 100 Kilometers

PACIFIC OCEAN

©PH

Applying Geographic Themes

Movement   The ability to move products has significant influence on a region's economic activity. How does movement along waterways influence the economic activity in the Great Lakes region?

populated land where the natural landscape has not been disturbed by people. Sometimes, however, the development of resources can greatly change the land in spite of low population density. Few people live in the Rocky Mountain region. But mining of the rich mineral deposits that lie under the surface of the mountains has radically altered the landscape in some places. Forestry is another activity that supports relatively few people. However, the harvesting or clearing of trees

for lumber or for papermaking often produces major changes in the physical environment. In some places, entire mountainsides have been stripped bare of vegetation.

Linking Population Density to Land Use   Compare the economic activity map on this page with the population density map. Much of the land in the United States and Canada is used for farming. Yet, because these farms are quite large, the families who own them live far from one another, so relatively few farmhouses dot the countryside. Agriculture significantly changes the natural landscape, however. Compare the map of natural vegetation on page 91 with the economic activity map. Notice some of the ways in which people have changed their natural environment. Forests have been cleared to provide pasture for livestock, and grasslands have been plowed under in order to plant crops.

Now find the manufacturing centers on the economic activity map. Notice how these centers correspond to the areas of greatest population density. Together, the United States and Canada produce about one third of the world's manufactured goods. Most people in these countries live in or near manufacturing centers.

Canada's industry and population are concentrated in the Great Lakes region and along the St. Lawrence River. **Hydroelectricity**, electricity that is generated by moving water, is used to supply power to many of these industries. The cities of Toronto and Hamilton on Lake Ontario are two of Canada's chief manufacturing centers. The nearby city of Detroit in the United States is also a major industrial center and is considered the world auto-manufacturing capital.

Movement of People and Goods

The United States and Canada have abundant natural resources that have helped both nations become industrial giants in the world market. Both countries are major trading partners of each other and as such share close economic ties.

The United States and Canada have long maintained friendly relations. No soldiers guard the border between the two countries, which is the longest undefended border on earth. People and goods pass easily from one country to the other.

Industry in the Great Lakes Region
Detroit, seen in the background, is a major industrial city on Lake Erie. Canada's industry and population are also concentrated in the Great Lakes region.

■■ SECTION **2** REVIEW ■■

Developing Vocabulary
1. Define: **a.** literacy **b.** standard of living **c.** hydroelectricity

Place Location
2. Where are the areas of greatest population density in the United States and Canada?

Reviewing Main Ideas
3. In what ways are the populations of the United States and Canada alike?
4. Which activities in the United States and Canada have produced the greatest changes in the landscape?
5. How is population density related to land use?

Critical Thinking
6. **Drawing Conclusions**   How has the physical geography of North America influenced links between the United States and Canada?

The United States and Canada

*The settlement of North America is a story of both
geographic barriers and opportunities.*

The geography and history of North America are interwoven like the threads of a tapestry. They must be viewed together for the complete picture to emerge. For example, geography provides some of the answers to the questions of how and why people first migrated to North America.

A Land Bridge from Asia

Geographers believe that during the last Ice Age, between 40,000 and 18,000 years ago, much of the world's water was frozen in great continental glaciers. Because so much water was frozen, the sea level dropped sharply, perhaps to as much as 300 feet (91 m) lower than it is today. These lower sea levels exposed a land bridge between Asia and North America where the Bering Strait exists today. If you look at the Bering Strait and the Bering Sea on the map on page 84, you will see the area that was once above sea level.

Most scientists think that people from Asia crossed the land bridge on foot in search of food. The descendants of these Asians, now known as Native Americans, spread across the continents of North and South America during a migration that lasted for thousands of years. Over time, each group developed its own language and culture.

European Migration to North America

Perhaps as early as the 1300s, French and Portuguese fishermen crossed the Atlantic to take advantage of the rich fishing grounds off the coast of Newfoundland, known as the Grand Banks. It is not certain how these fishermen learned of these rich fishing grounds. By the 1500s, many more European explorers, including Christopher Columbus, had sailed across the Atlantic in the hope of finding a western sea route to the riches of China. Instead they found a new world that they knew nothing about. They added what little information they could gather about the new land to their maps. The new world came to be called America.

Compared with the densely populated cities and towns of Europe, this new world seemed a land of overflowing abundance and wilderness. The continent held out to the adventurers all of the resources that were sorely lacking in Europe due to centuries of overuse: land, hardwood forests, and animals such as bears, turkeys, deer, fish, and beavers.

Such images of plenty were enough to pull thousands of European immigrants across the dangerous ocean to the shores of the unknown lands across the Atlantic Ocean. Spanish exploration resulted in the

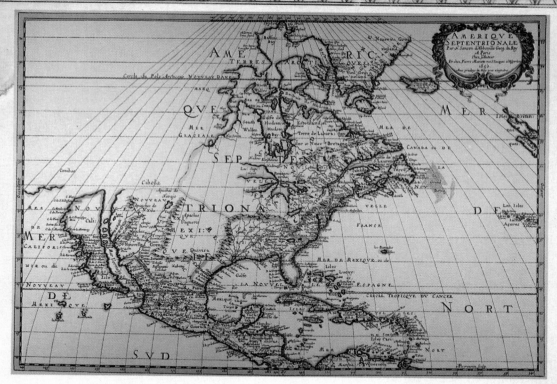

An Eastern Settler's View of North America
When Nicholas Sanson drew this map around 1650, few of the European settlers living along the east coast of North America had any idea how far the continent extended to the west.

settlement of what is today Florida and the southwest United States. Some of the first French and English colonists to arrive in North America during the 1600s and 1700s settled along the east coast of what became the United States and Canada.

Geography played a role in determining where the Europeans settled. French explorers traveled by boat up the St. Lawrence River, building settlements at Quebec and Montreal. English settlers along the coast also traveled inland along rivers until they reached the fall line. The fall line forms an imaginary boundary between the Appalachians and the coastal plain. It is the place where rivers and streams drop over many waterfalls and rapids as they pass from plateau to coastal plain. A string of major cities, including Montreal, Quebec; Baltimore, Maryland; Raleigh, North Carolina; and Co-

lumbia, South Carolina, are among the many cities and towns that grew from fall-line settlements.

Beyond the Appalachians

Few Europeans knew anything about western lands. One settler, Elias Pym Fordham, described North America as "nothing but an undulating surface of impenetrable forest." It was not until settlers braved the journey over the Appalachians in search of more land that they realized that they would find more than dense forest. Some of the first pioneers to do so followed Daniel Boone in 1769 into what is now the state of Kentucky. Thousands more soon followed.

In 1803, President Thomas Jefferson purchased more than 1 million square miles (1,609,000 sq km) of land from France. The purchase of this vast land, called the Louisiana Territory, doubled the area of the young United States and ensured United States control of the mouth of the Mississippi River at New Orleans. Farmers living west of the Appalachians depended on peaceful and free passage down the Mississippi to transport their crops to markets in the east. Jefferson also hoped that the Louisiana Purchase would encourage more easterners to move west and farm. And that they did.

By 1810 one out of every seven citizens of the United States lived west of the Appalachians. Just thirty years later, in 1840, that number had risen to one out of every three. Most settlers lived on farmland cleared in the forests of the Central Plains. Others ventured farther west to the grasslands of the Great Plains.

Crossing the Plains

The first explorers to see the grasslands of the Great Plains were not impressed. The country is like a bowl, declared one observer, so that when a man sits down, the horizon surrounds him all around at the distance of a musket shot. The interior of Canada and the United States contain vast stretches of grasslands. When Zebulon Pike explored the grasslands of Nebraska and Kansas in 1806 he left with a poor opinion of the region. He felt that pioneers would be better off to settle the eastern forests rather than try to cross the endless sea of grass. Those that did settle the grasslands soon discovered that the plains had rich soil, well suited to growing corn, wheat, and other grains.

Westward to the Pacific Coast

While some pioneers stayed to farm the plains, others traveled into unknown lands farther west. They sent back incredible stories about mountains that shone like silver, strange rock formations, and enormous brown and silver-streaked bears.

At the time of the Louisiana Purchase, no one knew exactly how far the lands of the territory extended. In fact, the Louisiana Purchase included almost half of the lands between the Mississippi and the Pacific. President Jefferson organized an expedition to explore the new lands and beyond. The trip was led by Meriwether Lewis and William Clark. Lewis and Clark set out from St. Louis, Missouri, in May 1804. More than one year later they saw the Rocky Mountains, reflecting not silver but the snow that capped them. They wrote,

Nothing can be imagined more tremendous than the frowning darkness of these rocks, which project over the river and menace us with destruction.

Despite these alarming remarks, the Lewis and Clark expedition crossed the Rockies, as well as the Cascades and coastal ranges farther west. In doing so, they opened up the American west to those hardy settlers who were willing to endure the dangers of grizzly bears and mountain crossings to gain more land. Throughout the Canadian and American West, rugged terrain and severe climate made life a challenge for pioneers.

Gold strikes in California in 1849 and in the Yukon—one of Canada's northern territories—in the late 1800s were a powerful force in luring adventurers westward. These adventurers were willing to endure tremendous hardship for the chance of finding gold. One writer described what he calls *Yukon fever*—the frenzied race for the wealth that gold would bring—as follows:

In my mind's eye I can see them. Men like ants, bent under their too-heavy load, plodding in lockstep up the frozen slope. . . . It was the first hurdle in a desperate race to the heart of the Yukon Territory. . . . The prize lay beneath the creeks. . . . The prize was gold.

As thousands of easterners headed west to make their fortune in gold, towns sprang up along trails through the western wilderness. Denver, for example, is located at the base of the Rockies on what was the most heavily traveled route into the mountains.

Westward Expansion

Thomas Otter's painting, "On the Road," shows a railroad and a wagon train, two of the most important modes of transportation used in settling and developing the western regions of North America.

Regions Linked by Railways

The west coast of the United States was very far from the major cities of the East and Midwest. Mountain ranges and rushing rivers that flowed north and south slowed overland travel. Lack of fast transportation was a severe problem for a young nation seeking to function as one United States.

Canada, too, was made up of different regions separated by thousands of miles. The westernmost province of British Columbia was almost completely blocked off from eastern cities like Toronto and Montreal by the immense wall of the Canadian Rockies.

Railroads provided the links that knit together the distant regions of both countries. In 1869 the last link of the transcontinental railroad in the United States was completed at Promontory, Utah. In 1885 Canada completed its transcontinental line from Montreal to Vancouver.

From the beginning, North American geography has provided both barriers and opportunities for the continent's settlers.

Tremendous motivation, skill, and hard work were required to overcome geographic barriers, including the Appalachians and the Rockies, and hundreds of years passed before the blank map of the earliest European settlers was filled in.

TAKING ANOTHER LOOK

1. How do most scientists think that people first migrated to North America?
2. What physical features discouraged settlers from westward migration?
3. Describe some of the resources that encouraged settlers to overcome the physical barriers to westward movement.

Critical Thinking

4. **Synthesizing Information** What do you consider to be today's new frontiers? What barriers currently exist to prevent people from reaching these frontiers? How might these barriers be overcome?

5
REVIEW

Section Summaries

SECTION 1 Land, Climate, and Vegetation
The United States and Canada share the same continent and have similar landform patterns. High mountains extend parallel to the western coast, low mountains stretch along the eastern coast, and vast plains form the interior. The high ridge of the Rocky Mountains forms a continental divide, separating rivers that flow west from those that flow east. Canada's climates are generally colder than those of the United States. Most of the natural vegetation of the two countries falls into four broad categories: tundra, forest, grassland, and desert scrub.

SECTION 2 Human Geography The United States has a much larger population than Canada. Many of the people in both countries enjoy a high standard of living compared with people in other countries of the world. Most live in cities and have ancestors who came from other countries. The majority of Canadians live in the southern part of their country near the United States–Canadian border. In the United States the greatest concentration of people is found in the northeast. Many of the manufacturing and population centers of the United States and Canada are located around the Great Lakes. The two countries have geographic, historic, cultural, and economic ties and are major trading partners of each other.

Vocabulary Development

Use each of the following words in a sentence that shows the word's meaning.

a. cordillera
b. province
c. bedrock
d. drainage basin
e. prairie
f. tributaries
g. literacy
h. standard of living
i. hydroelectricity
j. continental divide

Main Ideas

1. Describe two major landforms shared by the United States and Canada.
2. What are the Great Plains?
3. What is the Canadian Shield?
4. Explain why the Appalachian Mountains have lower elevations than the Rocky Mountains.
5. How do the Rocky Mountains influence the climate of Canada and the United States?
6. What are the four broad categories of vegetation found in most of the United States and Canada?
7. Give two reasons why some regions of Canada and the United States are less densely populated than others.
8. Why are the populations of both the United States and Canada described as culturally diverse?

Critical Thinking

1. **Making Comparisons** Name two similarities and two differences between the physical geography of the United States and Canada.
2. **Distinguishing False from Accurate Images** The descendants of the first people who lived in North America are called Native Americans. Explain why you think this name is or is not accurate.
3. **Recognizing Ideologies** What characteristics and values do you think early settlers needed in order to meet the challenges of North America's physical environment?

Practicing Skills

Using a Cross-Sectional Diagram Look again at the cross-sectional diagram on page 93 and answer the questions that follow.

1. Study the scales shown on the diagram. Every cross-sectional diagram requires two scales: one horizontal and one vertical. The horizontal scale, shown here in miles and kilometers, measures distance across the earth's surface. The vertical scale, shown in feet and meters, measures the height, or elevation, of the land above sea level. These scales are not equal. The vertical scale is much larger than the horizontal scale. As a result, changes in elevation appear more dramatic in the diagram than they are in real life. This difference is known as vertical exaggeration. (a) What distance in miles does one inch represent on the map's horizontal scale? (b) What distance in feet does one inch represent on the vertical scale? (c) What is the distance across the country?

2. Which are higher, the Central Plains or the Great Plains? Do you think that the higher elevation is obvious to the people living in that area?

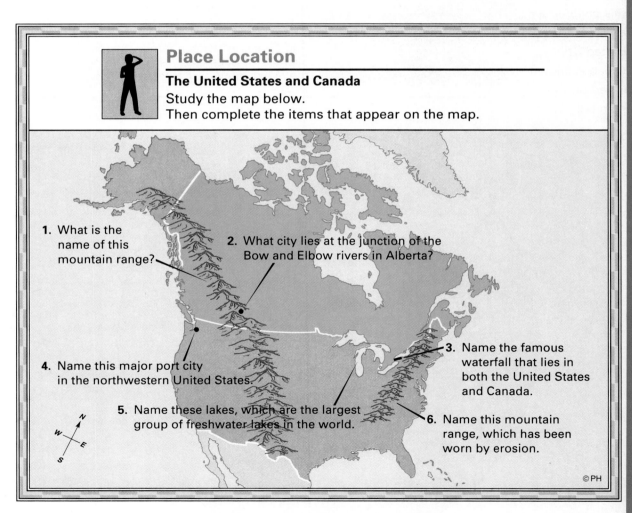

Place Location

The United States and Canada
Study the map below.
Then complete the items that appear on the map.

1. What is the name of this mountain range?

2. What city lies at the junction of the Bow and Elbow rivers in Alberta?

3. Name the famous waterfall that lies in both the United States and Canada.

4. Name this major port city in the northwestern United States.

5. Name these lakes, which are the largest group of freshwater lakes in the world.

6. Name this mountain range, which has been worn by erosion.

©PH

6

A Profile of the United States

Chapter Preview

Both sections in this chapter are about the United States, which is shown in red on the map below.

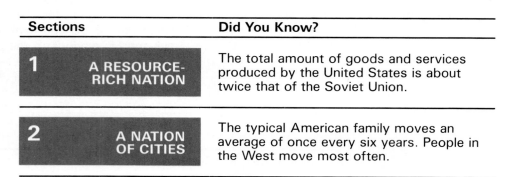

Sections	Did You Know?
1 **A RESOURCE-RICH NATION**	The total amount of goods and services produced by the United States is about twice that of the Soviet Union.
2 **A NATION OF CITIES**	The typical American family moves an average of once every six years. People in the West move most often.

Abundance from the Earth

The combination of moderate climate, fertile soil, modern technology, and hard work helps to produce large quantities of healthy crops such as these carrots, which were commercially grown on a United States farm.

1 A Resource-Rich Nation

Section Preview

Key Ideas

- Abundant natural resources, new technology, and a system of free enterprise led to the economic success of the United States.
- Transportation and communication systems provided links that were vital to the growth of American industries.

Key Terms

rugged individualism, free enterprise

Compared with most countries of the world, the United States is enormous. It is the world's fourth largest country in both area and population. A person could drive more than three thousand miles (4,800 km) from coast to coast and still not leave the United States mainland.

In addition to being large, the United States is wealthy. Today the United States ranks first in the world in the total value of its economic production. By 1990 the nation's gross national product (GNP) was higher than that of any other country in the world. The gross national product is a measure of the total value of all the goods and services produced by a country in a year.

How did the United States become such a wealthy country? At least four factors help answer this question: the hard work of the people who settled the country; the development of new technology; the country's political system; and the abundance of the country's natural resources.

Interaction: An Abundance of Natural Resources

"I think in all the world the like abundance is not to be found." These were the words of Arthur Barlowe, an English sea captain, shortly after his arrival in North America in 1584. The new continent seemed to offer an unbelievable degree of plenty, and with it the promise of wealth.

In 1832 Alexis de Tocqueville, a French visitor to the United States, made this comment about Americans:

Their eyes are fixed upon . . . [their] own march across these wilds, draining swamps, turning the course of rivers, peopling solitudes, and subduing nature.

With their "march across these wilds," Americans used their own hard work to turn the resources provided by nature such as soil, water, and timber, into useful, life-supporting materials.

Farming the Land One of the most abundant resources in the United States was the land itself. For much of the country's history, parcels of land were given to those who would promise to live on them for at least five years. In the 1700s much of the land that was given away or purchased at very low cost was used for farming. When the nation's first census was taken in 1790, more than three fourths of the people lived on farms. The prairie grasses of the Midwest and the Great Plains had some of the richest soils in the world. The Central Valley of California and the interior plateaus of the Pacific Northwest also had fertile soils.

Farming Technology By the 1800s, farmers had the ability to grow huge crops of grain, but without machines for large-scale harvesting they did not have the ability to turn crops into profit. A revolution in agriculture occurred in 1831 when Cyrus McCormick invented a mechanical reaper that was able to harvest the vast wheat fields of Ohio and Indiana.

Cyrus McCormick's mechanical reaper was only the beginning of tremendous improvements in farming methods and technologies. By 1990, American farmers were able to produce and harvest over 2 million bushels of wheat per year. About half of the wheat produced in 1990 was exported for sale outside the United States.

Clearing the Forests Forests were another of the United States' rich resources. No change in the natural landscape of the United States has been as striking as the clearing of vast tracts of forest over the past three hundred years. Many trees were first cut down to clear farm fields and pastures, and countless others were cut for lumber and wood products. After the forests of New England were cleared, loggers worked their way across the northern parts of Michigan, Wisconsin, and Minnesota before heading on to the forests of the Pacific Northwest. Each advance in lumbering technology, including machines for harvesting trees and sawmills, allowed the United States to produce and export more paper and other lumber products for building.

Wealth Beneath the Surface An abundance of mineral resources exists beneath the lush forests and rich soils of the United States. One of the country's most abundant minerals is coal, a solid fossil fuel that is used as a source of energy for industry, transportation, and homes.

Oil and natural gas are other fossil fuels found in the United States. They lie beneath the central and western plains, as well as beneath Alaska. In recent years, Americans became aware that these fuels, as well as coal, are nonrenewable. That is, they cannot

WORD ORIGIN

Wisconsin
The name Wisconsin most likely comes from the Native American word *misconsing*, which means "grassy place."

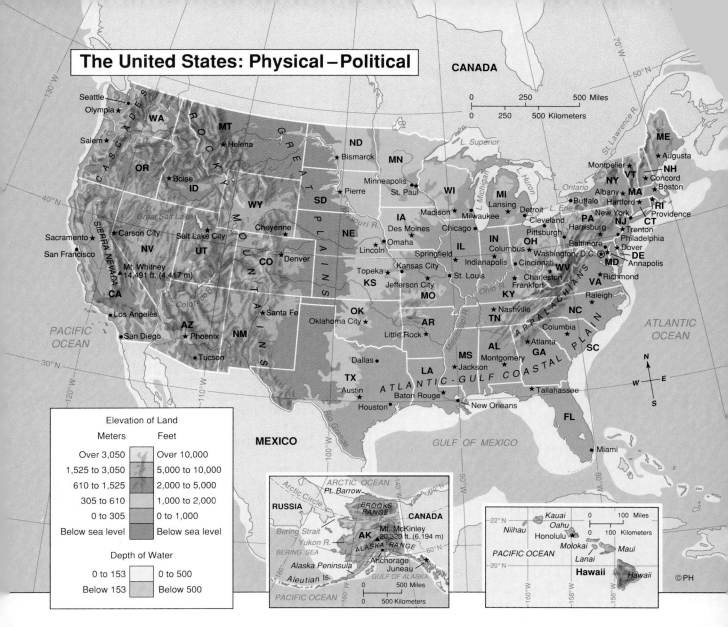

The United States: Physical–Political

CANADA

Elevation of Land

Meters		Feet
Over 3,050		Over 10,000
1,525 to 3,050		5,000 to 10,000
610 to 1,525		2,000 to 5,000
305 to 610		1,000 to 2,000
0 to 305		0 to 1,000
Below sea level		Below sea level

Depth of Water

0 to 153		0 to 500
Below 153		Below 500

Applying Geographic Themes

Location   The United States is made up of three land areas: the forty-eight contiguous states, Alaska, and Hawaii. Its lands extend from north of the Arctic Circle to south of the Tropic of Cancer. Which United States city is located at 30° N, 90° W?

ever be replaced once they are used. Because oil, natural gas, and coal are all extremely vital to the United States energy supply and economy, these precious resources require careful management to ensure that they are used sparingly.

Movement: Resources, Goods, and People

The United States could not have developed its resources without the development of new technologies for transportation. At each of the stages of

Moving by Rail
Workers, many of whom were Chinese immigrants, completed the first transcontinental railroad linking east and west in 1869.

development, transportation provided vital links that allowed producers to move raw materials to factories and finished goods to consumers. Wherever the new transportation routes went, they provided access to raw materials and new customers for farmers and factories.

In the early 1800s, transportation was faster on water than on land, so Americans built canals to make many places more accessible. For example, the Erie Canal, completed in 1825, linked the Hudson River in New York to the Great Lakes in the Midwest. After the canal opened, the trip between Albany and Buffalo in New York State took less than half the time of a wagon trip. And a barge could carry five times more freight than a wagon could take.

At roughly the same time, rivers became two-way, rather than one-way, highways. Unlike human-powered boats, which could only be easily used in the direction in which the river flowed, steamboats sped passengers up and down the nation's largest rivers. Steamboats also turned the Great Lakes into important transportation routes.

Steam-powered railroads later replaced steamboats as the most efficient means of transporting goods. At the start of the Civil War about 30,000 miles (48,270 km) of track had been laid across the United States. Workers laid the last rail that joined the east and west coasts of the country in 1869. By 1900, people and goods in nearly every settled part of the country were within easy reach of a railroad.

The next advance in transportation occurred in the early 1900s, when automobiles became popular. By the 1950s, hard-surfaced roads extended through much of the country. Today high-speed freeways permit people and goods to travel smoothly and rapidly between places.

Improving Communication

The industrial and economic growth of the United States was also closely tied to improved communications. Some people dreamed of faster methods of communicating. Samuel F. B. Morse found a way to make an electric current do the job. By 1848 the country had a telegraph network that stretched south from Maine to the Carolinas and west to St. Louis, Chicago, and Milwaukee. Newspapers talked about the "mystic band" that now held the nation together. The telegraph was the first major form of communication that did not require the transportation of people or paper. Its speed allowed Americans to transfer information in minutes instead of days. American businesses could now communicate more efficiently with the people who supplied raw materials and parts for their machines, as well as with their customers.

One entrepreneur saw a way to harness this new technology. Richard Sears, a telegraph operator in rural

Minnesota, began taking orders for watches by telegraph in the 1870s. Within a few years, his small business enterprise developed into one of the nation's largest businesses—Sears, Roebuck and Company. The company was taking telegraph orders and shipping products to customers throughout the nation.

New means of communication were the center of attention at the 1876 Philadelphia World's Fair. The star of the show was a small device invented that year by Alexander Graham Bell—the telephone. Starting in the 1870s, telephone systems operated in many large American cities. By 1915 a person could call from coast to coast. Today every American makes and receives a daily average of ten telephone calls.

The United States always has been a leader in using technology to harvest raw materials, turn them into products, and transport the finished goods to national and international markets. Production, transportation, and communication links have all been vital to the economic success of the country. So, too, has the political system of the United States.

Respecting the Value of the Individual

One of the most important shared values in the United States is the value of the individual—that is, the belief in individual equality, opportunity, and freedom. It was the notion that any hardworking individual—regardless of his or her wealth, social background, or religion—could find opportunity and success in the United States that drew many immigrants to the country. Many newcomers felt, as one Italian immigrant stated, that in America, "work was rewarded by abundance." The people of the United States have long praised the quality of rugged individualism—the willing-

ness of individuals to stand alone and struggle long and hard for something on the basis of their own personal resources and beliefs.

The government that was established and popularly approved by the people of the United States in 1789 reflected this value of individual liberty and freedom. The democratic ideas on which American government was based allow people to choose political leaders in open elections in which each person's vote counts equally. The United States also supports an economic system based on capitalism, or free enterprise, which allows individuals to own, operate, and profit from their own businesses.

SECTION 1 REVIEW

Developing Vocabulary
1. Define: **a.** rugged individualism **b.** free enterprise

Place Location
2. On which of the Great Lakes does Chicago border?

Reviewing Main Ideas
3. What were some of the abundant natural resources available to the people who came to the United States?
4. Describe three ways in which transportation and communication systems strengthened the nation's economy.
5. How has the United States' system of government contributed to the economic success of the country?

Critical Thinking
6. **Synthesizing Information** Free public education funded by the government has been a hallmark of American life. In what ways do you think education has contributed to the wealth of the country?

2 A Nation of Cities

Section Preview

Key Ideas

- The rate of an American city's growth depended on its location.
- Transportation, economy, and personal preferences affected the growth of United States cities.
- Cities of different sizes serve different functions and are related to one another in an urban hierarchy.

Key Terms

hierarchy, hinterland

A visitor to Vermont in the 1880s found a village nearly abandoned except for a few outlying farms. When he asked where all the villagers had gone, he was told they had left their village to find work in the factory towns and the large cities.

As the country's economy grew, its economy also changed—from an economy based primarily on local agriculture to one based on industry and manufacturing. Most recently service industries began to make up a larger share of the nation's economy. Service industries are businesses that are not directly related to manufacturing or gathering raw materials. Education, health care, banking, entertainment, transportation, and government are all service industries.

As these changes in the economy of the United States took place, life for men, women, and children living in American villages and towns was transformed. In fact, by 1890, many rural places became like the Vermont village described above whose people had left for new lives in the country's growing cities. Cities became the centers of transportation and production in the new industrial economy.

Location: The Growth of Cities

Once a nation dominated by rural settlers, the United States is now largely a nation of city dwellers. The country has more than two hundred and eighty metropolitan areas, nearly forty of them having more than one million people. A metropolitan area is a major city and its suburbs. How did all these metropolitan areas grow so large? Why have some grown faster than others? To answer these questions, keep in mind an old saying: "The value of any parcel of land is determined by three factors. The first is location. The second is location. And the third is location."

Every place on earth is said to be unique. A place is unique because of its absolute and relative location, its physical and cultural characteristics, and the ways in which it is connected to other places. As the United States economy changed, so did the circumstances of each village, town, and city within it. Three factors had an especially important impact on the growth of cities: changes in transportation, changes in economic activities, and changes in popular preferences.

Movement: The Impact of Transportation

For the first fifty years or so following American independence, sailing ships were the fastest and cheapest form of transportation. Cities functioned largely as places to carry on trade between the United States and Europe. All the major cities were busy Atlantic ports; the largest were Boston, New York, and Philadelphia.

Canals As the interior of the continent was developed, settlers came to rely on the country's abundant rivers to transport their crops. Many of the rivers the farmers used were tributaries of the Mississippi. New Orleans grew up at the mouth of the Mississippi on the Gulf of Mexico. The city flourished as trade goods from the Midwest emptied into its harbor and out to the world.

The people of many eastern cities realized that they, too, could benefit from more direct ties to the west. In the early 1800s, the governor of New York, DeWitt Clinton, came up with a daring plan to dig a 363-mile (584 km) canal from Lake Erie to the Hudson River. Western crops, instead of floating south to New Orleans, could be floated east through the Great Lakes into the canal and down the Hudson River to New York City. The plan was a success. The Erie Canal provided the best connection between the east coast and settlements near the Great Lakes.

New cities were established on the shores of the Great Lakes and also along the Ohio, Mississippi, and Missouri rivers. Buffalo, Cleveland, Detroit, and Chicago soon rivaled the older cities of the east. The success of the Erie Canal sparked a canal-building boom. By 1840 a vast canal system, stretching for 3,326 miles (5,352 km), linked the cities of the north and west.

Railroads   The same benefits from trade that motivated the building of canals also motivated the construction of railroads. The first successful railroads were built in the United States in 1830. By 1840, there were as many miles of railroad track as there were of canals. The great wave of railroad building followed about ten years later, creating the arteries that united the nation.

After the end of the Civil War in 1865, railroads became the country's most important form of transportation. Chicago, located centrally between the coasts, had the best location along the railroad network, so it became the largest city in the Midwest. Railroad tracks crossing the Great Plains and Rocky Mountains helped

Gateway to the West
A riverboat winds its way down the Mississippi River as sightseers enjoy St. Louis, Missouri at night. The Gateway Arch was built as a monument to the city's key location and role in the settlement of the West.

cities at both ends of these lines grow rapidly. New York acquired so many people and activities that it became the foremost metropolitan area in the United States.

Automobiles   Until the invention of the automobile, efficient travel was limited to waterways and railroads. But the automobile gave Americans a new freedom—they could go anywhere where there were roads.

The increased availability of automobiles and public transportation, such as trolleys and subways, made it possible for people to travel longer

WORD ORIGIN

Chicago
The name Chicago may come from an Algonquian word meaning "onion place," and first referred to a meadow where wild onions and garlic grew. In 1830 it became the name of the city.

distances to work. After World War II many businesses and people moved from the cities to the suburbs. A 1952 advertisement for Park Forest, a town about 40 miles (64 km) outside Chicago, lured people with these words: "Come out to Park Forest where small-town friendships grow—and you can still live so close to a big city." The scope of cities widened as suburbs grew.

The Impact of Personal Preferences

All of these advances in transportation allowed people more freedom to select where their businesses would operate and where they would live. Today many people choose locations that they feel offer the best possible surroundings. As a result, cities in the South and West, where winters are less severe than in the Northeast, have flourished. Because of the new industries along the Gulf coast, and its cultural attractions, New Orleans, which had declined when railroads replaced steamboat traffic, has grown in importance once again. At the same time, New York, Chicago, and other large centers have maintained their positions because they offer many jobs and varied activities.

Urban Hierarchies

Although today nearly 80 percent of all people in the United States live in metropolitan areas teeming with business and industry, about 20 percent continue to live in towns and villages so small that you might miss them if you blinked while passing through them. Regardless of how small or large each of these places is, they all play a specific role in the nation's economy.

The nation's urban places are often organized in a hierarchy, or rank, according to their function. Smaller places serve a limited area in limited ways, while larger cities provide other, wider-ranging functions.

The next time you're in a grocery store, find a can of peas or sweet corn. Where did these vegetables come from? There may be several answers to this question. The canned vegetables in your grocery store may have been grown in Minnesota, Mississippi, or California. Once they were perfectly ripe, they were harvested and trucked directly to a processing plant. There they were cleaned, tested, cooked, and canned within hours. Processing plants generally operate only a few weeks each year, so many workers have other jobs, tend children at home, or attend school at other times.

The can of vegetables may list the name of a major city. Obviously, the vegetables were not grown in a city. But the headquarters of the company that distributes the vegetables are most

Suburban Sprawl, Levittown, New York
During the 1950s, many Americans moved to suburban communities like this one. Each one of the 1,700 houses in Levittown was identical.

Small Town, Big City

The urban hierarchy plays a large part in Tina Miller's daily life. Tina is a ninth-grader from Palmer, Nebraska, a village of about 500 residents. Tina goes to school and church in Palmer, and her family gets bread and milk every few days at the small grocery store there. Once a week they drive about 20 miles (32 km) to Central City, a town with a population of roughly 3,000, to shop at a supermarket. When Tina visits the doctor, one of her parents drives her 30 miles (48 km) to Grand Island, a city with 33,000 residents. Tina may also go to Grand Island when she wants to buy new clothes, but she really prefers to make the nearly 100-mile (160-km) drive to Lincoln or Omaha, where the big shopping malls offer much greater variety. Occasionally, Tina and her family drive to Kansas City, Minneapolis–St. Paul, or Chicago, (shown at right), where they can watch a major-league baseball game.

1. Why do you think the shopping malls in Lincoln or Omaha are able to offer a great variety of clothing?
2. What role does the city or town where you live play in the urban hierarchy?

likely located in a large city. The corporate executives of the company have their offices there. The advertising campaigns and research also originate there. One hundred years ago, the peas or corn in that can would probably not have gone farther than the dinner table on the farm where they were grown. Because of advances in technology, canned vegetables, as well as farm-fresh produce, can be found in cities many miles from the nearest field.

Different terms describe urban places in each size category. Larger cities are called metropolises, and their

A Complex Economy
The floor of the New York Stock Exchange on Wall Street in New York's financial district hums with energy as traders invest their money in American companies by buying and selling shares of stock.

terlands. Like larger metropolises they engage in activities such as advanced medical care, fine art galleries, major-league sports teams, and sale of the highest quality clothing.

Cities like Des Moines, Albany, Nashville, and Tucson have a more limited range of activities and smaller hinterlands. These are places that usually have large shopping centers, daily newspapers, and computer software stores.

Some towns are small, and their service areas are quite limited. Few people outside the immediate area are familiar with these places. Such towns are home to automobile dealers, fast-food restaurants, and medium-sized supermarkets. Villages often have only small grocery stores. Post offices and video-rental stores may be present in some villages, but a general store often is the only business in the smallest hamlets.

Cities of similar size are not alike in all parts of the United States. They have distinct characteristics based in part on regional differences in the United States. In the next chapter you will read about the nation's four distinct regions.

hinterlands, the areas that they serve, are quite large. For some activities, the hinterland may be the entire United States and much of the rest of the world. For example, New York is the most important financial center in the Western Hemisphere. Chicago is the nation's leading agricultural market, where orders for millions of farm animals and billions of bushels of grain are made. Los Angeles is the world's leading film-production center.

Cities like Atlanta, Denver, Seattle, and Minneapolis–St. Paul are regional metropolises and have smaller hin-

---

![SECTION 2 REVIEW]

Developing Vocabulary
1. Define: **a.** hierarchy
   b. hinterland

Reviewing Main Ideas
2. How did railroads affect the growth of some United States cities?
3. Why have cities in the South and West flourished in recent decades?

Critical Thinking
4. **Drawing Conclusions** How does the urban hierarchy in the United States contribute to interdependence among people?

✓ Skills Check

- ☐ Social Studies
- ☑ Map and Globe
- ☐ Reading and Writing
- ☐ Critical Thinking

Reading a Time Zone Map

In 1884, an international conference developed the present system of twenty-four time zones. The International Date Line—located in the mid-Pacific Ocean at 180 degrees longitude—is the marker for where each day begins. So, if it is Sunday to the east of the line, it is Monday to the west. The following steps will help you to analyze a time zone map.

1. **Study the information on the map.** There are two sets of labels on this time zone map. At the top of the map is a sequence of hours labeled A.M. (before noon) and P.M. (after noon). The Prime Meridian at Greenwich, England, is labeled as "12:00 noon." Each colored band below the hour shares that hour's time. The sample hours shown express accurate readings only when the time is noon at the Prime Meridian. A second set of labels across the bottom of the map indicates the time difference in hours from the Prime Meridian. A value of +3 means a time three hours *later* than Greenwich time, and a value of −3 means three hours *earlier*. For example, if it is 2:00 P.M. in London, what time will it be in a zone bearing the label +7? If it is 3:00 P.M. in Cairo, what time will it be in a city six time zones to the west?

2. **Pinpoint your zone and compare the time difference with other cities.** Time zones make it easy to calculate global variations in time: If it is 2:00 P.M. in the time zone where you live, what time is it in Paris, France?

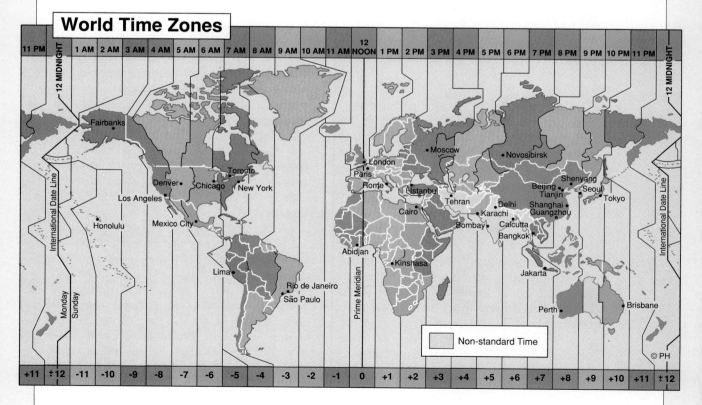

World Time Zones

Fighting the Drug War

The abuse of drugs is one of the most serious problems facing the United States today. In 1988, a survey of persons aged twelve years and older disclosed that more than fourteen million Americans had used illegal drugs in the previous month. In 1989, more than 57,000 Americans were admitted to emergency rooms because of cocaine- and heroin-related problems, and thousands of people died of drug overdoses.

Crime rates in the United States soared as drug use increased. More than two thirds of the males arrested in the nation's largest cities in early 1990 had used drugs within the previous three months. More than one person a day was murdered in Washington, D.C., during 1989 and 1990; police estimate that more than three quarters of these crimes resulted directly from drug use.

The Geography of Drug Production, Movement, and Use

Combating the illegal use of drugs requires a clear understanding of the geography of drug production, movement, and use. The most widely abused drugs—cocaine, heroin, and marijuana—originate as specific plants that grow in certain regions. Cocaine is derived from the leaves of coca trees, which thrive on the sunny, dry slopes of the Andes Mountains in northern South America. Most leaves are harvested in Peru and Bolivia and transported to Colombia to be processed into cocaine powder. In response to efforts to curb production farther south, drug traffickers recently began to grow more coca in Colombia itself.

Heroin is manufactured from the juice of the opium poppy, a flowering plant that grows in many warm regions. Heroin once came mostly from Turkey, then from Mexico. Crackdowns stifled production in those nations, however. Most of the heroin illegally shipped to the United States now comes from Pakistan, Afghanistan, and Iran in Southwest Asia and from Thailand, Myanmar, and Laos in Southeast Asia.

Marijuana comes from resins in hemp plants, which grow well in many tropical and temperate areas. Some marijuana used in the United States is grown illegally within this country, but most is shipped from the same areas of the world that produce cocaine and heroin.

All of these three major illegal drugs are shipped into the United States in a variety of ways. Some are smuggled in along with legitimate cargo on airplanes, ships, cars, and trucks. But large quantities of the drugs are transported on special flights that land at many of the thousands of private airstrips all over the country. Many imports are also hidden aboard ships that enter harbors along the Gulf, Atlantic, and Pacific coasts.

Drugs are used throughout the United States, but they are especially prevalent in metropolitan areas. This urban presence reflects two extremes. On the one hand, many users are residents of wealthy suburbs or prosperous city neighborhoods. On the other hand, drugs also are commonplace in the poorest communities, where many residents view participation in the drug trade as the only way to get money.

Combating the Illegal Use of Drugs

Authorities throughout the world try to combat the illegal use of drugs in many ways. In the United States, government efforts to intercept planes and boats carrying drugs have increased. In some areas, herbicides are sprayed on live plants. Elsewhere, farmers are encouraged to raise other crops. However, the prices they can receive for coca, opium, or marijuana are much higher than they can obtain for any other crops. As long as demand for illegal drugs in the United States remains high, drug traffickers will continue to find sources and routes for their shipments.

TAKING ANOTHER LOOK

1. In which South American country are most coca leaves processed into cocaine powder?
2. Why do you think there is so much violence associated with the illegal use of drugs?

Critical Thinking

3. **Identifying Alternatives** How do you think the United States government can best combat the illegal production, movement, and use of drugs? Why do you think this is the best method?

On Patrol *At many airports, officers use specially trained dogs to intercept the shipment of illegal drugs.*

CASE STUDY ON CURRENT ISSUES

117

6

REVIEW

Section Summaries

SECTION 1 A Resource-Rich Nation
Abundant resources, new technology, and a government supportive of free enterprise all helped to make the United States a wealthy nation. Transportation and communication systems were developed to link the nation's industries to both the raw materials and the consumers.

SECTION 2 A Nation of Cities
The United States has many large cities. Changes in transportation, economic activities, and popular preferences affected the growth of cities. Cities of different sizes are related to one another in an urban hierarchy. Small places function in limited ways for a limited area, while successively larger cities provide major services for an extensive surrounding area.

Vocabulary Development

Use each key term in a sentence that shows the meaning of the term.

a. hinterlands
b. hierarchy
c. rugged individualism
d. free enterprise
e. metropolitan area

Main Ideas

1. Name four factors that influenced the economic success of the United States.
2. Describe two types of technology that were developed to harvest or gain access to the nation's natural resources.
3. Why are fossil fuels important to the United States economy?

4. Why are transportation links important to the economic success of a country?
5. Name three factors that influenced the growth of cities in the United States.
6. How did transportation affect the growth of New York and Chicago?
7. Why do some cities grow and others decline in population over time?
8. Describe two ways the United States government reflects the value of the individual.
9. Why is having a wealth of natural resources not necessarily enough to make a country wealthy?

Critical Thinking

1. **Recognizing Ideologies** How did belief in individual worth influence the American political system?
2. **Drawing Conclusions** How do you think the free enterprise system contributed to the economic development of the United States?
3. **Perceiving Cause-Effect Relationships** How did technology change farming in the United States?
4. **Determining Relevance** How did changes in technology in the early 1900s influence the growth of cities in the United States?
5. **Predicting Consequences** Think about what has happened to American cities over the past two hundred years. What impact do you think the use of computers will have on where people will live and where they will work in the future?
6. **Recognizing Bias** Some people point out the problems that cities have while others point out the many benefits cities offer to people who live in them. What do you think are two of the problems and two of the benefits of city living?

Practicing Skills

1. **Reading a Time Zone Map** Time zone maps help us to picture activities occurring simultaneously across the world. Use the map on page 115 for these questions.

   a. Imagine you have a relative in Sydney, Australia, who writes to you asking to discuss her travel plans with you over the phone. "Please give me a call next Saturday around 2:00 P.M." she suggests. Use the time zone map to determine at what hour you should call from the United States.

   b. When it is 7:00 A.M. in New York, what time is it in Tokyo?
   c. When it is 5:00 P.M. in Moscow, what time is it in Bangkok?
   d. When it is noon in London, what time is it in Delhi?

2. **Reading a Population Density Map** Refer back to the population density map of the United States on page 95.
   a. What are some of the largest cities in the United States?
   b. Where are many of the large cities in the United States located? Why do you think this is true?

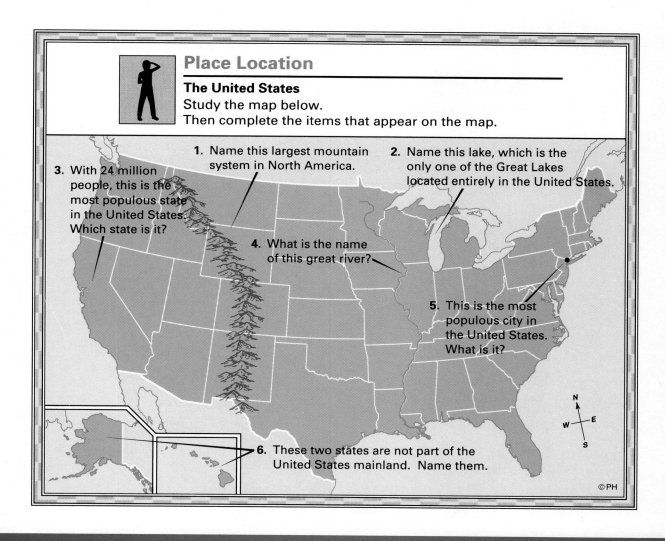

Place Location

The United States
Study the map below.
Then complete the items that appear on the map.

1. Name this largest mountain system in North America.

2. Name this lake, which is the only one of the Great Lakes located entirely in the United States.

3. With 24 million people, this is the most populous state in the United States. Which state is it?

4. What is the name of this great river?

5. This is the most populous city in the United States. What is it?

6. These two states are not part of the United States mainland. Name them.

© PH

7

Regions of the United States

Chapter Preview

Use the colors on the chart to the right and the map below to locate the four regions covered in this chapter.

Sections	Did You Know?
1 THE NORTHEAST	Maine is the only state that borders only one other state.
2 THE SOUTH	The Lake Pontchartrain Causeway in New Orleans, Louisiana, is the world's longest bridge. It is 24 miles (38 km) long.
3 THE MIDWEST	One hundred years ago 70 percent of all Americans were farmers. Today only 3 percent of the population are farmers.
4 THE WEST	The water in Utah's Great Salt Lake is four times as salty as any ocean.

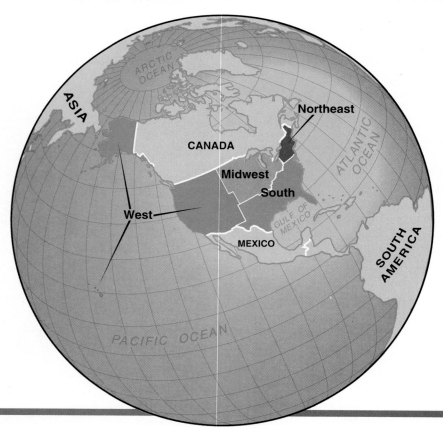

A Lighthouse in Portland, Maine
Some of the Northeast's most valuable resources are in its waters. Lighthouses situated all along the region's coast have guided trading ships and fishing boats safely into harbors. These harbors helped the Northeast to grow as a center of trade, commerce, and industry.

1 The Northeast

Section Preview

Key Ideas

- The Northeast's water bodies have been valuable resources throughout the region's history.
- The Northeast was one of the early sites of the Industrial Revolution.
- The Northeast has become one of the most highly urbanized and densely populated regions of the world.

Key Terms

suburb, megalopolis

When a group of American geographers visited the former Soviet Union in the late 1980s a Soviet geographer wanted to know more about the regions of the United States. So he talked with nine of the visiting geographers, spending at least an hour with each of them. At the end of the last talk, the Soviet smiled. "You Americans all have different ways to divide your country into regions," he declared, "and all of you are right!"

Can there be that many ways to define the regions of the United States? Indeed there can. People use the concept to identify places that have similar characteristics or close connections to one another. Regions may be defined historically; by the ways people live, work, and play; or by their political preference. As the maps in Chapter 5 show, landforms, climate, and vegetation all suggest different boundaries for North America's physical regions. The land use and population density maps suggest other divisions, based on human characteristics.

In this text, however, we look to the United States government for regional classifications. The government, for the purpose of collecting statistics, divides the country into four major regions: the Northeast, South, Midwest, and West. Look at the map on page 120 to identify these regions. The government's definition of these regions is based on a combination of physical, economic, cultural, and historical factors, many of which are examined in this chapter.

The Northeast at a Glance

Ogden Tanner, a writer whose ancestors settled in New England, a region that includes six states in the Northeast, wrote in the 1970s:

> *I think if I had to show someone New England only at one instant, in one time and place, it would have to be this: from a canoe suspended on a silver river, surrounded by the great, silent autumnal explosion of the trees. On the hills, the evergreens stand unchanging . . . Scattered in abstract patterns through their ranks, the deciduous trees . . . produce the glorious golds, oranges, reds, and purples . . . [of] a New England autumn . . .*

This brilliance is a result of the geography of the region. The unique combination of precipitation, type of soil, and the varieties of trees that thrive in the region give New England trees breathtaking fall colors. But the Northeast is far more than forests.

A visitor wanting a broad view of the Northeast might head for the craggy coast of Maine, New York's spectacular Niagara Falls, or the rolling farmlands of Pennsylvania. Still another distinctive feature of this region is its cities. Every year, millions of tourists flock to the Northeast just to explore such world-famous cities as New York, Boston, and Philadelphia.

Natural Resources of the Northeast

Compared to other regions of the United States, the Northeast has few natural resources. The region's soils are generally thin and rocky with steep hills that make farming difficult, keeping the size of most farms relatively small. The northern reaches of the Appalachian Mountains make some areas quite rugged. While the Appalachian area of Pennsylvania is rich in coal, the rest of the region is poorer in mineral resources.

The most valuable natural resources of the Northeast are in its waters. Since Colonial times, the rich Grand Banks in the North Atlantic Ocean have yielded large harvests of fish. The growth of the fishing industry and trade was aided by the region's jagged shoreline, which provides many excellent harbors. Throughout the 1700s and 1800s, these harbors helped the Northeast grow as a center of trade, commerce, and industry.

A Leader in Industry

The Northeast's many rivers, including the Connecticut and the Hudson, have been vital to its history and have helped the Northeast become a center of commerce. The same steep hills that hindered farming aided industrialists in the nineteenth century. The abundant precipitation, about 40 to 60 inches (100 to 150 cm) annually, combined with the hilly terrain, kept the rivers of the region flowing swiftly. Industrialists harnessed the power of these rivers by building water wheels, which converted the water power into machine power.

Throughout the 1800s—especially in Massachusetts, Rhode Island, and New Hampshire—factories powered by water sprang up at waterfalls along the region's many rivers. The factories manufactured shoes, cotton cloth, and other goods that were sold across the United States and shipped to markets around the world. The region's river valleys also served as trade routes, railroad routes, and modern highway routes for the Northeast. The rivers also provided the domestic water supply for the region.

Young people from the Northeastern countryside flocked to the factory towns to take industrial jobs. By the mid-1800s, European immigrants were streaming across the Atlantic to the Northeast. In 1840, about 80,000 Europeans immigrated to the United States; by 1850, the number skyrocketed to 370,000. Many of these immigrants hoped to earn a better living in one of the Northeast's many factories that dotted the landscape.

In addition to a landscape that favored industry and trade, the people who settled in the Northeast pioneered many new inventions that affect the way we live today. Notable Northeasterners include Eli Whitney, whose concept of interchangeable parts in 1798 paved the way for mass production. Thomas Edison, who invented the phonograph in 1877, the light bulb in 1879, and "talking pictures" in 1914, did most of his work in the Northeast. By the early 1900s, the Northeast held the position as the most productive manufacturing area in the world.

The Rise of the Megalopolis

Cities along the Atlantic coast first grew in importance as harbors of international trade and as centers of shipbuilding. As manufacturing grew, those cities attracted industries that needed a large supply of workers. Decade after decade, new industries sprang up—and the Northeast's cities grew in population. Throughout the 1800s and early 1900s, some of the

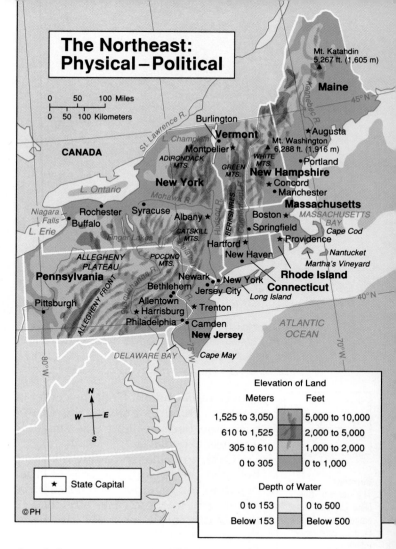

Applying Geographic Themes
1. Location Several ranges of the Appalachian system dominate the landscape of the Northeast. Name two of these ranges and the states where they are located.
2. Place Name three of the major rivers of the Northeast.

nation's most populous cities—New York, Boston, Philadelphia, Baltimore —were located in the Northeast.

Beginning in the late nineteenth century, **suburbs**—the mostly residential areas on the outer edges of a city—grew, too. Streetcars and commuter trains first carried workers from their homes in the suburbs to their jobs in the city. The rail lines resembled

A Multicultural Festival in New York City
New York City has a population of over seven million. Use the map on page 123 to describe New York City's relative location.

WORD ORIGIN

Vermont
The name of this New England state comes from the French, *verd mont,* which means "Green Mountain."

spokes radiating out from the center of a wheel. Suburbs grew at the end of each line because land was available and less expensive.

Over time, like spreading ink spots between the spokes, the coastal cities began to run together. The far suburbs of one city in some cases stretched to the suburbs of the next. By the 1960s, the area from Boston to Washington, D.C., had earned a new name: **megalopolis** (mehg uh LAH puh luhs), a word based on Greek roots meaning a very large city. Almost 50 million people now live in this megalopolis, one fifth of the entire United States population. The map on page 95 shows the population density of the Northeast.

The Problems of a Megalopolis

While the east coast megalopolis remains one of the dominant centers of American business and industry, it faces serious problems, too. After decades of steady expansion, its inhabitants now have serious concerns that the area might run short of water or of facilities for waste disposal.

Another problem facing some cities in the Northeast is the decline of population. Between 1980 and 1990, for example, the population of New York City decreased by about 900,000. As a result, the government of New York City collects fewer taxes from residents and businesses. Thus the city has less money to pay for basic services, such as street repairs, police protection, or garbage collection.

Yet the Northeast remains a vital area. New York City is still "the business capital of the world." New businesses and industries continue to locate in the Northeast. During the late 1980s, many highly technical and computer-related businesses opened their doors in the area. And less populous areas in Maine, New Hampshire, Vermont, upstate New York, and parts of Pennsylvania offer residents more agreeable natural environments in which to live.

■■■ SECTION **1** REVIEW ■■■

Developing Vocabulary
1. Define: **a.** suburbs **b.** megalopolis

Place Location
2. Identify four major cities in the Atlantic Coast megalopolis.

Reviewing Main Ideas
3. How have the Northeast's waters been important in its history?
4. What geographic factors have caused farming to be difficult in the Northeast?

Critical Thinking
5. **Drawing Conclusions** Overall, do you think the growth of the Boston-Washington megalopolis has had a mostly positive or mostly negative effect on the major cities of the Northeast? Support your answer with evidence.

2 The South

Section Preview

Key Ideas

- Warm climates and rich soils produce dense vegetation in many parts of the South.
- The South's climate and natural resources have shaped its economic development.
- Migration to the Sunbelt has led to rapid growth of Southern cities.

Key Terms

mangrove, bayou, Sunbelt

Ask an American where "the South" is and you're likely to be told it's the old Confederacy, the eleven states ranging from Texas to Virginia. Look at the map on page 126 and locate the nine remaining states considered to be part of the South today. Because of conflicts over economic and moral issues, which included tariffs and slavery, these states withdrew from the United States in 1861 and formed the Confederate States of America. Between 1861 and 1865, military forces from the Northeast and Midwest ultimately won a bloody civil war, and the South was reluctantly drawn back into the Union.

The South, as we are defining it here, includes the states of the old Confederacy, plus five others—Oklahoma, West Virginia, Kentucky, Maryland, Delaware—and the District of Columbia. As a region, it stands out from the rest of the country because of the humid subtropical climate and the lush mixed forests that are common in most of the region. One author recently captured the essence of the climate and vegetation in his description of a spring hike through Linville Gorge in the Blue Ridge Mountains of Virginia:

Whenever the trail led us away from the streamside and into the woods, we found ourselves walking through carpets of spring flowers dappled by the sun. Overhead the leaves of maple, oak, and birch trees, barely unfolding from their buds, let far more light penetrate to the forest floor than they would late in the summer.

Linking Climate to Vegetation

The South's location nearer to the Equator makes it warmer than other regions in the United States farther north. In addition, weather systems moving north from the Gulf of Mexico and the Caribbean Sea bring ample precipitation to most of the region. The coastal areas of Louisiana and Mississippi receive well over 60 inches (152 cm) of precipitation annually. Parts of Florida receive an average of 55 inches (140 cm) of rain per year.

In general, the farther west one moves within the South, the less the average annual precipitation. Oklahoma and western Texas have a warm, semiarid climate. Parts of Oklahoma average 30 inches (76 cm) of rain per year; El Paso, Texas, only 8 inches (20 cm). Such a climate supports the temperate grasslands known as prairie.

Most of the South, however, has a warm, wet climate that produces thick mixed forests of pine and other trees, or marshy stands of mangrove trees. Mangroves are tropical trees that grow in swampy ground along coastal areas. Other vegetation regions unique to the South include the bayous (BY oos)— marshy inlets of lakes and rivers—of Louisiana. In Florida, the Everglades—a large area of swampland covered in places with tall grass— provide a refuge for a wide variety of birds and animals. The South's wide variety of plant and animal life is due

> **WORD ORIGIN**
>
> **bayou**
> The Choctaw people, who originally lived in Louisiana, called the area's marshy inlets *bayuk*, which means "small stream."

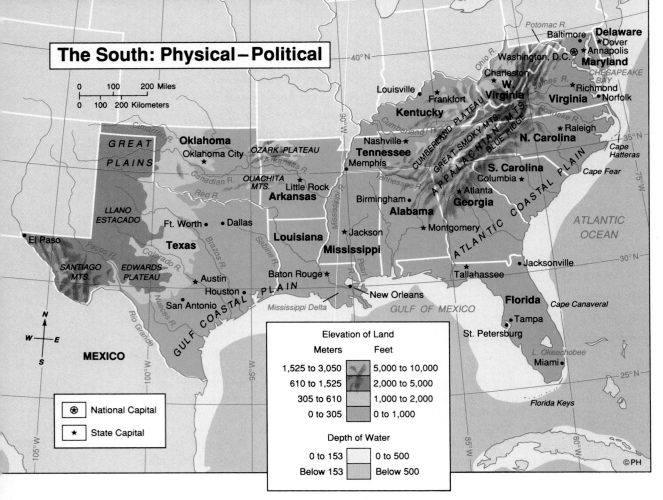

The South: Physical–Political

0   100   200 Miles

0   100   200 Kilometers

Elevation of Land

Meters		Feet
1,525 to 3,050		5,000 to 10,000
610 to 1,525		2,000 to 5,000
305 to 610		1,000 to 2,000
0 to 305		0 to 1,000

Depth of Water

0 to 153		0 to 500
Below 153		Below 500

⊗ National Capital

★ State Capital

Applying Geographic Themes

1. Place Which of the southern states is a peninsula?

2. Movement Which major river in the South has its mouth at New Orleans, Louisiana? Into which water body does this river flow?

not only to the warm, humid subtropical climate of most of the region, but also to the rich soils of the wide coastal plain.

Linking Climate, History, and Agriculture

The first permanent European settlements in the present-day United States were located in the South. In 1565 the Spanish settled in Florida.

As word spread about the South's rich soils and long growing season, more and more Europeans migrated to the region. Some built huge plantations and brought in slaves from Africa and the West Indies to do the work of raising tobacco, rice, or cotton. Today, farming remains an important part of the South's economy.

Despite its mostly fertile soil and mild climate, parts of the South have large areas where people live in bleak poverty. For example, a rural region called Appalachia, in the Appalachian Mountains, is one of the poorest areas in the United States. Its rocky soil and steep slopes make it an unproductive site for farming. At the same time, little new industry has located in the region.

DAILY LIFE

American Barbecue

Although fast-food restaurant chains now make it possible to eat the same foods anywhere in the country, many regional variations in foods and cooking remain. But no two areas of the United States serve and prepare their food in exactly the same way. Take barbecue, for example.

Barbecuers in eastern North Carolina flavor their meats with a vinegar-based sauce. One restaurant in Georgia feeds its fires with charcoal. Another popular spot in Missouri uses hickory.

Many barbecuers, like the one shown at the right, insist that cooking time also affects the taste. Says restaurant owner Warren Clark of Tioga, Texas, "other people say they cook their [meat] 18 or 20 hours. When we get there, we're not even started yet. We cook ours for three days."

1. How do tastes in food vary from place to place?
2. What are some foods that people in your part of the country enjoy?

Linking Resources to Industry

The traditional image of the South has been of a rural region, largely dependent on agriculture. But the South has long had a number of important industries, too.

In the 1880s, entrepreneurs built textile mills in the Piedmont section of the Carolinas on fast-moving streams that provided power. These mills were built on the Fall line—the imaginary line between the Appalachian Mountains and the Atlantic coastal plain. Many cities sprang up along the Fall line in both the Northeast and the South. Textile mills were built close to farms that grew cotton. Even today, textile mills in the Carolinas produce a variety of fabrics.

The oil industry in the South began in eastern Texas in 1901. Some of the United States' largest oil reserves are located in this region. By the 1960s and 1970s, that industry was bringing

The Oil Industry
Some of the United States' largest oil reserves are in the southern states of Texas, Oklahoma, and Louisiana. Oil workers in Oklahoma are shown above.

great wealth to the region. In the 1980s however, oil prices dropped sharply. This decline brought economic hardship to the oil-producing states of Texas, Oklahoma, and Louisiana.

A Changing Region

Until only recently, people often thought of the South as a slow-moving, slow-changing area. In the last few decades, that image has been shattered. As newspaper editor Joel Garreau said in 1981, "Change has become [the South's] most identifiable characteristic."

The Growth of Industry Beginning in the 1950s, both large and small industries began springing up in the South. Some were brand new, such as the space industry that mushroomed in the 1960s in Florida, Alabama, and Texas. But many industries were not

new; they simply moved south from northern cities. Within several years, this migration of business became a huge, steady wave.

A number of reasons make the South attractive to businesses. Industrial plants there were often newer, in better condition, and more efficient than those in the North. New factories could be built on land that was cheaper than in the crowded megalopolis of the Northeast. Because labor unions were much less common in the South, labor costs were usually lower.

Profitable Fisheries The Southern fishing industry is the largest in the nation. Both Atlantic and Gulf coast waters are rich in many kinds of fish.

Looking for job opportunities, thousands of people moved to the South. The band of southern states from the Carolinas to southern California became known as the Sunbelt. The Sunbelt region actually overlaps two other regions—the South and the West. By the 1970s, the Sunbelt's population was growing faster than any other population in the nation.

An Appealing Climate But business growth is not the only reason why the Sunbelt has thrived. Thanks to the South's mild climate, it has grown enormously as a retirement and tourism center. Beaches along the Gulf of Mexico and the southern Atlantic provide welcome relief from northern winters.

Southern Population: A Study in Contrasts

During the 1970s, the South's population increased in number more than any other region of the country—an increase of 7 million, or 20 percent. By the late 1980s, three of the largest cities in the nation were located in the South—Houston, Dallas, and San Antonio. Today the South remains the fastest-growing region in the country, especially the states of Florida and Texas.

A Varied Population Today, the South has a diverse population. Many white Southerners have ancestors who came from England, Scotland, or Ireland. With the rise of the Sunbelt and its increased job opportunities, more blacks, or African Americans, are moving into the South than are moving out of it. This reverses a century-long trend begun after the Civil War, during which blacks were migrating out of the South.

Another large segment of the Southern population is the hundreds of thousands of Hispanics who have moved there from Mexico and other Latin American countries. San Antonio, Texas, is the nation's ninth-largest city. It is also the first city in the United States with a Hispanic majority in its population.

Another large Hispanic group lives in southern Florida—the Cubans. Many Cubans have settled in the Miami area since 1960, after the communist takeover of their homeland. One area of Miami is populated by a Cuban majority. Called Little Havana, it is there that Cuban restaurants and Spanish-language television and radio stations reflect the heritage of the Cubans.

Many people of French ancestry live in Louisiana. The French settled the area in Colonial times and have made a lasting imprint on the region's culture. New Orleans, for example, is famous for its French cuisine.

Major Cities The South is home to many major cities. As you have read, New Orleans has a French flavor, while Cuban culture is a strong force in Miami. Other important urban areas are spread throughout the region.

Houston, Texas, is a large industrial and trading center. Much of the nation's new space exploration is monitored at the NASA headquarters located there. Dallas, about a four-hour drive north from Houston, is the heart of the Texas cattle industry. Houston is a center for the oil and banking industries. Atlanta, Georgia, once a major railroad center, is now a major airline hub.

The city of Washington is not located in any state, but rather in the District of Columbia. This district was carved from the states of Maryland and Virginia when it was chosen as the site for the nation's capital in 1790. Located on the shore of the Potomac River, Washington, D.C., was the first planned city in the nation. Because of its wide avenues, public buildings, and dramatic monuments, many people consider Washington to be one of the most beautiful cities in the world. As the capital, it is home to the nation's leaders and to hundreds of foreign diplomats.

SECTION 2 REVIEW

Developing Vocabulary
1. Define: **a.** mangrove **b.** bayou **c.** Sunbelt

Place Location
2. Locate El Paso and Houston, Texas on the map on page 126. Why do you think El Paso's climate is much drier than Houston's?

Reviewing Main Ideas
3. How does the climate of the South affect its vegetation?
4. Why has the South's population increased so much in recent years?
5. How has the South's economy changed since the 1950s?

Critical Thinking
6. **Making Comparisons** How does the recent industrial growth in the South compare to the industrial growth of the Northeast in the 1800s?

3 The Midwest

Section Preview

Key Ideas

- Agriculture is one distinctive characteristic of the Midwest.
- Because of variations in climate and soil, the Midwest produces a variety of crops and livestock.
- Industries grew up in the Midwest because resources and transportation were available.

Key Terms

silo, growing season, grain elevator, grain exchange

A Bountiful Harvest

The wheat fields of Nebraska testify to the abundance of food crops grown in the United States. Ten percent of the world's total wheat production is grown here.

The states of the Midwest share the center of the United States, but they vary greatly in physical landscape, climate, and economic activity. In the mid-1940s, British writer Graham Hutton met with a dozen Americans to discuss the region of the Midwest. Following the meeting, Hutton reported:

All of us agreed that there was a Midwest. Beyond that, agreement ceased. The region was defined by history, by zones of agriculture, by the [spread] of architecture, by the distribution of forms of local government, by political boundaries, by settlements and migrations, and so on.

Hutton and the Americans were right: the Midwest is a vastly diverse region without clear boundaries. The Midwest includes the lush, wooded hills of the Ozarks in Missouri; it also includes the barren, eroded Badlands of South Dakota. It is the vast blue-green of the Great Lakes and it is the acres of steel mills in the industrialized area around Gary, Indiana.

Yet one aspect most characterizes the Midwest: its broad, flat plains—home of the nation's most productive farms. A writer, Jack Schaefer, described the Midwestern plain in the mid-1950s as follows:

[It is a] natural grassland . . . settled, tamed, plowed, broken to harness. . . . As you push out into this prairie along any of the main east-west routes that slice across it, a sense of the land creeps in and grows until it dominates all else. The few farm buildings merge into their surroundings, natural objects in a natural world, and there is only the land, apparently limitless, serene, indifferent.

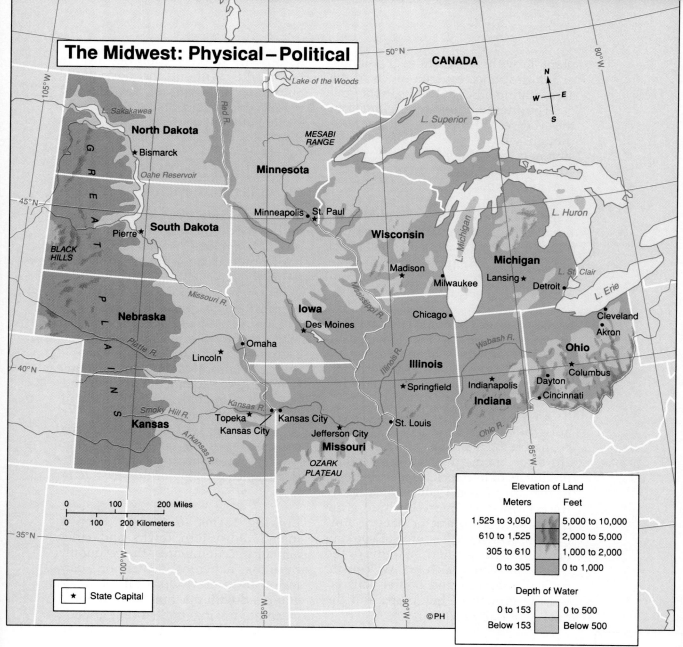

The Midwest: Physical–Political

CANADA

Lake of the Woods

50°N

L. Superior

L. Sakakawea

North Dakota

★ Bismarck

Oahe Reservoir

MESABI
RANGE

Minnesota

105°W

G R E A T

45°N

Minneapolis · St. Paul

South Dakota

Pierre ★

Wisconsin

L. Michigan

L. Huron

BLACK
HILLS

Madison
★

Michigan

Lansing ★

L. St. Clair

Milwaukee ·

Detroit ·

L. Erie

P L A I N S

Missouri R.

Nebraska

Iowa

Des Moines ·

Chicago ·

Cleveland ·
Akron

Platte R.

· Omaha

Wabash R.

Ohio

Lincoln ★

40°N

Illinois R.

Illinois

Indianapolis
★

Dayton ·

Columbus ·
★

Smoky Hill R.

★ Springfield

· Cincinnati

Indiana

Kansas R.

Topeka ★

Kansas City

Kansas City

Kansas

· St. Louis

Ohio R.

85°W

Arkansas R.

Jefferson City ★

Missouri

OZARK
PLATEAU

35°N

| 0 | 100 | 200 Miles |
| 0 | 100 | 200 Kilometers |

100°W

95°W

90°W

©PH

Elevation of Land

Meters		Feet
1,525 to 3,050		5,000 to 10,000
610 to 1,525		2,000 to 5,000
305 to 610		1,000 to 2,000
0 to 305		0 to 1,000

Depth of Water

| 0 to 153 | | 0 to 500 |
| Below 153 | | Below 500 |

★ State Capital

Applying Geographic Themes

1. Location What important industrial cities developed on the banks of the Great Lakes?

2. Place What landform dominates the southern region of Missouri?

The Agricultural Heartland

A drive along any of the numerous highways crisscrossing the Midwest reveals that farms unite this region. Mile upon mile of fields and pastures stretch as far as the eye can see, interrupted only by scattered farmhouses and **silos**—tall, round, airtight buildings for storing grain.

Most of the Midwest is relatively flat, and its soil is especially fertile. In many areas, ancient glaciers deposited mineral-rich materials that promote plant growth. Temperate grasslands, or prairies, once covered much of the area, adding other nutrients to the soil.

The humid continental climate of the Midwest also favors farming. Although winters can be bitterly cold, summers are generally long and hot. At least 20 inches (50 cm) of precipitation fall annually.

Regional Variations　Within the broad expanse of the Midwest are variations in climate and soil that affect farming. For example, eastern Ohio gets twice as much precipitation annually as does central South Dakota. In southern Kansas, the growing season —the average number of days between the last frost of spring and the first frost of fall—is more than 200 days long. Near the Canadian border, the growing season is less than 120 days long.

In the warmer, wetter parts of Illinois, Indiana, and Iowa, corn and soybeans are the major crops. These states are also among the nation's leading producers of livestock, especially hogs. In the drier Great Plains states to the west, farmers are more likely to grow grains such as wheat or oats or sunflowers, which are a source of cooking oil. Along the northern margins of the region in states such as Wisconsin, cooler conditions and poorer soils favor the growth of hay crops and the raising of dairy cattle.

The Nation's Breadbasket　Thanks to favorable natural conditions, Midwestern farms are among the most productive in the world. In recent years, for example, Iowa produced more corn, soybeans, and hogs than any other state in the nation. This output has earned the Midwest the nickname of "the nation's breadbasket."

Midwestern productivity is one factor responsible for the average American being well-fed. This remarkable productivity also allows the United States to export sizable amounts of

its produce to other countries around the world. Without the agricultural output of the farms of the Midwest region, the United States would be a far less affluent country.

The Changing Face of American Farms

In years past, American farms were mostly modest family enterprises, run by single families through long days of hard physical labor. Today, few such farms remain.

Farmers who hope to compete invest in the latest machines, which invariably carry high price tags. Those who can buy the machines find they can now cultivate much more land—if they can afford to buy more. Some farmers have expanded, but thousands have been unable to afford such investments and have "gone under."

While this change has occurred nationwide, it has affected the Midwest the most, because of the greater importance of agriculture to the region.

Linking Farms to Cities　Even in many Midwestern towns and cities, agriculture dominates the economy. Business activities center on dairies or grain elevators—tall buildings equipped with machinery for loading, cleaning, mixing, and storing grain.

Large Midwestern cities, too, are closely linked to the countryside. Some of the tallest office buildings in Minneapolis, Kansas City, and Omaha are homes to companies whose names appear on flour bags and feed sacks. Chicago radio stations broadcast frequent reports from the Mercantile Exchange and the Chicago Board of Trade. The Mercantile Exchange is the world's busiest market for eggs, hogs, cattle, and other farm products; the Board of Trade is the largest grain exchange—a place where buyers and sellers deal for grain.

Linking Industries to Resources

Partly because of its rich supply of natural resources, the Midwest's cities are also home to much heavy manufacturing. Most of the nation's iron ore comes from Minnesota, and sizable coal deposits are found in Indiana and Illinois. The steel industry developed on a large scale in western Pennsylvania. Availability of these minerals spurred development of steel mills in northwestern Indiana and in Ohio. The automobile industry grew up in the Detroit area largely because of the city's location near these steel-making centers.

Transportation and Industry

Look at the map on page 131. It shows that many Midwestern cities—Cleveland, Chicago, Minneapolis, St. Louis, Detroit, Omaha—are located on the Great Lakes or along major rivers. Water transportation aided the growth of heavy industries, such as the manufacture of automobiles and machinery. An average of 583 million tons of goods travel through the Mississippi River system each year.

By the 1880s, with the growth of America's railway system, thousands of railroad cars were pulling into Chicago every year. The trains brought millions of bushels of grain and millions of head of livestock from farms farther west. In Chicago, the grain was processed and the livestock slaughtered. The meat and grain were then shipped eastward by railroad.

The Heart of the Country

Toward the western edge of the Midwest, the climate grows drier, farming gives way to ranching, becomes less intensive, and distances between cities increase. Areas midwestern in climate and landscape blend into areas mostly western. On its eastern and southern boundaries, the Midwest blends similarly into its

Tornado Alley

More tornadoes form in the United States than in any other country in the world. Tornado Alley stretches from north central Texas to Kansas.

neighboring regions. The Midwest's central location truly makes it the crossroads of the United States.

SECTION 3 REVIEW

Developing Vocabulary
1. Define: **a.** silo **b.** growing season **c.** grain elevator **d.** grain exchange

Place Location
2. Explain three reasons why Chicago was ideally situated for growth.

Reviewing Main Ideas
3. What physical characteristics make the Midwest a productive farming region?
4. How were natural resources important in shaping the Midwest's industries?

Critical Thinking
5. **Synthesizing Information** Explain the relationship between the development of new farm machinery and the declining number of American farms.

✓ Skills Check

☐ Social Studies
☑ Map and Globe
☐ Reading and Writing
☐ Critical Thinking

Interpreting a Weather Map

The weather report on the television news gives you information about temperature, precipitation, and the movement of air masses. The map below shows the weather patterns in the continental United States on one day. Use the following steps to analyze the map.

1. **Evaluate the information pertaining to temperature.** The map legend uses seven colors to show temperature ranges for the country. After you have studied the key, answer the following questions. (a) What region of the country is the coldest? (b) What city is the coldest? (c) Name two cities that have temperatures in the 60s.

2. **Evaluate the information pertaining to precipitation.** Study the symbols used to show showers, rain, and snow. Then answer the following questions. (a) Name two states that are experiencing showers? (b) What state is experiencing rain?

3. **Evaluate the information pertaining to weather fronts.** A weather front is the boundary between two air masses of different temperature and humidity. There are three basic types of fronts: cold, warm, and stationary. The arrows and half-circles on this map indicate the direction in which the fronts and air masses are moving. Fronts often bring to an area sudden and dramatic changes in weather. A warm front is followed by warmer weather and a cold front by colder weather. Study the fronts shown on the map and then answer the following questions. (a) What two cities can expect colder weather in the days ahead? (b) Is Salt Lake City's weather likely to change in the next few days?

Weather Map

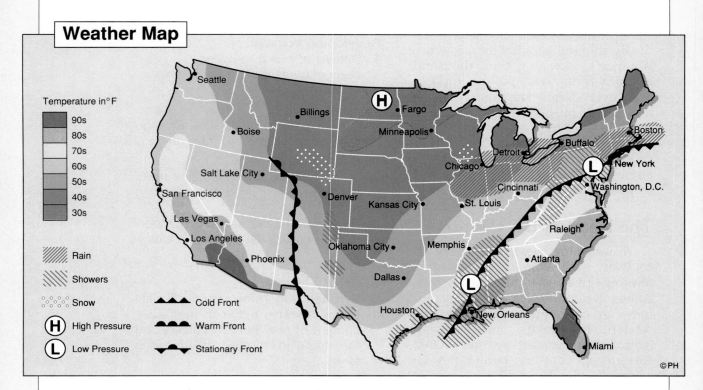

4 The West

Section Preview

Key Ideas

- The West's natural landscape is its most outstanding feature.
- Abundance or scarcity of water affects natural vegetation, economic activity, and population patterns in the West.

Key Term

aqueduct

Waikiki Beach, Hawaii
This seaside resort on Oahu Island is located in the southeast section of Honolulu. Diamond Head, an extinct volcano, is shown in the background.

Breathtaking natural landscape— this is the most memorable feature of much of the West. Towering snow-capped peaks rise throughout the Rocky Mountains. Rivers have carved spectacular canyons. Broad plains sweep on for hundreds of miles. Massive glaciers loom over icy Alaskan waters, while smoking volcanoes frequently spill red-hot lava over the Hawaiian land. The landscape of the West is varied and magnificent, but the physical characteristic that most affects the region is water.

Available Water

The abundance or scarcity of water is the major factor shaping the West's natural vegetation, economic activity, and population density. Looking again at the climate map on page 89, you will notice that most of the West has a semiarid or arid climate. San Diego, California, averages 9 inches (23 cm) of rain per year; Reno, Nevada, gets only 7 inches (18 cm). In dry areas such as these, the natural vegetation consists of short grasses, hardy shrubs, sagebrush, and cactus.

In contrast, the areas on the western side of the cordillera generally receive adequate rainfall and contain rich deciduous and coniferous forests at lower elevations. Along the northwest coast—in Seattle, Washington, for example, where rainfall averages 39 inches (99 cm) per year—lofty Douglas fir trees and giant redwoods thrive.

Hawaii and Alaska offer another contrast. Much of Hawaii has a wet and tropical climate and dense tropical rain forest vegetation. A world apart is northern Alaska's tundra—a dry, treeless plain that sprouts grasses and mosses only in summer, when the top layer of soil thaws.

Resources and the Economy

Beneath the jagged peaks of the Rocky Mountains and the Sierra Nevada lies an immense storehouse of minerals—gold, silver, uranium, and other metals. When this wealth was discovered in the mid-1800s, prospectors and settlers swarmed into the area. As towns grew up to provide the newcomers with necessary goods and services, the population of the West grew rapidly.

WORD ORIGIN

cordillera
The Spanish named the long mountain range that winds down North America for the word *cordilla,* which means "little rope." The Spanish word for rope is *cuerda.*

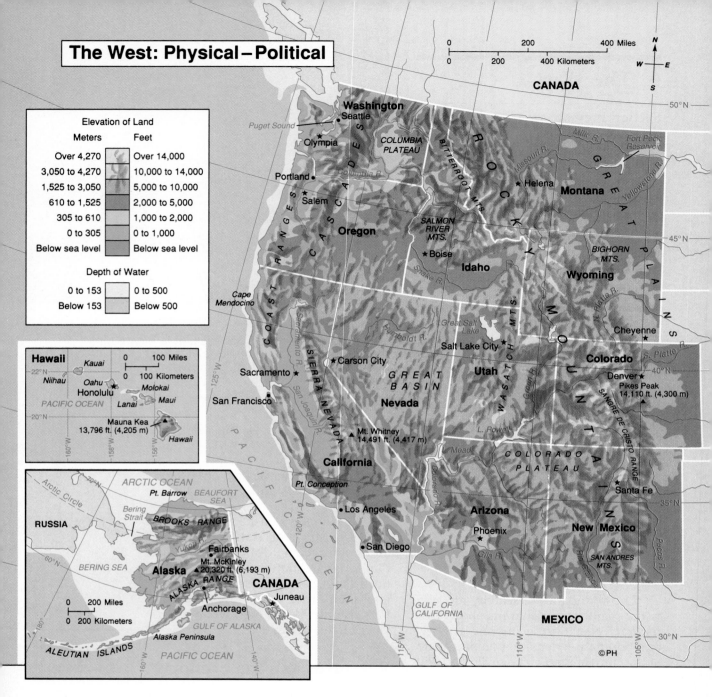

The West: Physical–Political

Elevation of Land

Meters		Feet
Over 4,270		Over 14,000
3,050 to 4,270		10,000 to 14,000
1,525 to 3,050		5,000 to 10,000
610 to 1,525		2,000 to 5,000
305 to 610		1,000 to 2,000
0 to 305		0 to 1,000
Below sea level		Below sea level

Depth of Water

0 to 153		0 to 500
Below 153		Below 500

CANADA

Washington
Seattle
Puget Sound
Olympia
COLUMBIA PLATEAU
BITTERROOT MTS.
Milk R.
Fort Peck Reservoir
Missouri R.
Helena
Montana
Portland
Columbia R.
Salem
Oregon
SALMON RIVER MTS.
Boise
Idaho
Snake R.
Yellowstone R.
BIGHORN MTS.
Wyoming
ROCKY MOUNTAINS
GREAT PLAINS
Cape Mendocino
COAST RANGES
CASCADES
Humboldt R.
Great Salt Lake
Salt Lake City
Carson City
Sacramento
San Francisco
GREAT BASIN
Nevada
Utah
WASATCH MTS.
N. Platte R.
S. Platte
Cheyenne
Colorado
Denver
Pikes Peak 14,110 ft. (4,300 m)
SIERRA NEVADA
San Joaquin R.
Mt. Whitney 14,491 ft. (4,417 m)
L. Powell
L. Mead
SANGRE DE CRISTO RANGE
California
Pt. Conception
Los Angeles
COLORADO PLATEAU
Colorado R.
Santa Fe
San Diego
Arizona
Phoenix
Gila R.
New Mexico
Pecos R.
SAN ANDRES MTS.
PACIFIC OCEAN
GULF OF CALIFORNIA
MEXICO
©PH

Hawaii

0	100 Miles
0	100 Kilometers

Kauai
Niihau
Oahu
Honolulu
Molokai
Lanai
Maui
PACIFIC OCEAN
Mauna Kea 13,796 ft. (4,205 m)
Hawaii

ARCTIC OCEAN
Arctic Circle
Pt. Barrow
BEAUFORT SEA
Bering Strait
BROOKS RANGE
RUSSIA
Yukon R.
Fairbanks
BERING SEA
Alaska
Mt. McKinley 20,320 ft. (6,193 m)
ALASKA RANGE
CANADA
Juneau
Anchorage

0	200 Miles
0	200 Kilometers

GULF OF ALASKA
Alaska Peninsula
ALEUTIAN ISLANDS
PACIFIC OCEAN

Applying Geographic Themes

1. Location Which two states do not border the rest of the United States?

2. Place Which major landform dominates the state of Nevada?

Deeper still within their rugged surface, Western lands also contain valuable deposits of natural gas and oil. The discovery of a major oil field near Prudhoe Bay, Alaska, in the 1960s led to the transformation of that state's economy. The Alaska pipeline, built in the 1970s, carries crude oil across the tundra south to Alaska's Prince William Sound.

The West's natural resources support two other important economic activities—forestry and fishing. About 40 percent of the nation's lumber is from the Pacific Northwest, while commercial fishing thrives in Alaska, Hawaii, and other Pacific Coast states.

The Growth of Western Cities

The completion of the first transcontinental railroad in 1869 spurred the growth of towns and cities along the ribbon of silvery track. In the 1880s the railroads lowered the fare between the Midwest and Los Angeles to only one dollar. Thousands jumped at the opportunity and moved out West. Because of the harsh landscape and climate, relatively few people settled in its countryside. Even today, a higher percentage of the West's population lives in cities.

Anchoring the southwest corner of the continental United States is the nation's second-largest city, Los Angeles. It began as a transportation center and by the 1920s the city was attracting new residents with the development of the airplane and movie industries.

The city has built projects to bring water in from hundreds of miles away. For example, more than half the water in the Owens River valley, 130 miles (209 km) north of Los Angeles, is brought into the city through a complicated system of **aqueducts**, large pipes that carry water over long distances. Such projects were needed for agriculture in the San Joaquin Valley and to support the growing population of Los Angeles.

Conquering Western Distances

The two outlying states of the Western region face distinct challenges in surmounting distances. Alaska is the largest state, but it is one of the least populated. Slightly more than 500,000 people live in an area larger than all of the Northeast. Very few roads pass through the rugged mountains—the Alaska Range and the Brooks Range—covering much of Alaska. Juneau, the state capital, can be reached only by boat or airplane. Even Anchorage, a city with more than 200,000 residents, has only two roads leading out of town.

Hawaii is made up of eight large islands and many smaller ones in the central Pacific Ocean. Hawaii is located more than 2,200 miles (3,540 km) from the mainland United States. Jet travel has made Hawaii a popular destination for tourists from North America and Asia. Until the recent development of communications satellites, Hawaiians relied on radios for news from the mainland. Technological improvements have helped shorten the distance between Hawaiians and the rest of the nation's people.

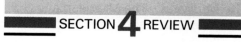

SECTION 4 REVIEW

Developing Vocabulary
1. Define: aqueduct

Place Location
2. Identify the states through which the Rocky Mountains pass.

Reviewing Main Ideas
3. How have farmers made dry areas of the West productive?
4. What is one way in which Los Angeles has provided enough water for its residents?

Critical Thinking
5. **Predicting Consequences** More and more tourists visit the West's national parks every year. What might be the consequences if this trend continues?

7

REVIEW

Section Summaries

SECTION 1   The Northeast   As one of the earliest sites of the Industrial Revolution, the Northeast became a world leader in commerce and industry, a position it still holds today. The area from Boston to Washington, D.C., is a region of high population density that is sometimes called a megalopolis. Although some cities are losing population to the Sunbelt, the Northeast remains a vital region.

SECTION 2   The South   Because of its warm climate and rich soil, most of the South has dense vegetation. Farming is the major economic activity, but recently the region, part of the so-called Sunbelt, has seen increased growth in business, industry, and population. The region's population is diverse, and many large cities are in the South.

SECTION 3   The Midwest   The Midwest, a diverse region, is sometimes called the heartland of the nation. Highly productive agriculture results from a humid continental climate, good soil, and a relatively long growing season, and is the distinguishing characteristic of this region. Many industries have developed in the Midwest because of available resources and a sound transportation system.

SECTION 4   The West   A beautiful, varied landscape is the distinguishing feature of the West. Water is the key factor in area development. Farmers use special techniques to raise crops in an arid climate. Industries are based on the region's natural resources. Los Angeles, the nation's second-largest city, faces problems associated with rapid growth. Alaska and Hawaii are unique regions that face the challenge of great distance from the other forty-eight states.

Vocabulary Development

Use each key term in a sentence that shows the meaning of the word.

a. suburb
b. megalopolis
c. mangrove
d. bayou
e. silo

f. growing season
g. grain elevator
h. grain exchange

Main Ideas

1. How did population migration help industry grow in the Northeast in the 1800s?
2. How did the Northeast's coastal cities become a megalopolis?
3. Why has the South grown so much in recent years?
4. Name five distinct population groups that live in the South.
5. Why is Washington, D.C., considered an international city?
6. Why is the Midwest so important to the rest of the nation?
7. How are the Midwest's cities linked to the surrounding farmlands?
8. How have various physical factors of the West shaped its development?
9. Name five mineral resources found in the West that aided its settlement.

Critical Thinking

1. **Formulating Questions**   San Diego, California, and Houston, Texas, are two cities that have grown very fast in recent years. Make up five questions that will help you discover what types of challenges face these two cities as they adjust to growth.

2. **Making Comparisons** In what ways is the Midwest like the Northeast? In what ways is it like the South?
3. **Distinguishing False from Accurate Images** Do the words *hot, dry,* and *empty* provide an accurate description of the West? Why or why not?
4. **Identifying Alternatives** Choose one basis—physical, economic, human, or historical—on which to divide the United States into regions different from the ones presented in this chapter. Give reasons that explain your choices.

Practicing Skills

Interpreting a Weather Map Use the weather map on page 134 to answer the following questions:
1. What are the temperatures in southern Texas and Florida on this particular day?
2. Describe the weather in Phoenix, Arizona.
3. Tell what the weather is like in Seattle, Washington.
4. What is the forecast for Salt Lake City, Utah, likely to include in the days ahead?

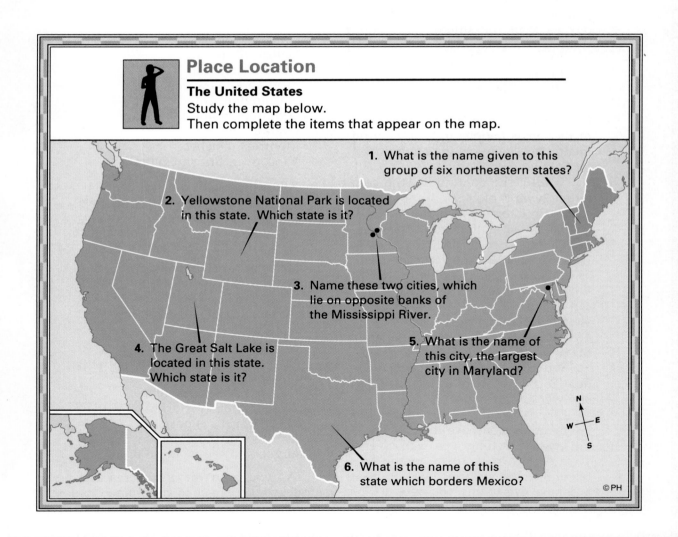

Place Location

The United States
Study the map below.
Then complete the items that appear on the map.

1. What is the name given to this group of six northeastern states?
2. Yellowstone National Park is located in this state. Which state is it?
3. Name these two cities, which lie on opposite banks of the Mississippi River.
4. The Great Salt Lake is located in this state. Which state is it?
5. What is the name of this city, the largest city in Maryland?
6. What is the name of this state which borders Mexico?

©PH

Place Location: WHERE ON EARTH?

The Next World's Fair

You're a television news reporter and you have just returned from Alaska where you've filmed a series on the petroleum industry. Today is your first day back in the studio. Your editor walks up and claps you on the shoulder. What she tells you isn't exactly good news.

"Remember the committee formed to pick the site of the next world's fair?" she asks. You nod. "The theme of the fair is people and the environment." You've made the environmental field your specialty.

"Well," she continues, "the committee won't officially announce its choice until tomorrow. But, thanks to a source, we've gotten our hands on a photocopy of their press release. We want to broadcast it on tonight's news as our lead story. The only problem is that the photocopy is smeared, and we can't read the name of the city. Our source has another copy, but he can't be reached for seven hours. We go on the air in three."

You guess what she expects you to do. "You want me to figure out which city will be the site of the fair?"

"That's right," your editor says. "Remember, you have three hours."

Losing no time, you walk over to your desk. On it are a computer, a telephone, an almanac, and an atlas. You have everything you need.

With a flick of a switch, you start the computer and call up your file on the world's fair committee. You recall that the committee had been quite secretive about the selection process. However, you had managed to learn that ten North American cities were still in the running. The list appears on your screen as follows:

San Francisco, California
Chicago, Illinois
Dallas, Texas
Atlanta, Georgia
New York, New York
Washington, D.C.
Toronto, Ontario
Ottawa, Ontario
Montreal, Quebec
Vancouver, British Columbia

Following this is a list of the committee's criteria for the site of the next world's fair. All the requirements have to be met.

Criteria for World's Fair Site

The world's fair city must:

1. not be a national capital.
2. be located on an ocean, a major lake, or a major river.
3. be located within the continent, not on either coast.
4. have one official language.
5. have previously hosted a world's fair or world's exposition.
6. be in a grassland region to allow fairgoers to see a prairie preserve.

As you scan the list, a research strategy takes shape in your mind. You can use the atlas and the almanac to check each city against the criteria. You should have no problem beating the

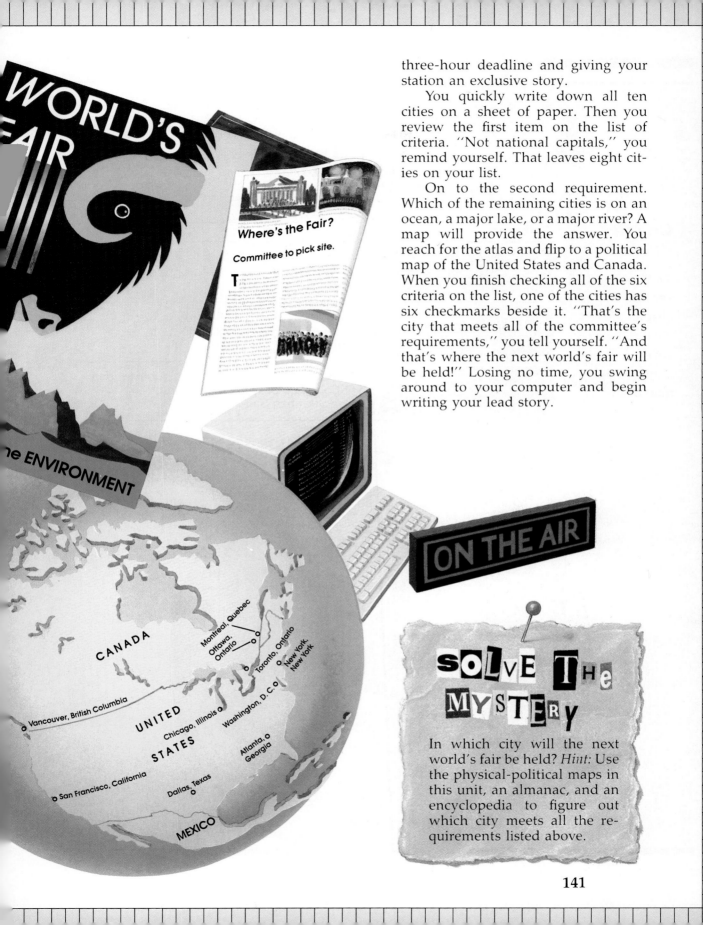

three-hour deadline and giving your station an exclusive story.

You quickly write down all ten cities on a sheet of paper. Then you review the first item on the list of criteria. "Not national capitals," you remind yourself. That leaves eight cities on your list.

On to the second requirement. Which of the remaining cities is on an ocean, a major lake, or a major river? A map will provide the answer. You reach for the atlas and flip to a political map of the United States and Canada. When you finish checking all of the six criteria on the list, one of the cities has six checkmarks beside it. "That's the city that meets all of the committee's requirements," you tell yourself. "And that's where the next world's fair will be held!" Losing no time, you swing around to your computer and begin writing your lead story.

WORLD'S FAIR

Where's the Fair?

Committee to pick site.

the ENVIRONMENT

ON THE AIR

SOLVE THE MYSTERY

In which city will the next world's fair be held? *Hint:* Use the physical-political maps in this unit, an almanac, and an encyclopedia to figure out which city meets all the requirements listed above.

Montreal, Quebec
Ottawa, Ontario
Toronto, Ontario
New York, New York
CANADA
Vancouver, British Columbia
UNITED STATES
Chicago, Illinois
Washington, D.C.
Atlanta, Georgia
San Francisco, California
Dallas, Texas
MEXICO

8

Canada

Chapter Preview

All three sections in this chapter are about Canada, which is shown in red on the map below.

Sections	Did You Know?
1 REGIONS OF CANADA	All the residents of Canada's Northwest Territories could fit into Toronto's domed stadium.
2 THE SEARCH FOR A NATIONAL IDENTITY	Canada is the largest country in North America, but it has only one tenth the population of the United States.
3 CANADA TODAY	Some scientists estimate that Alberta, one of Canada's provinces, has larger oil reserves than Saudi Arabia.

Banff National Park
Visitors gather at the shores of Lake Louise, in the Canadian province of Alberta. Use the map on page 145 to determine which mountains are shown in the background.

1 Regions of Canada

Section Preview

Key Ideas

- Canada can be divided into five distinct regions based on physical features, culture, and economy.
- Rugged landscapes and a cold climate have limited human interaction with the environment in many parts of Canada.
- Location has played a key role in the development of Quebec and Ontario as Canada's heartland.

Key Terms

maritime, lock

Canada is a vast land that covers most of the northern half of North America. Canada shares many similar physical characteristics with the United States. In other ways, however, Canada is quite different—it is a distinctive nation with its own unique opportunities and challenges.

Canada's ten provinces and two territories can be divided into five regions based on physical features, culture, and economy. As you read in Chapter 7, the regions of the United States overlap one another. The regions of Canada, however, are more distinct than those of the United States. Two reasons for this clear separation are the country's relatively small population and the structure of its government, which gives a great deal of power to the provinces.

The Atlantic Provinces

Tucked into the southeastern corner of Canada are the four Atlantic provinces of Newfoundland, Prince Edward Island, Nova Scotia, and New

Brunswick. Locate these provinces on the map on page 145. As the name of the region implies, all four provinces border on the Atlantic Ocean. The land in this region forms part of the Appalachian Mountains, which extend northward from New England in the United States. Hills covered with thick mixed and deciduous forest and rounded mountaintops highlight the landscape in most of the region. Thousands of lakes and small ponds dot the rugged terrain. As in New England, glaciers once moved across the area, leaving the soil thin and strewn with rocks.

The Atlantic provinces are often called "The Maritimes" because of their close ties with the sea. The word maritime means bordering on or related to the sea. The coastlines of these provinces are marked by hundreds of bays and inlets, providing excellent harbors for fishing fleets. Most residents of this region live along the coast.

The Atlantic provinces are the smallest of Canada's regions, including only about 5 percent of Canada's land and only about 10 percent of its people. Although small in area, the location of the Atlantic provinces has been key to Canada's settlement and development.

Historical Importance The shores of the Atlantic provinces were the first parts of Canada discovered by Europeans. Archaeological evidence indicates that the Vikings, probably under the leadership of explorer Leif Ericson, established settlements in Newfoundland around A.D. 1000.

Economic Activities Europeans are believed to have fished the Grand Banks area off the coasts of Newfoundland and Nova Scotia many years before Columbus arrived in the Americas in 1492. The Grand Banks remain among the world's richest fishing areas. Here, people are able to catch large quantities of cod, lobsters, herring, scallops, and other fish and shellfish. Nova Scotia earns more income from fishing than does any other Canadian province.

Forestry and fishing are important in the Maritimes. Some fruit, vegetable, and dairy farming takes place where the soil and local climate permit. The gentle, rolling plains and fertile soil of Prince Edward Island are particularly well suited to farming. Because it is a small island, and more of its land is close to the moderating influences of water, Prince Edward Island has a milder climate and a longer growing season than the mainland provinces, where farming is difficult.

In recent years, a large number of Maritime residents have found work in tourism and military or defense industries. Rugged coastlines and scenic hills make the region a popular vacation spot for many visitors. Despite the beauty that draws visitors, few natural resources other than the sea abound in the provinces. As a result, the Atlantic provinces are Canada's poorest income producing region.

The Great Lakes and St. Lawrence Provinces

In sharp contrast to the Atlantic provinces, the two provinces surrounding the Great Lakes and the St. Lawrence River are the core of Canada's population and economic activity. The large provinces of Quebec and Ontario make up the heartland of Canada. These provinces are distinguished by three distinct landscapes. The first is the Canadian Shield. It has poor soil and a cold climate but contains rich mineral deposits. Most of both Quebec and Ontario lies within the Canadian Shield. The second landscape is the Hudson Bay Lowlands—a flat, sparsely populated, swampy region between the Canadian Shield and Hudson Bay. The St. Lawrence Lowlands—third of

WORD ORIGIN

Ontario
The name Ontario comes from an Indian term meaning "shining waters." This name is most appropriate for the province because it contains about one fourth of the earth's total supply of fresh water.

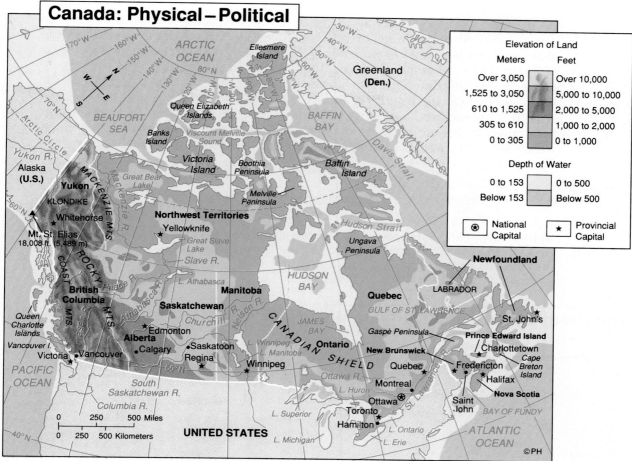

Canada: Physical–Political

Elevation of Land

Meters		Feet
Over 3,050		Over 10,000
1,525 to 3,050		5,000 to 10,000
610 to 1,525		2,000 to 5,000
305 to 610		1,000 to 2,000
0 to 305		0 to 1,000

Depth of Water

0 to 153		0 to 500
Below 153		Below 500

⊛ National Capital   ★ Provincial Capital

0   250   500 Miles
0   250   500 Kilometers

Applying Geographic Themes

1. Regions Canada's provinces and territories are political regions. In which province would you find the capital city of Ottowa? In which province would you find the city of Winnipeg?

2. Location Much of Canada is located in the far northern latitudes. Which lake lies north of the Arctic Circle in the Northwest Territories?

---

the landscapes—have rich soil and a relatively mild climate. Sixty percent of Canada's population lives in this region around the Great Lakes and the St. Lawrence River valley.

Location and Place: Characteristics of Ontario In addition to a central location and excellent waterways, Ontario has rich soil and mineral resources. Much of the land in the southeastern part of the province is used for farming, and it is here that most of the

province's people live. A network of cities has evolved in which a wide range of products—cars, food products, clothing, and building materials —are manufactured and distributed. Because of the province's location, industries based on processing minerals or manufacturing goods can easily ship their products to other parts of Canada and to the United States.

The St. Lawrence Seaway, which includes the Great Lakes, the St. Lawrence River, and several canals, has been called Canada's highway to the sea because of the volume of goods that travels its length. Lake Superior is 602 feet (183.5 m) above sea level. To make up for the differences in water levels, the seaway has a series of locks. A **lock** is an enclosed area on a canal that raises or lowers ships from one water level to another. Canada has taken advantage of the difference in height between the Great Lakes and sea level by constructing hydroelectric plants in several locations.

Toronto, Ontario's capital, is the largest metropolitan area in Canada. More than one third of Canada's largest companies now have their main offices in Toronto. This city also con-

---

Applying Geographic Themes

Movement   Early European explorers traveled inland along the St. Lawrence River from the Atlantic Ocean. The cross-sectional diagram shows the location of locks along the river. Why are these locks important to Canada's industries today?

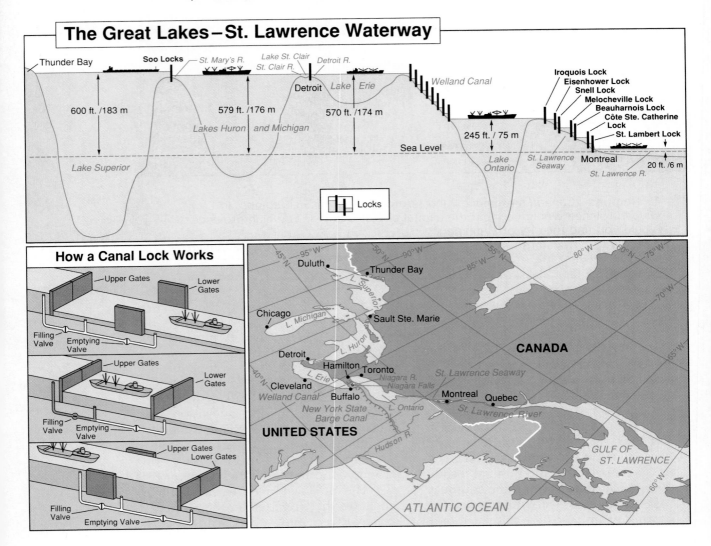

The Great Lakes—St. Lawrence Waterway

tains Canada's banking and financial center, as important to Canada as New York's Wall Street is to the United States. Ottawa, the national capital of Canada, is located on the Ottawa River in southeastern Ontario. The two cities of Ottawa and Hull, Quebec, which lie across the river from each other, make up Canada's fourth largest metropolitan area.

Characteristics of Quebec Although Quebec is Canada's largest province in terms of area, its population is not equally distributed. Most residents live in the cities in and around the St. Lawrence River valley. About nine tenths of the province of Quebec lies in the sparsely inhabited Canadian Shield to the northwest. Most of this region has remained a wilderness of forests, rivers, lakes, and streams. Treeless tundra with lichens and mosses covers the northern parts of Quebec.

The Appalachian Mountains rise gently along the southeastern border of the province. Both of these regions, the Canadian Shield and the southeast, are centers of mining and forestry. Farming remains an important activity in the fertile plains of the St. Lawrence Valley. But in recent decades, increasing numbers of Quebec's residents have been attracted instead to manufacturing and service jobs.

Quebec's largest city is Montreal, a beautiful metropolis at the Lachine Rapids of the St. Lawrence. Development that began when Montreal hosted "Expo '67" helped transform it from a quaint, provincial city into a bustling, modern center.

The capital of the province, also called Quebec, is the oldest city in Canada. It was founded in 1608 by Samuel de Champlain. The historic sites and European charm of Quebec make it a popular tourist attraction. The next section describes the province of Quebec's unique culture as the center of Canada's French population.

The French-Canadian City of Quebec
This street in the city of Quebec illustrates the charm, history, and French character of the province.

The Prairie Provinces

The provinces of Alberta, Manitoba, and Saskatchewan lie in southwestern Canada between the Rocky Mountains and the Canadian Shield. The Prairie provinces have long been associated with enormous rolling fields of wheat and other grains. One writer described the landscape looking "as if someone had taken a colossal pencil to the countryside and erased anything taller than a bush. . . ." Rather, the region presents a varied landscape:

. . . the prairies are more than crop-clad flatlands . . . there are scenic nuggets for the taking: crystal lakes hidden in groves of aspen, meandering rivers . . .

The Location of Cities   More than 75 percent of the people in the three Prairie provinces live in cities. The five largest cities in the Prairies are located on what were strategic points along the newly built railroad in the late 1800s. Winnipeg was established at an important river crossing as railroad tracks were laid from the east through the Canadian Shield. From Winnipeg, two rail lines were built to the West, each taking a different set of passes through the Rocky Mountains. The cities of Edmonton and Calgary in Alberta were established at points where each rail line headed into the mountains. Roughly midway between those cities and Winnipeg, the Saskatchewan cities of Saskatoon and Regina were founded as major service centers along the rail lines. Lacking both an inland river system and frequent precipitation, the Prairie provinces have been described as a region where "grains and trains dominate life."

A Changing Economy   The Prairie provinces provide most of Canada's grain and cattle. Wheat is the major crop grown. Most grain is exported, traveling by rail to ports on the Pacific Ocean, Great Lakes, or Hudson Bay.

Tourism is an important economic activity in many of the region's magnificent parks. The snowcapped Rocky Mountains of western Alberta have some of North America's most spectacular scenery. The discovery of oil and natural gas in Alberta provided a new source of wealth for the region. The oil industry also had a major effect on the growth of cities like Calgary and Edmonton.

British Columbia

The natural beauty of the Rockies stretches farther west into British Columbia. Canada's westernmost province is unlike any other region in Canada. The Inside Passage, a waterway between the long string of off-shore islands and the Coast Ranges of British Columbia, provides travelers with many scenic wonders:

> *Always the ultimate backdrop is rank on rank of mountains, some velvety green, some topped with snow, some populated with giant trees on the lower slopes but turning to sheer rock decorated with sheets of ageless ice. Here and there, like a silver thread in the mountain distance, is a glimpse of plunging river, disappearing into the green as if into a deep sponge.*

Notice on the map on page 145 that mountains of several ranges cover nearly all of British Columbia. As a result, more than four fifths of the province's residents live in or near the city of Vancouver.

Natural resources including salmon, forests, and minerals have helped British Columbia become one of Canada's wealthiest provinces. But for many residents and visitors, its cities are its most memorable attractions. Victoria, the capital, is located at the southeastern tip of Vancouver Island. It has the relaxed charm of a small British city, with manicured gardens that bloom year-round in the mild, wet marine West Coast climate.

Vancouver, the province's largest city, occupies a site by an excellent harbor. As Canada's major port on the Pacific Ocean, Vancouver grew rapidly during recent decades as trade with Asia increased. Immigration from Asia also increased Vancouver's population, while many Canadians from other provinces move to Vancouver when they retire because of its desirable climate and scenic beauty.

The Northern Territories

The northern 40 percent of Canada consists of the Yukon and the Northwest Territories. These cold, largely

Shop for Fun at the Mall

The West Edmonton Mall covers an area that is the equivalent of 108 United States football fields. Inside the gigantic complex are 836 stores, 110 restaurants, and 20 movie theaters. But the most exciting part of the mall is its amusement park. Roller coasters, a water slide, a hockey rink, a miniature golf course, and 47 other thrilling rides are available. Visitors may even cruise beneath the water of an artificial lake in one of the mall's four yellow submarines shown at right.

More than 100,000 people a day visit the mall. They are drawn to the wonder of a place where they can splash among machine-driven waves while snow blankets the frozen plains just outside the door. "We do not make money on the entertainment," admits one of the owners of the mall. "But it is the entertainment that brings in the people."

1. In which Canadian province is the West Edmonton Mall located?

2. Why did the developers of this mall need to make it an exciting place?

treeless lands are sparsely settled—together they are home to fewer than 1 percent of Canada's population. The residents of the Northwest Territories could fit into Toronto's domed stadium, with seats left over for one half of the Yukon's population. There are no large cities in the region.

Still, these lands have a stark and distinctive beauty that is awe-inspiring to those who are drawn to the far northern reaches of the continent.

A Changing Culture Many residents of the northern territories are native people who call themselves Inuit, a term that means "the people." Inuit, rather than Eskimo, is the name by which these people prefer to be known. The Inuit generally live north of the forests, while other Native American groups live farther south. Recently, a writer who traveled from one end of Canada to the other told of the Inuit's attitude toward the land.

Water Is a Precious Commodity
The Inuit villagers of this northern Canadian community help unload the water delivery truck. How does this photograph reflect some of the changes that have taken place in the Inuit way of life?

Of all Canadians, the Inuit have developed and maintained the closest relationship with the geography. They have a saying: "Our land is our life." Recognizing they are but one of the land's many elements—and certainly not the most important—the Inuit use the harsh geography to survive, as an astute judo student turns the momentum of an onrushing attacker to his own advantage.

Contact with persons of European ancestry has slowly changed the ways in which the Inuit live. Although seal hunting is still a major economic activity, modern Inuit hunters now use snowmobiles to cross the frozen lands

instead of dog sleds. Modern technology is used to overcome vast distances in other ways as well—Inuit children remain at home, taking classes transmitted by satellite over radio and television systems. Their teachers may be thousands of miles away.

Interaction: A Difficult Environment
The northern territories have rich deposits of many minerals. Gold, silver, copper, zinc, lead, iron ore, and uranium are found in the region. So are large reserves of petroleum and natural gas. Because the harsh climate and rugged terrain make it difficult to mine and transport these materials, however, many deposits have not been developed. In spite of the difficulties of life in the north, the people who reside there live with knowledge of the hardships and appreciation for the beauty and the bounty that the land offers.

■■■ SECTION **1** REVIEW ■■■

Developing Vocabulary
1. Define: **a.** maritime **b.** lock

Place Location
2. Which Canadian province lies to the west of Ontario?

Reviewing Main Ideas
3. Why are the Atlantic provinces often called "The Maritimes"?
4. How has the economy of the Prairie provinces changed in recent decades?
5. Name two characteristics of British Columbia that make it a desirable place to live.

Critical Thinking
6. **Making Comparisons** Besides the harsh climate, what factors do you think have limited the human impact on the lands of northern Canada?

2 The Search for a National Identity

Section Preview

Key Ideas

- Canada's history reflects conflict between two major cultures: English and French.
- Many Canadians continue to have strong ethnic and regional ties.
- The Canadian government supports the cultural diversity of its people while attempting to maintain a strong national unity.

Key Terms

separatism, secede

Like Canada's landscapes, the nation's population is extremely varied. Canada has come to define itself as a multicultural country—a mosaic of many pieces with varying colors and textures that coexist in an intricate design.

Unity is difficult to achieve, however, because the country is so vast and there are such great differences among the provinces and territories as well as among the people. Many Canadians in the eastern and western regions feel that the government does not show enough interest in their particular problems. The lack of unity is partly explained by Canada's historical development.

Historical Roots

Canada has had to struggle to develop a single national identity. One reason is that many of its people identify more strongly with regional and ethnic groups than with the nation as a whole. The two groups that still comprise most of the population are those of English and French ancestry. More than 40 percent of all Canadians have British ancestors; another 30 percent are of French descent.

Colonial Rivalries The French were the first to colonize land along the broad St. Lawrence River. Starting in the 1500s, fur traders operated along the St. Lawrence and its tributary rivers. Farmers also came to settle in the lowland valleys. An English group called the Hudson's Bay Company later established fur-trading posts north and west of the lands claimed by France.

Conflicts arose between French and English colonists. They competed with each other for the prosperous North American fur trade and clashed over land claims. In addition, warfare in Europe between Britain and France spilled over into North America. Between 1689 and 1763, British and French colonists fought four wars in North America. Finally, British troops defeated the French in the Battle of Quebec in 1759, and by 1763 France surrendered all of its empire in Canada. Britain then assumed control over the entire region.

Ties to Britain Canada remained under direct British rule until 1867 when the British passed a law creating the Dominion of Canada. This act gave Canadians control over their local affairs, while foreign and military decisions were still made by the British. Canada became a completely independent country in 1931, when the last British controls ended. Even today, however, Canada's symbolic ruler is the British monarch.

Conflict Between Two Cultures

When France lost Quebec to Britain in 1763, about 65,000 French colonists lived in that region. Since that time, Canada's French population has grown to more than 30 percent of the

WORD ORIGIN

Canada
The origin of the name Canada is uncertain. It may have originated from the word *kanata*—the Huron-Iroquois word for "village." Or it may have come from Portuguese sailors who, on visiting the Canadian shore in 1500, exclaimed, "Canada!" which means "Here, nothing!"

A Challenge to Unity

French-Canadian Separatists publicly call for support of their movement. In what ways do separatist groups threaten the future of Canada as a country?

country, the government continued to protect the rights of French-speaking citizens, and both English and French are official languages in Canada. However, only about 15 percent of Canadians speak both languages.

Promoting French Culture Many French Canadians today feel discriminated against by the English-speaking majority. They claim that they cannot get jobs in government or industry because they are of French descent.

The Quebecois (kay-beh-KWAH), Quebec's French-speaking citizens, consider themselves the guardians of French culture in Canada. Starting in the 1960s, many Quebecois began to press for changes that would assure the preservation of French culture. Some people favored separatism, that is, making Quebec an independent country.

In 1974 the government of Quebec made French the official language of the province. As a result, many English-speaking residents and businesses left Quebec, and the province suffered economically. The conflict between cultures grew when a political party dedicated to separatism was elected to office in Quebec in 1976.

Separatism itself has so far failed to gain the support of a majority of Quebec's voters. But the possibility that Quebec may secede, or withdraw, from the rest of Canada still exists.

Preserving a Balance Pierre Trudeau, who served as Canada's prime minister from 1968 to 1979 and from 1980 to 1984, worked to preserve a balance between the interests of the French-speaking and English-speaking Canadians. Trudeau grew up in Quebec. His father's family was French and his mother's family was British, and so he had an understanding of both cultures. Trudeau favored preserving French culture in Canada, but he opposed movements to separate Quebec from the rest of the nation.

country's total population. The great majority of French-speaking Canadians live in the province of Quebec.

In 1774 the British government passed laws to ensure that French Canadians, many of whom also live in Ontario, would be able to maintain their own language, laws, and culture. When Canada became an independent

A constitutional amendment, proposed in 1987, provided for Quebec to be recognized as a distinct society in Canada. The proposal, however, failed to win the approval of all provinces. So Canada continues to struggle with the question of how to keep French Canadians within a united Canada.

Place: Multicultural Characteristics of Canada

The multicultural nature of Canada's population is one of its most distinctive characteristics. Although most Canadians have British or French ancestors, many other groups are represented in the population.

Inuit and Native Americans had been living in what is now Canada for thousands of years before Europeans arrived. Today, most of Canada's 25,000 Inuit live in the territories and in northern areas of Newfoundland, Ontario, and Quebec. Canada has nearly 370,000 Native Americans, most of whom live on reservations.

Canada has welcomed immigrants from central and eastern Europe as well as other parts of the world. During the late 1970s, Canada accepted about 60,000 refugees from Cambodia, Laos, and Vietnam. Recently, people from other parts of South Asia and East Asia have settled in Canada, particularly in British Columbia.

Preserving Cultural Diversity Canadians celebrate their cultural heritages in annual festivals. The Highland Games in Nova Scotia preserve Scottish traditions of Highland dancing, bagpipe music, and caber tossing—a game in which contestants compete by tossing a heavy wooden pole. The Kitchener–Waterloo Oktoberfest in Ontario is one of the largest German festivals in the country. Manitoba holds a National Ukrainian Festival. The Calgary Exhibition and Stampede attracts nearly one million people annually. This famous festival includes a rodeo, a chuckwagon race, and other events that reflect the heritage of Canada's cattle ranchers, many of whom came from the United States in the late 1800s.

Uniting Canada's Regions Canada has been successful in uniting its regions and its people through transportation and communication links. The Canadian government supported private construction of a transcontinental railroad in the 1880s to unite British Columbia with the rest of the nation. Canada's modern leadership in telecommunications largely results from efforts to communicate with residents in its remote northern regions.

Establishing a distinct national identity is proving difficult to achieve. Responding to mass immigration, the Canadian national government adopted a policy of multiculturalism. This policy builds acceptance of cultural diversity by encouraging Canadians to be proud of their ethnic backgrounds.

SECTION 2 REVIEW

Developing Vocabulary
1. Define: **a.** separatism **b.** secede

Place Location
2. In which province is Calgary located?

Reviewing Main Ideas
3. What are the historical roots of the conflict between French-speaking and English-speaking Canadians?
4. Why is a Canadian national identity difficult to achieve?

Critical Thinking
5. **Making Comparisons** The multicultural characteristic of the United States is sometimes called a "melting pot." In Canada, it is often called a "mosaic." What do you think is the difference between these terms?

✓ Skills Check

☐ Social Studies
☐ Map and Globe
☐ Reading and Writing
☑ Critical Thinking

Identifying Central Issues

Part of your reading process should always involve thinking about the central issue the text is addressing. Sometimes the issue is clearly stated in the language of the text, but it may also be implied and not explicit.

Use the following steps to identify the central issues in the passage below.

1. **Find the topic of the passage.** Quickly scan the passage below. Watch for topic sentences or other key phrases that may summarize the content of the passage in question. Answer the following questions: (a) How would you state the topic being addressed in this paragraph? (b) What two aspects of this topic are considered?

2. **Determine the point of view.** The point of view may contribute bias to the passage. Be alert for strong language that seems to distort an argument in favor of one side or another. Also watch for statements lacking evidence to support them. Answer the following questions: (a) What is the point of view expressed in this paragraph? (b) How did you determine the point of view?

3. **Look for evidence supporting the key statements in the passage.** Answer these questions: (a) What evidence is given of cooperation between Canada and the United States during war? (b) What evidence is given of their peacetime cooperation?

4. **Identify the central issue.** By thinking back over what you have read, you should be able to identify the central issue in the passage. Answer the following questions: (a) How would you express the central issue here? (b) Which part of the central issue receives more attention in the passage?

Canada and the United States share the longest undefended border in the world. They share also the Great Lakes and the St. Lawrence Seaway, which the two countries developed as partners. During World War II, Canada and the United States formed a permanent joint defense board to protect the Atlantic and Pacific coasts. This joint effort resulted in a radar network called the Distant Early Warning System, or DEW Line, which stretches across northern Canada to detect hostile planes. Because they are neighbors, it is essential for the two countries to work together in war. Canada and America were allies during World War I, World War II, and the Korean War.

3 Canada Today

Section Preview

Key Ideas

- Canada's future, like its past, depends on overcoming the challenges of geography.
- Links between Canada and the United States have resulted in a peaceful, but uneven, relationship.
- Political and human resources have made Canada a leader in international cooperation.

Key Term

customs

The history of Canada has centered on the struggle to overcome a harsh environment. Canada has emerged from that struggle to become a prosperous nation. Its gross national product is among the top ten in the world. Its stable government and high standard of living attracted millions of immigrants in recent decades. Canada has developed a blend of cultures while at the same time becoming a leader in worldwide organizations.

Challenges and Opportunities

In spite of Canada's progress, the nation faces challenges as well as opportunities. Canada's future, like its past, largely depends on geography. The themes of movement and human-environment interaction are of basic importance in understanding Canada's future development.

Movement of Resources In the past, the development of extensive transportation systems was necessary for moving Canada's resources over great distances. Today, Canada must balance the opportunities of developing its northern resources with the challenges of protecting the fragile environment of the tundra. The use of heavy machinery to recover oil and other minerals in the Arctic causes permanent damage to the landscape. Construction of a pipeline above ground avoids the problem of disrupting the permafrost. But it creates barriers to the migration of caribou and other Arctic animals.

Movement of People In 1900 only about one third of Canada's people lived in urban areas. Today, 75 percent of the nation's people live in cities. Canada has more than twenty metropolitan areas with a population of 100,000 or more. Urbanization created challenges of providing housing and services, controlling pollution, and preventing overcrowding.

Like the United States, Canada has an urban hierarchy. Toronto, Montreal, and Vancouver are the major urban centers. Linked to these cities are regional centers such as Edmonton, Winnipeg, Halifax, Quebec, Hamilton, and Calgary. Smaller cities are in turn connected to these regional centers. This hierarchy provides the ties necessary for the movement of people, goods, services, and information throughout the country.

Movement: Links with the United States

The border between Canada and the United States is more than 4,000 miles (6,400 km) long. It is the longest unarmed border in the world. Travelers between the two countries must pay customs, tariffs or fees, charged by one country's government on goods people bring in from another country. But no fence exists along the Canadian–United States border.

Cultural Links As stated in Chapter 5, Canada has many links with the United States. Some of these ties are

WORD ORIGIN

Toronto
During the 1600s and 1700s Indians used the Toronto area as an overland route between Lake Ontario and Lake Huron. *Toronto* is a Huron Indian term meaning "meeting place."

cultural. People living close to the border can enjoy radio and television programs from stations in both countries. Professional baseball and hockey leagues include teams from both nations. Canadian football is a little different from the American game, but many players in the Canadian Football League come from the United States. And many Canadians are fans of National Football League teams.

Economic Links   Perhaps the most important links between Canada and the United States are economic. Canada buys nearly 25 percent of all United States exports, and the United States buys about 75 percent of Canadian exports.

In 1989 Canada and the United States signed the Free Trade Agreement (FTA), designed to drop export barriers and reduce customs tariffs over a ten-year period. Before the Free Trade Agreement, tariffs made the buying and selling of goods to other countries very expensive. But the agreement produced mixed reactions. On one hand, Canadian shoppers were able to take advantage of bargains in United States border cities such as Buffalo, New York, where prices are generally lower than in Canada. On the other hand, many Canadians blamed the FTA for plant closings and rising unemployment as major firms relocated south of the border, where costs were lower.

Supporters of the Free Trade Agreement pointed out that both countries would benefit in the long run. Canadian manufacturers have begun to modernize their operations. United States companies are investing in Canada, and Canadian firms are expanding southward. The end result is expected to provide 250,000 new jobs for Canadians and Americans.

An Uneven Relationship   Although there are many links between Canada and the United States, some Canadians are uncomfortable because the relationship between the two nations is so uneven. Pierre Trudeau once said of Canada's relationship with the United States, "It's like sleeping with an elephant; you lie awake waiting for it to turn over."

Canadians generally are far more aware of what is happening in the United States than Americans are about what goes on in Canada. Few Americans know who Canada's prime minister is or how Canadian policies may affect them.

Canada's location relative to the United States provides its people with great opportunities. At the same time, Canada still struggles to prevent its identity from being overshadowed by the United States. Mordecai Richler, a well-known Canadian writer, put it this way:

This country, . . . still blurry, is like a child's kaleidoscope that remains in urgent need of one more sharp twist of the barrel to bring everything into sharp focus. Making us whole. Something more than this continent's attic.

Links with the World

In contrast to the so-called superpowers, the United States and the former Soviet Union, Canada plays the role of a middle power in the global community of nations. Middle powers often join together to achieve their common goals. Because of its location, size, and multicultural population, Canada is very well suited to working with other nations.

The Importance of Location   Canada has a unique position with regard to other nations because of its location. With major ports on both the Atlantic and Pacific coasts, Canada has access to trade with Japan and other Asian countries as well as with Europe.

Canada's location and size also have given it an important role in defense alliances. It was a founding member of the North Atlantic Treaty Organization (NATO), formed in 1949 to defend North America and Western Europe against attack by the Soviet Union. The North American Aerospace Defense Command (NORAD) joins Canada with the United States in air and missile defense.

Membership in the Commonwealth

Canada maintains links with partners other than the United States through its membership in the British Commonwealth of Nations. This is a group of countries, mostly former colonies, that now have equality and independence under the symbolic protection of the British crown. Members of the Commonwealth often work together to promote better trade, health, and education in their countries.

As a member of the Commonwealth, Canada has links with developing countries in Asia and Africa. Membership in the Commonwealth also puts Canada in a favorable position with regard to trade with the European Economic Community, a group of Western European countries that have united their economic resources.

The Role of Peacekeeper Lester Pearson, Canada's prime minister from 1963 to 1968, once said:

> . . . the best defence of peace is not power, but the removal of the causes of war, and international agreements which will put peace on a stronger foundation than the terror of destruction.

Much of Canada's international policy has been based on Pearson's ideas. Canada has not developed any nuclear weapons, and it has taken an active part in promoting arms control and disarmament among other nations.

A City of the Future
Vancouver, located on the Pacific west coast, is one of the largest ports in Canada. How does its location favor further development today?

SECTION 3 REVIEW

Developing Vocabulary
1. Define: customs

Reviewing Main Ideas
2. Why is the relationship between Canada and the United States uneven?
3. How does Canada's location influence its role in international relations?

Critical Thinking
4. **Synthesizing Information** Why might Canada have advantages over the United States in attempting to mediate peaceful relations in the world?

8

REVIEW

Section Summaries

SECTION 1 Regions of Canada Canada is a vast land made up of ten provinces and two territories. Based on its physical features, culture, and economy, Canada can be divided into five distinct regions. Location has helped the development of some regions and has limited human interaction with the environment in other parts of the country.

SECTION 2 The Search for a National Identity Britain and France colonized Canada. Colonial conflicts between these two nations are reflected in the cultural conflicts between English-speaking and French-speaking Canadians today. Canada's multicultural people have strong ethnic and regional ties, which make it difficult to achieve a national identity. The Canadian government supports multiculturalism but encourages unity among its people.

SECTION 3 Canada Today Canada has become one of the world's most prosperous nations. It still faces the challenge of developing its northern resources while protecting the environment. Canada maintains a peaceful but uneven relationship with the United States. Economic and environmental problems concern both countries. Canada has become a leader in international cooperation by active involvement in world organizations.

Vocabulary Development

Use each key term in a sentence that shows the meaning of the term.

a. separatism
b. customs
c. maritime
d. secede
e. lock

Main Ideas

1. How did the Atlantic provinces play important roles in Canada's settlement and development?
2. What are the major economic activities in the area known as the Great Lakes–St. Lawrence provinces?
3. Why have the major cities of Alberta grown rapidly in recent decades?
4. How has contact with people of European ancestry changed the way of life of the Inuit?
5. Why do some people favor making Quebec an independent nation?
6. How have Canadians preserved their cultural diversity?
7. In what ways has the Canadian government worked to unite its citizens from different ethnic backgrounds?
8. What challenges does Canada face in developing its northern resources?
9. Why did the Free Trade Agreement produce mixed reactions among Canadians?
10. How has membership in the British Commonwealth of Nations benefited Canada and its people?

Critical Thinking

1. **Identifying Alternatives** What are some ways to make people in the United States more knowledgeable about Canada?
2. **Formulating Questions** What are five questions that you would ask before deciding to live in the Northwest Territories?
3. **Predicting Consequences** What consequences do you think would result if the Canadian government passed a law making English the only official language? Give reasons for your answer.

4. **Perceiving Cause-Effect Relationships** Why has Canada's commitment to multiculturalism made it difficult for the country to achieve a national identity?
5. **Testing Conclusions** How has Canada followed Lester Pearson's belief in the need to "put peace on a stronger foundation than the terror of destruction"?

Practicing Skills

Identifying Central Issues Use what you have learned in the skill lesson on page 154 to identify and describe the central issue in the following passage:

According to many Canadians, the United States has invested too heavily in Canada. Canadians want to work toward more control of their own economy, rather than depending on the United States. Presently, Canada's economy is vulnerable to conditions in the United States. Furthermore, the United States receives much of the profit from Canadian industries because it has controlling interests in many of them. These facts have led to disagreements among Canadians. . .

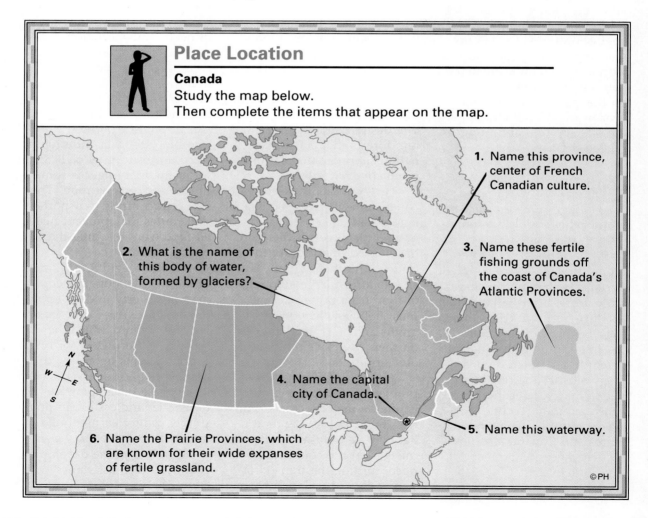

Place Location

Canada
Study the map below.
Then complete the items that appear on the map.

1. Name this province, center of French Canadian culture.

2. What is the name of this body of water, formed by glaciers?

3. Name these fertile fishing grounds off the coast of Canada's Atlantic Provinces.

4. Name the capital city of Canada.

5. Name this waterway.

6. Name the Prairie Provinces, which are known for their wide expanses of fertile grassland.

©PH

MAKING CONNECTIONS

•WHERE REGIONS MEET•

Food, Hunger, and Geography

For the past few decades, the farms of the world have generally succeeded in producing enough food to feed almost all of the planet's people. Yet as the map on the following page shows, this food is not reaching all the people in the world. In fact, an estimated 550 million people go hungry on this planet. Worse, each year nearly 15 million people die of starvation and hunger-related diseases.

Food Production

Hunger exists in every region of the world. It is found even in the United States and Canada, although most people in this region are well fed. Because both countries have large areas of fertile land, they produce far more crops than their people can consume.

In addition to having sufficient homegrown food supplies, most of the region's people have the economic resources to buy food from other countries. The United States and Canada also possess the necessary transportation to deliver that food all across the region.

Food and Hunger Around the World

Of course not all regions share the ability of Americans and Canadians to raise, distribute, and purchase food. Thus hunger is more widespread in other regions. One reason for this is that the earth's farmland is not evenly distributed around the globe. The deserts, mountains, and rain forests that cover many nations are not suitable for productive farming.

For example, the South American country of Bolivia is landlocked and has limited natural resources. It therefore has limited opportunity to trade with other regions. Unable to afford large amounts of food from other nations, most Bolivians live on the grains and starches that grow on their country's mountainous terrain.

In cases of famine, the problem of food distribution is often a major factor. Countries that can provide food relief may lie thousands of miles from the countries in need, turning the transportation of food over long distances into a crucial element in famine relief. Once food arrives in the region, often by airplane, the next challenge is transporting it from its point of arrival to the actual people in need. Transportation systems within famine-stricken countries are typically poor or nonexistent, making it often difficult for relief to reach its final destination. In addition, war or political turmoil can also block food distribution. In countries torn by civil war, for example, hunger is sometimes used as a weapon. The side that controls transportation routes may prevent needed food from reaching the opposing side.

Another significant group of the world's hungry people lives in poverty in developing nations. They may own land, but they cannot afford the seed, tools, fertilizers, or irrigation systems necessary to grow enough food. In the cities of these developing nations, the landless urban poor simply cannot earn enough money to buy food.

People in these developing countries need assistance with education, irrigation, health care, transportation systems, storage facilities, and distribution systems in order to be able to feed themselves. Several agencies of the United Nations, including the World Health Organization and the Food and Agriculture Organization, promote such projects in an effort to combat the world hunger problem. Through these agencies, regions connect with one another in their search for lasting solutions to world hunger and to provide relief when famine strikes.

The View from the U.S.A.

 The United States has supplied massive relief efforts for nations suffering from famine. The United States has also sought to relieve world hunger by offering technical aid and loans to farmers. In addition, Peace Corps volunteers assist people in developing countries with such projects as irrigation systems and the use of more advanced farming methods.

TAKING ANOTHER LOOK

1. What features of the United States and Canada help explain why many of the region's people are able to feed themselves?
2. Why are people around the world hungry in spite of sufficient food production?

Critical Thinking

3. **Demonstrating Reasoned Judgment** Make an argument for improving education as a means of attacking hunger.

Applying Geographic Themes

Regions   People require a certain number of calories in order to maintain good health. Scientists can estimate the general level of nutrition in a region by knowing the average number of calories each person consumes. In which regions are people generally well fed?

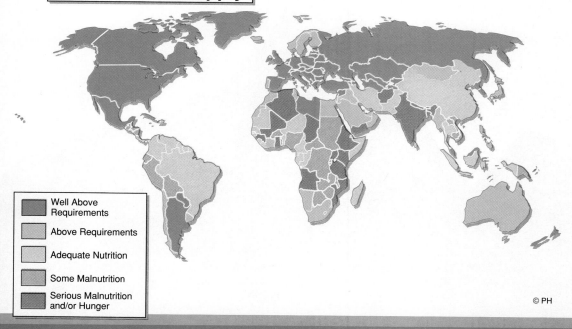

World Calorie Supply

Well Above Requirements

Above Requirements

Adequate Nutrition

Some Malnutrition

Serious Malnutrition and/or Hunger

© PH

UNIT 3

Latin America

CHAPTERS

Rain forests so thickly overgrown with tangled vines that a worker, swinging a machete and fighting the heat, can advance barely half a mile a day . . . gauchos in the saddle, gazing over a grassy plain that ends in mist at the horizon . . . bustling cities, filled to bursting with people . . .

From the northern border of Mexico to the far tip of South America, the region of Latin America is as varied as it is large. In this unit, you will discover how the landforms and vegetation, combined with the mix of people, have shaped the culture of Latin America.

Rio de Janeiro, Brazil ▶

An Indian weaver, Guatemala ▶

◀ Machu Picchu, Peru

163

9

Regional Atlas: Latin America

Chapter Preview

These three sections provide an overview of the countries of Latin America, shown in red and blue on the map below.

Sections	Did You Know?
1 MIDDLE AMERICA: LAND, CLIMATE, AND VEGETATION	The island group of the Bahamas includes more than seven hundred islands.
2 SOUTH AMERICA: LAND, CLIMATE, AND VEGETATION	The area covered by the Amazon River Basin is nearly as large as that of the contiguous United States.
3 HUMAN GEOGRAPHY	The Panama Canal allows ships to pass from the Atlantic to the Pacific without sailing around South America.

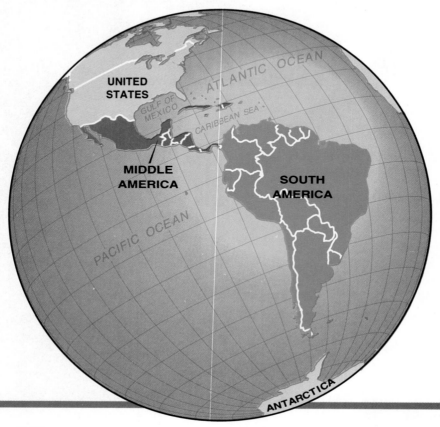

Mexico's Central Plateau

The Central Plateau runs north to south and is flanked by the country's eastern and western mountain ranges. Name the major mountain ranges between which this region is located.

1 Middle America: Land, Climate, and Vegetation

Section Preview

Key Ideas

- Mountains form the backbone of Middle America's mainland which stretches from Mexico to Panama.
- Climate and vegetation are strongly linked to elevation and distance from oceans.
- Landforms in the Caribbean differ from island to island.

Key Terms

mesa, cay, hurricane, tropical depression, tropical storm, storm surge, meteorologist

Imagine twisting the long, slender end of a funnel and you have an idea of how Middle America looks on a map. The mouth, or wide end, of the funnel to the north is Mexico, the largest country of Middle America. The narrow, slightly twisted tube of the funnel is Central America. To the west of this landmass lie the open waters of the Pacific Ocean. Off the eastern coast are the Gulf of Mexico and the blue-green Caribbean Sea, edged with many island nations. South of Central America lies the vast continent of South America.

Look at the globe on page 164 to locate this large region. Mexico, the seven countries of Central America, and the islands that lie like a string of pearls in the Caribbean Sea are shown in red. The twelve countries that make up the continent of South America appear as blue. Together, these many countries make up the region called Latin America—"Latin" because the major languages of the region, Spanish and Portuguese, are derived from the Latin language.

Place: The Land of Middle America

From north to south, mountain ranges run the 2,500-mile (4,023-km) length of Middle America and form the strong "backbone" of this long landmass. The mountains are interrupted only occasionally by plains, lowlands, and lakes.

Mountains Mexico is often called a country of mountains. One visitor said that the country seems to "lift itself in a mighty heave from the two oceans, with its . . . towering volcanic cones. . . ." Two mountain ranges dominate the landscape of northern Mexico. The eastern range is called the Sierra Madre Oriental (see EHR uh MAH dray aw ree EHNT l). In the west is the Sierra Madre Occidental (see EHR uh MAH dray ahk suh DENT l). Both names are Spanish. The first means "Eastern Mother Range," while the second means "Western Mother Range." Mexico's Madres mark one of the several places where the Western Hemisphere's west coast cordillera (KAWRD uhl EHR uh), or mountain range system, split apart. But 900 miles (1,448 km) south of the Mexican-United States border, in southern Mexico, the two chains connect again to form one high range. These highlands are called the Sierra Madre del Sur (see EHR uh MAH dray dehl sur), or "Mother Range of the South."

Continuing south, mountains form the central highlands of Central America. More than one hundred active volcanoes exist in the mountainous spine of Central America. In the last twenty years, more than forty fiery eruptions have occurred in the region, all of them along the Pacific coast.

Look at the map on page 167 to locate the chain of mountain ranges that stretches through Middle America's mainland.

Plateaus, Plains, Lakes, and Rivers Between the Sierra Madre Oriental and the Sierra Madre Occidental lies the plateau of Mexico, a huge, bumpy highland region that ranges in altitude from 6,000 to 8,000 feet (1,829 to 2,438 m). Although most of the people in Mexico live on the large central plateau, the mountains of the Madres and the mesas (MAY suhz) inland make east–west travel very difficult. A mesa is an isolated mountain with a flat top. The word *mesa* comes from the Spanish word for table.

Look again at the map on page 167. Notice the narrow, winding, green ribbons of lowland plains along both the Pacific and Caribbean coasts of Mexico and Central America. In many places, especially along the Pacific coastline, the plain is so narrow that it gets swallowed up by the forbidding, rocky walls of the cordillera. In places along the Caribbean coastline, however, the plain widens out into continuous patches of lowland. Mexico's Yucatán (yoo kah TAHN) Peninsula is one such place. The Yucatán is a thumblike peninsula that juts out into the sea and helps separate the Gulf of Mexico from the Caribbean Sea. This location attracts thousands of tourists each year. The Yucatán is the only place in Mexico where mountains do not dominate the landscape. Here, the mountains of the Sierras are too far away to even be seen.

Across the Gulf of California from Mexico proper lies another narrow peninsula 950 miles (1,206 km) long—the Baja (BAH hah) California. Its name, which means "Lower California," is deceiving because it is actually part of Mexico, not the state of California. Baja is made up of a series of mountains that stretch from north to south along its length.

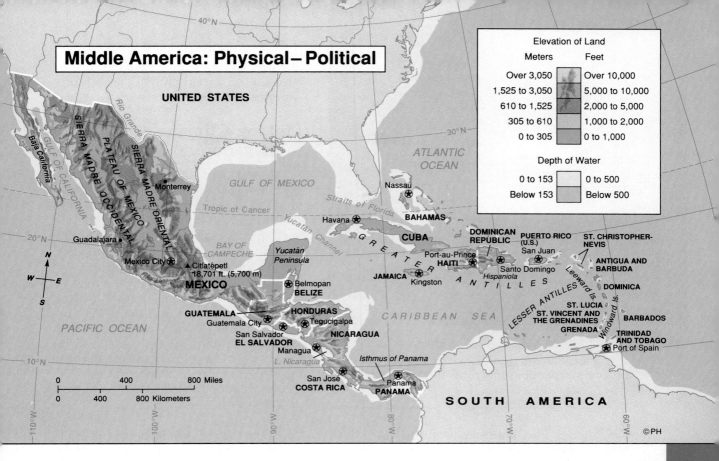

Middle America: Physical–Political

UNITED STATES

40°N

30°N

Rio Grande

SIERRA MADRE OCCIDENTAL

PLATEAU OF MEXICO

SIERRA MADRE ORIENTAL

Monterrey

GULF OF MEXICO

Tropic of Cancer

GULF OF CALIFORNIA

Baja California

20°N

Guadalajara

Mexico City

Citlatépetl
18,701 ft. (5,700 m)

MEXICO

BAY OF CAMPECHE

Yucatán Channel

Yucatán Peninsula

Belmopan
BELIZE

GUATEMALA

Guatemala City

HONDURAS

Tegucigalpa

San Salvador
EL SALVADOR

NICARAGUA

Managua

L. Nicaragua

PACIFIC OCEAN

San José
COSTA RICA

Panama
PANAMA

Isthmus of Panama

ATLANTIC OCEAN

Nassau

Straits of Florida

BAHAMAS

Havana

CUBA

GREATER

Port-au-Prince

HAITI

JAMAICA

Kingston

ANTILLES

Hispaniola

DOMINICAN REPUBLIC

Santo Domingo

PUERTO RICO
(U.S.)

San Juan

ST. CHRISTOPHER-NEVIS

ANTIGUA AND BARBUDA

Leeward Is.

DOMINICA

ST. LUCIA

LESSER ANTILLES

ST. VINCENT AND THE GRENADINES

GRENADA

Windward Is.

BARBADOS

TRINIDAD AND TOBAGO

Port of Spain

CARIBBEAN SEA

SOUTH AMERICA

N W E S

0   400   800 Miles
0   400   800 Kilometers

10°N

110°W 100°W 90°W 80°W 70°W 60°W

©PH

Elevation of Land

Meters		Feet
Over 3,050		Over 10,000
1,525 to 3,050		5,000 to 10,000
610 to 1,525		2,000 to 5,000
305 to 610		1,000 to 2,000
0 to 305		0 to 1,000

Depth of Water

0 to 153		0 to 500
Below 153		Below 500

Applying Geographic Themes

1. Regions Except for the narrow coastal strips, where is the largest lowland region in Mexico?

2. Place What are the names of the three main island groups in the Caribbean?

In addition to mountain ranges, plateaus, and plains, many lakes dot the land of Middle America. The largest of these is Lago de Nicaragua, or Lake Nicaragua, a large highland lake scattered with islands.

Because of its narrow shape, Middle America has no wide, deep navigable rivers. Instead, the streams of the region tumble swiftly from the central mountaintops to the narrow plains. They then quickly enter the oceans both east and west.

The Caribbean Islands

Along the floor of the Gulf of Mexico and the Caribbean Sea lies a vast mountain chain. This chain is the result of volcanic activity along the boundary between the Caribbean Plate and the North American Plate. In some places the very tops of these mountains peek above the waves, forming some of the Caribbean islands. Other islands of the Caribbean are cays (keez)—low-lying coral islands, formed over thousands of years from the accumulation of the rocklike skeletons of tiny sea animals.

The islands are divided into three main groups according to their location—the Bahamas, the Lesser Antilles, and the Greater Antilles (an TIHL eez). The Greater Antilles include the four largest islands of the region. Locate these three groups of islands on the map on this page.

Chapter 9, Section 1   **167**

Climate and Vegetation in Costa Rica
A tropical climate and lush vegetation create the beauty surrounding this rural settlement in Costa Rica.

of these factors is an important influence on the varied climates of Middle America.

Linking Climate to Elevation

If you laid the climate map of Middle America on page 169 over the physical map on page 167, you would see one of the major factors influencing the region's climates—Middle America's mountainous backbone. Because air cools as it rises in altitude, a variety of climates exist near one another on the region's mountainsides. Thus, the foot of a mountain may be hot and rainless, while a thousand feet up the mountainside, rain may fall in torrents. A few thousand feet higher, moisture may have been wrung out of the air completely, and cold, desert-like conditions may develop.

The landscape of the Caribbean islands varies from island to island. Cliffs, bluffs, and mountains jut up in many places throughout the islands. Overall, however, wind and surf have worn down and smoothed land surfaces into lowland plains. Most of the islands' coastlines have long stretches of white sandy beaches that meet the blue-green water of the Caribbean Sea.

On the Greater Antilles, mountains generally slope from the center of each island to coastal plains. The Lesser Antilles lie in an arc and include many islands with active and inactive volcanoes. These islands have seen some of the most devastating volcanic eruptions of the twentieth century. Elevation in the Lesser Antilles varies greatly and ranges from mountainous to relatively flat. A tangled chain of more than seven hundred small islands forms the Bahamas. These are mostly islands of coral and limestone.

Chapter 2 discussed how landforms, elevation, and distance from oceans affect a region's climate. Each

The Dry Lands One example of how elevation affects climate can be seen in northern Mexico. Arid and semiarid climates are the norm in northern Mexico because of the rain-blocking fortress of the Sierras on both coasts. Parts of northern Mexico receive less than 4 inches (10 cm) of precipitation each year. Farther south, however, some moist ocean winds squeeze through the mountains to drop welcome rains on the plateau of Mexico. In addition, the higher elevations of the plateau bring a pleasant climate all year round, with comfortably warm days and cool nights. In spite of its location south of the Tropic of Cancer, Mexico City, which is located on the plateau, enjoys very moderate temperatures. The average high temperature drops from 74°F (23°C) in July to 66°F (19°C) in January. This is because the city's altitude is 7,575 feet (2,309 m) above sea level. Perhaps it is also for this reason that the southern portion of the plateau is and has been the center of Mexico's population for thousands of years.

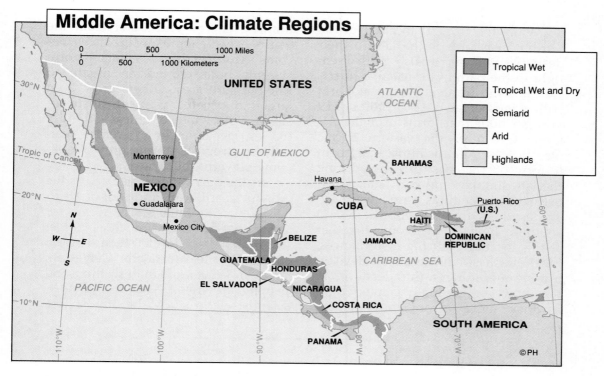

Middle America: Climate Regions

```
0        500          1000 Miles
0    500      1000 Kilometers
```

Legend:
- Tropical Wet
- Tropical Wet and Dry
- Semiarid
- Arid
- Highlands

UNITED STATES

ATLANTIC OCEAN

Tropic of Cancer

GULF OF MEXICO

Monterrey

MEXICO

Guadalajara

Mexico City

BAHAMAS

Havana

CUBA

Puerto Rico (U.S.)

HAITI

DOMINICAN REPUBLIC

JAMAICA

BELIZE

GUATEMALA

HONDURAS

EL SALVADOR

NICARAGUA

CARIBBEAN SEA

PACIFIC OCEAN

COSTA RICA

SOUTH AMERICA

PANAMA

©PH

Applying Geographic Themes

Regions Middle America is located south of the Tropic of Cancer. How does this affect the region's climates? Look back at the map on page 167 and explain why the region has such a variety of climates.

The Tropics of Middle America South of the Tropic of Cancer, including the lands from Southern Mexico to South America and all the islands of the Caribbean, tropical climates prevail. While several Central American towns at high elevations are known as "cities of eternal spring" because of their desirable mountaintop climates, the climates of other places usually depend on which side of the mountains the place is located. On the Pacific side of the region's mountains, a tropical wet and dry climate prevails with high temperatures all year and rainy and dry seasons. During the dry months of the year—usually December through April—the air may be dusty and the earth a dull brown. But during the rainy season—May through November—the land regains its lush, green look.

At low elevations near the Caribbean coast, a tropical wet climate exists. All year round, the air floats hot and humid. Northeast trade winds may bring as much as 100 inches (254 cm) of rain in a single year. Look at the climate map of Middle America on this page to find these two types of tropical climates.

The islands of the Caribbean enjoy a warm tropical climate all year, making them popular choices for vacationers. Summer temperatures average 80°F (27°C), but the northeast trade winds blowing over the ocean keep the days comfortable. Rainfall on the islands varies dramatically with elevation and exposure to winds. Sudden rains may fall daily on the northeast, or windward side of an island, while the leeward side is protected from rain-bearing winds by mountains.

While the tropical location of the Caribbean islands is the key to their desirable climate, it also puts them right in the path of destructive **hurricanes.** A hurricane is a tropical storm with winds of at least 74 miles (119 km) per hour that forms over the Atlantic Ocean or the eastern Pacific Ocean. Most Atlantic hurricanes begin to form off the coast of Africa during the late summer and early fall, when ocean temperatures are warmest. Trade winds from the northern and southern hemispheres come together at this location, causing disturbances. Most of these disturbances develop into no more than a thunderstorm. A few, however, gradually gain strength as they move across the Atlantic,

drawing energy and water from the warm ocean. Some disturbances become **tropical depressions**—storm systems built around an organized low-pressure area. Others grow even larger to become **tropical storms**—storms with winds of at least 39 miles (63 km) per hour. In an average year, only six tropical storms grow to become powerful hurricanes.

As the hurricanes sweep across the Atlantic, Caribbean, and Gulf of Mexico, they create huge **storm surges,** or waves that can rise as much as 25 feet (8 m) above sea level. When a storm surge of 20 feet (6 m) hit Galveston, Texas, in 1900, more than six thousand people drowned. **Meteorologists**—scientists who study the atmosphere

Applying Geographic Themes

1. Regions   What types of natural vegetation are found on the Yucatán Peninsula?

2. Interaction   Why do the farmlands of northern Mexico need extensive irrigation to grow healthy crops?

Middle America: Natural Vegetation

- Tropical Rain Forest
- Mixed Forest (Deciduous and coniferous)
- Coniferous
- Chaparral
- Tropical Grassland
- Temperate Grassland
- Desert Scrub

UNITED STATES

GULF OF MEXICO

Monterrey

MEXICO
Guadalajara
Mexico City

Havana

CUBA

JAMAICA

BELIZE
HONDURAS

GUATEMALA
San Salvador
EL SALVADOR

NICARAGUA
Managua

COSTA RICA

PANAMA
Panama City

PACIFIC OCEAN

ATLANTIC OCEAN

Tropic of Cancer

HAITI

DOMINICAN REPUBLIC

CARIBBEAN SEA

SOUTH AMERICA

30°N

20°N

10°N

110°W   100°W   90°W   80°W   70°W   60°W

N   W   E   S

0   250   500 Miles
0   250   500 Kilometers

©PH

and weather—at the National Hurricane Center in Florida can now use satellites and planes to track the path of tropical storms and hurricanes. By doing this they are able to warn people in the region to evacuate before storms and their surges reach land. Nevertheless, these violent storms can leave the islands without adequate power, water, communication, or food for weeks at a time.

Linking Climate to Vegetation

Compare the natural vegetation map of Middle America with the climate map on page 169. Northern Mexico's wide expanse of arid and semiarid regions on the climate map appear as desert scrub regions on the vegetation map. The rainfall in these regions is only enough to support twisted shrubs, cacti, and patches of coarse grass. The one exception to the desert scrub vegetation occurs on the hillsides of the Sierras, which are covered with both coniferous forest and mixed forest.

Stretching from the southern border of Mexico to South America, the vegetation is broad-leaved and luxurious because of tropical climates and heavy rainfall. Steamy, tropical rain forests blanket Central America. Look at the natural vegetation map to identify the variety of vegetation found on the Yucatán Peninsula.

The Caribbean islands also have tropical types of natural vegetation. On most islands, the leaves of trees and bushes are broad and do not fall seasonally. Palm trees and orange trees line beaches and coasts. Red poinsettia, hibiscus of vivid red, pink, and yellow, and many other colorful wildflowers reward the eye on all but the most arid islands. A tropical mix of perfumed breezes fills the air. On the eastern half of Hispaniola and all of Puerto Rico, tropical rain forests also appear.

Hurricane Destruction in the Caribbean
In September 1989, the eye of Hurricane Hugo passed directly over St. Croix. This devastating hurricane virtually destroyed the housing and economies of many Caribbean islands.

SECTION 1 REVIEW

Developing Vocabulary
1. Define: **a.** mesa **b.** cay
   c. hurricane **d.** tropical depression
   e. tropical storm **f.** storm surge
   g. meteorologist

Place Location
2. Which mountain range runs along Mexico's west coast?
3. What bodies of water are located off the east coast of Middle America?

Reviewing Main Ideas
4. Why does the southern half of Mexico receive more rainfall than the northern half?
5. What are the three main groups of islands in the Caribbean?

Critical Thinking
6. **Making Comparisons** How do the general climate and vegetation patterns of northern Middle America compare to those of southern Middle America?

2 South America: Land, Climate, and Vegetation

Section Preview

Key Ideas

- The Andes Mountains dramatically affect the climates and vegetation of the South American continent.
- The Amazon River Basin is an interior plain rich in tropical vegetation.
- Three grassy plains in South America provide an abundance of land suitable for agriculture.

Key Terms

timberline, canopy

South America possesses some of the world's most soaring mountains and widest plains. Some of the world's driest deserts and largest rain forests exist in South America. As you would expect on such a vast landmass, South America is a region of great variety and many contrasts.

Major Landform Regions

The physical geography of South America follows a pattern that is similar to that of North America. The west is lined with mountains of dizzying heights. The east is also mountainous, although more gently so, and in between, level plains smooth the interior.

Western Mountains The dimensions of the Andes range strain the imagination. From north to south, they extend 4,500 miles (7,240 km)—the world's longest unbroken range. At places in Peru and Bolivia the range is nearly 500 miles (804 km) wide. Some peaks of the Andes rise more than 20,000 feet (6,096 m) above sea level, making it the second-highest range in the

world. Only the Himalaya Mountains of South Asia are higher. Even the plateaus that lie between the steep slopes of the Andes range from 6,500 to 16,000 feet (1,981 to 4,877 m) above sea level.

Two Eastern Highlands Across the continent from the Andes, the comparatively small region of the Guiana Highlands swings across the northeast. The much larger expanse of the Brazilian Highlands to the south sprawls over one fourth of South America's total land area.

Compared to the Andes, the Guiana and Brazilian highlands are quite tame. Their greatest heights equal barely half those of the Andes. Their surfaces are more often gently rolling rather than steep and rocky.

Low Plains A usually narrow coastal plain edges the whole South American continent. One very long, thin plain strings along the Pacific coast. The southern end of this strip is often damp with fog and mist from the Pacific Ocean. Here, abundant rain falls throughout the year and forests thrive. In contrast, the midsection of this long coastal plain is the Atacama (ah tah KAH mah) Desert—the driest and one of the most lifeless places on earth. Winds blowing west across the cold waters of the Peru, or Humboldt, Current drop their moisture in the ocean. Thus only dry winds ever reach the land, where they create a desolate desert wasteland. At the convergence, or meeting point, of the Peru (Humboldt) and California current off the coasts of Peru and Ecuador, is one of the richest fishing regions.

Applying Geographic Themes

Regions Most of South America lies in the tropics. Which major mountain range has a greater effect on the climates and vegetation of the region than does latitude?

South America: Physical–Political

CARIBBEAN SEA

ATLANTIC OCEAN

Caracas

VENEZUELA

Georgetown

Paramaribo

GUYANA

Bogotá

SURINAME

Cayenne

GUIANA HIGHLANDS

FRENCH GUIANA (Fr.)

COLOMBIA

LLANOS

Equator

Quito

ECUADOR

Chimborazo
20,561 ft. (6,267 m)

Amazon R.

Madeira R.

Xingu R.

Tocantins R.

PERU

AMAZON BASIN

BRAZIL

Lima

Brasília

La Paz

BOLIVIA

BRAZILIAN HIGHLANDS

Sucre

PACIFIC OCEAN

GRAN CHACO

PARAGUAY

Paraná R.

São Paulo

Rio de Janeiro

Asunción

ANDES

Paraguay R.

Uruguay R.

Tropic of Capricorn

ARGENTINA

N
W   E
S

Aconcagua
22,831 ft. (6,959 m)

CHILE

URUGUAY

Montevideo

Santiago

Buenos Aires

PAMPAS

Río de la Plata

PATAGONIA

0   500   1000 Miles
0   500   1000 Kilometers

Falkland Islands
(Islas Malvinas)

Strait of Magellan
Tierra del Fuego
Cape Horn
Drake Passage

©PH

Elevation of Land	
Meters	Feet
Over 4,270	Over 14,000
3,050 to 4,270	10,000 to 14,000
1,525 to 3,050	5,000 to 10,000
610 to 1,525	2,000 to 5,000
305 to 610	1,000 to 2,000
0 to 305	0 to 1,000
Below sea level	Below sea level

Depth of Water	
0 to 153	0 to 500
Below 153	Below 500

Between the Andes and the highlands of the east, wide plains fill the center of the continent. In the north, a lowland region called the llanos (YAH nohs) straddles the border between Venezuela and Colombia. The llanos gets its name from the tall savanna grasses that grow on its vast plain. The Gran Chaco and the pampas make up two more large plains spread over Paraguay, Uruguay, and parts of Argentina. Like the llanos, the pampas gets its name from its vegetation of short prairie grasses.

South of Argentina's pampas is a dry plain and windswept plateau known as Patagonia. Patagonia is battered by the constant cold winds of the polar easterlies. Clouds never stop racing overhead. Dust and brush never stop crackling underfoot.

The Amazon Basin The equator cuts across the northern reaches of South America's largest plain—the Amazon Basin. Rising on the eastern slopes of the Andes, the Amazon River gathers the waters of countless tributaries and flows more than 4,000 miles (6,436 km) to the Atlantic Ocean. At the same time, the trade winds of the Atlantic Ocean, heavy with moisture, stream inland from the northeast and southeast, dropping more than 80 inches (203 cm) of rain each year. In some places showers occur almost daily. The Amazon constantly returns this water to the Atlantic. In its flow it carries more water and force than any other river in the world. When it finally reaches the coast, it gushes so much silt and water that the salty Atlantic waters are stained brown with fresh river water for 200 miles (322 km) beyond the river's mouth.

Linking Climate to Vegetation

Look at South America on the physical map on page 173. The continent runs from above 10°N latitude, north of the Equator, to nearly 60°S latitude, close to Antarctica. Most of the continent lies in the tropical zone between the Tropic of Cancer and the Tropic of Capricorn. Even so, the climates and vegetation are far from uniform.

The Variable Andes The north–south length of the Andes falls into a highlands category on the climate map. As discussed in Chapter 2, this means that temperatures and vegetation vary with altitude. In general, cooler temperatures and less vegetation are found at higher elevations. At very high elevations, the vegetation is known as Alpine tundra. Alpine tundra usually grows above the **timberline,** the boundary above which continuous forest vegetation cannot grow. Only plants that can survive cold temperatures, gusting winds, lack of precipitation, and short growing seasons grow in the Alpine tundra.

The highest altitudes of the Andes are in the midsection of the mountain chain. Mountaintop areas here are snow-covered and cold all year long. A climber in Argentina described these conditions:

At the crater's edge lies a log so heavy that modern climbers have reported they were unable to lift it. Perhaps those who tried were weakened by lack of oxygen and the incredible cold of these extreme altitudes, which I learned about firsthand. One day I flew . . . over Aconcagua's 22,834-foot summit, highest in the Americas. . . . The outside temperature was 31° below zero . . . Bernardo, . . . a radio weatherman, calculated that the wind dropped the chill factor to 100° below. . . . These killing temperatures have waylaid scores of Aconcagua climbers.

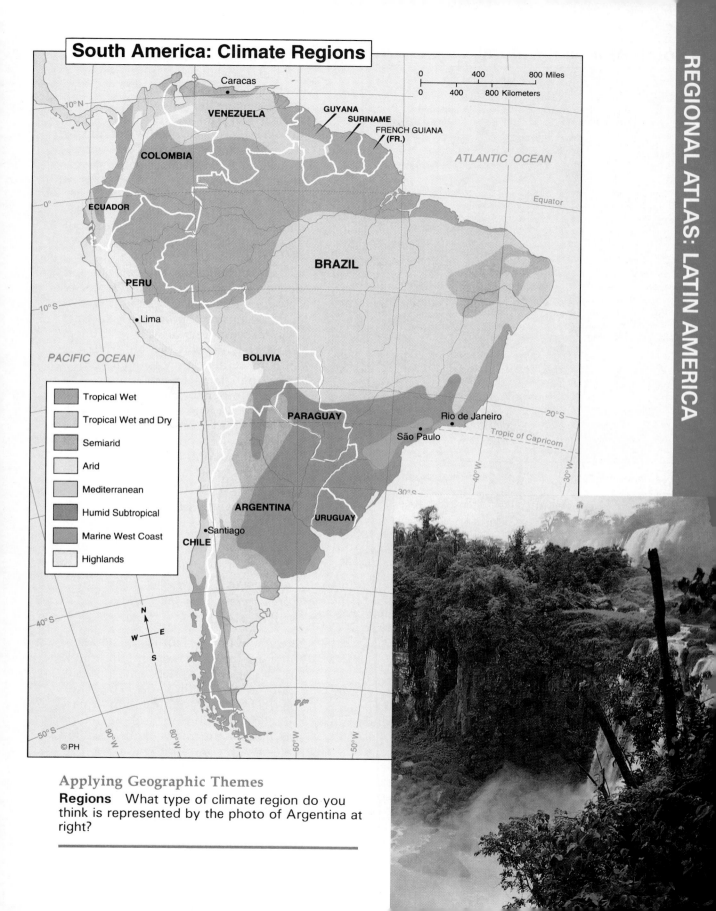

South America: Climate Regions

Caracas

VENEZUELA

GUYANA
SURINAME
FRENCH GUIANA
(FR.)

COLOMBIA

ATLANTIC OCEAN

ECUADOR

Equator

BRAZIL

PERU

• Lima

PACIFIC OCEAN

BOLIVIA

0   400   800 Miles
0   400   800 Kilometers

PARAGUAY

Rio de Janeiro

São Paulo

Tropic of Capricorn

ARGENTINA

URUGUAY

• Santiago

CHILE

Legend:
- Tropical Wet
- Tropical Wet and Dry
- Semiarid
- Arid
- Mediterranean
- Humid Subtropical
- Marine West Coast
- Highlands

N
W   E
S

10°N
0°
10°S
20°S
30°S
40°S
50°S

90°W   80°W   70°W   60°W   50°W   40°W   30°W

©PH

Applying Geographic Themes

Regions What type of climate region do you think is represented by the photo of Argentina at right?

South America: Natural Vegetation

Legend:
- Tropical Rain Forest
- Mixed Forest (Deciduous and coniferous)
- Chaparral
- Tropical Grassland
- Temperate Grassland
- Desert Scrub
- Desert
- Highlands (Vegetation varies with elevation)

Applying Geographic Themes

Regions Compare this map to the map on page 181. What type of vegetation is well suited to the raising of livestock, and where is it found?

Further north, however, the picture changes. Mountain temperatures there are warm, rains frequent, and rain forest growth thick and lush.

Pasturelike Plains The llanos, pampas, and western half of the Gran Chaco are the fertile grasslands of South America. Find these grassland regions on the climate map on page 169.

The llanos are tropical grasslands. Tropical grasslands are also known as savannas. Most savannas, like the llanos, lie near the Equator and have tropical wet and dry climates—long dry seasons and short rainy seasons—and warm temperatures year round. Because of their rainy seasons, savannas are covered with scattered trees as well as clumps of grass. For this reason, some people do not consider savannas to be grasslands at all, but rather transition zones between grasslands and forests.

Unlike the llanos, the pampas and the western half of the Gran Chaco are temperate grasslands. Temperate grasslands occur in regions with distinct seasonal variations of warm summers and cold winters. Cold air from the south or southwest occasionally blows across the plains in the winter. The violent thunderstorms and winds that follow are known as pamperos. Both long and short grasses thrive in these two regions, depending on the amount of rainfall. As one British traveler describes, the grasslands of the region are seemingly endless:

> On every side a sea of grass, grass, and more grass; "paja y cielo"— "grass and sky," as the natives of the country style their favorite landscape. Nothing to break the brown eternity of the Pampa.

The Amazonian Tropics The Amazon rain forest makes up the single largest and most varied mass of vegetation anywhere on earth. The Amazon Basin covers 2.3 million square miles (5.9 million sq km)—a region two thirds the size of the United States. With constantly warm temperatures of about 80°F (26°C) and at least 80 inches (203 cm) of rain every year, the growing season never ends.

Tangles of vines, grasses, and brush line the banks of the rain forest's rivers. This low growth soon leads into

Patagonia, Argentina
The cold, windswept desert region of Patagonia stretches through Chile and Argentina from the Andes to the Atlantic Ocean. It is the largest plateau in South America.

the dimly lit forest. Only thin shafts of sunlight struggle through the overhead canopy —the uppermost spreading branchy layer of a forest. Deprived of sunlight, the damp forest floor is bare and brown rather than green with low-growing plants. Closely spaced tree trunks ascend from the brown floor like pillars more than 100 feet (30 m) tall. Broad leaves mass together at the treetops and form a shady green "umbrella."

The rain forests of the Amazon Basin are the natural habitat of thousands of species of plants and animals, including the world's longest snake, the anaconda, which can stretch as long as 38 feet (11.6 m). The forest also harbors the world's biggest rodent, the ratlike capybara. Dagger-toothed piranha lurk within the waters of the Amazon, which, along with its tributaries, winds through the basin like the veins on the forest's glistening leaves.

Farther north and south from the Equator, the Amazon Basin's climate shifts to tropical wet and dry. The forest thins out, and an astonishing mix of low-growing plants appears among the trees. As you will read on pages 250 and 251, the impact of people's actions in the rain forests of the Amazon Basin are of interest and concern to people around the world.

SECTION 2 REVIEW

Developing Vocabulary
1. Define: **a.** timberline **b.** canopy

Place Location
2. In what country are the pampas of South America located?

Reviewing Main Ideas
3. What three major landforms enclose South America in the west and east?
4. How is the Atlantic Ocean affected by the tremendous amount of water and silt carried by the Amazon River?

Critical Thinking
5. **Making Comparisons**   Describe two ways in which the plain of the Amazon Basin differs from South America's other plains regions.

Chapter 9, Section 2   **177**

3 Human Geography

Section Preview

Key Ideas

- Native Americans, Europeans, Africans, and Asians make Latin America's population distinct.
- Spanish and Portuguese colonists have left their mark on Latin America's culture.
- Three ways of life are found in Latin America today: urban, rural, and traditional.

Key Terms

mestizo, dependency

One geographer described South America as "an empty continent surrounded by islands of people." The population density map on page 179 illustrates clearly what is meant by that description. People in South America tend to live in scattered, dense clusters near the coast rather than in the interior regions of the continent.

The people of Middle America, too, are concentrated in certain areas. Look again at the population density map to locate these areas of higher population density. Some are cool interior highlands, while others are coastal and island lowlands.

The People of Latin America

The entire region of Latin America is home to nearly 450 million people, almost twice as many as live in the United States. Latin Americans are descendants of Europeans, Africans, Asians, and Native Americans, or Indians. A great number are **mestizos** (meh STEE zohs)—people of mixed Spanish and Indian ancestry.

Native American Beginnings Long before the year A.D. 1000, the Maya of the Yucatán created a brilliant culture, advanced in farming, architecture, astronomy, and mathematics. By 1500, the even more widespread Aztec civilization dominated most of mainland Middle America. Ruins of the Aztec capital, Tenochtitlán (tay noch tee TLAHN), now lie buried beneath Mexico City, Mexico's modern capital. The ancient city once housed 250,000 residents. Its gleaming stone pyramids cast long shadows across a large central square.

At about the same time as the Aztec, the Inca of South America controlled an extensive empire in the Andes. Descendants of the Maya, Aztec, and Inca as well as other Indian groups still live in Latin America today.

The Spanish and Portuguese "In Spain there is nothing to compare." So said a Spanish soldier who gazed on

The Aztec

This mural by Mexican artist Diego Rivera shows the Aztec capital of Tenochtitlán, the ruins of which now lie beneath Mexico City.

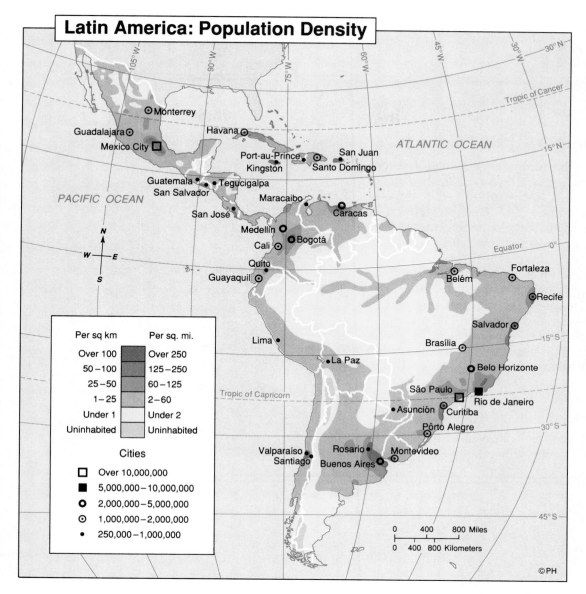

Latin America: Population Density

Per sq km	Per sq. mi.
Over 100	Over 250
50–100	125–250
25–50	60–125
1–25	2–60
Under 1	Under 2
Uninhabited	Uninhabited

Cities

□	Over 10,000,000
■	5,000,000–10,000,000
◉	2,000,000–5,000,000
⊙	1,000,000–2,000,000
•	250,000–1,000,000

Applying Geographic Themes

Regions As is true in many other parts of the world, the population density of Latin America is very unevenly distributed. Where are the regions of greatest population density located?

Tenochtitlán five hundred years ago. The soldier was only one among many Spaniards who immigrated to the region beginning around 1500. He, like others, was drawn to the region by stories of "El Dorado," a mythical city of gold and silver. By the mid-1500s, settlers from Portugal had arrived in South America and had settled in what is today Brazil.

The Spaniards and Portuguese exposed the Indians to diseases previously unknown in the Western Hemisphere. They forced the Indians to work their mines and plantations. Disease and hard labor killed thousands of

Where All Paths Cross

In hundreds of cities and towns throughout Latin America, the hub of daily life is the plaza—the graceful main square, like the one shown at right. The town's largest church, as well as shops, stores, and cafes often face onto the plaza.

Epic events in the lives of the villagers—rites of baptism, marriage, and death—take place at one end of the plaza, in the church. The plaza also serves as an informal meeting place. Lingering over coffee in the afternoon or visiting with friends on benches in the evening dusk are well-honored plaza traditions.

1. What buildings frame the plaza?
2. How is the plaza used by the people of the village?

Indians. Today, for example, only a small number of Indians still live in the Caribbean islands.

Over the span of three hundred years, Spanish and Portuguese colonists made lasting imprints on Latin America's culture. Spanish and Portuguese, both derived from Latin, became and remain the region's major languages. Roman Catholicism, spread by priests who arrived in the 1500s, became the strongest religion. Today, more than 90 percent of Latin America's people are Roman Catholic.

Spanish and Portuguese colonists also left their mark on the social structure of some parts of the region, including Mexico and Central America. The social structure in these places can be represented by a figure in the shape of a pyramid. A few landowning families of purely European descent became a dominant, elite upper class. Mestizo artisans and business people filled the middle, and large numbers of poor peasant farmers of mestizo, Indian, or black ancestry labored for all at the bottom.

Africans in Latin America   Other European nations, including France, Britain, and the Netherlands, joined the Spanish and the Portuguese in their scramble for colonies in Latin America. Where Indians were not present or had died of disease, the Europeans turned to Africa for laborers. Between the 1500s and the 1800s, as

Latin America: Economic Activity and Resources

Legend:

Forestry	⬤ Iron
Livestock Raising	+ Copper
Commercial Farming	+ Bauxite
Subsistence Farming	Gold
Manufacturing and Trade	Silver
Commercial Fishing	✽ Uranium
Little or No Activity	▲ Tin
Coal	▲ Lead
Petroleum	● Nickel
Hydroelectric Power	

PACIFIC OCEAN

ATLANTIC OCEAN

Tropic of Cancer

Tropic of Capricorn

Equator

©PH

Applying Geographic Themes

1. Regions   Which country has huge areas with no agricultural or manufacturing activity?

2. Interaction   Compare this map to the physical map on page 173 and explain what might impede development of economic activity in Bolivia.

many as ten million Africans were enslaved and brought to the Americas against their will. About half of them were sold in the Caribbean islands, another third in Brazil. Their descendants, freed generations ago, still live there in large numbers.

Independence and Immigration   As a result of the harsh rule of the Europeans, several independence movements developed in Latin America during the early 1800s. Despite independence, most countries continue to show influences of the European

Toussaint L'Ouverture

In 1791, a former slave named Toussaint L'Ouverture led the colonists of Haiti in their fight for independence from France.

---

countries that once ruled them. Today, the only mainland country that remains as a dependency of a foreign country is French Guiana. A **dependency** is a territorial unit under the rule of another country.

Living in Latin America

Indians built towns and large cities long before the arrival of the Europeans. But in their desire for gold and glory, the Spaniards destroyed the Indian cities and then built their own. The Spanish influence is still seen in the place names throughout the region, including the name *Argentina,* and in the design of many Latin American cities built around a central plaza.

Today's cities are home to well over half of all Latin Americans, and the cities' plazas are still the hubs, or hearts, of the cities. In Mexico, 70 percent of all people live in urban areas. Five of the largest metropolitan areas in the world are in Latin America. In Central America and the Caribbean, many rural people have moved to fast-growing cities in the hope of

WORD ORIGIN

Argentina
The name *Argentina* reflects the early Spanish explorers' hope to find silver in the region. The name is derived from the Latin *argentum,* meaning "silver."

finding factory jobs. Unfortunately, as you will read in Chapter 10, the number of people is often greater than the number of jobs.

The move to the cities is a relatively new trend. Until recently, Latin American society was more rural than urban. Wealthy members of the elite upper class owned huge tracts of farm and ranch land. They and the peasants who worked for them developed an agricultural way of life that is still practiced by many in the region today.

In a number of places, Latin Americans follow much older, traditional life-styles. Many Indians continue to follow their traditional ways of life— they speak their own languages and practice traditional religions. Indians living on Andean slopes in Peru and Bolivia dress, speak, and farm in the ways of their ancestors. Indians in Guatemala and western Mexico also observe traditional customs.

Today, the overall population of Latin America is growing rapidly. As you will read in the next four chapters, both the land and the people of the region face many challenges.

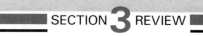

SECTION **3** REVIEW

Developing Vocabulary
1. Define: **a.** mestizo **b.** dependency

Place Location
2. Where did Tenochtitlán once stand?

Reviewing Main Ideas
3. Describe three ways in which the Spaniards influenced Latin American culture.
4. What three ways of life are practiced in the region today?

Critical Thinking
5. **Drawing Conclusions** What do you think might be two results of large numbers of people moving from a rural to an urban setting?

✔ Skills Check

☑ Social Studies
☐ Map and Globe
☐ Reading and Writing
☐ Critical Thinking

Reading Tables and Analyzing Statistics

Tables are often used to present a large amount of data or statistics. Use the following steps to read and interpret the table shown below.

1. **Determine what type of information is presented in the table.** This table presents statistics from five different categories for three South American countries: population, population density, per capita GNP, infant mortality, and life expectancy. GNP stands for gross national product, which means the total value of new goods and services produced in a country in a year. Per capita GNP is the number you get when you divide the GNP by the country's population. The per capita GNP shows what each person's income would be *if* the country's income were divided equally among all of its people. Because this is not often the case, per capita GNP figures can be misleading. A country's infant mortality rate is equal to the number of children who die before their first birthday, for every 1,000 live births. Life expectancy refers to the average number of years a person is likely to live. (a) What is the population of Bolivia? (b) What is the infant mortality rate for Argentina? (c) What is the life expectancy for people in Brazil?

2. **Find relationships among the figures.** Tables help to compare data. (a) How do the three countries rank in terms of per capita GNP? (b) How do the countries rank in terms of infant mortality? (c) What is the relationship between infant mortality and per capita GNP in the three countries?

3. **Use the data to draw conclusions.** Use the table to draw conclusions about the levels of population, wealth, and health care in these countries. For example, what is the relationship between a nation's wealth and its health services?

Population, Wealth, and Health Care for Three South American Countries			
Country	Argentina	Bolivia	Brazil
Population (in millions)	31.9	7.1	147.7
Population density (per square mile)	30	17	45
Per capita GNP (in U.S. dollars)	$2,370	$570	$2,020
Infant mortality rate (per 1,000 births)	29.7	110	63
Life expectancy at birth (in years as of 1989)	70	53	65

Source: Population Reference Bureau, Inc., World Population Data Sheet 1989

Latin America

*Natural resources in Latin America are one key
to understanding the region's Indian civilizations.*

Everywhere in the Americas, legends tell of the importance of corn. In Middle America, a story tells of the Creator giving to humans the first kernels of corn, then water, sun, and wind. The Indians planted the kernels, and the corn grew to become a garden. In a South American story, children walk in the rain forest. They drop kernels of corn along the path so that they not lose their way home. Elsewhere, legends describe the night sky as lit by burning, bright kernels of corn sent out from the hand of God.

The Use of Natural Resources

The Indians who passed along such stories were farmers who understood geography and the importance of natural resources. In addition to landforms and climate, the physical geography of a place includes its minerals, animal life, and vegetation.

People who live in places rich in resources have more to work with than people in places with scarce resources. Certain native plants in the Americas proved so useful that native people came to depend upon them as their main source of food. The cultivation of these plants was so successful that great cities and civilizations were able to develop. Corn in particular became the basis of the strong agricultural economies in the Americas. For the people of Latin America, corn, or maize, became the foundation of their diet and their way of life.

An Essential Plant Corn may also have been the oldest domesticated plant of the Western Hemisphere. About 10,800 years ago, an American Indian—probably a woman if traditions are a clue—gathered wild grass seeds in southern Mexico. After grinding the seeds, she could make a paste that the whole family ate with their fingers. Or she could bake the paste into a flat bread, like today's tortilla. The food had promise, so the woman returned again and again to the field of wild grass. Finally, she gathered the seeds of the hardiest plants with a different goal in mind. She took them home with her and started a garden.

Generation after generation, the children and grandchildren of the first gatherers repeated the planting ritual. Always they saved the hardiest seeds for the next crop. Slowly, the grass heads on which the seeds sprouted grew into tiny cobs, no bigger than a person's thumb. As time passed, corn plants became sturdier and larger, until the ear of corn we know today came into existence.

Other Native American Foods The natural vegetation of the Americas favored Native Americans with a number of other plants

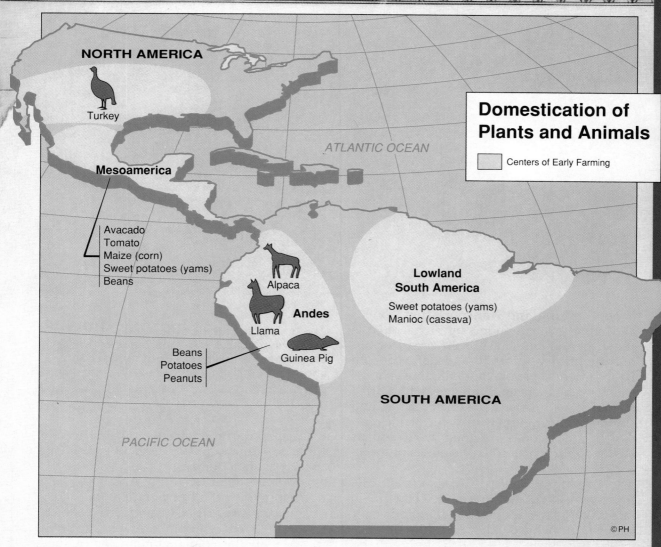

Domestication of Plants and Animals

☐ Centers of Early Farming

NORTH AMERICA

Turkey

ATLANTIC OCEAN

Mesoamerica

Avacado
Tomato
Maize (corn)
Sweet potatoes (yams)
Beans

Alpaca

Andes

Llama

Beans
Potatoes
Peanuts

Guinea Pig

**Lowland
South America**

Sweet potatoes (yams)
Manioc (cassava)

SOUTH AMERICA

PACIFIC OCEAN

©PH

Applying Geographic Themes

Interaction   The successful cultivation of certain plants in the Americas
enabled the people of the region to develop complex cultures, cities, and
empires. Which domestic plants originated in the Andes?

suitable for domestication. People worked
with these plants in much the same ways
they worked with wild maize. Thus people in
Middle America added the common bean,
squash, avocado, and chili pepper to their
diets. In addition to corn, these native foods
changed the nature of Native American life.
Permanent garden plots required stay-at-
home farmers.

The Establishment of Settlements

Many hunting groups gave up their nomadic ways and farmed in tiny hamlets and
villages. The formation of communities
spread farther afield. As settlers, these hunters not only learned how to grow crops, they
discovered the many uses for animals.

Food, Clothing, and Transportation Bones and other artifacts found by archaeologists show that animals have been a part of our human history for thousands of years. In Latin America, people hunted guinea pigs as well as goats, sheep, and pigs. They dried animal skins for use as blankets, tents, and clothing. They learned to value llamas, alpacas, and vicuñas for their wool and as beasts of burden as well as sources of food.

The Growth of Towns By about A.D. 150, some villages in Middle America had grown into towns. After another five hundred years, the first full-fledged cities in Middle America evolved. Farmers were not the only inhabitants of the towns and cities. Government and religious leaders, craftsworkers, tradespeople, and scholars also occupied the newly organized urban centers. Such specialization of labor and urban living are two parts of a highly organized culture known as a civilization. The major Middle American civilizations were those of the Olmec, the Maya, and the Aztec.

Farming in the Andes

What of other parts of the Western Hemisphere? How did agriculture affect them? The story is similar to that of Middle America. High in the Andes, under the dirt, people

discovered the tuber of a plant that thrived in the cold climate and thin, rocky soils. This tuber was the potato.

Native Americans of the Andes not only domesticated and improved the potato, they invented a way of preserving their harvests for future needs. At night they put the potatoes out in the cold mountain air so they would freeze. The next day, the moisture in the potatoes would slowly melt. The Native Americans squeezed the wetness out of the potatoes and set them out another night. After several days and nights of being processed in this way, the potatoes were freeze-dried and ready for storage. The Andean people could store the dried potatoes for up to six years. To revive the potatoes for cooking purposes, they merely soaked them in water before cooking them.

The Growth of Andean Settlements The potato dramatically affected the Andean peoples. It allowed them to take up farming, just as corn had done for people in Middle

America. Farming led to the establishment of villages, towns, cities, and in time a complex culture, or civilization with specialized workers. The potato, unknown to Europe prior to the 1500s, was brought back from South America and introduced as a food item.

Tropical vegetation in the lowlands of South America offered Native Americans still other possibilities for domesticated food crops. Cassava, also called manioc, became a basic source of food among farming settlements. Cassava comes from the starchy roots of many tropical shrubs, trees, and other plants. After some complicated processing, it can be made into a nutritious bread as well as other varieties of food.

The Exchange of Ideas As important as agriculture was, not everyone stayed at home and farmed. Traders and adventurers made their way to many different settlements. Pack animals, especially llamas, made such travel possible. The people of the Andes discovered that the llamas could travel long distances with little food or water, like another of their species—the camel. The Andean traders helped people throughout the Americas exchange ideas as well as goods.

Andean people started growing corn on the lower mountain slopes, where the air was warm and rain plentiful. Middle American farmers grew potatoes in their Sierra Madre highlands, where the air was too cool for corn. Where the climate was wet and hot enough, sweet potatoes and cassava filled out the agricultural menus of the American people.

The more varied agriculture became, the more balanced were the diets of the people —rich in starches, calories, vitamins, and proteins. Over time, healthier people led to a steady growth in population. By the time Columbus stumbled onto the "new world," most Native American peoples had practiced agriculture for dozens of generations. The Aztec and Inca were at their peak in terms of population, territory, and wealth. This "new world" that Columbus and other Spaniards stepped into, however, would have been much different without the benefits of its natural geography.

Potato Farming in the Andes
Like the ancient Incas, these Peruvian farmers rely on the potato as an important food crop.

TAKING ANOTHER LOOK

1. How did corn culture in Middle America and potato culture in the Andes contribute to the origins of civilization in the Americas?
2. How did climate help determine the places where certain foods originated?
3. What are some ways Native Americans processed or prepared foods?
4. In what ways does specialization of labor affect the structure and behavior patterns of a society?

Critical Thinking

5. **Distinguishing Between False and Accurate Images** Modern people sometimes believe that Native American peoples had nothing to offer the rest of the world. What facts about foods and agriculture argue against this false image?

9

REVIEW

Section Summaries

SECTION 1 Middle America: Land, Climate, and Vegetation The central plateau of Mexico, marked by broad arid and semiarid stretches, dominates northern Middle America. The Sierra Madre cordillera splits along both sides of the plateau and then comes together and extends southward through Central America. Elevated settings in southern Mexico and Central America have comfortably warm, moist conditions. Caribbean breezes help moderate tropical island environments, which include plains, assorted highlands, and lush vegetation.

SECTION 2 South America: Land, Climate, and Vegetation The Andes Mountains thread an unbroken chain through western South America. The Guiana Highlands and Brazilian Highlands form rolling uplands in the east. The llanos, Gran Chaco, and pampas are large, grassy interior plains. The Amazon Basin includes the world's largest tropical rain forest.

SECTION 3 Human Geography The culture of Latin America reflects the origins of its many inhabitants: Native American, European, African, and more recently, Asian. Today, Latin Americans follow either urban, rural, or traditional ways of life.

Vocabulary Development

Match the definitions with the terms below.

1. large waves caused by a storm
2. Latin American of mixed Native American and European ancestry
3. a scientist who studies the atmosphere and weather
4. a flat-topped, isolated mountain

5. a tropical storm with winds of at least 74 miles (119 km) per hour
6. a storm system built around an organized low-pressure area
7. the boundary above which continuous forest cannot grow
8. low-lying coral islands
9. a storm with winds of at least 39 miles (63 km) per hour
10. a territorial unit under the rule of another country
11. the uppermost spreading branchy layer of a forest

a. mesa
b. cay
c. hurricane
d. tropical depression
e. tropical storm
f. storm surge
g. meteorologist
h. timberline
i. canopy
j. dependency
k. mestizo

Main Ideas

1. Name three Latin American portions of the Western Hemisphere's "backbone."
2. What causes dry conditions in northern Mexico?
3. Name three groups of islands located in the Caribbean.
4. Why is the Atacama Desert one of the driest places on earth?
5. Why did Latin America's Indian population decrease after 1500?
6. Why do Brazil and the islands of the Caribbean have large African populations?

Critical Thinking

1. **Determining Relevance** European colonists built ports and cities along Latin American coasts to send plantation products and

other goods back across the Atlantic. What do you think this has to do with population density patterns in Latin America today?

2. **Identifying Central Issues** Why do you think a pyramid-shaped social structure developed in Latin America following the arrival of Europeans?

3. **Demonstrating Reasoned Judgment** Because both the Spanish and Portuguese languages derive from Latin, geographers have named all the Western Hemisphere's southern lands ''Latin America.'' Do you agree that this is a fair and accurate name for the region? Why or why not?

Practicing Skills

Reading Tables and Analyzing Statistics Refer to the table on page 183 to answer the following questions about three South American countries.

1. Which country has the smallest population?
2. Which country has the greatest population density?
3. How does the relative wealth of these countries relate to their infant mortality rates?
4. Explain the probable relationship between infant mortality and per capita GNP?

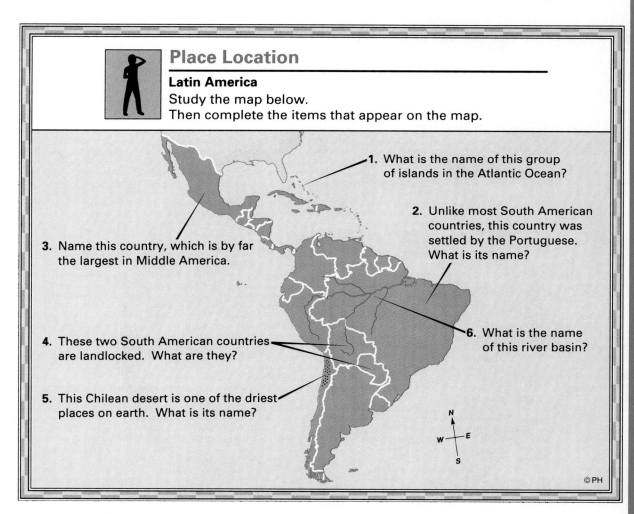

Place Location

Latin America
Study the map below.
Then complete the items that appear on the map.

1. What is the name of this group of islands in the Atlantic Ocean?

2. Unlike most South American countries, this country was settled by the Portuguese. What is its name?

3. Name this country, which is by far the largest in Middle America.

4. These two South American countries are landlocked. What are they?

5. This Chilean desert is one of the driest places on earth. What is its name?

6. What is the name of this river basin?

© PH

10

Mexico

Chapter Preview

Both of these sections provide an overview of Mexico, shown in red on the map below.

Sections	Did You Know?
1 REGIONS DEFINED BY MOUNTAINS	By the year 2000, Mexico City is expected to be the largest city in the world, with a population of 30 million.
2 A PLACE OF THREE CULTURES	The ancient Mayan civilization, which flourished from about A.D. 300 to 900, had the most accurate calendar of that era.

Mexico City

Mexico City retains its image as the age-old focus of Mexican life as it carries on the functions of a rapidly growing business and industrial center. Towering above the crowded capital city are ancient snow-capped volcanoes.

1 Regions Defined by Mountains

Section Preview

Key Ideas

- Mountains divide Mexico into five physical regions.
- Mexico's central plateau region is the heartland of the country.
- Within each of Mexico's coastal regions are a variety of contrasting physical characteristics.

Key Term

irrigation

A widely repeated story, perhaps true, perhaps legend, tells how Hernán Cortés, the Spanish explorer who conquered Mexico, was once asked by King Charles V of Spain to describe Mexico's physical features. As the story goes, Cortés answered the question by crumpling a piece of paper and throwing it down before the king. The crumpled paper represented what Cortés could not find words to describe: Mexico's mountains. High and rugged, they seemed to cover every part of the country.

Mountains do dominate Mexico's physical setting. The country's largest mountain range, the Sierra Madre Occidental, extends along the western coast. Mexico's second great mountain range, the Sierra Madre Oriental, extends along the eastern coast.

These two arms of Mexico's great mountain ranges divide the country into five physical regions. The largest region, the central plateau, lies between the Sierra Madres. West of the Sierra Madre Occidental is the Northern Pacific Coast, which includes the arm of the Baja California peninsula.

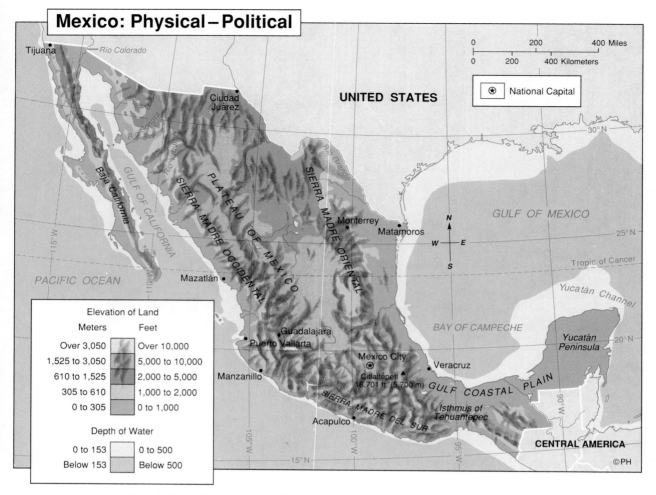

Mexico: Physical–Political

Tijuana
Rio Colorado
Ciudad Juárez
UNITED STATES

0   200   400 Miles
0   200   400 Kilometers

⊛ National Capital

30°N

Rio Sonora
Rio Grande

SIERRA MADRE OCCIDENTAL
PLATEAU OF MEXICO
SIERRA MADRE ORIENTAL

GULF OF CALIFORNIA
Baja California
PACIFIC OCEAN

Monterrey
Matamoros

GULF OF MEXICO

25°N

N
W — E
S

Tropic of Cancer

Mazatlán

Yucatán Channel

Guadalajara
Puerto Vallarta
Manzanillo

BAY OF CAMPECHE

Yucatán Peninsula

20°N

Mexico City
⊛
Citlaltépetl
18,701 ft. (5,700 m)

Veracruz

GULF COASTAL PLAIN

Acapulco

SIERRA MADRE DEL SUR

Isthmus of Tehuantepec

CENTRAL AMERICA

©PH

115°W   110°W   105°W   100°W   95°W   90°W

15°N

Elevation of Land

Meters		Feet
Over 3,050		Over 10,000
1,525 to 3,050		5,000 to 10,000
610 to 1,525		2,000 to 5,000
305 to 610		1,000 to 2,000
0 to 305		0 to 1,000

Depth of Water

0 to 153		0 to 500
Below 153		Below 500

Applying Geographic Themes

1. Location   Mexico is the northernmost country in Latin America. Because of their proximity, Mexico and the United States maintain close economic and cultural ties. What river forms two thirds of the Mexican-United States border?

2. Movement   Does movement between the two countries look difficult or easy?

East of the Sierra Madre Oriental lies the Gulf Coastal Plain, a lowland region that curves around the Gulf of Mexico. South of the central plateau lies the southern Pacific coast, a narrow coastal strip with the resort city of Acapulco forming its midpoint. Set apart from the rest of Mexico is the Yucatán Peninsula, which sticks out like a thumb northward into the Gulf of Mexico. Each region is shown on the map of Mexico above.

The Central Plateau: Mexico's Heartland Region

In addition to being the largest region in Mexico, the central plateau is the most important region. It is where most of Mexico's people live and where the country's major cities are located. The southern part of the plateau, nourished by rich soil and plenty of rainfall, makes up Mexico's best farmlands.

The central plateau is the most geologically unstable area of the country. The plate tectonics map on page 22 locates Mexico in the crossroads of four tectonic plates—the North American Plate, the Caribbean Plate, the Pacific Plate, and the Cocos Plate. Over time, as these plates have collided and slid against one another, they have pushed up mountains and have torn open the land with earthquakes. The mountains that form the southern edge of the plateau are studded with volcanoes, many of which are still active. Earthquakes continue to shake the land. In 1985, for example, an earthquake destroyed part of Mexico City, killing about twenty-five thousand people.

Why do so many people live in the midst of such danger? The answer lies in certain factors that make the central plateau an attractive place to live.

Climate counts as one reason why most Mexicans live in the central plateau. From north to south, the climate of the central plateau ranges from arid and semiarid to tropical wet and dry. As was described in Chapter 9, the Sierra Madres prevent rain from falling in northern Mexico. But farther south, moist ocean winds find their way through the mountains to bring rain to the southern end of the central plateau. Compare the annual precipitation map with the population map on this page. Note that population is densest in the southern part of Mexico.

Elevation is a key factor in the climate of the central plateau. The southern part of the central plateau sits in the tropics, but its climate is not hot because the high elevation, which

---

Applying Geographic Themes

Location Compare the two maps at the right. How does the amount of annual rainfall relate to the distribution of Mexico's population? Why do you think this is?

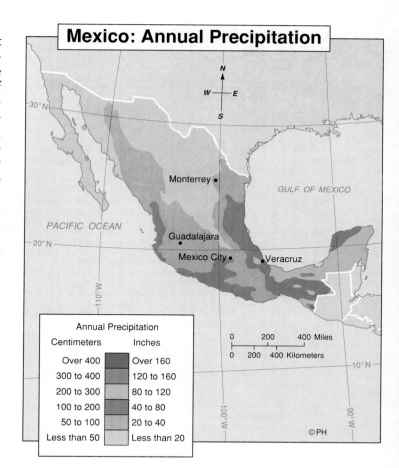

DAILY LIFE

Sunday in the Park

Located in Mexico City is a sloping, heavily wooded park, known as Chapultepec Park, where Mexicans go to relax. On a typical Sunday, more than half a million people visit the park, which is shown at right. They go rowing on one of its three lakes, they stroll through the zoo, or they listen to an open-air concert or play. People also visit the park to attend free classes in weaving, needlework, carpentry, and other trades.

Chapultepec Park is also the home of Mexico City's major museums including the National Museum of Anthropology. Other museums include the Museum of Colonial Art and the Museum of Modern Art.

1. What are some popular ways that residents of Mexico City like to use the park?
2. Why do you think so many people visit Chapultepec Park?

WORD ORIGIN

Tijuana
Tijuana, Mexico—originally, *El Rancho de Tia Juana,* or Aunt Jane's Ranch—grew during prohibition in the United States because alcoholic beverages were legal there.

averages about 7,000 feet (2,134 m), causes temperatures to be mild. Temperatures generally range between 50° and 80° F (10° and 27° C).

The Coastal Regions: A Study in Contrasts

On the other side of the mountains that frame Mexico's central plateau are three regions of mostly coastal plains. In many ways, these three regions are a study in contrasts. Notice on the map on page 192 that in each region, mountains dictate the width of the coastal plain. Along the Gulf coast and the northern Pacific coast, the coastal plain extends for about 80 miles (129 km) until it rises to meet the mountains. But along the southern Pacific coast, mountains crowd close to the coastline, leaving a coastal plain only about 15 miles (24 km) wide.

Northern Pacific Coast Dry, hot, and for the most part, thinly populated—describes Mexico's northern Pacific coast. The city of Tijuana (tee uh *WAH* nuh), just across the border from California, is one of Mexico's largest and fastest growing cities. This region can also be described, however, as having some of the best farmland in the country. A trio of widely spaced rivers, dammed and used for irrigation, make it possible for farmers to raise wheat, cotton, and other crops. These rivers are the Colorado, the Sonora, and the Yaqui [yah KEE]. Irrigation is the artificial watering of farmland, often by means of canals that draw water from reservoirs or rivers.

No such situation exists on the Baja California peninsula, which juts southward into the Pacific Ocean like a long, thin arm of mostly mountainous desert. An American who visited Baja California in 1989 described an overnight trip into Baja's desert one August in this way:

As we drove south . . . we passed the canteen back and forth in a kind of trance, lulled by heat waves rising off the pavement. I wiped dust from the little plastic thermometer I'd clipped to my bag; it read 110 degrees. The scene out our window was a no-man's-land of reddish volcanic mountains and scorched vegetation. Mars with cactus.

Southern Pacific Coast   The steep-sided Sierra Madre del Sur edges the narrow southern Pacific coast, providing few opportunities for farming. The natural setting of this region, however, favors another kind of economic activity—tourism. The resort cities and wave-washed beaches of Acapulco, Mazatlán (mahs uh TLAHN), and Puerto Vallarta (pwert o vuh YARH tuh), among others, draw thousands of visitors each year.

Gulf Coastal Plain   Like the southern Pacific coast, Mexico's Gulf coastal plain has key economic importance to the country. Along the plain and off-shore, beneath the waters of the Gulf of Mexico, lie deposits of petroleum and natural gas. These geological riches have made the Gulf coastal plain one of the world's major petroleum-producing regions.

The Yucatán Peninsula

The map on page 192 shows that the Yucatán is generally flat, forming a contrast to the mountains that cover much of Mexico. When rain falls on the peninsula, it passes through the surface of the land and drains underground. The Yucatán is covered by a layer of limestone (rock through which water passes freely). Underground erosion has carved out caverns where streams flow along the cavern floor. The water can be reached when the roof of a cavern collapses, creating a sinkhole which the ancient Mayans who lived in the Yucatán used as wells.

SECTION **1** REVIEW

Developing Vocabulary
1. Define: irrigation

Place Location
2. Where is Mexico's central plateau region located relative to the Sierra Madre Occidental and Sierra Madre Oriental?

Reviewing Main Ideas
3. How do climate and elevation help explain why most Mexicans live in the central plateau?
4. Which region of Mexico has a tropical rain forest?
5. Why are there no rivers in the Yucatán Peninsula?

Critical Thinking
6. **Demonstrating Reasoned Judgment**   Why do you think Mexico's tourist industry is centered on the southern Pacific coast rather than the Gulf coastal plain?

✓ Skills Check

☑ Social Studies
☐ Map and Globe
☐ Reading and Writing
☐ Critical Thinking

Interpreting Population Pyramids

A population pyramid is a type of bar graph that shows the percentages of males and females by age group in a given country. By its shape, a population pyramid also shows whether a population is growing or declining. If the pyramid is wide at its base, the population is growing. If the pyramid is narrow at its base, the population is declining. A graph with a nearly rectangular shape indicates a population that is neither growing nor declining.

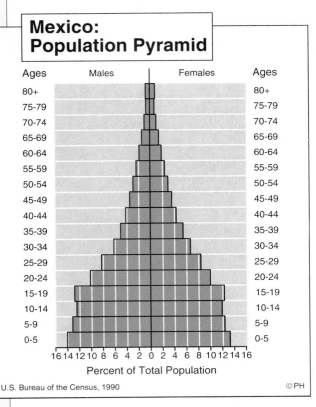

Mexico: Population Pyramid

Percent of Total Population

U.S. Bureau of the Census, 1990   © PH

Use the following steps to read and interpret the population pyramid below.

1. **Study the graph to become familiar with how information is presented.** Population pyramids present information in a certain way. The vertical axis shows age. The horizontal axis shows percentage of population. One side of the pyramid shows the male population. The other side shows the female population. Which side of the population pyramid for Mexico shows the male population?

2. **Practice reading the information shown in the population pyramid.** Each bar in the pyramid represents a percentage of people in a certain age range. For example, the bar at the bottom of the pyramid represents people age 4 and under. The horizontal axis shows that 9 percent of the people in this age group are males and 8.5 percent are females. Add the two numbers together to see that 17.5 percent of the total population is age 4 and under. Each full square represents 1 percent of the total population, so you could count the number of squares and get the same answer. Answer the following questions: (a) What percentage of Mexico's population is between ages 5 and 9? (b) What percentage of Mexico's population is between ages 30 and 34?

3. **Look for relationships among data.** A population pyramid is helpful in detecting characteristics of a population. For instance, the shape of the pyramid can tell you if a population is evenly divided among young and old or males and females. Answer the following questions: (a) Are the majority of people in Mexico under age 40? (b) Is the population of Mexico equally divided between males and females?

4. **Use the graphs to draw conclusions.** Population pyramids can provide clues that help you draw conclusions about a country's population. Answer the following questions: (a) Is the population of Mexico growing or declining? (b) How might demands for housing, food, and health care be affected by Mexico's population?

2 A Place of Three Cultures

Section Preview

Key Ideas

- Mexico today has roots in three cultures—ancient Indian, colonial Spanish, and modern Mexican.
- Spain conquered the Aztec civilization in 1521 and ruled Mexico as a colony for three hundred years.
- Since gaining independence, Mexico has become a democratic republic but is still working to improve social justice and economic opportunities.
- Most Mexicans live in urban areas.

Key Terms

hacienda, ejido, latifundio

In Mexico City sits a quiet square called the Plaza of the Three Cultures. If you stood on the plaza and gazed around, you would see evidence of three different Mexican "cultures," or traditions. Standing in the middle of the plaza are the ruins of an Aztec temple, which represents Mexico's Indian culture. The culture of the Spanish conquerors who destroyed the Aztecs is symbolized by a church built by the Spanish in 1609. Finally, Mexico's modern culture is represented by twin office buildings of glass and concrete.

These three "cultures"—the Indian, the Spanish, and the modern—are the human characteristics that make up modern Mexico. Like brightly colored cords braided together, they can be seen separately but they also overlap. The result is a mestizo nation keenly aware of the traditions of the past and the possibilities of the future.

From Aztec Empire to Spanish Colony

The greatest of Mexico's Indian civilizations was built by the Aztecs. By the early 1400s, their capital city of

Plaza of the Three Cultures, Mexico City

On this plaza, remains of centuries-old Indian cultures mingle with symbols of Mexico's Spanish heritage as well as symbols of modern Mexican culture.

Tenochtitlán (tay noh chit LAHN) was the center of an empire that spread over much of south-central Mexico. Tenochtitlán occupied an area that is now Mexico City. On the main square of Tenochtitlán—now the main square of Mexico City—stood great temples and palaces where Aztec kings lived. Before the Europeans arrived, the city was one of the largest in the world. Only four European cities—Paris, Venice, Naples, and Milan—had more people.

Into this bustling city, where every day about 100,000 Aztecs gathered in markets to trade goods, marched Hernán Cortés and six hundred Spanish soldiers. Cortés reached Tenochtitlán in 1519, after marching inland from Mexico's Gulf coast and making allies of the Aztecs' enemies along the way. With their help, Cortés destroyed the Aztec civilization within two years. Aztec warriors fought furiously, but when the final battle ended Tenochtitlán lay in ruins, flattened by Spanish cannon fire. The Spanish *conquistadors,* or conquerors, defeated the remaining Indian groups of Mexico. The territory won by Cortés became the colony of New Spain.

Four social classes emerged as the Spanish settled New Spain. At the top were the *peninsulares* (puh nin suh LAH rez). A *peninsular* was a Spanish person born in Spain. The second-highest group were the *criollos* (cree OH loz), or people of Spanish ancestry born in the New World. The mestizos ranked third, and the Indians ranked lowest. Over the next three hundred years, life in New Spain followed these social lines.

As in the other New World colonies set up by the Spanish, Indians provided the labor on *haciendas* (hah see EN duhs), large estates owned by the Spanish and often run as farms or cattle ranches. Both the *haciendas* and the Indian labor were granted to the *conquistadors* as rewards by the Spanish king. Under this *encomienda* system

low wages and constant debt unfortunately forced most Indians to live a slave-like existence.

The Search for Democratic Rule

By the early 1800s, the *encomienda* system was still in force. Change came in another way, however. The simmering anger the *criollos* felt for the *peninsulares* erupted into conflict. In 1810, a *criollo* priest named Miguel Hidalgo called for a rebellion against Spanish rule. His cry sparked a war of independence, which ended in 1821 when the independent nation of Mexico was established.

Mexico had achieved its independence, but not a democratic independence. The search for democracy lasted for more than one hundred years. During that time, Mexico experienced a series of political struggles between leaders who wished to establish democracy and leaders who ruled Mexico as dictators.

By the end of the nineteenth century, however, Mexico gained enough stability to attract large amounts of foreign capital and industry. Railroads were built, ranches were expanded, and Mexico's valuable oil reserves were developed. Such efforts to modernize the country mainly helped wealthy Mexicans become even wealthier. The gap between rich and poor from colonial times continued. Landlords still held on to the *haciendas,* on which Indian and mestizo peasants worked.

Change finally came in 1910. Peasant and middle-class Mexicans were able to stand up to the military dictator and the landlords who together controlled the country. The ten-year Mexican Revolution began. By the time the fighting ended in 1920, Mexico had a new president and a new constitution. The new government promised "land, bread, and justice for all."

Revolution, Germination **by Diego Rivera**
This mural, painted by Mexican artist Diego Rivera in 1927, shows Mexican peasants planting the seeds that led to the country's 1910 revolution.

The democratic republic established by the Mexican Revolution remains in place today. Like the United States government, the national government is headed by an elected president and congress. Unlike in the United States, however, one political party has had all the power. Called the Institutional Revolutionary Party, it has never lost a presidential election. Other political parties are gaining power, however, and one, the National Action Party, has won local elections in Mexico City.

The Mexicans of Today

Like no other country in Latin America, Mexico has worked to preserve both its Indian and Spanish heritages. Nearly all Mexicans use Spanish as their official language. People of purely Indian descent, however, often speak their ancient languages at home. The constitution grants freedom of religion, but nearly all Mexicans are Roman Catholics, like their former Spanish rulers.

Although Mexico has made great economic strides in modern times, the nation is still working to achieve social justice and create economic opportunities for more people. A small minority still holds much of the country's wealth. Next on the social ladder is a growing middle class, and ranking lowest is a huge number of poor people in rural and urban settings.

Patterns of Rural Life Rural life in Mexico usually involves working in agriculture. In 1910, nearly all Mexican land that could be used for farming was part of about eight thousand *haciendas.* After the revolution, however, the government began a program of buying out *hacienda* owners and breaking up their large holdings. The estates were then divided among landless peasants. The government still follows this land reform policy. To date, about half of the old *haciendas* have been broken up in this way.

The government has awarded most of the reclaimed farmland in the form of *ejidos* (ay HEE doz). An *ejido* is farmland owned collectively by members of a rural community. Many ejido farmers practice subsistence farming, raising only enough crops to meet the basic needs of one family.

Approximately one third of Mexican farms, however, are huge commercial farms owned by private individuals or farming companies. These commercial farms are called *latifundios*. Mexico's commercial farms and some *ejidos* raise cash crops such as corn, sugar cane, coffee, and fruit.

Many rural Mexicans never included in the *ejido* system try to coax crops from land unsuitable for farming. Regardless of where they farm, most farmers do not have enough money to buy the fertilizers and machines that could increase their harvests. Worst of all, there simply is not enough land to go around. Today three to four million rural families have no land or opportunities for work.

Many landless, jobless peasants become migrant workers. As such, they travel from place to place where extra workers are needed to cultivate or harvest crops or to work as day laborers. Many migrant workers also cross the Rio Grande River into the southwestern and western United States at harvest time. While some of these workers have permits to cross the border, others cross illegally, sometimes spending their life's savings to hire a guide to take them across the border. Every year, the United States Immigration and Naturalization Service arrests and sends home about one million people attempting to enter the United States illegally .

Patterns of Urban Life   Today, the real heart of modern Mexican culture beats in Mexico's urban areas. For many Mexicans, urban life means better economic opportunities than can be found in the countryside.

About 66 percent of Mexico's population is urban, with most urban dwellers crowding into Mexico City. With more than 18 million people, Mexico City is the largest urban area in the world.

Urban life in Mexico exists at four levels. A small upper class tends to be highly educated, well traveled, and politically powerful. A growing middle class includes government workers, bankers, lawyers, and owners of industries or businesses. Mexico's middle class members live in large apartments, drive the latest cars, and browse in shopping malls. Mexico's working class are generally skilled workers who maintain strong ties to traditional Mexican culture. They may live in adobe-block houses in older neighborhoods or in new worker apartment complexes. The lower class is estimated to be the largest of all social classes.

The Importance of Family Life   Whether urban or rural, rich or poor, life for most Mexicans revolves around the family. Mexicans regard the family as the foundation of their society. About 90 percent of Mexicans live with family members. The great majority of Mexican businesses are family businesses. And, in times of need or crisis, the family serves as the social security system and the bank. While economic and political forces have brought sweeping change to Mexico, families have provided a social anchor.

Major Economic Activities

While the family acts as social anchor for Mexico's people, a wide range of economic activities anchors Mexico's economy. Two of Mexico's most important economic activities are petroleum mining and tourism. Each has connections to the country's history and physical geography.

Mexico's petroleum industry is a result of some favorable geographic patterns. Great reserves of petroleum lie off Mexico's Gulf coast. Mexicans first discovered Gulf coast oil in 1901 near the city of Tampico. Additional discoveries have made Mexico the fourth-largest producer of oil in the world. Only the Soviet Union, the United States, and Saudi Arabia produce more oil. One petroleum worker described Mexico's off-shore oil reserves this way:

This is like Saudi Arabia. There they have sand. Here we have water. Underneath, though, we have one thing in common: lakes of oil.

Both the country's history and climate make tourism a leading industry in Mexico. National and foreign investments in tourism have exploded since Acapulco became a popular international tourist destination in the 1950s. Most foreigners visiting Mexico go only to the beaches, many of which line the southern Pacific coast and the Yucatán Peninsula. Mexico's ancient history also attracts visitors who marvel at the remains of Indian temples and cities.

Tourism ranks as an important source of income for Mexico. But this industry is important for another reason. Manufacturing has long contributed to Mexico's economy, with Mexico City serving as the country's leading industrial center. Manufacturing, however, also creates a heavy load of pollution. In Mexico City, polluted air from factories and cars tends to collect over the city because mountains trap the air on three sides. Mexicans regard tourism as a cleaner alternative. In fact, they call it the "smokeless industry."

The Lure of Mexico's Resorts
Tourism thrives because of vistas like this one on Mexico's beautiful Pacific coast.

SECTION 2 REVIEW

Developing Vocabulary
1. Define: **a.** *hacienda* **b.** *ejido*
   c. *latifundio*

Place Location
2. Where is Mexico's leading industrial center?

Reviewing Main Ideas
3. Name the Indian civilization in Mexico destroyed by the Spanish in the early 1500s.
4. How is Mexico's tourism industry related to the country's climate and history?

Critical Thinking
5. **Identifying Central Issues** What challenges will Mexico face if the number of people in urban areas continues to grow?

"Lost Cities" of Hope

Not long before the Mexican Revolution in 1910, a local artist painted a landscape of Mexico City. He correctly showed the city lying in a level farming valley surrounded by mountain slopes. An artist today would paint a very different picture. The mountains and the valley have endured, but the farms have disappeared. In their place stands a teeming metropolis surrounded by a spreading circle of urban fringe—Mexico's *ciudades perdidas*, or "lost cities"—one of the results of runaway urban growth.

Urban Growth

The ciudades perdidas are home to many of the one million people who migrate to Mexico City each year. Every day, an estimated 1,700 people move to Mexico City, wishing to trade the hard life of the peasant farmer for the hope of something better. In addition, one thousand babies are born in the city each day. At that rate, some experts predict that Mexico City will have between forty and fifty million residents by the year 2000. This astounding rate of growth, which might once have been a source of pride, is in some ways straining the city at its seams.

In Mexico City decent jobs for the often unskilled and illiterate newcomers are hard to find. Almost half of the city's new arrivals are "underemployed"—working but not earning enough money to support a family. Families are often forced to exist on the equivalent of only four dollars a day. A family's first home may be no more than a quickly built lean-to made of cardboard or flattened tin cans. So many homes have been built in this way that the new arrivals

have come to be known as *paracaidistas*, or parachutists—appearing as if from the sky.

Finding a Better Life

In the past, the government often treated the newcomers as illegal squatters on the land. Now, although the government officially discourages squatter settlements, it has done its part to make life more comfortable for the people of the lost cities. The government has improved schools and roads and has provided other services such as electricity. In addition, it often gives people a legal right to own their tiny lots after they have lived there for five years.

In spite of the odds against them, many residents of the lost cities live to tell stories with successful endings. They often find work in the "informal sector." That is, they may sell handmade craft items or perform a service such as auto repair. In this way they rely on their talents and hard work and are not dependent on scarce job opportunities. David Castro, for example, moved to the outskirts of Mexico City as a child. He finished elementary school and then earned enough money doing odd jobs to start a small lumber shop. A few years later, he borrowed money to start a cement business. By the time David was married, he was able to buy a house in a middle-class neighborhood inside the city.

Determined Hope

Many *paracaidistas* are far less fortunate than David Castro. As destitute as they may be, however, most residents of the lost cities would not consider returning to the countryside. Hope for their children's futures is the bright light in their lives. A *paracaidista* who recently arrived from the rugged mountain state of Oaxaca explains: "In Oaxaca, I went to school but I didn't really learn to read and write. But here, if my children go to school and if they wear better clothes, well, maybe it will be better for them."

Public schools in Mexico City are better than those in rural villages. *Paracaidistas* stand a chance of getting their children all the way through elementary school and possibly high school as well. Young people who finish school can usually find jobs as craft workers in factories or as white-collar workers in government or business offices. Like David Castro, they are able to move into Mexico's middle class.

One taxi driver lives with his family in a tarpaper shack. Yet, he shows confidence that his six children will have a brighter future when he says, "We thank God for letting us live here."

TAKING ANOTHER LOOK

1. Why do so many rural people move to Mexico City?
2. How has the Mexican government improved life for residents of the lost cities?
3. What is the greatest source of hope for many *paracaidistas?*

Critical Thinking

4. **Drawing Inferences** In what other ways might urban growth be causing Mexico City to "strain at the seams?"

Mexico's Urban Poor The residents of Mexico's "lost cities" work together and with the government to improve their living conditions.

10

REVIEW

Section Summaries

SECTION 1 Regions Defined by Mountains Two mountain ranges, the Sierra Madre Oriental and the Sierra Madre Occidental, dominate Mexico's physical setting and divide the country into five regions. The largest and most populous region is the Central Plateau where the country's major cities are located. High elevations keep the temperatures comfortable. Mexico has three coastal regions. The northern Pacific coast is dry, hot, and thinly populated. The resorts of the southern Pacific coast draw tourists. The Gulf coastal plain is a leading petroleum-producing region. A separate region is the flat Yucatán Peninsula.

SECTION 2 A Place of Three Cultures Mexico's human characteristics come from three "cultures," or traditions—the Indian, the Spanish, and the modern. Indian and Spanish cultures met in the 1500s, when Spain conquered the Aztec civilization and set up the colony of New Spain. Modern Mexico is a democratic republic still working to improve social justice and create economic opportunities for its people. The real heart of Mexico is in its urban areas, where about 66 percent of the population lives. The foundation of Mexican society is the family, which has served as an anchor of stability through sweeping economic and political changes.

Vocabulary Development

Match the definitions with the terms below.

1. farmland owned collectively by members of a rural community
2. a huge commercial farm owned by a private individual or farming company
3. the artificial watering of farmland
4. a large estate usually operated as a farm or cattle ranch

a. irrigation
b. *hacienda*
c. *ejido*
d. *latifundio*

Main Ideas

1. Describe the location of Mexico's central plateau.
2. What makes the central plateau region the most geologically unstable part of Mexico?
3. How does the elevation of the central plateau affect the region's climate?
4. Which of Mexico's coastal regions has a tropical rain forest?
5. Why does the Yucatán Peninsula have no rivers?
6. Describe the four social classes that made up New Spain.
7. What kind of government did the Mexican Revolution establish?
8. In what ways is the family the foundation of Mexican society?
9. Why is tourism an important industry in Mexico today?
10. Why does the majority of the Mexican population live in urban areas?
11. What economic gains were made between the Mexican war of independence and the Mexican Revolution? Which group of people mostly benefited from these gains?
12. How are farmers able to grow crops along Mexico's northern Pacific coast?

Critical Thinking

1. **Demonstrating Reasoned Judgment** Mexicans often say that the Mexican Revolution is still going on. Why?
2. **Predicting Consequences** The Mexican government is creating farms in its dry soil and tropical rain forests. How might this change Mexico?
3. **Expressing Problems Clearly** Tourism is an important source of Mexico's income, yet some people consider tourism an "irritant industry." Why?

Practicing Skills

1. **Interpreting a Population Pyramid** Refer to the population pyramid on page 196 to answer the following questions. (a) Developing nations have young populations. How does the pyramid show that Mexico has a young population? (b) If Mexico's population had an equal number of young and old, what shape would the pyramid be?
2. **Drawing Inferences** Mexico City is polluted and overcrowded. Why do you think rural Mexicans continue to move there?

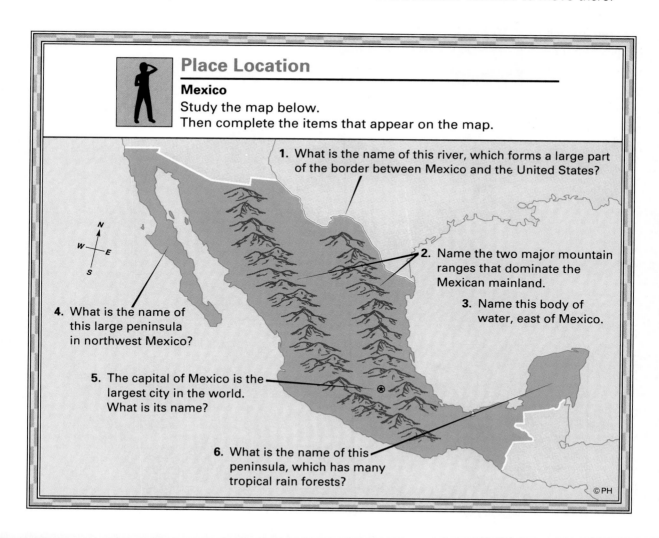

Place Location

Mexico
Study the map below.
Then complete the items that appear on the map.

1. What is the name of this river, which forms a large part of the border between Mexico and the United States?

2. Name the two major mountain ranges that dominate the Mexican mainland.

3. Name this body of water, east of Mexico.

4. What is the name of this large peninsula in northwest Mexico?

5. The capital of Mexico is the largest city in the world. What is its name?

6. What is the name of this peninsula, which has many tropical rain forests?

©PH

11

Central America and the Caribbean

Chapter Preview

Locate the countries covered in each of these sections by matching the colors on the right with those on the map below.

Sections	Did You Know?
1 CENTRAL AMERICA	Central America has one of the world's fastest growing populations. It will double in less than twenty-five years.
2 THE CARIBBEAN ISLANDS	When Columbus arrived on the island of San Salvador in the Bahamas, he believed he had reached an island near Japan or China.

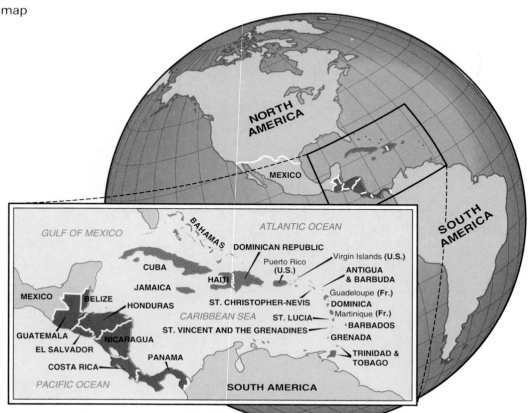

Market Day in a Guatemalan Village

Most Central American Indians live in the highland region and visit village and city markets to sell their produce and handmade textiles. Indians make up more than 50 percent of the population of Guatemala.

1 Central America

Section Preview

Key Ideas

- Central America is a place of diverse physical landscapes.
- Agriculture dominates the economy, and the vast majority of people are subsistence farmers.
- Extreme inequalities of income and power have caused violent political conflict.

Key Terms

isthmus, guerrilla

The small region of Central America connects the giant land masses of North America and South America like a curving, twisting vine reaching between two great trees in the rain forest. Central America is an **isthmus**, a narrow strip of land bordered on both sides by water and joining two larger bodies of land. As an isthmus, Central America forms a land bridge between two continents.

The Central American isthmus made movement between the Atlantic and Pacific oceans difficult until 1914. In that year, the Panama Canal opened, making it possible for ships to sail between the two oceans without having to travel around the tip of South America.

Seven countries occupy this narrow, curving strip of land. Beginning in the north, they are Guatemala, Belize, El Salvador, Honduras, Nicaragua, Costa Rica, and Panama. As the map on page 167 shows, they lie between Mexico and Colombia. The countries of Central America are small nations, with a combined land area only about one fourth the size of Mexico. Packed into this area is a physical and human landscape as diverse as the designs in traditional Indian clothing. This great diversity explains many of the challenges the region faces today.

A Place of Diverse Physical Landscapes

Naturalist Jonathan Evan Maslow captures the physical diversity of Central America in this description of Guatemala:

> *Up and down, round and round, the countryside never stayed the same more than a few miles at a stretch . . . Granite heights that looked clawed by blind and angry titans [giants] pitched into patches of lowland rain forest . . . It was like an entire continent stuffed as in an expertly packed suitcase into a country the size of Massachusetts.*

Much the same can be said about every country in Central America. Even a diverse landscape, however, can be divided into regions, and Central America is no exception. Three major landform regions make up Central America—the mountainous core, the Caribbean lowlands, and the Pacific coastal plain. Each landform region has its own climate.

The Mountainous Core As in Mexico, mountains cover much of Central America. Over the length of the isthmus, they tower up to 13,000 feet (3,963 m) above sea level. Because Central America's mountains are rugged and difficult to cross, they pose serious problems for transportation.

Central America's high elevations create two climate zones. Land that is 6,000 feet (1,830 m) above sea level is cold. Because of frequent frosts, few crops besides potatoes and barley can grow there. Elevations between 3,000 and 6,000 feet (915 and 1,830 m) have a year-round springlike climate, free of frosts but cool enough to grow corn, wheat, and coffee.

The Caribbean Lowlands On the eastern side of Central America, the mountainous core gives way to lowlands that edge the coast of the Caribbean Sea. The Caribbean lowlands have a tropical wet climate with year-round high temperatures and heavy rainfall. Most of its soil is not very fertile and can produce little besides dense rain forest vegetation.

The Pacific Coastal Plain Unlike the Caribbean coast, the Pacific coast has a tropical wet and dry climate with savanna, or grassland, vegetation. The difference in climate on the two coasts is caused by the winds that sweep from the northeast across the Caribbean toward Central America. These winds pick up moisture and dump it on the Caribbean coast and the eastern slopes of the mountains throughout the year. By contrast, the Pacific coast can depend on rain only in the summer. Volcanic activity and deposits of ash make Pacific coast soils fertile.

The Region's People

As the map on page 209 shows, Central America is home to several ethnic groups. Largely because the mountains blocked travel between areas, relatively little mixing of ethnic groups took place throughout Central America's history. Instead, at each location, one ethnic group tended to dominate.

Indians Those with the longest history in the region are the Indians. This category includes several different groups. Each Indian group has its own distinct history, culture, language, and traditional home.

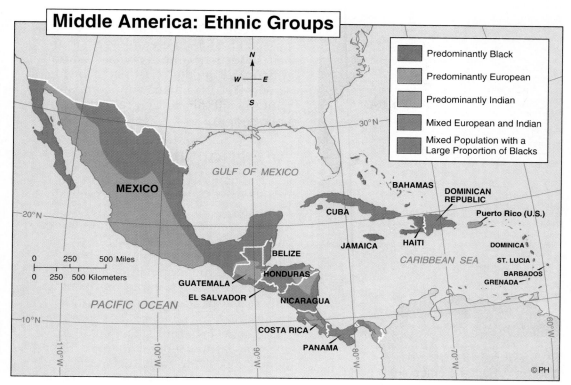

Middle America: Ethnic Groups

Legend:
- Predominantly Black
- Predominantly European
- Predominantly Indian
- Mixed European and Indian
- Mixed Population with a Large Proportion of Blacks

MEXICO • GULF OF MEXICO • BAHAMAS • DOMINICAN REPUBLIC • CUBA • Puerto Rico (U.S.) • JAMAICA • HAITI • DOMINICA • BELIZE • CARIBBEAN SEA • ST. LUCIA • HONDURAS • BARBADOS • GUATEMALA • GRENADA • EL SALVADOR • NICARAGUA • PACIFIC OCEAN • COSTA RICA • PANAMA

©PH

Applying Geographic Themes

Place European colonization and slavery both have left their mark on the human geography of this region. Some countries of Central America and the Caribbean are composed of one dominant ethnic group, while others have two or more ethnic groups coexisting as part of one country. Which countries are predominantly Indian?

The largest number of Central America's Indians live in Guatemala, where they make up half the population. Most of them are subsistence farmers.

Europeans and Mestizos Europeans have the second longest history in Central America, dating to the 1500s, when Spain first colonized the region. Because of this history, Spanish is the official language in six of the seven Central American countries. Belize, a former British colony, is the exception. Its primary language is English. The largest European settlement today is in Costa Rica, where 90 percent of the people are of European—mostly Spanish—descent.

A third group in Central America's population consists of the mestizos, people of mixed European and Indian background. Most people in El Salvador and Nicaragua are mestizos.

African-Americans Central America's Caribbean coast is home to many people of African descent. Some descended from African slaves who were shipped to Central America as early as the 1600s. But the largest numbers of African-Americans are the descendants of people who came to the region from the Caribbean Islands in the early 1900s. They came to work on Central American banana plantations or to help build the Panama Canal.

WORD ORIGIN

Panama
Native American place names often describe a region's physical characteristics. For example, *Panama* meant "abundance of fish" to the Native Americans who lived there before the Europeans arrived.

Work in a Tortilla Factory
Central America's cities are among the world's fastest growing. Many urban dwellers purchase their food rather than making it themselves.

The Extremes of Inequality

Most of the people of Central America fit into one of two categories —the wealthy and powerful or the poor and powerless. The powerful constitute only a tiny percentage of the population. Most of its members are wealthy plantation owners, predominantly European or mestizo, who also dominate their countries' governments and politics.

At least two thirds of all Central Americans are poor and powerless. Into this category fall millions of farmers with little or no land and laborers who earn low wages on plantations or in factories. Most of the region's Indians and African-Americans are in this group, as are many of the mestizos.

There is a small but highly important third category in Central America's social structure—a middle class.

This group includes small farmers who own land and some employees of urban industries and services. Central America's middle class is growing slowly. Yet it remains tiny compared with the millions of Central America's poor people.

An Agricultural Economy

The majority of Central America's people earn their living by farming. In Guatemala, El Salvador, Honduras, and Nicaragua, farming employs more than 50 percent of the people. In Belize, Costa Rica, and Panama, about 30 percent of the population works in agriculture. But farming in Central America can mean many different things.

At one extreme is the small subsistence farm. Here, with only their hands and a few basic tools, a family labors to grow enough corn, beans, or squash to stay alive. Most of the rural population of Central America lives on such farms.

At the other extreme is the large, fertile plantation owned by a wealthy family or a corporation. Plantation owners hire laborers at very low wages and bring in the newest machines, fertilizers, and pesticides to produce huge crops of coffee, bananas, or cotton. Most of these crops are not consumed in Central America but are shipped to the United States or Europe. Together, these three crops bring in 70 percent of Central America's profits from exports.

A Region of Political Turmoil

Imagine coffee beans ripening in the warm sun and workers reaching from ladders to cut clumps of bananas from the trees. Now imagine the crack of gunfire and the sound of soldiers scrambling through mountain forests. For years, both images have prevailed in Central America.

Armed conflict has troubled Central America for much of its history. The reasons for conflict vary, but a number of explanations apply to the region as a whole. The shortage of available farmland to meet the needs of the region's growing population has often sparked discontent throughout the region.

Adding to this problem, the land available for farming has been unevenly distributed. Finally, in recent years, with the exception of Nicaragua, governments in Central America have served the interests of the wealthy. Opponents of governments have risen up in armed rebellion, often fighting as guerrillas. A **guerrilla** is a member of an armed force that is not part of the regular army.

Nicaragua   Between 1979 and 1990, Nicaragua's government was controlled by the Sandinista Front for National Liberation, or the Sandinistas. The Sandinistas governed the country under a socialist system. Economic problems resulted, however, when government control of agriculture and industry caused lower production rates and a drop in exports.

More problems came when a group of Nicaraguans, including soldiers, tried to overthrow the Sandinistas. This group of guerrilla fighters, known as the contras, claimed that the Sandinistas were turning the country toward communism. The fighting that broke out strained Nicaragua's relations with the United States, which supported the contras with money and weapons. Fighting also caused thousands of deaths.

A measure of peace came to Nicaragua in 1990. More than a decade of Sandinista rule ended when a new president, Violeta Barrios de Chamorro (vee oh LEH tah BAH ree ohs duh chah MOH roh), took office in April 1990. For the first time in the country's history, power had been passed peacefully to a democratically elected government. Also in 1990, after about two years of negotiation, the contras agreed to lay down their weapons and return to civilian life.

El Salvador   Since 1980, El Salvador has been the scene of a civil war that has caused more than seventy thousand deaths. In an article in the *New York Times* on December 1, 1989, reporter Lindsey Gruson made this comment on the decade-long conflict:

The killing is fast becoming a personal and collective burden that threatens to make war a fixture in this country's political landscape. Each side . . . blames the other for furthering the fighting. But neither is ready to put away its guns.

The two sides Gruson referred to were the government, with its military troops, and rebel fighters armed by the Sandinistas and Cuba. The two sides have met a number of times to arrange a cease-fire, but progress has been painfully slow.

SECTION **1** REVIEW

Developing Vocabulary
1. Define: **a.** isthmus **b.** guerrilla

Place Location
2. Describe Central America's relative location.

Reviewing Main Ideas
3. What are Central America's three main landform regions?
4. What are the four main ethnic groups in the population?

Critical Thinking
5. **Checking Consistency**   What conditions in the countries of Central America might have had an impact on recent political conflicts there?

✔ Skills Check

☐ Social Studies
☐ Map and Globe
☐ Reading and Writing
☑ Critical Thinking

Comparing Two Points of View

In this chapter, you read that the Central American isthmus was finally broken through in 1914 with the opening of the Panama Canal. Two major figures in the building of the canal were United States President Theodore Roosevelt and Chief Engineer John Stevens. Roosevelt emphatically supported the building of the canal and did everything he could to bring about its creation. Stevens was chief engineer of the canal project from 1905 to 1907. As you will see in the following letters, each held a different point of view toward the building of the canal. Use the following steps to read and analyze their different points of view.

1. **Carefully study each source to gain an overall sense of the event it describes.** Read each passage and determine the event or topic being discussed. Answer the following questions: (a) What event does Roosevelt's letter describe? (b) What does Stevens mean by "the work"?

2. **Compare the points of view to identify their similarities and differences.** Determine how the two letters contradict or support each other. Answer the following questions: (a) Roosevelt called the canal "an epic feat." How did Stevens see the canal? (b) What other differences do you note between the two letters?

3. **Evaluate the validity and usefulness of the sources.** Study the language each writer used, to determine his opinions toward the building of the canal. Answer the following questions: (a) How would you describe Roosevelt's opinion of the building of the canal? (b) Would you say that Stevens held a differing opinion? Explain. (c) Would one source alone be as useful as both sources in understanding the story behind the Panama Canal? Explain.

From a letter from Roosevelt to his son, written in 1906 after the president had visited the canal site:

Now we have taken hold of the job. . . . There the huge steam shovels are hard at it; scooping huge masses of rock and gravel and dirt previously loosened by the drillers and dynamite blasters, loading it on trains which take it away . . . They are eating steadily into the mountain cutting it down and down. . . . With intense energy men and machines do their task . . . It is an epic feat, and one of immense significance.

From a letter to Roosevelt by John Stevens in early 1907:

The "honor" which is continually being held up as an incentive for being connected with this work, appeals to me but slightly. To me the canal is only a big ditch, and its great utility when completed, has never been so apparent to me, as it seems to be to others. Possibly I lack imagination. The work itself . . . on the whole, I do not like. . . . There has never been a day since my connection with this enterprise that I could not have gone back to the United States and occupied positions that to me, were far more satisfactory. Some of them, I would prefer to hold . . . than the Presidency of the United States.

2 The Caribbean Islands

Section Preview

Key Ideas

- The Caribbean islands are located in the Tropics and consist of three island groups.
- The people of the Caribbean are mostly black or of mixed race.
- Caribbean islanders have a tradition of migration among the islands and to areas outside the region.

Key Term

archipelago

Paradise Island, Nassau, Bahamas
The Caribbean's tropical location and pleasing climate make it a popular retreat for tourists from all over the world.

In the Dominican Republic, near the highway that links the city of Santo Domingo with its international airport, looms a huge billboard. It recently featured the sensational Dominican baseball player Pedro Guerrero as a spokesperson for a popular soft drink. Guerrero, however, plays baseball not for the Dominican Republic but for the United States. The billboard symbolizes much about the Caribbean. Like Pedro Guerrero, millions of talented people from the Caribbean have traveled to the United States and other countries in search of opportunity.

As you will learn in this section, the Caribbean is a beautiful region of forest-covered mountains, warm temperatures, and picture-perfect coasts. But the Caribbean is also a region that is struggling to develop its economy. And that is one reason why some of its people have left to find the opportunities the Caribbean cannot yet offer.

A Tropical Location

As was described in Chapter 9, three island groups make up the Caribbean islands. They are the Greater Antilles, the Bahamas, and the Lesser Antilles. Except for a number of islands in the Bahamas, all of the Caribbean islands are located just south of the Tropic of Cancer.

The Greater Antilles includes the four largest islands of the region—Cuba, Jamaica, Hispaniola (shared by the countries of Haiti and the Dominican Republic), and Puerto Rico. The nearly seven hundred islands of the Bahama **archipelago** (ar kih PEHL ih goh), or group of islands, lie northeast of Cuba. Most of the Lesser Antilles forms an archipelago that separates the Caribbean Sea from the Atlantic Ocean. A few other islands of the Lesser Antilles hug the coast of South America: Trinidad, Tobago, Aruba, and the Netherlands Antilles, which includes Bonaire and Curaçao.

Differences in Island Formation If you were to fly over the three groups of islands in the Caribbean, you would notice mountainous islands as well as islands with fairly level land. The difference in landforms is a result of the differing physical forces that shaped the Caribbean islands. The Greater Antilles and the islands of the Lesser Antilles off the coast of South America are the tops of underwater mountains pushed up from the ocean floor. The remaining islands in the Lesser Antilles are the tops of volcanoes whose repeated eruptions built up layers of volcanic soil.

The flat islands were created by the remains of once-living creatures called coral polyps. These tiny, soft-bodied creatures take in water and nutrients and release calcium carbonate, or limestone, to form a hard outer skeleton. This limestone skeleton is called coral. When sand and sediment build up on top of coral reefs, vegetation takes root. Eventually, an entire island is formed. All of the Bahamas are coral islands.

Because of their sandy soil, coral islands cannot support much agriculture. The mountainous volcanic islands have rich soil, but their slopes are quickly drained of nutrients and easily eroded. Thus, both mountainous islands and coral islands present challenges to agriculture.

A Climate Influenced by Sea and Wind As you read in the previous section, mountains influence the climate of Central America. The climate of the Caribbean islands, however, is affected more by sea and wind than by elevation. It was mentioned in Chapter 2 that nearness to water affects the climate of coastal areas. This geographic concept is at work every day in the Caribbean. As light breezes blow over the Caribbean Sea, they take on the temperature of the water beneath them. As these winds blow onshore, they moderate the temperature on the land. So, even though most of the Caribbean islands lie in the tropics, where the sun's rays are most direct, year-round temperatures only reach an average high of 80° F (27° C). But the humidity can be high.

Prevailing winds also affect amounts of rainfall in the Caribbean. On the northern and eastern windward side of the islands—the side facing the wind—rain can fall in torrents, reaching as much as 200 inches (508 cm) per year. On the leeward side, the side facing away from the wind, rainfall may amount to only 30 inches (76 cm) per year.

Ethnic Roots

A visitor to the Caribbean islands today would find little evidence of the original Indian inhabitants. European colonists arrived with Columbus in 1492; within a century, most of the Indians had vanished. Many died from the diseases of the foreigners, many from their cruelty.

The colonists needed laborers to work their plantations—usually growing sugar cane—so they brought hundreds of thousands of Africans to work as slaves on the islands that were then called the West Indies. Of the region's present population of thirty-four million, most descended from these slaves or from a mixture of Africans, Europeans, and native Indians.

The Caribbean islands also have a sizable Asian population. Most descended from East Indian and Chinese migrants who came voluntarily in the nineteenth century. By that time, slavery had been abolished in the Caribbean, so plantation owners sent half way around the world for replacements for their laborers.

Caribbean Nations Today

Today the Caribbean is a beehive of countries. Some of these countries are independent, including Cuba,

Making a Living in Haiti

This worker at a softball factory (above) is one of the few people able to find work in Haiti's commercial industries. In a poor country like Haiti, many people work at basket weaving and other cottage industries in order to make a living.

Haiti, the Dominican Republic, Barbados, Jamaica, the Bahamas, and Trinidad and Tobago (one country made up of two islands). About 90 percent of the Caribbean's population lives in these independent countries.

Many other Caribbean islands are still politically linked to European nations. The United States also has political connections with a number of Caribbean Islands. For example, Puerto Rico is a commonwealth of the United States. Its residents are American citizens. The United States government administers the U.S. Virgin Islands, whose residents are also American citizens. France claims Guadeloupe and Martinique. The Netherlands Antilles and Aruba are associated with the Netherlands but govern themselves. The British Virgin Islands, Cayman Islands, Montserrat, and several others remain colonies of the United Kingdom. They are governed according to British law.

Interaction: Making a Living from the Land

Today, the economies of the Caribbean islands revolve around agriculture. Thanks to the extremely fertile land, the islands produce much of the world's sugar, bananas, coconut, cocoa, rice, and cotton.

Many of the people of the Caribbean work in industries related to the region's agriculture—refining sugar, packaging coconut and rice products, and creating textile products. Still others work on the docks, shipping finished products to North America, Europe, or Northern Eurasia.

Because of their natural beauty, the islands are magnets for tourists from all over the world. Visitors flock here to relax and enjoy the tropical climate, lush green countryside, white sand beaches, and bright blue waters.

Yet while tourism thrives, the islands reap few benefits from this industry. Most of the hotels, airlines,

DAILY LIFE

The Breeze of Satire

At the edge of a village square on the island of Trinidad stands a man dressed in flashing jewelry and a straw hat. He is singing to a rhythmic beat, cleverly rhyming his words. Some listeners are clapping and singing along. Behind him, a small band plays a set of steel drums.

The man in the straw hat is a calypso singer like the one shown at right. A type of folk music that spread throughout the Caribbean from Trinidad, calypso features witty lyrics and clever satire. Its roots are in the songs of African slaves who worked on the plantations of Trinidad.

Once a calypso hit becomes popular, it circulates through the islands like a breeze. As West Indian poet Derek Walcott explained, "The song becomes common property. Children, old people are all singing that man's words. He's got an immediate audience that can comprise hundreds of thousands of people."

1. What is calypso?
2. How is the spread of calypso an example of the geographic theme of movement?

and cruise ships are owned by foreign corporations, not by people of the Caribbean islands. So the bulk of the profit ends up overseas. Local people hired for unskilled service jobs in the tourist industry are poorly paid and often face layoffs in the summer off-season. But since jobs are scarce, even these jobs are better than none.

Movement: In Search of Better Opportunities

Since the time of the first European colonization, Caribbean islanders have been active migrants. Most often, they have migrated in search of jobs. Sugar plantations have traditionally been the main employers on the islands, but the plantations' busy season lasts only four months. The other eight months are called the *tiempo muerto*—the dead season. During the dead season, idle workers pack their belongings and head for other islands or for Central America or the United States to find work. Whenever they receive a paycheck, they send some of the money to their families back home.

At the start of the twentieth century, many islanders found work in Panama, helping to build the Panama Canal for the United States government. Once the canal opened, most migrants returned to their homes, although many remained in Panama.

Starting in the 1940s large numbers of Puerto Ricans began moving to cities in the United States. A large percentage of these migrants have settled in New York City, where they have built a sizable and vibrant Hispanic community.

Movement away from the Caribbean has also been prompted by changes in government. For example, in 1959, Fidel Castro led a successful revolution to topple Cuba's corrupt, dictatorial government. In its place, Castro set up a communist government. Since that time, more than one million Cubans unhappy with the new order have emigrated to the United States. Many have settled in Florida.

Thousands of people from Haiti have also fled to the United States. Haiti is the poorest nation in the Western Hemisphere. Between 1957 and 1986, the country was ruled under a military dictatorship that did very little to improve the nation or raise living standards. In 1986 Haitians ousted their dictator, but the military quickly regained control of Haiti's government.

While the Caribbean islands have lost many people to migration, they have also experienced some benefits from it. The hundreds of millions of dollars that the migrants have sent home—not all of it from the United States—have helped reduce the burden of poverty throughout the Caribbean. With that money, the people remaining behind have bought consumer goods such as radios, televisions, and convenience foods. The resulting changes are so great that returning migrants are often amazed to find their island homes transformed. Their feelings of bewilderment are captured by Puerto Rican poet Tato Laviera, who wrote:

I fight for you, Puerto Rico, do you know that?
I defend your name, do you know that?
When I come to the island, I feel like a stranger, do you know that?

SECTION 2 REVIEW

Developing Vocabulary
1. Define: archipelago

Place Location
2. Which island group in the Caribbean includes the largest islands?

Reviewing Main Ideas
3. Describe how the three types of Caribbean islands were formed.
4. Why do the Caribbean islands have moderate temperatures?
5. Under what circumstances did Africans, East Indians, and Chinese people come to the Caribbean?
6. Why has migration become a feature of the human geography of the Caribbean islands?

Critical Thinking
7. **Identifying Alternatives** What changes to the tourism industry might bring more benefits to the people of the Caribbean islands? What barriers might there be to making those changes?

11

REVIEW

Section Summaries

SECTION 1   Central America The Central American isthmus, home of seven nations, contains rugged mountains, temperate middle-level elevations, and tropical lowlands. Its population includes four main groups—Indians, Europeans, mestizos, and African-Americans. Farming is the backbone of the economy, but most of the region's farmers are very poor. Rural poverty has caused much migration to the cities and has given rise to intense political conflict throughout the region.

SECTION 2   The Caribbean Islands The Caribbean islands include three island groups—the Greater Antilles, the Bahamas, and the Lesser Antilles. They share a tropical climate. Most of the region's people are descended from African slaves or from a mixture of African-American, European, and Indian ancestors. While the region relies on agriculture and a thriving tourist industry, widespread unemployment has led hundreds of thousands of people to migrate to other countries in search of work.

Vocabulary Development

1. Use each term in a sentence that shows the meaning of the term.
   a. isthmus
   b. guerrilla
   c. archipelago

Main Ideas

1. How can an isthmus be both a land bridge and a barrier to transportation?

2. In which physical regions are the tropical wet and dry climate zones found?
3. Which country in Central America has the highest population of European descent?
4. How can Central America's history as a Spanish colony be seen in language patterns today?
5. Describe the four groups of people that make up Central American society today.
6. How do the majority of the people in Central America make a living?
7. Which crops account for most of Central America's profits from exports?
8. Why might Nicaragua have greater political stability after 1990?
9. Describe how the Greater Antilles, the Lesser Antilles, and the Bahamas were formed.
10. How do the sea and the prevailing winds affect the climate of the Caribbean islands?
11. Name two Caribbean islands that are politically linked to the United States.
12. Why does tourism bring few benefits to the people of the Caribbean islands?
13. What factors have caused Caribbean islanders to migrate?

Critical Thinking

1. **Making Comparisons**   Compare and contrast the Indian populations of Central America and the Caribbean islands. What accounts for the differences between the two regions?
2. **Formulating Questions**   Migration from the Caribbean islands has had both positive and negative effects on the region. Make up three to five questions that would help you decide whether, on the whole, migration has helped or harmed the region.

3. **Formulating Questions** Since its revolution in 1959, Cuba has taken a very different course from other Caribbean lands. Think of four or five questions that you would want to ask in order to identify important differences between Cuba and other Caribbean lands.

4. **Checking Consistency** In travel brochures published in the United States, the Caribbean islands are often described as ''America's most beautiful playground.'' Is this image consistent with what you have read in this chapter? Why or why not?

Practicing Skills

Comparing Two Points of View Using the steps you learned on page 212 and what you have learned in Section 2 of this chapter, write a short paragraph that describes the Caribbean islands from a tourist's point of view. Next write a second paragraph that describes the islands from the point of view of a person who lives and works in the region. Then compare the differences and similarities in points of view in your two paragraphs.

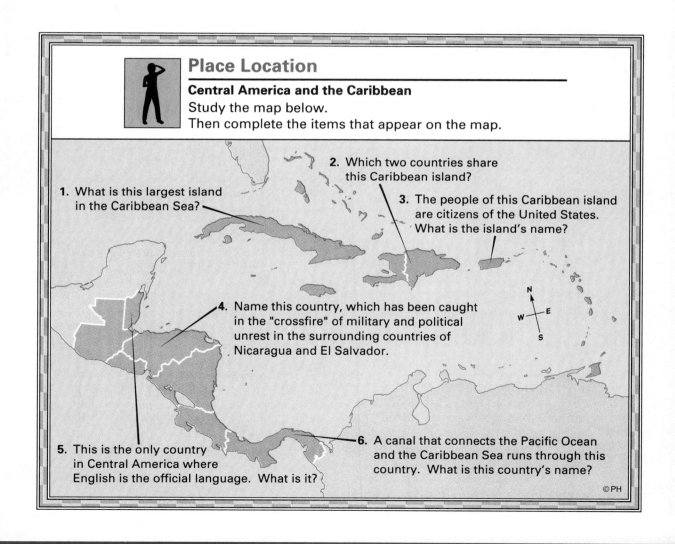

Place Location

Central America and the Caribbean
Study the map below.
Then complete the items that appear on the map.

1. What is this largest island in the Caribbean Sea?

2. Which two countries share this Caribbean island?

3. The people of this Caribbean island are citizens of the United States. What is the island's name?

4. Name this country, which has been caught in the "crossfire" of military and political unrest in the surrounding countries of Nicaragua and El Salvador.

5. This is the only country in Central America where English is the official language. What is it?

6. A canal that connects the Pacific Ocean and the Caribbean Sea runs through this country. What is this country's name?

© PH

The Vanished Expedition

You can't believe it. You're standing in a San Francisco flea market, and in your hand is the personal journal of the legendary Margaret Fitch.

Margaret Fitch! Her disappearance is one of modern archaeology's unsolved mysteries. Fitch, a world-famous archaeologist of the early twentieth century, possessed an unrivaled knowledge of Aztec and Incan cultures. Over a thirty-year span, she led a series of expeditions to remote Latin American sites. Her daring trek in 1925 was her last. Somewhere along the route, she and her party of nine vanished without a trace.

Well, not quite without a trace. You look down at the small journal in your hand. You gently open the cracked brown leather cover and carefully turn the pages. The ink has faded to a wispy brown. Some pages in the journal are dog-eared, creased, or torn. Other entries have been covered by dark stains. And many entries are missing altogether. In fact, only seven brief entries in the entire book are legible. Judging from the number of empty pages, the seventh entry seems to be the last one Fitch made on her final expedition.

As you study the journal, you become determined to find out where Fitch traveled and exactly where she disappeared. You'll follow the brief entries she jotted down. By using your detective skills and a few outside resources, you should be able to link each journal entry to a country.

You return to the first page of the leather-bound volume and read the first entry.

1. *March 20, 1925—We begin our expedition in the ancient Aztec capital of Tenochtitlán to search for relics. It is cool here at this high altitude, so I've brought warm clothing, along with my beloved hammer, brush, and chisel. . . .*

Tenochtitlán? Where exactly was that ancient city? You could look under "Aztec" in the encyclopedia to find out. With a growing sense of excitement, you quickly flip to the second entry in the journal:

2. *April 13—side trip to the Yucatán; hunch that Montezuma may have gone there and left his calling card; the members of my party (incl. Prof. Wilks, who imagines he knows the ancient Aztec language better than I do!) don't much care for the local dampness & heat, but we are pressing forward.*

It's obvious you're going to have to do some research to figure out a few of these entries. You know that an atlas and an encyclopedia will help you to understand the references to the Yucatán and Montezuma in this passage. You scan the next two entries:

3. *May 2—tents pitched in camp at isthmus, preparing for trek south; . . . how hard it must have been to dig Roosevelt's "big ditch"; heat is simply terrible.*

4. *May 17—My plan is to explore every modern country that the Incan empire once covered; right now we are at the northern tip of that empire . . .*

It's too bad that Fitch didn't have more explanations in her journal. For example, what—and where—is the "big ditch"? And what "isthmus" is Fitch referring to?

The last three entries in the journal don't give you any help at all:

5. *June 1—presently near Cuzco, we are traveling [remainder of page is lost]*

6. *July 22—The mountains seem never to end; breezes offer some relief; illness in our party slows progress. Today we reached a town with a name that means "sugar" in Spanish, to restock on—sugar!*

7. *August 3—since Santiago we have seen no one, though we tramp steadily southward along the coast of this narrow country; rain never ceases; but oh!—I have found what no one believed [remainder lost]*

You pay the flea market owner for your treasure, then hurry back home to begin research. Once you've linked up each entry with a country, you should be able to solve the mystery of Margaret Fitch's disappearance!

SOLVE THE MYSTERY

In what countries was Fitch when she wrote the above diary entries? *Hint:* An atlas and the encyclopedia will help you match each journal entry with the country it describes.

12

Brazil

Chapter Preview

Both sections in this chapter are about Brazil, which is shown in red on the map below.

Sections	Did You Know?
1 THE LAND AND ITS REGIONS	When viewed from the air, Brazil's capital city, Brasília, resembles the shape of a bow and arrow or an airplane.
2 BRAZIL'S QUEST FOR ECONOMIC GROWTH	Brazilians grow the fuel on which their automobiles run. Ethanol is an alcohol-based fuel that is made from sugarcane.

Carnival in Rio de Janeiro
The festival of Carnival, complete with costumed dancers, musicians, and street parades, takes place for four days each year before the Christian celebration of Lent.

1 The Land and Its Regions

Section Preview

Key Ideas

- Most people in Brazil's northeast region suffer extreme poverty.
- The southeast boasts many resources and two major cities.
- Brazil's capital city was built in the barren Brazilian Highlands.
- The Amazon River basin is Brazil's new frontier.

Key Terms

escarpment, *favela*

Brazil is the giant of South America. More than half the continent's people and land lie within its borders. Despite its large land area, Brazil contains only two major types of landforms—plains and plateaus. The plains consist of a fertile ribbon of lowlands, 10 to 30 miles (16 to 48 km) wide, that winds along the country's curving coastline. The immense Amazon River basin is also a plains region.

Behind the coastal plains lies a huge interior plateau that drops steeply to the lowlands beyond it. This drop forms an escarpment—a steep cliff that separates two level areas. The escarpment acted as a natural barrier to the interior in earlier centuries. As a result, much of the interior of Brazil is still undeveloped and very sparsely populated.

The Northeast Region

Portuguese colonists landed on the shores of northeastern Brazil in 1500. Through the 1500s, the Portuguese built large sugar plantations along the fertile coastal plain and built port cities from which to ship the valuable crop to Europe. Brazil became the world's major producer of sugar.

Over the next three hundred years, and until the practice was abolished in the late 1800s, Brazil's colonists brought more than three million Africans into slavery to work on the plantations. The folk tales, food, and religion of the northeast regions still reflect the people's African ancestry and cultural heritage.

Behind the northeast's coastal plain lies the *sertao* (ser TEYE oh), the Brazilians' name for the region's interior plateau. With a tropical wet and dry climate, the *sertao* often bakes through a year or more of drought, then suffers devastating rains. Yet

Applying Geographic Themes

1. Location Brazil is the largest country in South America, occupying nearly half of the entire continent. Yet 80 percent of its population live on only 10 percent of its land. Where are Brazil's major cities located?

2. Movement What physical characteristics serve to hinder the movement of Brazil's population away from the major cities?

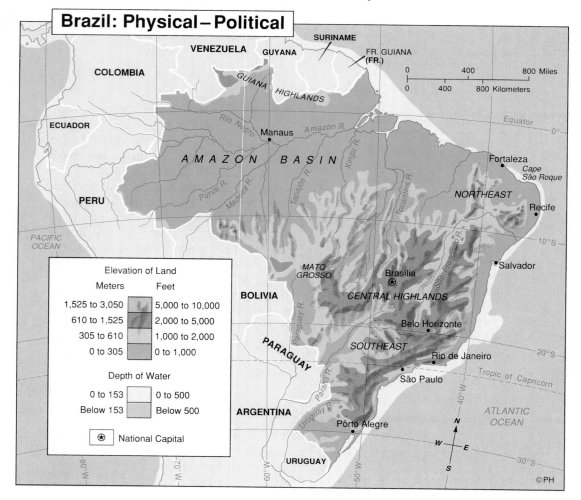

Brazil: Physical–Political

The Beach in Brazil

Brazilian beaches support a busy marketplace. People stroll up and down the sand selling straw hats, balloons, cold drinks, costume jewelry, and fruit-flavored ice cream. In addition, businesspeople often prefer to meet at the beach and consult with one another in the refreshing salt air.

About one third of Brazil's population live along the coast. For many of these people, the beach is a source of life itself. The clams and mussels they dig from the exposed mud flats at low tide yield an important source of protein. Whether directly or indirectly, the beaches of Brazil, like the one in Rio shown at right, nourish the nation.

1. What are some of the ways Brazilians use their beaches?

2. Explain this sentence: "Whether directly or indirectly, the beaches of Brazil nourish the nation."

many farmers in the region are able to feed their families by growing corn, raising chickens, and cattle.

The Southeast Region

Although the southeast is Brazil's smallest region, covering only 17 percent of the country's area, it is home to 40 percent of its population. Thanks to its mostly humid subtropical climate and its fertile soil, farmers can easily grow great quantities of crops such as cotton, sugar, and rice. Cacao beans grown in Brazil may end up in chocolate bars almost anywhere in the world!

But the southeast's biggest and most important crop is coffee. In the 1800s, thousands of newcomers—from other regions in Brazil, and from Portugal, Italy, Spain, even Japan—came to provide the huge numbers of laborers that coffee-growing requires. Today, Brazil is the world's "coffeepot, growing one fourth of the world's supply.

Despite the southeast's healthy agriculture, most of the region's population lives in or near São Paulo and Rio

Rio de Janeiro
Early Portuguese settlers named the city for the bay on which it stood. At the time, the bay was called Rio de Janeiro meaning "River of January." Historians believe a Portuguese explorer named the bay in 1503 for the month when he arrived there, thinking the bay was the mouth of a great river.

de Janeiro, Brazil's urban "gems." The mystique, beauty, and economic health of both Rio and São Paulo act as magnets for rural Brazilians whose economic opportunities are limited. Many, however, have become victims of low pay or unemployment, ending up in desolate slums called *favelas*. A journalist who recently visited Rio de Janeiro described the *favelas* there:

The houses [of the favelas], built illegally on hillsides or swampland, generally consist of wood planks, mud, tin cans, corrugated iron, and anything that comes to hand. Some cling to slopes so [steep] that the dwellings are in constant danger of being swept away in the heavy tropical rain storms that burst over the city.

Every Brazilian city has *favelas*. In recent years, however, the Brazilian government has taken steps to improve the situation. In some cases, *favelas* have been torn down and replaced by affordable public housing.

The Brazilian Highlands Region

North of the southeast region lie the Brazilian Highlands, the geographic heart of Brazil on the country's central plateau. The region focuses on Brasília, the nation's capital. Notice, on the map on page 224, how the location of Brasília contrasts with the locations of Brazil's other large cities. Forty years ago, Rio de Janeiro was the overcrowded capital of Brazil. But in 1955, hoping to boost development of the interior and to draw population away from the coastal cities, the national government decided to build a new capital city 600 miles (965 km) inland. Officially "inaugurated" in 1960, Brasília now has a population of one million.

The Amazon River Basin Region

Of Brazil's four major regions, the largest and the most mysterious is the Amazon River basin, swallowing up almost half of the country's area. Just about 10 percent of Brazil's population lives in the Amazon Basin, including about 200,000 Indians from 180 different tribes. This is, however, a small fraction of Brazil's original Indian population. At the time that the Portuguese arrived in the 1500s, between two and five million Indians were living within the country's present borders. Over the years many of Brazil's Indians have been killed by settlers and by diseases, that the settlers brought with them.

Today, the Indians of the Amazon Basin generally live in temporary settlements of as many as two hundred people, following a way of life that has remained mostly untouched by modern innovations. On pages 250 and 251 you will read more about economic development in the Amazon Basin.

■■■■■ SECTION **1** REVIEW ■■■■■

Developing Vocabulary
1. Define: **a.** escarpment **b.** *favela*

Place Location
2. Name two coastal cities in Brazil's northeast region.

Reviewing Main Ideas
3. What are Brazil's two biggest agricultural crops?
4. Why have Rio de Janeiro and São Paulo grown so much in recent years?

Critical Thinking
5. **Expressing Problems Clearly** Why do you think *favelas* appear in modern, newly built cities?

✔ Skills Check

☐ Social Studies
☐ Map and Globe
✔ Reading and Writing
☐ Critical Thinking

Identifying Main Ideas and Details

Authors of social studies books carefully organize their material in ways that help readers to understand it. One way that authors organize their writing is by grouping all the sentences that tell about a topic into paragraphs that have a main idea. The main idea is supported by several sentences that provide details. Every well-written paragraph has both a main idea that can be identified and supporting details. Use the following steps to identify the main ideas and details of the paragraphs below.

1. **Locate the topic sentence.** Every paragraph has a topic—the main idea on which the paragraph focuses. A paragraph's main idea is often stated in one sentence called the topic sentence. It is usually, but not always, the first or last sentence of the paragraph. (a) What is the topic sentence of paragraph 1? (b) What is the topic sentence of paragraph 2?

2. **State the main idea in your own words.** Some paragraphs have no topic sentences. In such cases, the main idea is implied rather than stated directly. When you read a paragraph without a topic sentence, try to state the main idea in your own words. Paragraph 3 has no topic sentence. Read paragraph 3 and state the implied main idea in your own words.

3. **Identify the supporting details.** Every paragraph contains details that support the main idea. These may appear anywhere in the paragraph. They usually take the form of examples that clarify the main idea. Read the paragraphs below. What are the supporting details in paragraph 3?

Paragraph 1
The histories of Brazil and the United States are similar in several important ways. Often the similar events occurred at different times in the past. But anyone familiar with the two countries will be struck by the things they have in common.

Paragraph 2
Take immigration, for example. Both countries have been destinations for immigrants from around the world. In Brazil as in the United States, remnants of the immigrants' cultures can still be seen. In Brazil's deep south, farms can be found whose houses and barns seem to have been transplanted from Germany and whose inhabitants still speak German. São Paulo has an area known as the "Japanese quarter."

Paragraph 3
Then there is the dark side of the histories of Brazil and the United States. Slavery was the cruel solution adopted by wealthy landowners when they needed labor to work their huge plantations. And when native Indians got in the way of pioneering newcomers, the Indians were often pushed aside—or killed.

2 Brazil's Quest for Economic Growth

Section Preview

Key Ideas

- In an attempt to reduce poverty, Brazil has developed new industries and encouraged settlement in the country's interior.
- Brazilian industry thrives, and thousands have migrated to the interior.
- These developments have improved the lives of some Brazilians but have harmed the environment and the Indians of the Amazon.

Key Term

ethanol

In the past fifty years, the government and people of Brazil have been struggling to modernize their economy. Although Brazil is rich in resources, its people do not receive equal shares of the nation's wealth. While a minority enjoys the luxuries that their wealth brings, a large majority struggles just to survive.

Brazil, like much of Latin America, is no longer a society of only rich and poor. The growth of industry and manufacturing has given rise to a middle class as people have been needed to manage and work in the factories and offices of Brazil. Likewise, as cities have grown, doctors, lawyers, teachers, government workers, and many others have moved in to fill the needs of a growing urban population.

A Wealthy Country with Much Poverty

Most of Brazil's poorest people live in the rural northeast or in the *favelas* of the region's big cities. Because of their impoverished lives and chronic malnutrition, the people of the northeast have an average life expectancy at birth of only forty-nine years—well below the rest of Brazil. Here, a family's average yearly income may be only one third the income of a similar family living in the southeast.

In *favelas*, poverty may be so intense that many parents cannot feed or house their children. Hungry, homeless children sometimes strike out on their own or with bands of other homeless children, seeking menial jobs or begging for a few coins with which to buy some scraps of food. Why does a country with so much natural wealth suffer from such widespread poverty?

Much of the country's poverty stems from conditions in agriculture, where one third of all Brazilian workers are employed. As an industry, Brazilian agriculture is productive and highly profitable. But the large plantations are owned by only a handful of families; the rest of the rural population owns only tiny farms—barely

The World's Coffee Pot
Filled with the beans of Brazil's coffee trees, bags such as these are exported to provide about one fourth of the world's total supply of coffee.

enough to keep one family alive—or owns no land at all. Those without land usually work on the large plantations, which pay very low wages. Thus, plantation workers may live in poverty even though they have steady jobs on extremely wealthy farms.

The unfavorable climate of the *sertao*, where many of Brazil's small farmers live, serves as another major cause of poverty. This interior region suffers from poor soil, scarce grazing lands, and uncertain rainfall. In addition, few farmers in the *sertao* can afford the farm machines that can help boost productivity.

Policies to Promote Economic Growth

Since the mid-1940s, the Brazilian government has undertaken several massive programs to ease the burden of poverty and produce more income for its people. These programs have had two major aims: to boost the growth of industry and to encourage settlement and development in the interior of the country.

Building an Industrial Base During the 1940s and early 1950s, the Brazilian government built the country's first steel mill and oil refinery. It also broke ground for the first in a series of huge dams that would produce the electricity on which industry could run. These dams were built where rivers dropped over the escarpment. To further encourage the growth of industry, the government established a bank that loaned money to people who wanted to start new businesses.

Manufacturing began to thrive in the 1950s with a tremendous growth in the automobile, chemical, and steel industries. As a result, within ten years millions of Brazilians began to move from rural to urban areas, seeking jobs in the new factories. Brazil's coastal cities, especially São Paulo, became crowded industrial centers.

Movement into Brazil's Interior

With São Paulo and Rio de Janeiro rapidly becoming teeming, overcrowded cities, Brazil's leaders realized the need for developing the country's vast, unused interior. The new capital city, Brasília, was "planted" in the Brazilian Highlands—600 miles (965 km) inland from the Atlantic coast—an oasis of shiny glass and gleaming steel amid barren rock. The city itself, when viewed from the air, resembles the shape of a bow and arrow or, some say, an airplane. Either way, the city's shape symbolizes movement or flight—the readiness of a country to take off and soar.

Brasília symbolized movement in another way as well. Because the country as a whole had few roads inside the coastal ribbon, the government began a massive road-building project with Brasília at its center. By the 1970s the country boasted thousands of miles of new roads, including one that stretched across the Amazon Basin for 2,700 miles (4,344 km).

To promote settlement in the north, the government gave away thousands of plots of land in the Amazon region, and thousands of mining or prospecting permits. Anxious to take advantage of the government's offers and begin new lives, great numbers of people relocated to this Brazilian frontier.

The Successes of Brazil's Development

To date, Brazil's development programs have had remarkable success. Manufacturing is now responsible for one third of Brazil's gross national product. Today, Brazil ranks among the world's leading industrial nations.

The high cost of imported oil in the 1970s led the country into the successful development of a new alcohol-based fuel called **ethanol.** Ethanol, sometimes called "gasohol," is made from Brazil's own sugarcane. In a

WORD ORIGIN

sertao
In Portuguese the word *sertão* means "interior," "midland part," or "heart of the country." The region of Brazil known as the Sertao is, in fact, an interior region of the country.

sense, Brazilian farmers are *growing* fuel. Brazil has broken free of its dependence on expensive foreign oil, for which it used to pay one billion dollars every year.

These and other industrial developments have directly affected Brazil's population. In 1940, two thirds of the work force were employed in agriculture. By 1980, one fourth were employed in manufacturing, construction, or mining. Almost half the labor force now work in service occupations —jobs in hotels and restaurants, retail stores, and government—that have sprung up as offshoots of the nation's industrial growth.

These new jobs usually pay better than most agricultural jobs. Brazil now has a small but growing middle class that is educated and skilled, something that scarcely existed before the 1940s. Yet much poverty still remains among the unemployed of the cities; it continues, too, in the agricultural northeast, which faces the same problems it faced fifty or more years ago.

High Technology in a New Industrial Age
Computers are one of many manufactured products that make Brazil an industrial leader. One fourth of all Brazilians work in factories.

Because of new roads and free land grants, the Brazilian Highlands and Amazon regions are buzzing with new settlers. Between 1970 and 1985, for example, more than one million people migrated to the Amazon region. Many of them were poor rural folk who went there to farm or to ranch.

Negative Effects of Development

While economic development has brought many positive changes to Brazil, it has also caused some unintentional changes. In the nation's big industrial cities, poverty has actually increased. Thousands of rural Brazilians have flocked to the cities in search of well-paying industrial jobs. Unfortunately, there have been more immigrants than jobs. As a result, the *favelas* have become an ever larger part of these cities. São Paulo, for example, saw its population double between 1970 and 1985, and the population of its *favelas* increased fifteen times.

Interaction: Changes in the Amazon Basin Along the Amazon, the challenges have been very different. After struggling to clear the forest to plant crops, settlers discovered that, because of the nearly constant rains that wash away nutrients, rain forest soil is really not fertile at all. It produces crops for only a few years. After that, settlers must plant elsewhere.

Land that has been exhausted by both the rains and by planting does not recover. Rather, it remains barren red clay, hard as brick. This gradual destruction of the rain forest has damaged the delicate ecology of the region and threatens to harm the ecology of the entire world. More information about the destruction of the rain forest is given on pages 250–251.

Uprooting the Amazon Indians As the rain forest has been opened by new roads and settlers, the Indians who have lived there for centuries have died at the hands of settlers or from disease. The government has set up reservations where those Indians who survive can live. But when Indians live in these reservations under government protection, they often lose their own culture. For example, many Indians have given up their language, religion, and style of living. Such change often marks the beginning of a tribe's disappearance, as Indians become assimilated into the country's majority culture.

Development of the Amazon has brought still more immediate dangers to the Indians. In spite of modern medicines, thousands of Indians have died from diseases against which they have no resistance. One tribe for example—the Parakana—had seven hundred to one thousand members in 1970. Ten years later when the region had been developed, only three hundred of them had survived. This is the result, one writer describes, "of living in permanent housing on unfamiliar land surrounded by people carrying unfamiliar germs." As two observers remarked: "There didn't seem to be anything you could call murder. Yet, the Indians were dying just the same."

Looking to the Future

The last half-century has brought Brazil much progress—and some serious new challenges. Yet there are reasons to expect the future to be brighter. The nation still has millions of acres of fertile land outside the Amazon region that could produce more food and a better living for much of its population. It has a rich culture and fine climate that can draw increasing numbers of tourists from other countries. Unlike some developing countries, Brazil has the potential to become a major world power. The people of Brazil are eager to enjoy their country's success in the coming years.

Learning the Tools of a New Culture
Resettlement has had both positive and negative results for the people of the Amazon like this Indian student.

SECTION 2 REVIEW

Developing Vocabulary
1. Define: ethanol

Place Location
2. Name the city located where the Rio Negro joins the Amazon River.

Reviewing Main Ideas
3. What were the two main parts of Brazil's program to develop the country?
4. Why has industrialization actually increased poverty in Brazil's cities?

Critical Thinking
5. **Analyzing Information** Name a specific group of people who have been helped by Brazil's economic development, and explain how the group has benefited.

12

REVIEW

Section Summaries

SECTION 1 The Land and Its Regions
Brazil is a huge land, well developed along its coast and little developed in the interior. Parts of its agriculture are extremely profitable, as are its mines. But many of Brazil's people live in extreme poverty. Most of Brazil's poor live in the *favelas* of the northeast's coastal cities, or in the dry *sertao*. The large cities of Rio de Janeiro and São Paulo are located in Brazil's southeast. Brasília, the new planned capital city, was built on the formerly barren Brazilian Highlands. A small Indian population lives in the Amazon Basin, as does a growing population of other Brazilians who seek to develop the region.

SECTION 2 Brazil's Quest for Economic Growth
In the past half-century, Brazil has developed new industries and encouraged settlement of the interior. These policies have helped to reduce poverty for some, but have caused other problems. Rural people in search of factory jobs have caused overcrowding in the coastal cities. Many live in city *favelas,* or slums. An increase of people in the interior has threatened to permanently damage the ecology of the rain forest. Settlers in the region are rapidly cutting down the rain forest in order to clear areas for farming. Brazil, however, has become a major industrial nation with a growing middle class. It is seeking to develop other land outside the delicate Amazon Basin.

Vocabulary Development

Use each term in a sentence that shows the meaning of the term.

a. *favela*   **b.** escarpment   **c.** ethanol

Main Ideas

1. Which landform in Brazil discouraged people from settling the region's vast interior?
2. How did the northeast region come to have many African influences in its own culture?
3. Why are so many people in the northeast so poor, when northeast sugar plantations are so productive?
4. What is the most important crop in Brazil's southeast region?
5. What geographic conditions make farms in the southeast so prosperous?
6. How has the development of ethanol had an effect on Brazil's economy?
7. Compare poverty in the *sertao* with poverty in the *favelas.*
8. What have been the successes of Brazil's policies for economic development?
9. Where and why did the Brazilian government build the new capital city of Brasília?
10. What have been some of the negative effects of settlement in the Amazon Basin?

Critical Thinking

1. **Testing Conclusions** Sometimes the total wealth of a nation is referred to as a ''pie,'' as in the following statement: When Brazil's leaders decided to attack the country's poverty, they decided to increase the size of the pie rather than divide up the pie more equally. What is meant by this statement? Is it supported by what you know about Brazil? Explain your answer.
2. **Identifying Alternatives** Suggest two or three different policies the Brazilian government might have tried in its attempt to reduce poverty. Discuss possible advantages and disadvantages of each policy.

3. **Determining Relevance** How do you think Brazil's sugar plantation owners might have benefited from the nation's economic development programs?
4. **Identifying Assumptions** Brazil's government did a number of things to encourage outsiders to settle and begin farming in the Amazon region. What assumptions about the region did these actions show?
5. **Recognizing Ideologies** In their attempts to reduce poverty, Brazil's leaders did little to change conditions in agriculture. Why do you suppose they took this approach?

Practicing Skills

Identifying Main Ideas and Details A well-written paragraph includes both a main idea and details. Read the paragraph on page 229 that begins with the heading "Building an Industrial Base." Write your answers to the following questions on a separate sheet of paper: What is the topic of this paragraph? What is its topic sentence? Give three supporting details from the paragraph that illustrate the meaning of the topic sentence.

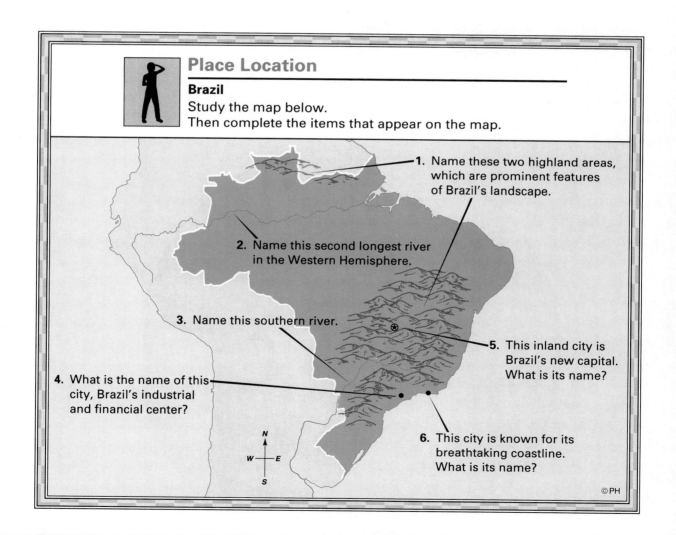

Place Location

Brazil
Study the map below.
Then complete the items that appear on the map.

1. Name these two highland areas, which are prominent features of Brazil's landscape.
2. Name this second longest river in the Western Hemisphere.
3. Name this southern river.
4. What is the name of this city, Brazil's industrial and financial center?
5. This inland city is Brazil's new capital. What is its name?
6. This city is known for its breathtaking coastline. What is its name?

© PH

13

Other Countries of South America

Chapter Preview

Locate the countries covered in these sections by matching the colors on the right with those on the map below.

Sections	Did You Know?
1 NORTHERN SOUTH AMERICA	Spaniards arriving in Colombia believed they had found El Dorado—a land of fabulous wealth and gold.
2 THE ANDEAN COUNTRIES	Indians living high in the Andes today have hearts that are as much as 20 percent larger than average.
3 SOUTHERN SOUTH AMERICA	Uruguay, one of the countries of southern South America, has ten times more sheep and cattle than people.

Canaima National Park, Venezuela

The world's highest waterfall, Angel Falls, drops 3,200 feet (970 m) into the Churun River, a tributary of the Orinoco. Look at the map on page 173. Into which body of water does the Orinoco flow?

1 Northern South America

Section Preview

Key Ideas
- Each of the Guianas reflects the European country that colonized it.
- Venezuela has a rich, oil-based economy that is rapidly diversifying.
- Colombia's economy is dependent upon a single crop.

Key Terms

mulatto, campesino

Grouped around Brazil like smaller paintings around a larger canvas are the twelve other countries of South America. These countries can be divided into the three regions shown on the globe on the facing page. The independent nations of Colombia, Venezuela, Guyana, and Suriname, together with the dependency of French Guiana, lie across the broad northern shoulder of South America.

The Guianas

Guyana, Suriname (SOOR ih NAHM), and French Guiana together are known as the Guianas. Although they share a tropical wet climate, vast stretches of rain forest, and a narrow coastal plain, their human geography makes them as different as three differently colored jewels set side by side. These differences are a reflection of each country's history and pattern of colonization. For example, Guyana's official language is English because the country was once a British colony. Dutch is the major language in Suriname, a colony of the Netherlands until 1975. And French is the official language of French Guiana.

The ethnic composition of the three Guianas is another example of the differences in their human geography. Guyana's two major ethnic groups are people of African and Asian descent. Africans were first brought to the country by Europeans to work as slaves on the sugar plantations. Asians from China, India, and Southeast Asia began arriving in the 1830s as replacements for the workers who left the plantations once slavery was abolished. Today, people of Asian descent make up about half of Guyana's population. People of African descent make up about 30 percent. Although a minority, people of African descent have the support of the country's other ethnic groups, and so dominate Guyana's government.

People of African and Asian descent also live in Suriname. But here only 10 percent of the population are of African descent, while just over 50 percent are of Asian descent. About 30 percent are mulattoes; most of the rest are American Indians. A mulatto is a person of mixed ancestry. The ethnic makeup in French Guiana is similar to that of Suriname, except that mulattoes are the largest ethnic group. People of European descent also live in French Guiana.

While their human geographies are very different, the three Guianas have similar economies because of their shared natural resources. Fishing boats harvest large quantities of fish and shrimp from the sea. In the lowlands, farmers grow sugar cane and rice. From the hills of Guyana and Suriname, miners extract bauxite, a mineral used in making aluminum. Guyana is one of the world's largest bauxite exporters.

Venezuela

To the west of Guyana lies the country of Venezuela. A shared boundary line is nearly all the two countries have in common. Fewer than one million people inhabit Guyana; nearly twenty times that number live in Venezuela. Guyana's annual per capita income of $457 makes it the poorest nation in South America; Venezuela, with an annual per capita income of $4,716, is the richest.

The cultures of Guyana and Venezuela also contrast sharply. While English is the official language of Guyana's mostly Asian and African population, the official language of Venezuela is Spanish, and its people are mainly mestizos or of European descent. While most Guyanans are Hindus or Muslims, the majority of Venezuelans are Roman Catholics.

Graph Skill

This graph shows the ethnic composition of several Latin American countries. Which country has a large Asian population?

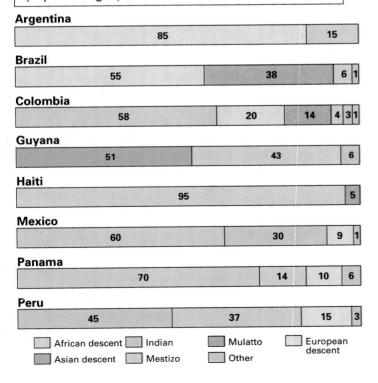

Major Ethnic Groups in Selected Latin American Countries
(in percentages)

Argentina: 85 | 15

Brazil: 55 | 38 | 6 | 1

Colombia: 58 | 20 | 14 | 4 | 3 | 1

Guyana: 51 | 43 | 6

Haiti: 95 | 5

Mexico: 60 | 30 | 9 | 1

Panama: 70 | 14 | 10 | 6

Peru: 45 | 37 | 15 | 3

Legend: African descent | Indian | Mulatto | European descent | Asian descent | Mestizo | Other

Landscapes Venezuela's land is varied—the area within its boundaries encompasses a narrow Caribbean coast, tropical rain forests, lowland plains, and impressive mountains.

The Andean Highlands Most of Venezuela's people live in fertile mountain valleys. The Andes Mountains rise powerfully in the country's northwest corner, overlooking a narrow coastal plain. A person sailing along the sultry Caribbean coastline never loses sight of the snowcapped Andes. Their extension, the Andean Highlands, consists of smaller mountain ranges, hills, and plateaus that stretch across northern Venezuela. Here in this region lies the capital city, Caracas, one of the world's most fascinating cities. One writer described Caracas as a place where "the present jostles the past." He continued:

Colorful murals, tiles, and mosaics enliven the faces of buildings. The effect is dazzling: the viewer, sated with brightness, sees the buildings vibrate against the purple splendor of the encircling hills.

Side by side with sidewalk cafes, universities, and busy department stores, however, are visions of poverty. As Rio de Janeiro has its *favellas,* so Caracas has its *ranchos,* or small shacks, where almost one third of the people live. In the last thirty years the government has launched massive programs to improve living conditions for the country's poor.

Waterfalls and Rain Forest Another mountain system, the Guiana Highlands, rises in southeastern Venezuela and covers nearly half of the country. Dense tropical rain forests choke much of the southern part of this region, where Venezuela borders Brazil.

Angel Falls, the world's highest waterfall, is located in the Guiana Highlands. As well as being nearly twenty times as high as Niagara Falls, Angel Falls differs from other waterfalls in an unusual respect. Instead of flowing over the top of a cliff, Angel Falls bursts out of caves below the cliff's edge to plunge more than 3,200 feet (970 meters) to the ground. South American legend has it that an angry Indian punched holes into the sides of the mountains to drain away the waters from a devastating flood.

The Sea of Grass Along both sides of the Orinoco River, which flows through central Venezuela, stretches the wide plain and tropical grassland, or savanna, region called the llanos. During the rainy season from April to December, the llanos become a shallow ocean. For the rest of the year, the hot sun of the dry season quickly burns the vegetation, and the soil becomes parched and cracked.

Elevation and Climate Venezuela lies within the tropics, but its varied climates depend more on elevation than distance from the Equator. In many mountainous areas of Latin America, people use their own language to describe the different climate zones that occur as elevation increases. The diagram on page 238 shows the Spanish terms used to describe these vertical climate regions. As you might expect, Venezuelan farmers grow different crops at different elevations due to variations in soil and climate. For example, coffee trees are ideally suited for growing in the *tierra templada* climate zone, where temperatures are relatively mild. In areas above 9,800 feet (2,987 m), which are common in the Andean Highlands, temperatures drop from "cold" to "freezing." This climate zone is known as the *tierra helada,* the frozen land of the Andes. At these soaring elevations, the mountains are covered permanently with ice and snowcaps that only grudgingly melt around their edges.

WORD ORIGIN

Venezuela
Venezuela means "Little Venice" in Spanish. The country was named by Spaniards who, when they first arrived, found Indians living on canals similar to those in Venice, Italy.

An Oil-Rich Region Venezuela's wealth can best be expressed in one word: oil. Two large beds of "liquid gold" lie in the Maracaibo Lowlands and the Lake Maracaibo region. A third oil field is located in the eastern part of the llanos.

Each year the Venezuelans pump about nine million barrels of oil out of the ground. This makes the country the fifth-largest oil producer in the world. Oil has brought great wealth to Venezuela. But economists estimate that at current production rates, the country's oil reserves may not last more than another twenty years. Wisely, Venezuela has reinvested a large share of its oil profits back into its economy—developing its bauxite and iron mines, building power plants, and setting up factories that will provide jobs when the oil wells run dry.

Colombia

Colombia—named after Christopher Columbus—is the only country in South America that borders both the Caribbean Sea and the Pacific Ocean. Its population of about thirty-two million makes it the third-largest country on the continent.

Land and Climate Like neighboring Venezuela, Colombia is a country of three physical regions—lowlands, mountains, and grassy plains, also called llanos. Nearly one third of Co-

---

Diagram Skill

In mountainous areas, climate varies greatly with elevation. What do Latin Americans call the region above 6,000 feet? Which crops are commonly grown in the *tierra templada?*

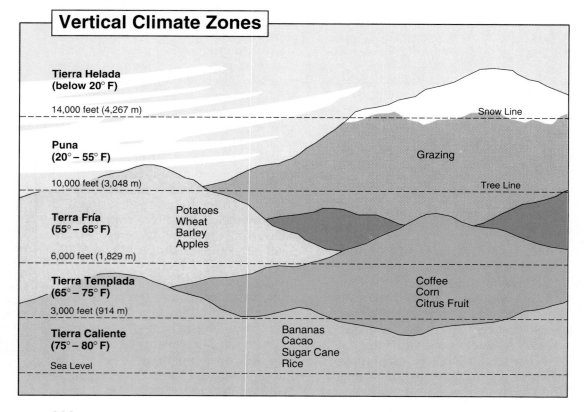

Vertical Climate Zones

Tierra Helada
(below 20° F)

14,000 feet (4,267 m)   Snow Line

Puna
(20° – 55° F)   Grazing

10,000 feet (3,048 m)   Tree Line

Terra Fría
(55° – 65° F)   Potatoes
Wheat
Barley
Apples

6,000 feet (1,829 m)

Tierra Templada
(65° – 75° F)   Coffee
Corn
Citrus Fruit

3,000 feet (914 m)

Tierra Caliente
(75° – 80° F)   Bananas
Cacao
Sugar Cane
Rice

Sea Level

lombia is covered by the Andes. About 75 percent of the country's people live in the fertile valleys between three cordilleras, or somewhat parallel mountain ranges, of the Andes. Bogota (BO guh TAH), Colombia's capital and largest city, lies in a basin of the Andes. Very few Colombians live in the llanos or in the forbidding rain forests which, combined, cover more than half of the country.

Single-Crop Dependence The variations in climate that occur in the mountains allow farmers to grow many different crops, but most of Colombia's farmers depend heavily on one crop. Second in the world only to Brazil, Colombia is renowned for its coffee, grown on more than 300,000 small farms. Most of the country's farmland is owned by a few wealthy individuals who rent small amounts of land to tenant farmers at high prices. Often the tenant farmers, called **campesinos,** are barely able to produce enough to feed their families, because they must focus their efforts on a cash crop, rather than on growing crops for food.

The government fears that if the world demand for coffee drops, or if the coffee trees are destroyed, the economy of the country will suffer. Officials are trying to reduce Colombia's dependence on a single cash crop and are encouraging the export of other farm products.

A Harmful Alternative As discussed in the case study in the previous unit, two products that have proved to be extremely profitable to a small minority of Colombia's people are causing international concern. Marijuana and cocaine, an illegal drug that is made from the leaves of the coca plant, are illegally exported from Colombia in huge quantities. Authorities estimate that the smuggling of illegal drugs brings twice as much money into Colombia as coffee does. The gov-

ernments of Colombia and the United States are working together to stop the flow of illegal drugs and the violence associated with the drug trade.

Other Challenges Colombia has had a stormy political history since it gained independence from Spain in the early 1800s. Continuing disputes between the country's two major political parties reached a head in the 1950s, when about two hundred thousand people were killed in a bloody civil war. In 1958, however, the two parties agreed to work together. Since that time, they have shared political offices and have tried to develop Colombia's economy.

Like many Latin American countries, Colombia struggles with the challenges that come with social inequality. A few people hold most of the country's wealth and power, while many suffer the results of extreme poverty. At times, antigovernment violence erupts among Colombia's least fortunate and angriest people.

SECTION 1 REVIEW

Developing Vocabulary
1. Define: **a.** mulatto **b.** campesino

Place Location
2. Name the river that flows from Venezuela's coast to its border with Brazil.

Reviewing Main Ideas
3. What are the advantages and the disadvantages of Venezuela's oil-based economy?
4. What is the problem of being a tenant farmer in Colombia?

Critical Thinking
5. **Perceiving Cause-Effect Relationships** Why do you think it is so difficult to stop the flow of illegal drugs from Colombia into the United States?

2 The Andean Countries

Section Preview

Key Ideas

- The Andes Mountains have helped to shape the economies of Ecuador, Peru, Bolivia, and Chile.
- Ethnic background influences the way in which most people in the Andean nations earn a living.

Key Terms

paramos, altiplano, selva

The Andes form the backbone of Ecuador, Peru, and Bolivia. They are the longest unbroken mountain chain in the world. The Andes soar higher than any of the world's mountain ranges except for the Himalayas. They have not only shaped the physical geography of this region but much of its economy as well.

A Mountainous Region

The Andes Mountains stretch more than 4,000 miles (6,400 km) from the Caribbean Sea to the southernmost tip of South America. Their rocky walls divide the Andean nations into three areas: coastal plain, highlands, and forest.

Coastal Plain As described in Chapter 9, a narrow plain stretches along the Pacific coast from Ecuador in the north to Chile in the south. At some points it is no more than a sandy beach at the foot of the mountains. In other places it reaches inland as far as 35 miles (55 km).

About half of the coastal plain covering all of Peru and the northern part of Chile is the Atacama Desert. The Atacama is so dry that archaeologists there have found many perfectly preserved relics from ancient times: colored textiles woven twenty-five hundred years ago, ancient mud-brick dwellings, and even mummies.

Both north and south of the Atacama, the coastal plain turns rainy. To the north, along the coast of Ecuador, lie oppressively hot and humid rain forests. To the south lies an area with a Mediterranean climate of hot, dry summers and mild, rainy winters.

The Highlands Inland from the coastal plain, the Andes rise skyward to incredible heights of nearly 23,000 feet (7,000 m). Between the towering peaks lies a series of highland valleys and plateaus. These flatter areas are known by different names depending on their location: *paramos* in Ecuador and *altiplano* in Peru and Bolivia.

Tropical Forests The Andes descend in the east to lowland regions. An incredible contrast exists between the mountains and the tropical land to their east. In Ecuador, Peru, and Bolivia, these regions are commonly known as *selva*, or forested regions. It is in the *selva* that the rain forests of the Amazon River basin begin. Jaguars, hummingbirds, monkeys, and toucans inhabit this region, but few people do.

A Wealth of Resources

People have always been drawn to the Andean Highlands because of the area's natural resources. The soil here is mostly rich and well suited for growing many crops, depending on the elevation. The mountains contain a wealth of gold, silver, tin, copper, and other minerals. At the same time, the mountains have often served as economic barriers, making trade among the Andean countries and with the outside world extremely difficult.

Vertical Trade One way the people of the Andes have met the challenge of the mountains is by vertical trade.

WORD ORIGIN

Chile
Meaning "end of the land," Chile was appropriately named by Indians who once lived on this long narrow strip on the west coast of South America.

In a typical Andean market town, such as Riobamba in central Ecuador, people from other villages meet to trade the crops that they are able to grow. Because different villages at different elevations grow crops specific to that climate zone, people trade "up" and "down." Tropical foods such as bananas and sugar, grown in the *tierra caliente*, may be traded to peasants for the potatoes and cabbages that they grow in the *tierra fría*. Highland cheese makers, coastal fishermen, and traveling peddlers all tend to meet in the Andean market town.

Living at High Altitudes Before the Spaniards arrived in the 1500s, the highlands were inhabited by various groups of Indians. Indians still make up between 40 and 70 percent of the populations of Bolivia, Ecuador, and Peru. The Andean Indians, who live at altitudes up to 17,000 feet (5,182 m) have developed physical characteristics, such as larger hearts and lungs, to make the most of the little oxygen that exists at those heights.

Ecuador

About 40 percent of the more than ten million Ecuadorians are of Indian descent. They speak Quechua (KECH wah), the language of the Incas and follow traditional Indian customs. They make their living as subsistence farmers in the highlands.

Another 40 percent of Ecuador's population are mestizos, who speak Spanish and generally live in the cities and towns of the highlands. In recent years many mestizos have migrated to the coastal plain where they work as tenant farmers on the plantations that grow bananas, cacao, coffee, and sugar for foreign markets. Other mestizos work in factories in Guayaquil (GWAH yah KEEL), the country's largest city.

In the 1960s, Ecuadorians discovered oil on the *selva* lowlands. In spite of the difficulty of transporting the oil

Plaza Independencia, Quito
The capital of Ecuador is in the Andes, about 9,350 feet (2,850 m) above sea level.

from the *selva* to the coast, petroleum became Ecuador's chief export. In 1987, however, an earthquake shattered the oil pipeline and sharply reduced the country's income.

People of European background make up only 10 percent of Ecuador's population. But they own the largest farms and factories and possess the greatest political influence.

Peru

Peru was the heart of the vast Inca Empire that met the Spanish conquistadors (kahn KEES tuh DAWRZ) in the early 1500s. The conquistadors destroyed the empire but never eliminated the Incas. Nearly 50 percent of Peru's population are Indians who speak either Quechua or Aymara (EYE muh RAH). As in Ecuador, most of Peru's Indians live in the highlands and make their living as subsistence farmers.

Most other Peruvians are mestizos. Like Ecuador's mestizos, they live in urban areas in or near the coastal

DAILY LIFE

Andean Farm Children

High in the mountains of South America, children who are eight and nine years old are treated as adults. They help their parents toil with the oxen in the fields, and they are expected to care for their younger brothers and sisters, too.

Such children live in small villages scattered through the Andes Mountains of Ecuador, Bolivia, and Peru. Their houses of rough stone are usually located several hours' walk from the nearest town. Andean farm families grow potatoes, wheat, and barley. Those at higher elevations herd alpacas and llamas for a living. Boys and girls, like the child shown, contribute in many ways. After checking their fields and animals, they may hike far into the mountains to bring grass for their cows or help with chores at home. Most of the time, the children themselves decide on the division of labor.

1. How do Andean children help their families?
2. What special challenges might a farm located high in the Andes face?

plain. For the most part, they work for low wages in factories that produce fish meal for feeding livestock, or on plantations that export cotton, sugar cane, and rice.

In Peru as in Ecuador, a minority of people of European descent control most of the country's wealth and are leaders in the government and in the army. In recent years many Asians have immigrated to Peru. In fact, in 1990 Alberto Fujimori, an ethnic Japanese citizen of Peru, was sworn in as Peru's President.

Bolivia

Landlocked Bolivia is similar to Ecuador and Peru but lacks their profitable coastal ports and factories.

Although Bolivia contains abundant mineral resources—especially tin—these are found high in the Andes and therefore are difficult to mine and bring to market. As a result, the country is very poor. Its annual per capita income is $590, way below that of Ecuador's at $1,428 and Peru's, which is $1,440.

As in Ecuador and Peru, the majority of Bolivia's people are Indians, mostly subsistence farmers who live in the highlands; most Bolivian mestizos work at unskilled, low-paying service jobs in the cities.

Chile: A Place of Geographic Contrasts

Chile's shape resembles a long, narrow ribbon. The country is 2,653 miles (4,200 km) long and at its narrowest point only 40 miles (64 km) wide.

About 77 percent of Chile's thirteen million people are mestizos. Another 20 percent are of European descent—mostly Spanish, British, and German. Unlike the other Andean nations, Chile has an almost nonexistent Indian population.

About three fourths of the Chilean people live in the Central Valley. This is a region of fertile river basins located between the Andes and the coastal cordilleras. Fruit and vegetables grow there in abundance. Chile's productive summer season occurs during the Northern Hemisphere's winter. Thus, most of the nation's farm products find their way to the produce sections of supermarkets in the United States and Europe.

The Central Valley also contains most of Chile's cities and factories. Santiago, the capital, is home to about one third of the country's total population. Many of the city's inhabitants are newcomers from the countryside, unskilled and illiterate. As a result, Santiago has a high unemployment rate and contains large slums.

A Return to Democracy
Chileans celebrate a vote to return power to a democratically elected president after being ruled by an iron-fisted military dictator since 1973.

■■■■ SECTION **2** REVIEW ■■■

Developing Vocabulary
1. Define: **a.** *paramos* **b.** *altiplano*
   c. *selva*

Place Location
2. Where do most of Chile's people live?

Reviewing Main Ideas
3. Why do Andean people engage in vertical trading?
4. How does the population of Chile differ from that of other Andean countries?
5. Why is Bolivia's economy significantly different from those of Peru and Ecuador?

Critical Thinking
6. **Drawing Conclusions** Large areas of South America are very sparsely populated. Why don't people settle these areas instead of moving to already overcrowded cities?

✓ Skills Check

☐ Social Studies
☐ Map and Globe
☐ Reading and Writing
☑ Critical Thinking

Determining Relevance

The study of geography often involves being able to see connections between the physical characteristics of a place and its human geography, people's lives. When thoughts or ideas are connected, or related to each other, we say that the thoughts and ideas are relevant to each other.

For example, the high elevation of the Andes is related to the dizziness felt by tourists traveling in that region. On the other hand, the fact that some places in the Andes are located at the Equator is *not* relevant when we discuss altitude-related dizziness. Learning to identify what is relevant and what is not when answering a question or solving a problem is an important skill in critical thinking.

Use the following steps to analyze the passages below.

1. **Identify the main ideas in the passage.** Read the passage to determine what the main ideas are. Answer the following questions: (a) What are the two main ideas expressed in Passage A? (b) How would you describe the main idea of Passage B?

2. **Look for connections among the ideas.** Ask yourself if the ideas are related to each other. Answer the following questions: (a) What connection is suggested in Passage A between the physical features of South America and human settlement, movement, and communication on that continent? (b) How are military rule and contemporary problems in Brazil linked in Passage B?

Passage A

In extreme contrast to the tropical rain forests of the Amazon, other parts of South America have cold deserts and dry, rugged mountains. The Atacama Desert of southern Peru and northern Chile is mostly barren and unable to support life. The Andes Mountains run about 4,000 miles (6,440 km) north to south along the western coast of South America. The Andes are a formidable barrier to transportation and communication. The southernmost parts of South America are close to the Antarctic and are thus very cold.

Passage B

Between 1956 and 1964, Brazil was a democracy. Then a military dictatorship ruled the nation from 1964 to 1985. Injustice in the form of censorship and political oppression marked those years. At the same time, Brazil became the most highly industrialized country in Latin America. In 1985, the military rulers agreed to free elections and a return to civilian rule. The elected rulers faced a massive foreign debt of over one hundred billion dollars. Overpopulation and poverty also present major problems.

3 Southern South America

Section Preview

Key Ideas

- Southern South America is a region of great physical diversity bound together by a river system.
- Uruguay and Argentina are two of the wealthiest and most urbanized nations of South America.

Key Term

gauchos

The three nations of southern South America—Uruguay, Paraguay, and Argentina—contrast sharply with the rest of the continent. Although they face economic problems, they are generally wealthier than the other South American nations. They also have more in common ethnically with Europe than with their neighbors.

Movement: A System of Rivers

Southern South America consists of regions with different terrains and climates, but it is bound together by a river system. Look at the map on page 173. The Río de la Plata is not really a river. It is an estuary made up of the flooded valleys of several rivers.

The Plata River system includes four major rivers that form national boundaries: the Uruguay, the Pilcomayo, the Paraguay, and the Parana. The Argentine capital of Buenos Aires (BWAY nuhs EYE RAYZ) is located at the place where the Uruguay and Parana rivers join. The Uruguayan capital of Montevideo (MAHN tuh vih DAY o) is located on the Río de la Plata. The river system provides a cheap and efficient way for people to ship goods and to move from place to place.

Varied Land Areas

The countries of southern South America fall into several physical regions, distinguished by varying landforms, climates, and vegetation.

The Andean Region   In the far west of Argentina, the Andes Mountains leap to their highest altitudes. The four highest mountains in the Western Hemisphere are located here, including Mount Aconcagua (ah kuhn KAH gwah), which towers 22,831 feet (6,959 m) above sea level. From this height, the Andes descend to lesser ranges, giving way at last to a region of gentle piedmont.

Tropical Lowlands   The Gran Chaco, meaning "great swamp," is a hot, lowland region of savanna and dense shrub that is shared by Paraguay, Argentina, Bolivia, and Brazil. Temperatures are mild and change little during the year. Rainfall, however, is seasonal

Montevideo, Río de la Plata, Uruguay

The Río de la Plata is actually a wide bay made up of several rivers. It provides a valuable means of transportation in the region.

A Gaucho at Work in Argentina
Argentina's gauchos herded cattle on the pampas until the 1800s. Today only a few continue to live and work on Argentina's interior ranches.

—summer rains turn the area into mud, while in winter the soil is dry and windblown.

Grasslands The pampas of Argentina and Uruguay are perhaps one of southern South America's best known features. These temperate grasslands, which stretch for hundreds of miles, were formerly home to the gauchos (GOW chos), the cowboys who herded cattle there. Today the pampas are Argentina's breadbasket, producing 80 percent of the nation's grain and 70 percent of its meat and meat products.

Cold Desert South of the pampas lies the windswept plateau of Patagonia. This desolate, dry, cold, and sometimes foggy plain is well suited for raising sheep and also contains rich deposits of oil and aluminum.

Paraguay

The western half of Paraguay consists of swampy chaco lands, while the eastern half is located on the Southern Brazilian Plateau. Although the country is landlocked, the Plata River system provides an outlet to the sea.

Almost all Paraguayans live in the eastern half of their country. Most are mestizos, and about one third of all Paraguayans live in urban areas, especially the capital city of Asuncion and its suburbs. Paraguayans generally speak Spanish, but many also speak Guarani, the language of the Indians of the region.

The Paraguayan economy is based on agriculture—mostly cotton, grains, and livestock. In 1984, Paraguay and Brazil cooperated in building a dam on the Parana River that is today the world's largest hydroelectric project. Paraguayans hope that cheap hydroelectric power from the dam will soon make up for their lack of mineral and other resources.

For thirty-five years, Paraguay was ruled by a military dictator, General Alfredo Stroessner. In 1989, however, discontented military officers ran Stroessner out of office, making Andres Rodriguez president. The new president surprised them by making the government more responsive to the needs of the people. Paraguay plans to hold its first democratic elections in 1993.

Uruguay

Uruguay's land is pampas and the nation's economy is based on its meat and meat-processing industries. Uruguayans devote about 75 percent of their land to livestock grazing and another 10 percent to raising grains to feed their cattle and sheep. The country's factories specialize in light industry, for example, preparing goods such as meat, wool, and leather for export. Uruguay produces no fuel or consumer goods and must import all it needs of these expensive products.

Even so, Uruguayans live fairly comfortably. Politically, however, the country has an unstable history. From

WORD ORIGIN

Uruguay
Uruguay meant "river of the painted bird" to the Indians who first dwelt there. The name probably comes from the brightly colored tropical birds found along the Rio de la Plata.

the early 1970s until 1984, Uruguay was led by a military dictator. In 1984 the army agreed to hold an open election, and Julio Maria Sanguinetti (HOO lee o mah REE ah sahn gwi NE tee) was chosen as president. In recent years the Uruguayans have staged many demonstrations to express their desire for freedom of the press and other democratic rights.

Argentina

Like their neighbors in Uruguay, most of Argentina's thirty-two million people have European ancestors, mostly Spanish and Italian. Twenty-eight million of them live in cities. More than ten million live in sprawling Buenos Aires and its suburbs.

Buenos Aires is a vibrant city. It looks toward Europe for its fashions, art, food, and decor. Busy factories produce goods for export, and the harbor is filled with freighters from many parts of the world. The capital dominates the country's political life. Like many of Latin America's cities, Buenos Aires acts as a magnet for poor rural people who are seeking jobs and a better way of life.

Until recently, Argentina's economic problems were made worse by its political problems. From the mid-1940s until 1983, Argentina was ruled by a series of military dictators. Some tried to develop strong ties with the country's labor unions. One of Argentina's best-known leaders was Juan Domingo Perón, who served as the country's president from 1946 to 1955. Perón was interested in developing Argentina's industry. He also wanted to distribute wealth more evenly among his people. Other leaders focused the government's efforts on assisting the wealthy and ignoring the problems of the poor and unskilled. But all of these leaders censored newspapers, closed down universities, and imprisoned those who objected to their policies.

Conditions during the 1970s were particularly bad. So many people were kidnapped by the military and never seen again that the period became known as the "dirty wars." Every Thursday afternoon a group of women marched in front of the presidential palace, carrying pictures of missing family members and demanding to know what had happened to them.

The military tried to give the appearance of progress by borrowing money from foreign banks to build dams, roads, and factories. When Argentina lost a war with Great Britain over the Falkland Islands off the country's southern Atlantic coast, the military faced disgrace. They agreed to the people's demands for open elections and Raul Alfonsin was elected president in 1983. He was succeeded in 1989 by another civilian, Carlos Saul Menem. As in the Andean nations, the future of democracy in Argentina depends on whether all of its people can achieve an adequate standard of living.

■■■■ SECTION 3 REVIEW ■■■■

Developing Vocabulary
1. Define: gauchos

Place Location
2. What capital cities are located on the Río de la Plata?

Reviewing Main Ideas
3. Name the four land areas of the southern South American countries.
4. Why are the pampas so valuable to Argentina and Uruguay?
5. How have the governments of the nations of southern South America changed in recent years?

Critical Thinking
6. **Determining Relevance** What does an adequate standard of living have to do with the future of democracy in southern South America?

13

REVIEW

Section Summaries

SECTION 1   Northern South America The northern South American countries are French Guiana, Suriname, Guyana, Venezuela, and Colombia. Although Guyana, Suriname, and French Guiana share many physical characteristics, their human geographies are very different. Their people speak a variety of languages depending on the European nation that colonized them. Venezuela's wealth is based on oil. Apart from Venezuela and Suriname, the region is poor.

SECTION 2   The Andean Countries The Andes Mountains divide the Andean countries —Ecuador, Peru, Bolivia, and Chile—into three regions: coastal plains, highlands, and forests. A vertical climate pattern is related to elevation. Indians and mestizos are mostly subsistence farmers or low-paid city workers. People overcame the barrier of the mountains by using a system of vertical trade. In recent years, the military rulers of the Andean nations have been replaced by elected civilians, who face the problem of improving their people's standard of living.

SECTION 3 Southern South America The three nations of southern South America —Paraguay, Uruguay, and Argentina—differ ethnically from their neighbors. Paraguay's population is mestizo, while the people of Uruguay and Argentina are mostly of European descent. The region possesses one of the most fertile grassland plains in the world, known as the pampas. Like nearly every other country on the continent, the nations of southern South America have endured military dictatorships. Although these have recently yielded to elected civilians, the new governments face many economic challenges.

Vocabulary Development

Match the following definitions with the terms that they define.

1. Andean Highlands in Ecuador
2. rural farmers
3. Andean Highlands in Bolivia and Peru
4. cowboy
5. forested regions
6. a person of mixed ancestry

a. campesinos   d. *selva*
b. *paramos*   e. gaucho
c. *altiplano*   f. mulatto

Main Ideas

1. How do the languages spoken in the three Guianas differ from those of the Andean countries?
2. What are the three geographical regions of most Andean countries?
3. Why is Venezuela trying to diversify its economy?
4. What is the danger of any country relying on one product?
5. Why do many farmers in South America find it difficult to earn a living?
6. What is the relationship between ethnic background and economic position in most South American countries?
7. What advantage does the Plata River system give to the countries of southern South America?
8. Describe the ethnic composition of Guyana, Suriname, and Ecuador.
9. Describe the population distribution or settlement patterns of Colombia.
10. Compare the pampas of Argentina and Uruguay to the region known as Patagonia in southern Argentina.

Critical Thinking

1. **Formulating Questions** Vertical climate zones are an important part of life in the Andes. Write one question about each of the zones whose answer would tell you more about life in the Andes.

2. **Distinguishing Fact from Opinion** Explain why you agree or disagree with this statement: "Many of the economic and political problems of South America stem not so much from the presence of natural wealth in a country as from the class structure of South American societies."

3. **Identifying Central Issues** Why do you think military dictatorships have been so common throughout South America?

Practicing Skills

Determining Relevance Using the steps you have learned on page 244, examine an editorial in your local newspaper. What is the main idea of the editorial? Is the evidence provided relevant? Can you extend the editorial's argument into related areas? Present your findings to the class.

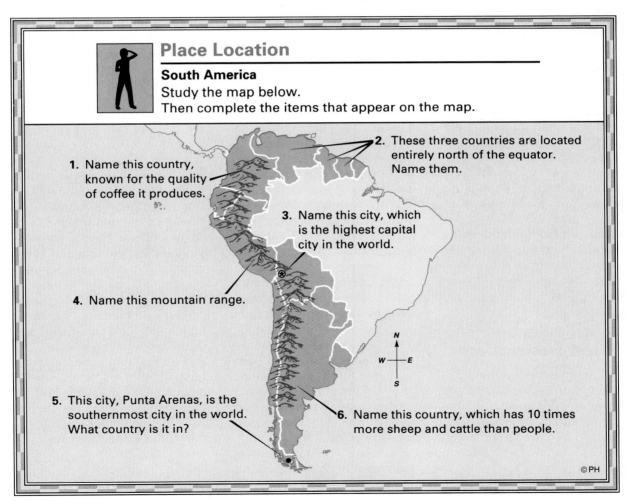

Place Location

South America
Study the map below.
Then complete the items that appear on the map.

1. Name this country, known for the quality of coffee it produces.

2. These three countries are located entirely north of the equator. Name them.

3. Name this city, which is the highest capital city in the world.

4. Name this mountain range.

5. This city, Punta Arenas, is the southernmost city in the world. What country is it in?

6. Name this country, which has 10 times more sheep and cattle than people.

©PH

MAKING
CONNECTIONS
·WHERE REGIONS MEET·

The Fate of the Tropical Rain Forests

How might the five senses be used to describe Brazil's tropical rain forest? One way would be to say that Brazil's tropical rain forest looks like a sea of green, sounds like a chorus of insects, smells like a fragrant flower, feels like the rough bark of a tree, and tastes like a cool drop of rain.

Today, however, an equally accurate way to describe rain forests in Brazil and other parts of the world is through the sound of chainsaws and the smell of smoke. People are cutting down the forests to make room for farms and to harvest the valuable wood of the forest trees. The destruction of the rain forest affects the whole world, including those countries that have no rain forests.

Brazil's Rain Forest Destruction

Brazil contains about 50 percent of the world's tropical forest. Until the early 1970s, the forest remained largely undisturbed, partly because there were few ways for people to reach the interior of the forest. But in the 1970s the Brazilian government began blazing roads into the interior in an effort to move people away from the overcrowded coast. Leaders hoped to provide more economic opportunity to their citizens and to develop a profitable resource in a nation troubled by debt.

Thousands of Brazilians have traveled these roads, seeking to escape poverty by establishing farms in areas of cleared forest. They find, however, that the rain forest soil supports crops for only a few years. When their crops fail, the farmers move deeper into the forest, leaving the land to ranchers who use it to graze cattle. The grazing further depletes the soil, exposing it to erosion. Thus within a few years, land once covered by awe-inspiring forest is reduced to useless wasteland.

Loggers and miners have also changed the face of Brazil's rain forest. Loggers strip the rain forest of trees to harvest valuable hardwoods for export. The machinery used by miners fouls the air, and the mercury they use to extract gold from sand pollutes rivers.

Worldwide Effects of Rain Forest Destruction

Brazilian officials, like those in many other nations with rain forests, contend that they must use rain forest resources to provide jobs and develop their economy. However, the rest of the world is now realizing that the destruction of rain forest has an impact far beyond the nation where it takes place.

Scientists estimate that no fewer than one out of every two species on our planet dwells in the rain forest. Many of these species have yet to be discovered. It is also estimated that one species of plant or animal life becomes extinct every day due to the cutting and burning. Scientists who create valuable chemicals from the rain forests' plant and animal life can only speculate about the cost of these extinctions to medicine and industry.

Another concern is that rain forest destruction worsens the quality of the world's air. The rain forest's enormous mass of green leaves helps purify the air by converting carbon dioxide into oxygen. As the forest shrinks, so does its ability to improve air quality. The burning of forests releases more carbon dioxide into the atmosphere, encouraging global warming.

The View from the U.S.A.

 American conservation groups have played a leading role in persuading the world of the importance of rain forests. Groups have tried a

number of tactics in recent years to balance the needs of developing nations' economies with those of the environment. One idea is called a "debt-for-nature swap." Conservation groups pay part of the national debt of a country in return for promises to spare rain forest. A Boston-based conservation group is seeking to change the economics of rain forest destruction through trying to create a market for nuts, fruits, and oils that can be regularly harvested from the rain forest without destroying it.

TAKING ANOTHER LOOK

1. How has Brazil used its rain forest to develop its economy?
2. In what ways is the earth affected by the rain forest destruction?

Critical Thinking

3. **Demonstrating Reasoned Judgment**   Some people have suggested boycotting hardwoods that come from rain forests. Do you think such tactics could help stop the rain forest destruction?

Applying Geographic Themes
Interaction   This cracked, dry soil that at one time supported rain forest vegetation is the result of leaching and erosion. On which continent is the damage or destruction of tropical forests most apparent?

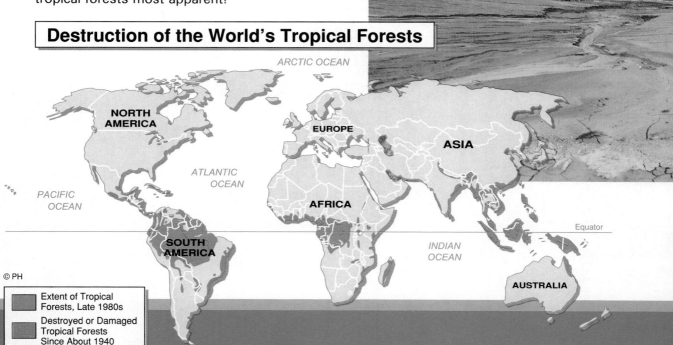

Destruction of the World's Tropical Forests

ARCTIC OCEAN

NORTH AMERICA

EUROPE

ASIA

ATLANTIC OCEAN

PACIFIC OCEAN

AFRICA

Equator

SOUTH AMERICA

INDIAN OCEAN

© PH

AUSTRALIA

Extent of Tropical Forests, Late 1980s

Destroyed or Damaged Tropical Forests Since About 1940

UNIT
4
Western Europe

CHAPTERS

From the rugged mountain peaks of Norway, clad year round in snow, to the flat olive groves and vineyards of Spain, warmed by the blazing sun, Western Europe is a puzzle of many pieces. Yet along towering mountain ridges, down rivers curving through fertile valleys, and across wide, choppy channels—all of the puzzle pieces come together.

Western Europe may be thought of as a puzzle of peninsulas, each reaching out farther and farther into the sea. In this unit, you will discover the peninsulas, plains, and mountains that have both benefited and hindered the people who live in this region.

Farmland in rural France ▲

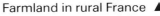

The Royal Guard, London ▶

◀ City street, Brugge, Belgium

14

Regional Atlas: Western Europe

Chapter Preview

Both sections of this chapter provide an overview of the countries of Western Europe, shown in red on the map below.

Sections	Did You Know?
1 **LAND, CLIMATE, AND VEGETATION**	No part of Western Europe is more than 300 miles (480 km) from the sea.
2 **HUMAN GEOGRAPHY**	Western Europe is less than half the size of the United States, yet it contains seventeen independent countries and six microstates.

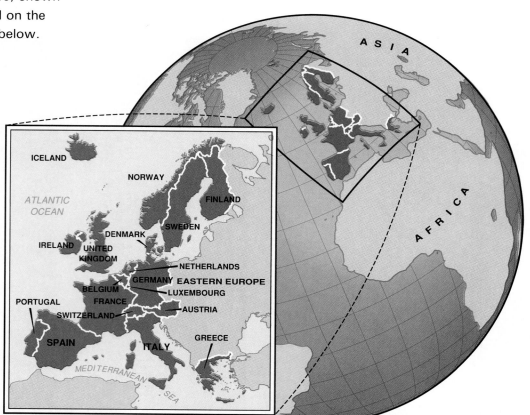

Southwest England

Because Europe is broken up into so many small and large peninsulas and islands, coastal scenes are a common feature of the landscape. This view shows the jagged, rocky coast of Southwest England.

1 Land, Climate, and Vegetation

Section Preview

Key Ideas

- Europe forms a large peninsula at the western end of a huge landmass called Eurasia.
- Mountains in the north and south of Western Europe border a central plain.
- Warm ocean currents produce a mild, moist climate in most of Western Europe.
- The Mediterranean climate region in the south has hot, dry summers and mild, moist winters.

Key Terms

summit, prevailing westerlies

Europe is a continent, but in some ways it is like a giant "peninsula of peninsulas," jutting out from one of the world's largest landmasses, Eurasia. Look at the map on page 257; notice that water surrounds much of Europe. What are some of these bodies of water?

A Peninsula of Peninsulas

Europe is sometimes called "a peninsula of peninsulas" because a number of smaller peninsulas dart out to the north, west, and south. Find, for example, the northern Scandinavian Peninsula. Three seas border it—the Baltic, North, and Norwegian seas. Denmark lies on a northern peninsula, too. Reaching southwest is the Iberian Peninsula, made up of Spain and Portugal. It is bounded by the Mediterranean Sea, the Atlantic Ocean, and the Bay of Biscay.

Chapter 14, Section 1   **255**

Now follow the jagged coastline of southern Europe. You will notice still other peninsulas, such as Italy, which stretches down like a boot into the Mediterranean Sea. Its toe points to Africa; its heel points to Greece, which is on a peninsula, too. Ancient scholars called Greece the "stepping stone to the world." Look at the map on page 257 to see what lands can be reached easily from Greece.

Mountains, Rivers, and Plains

Landforms in Europe vary greatly, from soaring mountains to flat plains. "Nowhere else is such a variety found within such small compass," wrote one European geographer in 1555.

Mountains   To visualize the landscape of Europe, think of a giant sandwich. Mountains spread along the region's northern and southern edges. In between lies a broad, rolling plain.

The northern mountains run from northern Finland, Sweden, and Norway through the western part of the British Isles. They formed more than 400 million years ago. Since then, they have been worn down into relatively low hills, much like the Appalachian Mountains in the United States. The highest peaks, in southern Norway, rise just over 8,000 feet (2,400 m).

The mountains in the southern part of Europe are higher and younger. Most formed about twenty-five million years ago when the tectonic plate containing Africa began folding into the Eurasian plate. These mountains form an almost solid barrier between the Mediterranean and lands to the north. The highest peaks are the Alps in Switzerland, Austria, France, and Italy. As the map shows, the summit, or highest point, of Mont Blanc rises 15,771 feet (4,807 m). In 1546, one frightened traveler described a journey through these peaks:

> The path goes up in the form of a snail shell or screw, with continuous turns and bends. . . . It is a narrow path and dangerous, . . . there are only [steep cliffs] and sheer drops to be seen. . . . I climbed the mountain . . . and quaked to my very heart and bones.

The steep cliffs and sheer drops described by the traveler were formed in part by glaciers. On at least four different occasions, glaciers have covered much of Western Europe. They have sculpted the mountains into spectacular peaks, ridges, and lakes. Today, people do not have to risk dangerous treks through glacial passes. Tunnels, allowing trains and motor vehicles to cut through the mountains, have considerably improved communications between countries.

Other young mountains in the south include the Apennines, Pindus, and Pyrenees. In this part of Western Europe, mountain building is still going on. In Italy, for example, volcanoes have erupted throughout history, destroying many towns and communities. Active volcanoes are evidence of tectonic activity—the plate movements that continually reshape the earth.

Applying Geographic Themes

1. Place   Which major mountain chain lies between Switzerland and Italy? Which plains region stretches across northern Europe?

2. Location   How might Switzerland's location help explain why it is a country with four official languages?

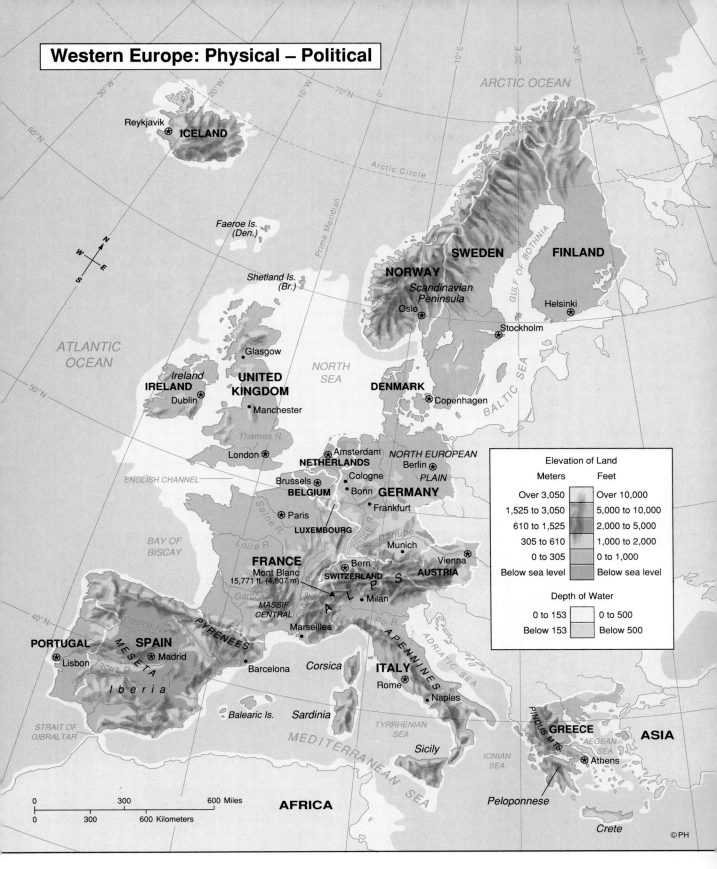

Western Europe: Physical – Political

ARCTIC OCEAN

Reykjavik

ICELAND

Arctic Circle

Faeroe Is. (Den.)

SWEDEN

NORWAY

FINLAND

Scandinavian Peninsula

Oslo

Helsinki

Shetland Is. (Br.)

Stockholm

GULF OF BOTHNIA

ATLANTIC OCEAN

Glasgow

NORTH SEA

DENMARK

Copenhagen

BALTIC SEA

Ireland

UNITED KINGDOM

IRELAND

Dublin

Manchester

Thames R.

London

Amsterdam

NETHERLANDS

NORTH EUROPEAN

Berlin

PLAIN

ENGLISH CHANNEL

Brussels

Cologne

BELGIUM

Bonn

GERMANY

Seine R.

Paris

Frankfurt

LUXEMBOURG

Rhine R.

Danube R.

BAY OF BISCAY

Loire R.

Munich

Vienna

FRANCE

Bern

AUSTRIA

Mont Blanc
15,771 ft. (4,807 m)

SWITZERLAND

Garonne R.

MASSIF CENTRAL

A L P S

Milan

Po R.

Marseilles

PORTUGAL

P Y R E N E E S

SPAIN

M E S E T A

Madrid

A P E N N I N E S

ADRIATIC SEA

Douro R.

Lisbon

Tagus R.

Barcelona

Corsica

ITALY

Rome

I b e r i a

Balearic Is.

Sardinia

Naples

STRAIT OF GIBRALTAR

TYRRHENIAN SEA

PINDUS MTS.

GREECE

ASIA

AEGEAN SEA

Athens

M E D I T E R R A N E A N   S E A

Sicily

IONIAN SEA

AFRICA

Peloponnese

Crete

Elevation of Land

Meters	Feet
Over 3,050	Over 10,000
1,525 to 3,050	5,000 to 10,000
610 to 1,525	2,000 to 5,000
305 to 610	1,000 to 2,000
0 to 305	0 to 1,000
Below sea level	Below sea level

Depth of Water

0 to 153	0 to 500
Below 153	Below 500

0   300   600 Miles

0   300   600 Kilometers

©PH

Rivers and Plains   Many rivers flow from the mountains and crisscross Western Europe. Throughout history, these rivers have carried much of the region's commerce. Two major rivers empty into the Mediterranean—the Po in Italy and the Rhône in France. Other rivers flow into waters to the north and west, including the Seine and Rhine rivers.

Some of Europe's great rivers cut across the North European Plain. This flat, fertile land stretches more than 1,200 miles (1,930 km) from France through Germany and into Eastern Europe and Northern Eurasia.

Other flat lands can be found on the Scandinavian Peninsula between southern Sweden and Finland. Here the soil is generally poor, however. During the last Ice Age, towering glaciers pushed south out of Scandinavia. They scraped off the topsoil and, when they melted about eight thousand years ago, deposited it on the North European Plain. As the glaciers retreated across Scandinavia, they left behind thousands of lakes.

Location and Climate

Compare the physical-political maps on pages 136 and 145 with the climate map of Western Europe on page 259. You'll notice that the following pairs of cities lie at about the same latitude: Calgary and London, and Anchorage and Stockholm.

All these cities are roughly the same distance from the equator. So you might expect them to have the same climate. But the European cities have milder climates than their North American counterparts. Why is there this difference in climates?

A City on a River
Paris was founded more than 2,000 years ago on an island on the Seine River. This location at the center of important land and water routes led to the growth of Paris into a major European capital.

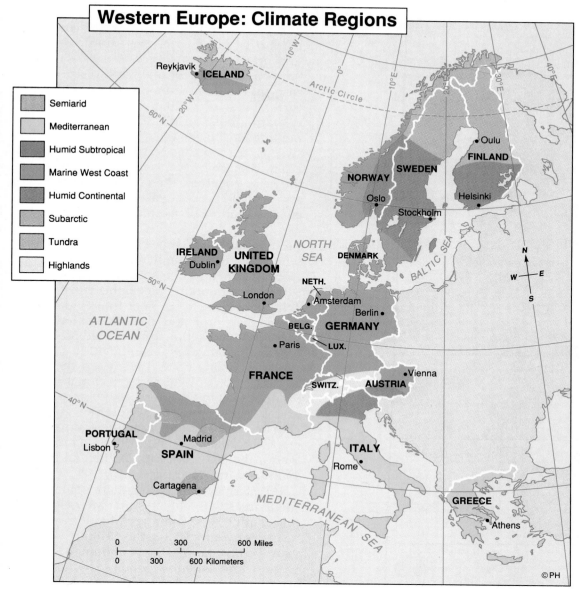

Western Europe: Climate Regions

Legend:
- Semiarid
- Mediterranean
- Humid Subtropical
- Marine West Coast
- Humid Continental
- Subarctic
- Tundra
- Highlands

©PH

Applying Geographic Themes

1. Regions What are the two most common climate types in the southern regions of Western Europe?

2. Movement Western Europe experiences less extreme climates than other parts of the world at similar northern latitudes. Why is this true?

Marine West Coast Climate Latitude is only one of the factors that determines a region's climate. The location of a place in relation to large bodies of water also plays a major role. Western Europe is a "peninsula of peninsulas." No point in the region is more than 300 miles (480 km) from the sea. Therefore, water has a strong influence on the region's climate.

Look back at the ocean currents map on page 34. Find the Gulf Stream, which edges past North America out of the Gulf of Mexico. This is part of a

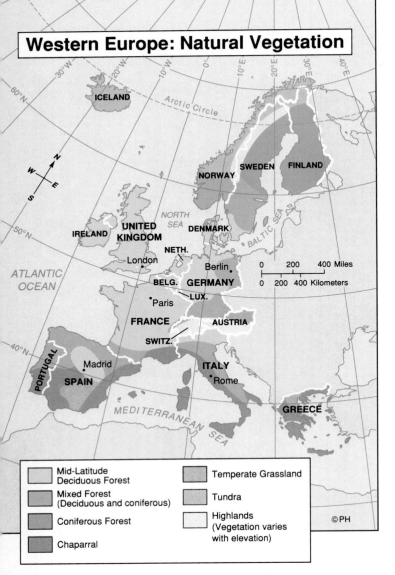

Western Europe: Natural Vegetation

ICELAND

Arctic Circle

NORWAY SWEDEN FINLAND

NORTH SEA

UNITED KINGDOM DENMARK

IRELAND

BALTIC SEA

London NETH.

Berlin

ATLANTIC OCEAN

BELG. GERMANY

LUX.

Paris

FRANCE AUSTRIA

SWITZ.

PORTUGAL

Madrid ITALY

SPAIN Rome

MEDITERRANEAN SEA

GREECE

0 200 400 Miles

0 200 400 Kilometers

Legend:
- Mid-Latitude Deciduous Forest
- Mixed Forest (Deciduous and coniferous)
- Coniferous Forest
- Chaparral
- Temperate Grassland
- Tundra
- Highlands (Vegetation varies with elevation)

©PH

Applying Geographic Themes

1. Interaction People have cut down most of the forests in Western Europe. Where are the largest areas of coniferous forest?

2. Place What type of vegetation is found at 60° N latitude in Western Europe?

Europe for much of the year. The winds that blow across the warm current are the **prevailing westerlies**—the constant flow of air from west to east in the temperate zones of the earth. These prevailing westerlies carry warm, moist air inland. Because no mountains block the moist air, it sweeps across the North European Plain, dropping rain at all seasons of the year. Together, the ocean currents and winds produce a warm, moist climate in most of Western Europe.

Mediterranean Climate A very different climate dominates the land along the Mediterranean Sea. Because southern mountains block the moist, Atlantic winds, in summer the region is relatively hot and dry. In winter, however, winds blow off the Mediterranean, bringing regular rainfall.

This climate pattern is known the world over as a Mediterranean climate. As discussed in Chapter 2, other parts of the world enjoy a Mediterranean climate, too. These include the southern California coastline in the United States, the coast of central Chile, and southwestern Australia.

Western Europe also contains some small areas with climates that are neither marine west coast nor Mediterranean. For example, on the Scandinavian Peninsula, northern mountains block warm Atlantic winds. So this area has a very dry, cold, subarctic climate. And just north of the Adriatic Sea, there is a small area with a humid subtropical climate.

Linking Vegetation and Climate

Compare the climate map on page 259 with the vegetation map on page 260. Note that bands of vegetation are closely related to the bands of climate. Most of the Mediterranean climate area was once covered with the broadleaf and mixed forests that grow well

large clockwise current that flows northeastward across the Atlantic. Known as the North Atlantic Drift, this current carries warm tropical waters out of the Caribbean Sea toward the coast of Europe.

Although the current cools slightly as it moves north, its warmth helps produce moderate weather in Western

in hot, dry summers. These forests included olive, cork, oak, and pine trees.

Forests also once blanketed the marine west coast climate area. Here, as deciduous trees dropped their leaves year after year and the leaves decayed on the ground, rich deposits of soil developed, making the land good for farming.

Needleleaf forests remain in the cold subarctic areas of Scandinavia. Farther north, tundra supports little vegetation other than grasses and mosses.

Interacting with the Land

Human beings have always interacted with and changed their environment. Over the years, people cut down most of the natural forests in Western Europe. They cleared land for farms and used the timber for fuel and building materials. Today, most land in Europe is used for farming and grazing. Pastures, orchards, vineyards, and gardens have replaced the forests of the past.

The rise of industry has also affected the region's natural vegetation. One of the few remaining woodlands is Germany's Black Forest region, a famous scenic area known for its dense forests on steep hillsides. A traveler in the early 1900s described its natural beauty:

Yet the setting in the heart of the pine woods is idyllic and nearby roars the waterfall. . . .The best view is from the top of a gorge. . . . When looking down, you behold a leaping torrent foaming wildly among the boulders.

Today, however, industrial pollution endangers the Black Forest and other forests. The next section tells more about human reshaping of the environment in Western Europe.

Lake Thun-Spiez, Switzerland
The elevation of the Alps has a great influence on climate and vegetation. What is the land on the lower slopes of these mountains being used for?

SECTION 1 REVIEW

Developing Vocabulary
1. Define: **a.** summit **b.** prevailing westerlies

Place Location
2. What is Western Europe's relation to Eurasia?

Reviewing Main Ideas
3. What are the major landforms in Western Europe?
4. How does the North Atlantic Drift affect Western Europe's climate?
5. Describe the climate found along the Mediterranean Sea.

Critical Thinking
6. **Perceiving Cause-Effect Relationships** How did human settlement of Western Europe change its physical geography?

✓ Skills Check

☐ Social Studies
☐ Map and Globe
☐ Reading and Writing
☑ Critical Thinking

Formulating Questions

An important part of the critical-thinking process is the ability to formulate questions while reading. Questions help us seek new information through inquiry. They help us gain insight and sharpen our understanding of any material we encounter.

To formulate good questions, keep in mind the question words used by reporters: *Who? What? When? Where? Why?* and *How?* The first four words help you gather the basic facts. The last two help you interpret them.

Use the following numbered steps to formulate questions which will enable you to understand and analyze the excerpt from the article printed below.

1. **Identify the topic.** To guide your reading, think of a question to identify the article's topic. Sometimes, you can do this by rephrasing the article's title in the form of a question. If the article does not have a title, first skim through it to get a sense of the general content; then form a question. Write a question to help you find the topic of the article on this page.

2. **Locate the important details.** As you read, look for details related to the main theme of the excerpt. Think of some questions a reporter might ask to find out the facts about Iceland.

3. **Identify a point of view.** Read carefully for an author's opinion on a subject. What question might you ask to help you look for the author's opinion on Iceland?

Iceland: Europe's Wonderland

Mountains, rivers, geysers, steam holes, volcanoes, and Europe's largest glacier — you can find all this and more in Iceland. This island nation, about the size of Kentucky, is packed with natural wonders.

Iceland is a land of contrasts. In its northeastern corner, cracked beds of lava are so like the moon's surface that American astronauts have practiced walking and driving a lunar rover there. In the southeastern corner, a glacier covers some 3,200 square miles (8,290 sq km). It is so big that it can be seen completely only from the air.

Beneath Iceland's surface are beds of molten rock. Volcanic activity has created small offshore islands from time to time. It also sends lava spewing up beneath the glacier. Huge water domes appear in the ice and explode into geysers. In Iceland you can see geological forces that shaped the landscape of Western Europe.

2 Human Geography

Section Preview

Key Ideas

- Western Europe is one of the most densely populated areas of the world.
- Deposits of coal and iron and access to sea routes helped Western Europe to industrialize.
- Since World War II, Europeans have cooperated to overcome past conflicts among their nations.

Key Terms

multilingual, minority

Western Europe is only about half the size of the continental United States. But more than 375 million people live within Western Europe. This means that Western Europe is three times as densely populated as the United States.

To get an idea of how these statistics affect daily life, turn to the map of the southern United States on page 126. Suppose you set out from El Paso, Texas, and drove east at 55 miles per hour. After fifteen hours, you would still be in Texas.

Now look at the map on page 257, and imagine making a trip of the same length between Milan, Italy, and Amsterdam, the Netherlands. After fifteen hours, you would have passed through six countries! Their populations total more than twelve times that of Texas, and each has a different language and history.

Linking Population Density and Land Use

Western Europe occupies less than 2 percent of the world's landmass. Yet it has one of the highest population densities in the world. For example,

the Netherlands has an average of 920 people living in each square mile.

To get a visual picture of population densities in Western Europe, look at the map on page 264. Notice that two areas are especially dense. The first extends from central England into France, across the North European Plain, and into Eastern Europe. The second runs south along the Rhine River valley into southeastern France and Italy.

Housing along Amsterdam's Canals

Because the Dutch must eke out every inch of land, they build their houses narrow and tall. Steep staircases wind their way from floor to floor and wooden posts near the roofline outside are used to hoist furniture to the upper floors.

Urban and Industrial Areas Two thirds of Western Europe's people live in four nations: the United Kingdom, France, Germany, and Italy. In these countries people are concentrated in large metropolitan areas. Using the map on page 264, identify those Western European cities with populations of over one million people.

Population growth and the distribution of natural resources have gone hand-in-hand in Western Europe.

Applying Geographic Themes

Regions Western Europe is one of the most densely populated regions on the earth, but its population is unevenly distributed. After comparing this map to the map on page 257, name two physical features that explain the low population density in some regions. Where are the areas of greatest population density in Western Europe?

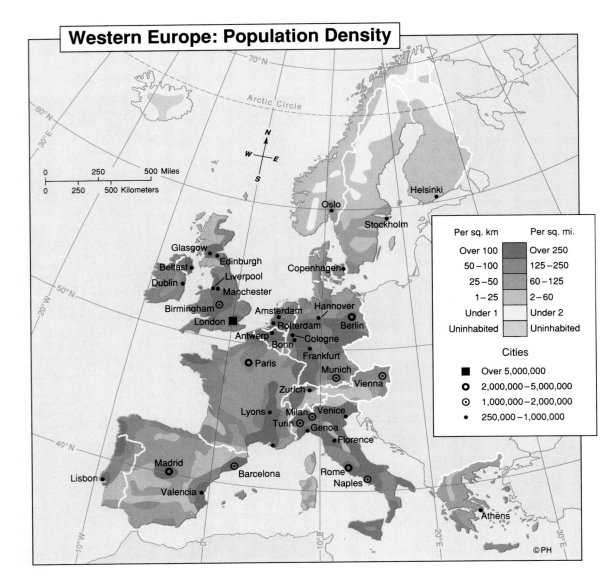

Western Europe: Population Density

Per sq. km	Per sq. mi.
Over 100	Over 250
50–100	125–250
25–50	60–125
1–25	2–60
Under 1	Under 2
Uninhabited	Uninhabited

Cities

■ Over 5,000,000
◉ 2,000,000–5,000,000
⊙ 1,000,000–2,000,000
• 250,000–1,000,000

©PH

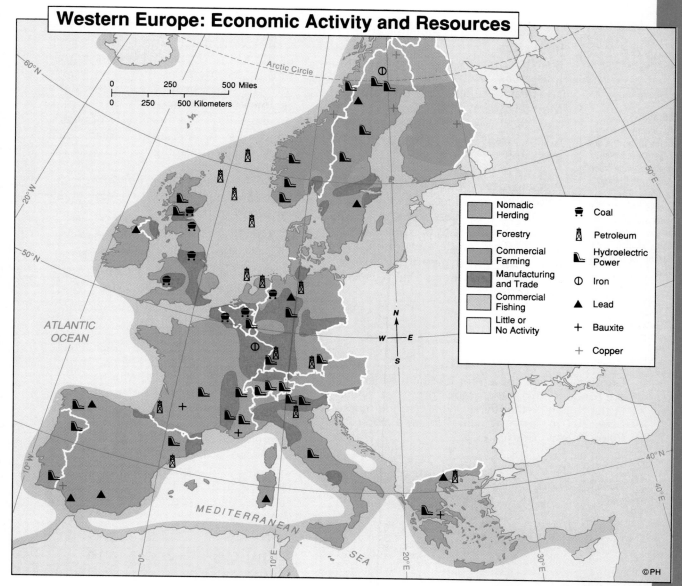

Western Europe: Economic Activity and Resources

Legend:

- Nomadic Herding
- Forestry
- Commercial Farming
- Manufacturing and Trade
- Commercial Fishing
- Little or No Activity
- Coal
- Petroleum
- Hydroelectric Power
- Iron
- Lead
- Bauxite
- Copper

Applying Geographic Themes

1. Interaction How have natural resources influenced the way many Scandinavians make their living?

2. Regions What natural resources do you think have brought economic independence to some of the countries bordering the North Sea? Why is this true?

When the Industrial Revolution began to transform Western Europe from an agricultural to an industrial society toward the end of the eighteenth century, factories and mills sprang up near deposits of coal and iron ore. These resources are abundant in population centers such as Manchester, Scheffield, and Hamburg. During the 1800s, therefore, these cities grew quickly as people left farming areas for work in manufacturing or mining.

Access to sea transportation also influences urban and industrial growth. Look again at the most populous areas on the map. Many of them are near excellent seaports along the Mediterranean and the Atlantic. Today, most Western Europeans live in urban areas with strong industrial economies.

Agricultural Areas   Farming areas generally remain less densely populated. Modern farmers use sophisticated farming methods to produce high crop yields. The moderate temperatures, fertile soils, and plentiful rains make the North European Plain especially bountiful.

Agriculture along the Rhine River

Germany is well known for its Rhine wines, which are produced from the grapes grown at vineyards in the region of the Rhine River. Use the map on page 257 to name two cities located along the Rhine River in Germany.

Some of the richest lands can be found along the mouths of rivers, such as the Rhine. These rivers drop alluvium, or sediment, as they flow slowly into the sea. Locate the Rhine's delta region in the Netherlands on the map on page 257. For hundreds of years, windmills have helped create rich farmland by pumping water out of the low delta areas.

A nineteenth-century author described Dutch efforts to create more farmland in this way:

> The enemy from which they had to wrest the land was triple: the sea, the lake, and the rivers. They drained the lake, drove back the sea, and actually imprisoned the rivers.

Europeans still use these techniques today, though electric pumps have replaced the windmills.

In less populated areas along the edges of Western Europe—in Scandinavia, Ireland, Spain, and Greece—farmers use land for grazing or reserve it for forests. Despite the region's high demand for agricultural products, farmers manage to produce enough surpluses for export to other parts of the world.

Conflicts, Diversity, and Cooperation

Cooperation among the nations of Western Europe is new. For hundreds of years, nations used warfare to redraw the map of Europe. Nation conquered nation, and territory shifted hands. At various times, people in what are now called Scandinavia, Germany, France, and Italy have claimed most of Western Europe as their own.

In this century, European power struggles set off two world wars. Both resulted in new political boundaries for the region. Post–World War II settlements created most of the boundaries on the political map on page 257.

DAILY LIFE

Flying by Rail

Western Europe's compact size has encouraged the growth of a rail network that is heavily used by commuters and vacationers alike.

European trains are often the fastest way to go. This is particularly true in France, a world leader in the development of high-speed rail systems. Parisians heading for Le Mans can do so at an average speed of 137 miles (218 km) per hour—the fastest regularly scheduled train service in the world.

1. Why are trains an efficient means of transit in Western Europe?

2. What are some economic advantages of travel by rail?

For many years after World War II, European boundaries reflected two rival political camps—a Soviet bloc of nations in Eastern Europe and a democratic bloc of nations in Western Europe. But these divisions have begun to break down as the Soviet-controlled communist governments in the Eastern bloc finally collapsed. In 1990, the Soviet Union agreed to the reunification of East and West Germany. One German leader declared:

For myself and millions of people, Germans included, it is a dream come true to see German unity and European unity taking place together.

Two Major Language Groups Centuries of movement and interaction between different groups in Western Europe have produced a multilingual region, where many different languages are spoken and understood. In Switzerland alone, for example, official languages include Swissdeutsch, French, and Italian. In Belgium, people speak Flemish and French.

Many of the languages spoken in Western Europe fall into two major groups. Portuguese, Spanish, French, and Italian evolved from Latin, the language of the Romans who dominated Europe two thousand years ago. The languages in this group are known as Romance languages.

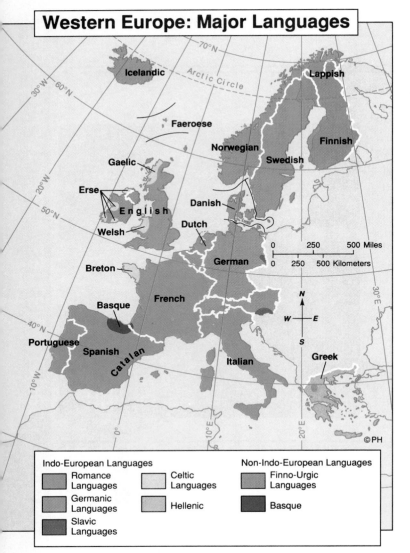

Western Europe: Major Languages

Icelandic

Faeroese

Lappish

Norwegian

Finnish

Swedish

Gaelic

Erse

English

Danish

Welsh

Dutch

Breton

German

French

Basque

Portuguese

Spanish

Catalan

Italian

Greek

©PH

0   250   500 Miles
0   250   500 Kilometers

N
W — E
S

Indo-European Languages

- Romance Languages
- Germanic Languages
- Slavic Languages
- Celtic Languages
- Hellenic

Non-Indo-European Languages

- Finno-Urgic Languages
- Basque

Applying Geographic Themes

Movement Language helps scholars trace the history and migration of a culture. Which group is linguistically different from its French and Spanish neighbors? In which four regions do people speak a Celtic language?

various Germanic tribes that overran Europe after the fall of Rome, around A.D. 476. Both Germanic and Romance languages belong to the Indo-European language family.

A few of the languages spoken in Western Europe have still other roots. Finnish is derived from a language spoken in Central Asia. Some people in the western parts of the British Isles and just across the English Channel in France speak languages handed down from the Celts, a people who occupied the region before the Romans arrived.

A Shared Belief in Christianity Most people in Western Europe follow one of three Christian faiths. In Ireland, France, Spain, Italy, Portugal, Austria, and Belgium the majority belong to the Roman Catholic Church. Most other Western European countries—Switzerland, Scandinavia, northern Germany, the Netherlands, and the United Kingdom—are largely Protestant. Greeks belong to the Eastern Orthodox Church. There are also Jewish and Islamic minorities in the countries of Western Europe.

A shared belief in Christianity has not meant religious unity in Western Europe. Many wars have resulted from conflicts among Christian factions. Locate Northern Ireland, a part of the United Kingdom, on the map on page 260. To this day, a struggle continues there between the Protestant majority and the Roman Catholic minority. A minority is a group that has fewer members than the major ethnic, national, or religious group in an area.

Increased Cooperation As Western Europe heads toward the year 2000, the region shows more signs of pooling its resources. Since 1949, Western Europe has banded with the United States and Canada in a mutual defense alliance called the North Atlantic Treaty Organization (NATO). NATO was set up to protect Western Europe from attack by the Soviet Union.

Most of the other nations in Western Europe speak Germanic languages. For example, Swedish, Danish, Norwegian, Icelandic, Dutch, and Flemish are all Germanic languages. The languages in this group are derived from the dialects spoken by the

Applying Geographic Themes

Regions The Roman Catholic Church is prominent in many countries of Europe. Where are the largest areas of Protestantism in Europe today? Many cathedrals are built with stained-glass windows such as the one above from the cathedral in Chartres, France.

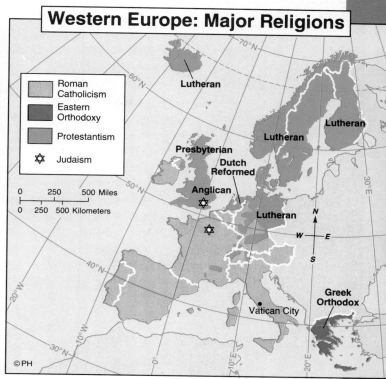

Western Europe: Major Religions

- Roman Catholicism
- Eastern Orthodoxy
- Protestantism
- ✡ Judaism

Lutheran
Lutheran
Lutheran
Presbyterian
Dutch Reformed
Anglican
Lutheran
Greek Orthodox
Vatican City

0   250   500 Miles
0   250 500 Kilometers

©PH

Beginning in the 1950s, some Western European countries formed a "common market" for their mutual economic benefit. For example, countries that were members of the Common Market agreed not to tax goods passing from one country to another. They hoped that this arrangement would encourage a greater amount of trade among countries. Member countries also worked together to coordinate their transportation and banking systems. Later on, other countries joined this group, known as the European Economic Community. This community of European nations set the year 1992 as a target date to lower barriers to trade and travel and, possibly, to set up a common monetary system. You will read more about the cooperative efforts of the European Community in the Case Study on pages 312 and 313. Western Europeans hope that a united Europe will make the region a world power in the twenty-first century.

SECTION 2 REVIEW

Developing Vocabulary
1. Define: **a.** multilingual **b.** minority

Place Location
2. Identify countries that lie in the two most populous corridors in Western Europe.

Reviewing Main Ideas
3. How does the population density of Western Europe differ from that of the United States?
4. How did industrialization alter population patterns in Western Europe?
5. How have Western European nations cooperated since World War II?

Critical Thinking
6. **Predicting Consequences** How do you think the European Community will affect the position of the United States in world affairs?

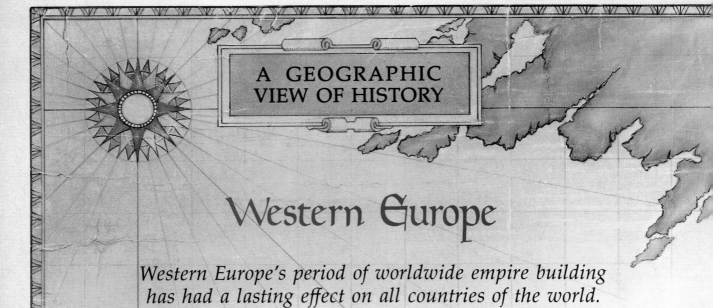

A GEOGRAPHIC VIEW OF HISTORY

Western Europe

Western Europe's period of worldwide empire building has had a lasting effect on all countries of the world.

A popular saying during the 1800s stated that "The sun never sets on the British empire." By that time, British colonies spanned the globe. There was indeed no time of the day when the sun did not shine on the British flag, whether in China, Africa, Europe, or the Americas.

Great Britain had not been alone in its quest for empire. Other Western European nations sought far-flung empires, too. Because of this, some historians have called the nineteenth century "The Age of European Dominance."

Western Europe has had a tremendous impact upon the rest of the world. Although they have lost their great empires, Europeans have nevertheless left their cultural mark on every continent. The official languages in many parts of Africa, Asia, and the Americas reflect a history of European dominance over them.

Economic Activity

Historians disagree about why Western European countries entered an age of expansion. Some say they sought to expand their wealth and add glory to their nations. Others mention the desire of popes and rulers to spread the Christian faith. Still others cite political competition among nations. More territory meant more power. However, it is probable that no single reason can explain Europe's dominance in the world.

Geography is what first pushed Western Europeans away from their shores, as their rulers realized the tremendous profits that might be earned through exploration and conquest. As one historian put it:

The motives of the sea-fancying Europeans may have seemed . . . varied. . . . Yet . . . among the three elements . . . summing up colonial objectives—God, gold, and glory—gold would stand out by far.

The explorers themselves seemed to confirm this view. "We Spaniards," said one conqueror of the Americas, "suffer from a disease that only gold can cure." To seek relief from their illness, Europeans sought land after land to conquer.

European "Discovery" of the World

In the tenth century A.D., long wooden ships began edging westward into the Atlantic. The heads of fearsome monsters decorated the prows; brightly colored shields hung off the sides. "The ships alone," wrote one observer, "would have terrified any enemy."

These were the Viking "dragonships." Viking warriors used them to conquer large parts of Western Europe and to sail into uncharted waters. They built colonies in Iceland and Greenland. Then, in the year 1000, the Vikings set up a short-lived colony in North America.

Western Europe naturally pointed people to sea. Its navigators searched for new ways to improve their skills. By the 1400s, they had made many improvements upon the Viking longboats. The compass helped sailors determine direction. Square-rigged sails fared better than the Viking ships by increasing the amount of wind that could be harnessed. More important, improvements in ship design even allowed sailors to sail into the wind.

Conquest of the Americas On October 12, 1492, a sailor aboard a ship commanded by Christopher Columbus shouted out *Tierra! Tierra!* (Land! Land!). Columbus did not know it, but these words foreshadowed the European discovery of two continents— North and South America. As Columbus landed on islands in the Caribbean, he claimed them without regard to the native people living there. Wrote Columbus to the king and queen of Spain:

> *I found many islands inhabited by men without number, all of which I took possession for our most fortunate king, . . . [with] no one objecting.*

This pattern was followed by other European explorers and conquerors in the Americas. While European discoveries created many opportunities for the people of Western Europe, native people in those lands lost control of their homelands and faced many hardships. Within one hundred years, only about 10 percent of the native populations of North and South America remained alive. They fell victim to war, forced labor, or European diseases against which they had no immunity.

The contact between the Americas and Western Europe brought about far-reaching changes. Europeans learned to grow plants

Christopher Columbus
Columbus traveled west from Spain in 1492 hoping to reach the riches of East Asia. His voyage to the Americas began a period of extensive European domination in the Western Hemisphere.

from the Americas such as potatoes, tomatoes, and corn. People in the Americas, in turn, planted European grains and raised new livestock, including pigs, sheep, horses, and cattle.

Early Colonial Empires Portugal and Spain took an early lead in empire building. Portuguese navigators found the first all-water routes to the rich spice islands of Southeast Asia. They sailed around Africa and headed east to China, setting up colonies in seaports all along the trade routes. These colonies provided stops for trading ships and protected the ships from attacks. They also gave Portuguese settlers a place to build new lives and personal fortunes.

Taking control over a dependent area or people is called colonialism. Over time, colonialism became a way of building worldwide empires. Following Portugal's example, other nations joined the race for colonies.

Spain, too, built its empire in the Americas. At first, settlers were restless. As one historian expressed it, their purpose was to find a shortcut to "legendary treasures of gold," and their intent was to stay only long enough to make themselves rich. But as time passed, the Spaniards stayed on to settle and govern in the Americas and raise families there as well.

Colonialism Spreads By the 1600s, other Western European nations had become colonial powers, too. The Netherlands, France, and England joined Portugal and Spain in building colonies in Africa, Asia, and the islands and coasts of the Americas. Colonies supplied the home countries with gold and silver, slaves for forced labor, tea, coffee, sugar, and other spices.

The nations of Western Europe used superior military technology and equipment, such as rifles and cannons, to subdue and control native people. But these weapons cost a great deal of money. So did the vast navies needed to protect a nation's colonial holdings. In the 1600s, for example, the Dutch and Portuguese battled for the control of colonies in Brazil and on the west coast of Africa. Such colonial wars often drained a nation's treasury and eventually weakened the economy of the home countries.

The Industrial Revolution

When the Industrial Revolution took place in Western Europe, the purpose of empire building changed. Industrial nations needed raw materials for their factories and markets for their manufactured goods.

By the 1800s, Western Europeans wanted more than simple trading posts. In the days of colonialism, settlers often stayed on the coast near seaports. As European nations industrialized, however, they began to establish mines and plantations in the interior of their colonies. They used military force to take more and more land from the native people. Over time, the home countries exercised greater control over their "overseas provinces." They set up governments, and the settlers became the ruling class.

This period of empire building became known as the age of imperialism. In the 1800s, the economies of Western Europe depended increasingly upon the export of manufactured goods. Colonies became the buyers. In 1850, for example, India supplied Britain with most of its raw cotton. But it also bought one quarter of all British manufactured cloth.

The Movement of People Western Europe's population boomed during the nineteenth century. In fact it more than doubled, from less than 200 million to more than 400 million. At the same time, living conditions in the cities and factories declined. In some cases, the advent of machines threw people out of work. Unemployment became a huge problem in industrial cities, such as Manchester, England.

In the mid-1840s, a blight killed most of the potato crop in Ireland. A great famine swept through the countryside. Around the same time, political turmoil rocked all of Europe. All over Western Europe, people looked for ways to rebuild their lives.

The solution for many was found in Western Europe's overseas colonies. People moved by the millions to the Americas, Australia, and New Zealand. Here immigrants found jobs or built farms on unsettled lands. Many European immigrants joined with the flood of United States citizens who pushed toward the western frontier to build farms on the sweeping Great Plains.

Success stories lured wave after wave of immigrants away from Europe's shores. In the late 1800s, for example, German, Swiss, and Spanish immigrants arrived in Argentina in large numbers. They worked and sent back money to their families in Europe. Explained one writer:

You work long hours and your family works and the harvest comes and you remit [sent money home] and Europe knows that all is well. Then the immigrant stream begins to swell. There is no propaganda for immigration like . . . prosperity.

The World About 1750

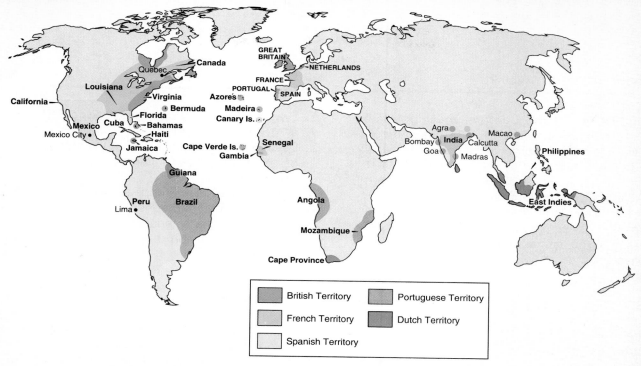

Legend:
- British Territory
- French Territory
- Spanish Territory
- Portuguese Territory
- Dutch Territory

Applying Geographic Themes

Movement Few places on the earth have remained untouched by the domination of the technologically advanced countries of Europe and the United States. Which two European countries dominated South America?

Expansion of European Power European expansion—from the days of the Vikings to today—has spread Western European technology and culture throughout the world. During the late 1800s, for example, European business owners brought their knowledge of the Industrial Revolution to other lands. By 1885, five hundred of the seven hundred industries in São Paolo, Brazil, were owned by European immigrants or descendants.

In time, a spirit of nationalism swept through these European colonies, and they won independence. But systems of government, religious practices, and other customs still bear witness to European dominance.

In the 1990s, Western Europeans continue to emigrate. However, the great waves of the past have slowed to a trickle. Western European nations exert their influence in other ways, as members of a global community of independent nations.

TAKING ANOTHER LOOK

1. What factors led Western European countries to expand beyond their borders?
2. How did colonies suffer as a result of European empire building? How did they benefit?

Critical Thinking

3. **Demonstrating Reasoned Judgment** ''The Industrial Revolution was Western Europe's greatest contribution to the world.'' Attack or defend this statement based on information in this section.

14

REVIEW

Section Summaries

SECTION 1   Land, Climate, and Vegetation
Old mountains in the north and younger mountains in the south border a plain called the North European Plain. The climate of most of Western Europe is mild and moist because of warm Atlantic Ocean currents and prevailing westerly winds. The Mediterranean climate in the south has hot, dry summers and mild, moist winters. Many of the natural forests in Western Europe have been cut down to make room for farmland.

SECTION 2   Human Geography   Western Europe is one of the most densely populated areas in the world. It also has one of the strongest industrial economies. Farming on the North European Plain must be very productive to grow food for much of the region. Most of the many languages are either Germanic or Romance languages, and most of the people belong to either the Roman Catholic or Protestant branches of the Christian religion. The countries in Western Europe have a very long history of wars among themselves but they are now beginning to cooperate politically and economically.

Vocabulary Development

Match the definitions with the terms below.

1. the highest point of a mountain
2. a group that is smaller than the major ethnic, national, or religious group in a particular area
3. flow of air from west to east in the temperate zones
4. speak and understand many languages

a. multilingual   c. summit
b. minority   d. prevailing westerlies

Main Ideas

1. Why is Western Europe sometimes called a ''peninsula of peninsulas''?
2. Describe the location and give the names of some of Western Europe's younger mountains.
3. Describe the North European Plain.
4. What are Western Europe's most distinctive physical characteristics?
5. How has population affected the natural vegetation?
6. Why are climate regions in Western Europe generally milder than climate regions at similar latitudes in other parts of the world?
7. Describe the location of two heavily populated regions of Western Europe.
8. What are the two major language families in Western Europe?
9. How did the Industrial Revolution affect population distribution?
10. To what extent do Western Europeans share language and religion?
11. How have Western European countries begun to cooperate since World War II?

Critical Thinking

1. **Determining Relevance**   How has Western Europe's dense population affected farming methods on the North European Plain?
2. **Expressing Problems Clearly**   Changes caused by the Industrial Revolution made things easier and more profitable for some people, but caused problems for others. Explain this statement.
3. **Drawing Conclusions**   How has the location of Western Europe affected its role in world political and economic affairs?

Practicing Skills

1. **Formulating Questions** Find the section in the Geographic View of History on pages 272–273 under the large boldfaced heading "The Industrial Revolution." Then write three questions about that section the answers to which would give you a greater understanding of the topic.

2. **Interpreting an Economic Activity Map** Refer to the Economic Activity and Resources map on page 265 to answer the following questions:

   a. Where are the largest coal deposits in Western Europe located?
   b. Where are the largest regions of manufacturing and trade?
   c. Compare the Economic Activity map to the physical map on page 257. What do most of the regions with hydroelectric power have in common in terms of their physical landscape?
   d. What valuable resource can be found in the North Sea?
   e. What are the major activities in Spain?

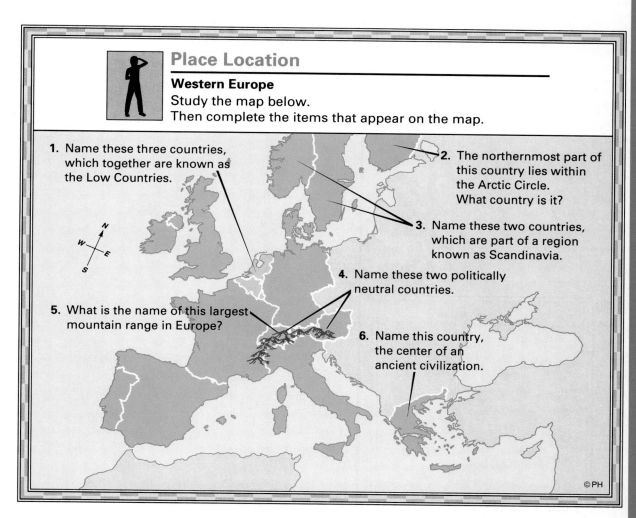

Place Location

Western Europe
Study the map below.
Then complete the items that appear on the map.

1. Name these three countries, which together are known as the Low Countries.

2. The northernmost part of this country lies within the Arctic Circle. What country is it?

3. Name these two countries, which are part of a region known as Scandinavia.

4. Name these two politically neutral countries.

5. What is the name of this largest mountain range in Europe?

6. Name this country, the center of an ancient civilization.

© PH

15

The British Isles and Nordic Nations

Chapter Preview

Locate the countries covered in each section by matching the colors on the right with those on the map below.

Sections		Did You Know?
1	ENGLAND	No part of England is more than 65 miles (105 km) from the sea.
2	SCOTLAND AND WALES	There have been at least 3,000 recorded sightings of the Loch Ness monster, said to inhabit a lake, or loch, in Scotland.
3	THE TWO IRELANDS	Patrick, the patron saint of Ireland, was born in either Wales, Scotland, England, or France, not in Ireland.
4	THE NORDIC NATIONS	No part of Denmark is very high above sea level, so there is a saying that you can see the entire country if you stand on a box.

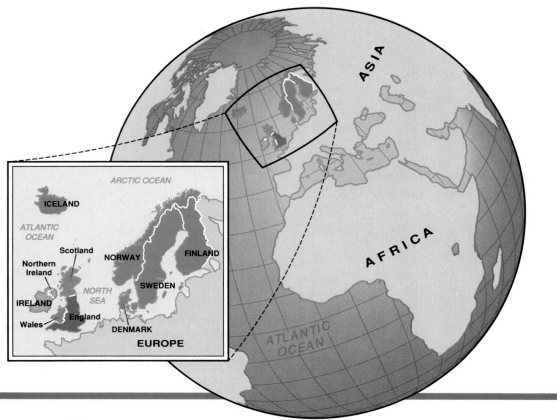

The English Countryside
Elizabethan cottages, wildflower gardens, and medieval churches draw thousands of admirers to England annually. This village is in the Cotswolds, in the southwestern county of Gloucestershire.

1 England

Section Preview

Key Ideas

- The United Kingdom includes the subdivisions of England, Scotland, Wales, and Northern Ireland.
- England is the most densely populated area in the United Kingdom.
- Britain's location made it a center of Atlantic exploration and trade.
- Britain possessed the human and natural resources to fuel the Industrial Revolution.

Key Terms

estuary, ore

The British Isles is the name given to the more than five thousand islands clustered off the northwest coast of Europe. The largest island in the British Isles—and in all of Europe—is Great Britain. The second largest is Ireland.

The island of Great Britain is about the size of the state of Minnesota. It comprises three formerly independent countries: England, Scotland, and Wales. Together with Northern Ireland, these form the United Kingdom of Great Britain and Northern Ireland, or simply the United Kingdom.

The core of the United Kingdom is England. Notice from the population map on page 264 that England is the most densely populated area in the British Isles. Nearly 80 percent of the region's population live here.

England's Rural Landscape

"Our England is a garden," declared English poet Rudyard Kipling in the late 1800s. Kipling was describing rural England with its rolling meadows, peaceful rivers, and neat farms.

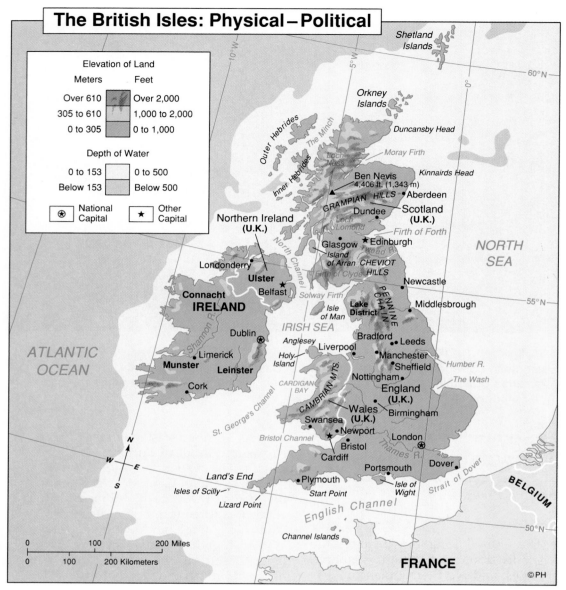

The British Isles: Physical–Political

Elevation of Land

Meters		Feet
Over 610		Over 2,000
305 to 610		1,000 to 2,000
0 to 305		0 to 1,000

Depth of Water

0 to 153		0 to 500
Below 153		Below 500

⊛ National Capital   ★ Other Capital

Shetland Islands

Orkney Islands

Duncansby Head

Outer Hebrides

The Minch

Loch Ness

Moray Firth

Ben Nevis 4,406 ft. (1,343 m)

Kinnairds Head

GRAMPIAN HILLS · Aberdeen

Inner Hebrides

Dundee · Scotland (U.K.)

Loch Lomond

Firth of Forth

NORTH SEA

Northern Ireland (U.K.)

Glasgow ★ Edinburgh

Island of Arran

Tweed R.

CHEVIOT HILLS

Londonderry

North Channel

Firth of Clyde

Newcastle

Ulster ★

Belfast

Solway Firth

Isle of Man

PENNINE CHAIN

Middlesbrough

Connacht

IRELAND

Lake District

55°N

Bradford · Leeds

Dublin ⊛

Anglesey

Liverpool

Manchester

IRISH SEA

· Limerick

Holy Island

Sheffield

Humber R.

Munster

Leinster

Nottingham ·

The Wash

ATLANTIC OCEAN

· Cork

CARDIGAN BAY

CAMBRIAN MTS.

England (U.K.)

St. George's Channel

Wales (U.K.)

Birmingham

Swansea

Newport

London ⊛

Bristol Channel

★ Bristol

Thames R.

Cardiff

Land's End

Portsmouth

Dover

N
W — E
S

Isles of Scilly

Start Point

Isle of Wight

Strait of Dover

BELGIUM

Lizard Point

English Channel

50°N

Channel Islands

0   100   200 Miles
0   100   200 Kilometers

FRANCE

©PH

60°N
10°W
5°W
0°

Applying Geographic Themes

1. Regions   The British Isles have clearly defined regions of highlands and lowlands. Where are the Cambrian Mountains located?

2. Location   Which body of water lies east of the islands?

By this time, however, some parts of England had grown into busy industrial centers. Even England's rural landscape was far more varied than Kipling's garden suggested.

What does a typical English landscape look like? Most people in England would answer this question by mentioning three strikingly different areas: the Highlands, Midlands, and Lowlands.

The Highlands are a band of hilly lands facing the Atlantic. As the map on this page shows, they run west of a line stretching roughly from Newcastle in the northeast to Plymouth in the southwest. Older and harder rocks in this region have been worn down by centuries of weathering. Even so, some peaks rise to 3,000 feet (915 m), and the land is difficult to farm.

A short distance to the southeast are the Midlands. This area presents a sharp contrast to the less populated Highlands. Here lie the thick veins of coal that fired the country's Industrial Revolution. Factory towns such as Birmingham, Manchester, and Stoke-on-Trent still darken the air with fumes from their mills. Some of England's highest population densities are in the Midlands.

To the south and east are the rolling Lowlands. The land slopes gently toward the English Channel, and elevations rarely top 1,000 feet (305 m). Younger, softer rocks lie beneath the land's surface. Because these rocks break up easily, soil in the Lowlands tends to be fertile.

The Lowlands provide England with some of its most productive farms. Farmers grow wheat, vegetables, and other crops on small plots. They set aside larger parcels of land for pasture. The cool, moist weather of England's marine west coast climate is perfect for raising sheep and dairy and beef cattle.

England's Urban Landscape

Even before industrialization, England's fertile soil and favorable climate allowed farmers to produce surplus goods for export. Trade within England and with Europe fostered the growth of cities along many inland rivers and along the coast. Of these, London was the most important. Why did London become one of the greatest commercial cities in the world? The answer can be found in one of the five geographic themes—location.

London's Relative Location The map on page 278 shows that London is about 70 miles (113 km) from the

The Tower Bridge, London
More than a dozen bridges span the Thames River in London, but none are more dramatic than the Tower Bridge at night.

continent of Europe. But Dover is even closer. So why isn't Dover the English capital of trade? London possesses a key advantage over Dover and other coastal ports: it is located on the Thames (TEHMZ) River.

Since the Thames Valley formed, the water level of the ocean has risen and created an estuary, a flooded river valley at the wide mouth of a river. Water levels in an estuary rise and fall with the tides. For centuries, ocean-going ships have sailed up the Thames to reach London. Use the map on page 278 to locate inland cities that are within easy reach of London.

Even in the 1500s, London was a bustling port. One writer described the waterfront in this way:

> . . . a forest of masts . . . Huge square rigged ships lay side by side, surrounded by barges and small craft. . . . The boats had to fight to their landing places. . . .

DAILY LIFE

The Postbuses of Great Britain

In the British countryside, marked by small villages and isolated farms, the Post Office helps knit people together with more than just mail service. The United Kingdom maintains a network of vans called "Postbuses." The Postbuses offer an inexpensive transportation and delivery service to rural residents who might otherwise remain isolated.

These large, boxy vans have room for up to eleven passengers. People can pile aboard whenever the driver stops to deliver or pick up mail. People use the Postbuses to go shopping in town, get medical care, reach their jobs, or simply visit friends. They can travel on 36 rural routes in England and Wales, and 138 in Scotland.

1. What advantages does the Postbus system offer to rural residents?
2. Why do you think the government offers this service to rural residents?

Changes in Relative Location The port of London grew rapidly in the 1500s because of changes in the patterns of world settlement and trade. Before 1500, the Mediterranean Sea was the center of trade. Few sailors dared venture into the Atlantic. London remained somewhat isolated on the far edge of European trade. Starting in the 1500s, however, improved ships and navigation devices allowed Europeans to push westward across the Atlantic. Great Britain's strategic, central location on the Atlantic was ideal for trade. So, as trade along the Atlantic increased, Britain's relative location improved.

Workshop of the World

In the 1500s, Britain shipped mostly the products of its farms. But within its small area, the island nation had the resources to fuel the start of the Industrial Revolution—a revolution that would change the world.

As shipowners and merchants earned profits from trade, they looked for new ways to invest their money. Britain had far-flung colonies and a growing population to produce goods to sell to these colonies. Wealthy business owners built factories for the production of manufactured goods. As ships plied the oceans loaded with British goods, Britain became known as the "workshop of the world."

The Rise of Heavy Industry Some of the earliest technological advances of the Industrial Revolution were used in factories that produced textiles, or cloth. At first, British manufacturers used water power to run spinning machines. But they later switched to coal as a source of power.

The mining of coal became a big business in itself. Find the Pennine Mountains on the map on page 278. Major coal fields lay along the edges of this range, as well as in the northeast near the city of Newcastle.

Britain also possessed large reserves of iron **ore,** or rocky material containing a valuable mineral. Over time, inventors improved methods of melting iron ore and using it in the production of steel. New manufacturing centers such as Birmingham, Sheffield, and Newcastle sprang up alongside the huge deposits of coal and iron.

The Industrial Revolution brought wealth to Britain, but the factories and mines also changed the English landscape. English poet Robert Blake condemned the "dark, Satanic mills" for destroying "England's green and pleasant land." A visitor to Birmingham in the early 1800s described the effect of industry.

The noise of Birmingham is beyond description; the hammers seem never to be at rest. The filth is sickening. . . . I feel as if my throat wanted sweeping like an English chimney.

Challenges to British Industry Britain's plentiful supply of raw materials and position astride major sea routes made it the world's industrial leader for many years. But in the late 1800s, Britain was challenged by two new industrial powers—Germany and the United States. By 1900, the United States was making as much steel as the United Kingdom; Germany was close behind.

In recent years, British industry has fallen upon hard times. Britain is no longer the world's leading exporter, the "workshop of the world." Now it imports many goods from the United States, Germany, and Japan. To offset the losses of heavy industry, the British government has encouraged the development of service industries, such as tourism, finance, and data processing. Britain also still relies on shipping profits. In the 1990s, Britain continues to have one of the largest merchant marines afloat. London, Liverpool, Manchester, and other British ports are among Europe's leading outlets to the Atlantic.

SECTION 1 REVIEW

Developing Vocabulary
1. Define: **a.** estuary **b.** ore

Place Location
2. Where is Manchester located in relation to London?

Reviewing Main Ideas
3. Describe the Highlands, Midlands, and Lowlands of England.
4. What resources fueled Britain's Industrial Revolution?

Critical Thinking
5. **Perceiving Cause-Effect Relationships** How did Britain's relative position in the world change in the 1500s? How would you describe its relative location today?

2 Scotland and Wales

Section Preview

Key Ideas

- Scotland and Wales have kept separate cultural identities from England.
- Scotland and Wales are each divided physically into highlands and lowlands.
- Older industries in Scotland and Wales have fared poorly.

Key Terms

moor, bog, glen

The Highland Games, Isle of Mull
The plaid, or tartan, worn by these dancers shows their clan.

An English writer once remarked, "[A Scot] is British, yes, and he will sing 'There will always be an England' but he murmurs to himself, 'As long as Scotland is there.'" This story reveals something about how the Scots view their relationship with England. The two nations have been bound politically for almost three hundred years. But Scotland has always kept its own identity apart from England.

The same can be said of Wales, which has been united with England since the late 1200s. Like the people of Scotland, the Welsh have maintained their own culture.

Interaction: The Scottish Landscape

Scotland occupies more than one third of the land area in the United Kingdom, but less than 10 percent of the nation's population live there. The landscape is rugged and bears the marks of the heavy glaciers that moved across the northern part of Great Britain during the last Ice Age.

The map on page 278 shows that the Cheviot Hills and the River Tweed separate Scotland from England. Scotland itself is divided into three regions —the northern Highlands, the central Lowlands, and the southern Uplands.

The Highlands When people hear the name Scotland, they imagine kilted Highlanders playing bagpipes, or they think of the legendary Loch Ness monster that supposedly lives in one of the region's lochs (LAHKHS), as lakes are called in Scotland. The Highland region is really a large, high plateau. The Grampian Mountains cut across it with peaks reaching past 4,000 feet (1,219 m).

Much of the Highlands are covered with moors—broad, treeless rolling plains. The moors are covered with bogs—areas of wet, spongy ground. Steady winds blowing off the Atlantic

Ocean bring abundant rainfall to the moors. But the dampness of the soil limits plant growth to grasses and low shrubs such as purple heather.

The Highlands also have many lochs, which were carved out by retreating glaciers. The most famous, Loch Ness, is located about 90 miles (145 km) west of Aberdeen. According to legend, a dinosaur-like monster nicknamed Nessie lives here. Although little hard evidence supports Nessie's existence, people from around the world come to this remote area to search for the legendary monster.

The land, water, and climate of the Highlands are well suited to the region's economies of fishing and sheep herding. A few people produce a type of handwoven woolen cloth known as tweed. This Scottish home industry has continued over hundreds of years, in sharp contrast to the factory production of textiles in England.

The Central Lowlands Between the hilly border with England and the Highlands runs a long lowland region. Nearly 75 percent of Scotland's people live in this region, which stretches between Glasgow and Edinburgh (EHD ihn BUHR oh).

Industry came to the Scottish central Lowlands in the early 1800s. The Clyde River near Glasgow grew into a huge shipbuilding center. Ships in various stages of construction crowded the docks along the river in the early 1900s. The Clyde shipbuilders played a major role in establishing the United Kingdom as the world's leading naval power. Explained one observer:

Through all the transitions—wood to iron, iron to steel, paddle to . . . turbine engines—Clyde shipbuilders have been to the front with [ideal] ships.

Since the mid-1900s, however, heavy industries in Scotland have fallen on hard times. Old factory centers

The Highland Moors
No trees grow on the Scottish Highlands, but tiny, purplish-pink flowers called heather flourish on the damp moors.

such as Glasgow have declined. The loss of jobs has caused more than one third of Glasgow's residents to leave since 1960.

The Southern Uplands Closest to the English border, the Southern Uplands is primarily a sheep-raising region. The Tweed River valley woolen mills are kept well supplied with wool by area farmers. Medieval abbeys and low, hilly landscapes draw many visitors to the region.

Scotland Today

New industries are slowly taking the place of mining, steel making, and shipbuilding. Oil discoveries in the North Sea off the northeastern shore of Scotland have helped some cities such as Aberdeen. Computer and electronic businesses have also developed along the Clyde and Tweed rivers. Some people call the Clyde Valley

WORD ORIGIN

moor
The word *moor* stems from the Old German word *meri* meaning "sea," and from the Old English word *mor* meaning "boggy swamp." A moor is indeed a "sea" of open, rolling, boggy land where grasses and heather grow.

"Silicon Glen," after the area in California known as "Silicon Valley." A glen is a narrow valley.

Although united with England, Scotland has retained its own culture. When the Scottish and English parliaments were united through the Act of Union in 1707, Scotland kept important trading and political rights. Scotland still has its own system of laws and its own system of education. Many Scots also remained members of the Presbyterian Church, rather than joining the Church of England. A small minority of Scottish people even talk of once again becoming a separate country. As one Scottish patriot explained:

You can tell me that Scotland is part of the United Kingdom, and I will tell you that is the truth but not the whole truth. . . . You see, national boundaries are not simply a matter of geographical frontiers. It's culture we're talking about, a set of national characteristics. These Scotland has retained.

Wales: A Country Apart

A similar spirit of pride and independence burns in Wales. The Welsh are quick to point out that Wales, like Scotland, is a country, not a British state. It has its own capital city, postage stamps, national flag, and language.

However, Wales is strongly influenced by its powerful neighbor England, by which it was conquered in 1284. Since that time, the seat of Welsh government and law has been in London. As a symbol of England's authority over Wales, the heir to the English throne has been traditionally called the Prince or Princess of Wales.

The Welsh Landscape Wales is really a peninsula. It is about the size of the state of Massachusetts and has a landscape similar to Scotland. On the map on page 278 you can find a highland area to the north of Wales and lowlands running along the southern coast near Cardiff. Locate the Cambrian Mountains in the center of Wales.

Wales enjoys a marine west coast climate like the rest of Great Britain. However, the rain-carrying winds from the Atlantic pass over Wales before reaching England. So Wales usually receives even more rain than southern England.

A Separate Language Since the 1500s, Welsh representatives have sat in Parliament, and some have risen to high office in the British government, including that of Prime Minister. Even so, the Welsh have fought for cultural independence.

A Scottish City on the Clyde
Like many industrial cities, Glasgow's growth was due to its proximity to a major water route.

One of the keys to Welsh culture is language. Wales is actually a bilingual country. Its 2.8 million people speak English, but nearly 20 percent still speak Welsh as their first language.

Welsh is a language handed down from the Celtic peoples who lived in Wales for thousands of years. It is spoken mainly in the mountains of northern Wales, where people fish, tend livestock, or produce crafts by hand. In the 1980s, Welsh patriots won the right to broadcast television programs entirely in Welsh.

Turning Around the Welsh Economy

The economic history of Wales is similar to that of England and Scotland. In the late 1800s and early 1900s, industry and coal mining changed the landscape of southern Wales. Mines in the Rhondda Valley, situated just north of Cardiff, became some of Britain's biggest coal producers.

By the mid-1900s, however, heavy industries in Wales had fallen behind in technology. A writer described the Welsh economy in 1945:

Steel sheets were still being hand-dipped in molten tin to make tin plate. . . . [Miners] still picked coal by hand from two-foot seams. . . . In many places, things like bread, milk, and coal were still delivered by horse and cart.

By the 1980s, most of the coal mines in the Rhondda Valley had closed. Unemployment rates soared, and less than one half of the students leaving high school found jobs. Oil companies built petroleum refineries in Wales. But economic conditions were dim. Welsh representatives in Parliament argued for funding that would help to start some new service

Welsh Coal Miners
Since the Industrial Revolution, England has depended on coal from Wales for fuel.

industries in Wales, including tourism for people interested in seeing the traditional Welsh way of life.

SECTION 2 REVIEW

Developing Vocabulary
1. Define: **a.** moor **b.** bog **c.** glen

Place Location
2. Where is Edinburgh located in relation to Glasgow?

Reviewing Main Ideas
3. Describe the landscape of Scotland.
4. Why have heavy industries such as steel making and coal mining fared badly in Scotland and Wales in recent times?

Critical Thinking
5. **Drawing Conclusions** Why do you think survival of the Welsh language goes hand in hand with a spirit of nationalism in Wales?

✔ Skills Check

☑ Social Studies
☐ Map and Globe
☐ Reading and Writing
☐ Critical Thinking

Using Time Lines

A time line helps you place events in chronological order. Study the time line below, then practice using a time line by following these steps.

1. **Identify the time period covered by the time line.** Study the entire time line to learn the span of history that it covers. (a) What is the earliest date shown on the time line? (b) What is the latest date shown?

2. **Determine how the time line has been divided.** Most time lines are divided into equal intervals, or units, of time. Into what intervals is this time line divided?

3. **Study the time line to see how events are related.** A time line helps you to see relationships between events. (a) What events in the 1700s help explain how the Industrial Revolution started in Britain? (b) What events on the time line help show the effect of industry on children and other workers?

The Industrial Revolution in England

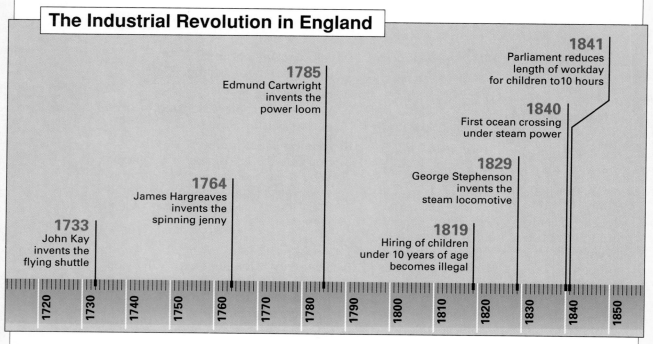

1733 John Kay invents the flying shuttle

1764 James Hargreaves invents the spinning jenny

1785 Edmund Cartwright invents the power loom

1819 Hiring of children under 10 years of age becomes illegal

1829 George Stephenson invents the steam locomotive

1840 First ocean crossing under steam power

1841 Parliament reduces length of workday for children to 10 hours

1720 1730 1740 1750 1760 1770 1780 1790 1800 1810 1820 1830 1840 1850

3 The Two Irelands

Section Preview

Key Ideas

- The island of Ireland is divided into two political regions: Northern Ireland, which is part of the United Kingdom, and the independent Republic of Ireland.
- Religious and political conflicts begun in the 1500s still trouble the Irish today.

Key Terms

peat, annex

Ireland is divided. These three words say a lot about the island. It is divided politically into two parts: Northern Ireland and the Republic of Ireland. Ireland is also divided religiously between Protestants and Catholics. Finally, Ireland is divided culturally between the descendants of native Celtic peoples and the descendants of English and Scottish immigrants.

Place: The Landscape of the Emerald Isle

The divisions in Ireland are not visible immediately. Most people flying into Ireland are struck by the island's peaceful appearance. Ireland's moist marine west coast climate keeps vegetation a brilliant green for most of the year. From the air, Ireland lives up to its nickname: "The Emerald Isle."

The island itself looks like a huge bowl. Hills ring most of the coastline, while the middle of the island is a plain that drains into the River Shannon. About one sixth of the island is covered by **peat**, a spongy material containing waterlogged mosses and plants. As the plant material slowly decays, it forms a thick layer of damp, brownish matter.

Because Ireland has few forests, farmers have learned to cut and dry blocks of peat as fuel for cooking and heating. The Republic of Ireland recently developed a method for using peat in power plants. These plants now produce one quarter of the nation's electricity.

A Region Torn by Conflict

Despite a peaceful physical appearance, Ireland's history has been shaped by invasions and wars. Celtic tribes from Europe first settled Ireland around 400 B.C. They repeatedly had to defend themselves against Viking raids, lasting from roughly A.D. 800 to 1016. The next challenge came from England.

Dingle, Southwest Ireland

Most Irish people still live off the land and in cottages as simple as the landscape. The stone fencing is often hundreds of years old.

Ireland
In the ancient Celtic language of Gaelic, Ireland is called Éire (pronounced AIR uh). The Gaelic phrase *Eireann go bragh* means "Ireland forever."

In 1066, Norman invaders from France conquered England. Some of these conquerors set their sights on Ireland. They took over large tracts of land in Ireland and built castles to protect themselves. The Normans tried to take firm control of the Celts. They forbade marriage between Normans and Celts, banned use of the Celtic language, known as Gaelic (GAY lik), and even outlawed Celtic harp music.

Fearing the growing power of these Norman lords in Ireland, King Henry II of England declared himself Lord of Ireland in 1172. He did not have enough power to force Norman lords to obey him. But English rulers who followed Henry held on to the title and began thinking of Ireland as a possession.

Religious Conflicts In the 1500s, religious turmoil in Europe increased conflicts between England and Ireland. Until this time, the Roman Catholic church, headed by the Pope in Rome, had directed religious affairs in much of Western Europe. But as kings became more powerful, they competed with the Pope for their people's loyalties.

In the early 1500s, groups formed in Europe tried to change some of the Church's practices and started a reform movement known as the Reformation. A split soon developed between the Roman Catholics and the Protestants, the name given to those Christians who protested Church policies.

In 1534, Henry VIII of England broke away from the Roman Catholic church and founded the Church of England. He made himself head of the church and moved to strengthen his royal power. He changed his title from Lord of Ireland to King of Ireland.

Henry did not try to force the Irish to give up the Roman Catholic religion. But some of his descendants tried to. Most of the Irish remained strongly Roman Catholic, however, and fought bitter battles against the English. When the English eventually won, they imposed harsh laws on the Irish and gave away large parcels of Irish land to Protestant settlers from Great Britain.

The divisions in Ireland soon became economic as well as religious. The Protestant minority controlled much of the wealth, while the defeated Irish Catholics fell into poverty.

Lasting Bitterness The British policy toward Ireland left a legacy of bitterness and hatred. In 1798, the French supported a rebellion in Ireland. The United Kingdom responded in 1801 by annexing Ireland, or formally adding it to its territory.

Movement for Independence Many Irish continued to press for independence throughout the nineteenth century. Fierce rebellions between 1916 and 1921 led officials in the United Kingdom and Ireland to divide the island into two parts. The six northeastern counties remained part of the United Kingdom, but the rest of Ireland became a free state under British supervision. This free state declared its total independence as the Republic of Ireland in 1949. Independence did not end political turmoil on the island.

Roughly two thirds of the 1.7 million people who live in Northern Ireland today are Protestant; the rest are Catholic. Most Catholics support the reunification of Ireland, while most Protestants oppose it.

Both Protestant and Catholic extremists have used violence to try to win control of Northern Ireland. In the late 1960s, Great Britain sent troops there to protect the Catholic population from Protestant extremists. Now Catholic extremists see these troops as enemies. So in the 1990s, the governments of Ireland and the United Kingdom continue to struggle with a problem begun in the 1500s.

Changing Economic Patterns

The poverty of Ireland's early years still troubles it today. Find the per capita GNP for the Republic of Ireland in the table on pages 751–761, and compare it with those of the United Kingdom, France, and Sweden. As you can see, the Republic of Ireland ranks low in national wealth.

The Potato Famine Ireland never industrialized as did most of Great Britain. Instead, it depended heavily on the harvesting of potatoes, which thrived in Ireland's moist climate and crumbly soil. However, in the 1840s, a severe blight, or disease, wiped out Ireland's potato crop and created massive famine. The island's population dropped from roughly 6.5 million in 1841 to slightly more than 3 million by 1900. Nearly one million died of hunger and disease. Many more emigrated to other lands. During the 1900s, Ireland sought to rebuild. Today farmers rely more on beef and dairy cattle.

Putting Advantages of Place and Location to Work In the late 1980s and early 1990s, the Republic of Ireland has attempted to make better use of its location. As the world map on pages 762–763 shows, the Republic of Ireland is centrally located between Europe and North America. The Irish have taken advantage of this location by upgrading Shannon Airport near Limerick in the west. Many transatlantic flights now refuel there; this activity in turn attracts other industrial operations. The Irish have worked hard to attract tourists, too. As journalist Flora Lewis explained:

> The Irish came to appreciate the economic value of their rolling hills and quiet valleys. . . . The geography that was a heartache can be seen now as a blessed rentable refuge to people from the hectic outside world.

A Legacy of Strife in Northern Ireland
Conflict between Protestants and Catholics has brought violence to Northern Ireland.

SECTION 3 REVIEW

Developing Vocabulary
1. Define: **a.** peat **b.** annex

Place Location
2. Identify the two parts of Ireland.

Reviewing Main Ideas
3. How have the Irish turned their poor drainage system to their advantage?
4. How did religious and cultural conflicts emerge in Ireland between 1100 and 1700?
5. How do the Irish hope to rebuild their economy?

Critical Thinking
6. **Analyzing Information** Tell whether you agree or disagree with the following statement; support your answer with evidence from the text: "Protestants and Catholics in Northern Ireland both feel threatened by the past and both are reluctant to negotiate."

4 The Nordic Nations

A Coastal Village in Norway

Flooded glacial valleys with deep water and steep mountain walls are known as fjords. Many of Norway's villages are located in coastal fjords.

The people of northern Europe call their land Norden, from an ancient word meaning "Northlands." Norden includes five independent nations: Norway, Sweden, Finland, Denmark, and Iceland. These nations, called the Nordic nations by English-speaking peoples, are unified as a region by location and strong cultural bonds.

A Northern Location

Norden is identified as a region in part by its location in the northern latitudes. As the world map on pages 762-763 shows, the Nordic nations occupy a latitude comparable to Alaska. Parts of some Nordic nations reach into the Arctic Circle and the polar zones.

A Varied Landscape Norden is not a single expanse of land. It is a collection of peninsulas and islands separated by seas, gulfs, and oceans. The most continuous land masses are the Scandinavian and Jutland peninsulas. Find these two peninsulas on the map on pages 764-765.

The terrain varies dramatically throughout the Nordic nations. Denmark is so flat that its highest point is less that 600 feet (183 m), while Norway is one of the most mountainous nations in Europe. Because of this difference, Norwegians sometimes call their Danish neighbors "flatlanders."

The Effect of Glaciers Much of the landscape on the Scandinavian Peninsula is the product of the last Ice Age. Huge glaciers carved out thousands of lakes across the peninsula. They also removed topsoil and other materials and deposited them in Denmark and other parts of Western Europe. Much of the soil in Scandinavia today remains rocky and difficult to farm.

Notice the jagged coastlines along the Scandinavian Peninsula on the map on pages 764-765. When the glaciers advanced, they carved out deep

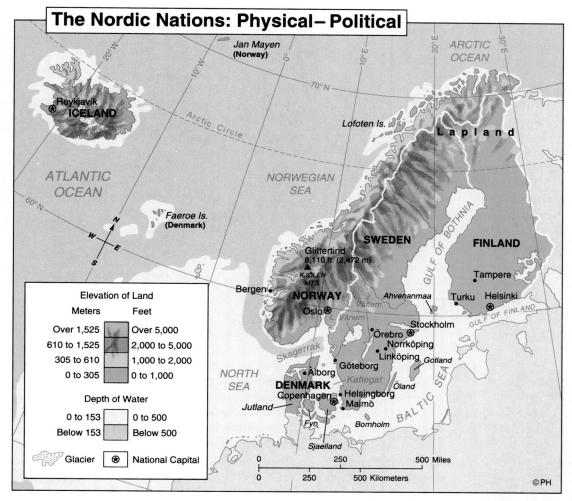

The Nordic Nations: Physical–Political

Jan Mayen (Norway)

ARCTIC OCEAN

Reykjavík
ICELAND

Lofoten Is.

Lapland

ATLANTIC OCEAN

NORWEGIAN SEA

Faeroe Is. (Denmark)

GULF OF BOTHNIA

SWEDEN

FINLAND

Glittertind 8,110 ft. (2,472 m)
KJØLEN MTS.
Bergen

Tampere

Ahvenanmaa

Turku   Helsinki

NORWAY

Vättern

Oslo

L. Vänem

GULF OF FINLAND

Stockholm

Örebro

Norrköping

Linköping

Gotland

Skagerrak

Göteborg

NORTH SEA

Ålborg

Kattegat

Öland

DENMARK

Copenhagen

Helsingborg

Malmö

BALTIC SEA

Jutland

Fyn

Bornholm

Sjaelland

Elevation of Land

Meters	Feet
Over 1,525	Over 5,000
610 to 1,525	2,000 to 5,000
305 to 610	1,000 to 2,000
0 to 305	0 to 1,000

Depth of Water

0 to 153	0 to 500
Below 153	Below 500

Glacier   ⊛ National Capital

0   250   500 Miles
0   250   500 Kilometers

©PH

Applying Geographic Themes

1. Location   Which body of water separates the Scandinavian Peninsula from the rest of Europe to the south?

2. Place   Which Nordic country has the highest overall elevations? What is the capital of Iceland?

glacial valleys along the coasts. When the glaciers melted, water filled the valleys, creating flooded glacial valleys known as **fjords** (fee YAWRDZ). Some fjords are so deep that ocean-going ships can sail into them. Most have such steep walls that even mountain climbers find them difficult to scale.

Interaction: Iceland's Land of Fire and Ice   The most unusual Nordic nation is Iceland. It is the product of

volcanic action along the North American and Eurasian tectonic plates. Movement along these plates permits molten rock from the earth's crust to bubble upward, creating land.

In Iceland, volcanoes and glaciers exist side by side. Icelanders call their island a land of fire and ice. They have learned to take advantage of the island's particular geology to produce geothermal energy, or energy produced from the heat of the earth's

aurora borealis
Roman mythology
named the arches of
light that appear at
night in Arctic skies
Aurora Borealis. Au-
rora was the god-
dess of dawn, Bore-
as was the god of
the north wind.

interior. Today, geothermal energy accounts for a large share of the power used for heat and electricity in Iceland.

Long Winters and Short Summers
"Winter is the element for which we are born," declared a Finnish historian in the 1800s. And certainly winter is the one element that has deeply affected the lives of most Nordic peoples.

Norden's location to the far north results in long winters and short summers. It also affects the number of hours the sun shines in each season. In midwinter, the sun may shine only two or three hours a day. In midsummer, it shines for more than twenty hours.

In the depths of winter, much of the land and water in the Nordic nations freezes. It is the time when the aurora borealis, or northern lights, shine most brightly. These lights appear when atomic particles from the sun, attracted by the magnetic fields of the North Pole, break through the northern atmosphere. The particles create white, red, and blue lights.

The start of summer is a public holiday in most Nordic nations. Darkness gives way to the long-awaited light, and people celebrate the return of the "midnight sun." In the northernmost territories, the sun never really sets for several weeks in midsummer. People call the long twilight hours of evening the "white nights."

The Effect of the Ocean on Climate
Despite the length of winter, the climates in much of Norden are milder than you might expect. As is evident from the climate map on page 259, most of Iceland, all of Denmark, the west coast of Norway, and southern Sweden have mild marine west coast climates. This is because the warm currents of the North Atlantic Drift moderate the weather and keep the coast free of ice.

The coldest areas in Norden lie just east of a mountain chain that runs northeast to southwest through Norway. This range prevents the warm, moist ocean winds from reaching the rest of the Scandinavian Peninsula.

Shared Cultural Bonds

More than climate and location bind the Nordic nations into a region. They also have strong cultural ties.

Similar Historic Roots   The Nordic nations have similar histories. From around A.D. 800 to 1050, Vikings sailed out of the fjords and inlets of

A Folk Festival in Denmark
These dancers are forming a circle around the musicians at a folk festival in Ribe, a Danish town.

southern Scandinavia to raid much of Western Europe. The Vikings were more than warriors—they were traders, colonizers, and explorers who left their mark on world history.

The Nordic nations were also united at times. Queen Margrethe of Denmark joined the five lands under one crown in 1397. The union ended in 1523 when Sweden withdrew. But Denmark, Norway, and Iceland remained united for several centuries more. Sweden and Finland were united until the early 1800s, when Sweden ceded Finland to Russia.

Religion, too, unites the Nordic people rather than dividing them, as in the British Isles. Most Nordic peoples belong to the Lutheran Church, first established during the Reformation.

With the exception of Finland, Nordic languages derive from common roots. Even Finland is bilingual, and most Finns have a working knowledge of Swedish, Finland's second language. In addition, Nordic schools require students to learn English, which helps bridge any linguistic differences.

Democratic Governments and Mixed Economies Nordic countries share certain political and economic beliefs. All five of the Nordic nations are democracies, and their economic systems are mixed economies. That is, they practice a mixture of free enterprise and socialism.

Most businesses in the Nordic countries operate much as they do in the United States. But the Nordic governments guarantee certain goods and services to everyone and operate some industries that are run privately in the United States. For example, Denmark and Sweden have state-run day-care centers and state-supported medical care.

As a rule, the Nordic nations follow a neutral course in foreign affairs. Currently, Norway refuses to open its excellent harbors for military use. It also forbids the storage of nuclear weapons on its territory. Denmark and Sweden actively promote peaceful solutions to international crises.

Sound Economies

Compared with other regions of the world, the Nordic nations have sound economies. They derive their wealth from varied sources. Denmark and Sweden have enough flat land and a mild climate suitable for agriculture. Denmark uses almost three quarters of its land for farming, and in the late 1980s produced more than four times the amount of food needed to feed its people. Fishing is also an important economic activity for the peninsula and island nations of northern Europe. The Norwegians, in particular, look to the sea. They compare it to farmland and call their offshore waters "The Blue Meadow." The region also profits from high-grade ores and from vast expanses of forest.

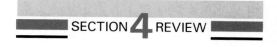

SECTION **4** REVIEW

Developing Vocabulary
1. Define: fjord

Place Location
2. Which Nordic nations border Finland?

Reviewing Main Ideas
3. Name at least three cultural bonds that help to link the nations of northern Europe.
4. Describe the relationship between the government and the economy in the Nordic nations.

Critical Thinking
5. **Demonstrating Reasoned Judgment** Give evidence that would support this statement: "Because of their physical geography, the Nordic nations rely more heavily on shipping than any other region in Europe."

15

REVIEW

Section Summaries

SECTION 1 England England comprises three strikingly different regions: the Highlands to the north and west, the central Midlands, and the Lowlands to the south and east. For centuries, London's location on the Thames River has made it a world trading center. Britain's rich human and natural resources fueled the Industrial Revolution.

SECTION 2 Scotland and Wales Although part of the United Kingdom, Scotland and Wales retain their cultural identities. In both regions the rugged highlands remain rural while the lowlands are urban and industrial.

SECTION 3 The Two Irelands Religious conflicts begun in the 1500s continue to divide Northern Ireland, which is part of the United Kingdom, and the Republic of Ireland. The climate allows vegetation to remain brilliant green for most of the year. However, poor drainage hinders agriculture. The Republic of Ireland hopes that its location will attract tourists in the 1990s.

SECTION 4 The Nordic Nations Denmark, Norway, Sweden, Finland, and Iceland share a northern location, a common religion, related languages, and similar political beliefs. Their landscape and resources are varied and have encouraged strong, diverse economies.

Vocabulary Development

Match the definitions with the terms below.

1. a flooded glacial valley
2. a flooded valley near the mouth of a river where tides change the water level
3. rocky material that contains a mineral
4. formally add to a country's territory
5. damp, brownish dirt composed of decaying mosses and plants
6. marshes
7. a broad, treeless plain
8. a narrow valley

a. ore	**e.** moor
b. peat	**f.** bog
c. fjord	**g.** annex
d. estuary	**h.** glen

Main Ideas

1. What are the four main areas included in the United Kingdom?
2. Given its northern latitudes, why is the climate of northern Europe unusual?
3. What impact did the Industrial Revolution have on Britain?
4. Describe the mixed economies of the Nordic nations.
5. What are the causes of the conflict in Ireland?
6. Describe the landscape of the Scottish Highlands.

Critical Thinking

1. **Making Comparisons** Compare the current economies in the British Isles to those in the Nordic nations.
2. **Synthesizing Information** What generalizations can you make about northern Europe's climate and vegetation by analyzing many of the key terms for this chapter?
3. **Identifying Central Issues** How has the maritime setting of northern Europe influenced the economies of the countries of the region?

4. **Demonstrating Reasoned Judgment** Describe four ways in which the people of northern Europe have interacted with their natural environments to produce energy for heating their homes. Which of these examples of interaction do you think is the least harmful to the environment?

5. **Predicting Consequences** Some demographers estimate that Catholics will outnumber Protestants in Northern Ireland by the early twenty-first century. What might be the political implications of such a trend?

Practicing Skills

Using Time Lines Look again at the time line on page 286. Then answer the following questions.

1. How many years passed between the invention of the steam locomotive and the first steam-powered ocean crossing?
2. What do the events on the time line regarding child labor imply about the impact of the Industrial Revolution on people's lives?

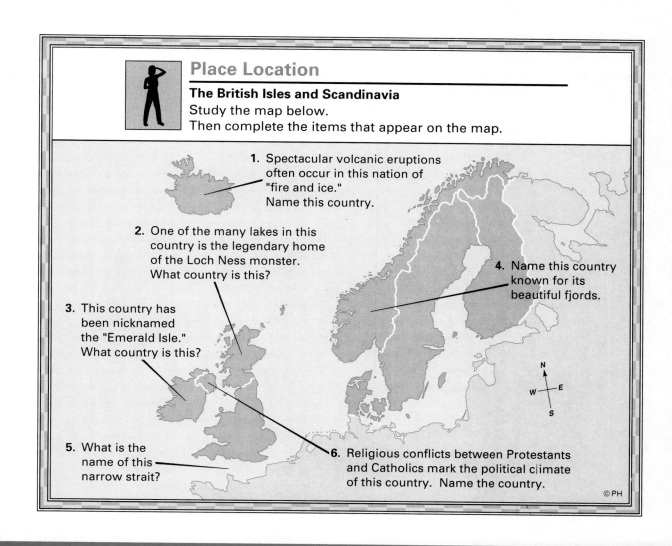

Place Location

The British Isles and Scandinavia
Study the map below.
Then complete the items that appear on the map.

1. Spectacular volcanic eruptions often occur in this nation of "fire and ice." Name this country.

2. One of the many lakes in this country is the legendary home of the Loch Ness monster. What country is this?

3. This country has been nicknamed the "Emerald Isle." What country is this?

4. Name this country known for its beautiful fjords.

5. What is the name of this narrow strait?

6. Religious conflicts between Protestants and Catholics mark the political climate of this country. Name the country.

©PH

16

Central Europe

Chapter Preview

Locate the countries covered in these sections by matching the colors on the right with the colors on the map below.

Sections	Did You Know?
1 FRANCE	The chateau of Ussé on the Indre River in France's Loire Valley was the inspiration for the story "Sleeping Beauty."
2 GERMANY	The birthday custom of having a cake topped by lighted candles began with the thirteenth century German *kinderfeste.*
3 THE BENELUX COUNTRIES	The Belgians, not the French, invented French fries and called them *patates frites.*
4 SWITZERLAND AND AUSTRIA	Every Swiss citizen devours an average of twenty-two pounds of chocolate each year. Americans eat only two to ten pounds of it.

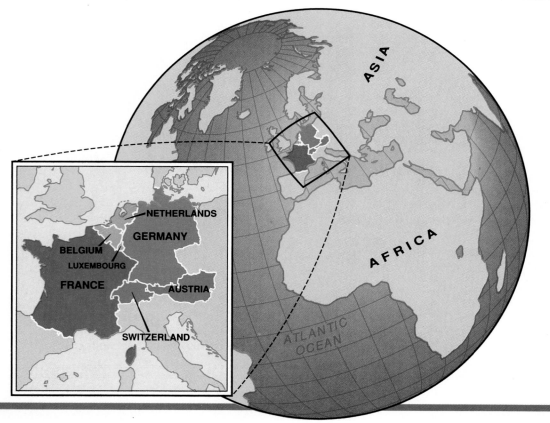

City of Light
Brightly lit boulevards line the Seine River that flows through Paris, France's City of Light. The Eiffel Tower, standing at left, has been a Parisian landmark since Alexandre Gustave Eiffel designed it for the 1889 World's Fair.

1 France

Section Preview

Key Ideas

- France has a great variety of physical and economic regions.
- History, language, and culture combined have created a distinct French identity.
- The French have a strong economy.

Key Term

dialect

The map of France on page 296 shows why some people might think of a hexagon when they look at the outline of France. If you smooth out the zigs and zags of France's borders, you will see that the country is roughly six-sided. Water borders three of the sides. Mountains form barriers on two other sides. Only in the northeast do low hills and plains provide easy passage from France into neighboring countries.

For nearly five hundred years, the boundaries of France have remained almost unchanged. Over the centuries, the French have established a strong national identity. As journalist Flora Lewis observed, "The French have no problems of identity. They know who they are and can't imagine wanting to be like anybody else."

A Country of Varied Regions

Even while maintaining a strong national identity, historic cultural and regional economic regions exist within France. The people of each of France's regions, from Normandy and Brittany in the north to Provence (pruh VAHNS) in the south, maintain their own traditions and way of life. From rich farming areas to huge, urban manufacturing and commercial centers, the different regions of France also contribute to the country's strong economy.

Applying Geographic Themes

Movement The six-sided country of France is bordered by three major bodies of water and by seven other countries. Which river forms part of the border between France and Germany?

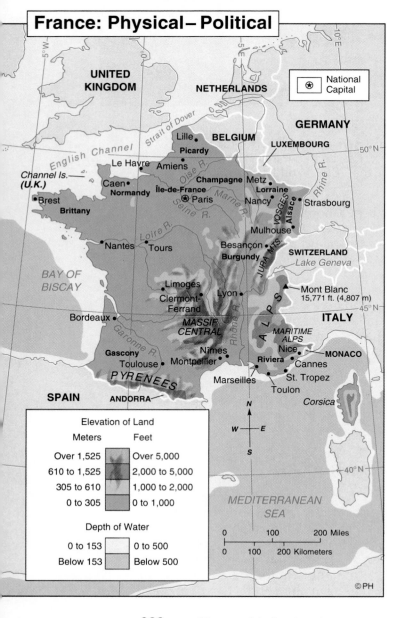

France: Physical– Political

UNITED KINGDOM
NETHERLANDS
National Capital
Strait of Dover
GERMANY
LUXEMBOURG
BELGIUM
Lille
Picardy
50°N
English Channel
Le Havre
Amiens
Oise R.
Metz
Channel Is. (U.K.)
Caen
Champagne
Lorraine
Normandy
Île-de-France
Nancy
Rhine R.
Brest
Marne R.
Paris
Strasbourg
Brittany
Seine R.
VOSGES MTS.
Alsace
Loire R.
Mulhouse
Nantes
Tours
Besançon
JURA MTS.
BAY OF BISCAY
Burgundy
SWITZERLAND
Lake Geneva
Limoges
Lyon
Mont Blanc 15,771 ft. (4,807 m)
Clermont-Ferrand
45°N
Bordeaux
Rhône R.
ITALY
MASSIF CENTRAL
A L P S
MARITIME ALPS
Garonne R.
Nice
Gascony
Nîmes
MONACO
Toulouse
Montpellier
Riviera
Cannes
PYRENEES
Marseilles
St. Tropez
Toulon
SPAIN
ANDORRA
Corsica

N
W — E
S
40°N

MEDITERRANEAN SEA

Elevation of Land

Meters	Feet
Over 1,525	Over 5,000
610 to 1,525	2,000 to 5,000
305 to 610	1,000 to 2,000
0 to 305	0 to 1,000

Depth of Water

0 to 153	0 to 500
Below 153	Below 500

0   100   200 Miles
0   100   200 Kilometers

©PH

Agriculture and Industry in Northern France In the interior of northern France lies the Paris Basin, a part of the North European Plain which stretches across northern Europe. The Paris Basin is a large, flat, circular area drained by the Seine (SEN) and other rivers.

In the center of the Paris Basin, on the banks of the Seine River, lies Paris, the economic, political, and cultural capital of France. Paris and its surrounding area form France's chief manufacturing center. Raw materials shipped here from other parts of France and from other countries are turned into finished products.

The city of Lille (LEEL), north of Paris, is another important industrial center. Coal is mined on the nearby Belgian border. In the late 1800s, the availability of coal as a source of fuel attracted many industries to this area. Since then, steel mills, textile factories, and chemical plants in Lille and the surrounding cities and towns have provided jobs for many people. However, like industrial cities in Great Britain, the area is now facing economic problems. Unemployment is high, and many people are moving closer to Paris and to other parts of France to make a living.

The Vineyards of Southwestern France Away from northern France toward the south of the country, the climate changes. The air becomes warmer, the soil drier. It is in these conditions that the grapes used in making French wines thrive. Grapes are grown for wine making in many parts of France. However, the region around the busy seaport of Bordeaux (bor DOH) in southwestern France has the reputation for producing the best wines. The town of Bordeaux has given its name to the whole wine crop of the region. The importance of the region's physical geography to wine production is explained by Baron Geoffrey de Luze, who owns vineyards near Bordeaux.

It's a combination of sun, . . . just the right amount of rainfall and no frost, and . . . the miserable soil. . . . It's true. You'll notice how stony and poor the soil is here . . . When the soil is rich, the production of grapes is large. So the individual grapes draw less concentration of the good things in the earth and from the sun. You'll find that the most refined wines come from the poorest soil. With fewer fruits and more sun, one arrives at unbelievably good grapes.

Farming and Tourism in South Central and Southeastern France East of Bordeaux lie two huge mountainous areas—the Massif Central (ma SEEF sahn TRAHL) and the Alps. Dividing these two rugged regions is the Rhone River. The Massif Central lies to the west of the Rhone. This ancient mountain system, which forms one sixth of France's land area, was ruptured by new volcanoes when the Alps were created thousands of years later to the east. As a result, the landscape is a mixture of older peaks worn flat by time and newer, sharper peaks that are not yet eroded. Because much of the soil in the Massif Central is too poor to grow crops, raising livestock is the most important economic activity.

East of the Rhone River are the Alps—a mighty barrier of mountains that provide some of the most spectacular scenery in Europe. Unlike the Massif Central, the Alps are a long range of towering, snowcapped mountains. Mont Blanc, the tallest peak in the Alps, rises 15,771 feet (4,807 meters) above sea level.

For centuries, the Alps made movement between France and Italy difficult. In 1787, Horace de Saussure, a naturalist and physicist, climbed to the top of Mont Blanc. He wrote, "Someday, a carriage road will be built under Mont Blanc, uniting the two

French Vineyards
Vineyards like this are a common sight in the south of France. Each region produces grapes that give every wine its own special flavor.

valleys. . . ." His vision took 178 years to become reality. In 1965, engineers dug a highway tunnel through Mont Blanc, which straddles the border between France and Italy.

The Alps are known worldwide for their fashionable and challenging ski resorts. During the summer, a magnificent array of alpine wildflowers on the mountain slopes delights the eyes of hikers and climbers.

Work and Play Along the Mediterranean Nestled between the Alps and the Mediterranean Sea in southeastern France is a thin strip of low-lying coastal land. This area, known as the Riviera, attracts millions of tourists each year. The warm climate is ideal

The Daily Catch
Commercial fishing workers bring in a huge yearly catch working off the French coasts. This boat is in port at Cannes on the Mediterranean Sea.

for sunbathing on the region's beautiful beaches and swimming in the sea. The Riviera is also known as the Côte D'Azur—the Azure Coast—for the splendid symphony of blue created by the sky, the sea, and the local flower, lavender. Many people also visit the lively resort cities of Cannes (KAHN), Nice (NEES), and St. Tropez (SAN troh PAY).

The Mediterranean coast is not all play, however. The port of Marseille (mar SAY) is the busiest seaport in France and the second most active in all of Western Europe. Here, ocean-going tankers filled with petroleum from the Middle East and North Africa wait to unload. The petroleum is processed at large oil refineries along the coast. Many French exports, including wine, electronic goods, and chemicals, are shipped from Marseille to other countries.

Mining and Industry in the East
North of the Mediterranean coast and over the peaks of the Alps and Jura

Mountains lies the Rhine Valley. Here the Rhine River, Europe's busiest waterway, forms part of France's border with Germany. Two resource-rich provinces in the Rhine Valley, Alsace (al SAS) and Lorraine, have changed hands during conflict between France and Germany many times. Lorraine has France's largest deposits of iron ore. Nearby, coal is mined. Strasbourg, France's major port on the Rhine, is located in Alsace.

Place: France's History, Language, and Culture

Referring to France's great diversity, former French President Charles de Gaulle once asked, "How can you govern a nation that makes 365 kinds of cheeses?" Despite having the kinds of cultural and economic differences that have often caused other countries to break apart, France is a highly unified country.

A Long History of Unity Modern France was known as Gaul when the Romans conquered it in the first century B.C. For nearly five hundred years, the area prospered under the Romans. The Gauls, the native people of the area, were strongly influenced by the Romans, adopting the Latin language and Christian religion.

As the Roman empire declined, the Franks, who came from the area that is today Germany, conquered the region. The Franks gave France its name. They also gave the country one of the most famous conquerors of all times— Charlemagne (SHAR luh mayn). He became king of the Franks in A.D. 768. By the time Charlemagne died in 814, he controlled a huge empire, known as the Holy Roman Empire, that included much of Western Europe.

Charlemagne's empire fell apart after his death. By the tenth century, most of the power lay in the hands of the nobles who controlled land in the kingdom. In 987, these nobles chose

Hugh Capet (HUE ka PAY), the ruler of Paris and the lands around it, as their new king.

Under Hugh Capet and his heirs, the monarchy grew strong. The lands ruled by the various nobles were united under one leader—the king. Gradually, the ruling monarchs of France expanded the kingdom's boundaries until, by 1589, they were almost the same as those of modern France. For the next two hundred years, French kings exercised absolute control over their lands. Then, in 1789, the monarchy came to a violent and bloody end during the French Revolution.

Ever since, France's political history has had many ups and downs. After the revolution, a republic run by representatives of the people was established. Since then, France has had several different forms of government. Napoleon Bonaparte and his nephew, Louis-Napoleon, both established empires in the nineteenth century. Three times since 1870, German armies have overrun northern France. The last two invasions, during World War I and World War II, were repelled with help from other countries, including the United States.

Despite this turmoil, the people of France have maintained a strong sense of national identity. One reason for this is their belief in the historical unity of France. Language and culture have also played important roles in creating the French identity.

One Country, One Language Before the 1500s, the language that is now called French was spoken only in and around Paris. In the reign of Hugh Capet, Paris became the political capital of France. As the French kings expanded their control over France, they decreed that the language of Paris become the language of all the lands they ruled.

Several other languages, for example, German, Basque, and Breton, are still spoken in parts of France, as are

The French Revolution

In 1789 the French monarchy came to a violent and bloody end during the French Revolution. This French etching shows the women of Paris marching toward Versailles, the residence of Louis XVI located just outside the city.

several dialects—variations of a language that are unique to a region or community. French, however, is the national language, and it is carefully controlled. New French words are published in official dictionaries only if they are approved by the French Academy. This body, which is symbolic of French cultural pride, was established in 1635 to preserve the purity of the French language.

A Strong Cultural Identity The French also take enormous pride in their intellectual and artistic achievements. Among their greatest heroes are philosophers like René Descartes (ruh NAY day CART), Voltaire, and Jean-Paul Sartre (ZHAHN PAUL SART

Artists at Work
Easels line the streets of Montmartre, an area of Paris that is well known as an artist's colony.

France Today

Following World War II, the French government established national planning programs to modernize the economy and encourage more balanced growth among France's regions. It also reached out to its Western European neighbors to form new trade agreements. Today, because of these changes, France ranks among the leading exporters of the world.

France's prosperity has helped to make it powerful. As a wealthy nation, France gives military and economic aid to many countries. It is a leader in the development of sophisticated defense systems and it is an influential voice in world politics.

But, like all countries, France has challenges within its borders. Not everyone enjoys a high standard of living. Rents in cities like Paris are skyrocketing. Many people are unable to afford the rising housing costs and are forced to live in run-down buildings. Elderly and foreign residents are particularly affected.

ruh). Many of the world's most famous painters have been French, including Claude Monet (mo NAY) and Pierre Auguste Renoir (PYER aw GOOST ruhn WAHR). And France continues to be a world leader in fashion design. Yves Saint Laurent (EEV SAN law RAHN) and other French designers create clothes that influence fashion all over the world.

For centuries the beautiful city of Paris has been the cultural center of France. The city's atmosphere of freedom has attracted artists and intellectuals from all over the world. Many important developments in the arts and literature can be traced to the studios of artists and writers living in Paris. Today the city's art galleries and museums, including the famous Louvre (LOOV ruh), celebrate the achievements of these artists. Theater, ballet, opera, and music performances also add to the city's artistic flair.

WORD ORIGIN

Paris
Ancient Romans named Paris *Lutetia Parisianorum,* "mud flats of the Parisi." Later, the reference to mud flats was dropped, but the name of the Parisi people remained.

■■■ SECTION **1** REVIEW ■■■

Developing Vocabulary
1. Define: dialect

Place Location
2. What countries border France to the east?

Reviewing Main Ideas
3. Where is France's manufacturing center?
4. How has language helped to unite France?
5. Why do the French have a strong sense of national identity?

Critical Thinking
6. **Drawing Inferences** How do you think France's diverse regions contributed to its economic prosperity?

2 Germany

Section Preview

Key Ideas

- Germany was reunited in October 1990.
- Germany is Europe's leading industrial country.
- Germany is one of the world's leading economic powers.

Key Terms

inflation, lignite

No Longer Divided

Germans participated in tearing down the Berlin Wall that divided east from west for twenty-eight years. Today, Germany is a united country.

On Thursday night, November 9, 1989, thousands of East and West Berliners gathered along the Berlin Wall. Just hours earlier, the East German government had announced that the borders between East and West Germany would be opened. As one reporter wrote:

> *They seemed to be drawn by the sense that . . . the barrier of concrete and steel that had figured so prominently in the history of this city and the world, might soon be relegated to history. Some came with hammers and chisels, others with guitars, most with cameras.*

People all over the world were moved to tears of joy as the wall that separated east from west was finally torn down.

Germany's Struggle for Unity

For most of the years since the end of World War II, East and West Germany watched each other cautiously across the Berlin Wall. The 99-mile-long wall was built in 1961 by the communist East German government to keep its citizens from escaping to West Germany. The wall created a physical boundary between two differing political regions.

Since it was built, the Berlin Wall came to symbolize the division of Germany that took place after World War II. Less than a year after the wall finally opened, Germany was reunited. However, Germany's history as a nation is comparatively recent.

Divided German States The area that is now Germany was once part of Charlemagne's great empire. After Charlemagne's death, Germany broke up into many small, independent political units. Princes, dukes, counts, and bishops all ruled their own domains. Many cities were free states. Often there was bitter rivalry and fighting among all these states.

During the 1500s, a movement called the Protestant Reformation divided the German states even further. The Reformation was led by Martin

Money to Burn

This woman is using German currency to light her stove. During the economic collapse in the 1920s, German currency had little worth.

Luther, a German monk. Luther objected to many of the practices and teachings of the Roman Catholic church. In the early 1600s, the Reformation sparked thirty years of warfare between Protestants and Catholics throughout Germany and other parts of Central Europe.

Starting in the late 1700s, the state of Prussia in what is now eastern Germany gained power in the region. Under Prussia's influence, many German states began to merge with one another to form a single confederation. Germany's power increased and in 1871, Germany defeated France in the Franco-Prussian War. The leader of Prussia was crowned head of the new German Empire. German states that until then had remained independent agreed to join a united German nation.

United Germany's Defeats In 1882 Germany joined with Austria-Hungary and Italy to form a military alliance known as the Triple Alliance.

Between 1914 and 1918, Germany, Austria-Hungary, and other countries fought against France, Russia, the United Kingdom, the United States, and other allies in World War I.

When the war ended, Germany was defeated. According to the terms of the treaty following the war, Germany had to pay the victors huge sums of money for war damages. As a result, Germany suffered economically. The economy collapsed in the early 1920s when **inflation,** or sharply rising prices, ruined the value of Germany's currency. In 1929, a worldwide economic depression left millions of Germans without jobs.

In the early 1930s, Adolf Hitler and his Nazi Party came to power in Germany. Hitler promised to restore Germany's past glory and to improve the economy. He blamed the Jews and other people whom he considered to be racially inferior for all of Germany's problems.

In 1939 Germany invaded Poland and World War II began. During the war Hitler had millions of Jews, Poles, Gypsies, Slavs, and other people killed in concentration camps. Finally, in 1945 Germany was defeated by the allied countries—the United States, the United Kingdom, France, and the Soviet Union.

One People, Two Countries Following the war, tensions grew between the Western Allies and the Soviet Union concerning Germany's future. Western leaders responded by establishing the democratic country of the Federal Republic of Germany—West Germany—in 1949. Soon the Soviet Union set up the Communist German Democratic Republic—East Germany. The former capital of Germany, Berlin, was also divided. Although Berlin was located within East Germany, American, British, and French forces remained in the western half of the city, which became part of the Federal Republic.

The countries remained separate for nearly forty years. Then, in late 1989, a wave of democratic demonstrations swept through Eastern Europe and overturned East Germany's Communist government. Soon the new East German government announced that it would open the country's borders. Celebrations in East and West Berlin were especially joyous. Within weeks large sections of the Berlin Wall, the symbol of divided Germany, were destroyed. On October 3, 1990, East and West Germany were officially reunited.

A Mosaic of Regions

The physical regions of Germany are varied, but the differences between regions are not as dramatic as they are in France. As journalist Flora Lewis observed:

It is a [mild] land, brisk but bright along the North Sea coast, heaving gently above green valleys to the majestic Bavarian Alps. The mighty Rhine, one of Europe's oldest, most traveled highways, is still a great commercial lifeline.

Germany's land can be divided into three bands that extend across the country. The high, rugged land in the south turns into hills, low mountains, and tall plateaus in central Germany before leveling off into the flat lands of the north.

Germany has a generally mild climate, largely because of the influence of the North Sea. Away from the sea in southern areas of the country, a humid continental climate prevails, causing colder winters and warmer summers. But even in January, temperatures are usually above freezing. Cold winds from the east, however, may bring a sharp drop in temperature for short periods. In July, temperatures throughout the country average a comfortable 70°F (21°C).

Plains, Rivers, and Cities in Northern Germany Northern Germany is covered by the North German Plain, which is a part of the North European Plain. For hundreds of miles flat, sandy plains spread out over northern Germany until they reach the North and Baltic seas. Wide rivers flow north out of the southern highlands across the plains to the sea.

Applying Geographic Themes

Movement A string of urban areas lies along the Rhine River and its tributaries in western Germany. Name three of these cities.

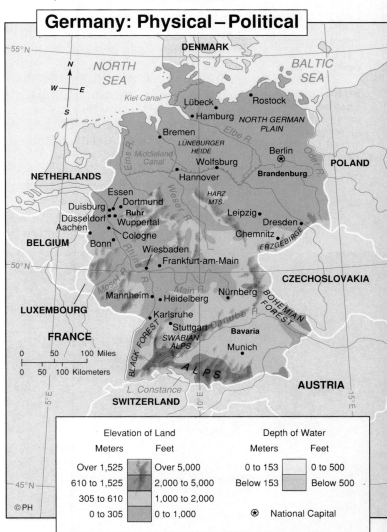

Although much of the land in the plains is farmed, manufacturing and trade are also important economic activities. Hamburg, Germany's largest port and second-largest city, is built around a harbor where the Elbe River flows into the North Sea. Since the end of the Middle Ages, Hamburg has been a leading center of trade. Today, the city's historic core gives it a special character. However, its "old" buildings are actually quite new. Most of the old structures were destroyed in massive bombing raids during World War II and were rebuilt after the war.

Another German port, Rostock, is also a tribute to German achievement after World War II. When East Germany cut its connections with West Germany, it lost access to West German ports on the North Sea. Because the East Germans needed an outlet to the sea for shipping, they dug a new harbor at Rostock, creating Germany's only major port on the Baltic Sea.

Berlin, Germany's capital and its largest city, is an inland city. Although it was badly damaged during World War II, both East and West Germany spent a great deal of money to rebuild the parts of Berlin that they controlled. Today, Berlin is once again the prosperous capital of a united Germany.

Rich Resources and Industry in Central Germany The rugged central part of Germany has some of the most beautiful scenery in the country. Two major rivers, the Rhine and the Oder, flow through this region. In some places, the Rhine cuts steep gorges through the mountains. In others, the rivers run beside villages that were built centuries ago.

In addition to its idyllic setting, this region of Germany is one of the most important industrial centers in the world. In the 1800s, huge coal deposits were found near the Ruhr (ROOR) River. With plenty of available fuel, the Ruhr Valley became Germany's first industrial center.

Today, the Ruhr Valley produces most of Germany's iron and steel. It also has important chemical and textile industries. Over eight million people live in the large cities of Duisburg (DOOS boorg), Essen, Bochum (BO khuhm), and Dortmund and the smaller cities and towns in the area that form one huge metropolis. Germans now refer to the entire region as *Ruhrstadt* (Ruhr City).

In the eastern part of central Germany is another large industrial region. Steel, machinery, automobiles, and textiles are produced in the factories of cities such as Leipzig and Dresden and in the surrounding area. Power for the factories comes mostly from lignite, a soft, brown coal. The lignite is easy to mine. However, it pollutes the air heavily.

Not everyone who lives in central Germany lives in a big, industrial city. Many people live in cities such as Frankfurt, Germany's banking center, and Heidelberg (HY duhl berg), the site of a world-famous university. Others live on farms located in the

WORD ORIGIN

lignite
The coal called *lignite* still has the brown look of decayed wood. Hence, its name comes from the Latin *lignum*, "wood."

Hamburg's Harbor on the North Sea
Although destroyed by bombs during World War II, Hamburg is a center of modern architecture.

southern part of central Germany. Here the soil is more fertile than anywhere else in the country.

Scenic Southern Germany Along Germany's southern border lie the Bavarian Alps. These mountains have many high peaks that are capped with snow all year round. In the winter, the resorts in these mountains are filled with skiers. North of the Alps, the land is less mountainous. The Rhine and Danube rivers flow through this hilly land. Hikers enjoy spectacular scenery as these rivers wind their way through the rugged hills and thick, dark forests of firs and spruce.

The largest city in southern Germany is Munich (MUE nikh). Since World War II, Munich has developed as Germany's cultural center. Theaters and museums that were destroyed during the war have been renovated. Even damaged paintings and sculptures have been restored and are once again exhibited.

Germany in the World Today

After World War II, helped by massive American aid, the Germans worked hard to rebuild their shattered economies. In 1991 Germany was the leading industrial country in Western Europe and it ranked fourth in the world, after the United States, Japan, and the Soviet Union, in the amount of goods it produced. It is one of the strongest economic powers in the world.

German reunification presents the challenge of merging two separate countries. It also promises to increase prosperity. Germany can maintain the strong economic ties it has with Northern Eurasia and other Eastern European countries, and it is a leading member of the European Community. Germany will therefore have increasing markets in which to sell its valuable products.

Automobile Assembly Line
Much of Germany's steel is made into quality cars, making it one of the world's leading car manufacturers, after the United States and Japan.

SECTION **2** REVIEW

Developing Vocabulary
1. Define **a.** inflation **b.** lignite

Place Location
2. Name the countries that border Germany on the east.

Reviewing Main Ideas
3. Why was Germany divided after 1945?
4. Why is the Ruhr Valley important to Germany's economy?

Critical Thinking
5. **Drawing Conclusions** Why do you think some European countries, such as France, might be uneasy now that Germany is united?

✔ Skills Check

☐ Social Studies
☐ Map and Globe
☐ Reading and Writing
☑ Critical Thinking

Distinguishing Fact from Opinion

To determine the soundness of an author's ideas, you need to be able to distinguish between fact and opinion. This ability allows you to reach your own conclusions about issues and events. Use the following steps to practice this skill.

1. **Determine which statements are based on facts.** A fact can be proven by checking other sources. It does not include someone's own values or opinions. Read statements A through G below. Answer the following questions: (a) Which statements are based solely on facts? (b) List two sources you could use to check that these statements are true.

2. **Determine which of the statements are opinions.** An opinion states a person's belief or feeling about a subject. It usually cannot be proven even if it is a widely held opinion. Study statements A through G again. Answer the following questions: (a) Which of the statements obviously include someone's opinion of Otto von Bismarck? (b) Which words in each of the opinion statements indicate that the statement is an opinion?

Statements

A. Otto von Bismarck was born in 1815.

B. Bismarck was truly an extraordinary human being.

C. As leader of Prussia, he won a series of wars, including the Danish War of 1864 and the Austro-Prussian War of 1866.

D. He was a very capable general.

E. Few people have ever exercised power more ruthlessly than Bismarck.

F. Bismarck once said, "Not by speeches and resolutions of majorities are the great questions of time decided upon—but by blood and iron."

G. Bismarck, tall and handsome, was a leader in the move for German unification in the nineteenth century.

3 The Benelux Countries

Section Preview

Key Ideas

- The Dutch have reclaimed over half of the Netherlands from the sea.
- Two distinct ethnic groups make up Belgium's population.
- Luxembourg is a small country with a high standard of living.

Key Term

polder

Crowded together in northwestern Europe are three small countries—Belgium, the Netherlands, and Luxembourg. From the first letters of their names, together these countries are known as the Benelux countries. The Benelux countries are also called the Low Countries because so much of their land is low and flat. Their combined land area makes them just a little larger than the state of West Virginia. But their combined population of 25.3 million people is almost as large as Canada's. The combination of a small land area and a large number of people makes these countries the most densely populated in Europe.

Interaction: Claiming Land from the Sea

"God made the world, but the Dutch made Holland [the Netherlands]," commented French philosopher René Descartes. In few places is the result of human interaction with the environment more evident than in the Netherlands. The map on page 310 shows that the entire western side of the country is bordered by the North Sea. In many places, inlets of the sea reach deep into the Netherlands. In fact, much of the low, flat land you see on the map was once covered by water. This is because the Dutch have created almost half of their country's land by reclaiming it from the sea, or from lakes and swamps. A Netherlander stated the national goal of his country in one sentence, "It is to possess land where water wants to be."

Technology Helps the Dutch to Create More Land Over two thousand years ago, people living in the area that is now the Netherlands began to create land on which to live and farm. They built low mounds and surrounded them with stone walls to make dry islands from the watery marshlands near the coast. When the Romans conquered the area, they constructed sophisticated dikes, or dams, to hold back the water.

Gradually, the Dutch became even more skillful at creating new land. They circled a piece of land with dikes and then pumped the water out into

Fields of Flowers

Tulips of all colors carpet the countryside in the Netherlands every April and May.

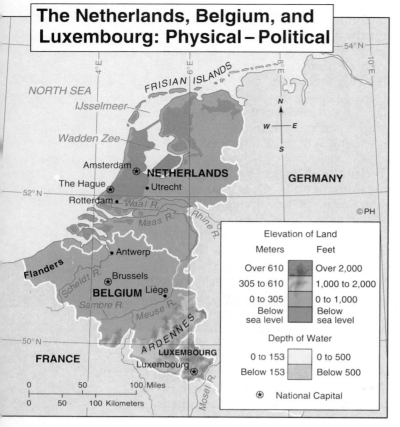

The Netherlands, Belgium, and Luxembourg: Physical–Political

NORTH SEA
FRISIAN ISLANDS
IJsselmeer
Wadden Zee
Amsterdam
NETHERLANDS
The Hague
Utrecht
Rotterdam
Waal R.
Maas R.
Rhine R.
GERMANY
©PH
Antwerp
Flanders
Scheldt R.
Brussels
BELGIUM Liège
Sambre R.
Meuse R.
ARDENNES
LUXEMBOURG
Luxembourg
FRANCE
Mosel R.
54°N
52°N
50°N
4°E
6°E
8°E
10°E

N
W E
S

0   50   100 Miles
0   50   100 Kilometers

Elevation of Land

Meters		Feet
Over 610		Over 2,000
305 to 610		1,000 to 2,000
0 to 305		0 to 1,000
Below sea level		Below sea level

Depth of Water

0 to 153		0 to 500
Below 153		Below 500

⊛ National Capital

Applying Geographic Themes

1. Movement Three of the world's busiest ports lie within a distance of about 80 miles (130 km): Rotterdam, Amsterdam, and Antwerp. On which river is Antwerp located?

2. Regions Where is the wooded plateau region of Ardennes located?

Amsterdam
Amsterdam, the capital of the Netherlands, takes its name from a thirteenth century dam on the Amstel River.

canals. The Dutch call land reclaimed from the sea in this way a **polder**. Beginning in the 1200s, the Dutch used windmills to power the pumps that removed water from the land. A great deal of this new land is used for farming, but cities have also been built on some of the land.

Much of the land in the Netherlands has an unusual appearance because of the unique way in which it was created. Almost one third of the country is below sea level. Standing in a polder field, one often looks up to see ships passing by in canals that run alongside the land.

Making Good Use of Valuable Land

Despite the great success the Dutch have had in land reclamation, land is still scarce in the Netherlands. The table beginning on page 751 shows that the Netherlands has an extremely high population density. With so many people living in such a small area, the Dutch have learned to make the best possible use of their land.

Dutch farmers cannot afford to waste any of the Netherlands' farmland. Over half of the land is used for dairy farming, but dairy farmers also grow crops. Throughout the Netherlands, farmers fertilize the soil heavily and use modern agricultural methods.

Government leaders are devoting special attention to preserving the country's farmlands. The cities of The Hague (HAYG), Rotterdam, Amsterdam, and Utrecht (YOO trekht) lie near one another. They form one huge arc-shaped metropolis that the Dutch call Randstad, or ring city. This part of the Netherlands is the most densely populated. The government is trying to prevent this area from expanding into nearby rural areas.

Advantages of a Seaside Location

The Dutch have also learned to make good use of their location on the North Sea. Rotterdam and Amsterdam are both important ports for foreign trade. Rotterdam is the busiest port in terms of the quantity of cargo that it handles. Because it is situated near the mouth of Europe's largest inland waterway, it serves as a link between much of Europe and the world.

Belgium: Two Peoples, One Place

The people who inhabit the modern country of Belgium are an uneasy mix. About 30 percent of all Belgians speak French and call themselves Walloons. About 55 percent speak Flemish, a language close to Dutch.

After Belgium gained independence from the Netherlands in 1830, relations between the Walloons and the Flemings grew tense. French, the language of the Walloons, was the country's only official language. Most government leaders spoke French, and all Belgian universities used French. As a result, the Flemings, who spoke Dutch, were prevented from fully participating in Belgian life. They could not hold government positions or enter professions in which a university education was needed.

Yet the Flemings made up a large part of the population. They wanted the same cultural and economic rights that the Walloons enjoyed. To resolve the conflict, the Belgian government made Flemish an official language in the late 1800s. The conflict between the two groups continues, however. Both have strong political parties.

Luxembourg

Luxembourg covers only 998 square miles (2,586 sq km), an area smaller than the state of Rhode Island. Despite its small size, Luxembourg has managed to endure for more than one thousand years. The tiny country lies where Germany, France, and Belgium meet. Although Luxembourg has close cultural ties to all of these countries, it has maintained an independent spirit. Like Belgium, the Luxembourgers have two languages. French is the official one, but many people also speak a German dialect.

Luxembourg has one of the highest standards of living in Europe. One of the most important economic activities is the manufacture of steel from the iron ore mined in the northern hills. Luxembourg is a member of the European Community, and it trades most of its steel with other members. Banking and other financial services are also important economic activities in Luxembourg.

One of Belgium's Major Crops
This flax will be used to manufacture rope, thread, linseed oil, and linen fabric. Belgian linens and lace are considered the world's finest.

SECTION 3 REVIEW

Developing Vocabulary
1. Define: polder

Place Location
2. Locate and then name the capital of Belgium.

Reviewing Main Ideas
3. Describe how the Dutch reclaim land from the sea.
4. Which two ethnic groups make up most of Belgium's population?
5. What is Luxembourg's most important economic activity?

Critical Thinking
6. **Perceiving Cause-Effect Relationships** How did technological advances help the Netherlands to develop as a country?

The European Community

The television commercial opens with a scrawny French boxer standing in the ring facing two huge opponents— an American football player and a Japanese sumo wrestler. Before viewers have time to gasp at the thought of this poor little fighter being beaten by the two giants, eleven of his friends come to his rescue. Together the twelve turn back the two Goliaths.

This comic advertisement was created to promote the European Community, or European Common Market as it is sometimes called. The European Community is a group of twelve Western European nations that have joined together to form a single market in order to compete economically with the United States, Japan and other countries.

A History of Success

In 1950, France and West Germany took the first step toward economic unity when they agreed to share their coal and steel resources. Soon Italy, Belgium, Luxembourg, and the Netherlands joined them. Together, these six nations established the European Coal and Steel Community (ECSC).

The ECSC was so successful that in 1957 member countries established two more organizations: the European Atomic Energy Commission (Euratom) and the European Economic Community (EEC). Euratom enabled countries to produce nuclear energy for peaceful uses. The EEC was founded to create a common market for all of the member countries. Together the ECSC, Euratom,

The European Community

Share of Exports in World Trade*

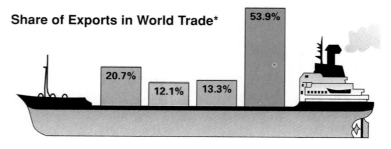

20.7% 12.1% 13.3% 53.9%

Share of Imports in World Trade*

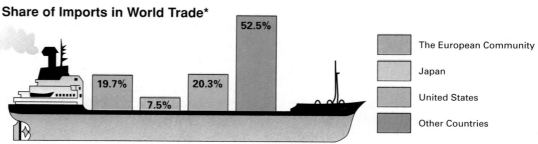

19.7% 7.5% 20.3% 52.5%

*Excluding trade within the European Community

Graph Skill *The combined imports and exports of the European Community form a large percentage of the world's trade. How do European Community exports compare to those of the United States?*

- The European Community
- Japan
- United States
- Other Countries

and the EEC form the European Community. Working together, the countries of the European Community have made great economic progress.

Movement: The Removal of Trade Barriers

With over 350 million citizens, the European Community is a leading economic power. It produces more goods than the United States and is the largest exporter in the world.

In its short history, the European Community has achieved significant accomplishments. Many of the barriers to the free movement of goods and services among member countries have been eliminated. Tariffs on goods traded among members have been abolished.

A United States of Europe?

As the European Community approaches the twenty-first century, plans for even greater European unity are under way. The community has agreed that borders between member countries will be practically eliminated by 1992. People, goods, services, and money will be able to move freely between member countries in much the same way as they do between states in the United States. Other changes have occurred, too. For example, Europeans now elect delegates to a European Parliament, and the parliament selects ministers to administer new laws. Europeans also have established a European monetary unit to protect the currencies of various countries from economic ups and downs. As the French advertisement clearly points out, a united Europe is a strong Europe.

TAKING ANOTHER LOOK

1. What is the European Community?
2. Why is the European Community working to increase unity among its members?

Critical Thinking

3. **Predicting Consequences** What might be the economic and political effects of a united Europe on the United States?

An Economic Union The member countries of the European Community hope to gain a stronger position in world trade by joining resources. What historical and cultural factors must they overcome before they can unify?

4 Switzerland and Austria

Section Preview

Key Ideas

- Switzerland is a federation of people with different cultural traditions.
- The Swiss have a high standard of living.
- Austria has a short history as an independent country.

Key Term

canton

The Alps tower high above sea level in the two small countries of Switzerland and Austria; they cover more than half of each country's land area. Both countries are neutral. Neither is a member of the European Community or NATO. Despite these similarities, Switzerland and Austria are strikingly different.

Switzerland: A Place of Diversity

The Swiss have many different names for their country. Switzerland is the name used by people who speak English, but it is not used by most residents of Switzerland. Schweiz

Applying Geographic Themes

1. Location Describe Vienna's location in three different ways. Where is Austria located in relation to Czechoslovakia?

2. Regions Which region of Austria is the most mountainous?

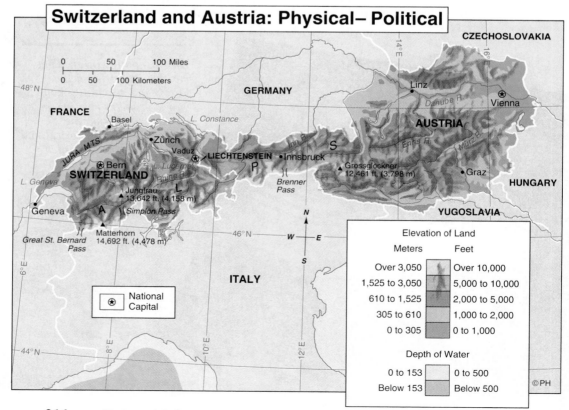

Switzerland and Austria: Physical– Political

DAILY LIFE

A Slower Pace

There is a side of European daily life that is commonly seen in the region's secluded gardens, sidewalk cafés, and especially in its softly lit coffeehouses where people take time to enjoy life's simple pleasures. In Austria and most other countries of Europe, it is common for a person to linger unhurried in the relaxed atmosphere of a coffeehouse sipping a cup of coffee, talking to friends, writing a letter, or reading a newspaper. In Austria, famous for its coffeehouses like the one shown at right, the tradition is an old one. In 1683 Turkish invaders who had been camped outside Vienna in silken tents were finally driven away. As legend has it, they left behind piles of coffee beans. Austria's first coffeehouse opened shortly thereafter.

1. Describe the tradition of European coffeehouses.

2. What does the tradition of coffeehouses imply about the quality of life in Europe?

(SHVYTS) is the term used by the Swiss who speak German—about 65 percent of the population. Suisse (SWEES) is the name used by the 20 percent of the Swiss who speak French. And those who speak Italian refer to their country as Svizzera (SVEE tsay rah). In addition, a very small number of people speak a dialect called Romansch.

So what is the official name of Switzerland? Although these names— Schweiz, Suisse, and Svizzera—are the popular names given to the country, its official name is *Confederatione Helvetica*. The name is in Latin, but we know it as the Swiss Confederation.

For over seven hundred years, the Swiss have bound people from different cultural traditions into a proud, prosperous, independent country. Yet these various cultural groups have maintained their distinctive identities. They have also maintained much of their political autonomy, or independence.

Uniting Together for Defense

Switzerland was formed in 1291 when leaders of three small cantons, or states, banded together to defend their freedom against an Austrian emperor. Together these united cantons formed the Swiss Confederation. They fought several wars of independence against the Austrians. Attracted by the growing strength of the confederation, other cantons began to join. By 1513, thirteen cantons belonged to the confederation.

In 1515, after a defeat in Italy, Switzerland decided never again to fight in a foreign war. It hoped to avoid any more losses in the bloody battles that were going on around it. However, in 1798 Napoleon's armies occupied Switzerland. When Napoleon's forces were finally defeated, the countries of Europe formally recognized Switzerland as a neutral country. They guaranteed that no other European country would invade its borders. Since that time, Switzerland has not taken sides in conflicts between other countries.

The Cantons Maintain Their Strong Identities Today, twenty-six cantons make up Switzerland. These cantons differ from one another greatly in language, religion, customs, and the ways in which people make a living. The people of each canton are proud of their particular way of life, and they work hard to preserve it.

The cantons have a great deal of control over their own affairs. This tradition dates back to the early history of the Swiss Confederation. At that time each canton governed itself as a separate country. Even today, any law passed by the national government must be ratified by popular vote if enough Swiss citizens so request. In this way, the people can accept or veto the law.

A Prosperous Economy The independent spirit of the inhabitants of the cantons exists alongside their strong feelings of national unity. This, together with Switzerland's neutrality, has helped the country to thrive. The Swiss enjoy one of the highest standards of living in the world. Although Switzerland has few natural resources, it has developed specialized economic activities that are highly profitable.

Dairy farming is the most important form of agriculture, because there is little flat land on which to grow crops. Cattle are driven to high mountain pastures in the summer. In the winter they are brought down to valleys where they are more protected from the weather. Since milk does not stay fresh for long, most of it is turned into processed products like milk chocolate and cheeses that are shipped around the world.

Alpine Village
The Alps almost overpower this rural village as they do almost any vista in Switzerland.

Switzerland is famous for certain types of manufacturing. The country has none of the mineral resources, such as iron ore, coal, or petroleum, needed for heavy industry. These materials are expensive to import. So the Swiss specialize in making products that do not require many materials or costly transportation, but instead rely on skilled labor. Since the late 1600s, Swiss jewelers have produced watches known the world over for their accuracy. Switzerland produces very high-quality tools, including microscopes and measuring and cutting tools. Today, they are also world leaders in the development of new medicines.

Banking is an important service industry in Switzerland. Switzerland is seen as a safe place to keep money because of the country's neutrality. People from many countries deposit their money in banks in Zurich, Geneva, and other Swiss cities.

Tourism is also very important to Switzerland's economy. Switzerland's landscapes are among the most spectacular in the world. Many come to ski at resorts such as Zermatt and St. Moritz in the snowy Alps. Others come to hike, climb mountains, or simply to enjoy Switzerland's scenery.

Austria: One of Europe's Newest Countries

In contrast to Switzerland, Austria is a new country. Until the end of World War I it was part of a larger empire, the Austro-Hungarian empire. In the late 1800s, this empire controlled parts of Italy, Romania, Czechoslovakia, and Yugoslavia. However, in World War I the Austro-Hungarian empire collapsed. Austria and Hungary were separated into independent countries as a result. Much of the land they controlled was taken to form new Eastern European countries.

Since modern Austria is much smaller than the empire that preceded it, one of its biggest challenges has

been to rebuild itself as a country within its new boundaries. In response to this challenge, business and political leaders have redirected the country's economy.

Austria has used Switzerland as a model for its economic renewal. Like Switzerland, Austria has created specialized industries. Much of its economic activity centers on the manufacturing of machine tools, chemicals, and textiles. Dairy farming is the most important agricultural activity. However, unlike Switzerland, Austria has some mineral resources. Deposits of iron ore are mined in the eastern Alps and then processed into iron and steel products.

Vienna, the country's capital, has also had to adapt to its changing role in history. Once the political and cultural center of the Austro-Hungarian empire, Vienna had two million residents in 1910. It was one of the world's six largest cities. Today its population is only one and one-half million. But other Austrian cities are growing. Modern industries that find Vienna too congested prefer to locate in smaller cities like Graz, Linz, or Innsbruck.

SECTION 4 REVIEW

Developing Vocabulary
1. Define: canton

Place Location
2. Which countries border Switzerland?

Reviewing Main Ideas
3. Why does Switzerland have three official languages?
4. What are Austria's most important economic activities?

Critical Thinking
5. **Identifying Central Issues** What factors have helped Switzerland to become a prosperous country?

16

REVIEW

Chapter Summaries

SECTION 1 France France is a country of many varied physical regions. The way people use the land to make a living differs greatly from place to place—from rich farming areas to huge urban centers. Despite this, a shared history, language, and culture has given the French a deep sense of national unity. Today, France is a prosperous country and a leading world power.

SECTION 2 Germany The physical regions of Germany can be divided into three wide bands—rugged highlands, gentle hills, and flat lowlands. As a result of World War II, Germany was divided into two separate countries— East Germany and West Germany. In 1990, Germany was reunited. The country is heavily industrialized. It is one of the world's leading economic powers.

SECTION 3 The Benelux Countries Belgium, the Netherlands, and Luxembourg are called the Benelux countries. All three countries are densely populated. The Dutch have actually increased the size of their country by reclaiming land from the sea. In Belgium, relations between the country's two main ethnic groups— the Walloons and the Flemings—are tense. The people of Luxembourg, the smallest Benelux country, lead a peaceful and prosperous life.

SECTION 4 Switzerland and Austria Switzerland is a federation of peoples from different cultural backgrounds that dates back to 1291. Today, four languages are spoken in the country. As a neutral country, Switzerland does not take sides in conflicts between other countries. Austria was part of the huge Austro-Hungarian Empire until 1918. One of Europe's newest countries, Austria, has used prosperous Switzerland as a model for some aspects of its economic development.

Vocabulary Development

Match the following definitions with the terms below.

1. a political division within Switzerland
2. a soft, brown coal
3. sharply rising prices
4. low land drained of water and surrounded by dikes
5. a regional variation of a language

   a. inflation
   b. dialect
   c. polder
   d. canton
   e. lignite

Reviewing Main Ideas

1. How have history, language, and culture combined to give the French a strong sense of identity?
2. Describe two cultural contributions that France has made to the world.
3. Describe Germany's economic activities.
4. Why were Hamburg and many other cities rebuilt to look old after World War II?
5. Why is land scarce in the Benelux countries?
6. What are the two major ethnic groups in Belgium.
7. Why is Switzerland politically unique among the world's countries?
8. Explain why four languages are spoken in Switzerland.
9. What positive impact do the Alps have on France?

Critical Thinking

1. **Identifying Central Issues** What impact did the defeat in World War II have on Germany?
2. **Drawing Inferences** How do you think the nature of Dutch land influences the way the land is used?
3. **Perceiving Cause-Effect Relationships** How has Switzerland's neutrality helped it to become prosperous?
4. **Making Comparisons** How do France and Switzerland differ in the role that language plays in their identities?

Practicing Skills

Distinguishing Fact from Opinion
Generally, an opinion is more convincing when its author gives facts to support it. Answer the question: What opinion is expressed in the paragraph below?

Paragraph
Bismarck couldn't wait to go to war with France. He knew that war would help unification. It also would be popular with the German people, which would encourage other states to join the North German Confederation.

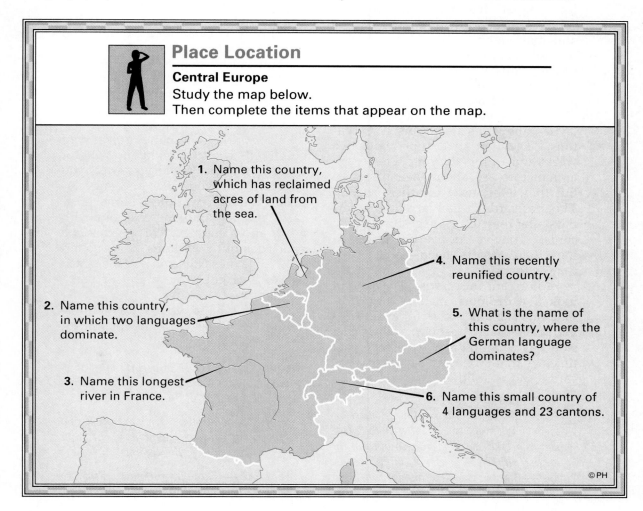

Place Location

Central Europe
Study the map below.
Then complete the items that appear on the map.

1. Name this country, which has reclaimed acres of land from the sea.
2. Name this country, in which two languages dominate.
3. Name this longest river in France.
4. Name this recently reunified country.
5. What is the name of this country, where the German language dominates?
6. Name this small country of 4 languages and 23 cantons.

© PH

Place Location: WHERE ON EARTH?

Postcards from Europe

"They're coming from the Wiltons," says Phil Jaynes. He hands you some postcards. "Four, so far."

You are a young reporter for the local paper. Just this morning Phil called your paper with a news tip. He and his neighbors have been getting mysterious messages in the mail. Your editor wants you to check out Phil's story, adding that it might be a good item for the feature section.

So here you are, talking to Phil Jaynes about his neighbors, the Wiltons. "They left for Europe about two weeks ago," Phil tells you. "I never got a chance to ask about their itinerary, but they'll be away for three weeks. The postcards began arriving a few days after they left." He gestures at the cards. "They've got everybody guessing where they've been—and where they're going."

You scan the cards. They don't look like ordinary picture postcards. One side is blank except for an address and a smudged postmark. The other side has a handwritten message, but no clues that give away a location. The cards read as follows:

Sept. 12—*Hi, everybody! Guess where we are? That's right, we want you to guess. Here's a clue: we've spent two days in the largest city of this island country. Must dash—Big Ben says it's time for lunch!*

Sept. 14—*What country are we in now? A really beautiful one. We sailed through a fjord and hiked on a glacier yesterday. Now we're in the capital, once named after King Christian IV.*

Sept. 18—*Traveled south and arrived today in Hans Christian Andersen's home town. Weather here a bit rainy, but we're having too much fun to care!*

"Pretty clever idea, don't you think?" asks Jaynes.

You agree, pointing to the postmarks on each card. "They've even smudged the postmarks, so there's no help there. I'll bet there's a mail clerk in each country who's in on this."

Back at the newspaper office, your editor tells you to sit tight and wait for more postcards. While you wait, you get busy with your atlas and encyclopedia. You read over the Wiltons' postcards and begin linking each one with a city and a country. This gives you a good idea of the Wiltons' route so far.

Phil Jaynes calls you early the next week. "I've got some more for you," he reports.

Pencil in hand, you jot down information as Phil reads the postcards. They are as follows:

Sept. 22—*Hot dog! Transportation hub of this city—on Main River—is huge. City's airport is largest in Europe!*

Sept. 23—*Now have firsthand proof that Jungfrau is loveliest mountain in the Alps. Right now we're in the capital.*

Sept. 24—*Through Brenner Pass to Europe's leading cultural center. Breathtaking city! Also home of Johann Strauss and the waltz.*

320

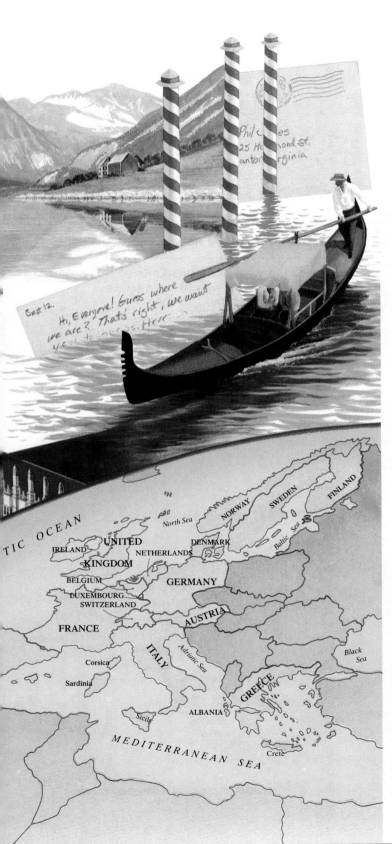

As you hang up, you lean back to study the messages. The encyclopedia can help you locate the land of Strauss, the waltz, and Europe's cultural center. Then you can check the atlas for the Alps, the Brenner Pass, and the Main River.

When Jaynes calls the next week, you're ready for more clues. He reads you three final cards.

Sept. 26—Gondolas are the way to go! Most of the people living here use motorboats, though. Right now we're taking a break in St. Mark's Square.

Oct. 1—We're starting the last leg of our trip. Good thing! We're on our last legs! We sailed between Corsica and Sardinia to this Mediterranean port in Europe's third largest country. We're about fifty miles south of the French border, but we're not in France!

Oct. 3—Spending our last day in westernmost country of continental Europe. We fly home from its capital tonight. See you soon!

As soon as Jaynes hangs up, you head for the dictionary to check *gondola.* You also reach for your atlas and encyclopedia. In just a short while, you will have solved the mystery of the Wiltons' postcards and will start writing your feature article!

SOLVE THE MYSTERY

From what city and country did the Wiltons write each postcard? *Hint:* Use an atlas and a set of encyclopedias to help you find the answers. A dictionary might help, too.

321

17

Mediterranean Europe

Chapter Preview

Locate the countries covered in these sections by matching the colors on the right with the colors on the map below.

Sections	Did You Know?
1 SPAIN AND PORTUGAL	In 1488, the Portuguese explorer Bartolomeu Dias was the first European to sail around the southern tip of Africa.
2 ITALY	The Vatican, a walled city within Rome, is an independent state with its own railway, currency, and post office.
3 GREECE	The first Olympic Games were held in honor of the god Zeus. The modern Olympics are based on these festivals.

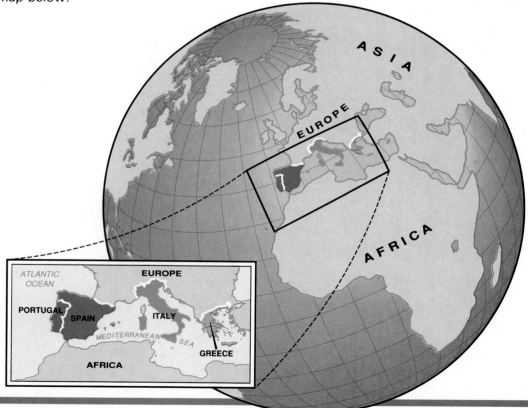

A Remembrance of Times Past

The Spanish landscape is dotted with reminders of its proud past. A medieval castle that once withstood invaders who braved the Pyrenees overlooks the dry northeastern region of Aragon.

1 Spain and Portugal

Section Preview

Key Ideas

- The physical characteristics of the Iberian Peninsula isolate it from the rest of Europe.
- Spain's economy is moving away from agriculture into other areas.
- Minority groups in Spain are striving for greater independence.
- Portugal is a small country with a history of overseas trade.

Key Term

navigable

As the globe on the facing page shows, the Iberian Peninsula dangles off the southwestern edge of Europe, separating the waters of the Mediterranean from the Atlantic Ocean. Two countries dominate the peninsula, Spain and Portugal. Spain covers most of the peninsula; Portugal occupies about one sixth of the land.

Looking at their locations on the map, you might guess that Spain and Portugal are closely tied to the rest of Europe. But location isn't always what it appears. The French emperor Napoleon once said, "Europe ends at the Pyrenees (PIHR uh neez)"—the mountains that divide the peninsula from the rest of Europe. Poet W. H. Auden described Spain as "a fragment nipped off from hot Africa, soldered so crudely to inventive Europe." Why would people think of Spain as isolated from the rest of Europe when it so clearly is a physical part of it? The answer is revealed in the histories of Spain and Portugal and in the distinct characteristics of the two countries.

Spain: A Unique Place in Europe

A castle appears on Spain's coat-of-arms and also on its flag. The castle is a symbol both of Spain's history and of its physical setting. In the historical sense, the castle represents Castile (cas TEEL). Castile was one of the many kingdoms that banded together in the late 1400s to expel the Muslim Moors who had ruled Spain for more than seven hundred years. The symbol also stands for the many beautiful castles that still dot the Spanish countryside.

Applying Geographic Themes

1. Regions What natural boundary separates Spain from France?
2. Place Where are the Cantabrian Mountains located?

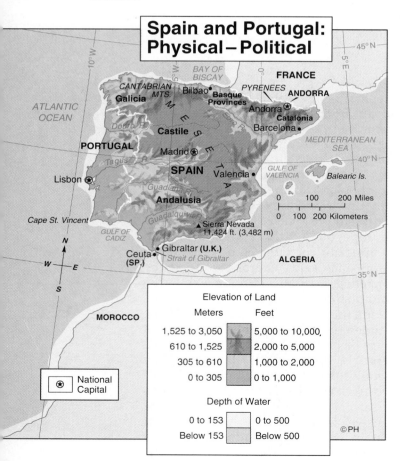

Spain and Portugal: Physical–Political

Elevation of Land

Meters		Feet
1,525 to 3,050		5,000 to 10,000
610 to 1,525		2,000 to 5,000
305 to 610		1,000 to 2,000
0 to 305		0 to 1,000

Depth of Water

0 to 153		0 to 500
Below 153		Below 500

⊛ National Capital

©PH

These fortresses are reminders of hundreds of years of war that are part of Spain's history.

In a geographical sense, Spain itself is like a well-guarded castle. The Pyrenees Mountains block easy passage across the nation's only land border with the rest of Europe. Approaches by water are no easier. Steep cliffs rise directly from the water along large stretches of the coastline. Elsewhere, coastal plains are very narrow.

Rising from the slender margins of coastal plains are the high plateaus that form most of Spain. The plateau of central Spain is known as the Meseta (me SAY tuh), a word that means "little table." Several large rivers flow across the Meseta and between the few mountain ranges that divide the plateau. Only one of these rivers is navigable; that is, deep and wide enough to allow ships to pass. Dangerous rapids make all other rivers unnavigable.

The Effect of Elevation on Climate

Almost all of Spain has a typical Mediterranean climate of mild, rainy winters and hot, dry summers. Spain's elevation also has a strong influence on its climate. Moist Atlantic winds rising over the Cantabrian (can TAH bree uhn) Mountains along the northern coast drop ample rain for farmers to raise corn and cattle there. The Meseta in the interior, however, is in the rain shadow of the mountains. This region is much drier. Farmers in the Meseta grow wheat or barley, using dry-land farming methods that leave land unplanted every one or two years to gather moisture. Sheep and goats graze on slopes too steep or dry for growing crops.

Parts of southeastern Spain are much drier than the rest of the country, making them semiarid. The winds that usually blow over this area come from Africa and carry little moisture. Citrus and olive trees grow well on eastern coastal plains near cities such as Valencia and Barcelona, where irrigation systems provide water that the climate does not.

Spain's Economy

For much of its history, Spain has been an agricultural nation. As late as the 1940s, more than half of all workers were farmers. Because of the climate, however, farming can be difficult in Spain. In recent decades, the Spanish have used their natural resources to build new industries. One major industrial center is in the north, around the city of Bilbao. Local iron ore provides material for producing steel and other products. Barcelona, the nation's largest port, is also a major industrial center.

The Capital's Central Location

Spain's largest city, Madrid, is located near the center of the country. King Philip II made this city Spain's capital in 1561. One legend suggests that the king selected this site on the Meseta because its dry climate eased the pain of his gout, a disease that causes painful joints. Historical geographers give another reason for the capital's status: its central location. This factor allowed Philip and later Spanish rulers to have strong control over people and resources in all parts of the nation.

Over the years such central control grew easier as Madrid became the hub of new transportation routes. As a result, the city prospered by tapping the wealth of other Spanish regions. An old Spanish saying suggested, "Everyone works for Madrid, and Madrid works for no one."

In recent decades, the Spanish have built newer industries in the area around Madrid. Many migrants from poorer farming areas have moved to the city and surrounding area. As a result, the metropolis now has more than four million residents. But it also has problems, including heavy traffic and air pollution.

Spain's Second City
Barcelona, the capital of Catalonia on the southeast coast, is Spain's industrial center. Inhabitants speak Catalan as well as Spanish.

The Regions of Spain

Despite nearly five hundred years of central control, Spain's regions hold on to their strong independent identities. Writer V. S. Pritchett said this about the Spanish people's strong regional ties:

> They are rooted in their region, even nowadays. . . . They are Basques, Catalans, Galicians, Castilians, . . . and so on, before they are Spaniards; . . .

Of all regions, the most striking example of independent identity may be the Basque (BASK) people of northeastern Spain. The Basques number fewer than one million people in Spain, yet they inhabit one of its richest areas.

The Basque language, Euskera, is not related to any other European language and is difficult to learn. A Spanish story tells of a person who "spent seven years learning it and in the end knew only three words."

The Basques have always wanted to maintain their cultural identity apart from the rest of Spain. As a result they have been persecuted by many Spanish leaders. Some Basques demand nothing less than total independence. A few of these separatists have engaged in violent acts against the central government.

Tensions are less severe in Catalonia, the region surrounding Barcelona. However, pressures for greater use of the Catalan language, a mixture of French and Spanish, are evident in this region, too. Other parts of Spain are also asking for greater local control.

Portugal: Links to the Sea

English professor and novelist Frank Tuohy explained the differences he saw between Spain and Portugal this way:

> *Spain is like a novel with half a dozen chapters; Portugal is a short story. A compact country, with variety in a limited space, one small village church will commemorate six centuries of history and three golden ages of architecture.*

Portugal is a nation about the size of the state of Indiana but with twice as many residents. It shares a landmass with Spain, but it is quite different from its larger neighbor. The northeastern corner of the country is mountainous, but the land slopes gently toward the ocean. At least 20 inches (50 cm) of rain falls each year in most parts of the country.

The rainfall favors farming. Grains such as wheat, corn, and barley grow on flatter lands. The oil that is pressed from the olives that flourish in the south is a major export. So is the port wine produced in northern valleys near the city of Oporto. The bark from some varieties of oak trees that grow well in southern Portugal yields cork, making cork and cork products major exports, too.

A History of Exploration Portugal is small, but it has had a large impact on world affairs. It emerged as an independent nation in 1143 when rulers of the area around Oporto defeated the Moors. Portugal quickly became a trading nation. Portugal's capital, Lisbon, became the leading port of the new nation.

The Basques

The Basques, who live in northern Spain, have their own language and traditions. Many would also like political independence.

In the fifteenth century, Spain sent Christopher Columbus and other sailors across the Atlantic to find routes to China and India. At the same time, Portugal explored new sea routes to East Asia around Africa. As expeditions reached farther around Africa, Portugal established many trading colonies. When both Spain and Portugal expanded their colonial empires into South America, conflicts arose over the division of land. In 1494 the two countries signed a treaty and Portugal gained control of large parts of Africa and Brazil; Spain claimed most of the rest of Latin America.

Independence in Africa The empires of Portugal and Spain shrank in the early 1800s as many colonies gained their independence. Not until 1975 did the Portuguese grant independence to their African colonies. Since that time, more than six hundred thousand people from the former African colonies have immigrated to Portugal, seeking greater opportunities. As a result, Portugal's population has become more diverse.

Banker Antonio Vasco de Mello observed about the old Portugal, "We didn't know if we were a small European country with big African holdings or a big African country with a foothold in Europe." When Portugal gave its colonies their freedom, the country clearly turned back to Europe. Like Spain, Portugal joined the European Economic Community, also known as the Common Market, in 1986.

Portugal's economy remains heavily agricultural, but many farmers have been drawn to work in new factories in spite of the low wages that they earn there. The nation is working to increase its literacy rate, but more than 15 percent of its people still can't read or write. Portugal will have to wrestle with these issues in the coming decades if it hopes to regain its former position as a world economic power.

Portugal's Industrial and Cultural Center
Like most other capital cities, Lisbon grew in importance because of its proximity to an important water route.

SECTION 1 REVIEW

Developing Vocabulary
1. Define: navigable

Place Location
2. Where is Lisbon located in relation to Barcelona?

Reviewing Main Ideas
3. How do Spain's mountains affect its climate?
4. What is the meaning of "everyone works for Madrid, and Madrid works for no one"?
5. How did a small country like Portugal come to have a great influence on the world from the 1400s to the 1800s?

Critical Thinking
6. **Predicting Consequences** Why might the Spanish government be reluctant to grant the Basques their independence?

2 Italy

Section Preview

Key Ideas

- Italy has a mountainous terrain, but agriculture is still important to the economy.
- Many Italians have migrated to the industrial north to find employment in factories.

Key Term

Renaissance

Applying Geographic Themes

Place   What landform separates Italy from its northern neighbors? Name two islands that are part of Italy.

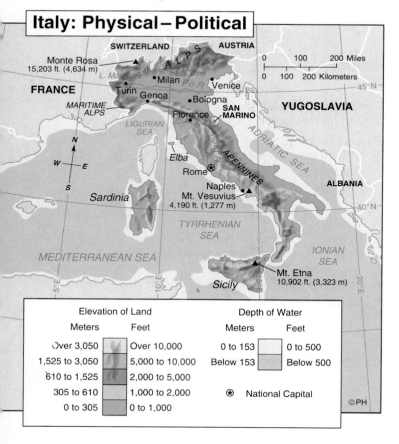

Italy: Physical–Political

Elevation of Land		Depth of Water	
Meters	Feet	Meters	Feet
Over 3,050	Over 10,000	0 to 153	0 to 500
1,525 to 3,050	5,000 to 10,000	Below 153	Below 500
610 to 1,525	2,000 to 5,000		
305 to 610	1,000 to 2,000	⊛ National Capital	
0 to 305	0 to 1,000		

©PH

Italy has perhaps the best-known outline of any country in the world. Most people suggest that Italy looks like a giant boot ready to kick the triangular "rock" of Sicily across the Mediterranean Sea.

Struggling Farms and Booming Factories

Italy's boot is formed around the Apennine Mountains. This young and seismically active range begins in the northwest and arcs its way down the peninsula into Sicily. The Apennines are not tremendously high; no peak is higher than 10,000 feet (3,000 m) above sea level. But they and other highlands cover much of the Italian peninsula, leaving narrow coastal plains as the country's only flat land. The southern toe of Italy and the island of Sicily have been sites of historic and recent volcanic eruptions. Sicily's Mount Etna erupted most recently in 1981.

Climate and Vegetation   The Alps run from east to west along the entire northern boundary of Italy. Although their passes allow easy human travel, their tall peaks block much of the moisture that the prevailing westerlies carry from the North Atlantic into Western Europe. As a result, Italy's climate south of the Alps is Mediterranean—hot and dry in summer and mild and wet in winter.

Trees that once covered many hillsides have been cleared for space and fuel over the centuries. Only scrub vegetation remains. In addition, large volumes of soil have eroded through overgrazing by goats and sheep.

In spite of the scarcity of flat land and the dry climate, until recently Italy relied heavily on agriculture. As late as 1960, more than one third of the population lived and worked on farms. Today, however, less than 10 percent of the Italy's work force is agricultural.

Overpopulation Italy is just about the same size as Florida and Georgia combined. But with a population of about fifty-eight million people, Italy contains four times as many people as these two states. In addition, people cannot make their homes on the rocky mountain peaks that dominate much of Italy's landscape. Therefore, the populated area is very crowded.

In the early 1900s, many Italians were forced to move because its small amount of farmland could not support the population. Unemployment in rural areas is still high, especially in southern Italy. Since World War II, many workers have migrated from the poor southern regions to the northern provinces of Lombardy and Piedmont to find jobs in factories.

A Growing Industrial Economy The Italian government has encouraged the development of new factories and services in recent years. But most economic growth has resulted from private business. Automobiles, home appliances, and other metal goods have been the most successful products. These industries have boosted Italy's steel industry and helped the growth of many smaller factories that supply parts and machines.

The success of modern Italian industry shows that a country can turn geographic disadvantages into opportunities. Before the 1950s, Italy was largely agricultural and relatively poor. But the country worked hard to help form the European Economic Community (EEC). Once the EEC began to operate, Italy could reach a much larger and richer market. Because Italy was poor, its workers could work for lower wages than workers in other European countries. Italian goods therefore could be sold at lower prices, and Italian industries boomed.

Italian creativity also played a role in the country's industrial boom. The Italian businesses developed many new styles, designs, and methods for making their products. These improvements made Italian products, such as sleek home furnishings and high-fashion clothing, more attractive to foreign markets.

Region: One Nation from Many Parts

When you think of Italy's history, you may think of the Roman empire. But after that empire ended in the fifth century, the Italian peninsula became a changing patchwork of separate political units. Over the next thirteen hundred years, many Italian cities operated as independent states. Kingdoms grew and declined. As the influence of Christianity spread, the Roman Catholic church gained control over large amounts of land.

It was not until 1861 that states in the northern part of the peninsula joined together to form the country of Italy. Within a decade, the entire peninsula was united. During the twentieth century, a united Italy has survived two world wars and many periods of change in national government.

Regional Italy Italy's survival as a unified nation is especially impressive because of the striking differences that exist among its many regions. Although each of its smaller regions has a distinct local character, Italy may be roughly divided into three large regions. Northern, central, and southern Italy are as different from one another as are the green, white, and red colors of the country's flag.

Northern Italy The country's northern region often is called European Italy. The provinces in this region are located closest to the rest of Europe, and they resemble central European nations more than do other Italian provinces.

The heart of northern Italy is the lush Po River valley. This broad, well-watered plain lies between the Alps

Italy's Breadbasket
The Po Valley is Italy's richest agricultural region. How might climate and landforms contribute to its prosperity?

of raw materials. Today hydroelectricity from the Alps powers many factories. The industrial growth of the Po valley has made Genoa a thriving port city.

Other parts of the northern region also are prosperous. In winter vacationers flock to the fashionable ski resorts in the Italian Alps. In summer, tourists are attracted to the area's splendid lakes. Dairy farms, like those in nearby Switzerland, are very productive and profitable.

Frequent flooding in the area around Venice has stunted its agricultural and industrial growth. Flooding in Venice is described in more detail on page 331. But Venice remains a popular destination for tourists. The intricate network of canals that serve as streets enchant visitors, as do the mammoth palaces built in the late Middle Ages, when Venetian traders conducted business throughout the world.

Central Italy Central Italy consists of Rome and the surrounding regions, which were once controlled by the Roman Catholic church. National leaders selected Rome as the capital when Italy was unified in the late 1800s for two reasons. First, its location was central. Second, it had been the capital of the Roman empire, and its history symbolized the glory that the Italians hoped to restore to their new nation. Visitors to Rome can see the ruins of two impressive buildings from the time of the Roman Empire. Remains of the Colosseum—Rome's largest stadium—and the Forum—a public meeting place—still stand.

Rome is a bustling city unlike any other in the world. American novelist Michael Menshaw described the flurry of activity one sees in Rome like this:

Many streets are as narrow as hallways, as steep as staircases, as dim and cool as cellars. Yet even where these cramped passages

and the Apennines. Rivers flowing out of the mountains provide even more water. Since drainage was improved in the Middle Ages, the valley has been Italy's most productive agricultural area. Wheat and rice are the most important crops.

The Po valley is even more important now as an industrial center. About two thirds of Italy's factory products are made in the region. Early industrial development focused on the cities of Milan and Turin. These vital trade centers were also located near sources

Flooding in Venice

Few places on earth are more prone to flooding than Venice, shown at right. In recent decades, up to two hundred floods a year have washed through this elegant Italian city.

Residents are getting used to the routine. Whenever high water threatens, a citywide siren sounds. People who live in ground-floor apartments then race to put their furniture up on blocks. Shops in low areas are forced to close.

Why does it flood so much here? Mineral deposits from the very salty waters of the Adriatic Sea are slowly building up, and the entire Adriatic basin is sinking under the weight. Venice is sinking along with it, allowing more and more water to gush over the city's streets each year. Plans to construct huge gates to hold back the Adriatic flow are currently being considered.

1. Why does Venice have so much flooding?
2. What impact does the flooding have on the daily lives of those who live in the city?

---

open into broad avenues and room-like piazzas [open squares] full of people, Romans maintain their inalienable right to do outdoors anything they might do at home. . . . Romans simply like to do things together; they enjoy sharing with the world the endless wonder they take in themselves and in one another.

Set within central Rome is an area measuring less than one square mile (2.6 sq km) known as Vatican City. This small tract serves as the world headquarters of the Roman Catholic church. St. Peter's Cathedral and the Vatican Museums are the main structures in the Vatican. Fewer than one thousand people live in Vatican City, but the district swells with visitors daily throughout the year.

Southern Italy The southern region of Italy is known as the *Mezzogiorno* (MET so ZHOR no) and includes Sicily and Sardinia. The name means midday and points out one of the region's most noted features: its intense noontime sun. Poor roads used to make travel difficult in this area, but new freeways now bring it closer to the rest of the nation. Agriculture is not profitable here because too many people live on too little arable land. Some heavy industries located here in the early years after World War II, but they have suffered in recent decades. As a result, many southerners have migrated to northern Italy.

Other southern Italians have moved to Naples, the largest city in the region. This port city suffers from some of the worst poverty in Europe. The number of available jobs cannot keep up with the number of people who wish to work. The government hopes that as Italy's economy develops within the European Community, more jobs will be available.

SECTION 2 REVIEW

Developing Vocabulary
1. Define: Renaissance

Place Location
2. Why does Rome's location make it a good choice for Italy's capital?

Reviewing Main Ideas
3. How do Italy's landforms affect the density of its population?
4. How has Italy developed as an industrial nation?
5. Why have many southern Italians migrated to northern Italy?

Critical Thinking
6. **Analyzing Information** What are some of the advantages to Italy of having such diverse regions? What are some disadvantages?

A Country Within a Capital
St. Peter's Basilica, with its famous dome designed by Michelangelo, is the showpiece of a tiny, independent state within the city of Rome known as Vatican City. The Vatican is the headquarters of the Roman Catholic Church.

Rome is not the only major city in central Italy. Bologna is a leading agricultural center nicknamed "The Fat" because of its wonderful variety of foods. Florence is a cultural center made famous by Michelangelo and other Italian painters during the Renaissance—a great rebirth of art and learning that started in Italy in the 1300s and spread throughout Europe.

✓ Skills Check

✓ Social Studies
☐ Map and Globe
☐ Reading and Writing
☐ Critical Thinking

Analyzing Line and Bar Graphs

Line and bar graphs are useful ways to present information visually and condense large amounts of data. Graphs allow us to see the relationships between two or more sets of data, and to discern trends.

Use the following steps to read and analyze the line and bar graphs below.

1. **Identify the kind of information presented in the graph.** Line and bar graphs are useful for showing different types of information. Answer the following questions: (a) What subject does the line graph portray? (b) What do the numbers on the vertical axis (the left side) of the line graph represent? (c) What do the numbers on the horizontal axis (along the bottom) of the line graph represent? (d) What does the bar graph portray? (e) What does the key for the bar graph tell you?

2. **Practice reading the information shown in the graphs.** Line graphs often show changes or trends over time whereas bar graphs allow you to compare data. Answer the following questions: (a) Which of the countries shown on the line graph had the greatest population growth between 1960 and 1970? (b) Which country experienced a population decline during these same years? (c) Which country shown on the bar graph imported the most goods?

3. **Look for relationships among the data.** In addition to reading individual pieces of data on the graphs you can also study trends and compare data. Answer the following questions: (a) How much more than Spain did Italy import from 1986 to 1987? (b) During what ten-year period were the rates of population growth of Portugal, Spain, and Mexico roughly the same?

Total Population
(in millions)

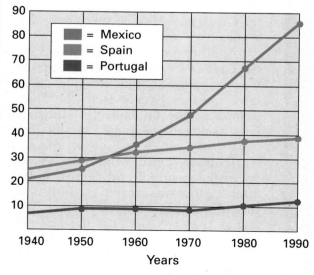

Source: The Stateman's Year-Book

Total Imports
(in billions of U.S. dollars)

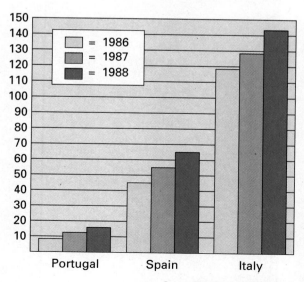

Source: The Stateman's Year-Book

3 Greece

Section Preview

Key Ideas

- Greece has felt the influence of both Western and Eastern cultures.
- Greece is a mountainous land with a largely agricultural economy.
- Greece has relied on the sea for trade and contact with its many islands throughout its history.

Key Terms

graben, tsunami

Many countries fit neatly into regional groups. Mexico, for example, is clearly part of Latin America. Others lie in a gray area between distinct regions. Greece is one of the latter

A Proud Heritage

As in ancient times, Greek villagers make the best of their sparse landscape by raising livestock and growing olives and citrus fruit.

countries. There are several reasons to consider Greece as part of Mediterranean Europe. First, Greece has strong geographical and historical ties to the Mediterranean. Second, Greece is now a member of both the EEC and NATO. And third, Greece is the birthplace of a culture that reached full expression in Western Europe.

But Greece bears the imprint of other regions, too. As the map on page 335 shows, its nearest neighbors are not in Western Europe. Greece shares its northern border with the Eastern European nations of Albania, Yugoslavia, and Bulgaria, while its eastern boundary meets the Middle Eastern nation of Turkey.

Place: A Rugged Land

The land area of Greece includes about 1,450 islands. Its northern mountains are older extensions of the Dinaric Alps, which form the mountainous backbone of the Balkan nations. Southern Greece is the product of tectonic forces—the place where the Eurasian tectonic plate meets the African plate. Major faults here thrust some lands higher and caused others to sink. Grabens, areas of land that have dropped down between faults, were flooded. The Aegean Sea to the east of the Greek mainland occupies one such graben.

Another graben was flooded by the Gulf of Corinth. As shown by the map on page 335, this thin inlet separates most of Greece from the Peloponnesus (pel uh puh NEE suhs), a large peninsula of rugged mountains.

Agriculture amid Mountain Peaks

A Greek legend tells how one country after another was created by fertile soil that was sifted through a strainer. The stones that were left in the strainer were thrown away. According to the legend, these stones became Greece—a country covered by mountains and rocky soil.

The stony mountains of Greece are fairly high. The tallest peak, Mount Olympus, rises 9,570 feet (2,920 m), and many areas have elevations over 3,000 feet (915 m). Parallel ranges make travel difficult in many parts of the country. Narrow coastal plains, however, provide flat areas on which wheat and other grains are grown. Here, olive and citrus groves also abound. Because Greece has a Mediterranean climate with dry summers as well as poor irrigation methods, it is difficult to produce fruit that is always in good condition. Greek agricultural products therefore do not sell as well as those from other countries.

On the more rugged slopes where growing crops is not possible, farmers graze sheep and goats. As in other Mediterranean nations, however, these animals have destroyed natural forests, leaving a scrubby vegetation that does little to prevent soil erosion.

Athens Although the largest coastal plains are in northern Greece, the most populous area is a part of Greece known as Sterea Hellas. Here Greece's capital, Athens, is located. Although most foreigners think of Athens as an ancient city, it has grown up mainly within the last hundred years. Modern Athens is one of the youngest capital cities in Europe. Even Washington, D.C., is older. But the monuments of Athens have stood on the hill known as the Acropolis for thousands of years. A recent visitor to Athens wrote:

If you ask anyone who has visited Athens to describe his most magical memory of the city, you can be virtually certain that the Acropolis . . . will figure in his reply. The answer may be a recollection of . . . a view of the 2,400-year-old Parthenon soaring triumphantly against the night sky; . . . [or] seeing one of the most stunning

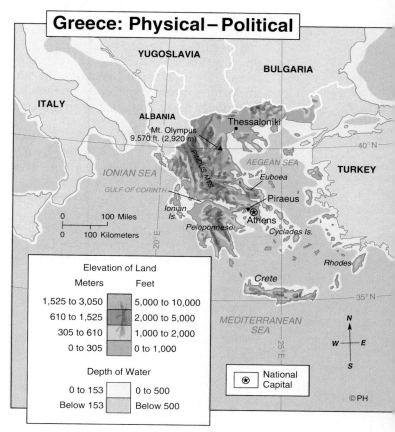

Greece: Physical–Political

Elevation of Land

Meters		Feet
1,525 to 3,050		5,000 to 10,000
610 to 1,525		2,000 to 5,000
305 to 610		1,000 to 2,000
0 to 305		0 to 1,000

Depth of Water

0 to 153		0 to 500
Below 153		Below 500

National Capital

©PH

Applying Geographic Themes

1. Place Greece is composed of many island groups and peninsulas. Name two of the larger islands.

2. Location Where is Greece located in relation to the Ionian Sea?

sights on earth: the Aegean sunset turning the white marble of the Parthenon to brilliant gold.

The homes of more than three million people surround the Acropolis. Almost one third of Greece's total population live in this crowded city. Modern apartments and houses line crowded city streets, as do new office buildings and a dazzling array of stores, taverns, and restaurants. And the residents of Athens daily endure chaotic traffic jams—a hallmark of this modern city. The downtown streets are so choked by traffic that

WORD ORIGIN

Athens
Many place names in Mediterranean Europe grew out of the myths of ancient Greece. The city of Athens was named *Athenai* in honor of Athena, the Greek goddess of wisdom.

walking is faster than driving. As one visitor noted, "There is only one proven solution to Athens's traffic problems: live television coverage of an important international soccer match. Whenever that happens, the streets are deserted."

Focus on the Sea   Just 5 miles (8 km) to the south Athens merges with the seaside suburb of Piraeus (py REE us), Greece's largest port. The harbor there has grown steadily in importance in the twentieth century. As one might expect in a nation where no point is more than 85 miles (135 km) from the sea, Greece relies heavily on trade over the waters. It has one of the world's largest commercial fleets, and shipbuilding is an important industry. Other industries also have located near the docks of Piraeus, taking advantage of low transportation costs for imported raw materials and exported manufactured goods.

Greece also focuses on the sea to maintain contact with the many islands that it claims. Many of these are in the Aegean Sea, although the largest, Crete, is south of the mainland in the Mediterranean Sea. About two thirds of the islands are too small and rocky to support permanent residents. But there are more than four hundred islands on which people do live, many of them making a living from fishing. As elsewhere in Greece, tourism also continues to grow as a major economic activity. Visitors from around the world seek the sun, sparkling water, and gleaming beaches of the Greek islands.

The Mystery of Crete

Part of Greece's appeal to visitors lies in the richness of its history. One historical mystery surrounding the island of Crete puzzled archaeologists for decades. In recent years, geology has provided part of the solution to this mystery.

About thirty-five thousand years ago, the Mediterranean island of Crete was the center of Greece's flourishing Bronze Age culture. This culture was called Minoan after Minos, a legendary king of Crete. Because they were expert shipbuilders, the Minoans were able to travel and conduct trade throughout the Aegean Sea and as far away as the British Isles. They were so secure in their dominance over the seas that their beautiful cities needed no walls to protect them.

Then, around 1500 B.C., the glorious Minoan culture fell into a rapid decline. Some scholars believed that warlike people from the Greek mainland attacked and destroyed Crete. Others thought that an earthquake had demolished the island. But Greek archaeologist Spyridon Marinatos did not believe that these explanations were complete. In 1932 he uncovered evidence that added another piece to the puzzle.

Excavating near the ancient harbor town of Amnisos in Crete, Marinatos noticed that many huge stone blocks had been uprooted from their foundations and strewn toward the sea. "Some terrific power," he reported ". . . had swept them from their original places." Marinatos thought that this power might have been a volcanic eruption on the island of Thera, located 70 miles (110 km) from Crete.

In the 1960s, he and his team found evidence to support his theory. On Thera, his team unearthed an entire town that had been buried in white volcanic ash. In the ruins were pieces of Cretan pottery from the period 1520 to 1500 B.C.

Marinatos reasoned that the awesome eruption on Thera had caused tsunamis (TSOO nahm eez), or tidal waves, as high as 300 feet (90 m) to crash over neighboring islands, including Crete. The tsunamis and earthquakes that accompanied Thera's eruption destroyed many of Crete's lovely palaces and other buildings.

A History of Varied Influences

Discussing the gray area to which some say Greece belongs, geographer T.R.B. Dicks observed:

> Many would argue that the Greeks are a curious mixture of eastern and western. . . . It is in the towns and cities that western influence is most marked, but even in Athens the colours of the Orient [East] are strongly represented.

Greece as a Western Nation Greece may be considered a Western nation in part because Western culture has so many of its roots in ancient Greece. One rich example is found in Homer's great epic poems, the *Iliad* and the *Odyssey,* written during the eighth century B.C. These poems were based on stories about the Trojan War and the fall of Troy. But, to the ancient Greeks, they were more than inspiring stories of heroism. They also provided a guide for moral behavior and were the cornerstone of Greek education.

Foreign Influences on Greek Culture While the influence of ancient Greek culture was spreading through Western Europe, other cultures were putting their stamp on Greece. Usually this influence was the result of military conquest. From the second century B.C. to the fifth century A.D., Greece was part of the Roman empire. As the Roman empire declined, Greece became an important part of the Byzantine empire, centered in what is now Istanbul, Turkey. For the next thousand years, Greece was invaded from all directions, over land and water. Slavs, Albanians, and Bulgarians came from the north. Arabs swept in from the south. Normans and Venetians attacked from the west. In 1453, Turks conquered Constantinople, now the city of Istanbul, and ruled Greece for almost four centuries. Finally, after a

An Aegean Paradise
The whitewashed houses of Thera are a familiar sight on many of the Aegean Islands.

ten-year rebellion, the modern state of Greece gained its independence in 1830.

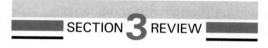

SECTION 3 REVIEW

Developing Vocabulary
1. Define: **a.** graben **b.** tsunami

Place Location
2. Name two Greek islands in the Aegean Sea.

Reviewing Main Ideas
3. Describe the major landforms that are found throughout Greece.
4. What physical characteristic has made Greece so vulnerable to foreign invasion?

Critical Thinking
5. **Synthesizing Information** Do you think that centuries of foreign rule weakened or strengthened the Greek national identity? Give reasons for your answer.

17

REVIEW

Section Summaries

SECTION 1 Spain and Portugal The location and geography of the Iberian Peninsula has historically kept Spain and Portugal remote from the rest of Europe. The Meseta in central Spain is dry but does support farming. Madrid, Spain's capital, is located near the center of the country. Spain's regions have strong individual identities; many Basques in northern Spain want independence for their region. Catalans, who live in the region near Barcelona, are demanding greater use of their own language. Portugal is a small country that has used the sea as its trading passport to the world. Its climate supports many farming activities.

SECTION 2 Italy Italy's mountainous terrain makes farming difficult; industry has become more important in recent years. Many people are crowded onto Italy's limited habitable land. Italy can be divided into three large regions. Northern Italy is quite European and has the most industry. Central Italy contains Rome, the nation's capital. Southern Italy is deprived economically, and unemployment has led many southerners to move north to find work in factories.

SECTION 3 Greece Because of its location, Greece has evolved into a mixture of Eastern and Western cultures. It is a mountainous land, but farming remains important to the country's economy. Athens, Greece's capital, has ancient ruins alongside modern buildings. The sea has always been important for Greek trade and cultural exchange and has made the country vulnerable to invasion. Ancient Greek culture has had a strong influence on Western civilization. Greece has been influenced by other cultures but has maintained a strong sense of its own identity.

Vocabulary Development

Match the following definitions with the terms they define below.

1. areas of land that have dropped down between faults
2. a rebirth of art and learning during the 1300s
3. tidal waves
4. deep and wide enough to allow ships to pass
   a. graben
   b. Renaissance
   c. navigable
   d. tsunami

Main Ideas

1. Where is the Iberian Peninsula located in relation to Europe and Africa?
2. Where is the Meseta, and what type of climate does the region experience?
3. Why have Spain and Portugal seemed isolated from the rest of Europe?
4. Name one characteristic of the Basque region that makes it unique.
5. How has Portugal taken advantage of its location on the Atlantic Ocean?
6. How do the Alps influence the climate of northern Italy?
7. Where is the Po valley located, and why is it an important region?
8. How is the economy of Italy changing?
9. Name two ways in which northern Italy differs from southern Italy.
10. Why are the Mediterranean and Aegean seas so important to Greece?
11. Why has Greece felt the influence of Eastern cultures?

Critical Thinking

1. **Demonstrating Reasoned Judgment** Explain why you agree or disagree with the following statement: ''The Spanish government should allow the Basques to form an independent nation.''
2. **Making Comparisons** Compare the physical and human geography of Spain and Italy. How are they alike? How do they differ?
3. **Drawing Inferences** How did Portugal come to have colonies in Africa?

4. **Analyzing Information** The Greek philosopher Plato mentioned a ''lost continent'' of Atlantis, thought to lie beneath the sea. What events in Greek history could have given rise to such a theory?

Practicing Skills

Analyzing Line and Bar Graphs Look again at the line graph on page 333. In approximately what year did the population of Mexico equal that of Spain? How did Spain's population compare to Mexico's by 1990?

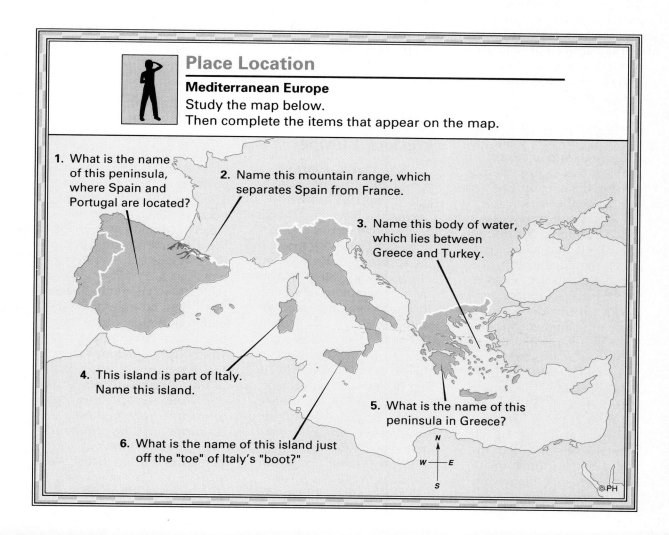

Place Location

Mediterranean Europe
Study the map below.
Then complete the items that appear on the map.

1. What is the name of this peninsula, where Spain and Portugal are located?

2. Name this mountain range, which separates Spain from France.

3. Name this body of water, which lies between Greece and Turkey.

4. This island is part of Italy. Name this island.

5. What is the name of this peninsula in Greece?

6. What is the name of this island just off the "toe" of Italy's "boot?"

©PH

An Acid Rainfall

The evidence of the damage is alarming: lakes and streams that once teemed with plants, fish, and other animals lie barren and still, empty of most aquatic life. Some of the world's greatest art and architectural treasures—from Mayan ruins in Mexico to Renaissance statues in Italy—deteriorate at a rate that shocks historians and art lovers. Scientists and politicians struggle to find answers to the problem that is suspected of causing this widespread damage: acid rain.

Acid rain forms as a result of the burning of fossil fuels—coal, oil, and natural gas. When these fuels burn, certain pollutants are released into the air. As factory smokestacks, cars, trucks, and other types of equipment spew these pollutants into the atmosphere, they combine with water vapor to form acid-laden chemicals. These acidic chemicals then fall to earth in the form of precipitation known as acid rain. The abnormal levels of acid in the rain, scientists think, then wreak havoc on plants, animals, and buildings.

The issue of acid rain connects all regions of the world. Because wind currents and precipitation patterns recognize no political boundaries, the acid rain created by one region may actually fall in another region. The spread of acid rain has therefore become a global concern.

A Tree Dies in Western Europe

Acid rain was first observed about twenty years ago. In the 1970s, scientists in Germany's Black Forest began to observe a disturbing phenomenon: needles on some of the spruce trees were yellowing and falling off. This was the first symptom of an illness that soon spread to other trees in the area.

As the trees sickened, their growth was stunted and their roots shriveled. Insects and disease took advantage of weakened plants and caused further damage. The culprit in this disaster? Scientists suspected acid rain.

Today, about one third of the trees in the Black Forest have been affected by acid rain; forest areas in other Western European countries have also been affected. This damage is causing Europeans serious concern. The destruction of aquatic environments, buildings, and artwork is also a major concern for Europeans.

A Global Problem

As the map on the following page shows, acid rain presents a problem that extends far beyond Western Europe. The problem is widespread because the chemicals that create acid rain have no respect for international boundaries. Winds and weather systems can carry airborne pollutants five hundred, even one thousand miles from the nation or region where they originated. For example, winds carry pollutants from the coal-burning power plants in the United Kingdom across the North Sea to Norway and Sweden. Forests in these countries are now suffering the withering effects of acid rainfall as a result. Scientists suspect that, in some cases, the region producing the harmful pollutants may be less affected by acid rain than other areas nearby.

The international character of the acid rain problem has prompted many nations around the world to join together in an effort to combat it. Over two dozen nations have so far agreed to cut sulfur dioxide emissions—a major component of acid rain—by 30 percent in the 1990s.

The View from the U.S.A.

 The map below indicates that much of the eastern portion of the United States and Canada is affected by acid rain.

Most of this pollution is generated in the United States, much of it in the Midwest, and then carried by winds to Canada. As a result, acid rain is damaging nearly half of the lakes and forests in eastern Canada.

For more than a decade the two countries have been discussing how to solve the problem. Scientists and political decision makers in the United States disagree on how to address the causes or effects of acid rain. Many people urge quick action, while supporters of industry continue searching for less costly ways to reduce harmful emissions. President Bush has pledged to reduce acid rain in

an effort to soothe strained relations with Canada. Lawmakers in the United States are now struggling to meet this pledge without destroying the industries that contribute to the problem.

TAKING ANOTHER LOOK

1. What are the causes and the effects of acid rain?
2. Why is it important for different countries to work together to reduce emissions that cause acid rain?

Critical Thinking
3. **Predicting Consequences** What impact will the continued industrial development of nations around the world have on acid rain?

Applying Geographic Themes

Interaction Compare this map with the map on page 72. What economic activity do all of the regions of acid rain have in common?

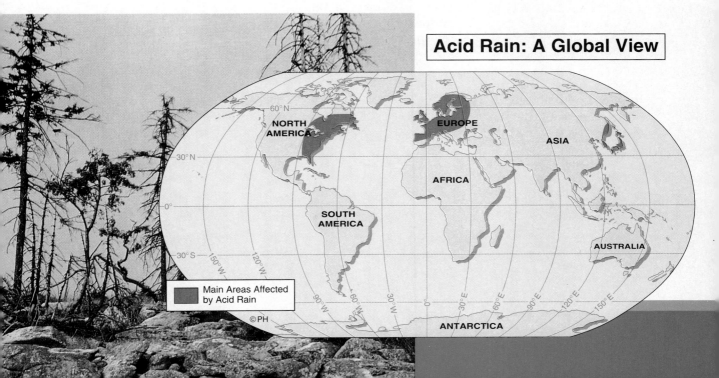

Acid Rain: A Global View

Main Areas Affected by Acid Rain

©PH

Eastern Europe

Wide plains, high mountain ranges, ancient cities, small farms, rich coal mines . . . climates that range from freezing in the north to hot and humid in the south . . . the Danube River winding its way to the swampy delta at the Black Sea . . .

Long dominated by the rise and fall of empires, Eastern Europe is a region in flux. For centuries its varied landforms have been a key factor in the movement of people and armies. In this unit, you will discover how the people of Eastern Europe are using the resources of the region to adapt to the growth of democracy.

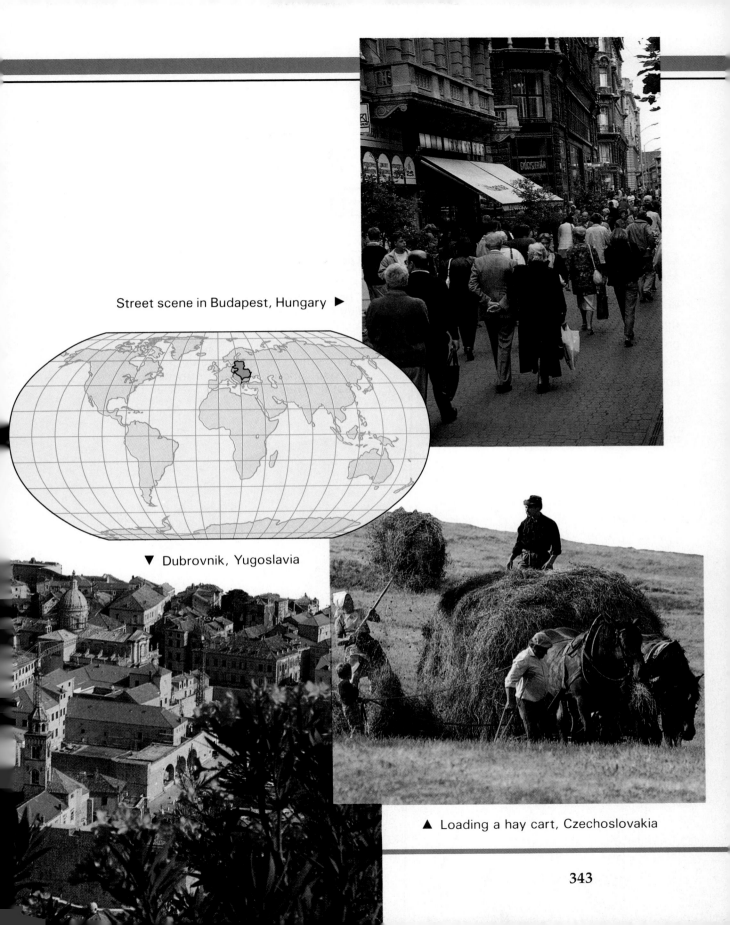

Street scene in Budapest, Hungary ▶

▼ Dubrovnik, Yugoslavia

▲ Loading a hay cart, Czechoslovakia

343

Regional Atlas: Eastern Europe

Chapter Preview

Both sections in this chapter provide an overview of the countries of Eastern Europe, shown in red on the map below.

Sections	Did You Know?
1 LAND, CLIMATE, AND VEGETATION	The Danube River runs through, or borders, five Eastern European countries.
2 HUMAN GEOGRAPHY	Slavic ethnic groups make up the majority of the population between Germany in the west and Russia's Pacific coast in the east.

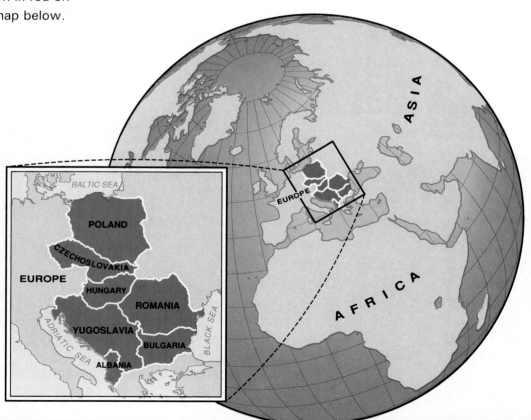

The Chain Bridge, Budapest, Hungary
Budapest straddles both banks of the Danube River. After World War II, Hungarians rebuilt many of the buildings and bridges of this war-torn capital city in their original style.

1 Land, Climate, and Vegetation

Section Preview

Key Ideas

- The many nations of Eastern Europe form a corridor between Europe and Asia.
- Eastern Europe can be divided into four landform regions.

Key Term

karst

The term *Eastern Europe* is a product of history. It came into use at the end of World War II to describe nations under the control of the Soviet Union. As Czech author Milan Kundera explained: "After 1945, . . . several nations that had always considered themselves to be Western woke up to discover that they were now in the East."

A Region of Political Change

Until 1990, Eastern Europe included eight nations: Albania, Bulgaria, Czechoslovakia, East Germany, Hungary, Poland, Romania, and Yugoslavia. In 1990, however, East and West Germany were reunited. Without any conflict, East Germany was taken off the map of Eastern Europe.

Boundaries in the area comprising Eastern Europe have changed many times throughout history. In 1921, one geographer called it the "zone of political change."

One reason for Eastern Europe's changing boundaries is its strategic location. Eastern Europe connects the European peninsula with the rest of Eurasia. For thousands of years, the region was a corridor for armies.

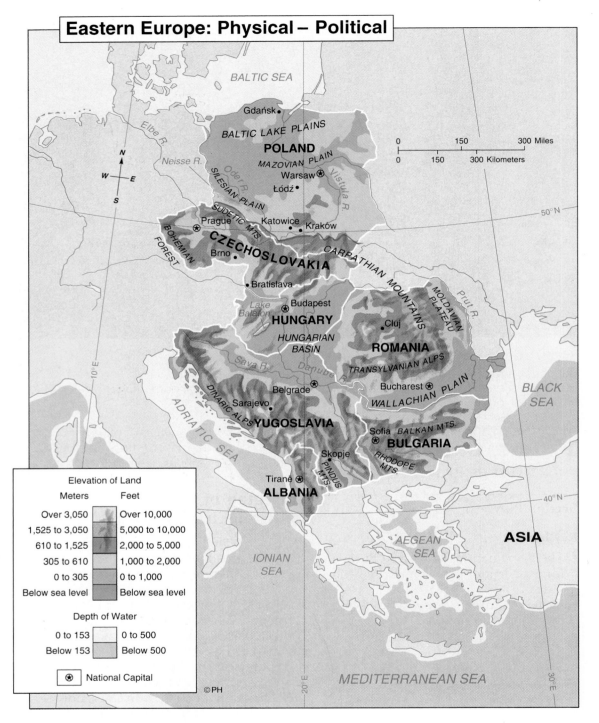

Eastern Europe: Physical – Political

BALTIC SEA

Gdańsk

BALTIC LAKE PLAINS

POLAND

MAZOVIAN PLAIN

Warsaw ⊛

Łódź

Elbe R.

Neisse R.

Oder R.

SILESIAN PLAIN

Vistula R.

SUDETIC MTS.

Prague • Katowice

Kraków

CZECHOSLOVAKIA

Brno •

BOHEMIAN FOREST

CARPATHIAN MOUNTAINS

• Bratislava

Lake Balaton

⊛ Budapest

HUNGARY

HUNGARIAN BASIN

Cluj •

MOLDAVIAN PLATEAU

Prut R.

ROMANIA

Sava R.

Danube R.

TRANSYLVANIAN ALPS

Bucharest ⊛

WALLACHIAN PLAIN

DINARIC ALPS

Belgrade •

Sarajevo •

YUGOSLAVIA

ADRIATIC SEA

Skopje •

PINDUS MTS.

Sofia BALKAN MTS.

⊛ **BULGARIA**

RHODOPE MTS.

Tiranë ⊛

ALBANIA

BLACK SEA

ASIA

AEGEAN SEA

IONIAN SEA

MEDITERRANEAN SEA

50° N

40° N

10° E

20° E

30° E

N / W–E / S

| 0 | 150 | 300 Miles |
| 0 | 150 | 300 Kilometers |

Elevation of Land

Meters		Feet
Over 3,050		Over 10,000
1,525 to 3,050		5,000 to 10,000
610 to 1,525		2,000 to 5,000
305 to 610		1,000 to 2,000
0 to 305		0 to 1,000
Below sea level		Below sea level

Depth of Water

| 0 to 153 | | 0 to 500 |
| Below 153 | | Below 500 |

⊛ National Capital

©PH

Applying Geographic Themes

1. Location   The Danube River is one of Eastern Europe's most valuable natural resources. Into which body of water does it empty after going from western Germany through many Eastern European countries?

2. Movement   Which two important Eastern European plains are separated by mountains but linked by the Danube?

The Landscape of Eastern Europe

The eight Eastern European nations cover roughly the same area as the states of Oregon, Washington, Idaho, Montana, and Wyoming; or about 490,000 square miles (1.27 million sq km). Despite its relatively small size, Eastern Europe has a remarkably varied landscape. Four broad bands of distinct landforms cross the region from east to west. Many of these landforms begin in Western Europe and extend into Northern Eurasia.

The North European Plain   The North European Plain forms the northernmost landform region. It stretches from the Atlantic Ocean in the west to the Ural Mountains in the east. Throughout history, many cultural groups have battled for control of the plain. Today most of Poland occupies the area on the Vistula River and its tributaries that flow across the plain. "We shall cross the Vistula," declares the Polish national anthem. "We shall remain Poles."

The Carpathian Mountains   A system of zigzagging mountains known as the Carpathians makes up the next landform region. As the map on page 346 shows, these mountains curve in a huge crescent from southern Poland, through Czechoslovakia, and into northern Romania. In Romania, part of the range is called the Transylvanian Alps, home of the legendary Count Dracula.

The Hungarian Basin   The Danube River dominates the third landform region. Locate the Danube on the map on page 346. As you trace the length of the river, keep in mind some of the names given to it: "river of destiny," "dustless road," and "lifeline."

The Danube winds its way from west to east across a huge plain covering most of Hungary and parts of western Romania and northeastern Yugoslavia. Because Hungary lies at the center of this plain, the area is known as the Hungarian Basin. Hungarians call it the *pustza* [POOSH tza] —an area once covered with sweeping grasslands. Today, most of the *pustza* is farmland. Only a small stretch of land, called the *Hortobaby*, is preserved as prairie.

In trying to capture the spirit of the lands along the Danube, one journalist wrote:

> *The Danube is Eastern Europe's great, throbbing artery. Its valley and plain . . . has for centuries provided a route to migrating peoples. . . . No other river in Europe . . . flows through as many nations . . . or echoes to as many languages.*

A Hydroelectric Dam in Romania
Dams along the Danube River provide hydroelectricity for many industries.

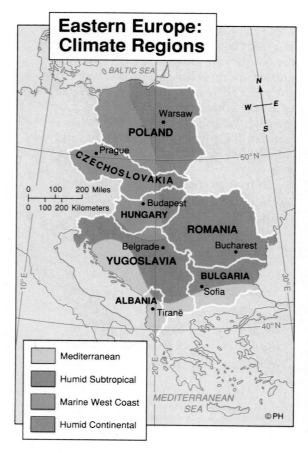

Eastern Europe: Climate Regions

BALTIC SEA

- Warsaw
POLAND
- Prague
CZECHOSLOVAKIA
50° N
HUNGARY • Budapest
ROMANIA
Belgrade • • Bucharest
YUGOSLAVIA
BULGARIA
• Sofia
ALBANIA
• Tiranë

0   100   200 Miles
0   100 200 Kilometers

10° E
20° E
30° E
40° N

MEDITERRANEAN SEA

©PH

Mediterranean

Humid Subtropical

Marine West Coast

Humid Continental

Applying Geographic Themes
Interaction   This Yugoslavian farming scene is located in a Mediterranean climate region. What other type of climate region is found in Yugoslavia?

WORD ORIGIN

Adriatic Sea
The Adriatic Sea takes its name from the Italian city of Adria. In ancient times, Adria was a flourishing port. Today it is fourteen miles inland due to silt buildup at the Po River delta.

The Balkan Peninsula   South of the Danube is Eastern Europe's fourth topographic region—the Balkan Peninsula. Four Eastern European nations are located on the peninsula: Romania, Bulgaria, Yugoslavia, and Albania. The Western European nation of Greece lies at the tip.

From a satellite view, the Balkan Peninsula looks like a tangled web of mountain ranges and valleys. The Balkan mountains have acted as walls against invasions. But they have also been an obstacle to unity. Today, dozens of languages are still spoken on the peninsula.

The Dinaric Alps are one of the peninsula's most imposing mountain ranges. These mountains run along much of Yugoslavia's Adriatic coast before arching up into Bulgaria and Romania. Most of the rock formations in the Dinaric Alps are made of karst, a soft limestone that is easily dissolved by water. Over time, the forces of erosion have sculpted the karst into spectacular jagged white peaks. One writer compared them to "raging white seas turned to stone."

Climate and Vegetation

The climate of Eastern Europe reflects a transition between the moderate marine climates of Western Europe and the harsher continental climate of

Yugoslavia's Adriatic Coast
The coastal town of Makarska faces the Adriatic Sea. Look at the physical-political map on page 346 and name the major mountain range separating the Adriatic coast from the rest of the country.

Northern Eurasia. Lands in the northeast are generally cooler and drier than lands on the Balkan Peninsula, where a Mediterranean climate prevails. Ice hampers activity on the rivers on the northern plains for two to three months each year.

The air masses that sweep over most of Western Europe also reach Eastern Europe. But they lose moisture as they travel over land. So Eastern Europe has drier seasons than Western Europe. The driest areas lie in the Hungarian Basin, south of the Carpathians.

Thousands of years ago, thick forests blanketed the plains of Eastern Europe. However, settlers eventually cleared most of these forests for farmland. Almost half the land in Eastern Europe—about 45 percent—is used for planting crops or grazing animals.

SECTION **1** REVIEW

Developing Vocabulary
1. Define: karst

Place Location
2. Where is the Adriatic Sea located?

Reviewing Main Ideas
3. Why has Eastern Europe's location made it a target for many invasions?
4. Name the four landform regions that stretch across Eastern Europe.

Critical Thinking
5. **Drawing Conclusions** Some geographers have called Eastern Europe one of the world's great crossroads. What geographic facts support this conclusion?

Chapter 18, Section 1   **349**

✔ Skills Check

☐ Social Studies
☐ Map and Globe
☐ Reading and Writing
☑ Critical Thinking

Identifying Assumptions

Popular journalism can be an excellent source of information about public attitudes toward historical events. Articles in weekly news magazines commonly present a blend of facts, quotations, and analysis that conveys a point of view. To determine the validity of an article's point of view you must be able to identify and evaluate its assumptions—the ideas that the writer takes for granted without offering additional facts to support them. Use the following steps to analyze the assumptions contained in the article excerpt below.

1. **Read the excerpt carefully to determine its topic and its approach.** Identify the subject with which the article is concerned. Answer the following questions: (a) With what subject does the article excerpt deal? (b) What approach does the article take toward its subject?

2. **Define the article's point of view.** Determine if the author is presenting a particular point of view. Answer the following questions: (a) How would you describe the tone of the article? Is it hopeful or despairing, with regard to the future of Eastern Europe? (b) What evidence does the article offer in support of this view?

3. **Identify the assumptions upon which the article is based and decide if they are valid.** Decide if the assumptions are supported by factual evidence. Answer the following questions: (a) Which assumption is plainly stated in the article? (b) What unstated assumptions can you identify? (c) How could you find out if the article's assumptions are valid?

As an ideological earthquake rocks the Soviet empire, fracturing the social, political and economic arrangements that have guided East bloc relations since 1945, the first impulse is to check its force on the Richter scale. But the next task, the part where the debris must be cleared away and planners must construct something new, has not been addressed . . .

The current pace of change in Eastern Europe, coupled with a global impulse toward interdependence, suggests that economic integration between East and West is inevitable. It is easy to imagine the formation of pan-European institutions. As those efforts gain strength, a gradual demilitarization might follow. "The Warsaw Pact will put more emphasis on political coordination and less on defense and military issues," predicts a U.S. State Department official.

Such cooperation assumes that the East European experiment will not suffer a sudden reversal, exploding in crackdowns, nationalist upsurges or anarchy. A return to the old orthodoxies and iron-fisted Soviet control might follow, but in the present climate, that is all but impossible to imagine.

From "There Goes the Bloc," *Time*, November 6, 1989

2 Human Geography

Section Preview

Key Ideas

- Eastern Europe has a complex mix of ethnic groups, but roughly two thirds of the region's people are descended from the Slavs.
- Most Eastern Europeans today live in urban areas.
- By 1990, Eastern Europeans had begun to move away from governments based on communism.

Key Term

multiethnic

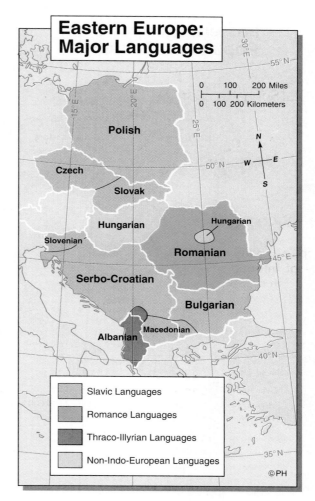

Eastern Europe: Major Languages

Legend:
- Slavic Languages
- Romance Languages
- Thraco-Illyrian Languages
- Non-Indo-European Languages

©PH

The 1984 Nobel-prize winning poet from Czechoslovakia, Jaroslav Seifert, wrote: "For us, there is no Eastern Europe. It is a collection of countries. . . . You should not see us as a single entity." Seifert's words reflect the strong national identities felt by most Eastern Europeans today.

A Region of Cultural Diversity

The many languages spoken in Eastern Europe are the best indication of the region's great cultural diversity. Polish, Czech, Slovak, Hungarian—these are but some of the languages that can be heard on a journey between the Baltic and Black seas.

Because conquering nations have redrawn the map of Eastern Europe so many times, more than one language is often spoken within the borders of a single nation. Czechoslovakia, for example, includes two main cultural groups—the Czechs and the Slovaks—each with its own language. Like Czechoslovakia, most Eastern European nations are **multiethnic,** or composed of many ethnic groups.

Applying Geographic Themes

Place   Most people in this region speak Slavic languages. In which Eastern European country is a non-Indo-European language spoken?

Movement: Shifting Populations

The languages spoken in Eastern Europe reflect the movement of many different peoples into the region over thousands of years. Today, roughly two thirds of the people in Eastern Europe are descended from the Slavs.

The Slavs   The Slavs were not the earliest inhabitants of Eastern Europe. Ancient Celts, the group of people

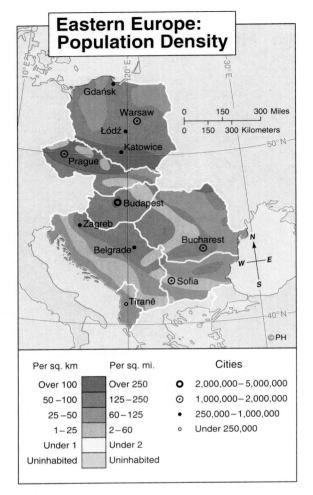

Eastern Europe: Population Density

Per sq. km	Per sq. mi.	Cities	
Over 100	Over 250	◉	2,000,000–5,000,000
50–100	125–250	⊙	1,000,000–2,000,000
25–50	60–125	•	250,000–1,000,000
1–25	2–60	○	Under 250,000
Under 1	Under 2		
Uninhabited	Uninhabited		

Applying Geographic Themes

Place Which country in Eastern Europe has no city with more than 250,000 people? Which city, with more than 2,000,000 people, is the largest on the map?

who eventually settled Ireland, once roamed the banks of the Danube. Roman legions occupied half the region, too. The Goths and Huns, whose invasions helped break the power of Rome, also swept across Eastern Europe. But the Slavs were the people who stayed and settled the land.

The Slavs probably arrived at the Carpathian Mountains around the second or third century A.D. Over time, they moved north to the Baltic, south to Greece, west to Germany, and east to Russia.

Once they were settled, mountain ranges separated the Slavs from each other. Each group gradually developed its own culture. Today Eastern Europe's many Slavic ethnic groups include the Poles, Czechs, Slovaks, Serbs, Croats, and Slovenes.

The Magyars "I profess with pride, both here and abroad, that we are a people of Asian origin!" So said one Hungarian of his ancestors, the Magyars. The exact origins of the Magyars remain unknown, but they most likely came from central Asia.

According to Hungarian tradition, the Magyars, who had moved into the Russian steppes, invaded the Hungarian Basin from Russia in A.D. 895. Expert horseback riders, they were able to claim the land by crushing opposition quickly. The Magyars quickly formed a representative form of government. In 902, Prince Arpad, a hero of the Hungarian people, set up a national assembly. One Hungarian historian boasted, "We had a parliament before we had chairs."

Other Ethnic Groups During the seventh century, a nomadic people known as the Bulgars settled in the Balkan Peninsula near the Black Sea. These people, together with the Slavs who already lived there, are the ancestors of today's Bulgarians.

Modern Albanians like to trace their history to people who lived on the Balkan Peninsula long before the arrival of the Slavs, Magyars, and Bulgars. Romanians link themselves to the ancient Romans, who ruled the region from the second century B.C. to the fifth century A.D.

Other ethnic groups also moved into Eastern Europe. Descendants of conquerors—Germans, Turks, and others—remained in Eastern Europe long after empires fell. Gypsies and Jews came from many different lands. Both of these peoples faced long histories of discrimination.

Preserving Traditions

Ethnic groups in Eastern Europe are proud of their heritage and traditions. These couples from Poland (right) and Yugoslavia (left) are wearing their traditional wedding costumes.

The Economy of Eastern Europe

At the end of World War II, the Soviet Union helped to set up Communist governments throughout Eastern Europe. The Soviets hoped to establish a buffer zone of Communist countries between themselves and Western Europe. Communism is a political theory based on the ideas of Karl Marx that calls for a society without social classes and with common ownership of all economic resources such as factories, mines, and farmland. In practice, however, communism in this region has meant totalitarianism, a system in which government has complete control over people's lives.

From Farms to Factories   The Communist governments in Eastern Europe worked to create more industry in the region. They took advantage of the rich reserves of coal, iron ore, zinc, and mercury in the Carpathians.

The switch from agriculture to industry changed the population distribution in Eastern Europe. Before World War II, much of Eastern Europe had been rural. As industry in the region grew, many people moved from farms to cities in search of factory jobs. Compare the population map on page 352 with the economic activity map on page 355. Notice that high population densities and urban land use go hand in hand. Currently, most Eastern Europeans live in urban areas.

The Shock of Freedom

In country after country, Eastern Europeans have ended Communist rule. "In forty-five years it did not succeed economically, socially, or morally," explained one Romanian citizen. "It was the system that produced dictatorship."

Today many Eastern Europeans are beginning to feel the pressure of reform in their daily lives. "We're not used to thinking for ourselves," remarked one woman. "It's very difficult learning to be free." Shopkeepers, like the ones pictured, who once ran state-controlled stores fear for the future. "We realize if the shop is not bringing in profits, we'll be out of jobs ourselves." The question that most troubles Eastern Europeans was best phrased by a Czech student: "How will we learn to take care of each other?"

1. Why do you think so many Eastern Europeans finally rejected communism?

2. Why do you think change is difficult for many people in Eastern Europe?

Economic Shortages The heavily industrialized countries of Poland, Czechoslovakia, and Hungary are the wealthiest areas in the region today. Even wealthy areas, however, are not rich by Western European standards. As the table on pages 751–761 shows, per capita incomes in Eastern Europe are less than half of those in most Western European countries. Although the standard of living varies somewhat from nation to nation, almost all of the people who live in Eastern Europe cities face severe housing shortages. Families lucky enough to find housing of their own live in simple apartments of two or three rooms.

Other shortages abound. Few people own televisions, electric kitchen appliances, or automobiles. Even basic goods—food, soap, shoes, and other

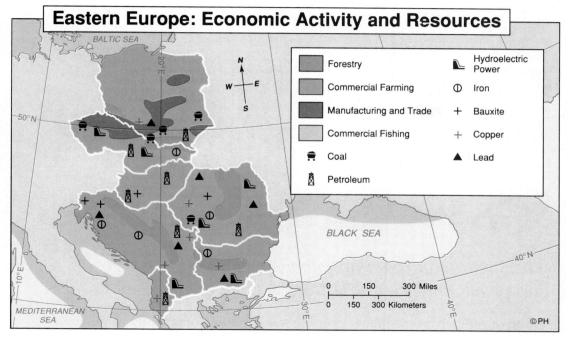

Eastern Europe: Economic Activity and Resources

Forestry

Commercial Farming

Manufacturing and Trade

Commercial Fishing

Coal

Petroleum

Hydroelectric Power

Iron

Bauxite

Copper

Lead

BALTIC SEA

50°N

BLACK SEA

40°N

MEDITERRANEAN SEA

0   150   300 Miles

0   150   300 Kilometers

©PH

Applying Geographic Themes

Interaction Which of the Eastern European countries shown on this map have deposits of coal? Describe two resources shown on this map that aid industrial development in the region.

clothing—are in short supply. People stand in long lines waiting to buy meat and baked goods only to find stores sold out by the end of the day. Shortages also keep prices high.

In rural areas, people live in small brick or wooden cottages. Some cottages are painted with brightly colored decorations. However, many lack such conveniences as indoor plumbing and central heating.

Challenging Communism

By 1990, reform movements had swelled up throughout Eastern Europe. Poland, Czechoslovakia, and Hungary demanded freely elected governments and private ownership of farms and factories. Independence movements also took shape in Albania, Bulgaria, Romania, and Yugoslavia. "We're in the middle of a revolution," commented one Eastern

European in 1990, "and nowhere near the end of it." As this remark implies, Eastern Europeans are clearly trying to set a new course for the future.

SECTION **2** REVIEW

Developing Vocabulary
1. Define: multiethnic

Reviewing Main Ideas
2. What ethnic groups settled in Eastern Europe?
3. How has the rise of industry affected the population distribution in Eastern Europe?

Critical Thinking
4. **Determining Relevance** How might the low standard of living in Eastern Europe have encouraged opposition to communism?

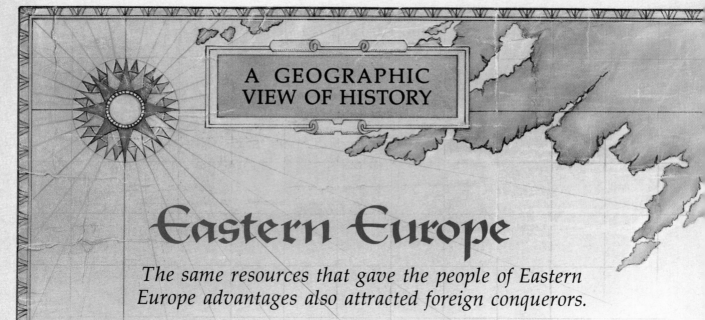

Eastern Europe

The same resources that gave the people of Eastern Europe advantages also attracted foreign conquerors.

One day a group of campers hiked along a trail until they found a spot where three trails met. "What a great spot to pitch our tent," said one camper. "We have ready access to everything we need. One trail goes down to the lake, one to the stream, and one to the berry bushes." So they pitched their tent for the night.

These campers are like the Eastern Europeans. Their region's central location gave them certain advantages. They easily communicated and traded with Western Europe, central Asia, and the Middle East. They also possessed valuable natural resources, especially rich farmland.

But these same resources that have given the people of Eastern Europe so many advantages also attracted conquerors from the east, west, and south. The region's valuable trade routes—the Danube River and the North European Plain—made it temptingly easy for foreign armies to march into Eastern Europe.

Rule by Rome and Byzantium

At the start of the first century A.D., Roman emperors set their sights east. They sent Roman legions to conquer much of modern Bulgaria, Romania, and Albania. By the end of the century, Rome controlled most of the lands south of the Danube.

In A.D. 395, the Roman Empire split into a western empire and an eastern empire. Two rival capitals emerged, Rome in the west and Byzantium in the east. The Christian church split, too. In the west, Christians followed Roman Catholicism, while in the east they practiced Eastern Orthodoxy.

Both Rome and Byzantium (later renamed Constantinople and now known as Istanbul) influenced cultures in Eastern Europe. Today the Roman Catholic church remains strong in Poland, Czechoslovakia, and Hungary.

The eastern part of the Roman Empire became known as the Byzantine Empire. The Byzantines were heavily influenced by cultures along the Mediterranean and Black seas, especially the Greek and Turkish cultures. From about 500 to 1500, the Byzantines spread their influence throughout the Balkan Peninsula. Byzantine missionaries introduced Eastern Orthodoxy and the Cyrillic alphabet, a variation of the Greek alphabet. This alphabet is still used now in Bulgaria, Russia, and parts of Yugoslavia.

Waves of Invaders

"When some hordes depart others immediately appear," lamented one Polish writer of his homeland. Such was the case when the mighty Roman Empire crumbled.

For hundreds of years, waves of invaders—Huns out of Asia, Vikings out of Scandinavia, and Goths out of Western Europe—swept across Eastern Europe. Some groups departed; others, such as the Slavs and Magyars, stayed.

In the 1300s, the Ottoman Turks conquered Constantinople. From here, they pressed into the Balkan Peninsula and other parts of Eastern Europe. The Ottoman Turks brought yet another layer of culture to an already culturally diverse region. The Turks came from southwest Asia. But as they spread west, they introduced many Middle Eastern traditions, including the Islamic religion.

Ottoman power began to decline in the late 1500s, but the Turks managed to keep hold of some regions. Most parts of modern Yugoslavia were not removed from Turkish control until the late 1800s, and Albania did not win independence until 1912.

Dividing Up Eastern Europe

The decline of the Ottoman Empire coincided with the rise of three new empires: the Hapsburg dynasty of Austria, the Prussian Empire in Germany, and the Russian Empire. Eastern Europe's location and resources whetted the appetites of these three land-hungry neighbors. By the eighteenth century Prussian leader Frederick the Great boasted that it would be "consumed leaf by leaf, like an artichoke."

The Hapsburgs had already plucked the first leaf. In the early 1500s, they took control of much of present-day Czechoslovakia. By the mid-1700s, they controlled Hungary and parts of modern Yugoslavia and Albania. In 1867, Austria joined with Hungary to form the powerful Austro-Hungarian Empire.

As Austria took its share of Eastern Europe, the Russians and Prussians did the same. By 1812 the Russians had conquered lands up to the mouth of the Danube River. Meanwhile, the Prussians slowly expanded their bases along the Baltic coast.

In the late 1700s, the three empires finally converged on the plains of Poland. In a series of agreements, they simply divided

Sarajevo, Yugoslavia
An Islamic mosque towers over an old Turkish market and a town that was once part of the Ottoman Empire.

Poland up among themselves. From 1795 to 1918, Poland vanished from the map of Europe.

World War I

In the 1800s, ethnic groups in Eastern Europe began to demand independence and call for kingdoms of their own. Meanwhile, imperial powers in the rest of Europe were competing for worldwide empires. To protect their holdings, some of these nations gradually formed alliances. One alliance included Germany, Austria-Hungary, and Italy. Another included Russia, Great Britain, and France.

In early 1914, United States President Woodrow Wilson sent an aide to investigate the situation in Europe. The aide compared the continent to a powder keg. He told the president: "It only requires a spark to set the whole thing off."

The spark came on June 28, 1914. On that day, a young Bosnian shot and killed Archduke Francis Ferdinand, heir to the throne of Austria-Hungary, and his wife, Sophie. The assassinations took place in Sarajevo [sa ra YAY vo], a city in Yugoslavia.

Austria-Hungary responded by attacking Serbia, which supported Bosnia. Because of the existing European alliances, other nations immediately were dragged into history's first world war. The fighting caused massive destruction all over Europe. It also led to the defeat of the Central Powers: Germany, Austria-Hungary, Bulgaria, and Turkey.

The victorious Allies—France, Great Britain, Italy, and the United States—remapped Europe. Hungary gained its independence. Poland reappeared. Two new nations—Czechoslovakia and Yugoslavia—emerged for the first time. The Allies tried to prevent future wars by organizing ethnic groups into independent countries. But satisfying so many groups proved an impossible task. Some of the new national boundaries cut through ethnic territories. Other boundaries drew together former enemies. In Yugoslavia, for example, Serbs, Croats, and Slovenes found themselves forcibly united.

World War II

Following World War I the quest for empire continued. Japan marched into China and Korea. Italy seized Ethiopia. And in the late 1930s, Nazi Germany's leader, Adolf Hitler, ordered troops into Czechoslovakia and Poland. So began World War II.

In 1939 Germany and the Soviet Union made a pact not to attack each other. Hitler's armies then swept across Eastern Europe. Hitler also enacted a plan to destroy Europe's Jews. This Holocaust, or great destruction, claimed the lives of twelve million people including six million Jews, two million Gypsies, and four million other political prisoners.

In 1941 Hitler broke his promise and turned his armies on the Soviet Union. Soviet forces then joined with troops from Great Britain, France and, soon, the United States

to break Germany's power. The Soviets attacked from the east. The other nations attacked from the west.

When the war ended, the Soviet Union stayed in Eastern Europe. Soviet leaders wanted to create a buffer zone of Communist countries between themselves and the West. In 1946, in Fulton, Missouri, British Prime Minister Winston Churchill spoke these words in a speech to the world.

From Stettin in the Baltic to Trieste in the Adriatic, an iron curtain has descended across the Continent. Behind that line lie all the capitals of the ancient states in central and eastern Europe. Warsaw, Berlin, Prague, Vienna, Budapest, Belgrade, Bucharest, and Sofia, all these famous cities and the populations around them lie in what I might call the Soviet sphere.

The Invasion of Poland, 1939

When Hitler's German army invaded Poland, the outraged Allies immediately declared war on Germany, and World War II began.

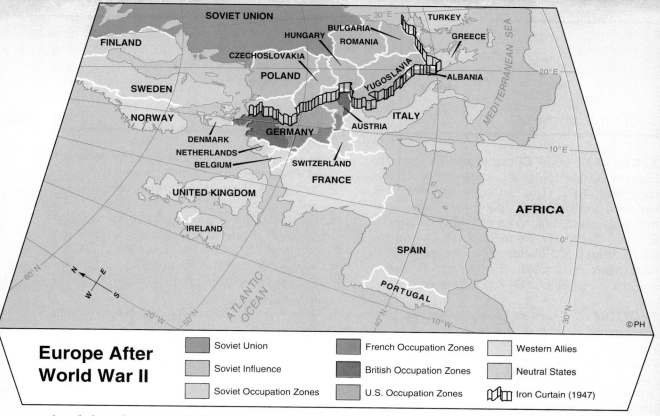

Europe After World War II

▇ Soviet Union	▇ French Occupation Zones	▢ Western Allies
▇ Soviet Influence	▇ British Occupation Zones	▢ Neutral States
▢ Soviet Occupation Zones	▇ U.S. Occupation Zones	▧ Iron Curtain (1947)

©PH

Applying Geographic Themes

Movement After World War II the Soviet Union extended its communist sphere of influence over Eastern Europe. Which present-day country was split in two by the Communist-imposed Iron Curtain?

Eastern Europe Today

In the postwar years, a "cold war," or a war of conflicting ideas fought without bullets, was waged between the Soviet Union and Western nations led by the United States. The Soviets responded to the threatening situation by tightening their control on Eastern Europe's Communist-backed governments. Rebellions flared in Hungary in 1956 and Czechoslovakia in 1968. But Soviet tanks crushed both uprisings.

Tensions in Eastern Europe lessened during the 1980s. The Soviet Union's government focused on its own problems, leaving Eastern Europeans freer to determine their own fates. In 1989 Communist governments were ousted in most countries and Eastern Europe's relationship with the rest of the world changed dramatically.

TAKING ANOTHER LOOK

1. What foreign powers have exercised political control over Eastern Europe?
2. How did efforts by European powers to dominate Eastern Europe help lead to two world wars?

Critical Thinking

3. **Identifying Central Issues** What are some of the key problems that Eastern European leaders might face in the 1990s?

18

REVIEW

Section Summaries

SECTION 1 Land, Climate, and Vegetation
Eastern Europe forms a corridor between Europe and Asia. This strategic location has exposed the region to many invasions. Four major landform regions stretch across Eastern Europe: the North European Plain, which includes most of Poland; the crescent-shaped Carpathian mountains; the Hungarian Basin; and the mountainous Balkan Peninsula.

SECTION 2 Human Geography Eastern Europeans trace their roots to a variety of people, including the Slavs, Magyars, and Bulgars. Because of isolation caused by Eastern Europe's many mountains, dozens of ethnic groups have emerged over time. Roughly two thirds of these groups are descended from the Slavs. Today, most Eastern Europeans live in urban industrial areas. Under Communist rule, Eastern Europe experienced many economic shortages. In 1989, the Soviet Union loosened its hold on Eastern Europe, and the region took the first steps toward building new, independent governments.

Vocabulary Development

Match the definitions with the terms below.

1. soft limestone easily eroded by wind and water
2. composed of many ethnic groups
3. calls for a classless society and common ownership of all economic resources
4. system in which government has complete control over people's lives

a. communism   c. totalitarianism
b. karst   d. multiethnic

Main Ideas

1. How has location affected the development of Eastern Europe?
2. What are the historic origins of the term *Eastern Europe*?
3. How has the mountainous topography in parts of Eastern Europe encouraged the formation of many ethnic groups?
4. How does the standard of living in Eastern Europe compare with the standard of living in Western Europe?
5. How have Eastern Europeans responded to the loosening of Communist ties?
6. Why is the Danube River sometimes referred to as Eastern Europe's "dustless road?"
7. What effect did the switch from agriculture to industry have on Eastern Europe's population?
8. What effect did the "cold war" between the Soviet Union and Western nations have on the countries of Eastern Europe?

Critical Thinking

1. **Drawing Conclusions** Eastern Europe lies astride several seas. How might this have influenced Soviet policy toward the region?
2. **Identifying Assumptions** Under Soviet domination, travel outside Eastern Europe was tightly controlled and often forbidden. What assumptions can you draw about Soviet control from this practice?
3. **Drawing Inferences** What types of challenges do you think the people of Eastern Europe will face in trying to carry out political and economic reforms over the next several years?

Practicing Skills

1. **Identifying Assumptions** Select an editorial from your local newspaper. Using the steps you learned on page 350, analyze the assumptions in the editorial and decide if they are valid. Report your findings to the class.

2. **Distinguishing Fact from Opinion** Read the paragraph in the next column to find two statements of fact and two statements of opinion. Refer to page 308 for information on identifying facts and opinions.

Hungarian Communist party leaders, some of whom are among the most creative politicians in Eastern Europe, helped stir up the rage that brought down the Berlin Wall. They allowed East Germans to flee west through Hungary. Since then, they have pardoned almost every Hungarian ever punished for political crimes, ordered officials out of luxury homes, and pulled down the Red Star from the parliament building. But too much blood has flowed under the bridge, and Hungarian voters don't appear to be in a forgiving mood. Party membership has collapsed, falling from 700,000 to 50,000.

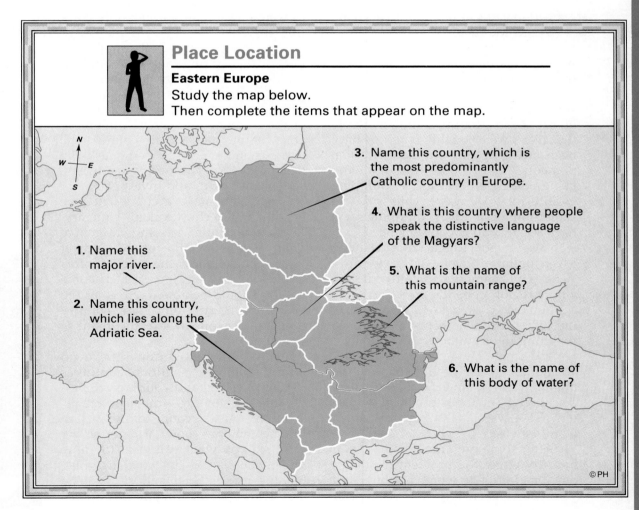

Place Location

Eastern Europe
Study the map below.
Then complete the items that appear on the map.

1. Name this major river.

2. Name this country, which lies along the Adriatic Sea.

3. Name this country, which is the most predominantly Catholic country in Europe.

4. What is this country where people speak the distinctive language of the Magyars?

5. What is the name of this mountain range?

6. What is the name of this body of water?

©PH

The Danube Polluter

You are a member of a United States team of environmental scientists based in Eastern Europe. You have been called to an emergency meeting at headquarters in Belgrade, Yugoslavia, where the head of your organization is waiting in a conference room. As you walk in and take a seat, you see that his expression is tense.

"Terrible things are happening somewhere along the Danube River," he says. "For the third straight week, the river's water quality readings have been swinging into the danger zone. Someone is breaking the law and using the Danube as a dumping ground for toxic waste. Unless we do something fast, that dumping will have some serious consequences."

"We've got to get that Danube River polluter!" your boss declares. "The environmental health of the whole region depends on it."

You ask your boss to give you all the information he can. "Whoever is polluting the Danube is very sneaky," he tells you. "The toxic chemicals are always dumped at night to conceal the point of release. And the chemicals are discharged in an unpredictable pattern." He warns, "Nailing this down won't be easy."

After some preliminary study, you have to agree: the problem's a tough one. In addition to what your boss told you, you've been able to discover only two more facts. From the type of chemicals being dumped, the violator is likely to be a large industrial plant. Also, the pollution is occurring somewhere along the Danube River's course through five countries: Czechoslovakia, Hungary, Yugoslavia, Bulgaria, and Romania. What a problem: almost a thousand miles of riverbank to watch! How can you possibly crack this puzzle?

You sit at your desk, tapping a pencil and gazing out the window. Suddenly an idea hits you. You have political connections that could be helpful. If you can tap your network of government contacts you might be able to find out which country the plant is in. Then you could ask the help of the local authorities to move in, shut them down, and investigate.

First, you make a chart with five headings: Czechoslovakia, Hungary, Yugoslavia, Bulgaria, and Romania. Then you get on the phone and let your contacts know the kind of information you are seeking. As facts come in, you should be able to cross off the possibilities until only one country remains: the one where the plant is operating.

Three days pass. You receive no information, the pollution continues to foul the river, and your boss fumes. On the fourth day—the first breakthrough! Alex, your informant in Prague, phones you. You know he can't tell you what you want to know directly: it would be too dangerous. But his clue is brief and clear. "The polluter is operating in one of the three smallest countries in land area on your list," he says.

Baltic Sea

Gdańsk
Szczecin
POLAND
Poznań
WESTERN
Łódź • Warsaw
Wrocław • Lublin
EUROPE
Prague • Kraków
CZECHOSLOVAKIA
Ostrava
Brno
Bratislava • Miskolc
Debrecen
HUNGARY
Ljubljana • Budapest
Zagreb • Szeged
Cluj
Rijeka • Timişoara
ROMANIA
YUGOSLAVIA
Danube
Ploeşti
Belgrade
Bucharest
Constanţa
Sarajevo • River
Adriatic Sea
Pleven • Varna
BULGARIA
Black
Dubrovnik • Sea
Sofia
Plovdiv
Skopje
Tiranë
ASIA
ALBANIA
MINOR

YUGOSLAVIA
BULGARIA
ROMANIA
HUNGARY
CZECHOSLOVAKIA

You quickly consult an almanac, jotting on your chart the land areas of each country. "Aha!" you say to yourself, crossing two nations off your list. "That helps some."

The second clue, four days later, comes during a midnight phone call to your home from a source in Bucharest, Romania. "I can say little," she begins. "Only this much—the total population of the country you seek is under fifteen million." Her clue is followed by a click and a buzzing sound.

The field is narrowing. After more research, you draw a neat line through another name on your list. Only two countries are left.

The final tip, also from Bucharest, arrives by mail. It is typewritten and unsigned, and, like the other clues, brief and clear. "The offender's flag has three stripes, the middle one white."

That does it! You know in a flash that you'll be able to find the Danube polluter now. The sources who contacted you can be trusted, and you've done your homework properly. In ten minutes you'll know the name of the country where the culprit is hiding. The local authorities will help you do the rest.

SOLVE THE MYSTERY

From which country is the Danube River being polluted? *Hint:* Use an atlas, almanac, or set of encyclopedias. As you check out each clue, eliminate choices until you are left with one country.

CHAPTER

19

The Countries of Eastern Europe

Chapter Preview

Locate the countries of Eastern Europe covered in these sections by matching the colors on the right with those on the map below.

Sections	Did You Know?
1 **POLAND**	Famous Poles include writer Joseph Conrad, composer Frederic Chopin, and scientists Nicolaus Copernicus and Marie Curie.
2 **CZECHOSLOVAKIA AND HUNGARY**	Until 1873, Hungary's capital, Budapest, was two separate cities, Buda and Pest, that lay on opposite banks of the Danube River.
3 **THE BALKAN NATIONS**	Yugoslavia has an extremely diverse population made up of about twenty-five major ethnic groups.

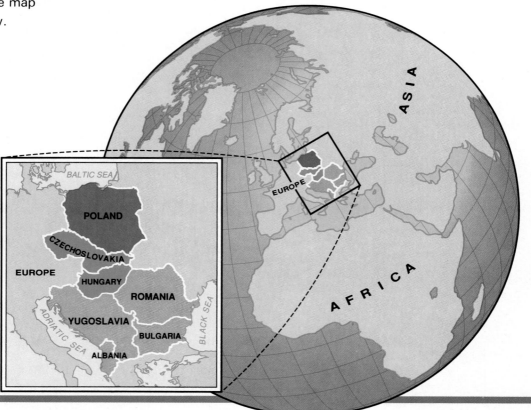

Eastern Europe's "Field Country"
Nearly all of Poland is covered by the broad expanses of the North European Plain. In spite of open fields and fertile soils, farm productivity in Poland is hampered by a lack of modern machinery.

1 Poland

Section Preview

Key Ideas

- The Poles have clung to their culture through nearly two centuries of foreign domination.
- Attachment to the land and belief in the Roman Catholic religion have helped shape the Polish character.
- Today Poles are seeking to set up a free-market economy.

Key Term

ghetto

"If you cannot prevent your enemies from swallowing you, at least prevent them from digesting you." This was the advice that French philosopher Jean-Jacques Rousseau offered to the Poles in the late 1700s.

For more than two centuries, Poles have followed Rousseau's advice. Although they have seen their nation "swallowed" many times, the Poles have refused to allow foreign nations to "digest" the Polish national identity. For example, although the Soviet Union controlled Poland in the years following World War II, the Polish people launched an independence movement during the 1980s to reclaim their identity as a nation.

Place: The "Field Country"

One factor that has helped the Poles, and many other nationalities, retain their identity as a people is their attachment to the land. Most of Poland is covered by the North European Plain. Forests once covered the flat lands, but the trees were cut long ago to create farmland. Today, more than four fifths of Poland is open fields.

Although most of Poland's soil is fertile, it tends to become poor and sandy in the east and northeast. In the northeast, around the Baltic Sea, thousands of wooded lakes break up the landscape. In these forests, boar, elk, wolves, and bison can still be seen.

Poland has valuable industrial resources, too. In the Carpathian Mountain region of the south, large deposits of coal, sulfur, and copper have been found. However, Poland depends upon other countries for two vital minerals—iron ore and petroleum.

A Polish Nation

Today nearly ninety-five out of every one hundred people who live in Poland are Roman Catholic. But this was not always the case. Before World War II, Poland was a multiethnic nation. However, Nazi occupation and Soviet control changed this.

The Holocaust   Today no more than nine thousand Jews live in all of Poland—a nation once home to more than 3.5 million Jews. Almost all of Poland's Jews were destroyed at the hands of the Nazis during World War II. The Nazis sealed off Jewish ghettos in Polish cities such as Warsaw. A ghetto is an area of a city where a racial minority is forced to live. When Jews in the Warsaw ghetto rebelled, the Nazis slaughtered all the people remaining in the ghetto and burned it to the ground.

The Nazis also set up six of their infamous concentration camps, or prison camps, in Poland. Here people from many nations suffered or were brutally murdered, but the majority of those who lost their lives were Poles. By the war's end, roughly six million Poles were killed in concentration camps, about half of them Jews. This great destruction of human life together with the deaths of six million additional Europeans has become known as the Holocaust.

Fleeing Soviet Control   Following the war, the Soviet Union took over lands in eastern Poland. The Soviets then expanded Poland's western border into what had once been Germany. Millions of Poles fled from lands swallowed up by the Soviet Union. Germans living in lands given to Poland also fled. Today, therefore, nearly everyone in Poland is Polish.

Applying Geographic Themes

1. Movement   Judging from its location, why do you think Gdansk is an important Polish city?

2. Location   Describe the location of Warsaw.

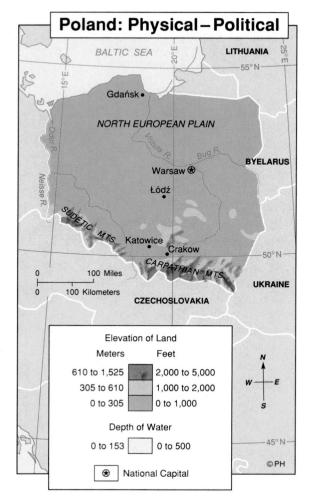

Poland: Physical–Political

Elevation of Land

Meters		Feet
610 to 1,525		2,000 to 5,000
305 to 610		1,000 to 2,000
0 to 305		0 to 1,000

Depth of Water

0 to 153		0 to 500

⊛ National Capital

The Influence of the Church

The Roman Catholic church has played an important role in shaping Polish culture. Even when there was officially no Polish nation, churches continued to unify the Poles.

When the Soviet Union took effective control of Poland, it tried to ban religious worship there. However, as in the past, priests took the lead in opposing foreign rule. Church leaders in Poland worked out an uneasy compromise with Communist leaders that allowed Catholic churches to remain open.

In 1978, a Polish-born bishop became Pope John Paul II, leader of the Roman Catholic church. When the Pope visited his homeland in 1987, he angered Communist officials by voicing support for a Polish workers' union known as Solidarity.

Solidarity

Solidarity gained worldwide recognition in August 1980. That summer, an extraordinary strike by shipyard workers in the Baltic port of Gdansk gave birth to Poland's first independent workers' union.

Solidarity pressed for economic reform and greater personal freedom. Its leader, Lech Walesa (LEK vah LEN sah) boldly declared:

> If you look at what we Poles have in our pockets and in our shops, then communism has done very little for us. . . . They wanted us to be afraid of tanks and guns, and instead we don't fear them at all.

In December 1981, Poland's Communist government outlawed Solidarity. However, the movement did not die. Support for Solidarity grew in the late 1980s when the Soviet Union loosened its grip on Poland. In the spring of 1990, Poland held its first free elections in more than forty years. The voters elected leaders of Solidar-

Solidarity

Solidarity leader and recently elected president Lech Walesa and Polish workers seek to lead Poland into the world's free-market economy.

ity, and rejected many of the Communist leaders running for reelection. The new leaders immediately began the huge task of shifting Poland's government-controlled economy to one based on a free market.

■■■■ SECTION **1** REVIEW ■■■■

Developing Vocabulary
1. Define: ghetto

Place Location
2. Where is Gdansk located?

Reviewing Main Ideas
3. What cultural traits have helped define the Polish people as a nation?

Critical Thinking
4. **Perceiving Cause-Effect Relationships** How do you think the economic changes in Poland will affect workers?

Belt Tightening in Poland

"There's no country in the world in a state of crisis, with the economy as bad as here," said a Communist official in 1988. This was an amazing thing for a Communist leader to tell a Western reporter. But in 1988 Poles openly discussed their desire for a change in the government. "They [the Communists] were wrong on all scores," a newspaper editor told the same reporter. "Instead of a socialist [society], they've created a monstrosity."

The 1990 Revolution

In 1990, the Poles successfully ousted the Communists from power. "The Communist regime," explained one voter, "could never really reform itself from within." Poles in Warsaw celebrated victory by hanging a wreath from the door of the headquarters of the Communist party.

In early 1990, Poles had high expectations for the future. But democracy was no quick cure for Poland's economic ills. The economic decline under communism had been too steady and too extensive. The new government faces pressing economic problems. Farmers in Poland still farm small plots of land. Modern farm equipment and feed for cattle and hogs are in short supply. So agricultural production remains low, and Poland must import food.

Poland has many industries. And these industries bring prosperity to many parts of Poland. However, some regions, particularly northern and central Poland, need to develop new industries. The development of these new industries is one challenge facing Poland as it begins its switch to a free-market economy.

With the end of communism, goods suddenly filled stores. However, few workers could afford the prices, as financial support from the government, which had kept prices low, also ended. In 1989, an average Polish family spent 30 percent of its income on food. By late 1990, that same family spent 60 percent of its income on food.

Other costs skyrocketed, too. Rents in Warsaw quadrupled. Gasoline suddenly cost five times as much, and owning a car became a luxury few could afford.

Shipworkers in Gdansk Shipworkers were among the first to oppose Communist rule.

Restless Workers and Soup Kitchens

By the end of 1990, the number of unemployed topped one million. Much of the joblessness resulted as old, outdated factories closed. Without support from the government, these factories could not survive the shift to a free market. They made too little money to stay in business.

To help the unemployed, the Poles set up soup kitchens where people could receive food free of charge. Workers demanded higher wages. Shipworkers in Gdansk, the birthplace of Solidarity, once again talked of strikes.

Hopes for the Future

Poland's deep economic troubles promise to test the nation's commitment to democracy. But for the first time in many years, the Poles have shaken themselves free of foreign domination. ''Yes, this is real,'' said one Pole. ''There's no turning back. We are free people!'' To maintain their newfound freedom, most Poles are willing to tighten their belts and wait for capitalism to take root.

A manager at an automobile plant noted changes as early as late 1990. ''The reforms were needed,'' said the manager. ''At the moment the threat of unemployment appeared, we had a 20 percent turnaround in productivity.'' Polish officials are hoping for similar success stories in the upcoming years.

TAKING ANOTHER LOOK

1. How has the removal of government controls affected the Polish economy?
2. Why has the switch to a free-market economy forced many factories and businesses to close?

Critical Thinking

3. **Determining Relevance** How might Poland's history help Poles withstand the economic hardships of the 1990s?

Shopping in Poland *High costs make fresh fruits, vegetables, and cut flowers luxuries that many Poles cannot afford.*

✔ Skills Check

- ☐ Social Studies
- ☐ Map and Globe
- ☑ Reading and Writing
- ☐ Critical Thinking

Writing Effective Paragraphs

There are many different ways to write an effective paragraph. However, all well-written paragraphs have three elements. The first of these elements is unity. Unity means that all of the sentences in the paragraph are closely related to the main idea. The main idea may be stated in a topic sentence, or it may be only implied. In either case, every sentence in the paragraph will support the main idea.

A second element of an effective paragraph is development. This means that the main idea is explained clearly so that the reader can understand it. It also means that enough details or examples are included to make the main idea believable

The third element of an effective paragraph is coherence. Coherence means that all of the ideas in the paragraph are organized logically so that the paragraph reads smoothly and the points it makes are clear.

Use the following steps to arrange the sentences below into a unified, coherent, and well-developed paragraph.

1. **Identify the main idea you want to express.** A topic sentence both expresses the main idea and defines the scope of a paragraph. It should be neither too broad nor too narrow. Answer the following questions: (a) Which of the sentences below would make the best topic sentence? (b) Why is this sentence the best choice?

2. **Arrange the topic sentence and supporting statements in logical order.** Supporting statements contain facts or ideas related to the main idea. They should be arranged carefully to make sense and read smoothly. Answer the following questions: (a) Should the topic sentence come first? (b) In what order should the supporting statements below be organized? (c) Why does this order make the most sense?

3. **Use transition words for clarity.** Transition words such as *although* and *therefore* are devices that help to hold a paragraph together. After forming a paragraph using the statements, answer the following question: What are two transition words in the paragraph you have formed?

Statements

A. On September 3, 1939, Britain and France declared war on Germany.
B. Confident of his support and his strength in Eastern Europe, Hitler invaded Poland on September 1, 1939.
C. It took a long and costly war to defeat him.
D. Hitler had grown strong, however.
E. World War II had begun.
F. He believed that Britain and France would go along with him again, but this time he was mistaken.
G. Some thought the conflict would end in a matter of weeks.

370

2 Czechoslovakia and Hungary

Section Preview

Key Ideas

- Czechoslovakia has two main ethnic groups: Czechs and Slovaks.
- Hungary has both agricultural and industrial strength.

Key Term

collective farm

Bratislava, Czechoslovakia
These young people are gathered around a kiosk, or pillar with public information, in the city of Bratislava.

"A Return to Europe!" Czech President Vaclav Havel (VAH tsluhv HAH vuhl) chose these words as one of his campaign slogans in 1989. Today, people visiting Havel's offices are warned, "Don't call the Czechs Eastern Europeans."

Similar talk is being heard in Hungary. In 1990, a Catholic priest who had fled Communist Hungary in 1947 returned to address the nation's new, independent Parliament. He said, "Implanted in every Hungarian head and heart must be the knowledge that we belong to Europe."

Czechoslovakia and Hungary differ in many ways. But they share traditionally Western outlooks. In 1990, after more than forty years of Communist control, both rejoined the community of free European nations.

The Landscape of Czechoslovakia

Czechoslovakia occupies an area about the size of the state of New York. It has few flat areas, except the plains that stretch along the Elbe and Danube rivers. The landscape is dominated by rolling hills and plateaus. Mountainous ridges define the country's boundaries to the north, northwest, and southwest.

More than fifteen million people live in Czechoslovakia. It is a multi-ethnic nation, but the two dominant groups are Czechs and Slovaks.

Two Peoples United

In the final days of World War I, the Allies approved the Czechoslovak National Council's plan to unite the Czechs and Slovaks into a single republic. Czechoslovakia came into being on October 28, 1918.

Many Germans lived in Czechoslovakia, too. Hitler eventually used this fact as an excuse to invade the republic. After the Soviets liberated Czechoslovakia in 1945, they backed a Communist takeover. From 1948 to 1989, Communists ran the nation.

In 1989, Czechoslovakia held its first democratic elections in more than forty years. Vaclav Havel—poet, philosopher, and anticommunist—was elected President.

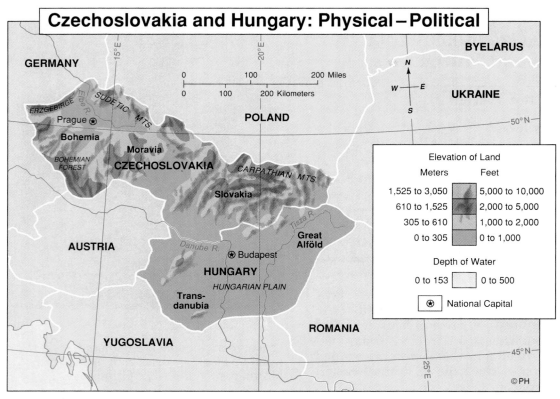

Czechoslovakia and Hungary: Physical–Political

GERMANY

BYELARUS

UKRAINE

POLAND

Prague

Bohemia

BOHEMIAN FOREST

Moravia

CZECHOSLOVAKIA

Slovakia

CARPATHIAN MTS.

AUSTRIA

Danube R.

Budapest

Great Alföld

HUNGARY

HUNGARIAN PLAIN

Trans-danubia

ROMANIA

YUGOSLAVIA

SUDETIC MTS.

Elbe R.

ERZGEBIRGE

Tisza R.

0   100   200 Miles

0   100   200 Kilometers

Elevation of Land		
Meters		Feet
1,525 to 3,050		5,000 to 10,000
610 to 1,525		2,000 to 5,000
305 to 610		1,000 to 2,000
0 to 305		0 to 1,000

Depth of Water		
0 to 153		0 to 500

National Capital

©PH

Applying Geographic Themes

Place   Compare this map with the map on page 366. How do the physical characteristics of Czechoslovakia differ from those of Poland?

Interaction: Saving the Environment

One of the biggest problems facing Czechoslovakia, said Havel, was saving the environment. "We have laid waste to our soil and the rivers and the forests, and we have the worst environment in the whole of Europe today."

Experts predicted that some 60 percent of the nation's forests would be destroyed by acid rain and industrial pollution by the year 2000. The experts also noted that water and polluted air had become major health problems. One Czech scientist stated: "If you go to the doctor with a sore throat, cough, or a headache, the [doctor] . . . will tell you, 'You must have opened a window last night.'"

The Czech Economy: Three Regions

The industries that threaten the Czech environment are also the basis of the Czech economy. When Czechoslovakia was created in 1918, it inherited a wealth of existing industries from the Austro-Hungarian empire to which it had belonged. More than 40 percent of the old empire's labor force worked in Bohemia, Moravia, and Slovakia—the three regions that today form Czechoslovakia.

Bohemia   Bohemia is the western region and industrial heart of Czechoslovakia. Mines in the north produce coal, iron ore, copper, and lead. Bohemia also has deposits of quartz, which is used to make glass.

More than one hundred church steeples rise above the city of Prague, the Czech capital, as a visual reminder of the region's Roman Catholic heritage. Equally visible is the dense, blue-gray haze that covers the city as a result of the industry concentrated in the area. Chemical fumes pour from the chimneys of coal-burning plants.

Moravia   Moravia is the central region of Czechoslovakia. Coal mines and coal-burning industrial plants have led Czechs to nickname the northern part of Moravia "the Black Country."

Moravia has some of the oldest factories in Czechoslovakia. The region greatly profited from the early arrival of the Industrial Revolution. And it continued to prosper under Communist rule.

Moravia's old coal and steel industries now face an uncertain future. In the 1990s, free trade returned to Czechoslovakia. In the past, the Soviet Union accounted for more than 50 percent of the nation's trade. Czech leaders are now faced with making Moravia's industries competitive in the world market. "Whole sectors of industry are producing things in which no one is interested," said Vaclav Havel in his first speech as President. "Our outdated economy is squandering energy, of which we are in short supply."

Slovakia   Slovakia is Czechoslovakia's eastern region. Cradled within an arch formed by the Carpathian Mountains, its lands unfold from rugged peaks in the north to the fertile plains of the Danube in the south.

Slovakia is the most agricultural part of Czechoslovakia. In 1948, the Communist government ended private ownership of farms and set up state-owned collective farms. On a collective farm, workers were paid by the government and shared the profits from their produce.

Not until the 1960s did the number of industrial workers in Slovakia equal the number of agricultural workers. Because Slovakia was the poorest of the country's three regions, the Communist government built new plants there. Higher wages and paid vacations lured many Slovaks off the land and into the factories. Today, only about 15 percent of the region's people till the land.

Ethnic Tensions

The Czechoslovakian government built the factories in part to reduce tensions between Czechs and Slovaks. Slovaks number five million in a nation of fifteen million. They jealously guard their rights as a minority. Like the United States Congress, Czechoslovakia's lawmaking body, the Federal Assembly, has two houses. In the

WORD ORIGIN

Slovakia
The Czech region of Slovakia, "land of the Slavs," stems from the word *slave*. In the Middle Ages, the Slav people were enslaved or destroyed during Germany's eastward expansion.

An Industrial Economy
Industry, not agriculture, is the basis of the Czech economy. These women are soldering components in a Czech television factory.

DAILY LIFE

Learning the Truth

Czech teachers write the words "Prague Spring" on the chalkboard. They tell students about attempts to win reform in the spring of 1968 and about brutal Soviet arrests. Hungarian teachers, like the teacher at right, openly discuss the 1956 uprising. They no longer call Soviet intervention "brotherly assistance." They call it repression. Students who normally fidget listen intently.

After years of government censorship, teachers are speaking freely. "At last we can tell the truth to children," said a teacher in Prague. What is the truth? One student put it this way: "Communism made mistakes."

1. How has education changed in Eastern Europe?
2. Why do you think control of education was important to Communist leaders?

---

lower house, representation is by population. Czechs, therefore, hold the majority of seats. In the upper house, however, representation is by political subdivision. Since Slovakia covers one half of the country, it receives the same representation as Bohemia and Moravia combined.

Hungary: A Thousand-Year History

Unlike Czechoslovakia, Hungary has no ethnic divisions. About 95 percent of Hungarians are descended from the Magyars who settled the area in the late 800s.

The Hungarians also claim a longer history than Czechoslovakia. They date the start of their nation at 1001, the year the Pope crowned St. Stephen king. St. Stephen is still one of Hungary's most revered rulers.

The Roman Catholic faith and fierce patriotism have guided much of Hungary's history. The Hungarians revolted against their Austrian rulers, and later against their Communist rulers in 1956. In 1990, Hungarian voters freely elected their first non-Communist government since 1947. As one member of the new Parliament triumphantly declared, "The Magyar nation has been preserved!"

The Hungarian Landscape

Hungarians still speak a very different language from their Slavic neighbors. But this does not mean that they have been culturally isolated. Throughout their history, the Hungarians have actively traded with all parts of Europe. They also commanded an empire that included many Slavic groups. Hungary was created along with Austria, Yugoslavia, Czechoslovakia, and Poland at the end of World War I, when the Austro-Hungarian empire was broken up.

Today, Hungary is about the size of the state of Indiana. The Danube River divides it into two parts. The eastern half consists of a broad plain known as the Great Alfold. This region's fertile soil has led Hungary to be called "the breadbasket of Europe."

The western half of Hungary has more hills. Because this region lies west of the Danube, it is known as Transdanubia, or "across the Danube." This region is an area of plateaus, hills, and valleys. It contains enough deposits of bauxite, coal, and iron ore to support Hungary's aluminum and steel industries.

Seeds of Free Enterprise

When newly elected officials walked into Parliament in October 1990, an American reporter asked one member to describe Hungary's biggest problem. "The past forty years," he replied.

Like other nations in the region switching from an economy controlled by the government to one based on a free market, Hungary faces many economic difficulties. But in many ways the Hungarians are better prepared for the change than some of their neighbors. The collective farms set up by the Communists were successful. But so were small, privately held plots.

In 1989, about 6.5 percent of Hungary's farmland belonged to private individuals. These plots produced just under one third of the nation's entire agricultural output and slightly more than one third of its meat. The new democratic government is allowing both collective and private ownership to continue, with an emphasis on individual production.

Free enterprise never died in Hungary. Even under communism, people continued to buy goods or services produced by privately owned businesses. Often Hungarians gave more time to their small, privately run businesses than to the state-run businesses. "We are socialists in the morning and capitalists in the afternoon," declared one Hungarian joke.

In the 1970s, a mathematics professor named Erno Rubik invented a cube-based puzzle to help his students learn. Rubik's cube made millions of dollars on the world market, and its inventor became rich. Officials hope that such initiative will help Hungary compete successfully in world markets in the 1990s.

SECTION 2 REVIEW

Developing Vocabulary
1. Define: collective farm

Place Location
2. What are the three regions that form Czechoslovakia and where are they located?

Reviewing Main Ideas
3. In what ways does Czechoslovakia's government reflect its multiethnic population?
4. Why do Hungarian leaders think their country will do well competing in the free market?

Critical Thinking
5. **Making Comparisons** How do the human geographies of Czechoslovakia and Hungary differ?

3 The Balkan Nations

Section Preview

Key Ideas

- The Balkan nations share a history of internal division and foreign domination.
- Yugoslavia contains many ethnic groups.
- Democratic revolutions have shaken Communist control of the Balkan Peninsula.

Key Term

nonaligned

Sarajevo, Yugoslavia

Located in the scenic Dinaric Alps, Sarajevo was the site of the 1984 Winter Olympics and is known for its mosques—Muslim houses of worship.

In 1918, a new term crept into the English language: *balkanize*. The word *balkanize* means to break up into small, mutually hostile political units, as occurred in the Balkans after World War I. The term grew out of the complex cultural and political geography of the Balkan Peninsula.

Perhaps the one thing that the Balkans share is their historical experience. The peoples of this region have all known the ordeal of foreign domination. For five hundred years they were ruled by the Turks, whose influence can be seen to this day.

Today the Balkan Peninsula is divided into five nations, including Greece. Four of these nations—Yugoslavia, Albania, Romania, and Bulgaria—fell under communism after 1948. In the late 1980s and early 1990s, anticommunist rebellions swept the peninsula. However, the march toward democracy has been slow and sometimes violent.

Yugoslavia: Divided Regions

"We're all supposed to be Yugoslavs. But scratch one of us, and you'll find a Serb or Croat or something else." This comment by a Croat lawyer provides a clue to the human geography of Yugoslavia. Yugoslavia contains twenty-four million people. But they are divided among twenty-four ethnic groups and three religions—Roman Catholicism, Eastern Orthodoxy, and Islam. Both the Latin and Cyrillic alphabets are used in Yugoslavia, which is composed of six republics. These republics are: Serbia, Croatia, Slovenia, Montenegro, Bosnia-Herzegovina, and Macedonia.

Yugoslavia's physical geography is equally varied. To the north stretches the Danube River and the Hungarian Basin. But, as discussed in Chapter 18, mountains, which cover nearly 70 percent of the land, are the region's major landform.

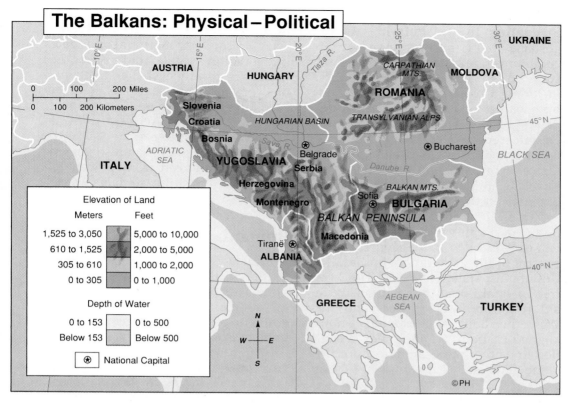

The Balkans: Physical–Political

AUSTRIA
HUNGARY
UKRAINE
CARPATHIAN MTS.
MOLDOVA
ROMANIA
Slovenia
Croatia
HUNGARIAN BASIN
TRANSYLVANIAN ALPS
45°N
Bosnia
Sava R.
Bucharest
ADRIATIC SEA
YUGOSLAVIA
Belgrade
BLACK SEA
ITALY
Serbia
Danube R.
Herzegovina
BALKAN MTS.
Sofia
Montenegro
BULGARIA
BALKAN PENINSULA
Macedonia
Tiranë
ALBANIA
40°N
GREECE
AEGEAN SEA
TURKEY

0   100   200 Miles
0   100   200 Kilometers

Elevation of Land

Meters	Feet
1,525 to 3,050	5,000 to 10,000
610 to 1,525	2,000 to 5,000
305 to 610	1,000 to 2,000
0 to 305	0 to 1,000

Depth of Water

0 to 153	0 to 500
Below 153	Below 500

⊛ National Capital

N W E S

©PH

Applying Geographic Themes

1. Place The mountains and forests of the Balkan countries have served to create pockets that have isolated the people of the region. What mountain system dominates Yugoslavia? What mountains dominate central Romania?

2. Location What river forms the border between Romania and Bulgaria?

A Nation Divided Yugoslavia means "Land of the Southern Slavs." But a common Slavic ancestry does not mean unity. In fact it means twenty-four fiercely independent ethnic groups living in an area roughly the size of the state of Wyoming.

The two largest groups are the Serbs and the Croats. Serbs make up about 40 percent of the nation's population. Most live in Serbia, home of the Yugoslav capital of Belgrade. Croats account for about 20 percent of the population. They dominate Croatia, which forms one of Yugoslavia's most prosperous industrial centers.

Differences between the Serbs and Croats highlight divisions among the Yugoslavs. Both groups are descended from the same early Slavic peoples that settled the region. Their spoken languages are nearly identical. But the Serbs practice Eastern Orthodoxy and write in the Cyrillic alphabet. The Croats, on the other hand, practice Roman Catholicism and write in the Latin alphabet.

Other ethnic groups include the Slovenes, Montenegrins, Hungarians, Bosnians, Macedonians, and Albanians. The Albanians, who number nearly five hundred thousand, do not have their own province. Most members of the Albanian ethnic group follow Islam and live in the southern portion of Serbia. Violent clashes have taken place between the Serbs and Albanians.

WORD ORIGIN

Cyrillic
St. Cyril, a ninth-century Greek missionary, invented the Cyrillic alphabet, now used in Russian, Bulgarian, and other languages of Northern Eurasia.

Struggling Toward Reform As Yugoslavia attempts to move away from communism, ethnic divisions have resurfaced. In 1991, several republics, including Croatia, Slovenia, and Bosnia-Herzegovina declared their independence and a civil war broke out. Despite repeated cease-fire attempts, fighting took place between Croatian forces on one side and the Serbian-led Yugoslav military and Serbia-backed guerillas on the other.

This situation has ended a long period of stability in Yugoslavia. At the end of World War II, Marshal Tito took control of the government and built a Communist state. However, Tito kept free of total Soviet domination. With aid from both the Soviets and the West, he built a prosperous, nonaligned nation. Nonaligned means tied to neither the Communist nor the democratic bloc of nations.

Albania

Some have called Albania "Europe's hermit." Since the late 1970s the country has turned its back on its neighbors.

Albania has paid a price for its isolation. It is one of the poorest nations in Eastern Europe, and few of its workers earn more than eighty dollars a month. Albania has many mineral resources, including chromium, copper, and oil. But its isolation has made it difficult to export these goods.

In 1990, Albania's leaders cautiously tried to form new ties with the rest of the world. The nation's Muslims, who make up 70 percent of the population, and members of other Christian churches pressured the government to allow them to practice their religions freely. But calls for reform met with stiff opposition from the Communist party.

In the summer of 1990, more than four thousand Albanians fled the nation. News of anticommunist revolutions elsewhere encouraged them to make the dangerous leap to freedom. They crashed through embassy gates in the nation's capital and asked other nations to give them shelter. Explained one twenty-four-year-old refugee: "We don't want these Communists. We want freedom. We want to think. We want enough to eat."

In 1991, the Communist government was removed from power, but the country's future remains uncertain.

Romania

Romania possesses broad plains of fertile soil along the Danube River and mineral-rich foothills in the Carpathian Mountains. But, as in Albania, government policies have left the nation impoverished.

After World War II, Romania's government was heavily influenced by the Soviet Union. In 1965 a brutal dictator, Nicolae Ceausescu (NIH kah ly chow SHESS koo) took control.

Ceausescu used Romania's wealth to enrich himself. He generally ignored improvements in agriculture, industry, and mining. Energy became so scarce that television shows aired only two nights a week.

Under Ceausescu, owning a typewriter was a criminal offense. All writing and art had to glorify Ceausescu and communism. But Ceausescu could not stop the 1989 democratic revolutions from reaching his nation. In 1990, he tried to flee Romania, but rebels caught and executed him.

A multiparty system emerged in Romania. However, disputes among the parties broke into open riots. Many observers feared the rise of another dictatorship. Yet some leaders expressed hope for the future. Said one party leader:

In forty-five years, [communism] did not succeed economically, socially, or morally. . . . It was a system which produced dictator-

ship. There is no way to reform except abolish the system along with the dictatorship. Our great opportunity . . . is to start with a clean sheet.

Bulgaria

Bulgaria borders the Black Sea to the east and Turkey and Greece to the south. During a long period of Turkish rule, the Russians were among the strongest supporters of native Bulgarians. Both peoples shared Slavic origins and a belief in Eastern Orthodoxy. These ties led Bulgarians to view Soviet control following the end of World War II more favorably than did other Balkan nations.

The Danube plains that run through Bulgaria provide the region with much fertile soil. The nation's southern location along the Black Sea also gives it a long growing season. Bulgaria is called "the garden of Eastern Europe." But under Soviet domination most of its produce went to the Soviet Union. Some Bulgarians complained that this kept their nation poor. Still, Bulgaria remained one of the Soviets' most loyal allies.

When the Soviets loosened controls on Bulgaria, the nation held free elections. Communists claimed 40 percent of the vote in 1990, thereby guaranteeing their party a strong voice in Bulgaria's government. One supporter of democracy explained the situation Bulgaria faced in the early 1990s.

We're in the middle of a revolution and nowhere near the end of it. The second stage of the revolution [building a new government] . . . is a bigger challenge than the first. . . . If nothing is done soon to cement the gains we have made a new form of totalitarianism lies ahead.

A Democratic Revolution in Romania—1989

In a newly democratic Romania, citizens display their country's flag. The symbols of the country's former communist regime have been cut away.

SECTION **3** REVIEW

Developing Vocabulary
1. Define: nonaligned

Place Location
2. Describe Bulgaria's location.

Reviewing Main Ideas
3. What historical experiences have the Balkan nations shared?
4. How do ethnic divisions influence government in Yugoslavia?

Critical Thinking
5. **Drawing Inferences** What do you think are some of the possible sources of conflict among Yugoslavia's ethnic groups?

19

REVIEW

Section Summaries

SECTION 1 Poland Poland has valuable agricultural and industrial resources. The Polish people have clung to their culture through nearly two hundred years of foreign domination. Two keys to the Polish national identity are attachment to the land and to the Roman Catholic religion. Nazi destruction of the Jews during the Holocaust and Soviet remapping of Polish borders after World War II dramatically reshaped Poland's human geography. Today 95 percent of the people living in Poland are Roman Catholic. During the 1980s Solidarity pressed for economic reform and personal freedom. In 1989, the Poles took the first steps toward building a democratic republic.

SECTION 2 Czechoslovakia and Hungary Both of these countries rebelled against their Communist governments in the late 1980s. There are tensions between the two main ethnic groups within Czechoslovakia—the Czechs and Slovaks. Almost all Hungarians, on the other hand, are descended from the Magyars who first settled the region. Both Czechoslovakia and Hungary face difficult economic periods as they try to transform their Communist, state-controlled economies into free-market systems.

SECTION 3 The Balkan Nations Internal ethnic divisions and long periods of foreign domination have troubled the people of the Balkan Peninsula for centuries. Five hundred years of Turkish rule has left a strong influence on the region. Today Balkan countries are struggling on a middle course between the Communist governments of the past and the democratic governments that some people hope to build.

Vocabulary Development

Use each term in a sentence that shows the meaning of the term.

a. ghetto
b. collective farm
c. nonaligned

Main Ideas

1. How has repeated territorial conquest of Poland affected its human geography?
2. Why has the Roman Catholic church been a strong unifying force in Poland?
3. How did the rise of Solidarity reflect Polish nationalism?
4. How have ethnic divisions within Czechoslovakia helped shape its government?
5. How have history and religion influenced the national character of Hungary?
6. How did Communist rule change the agriculture and industry in Czechoslovakia and Hungary?
7. Why did the political remapping of the Balkan Peninsula after World War I intensify ethnic rivalries?
8. How has each of the following nations responded to the loosening of Soviet ties: Yugoslavia, Albania, Romania, Bulgaria?

Critical Thinking

1. **Formulating Questions** Not all Eastern European nations have the same attitude toward government. What questions would you ask leaders in Poland, Yugoslavia, and Albania to determine the type of government they plan for the 1990s?

2. **Recognizing Bias** Many people in Eastern Europe believe that use of the term "Eastern" reflects a bias on the part of the United States and the rest of Europe. Do you agree? Why or why not?

3. **Identifying Alternatives** What economic alternatives are available to people living in Czechoslovakia and Hungary during the 1990s? How do these alternatives differ from the ones they had in the 1950s?

4. **Drawing Conclusions** Why might Czechoslovakia need international cooperation to save its environment?

Practicing Skills

Writing Effective Paragraphs Using the steps you have learned on page 370, write a paragraph giving your opinion on a recent issue in your school or community. Be sure to create an effective topic sentence and to use clearly organized supporting statements and transition words where needed. Finally, exchange your paragraph with that of another class member and analyze each other's paragraphs for unity and coherence.

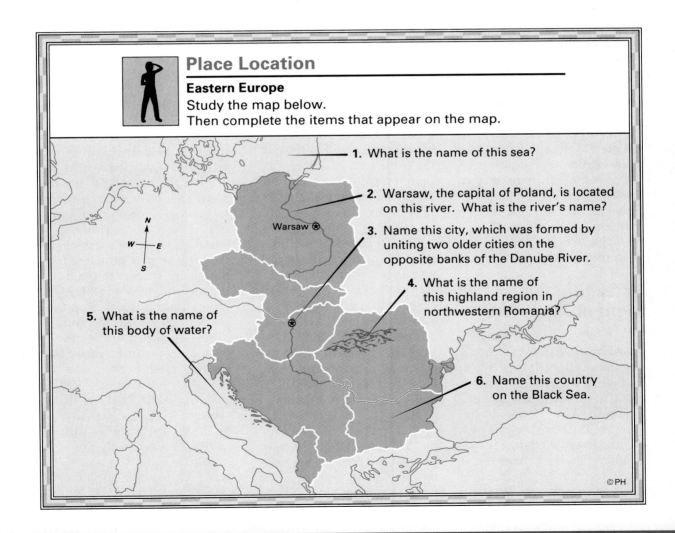

Place Location

Eastern Europe
Study the map below.
Then complete the items that appear on the map.

1. What is the name of this sea?

2. Warsaw, the capital of Poland, is located on this river. What is the river's name?

3. Name this city, which was formed by uniting two older cities on the opposite banks of the Danube River.

4. What is the name of this highland region in northwestern Romania?

5. What is the name of this body of water?

6. Name this country on the Black Sea.

Warsaw

©PH

MAKING
CONNECTIONS
·WHERE REGIONS MEET·

World Communication Links

It is not hard to imagine the scene: A family, sitting around a table in their cramped apartment, listening to the shocking, almost unbelievable news coming over their radio. It is news of a sort of revolution taking place in a far-away nation. Yet it is news that promises to shake the foundations of their own nation. As they listen to the events unfolding in the Soviet Union and hear the changes in the structure and power of the government there, they instantly recognize the opportunity in their own nation. For this family, the radio carries a message of hope that one day soon they will be free.

Spreading the Word of Democracy

The shock waves that transformed Eastern Europe's Communist governments began when Mikhail Gorbachev came to power in the Soviet Union in March 1985. Gorbachev instituted many policies of fundamental change in his country. These changes eventually led many Eastern European nations to seek similar

and, in many cases, more far-reaching reforms.

One of the most important factors behind the democratic movement in Eastern Europe was communication. The media—television, newspapers, and radio—were instrumental in keeping the message of freedom alive and moving in Eastern Europe. Voice of America (VOA), Radio Free Europe, Radio Liberty, and the World Service of the British Broadcasting Corporation (BBC) kept Eastern Europeans from isolation and provided Western views of events. East Germans received television broadcasts from West Germany. Satellite dishes in Hungarian villages transmitted television broadcasts from the West. Copying and fax machines helped disseminate information.

Citizens heard of the first reforms in Poland and Hungary—and learned that the Soviet Union was not rushing in to crush the spirit of democracy and freedom as it had in the 1950s and 1960s. People became inspired by the reality that the changes in the Soviet Union were genuine. When

Lech Walesa, leader of Poland's Solidarity, was asked if these radio broadcasts had been important to the Polish democracy movement, he responded: "Would there be earth without the sun?"

The World Beyond Europe

The electronic communications signals that carried news of the dramatic events in Eastern Europe traveled far beyond the borders of that region. Today's communications technology ensures that news of events in one part of the world can quickly reach even the most remote corners of the globe. Thanks to radio broadcasts, people in Mongolia—a country located on the border between China and Russia—learned of the new developments in Eastern Europe. "I listen to the radio—BBC, Voice of America, Radio Moscow—and I hear of the changes in Poland," a shepherd told a reporter in early 1990. "I feel strongly that Mongolia must find a new way." Eastern Europe's message of democracy was apparently heard by other Mongolians as well; in mid-1990 Mongolia became the first Communist country in Asia to hold free elections.

The 1989 student-led pro-democracy movement in China also derived strength and inspiration from the news from Eastern Europe. Listening to radio reports of the VOA, BBC, and other communications networks, the students pursued their demonstrations in Beijing. Their movement was brutally crushed by the Chinese government. Yet many students insist that the message of change that inspired them remains alive.

The View from the U.S.A.

 No nation in the world demonstrates a greater appreciation for the power of communications than the United States. The government has a deep commitment to providing news and information to people living under censorship. VOA programming, broadcast in forty-two languages, aims to serve as a reliable source of news and to describe and discuss United States policies. VOA draws 120 million listeners in Europe, Latin America, the Middle East, and Asia.

TAKING ANOTHER LOOK

1. Name one major factor that contributed to the spread of developments in the Soviet Union to Eastern Europe.
2. Why is the United States committed to providing radio broadcasts to nations around the world?

Critical Thinking

3. **Predicting Consequences** How might have events in Eastern Europe and around the world been different if the Soviet Union had intervened militarily in an Eastern European nation in 1989?

Communication Links End Isolation

Even in remote corners of the world, satellite dishes enable people to receive information quickly—like the news of Germany's reunification.

UNIT
6
Northern Eurasia

CHAPTERS

A huge expanse of northern land stretching from the Baltic Sea in the west to the Pacific Ocean in the east . . . a region of tumultuous change. This is Northern Eurasia.

When central power faltered in 1991, the republics that formed the Soviet Union became independent nations. Some of these nations worked to develop a commonwealth. In this unit you will discover how the people of these independent countries have made use of the region's climates, vegetation, and resources.

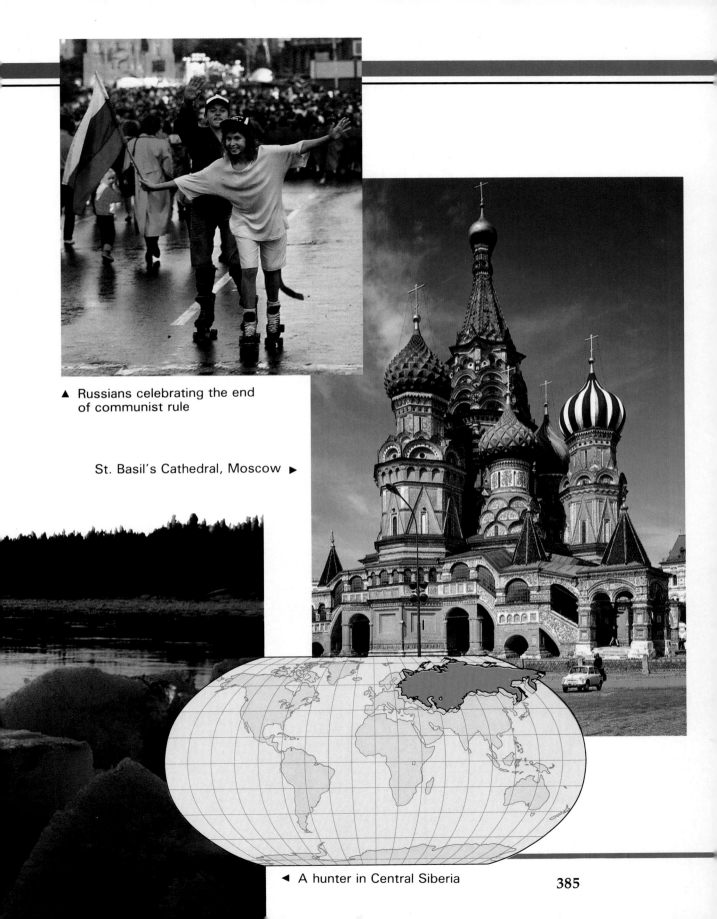

▲ Russians celebrating the end of communist rule

St. Basil's Cathedral, Moscow ▶

◀ A hunter in Central Siberia

385

20

Regional Atlas: Northern Eurasia

Chapter Preview	Sections	Did You Know?
Both sections in this chapter are about Northern Eurasia, shown in red on the map below.	**1** LAND, CLIMATE, AND VEGETATION	The Bering Strait separates Northern Eurasia and the United States by only 3 miles (4.8 km) of water.
	2 HUMAN GEOGRAPHY	More than two thirds of Northern Eurasia's people live in cities, 25 of which have populations of one million or more.

Harvesting Wheat in Kazakhstan
More than one quarter of the land in Northern Eurasia is used for agriculture. Because of inefficient farming methods and poor systems for distributing goods, however, nations in the region must import wheat to feed their people.

1 Land, Climate, and Vegetation

Section Preview

Key Ideas

- Northern Eurasia has plains in the west, becomes more mountainous toward the east, and contains varied landforms on its southern border.
- The climate regions of Northern Eurasia closely correspond to its physical regions.
- Northern Eurasia contains vast areas of tundra, forest, steppe, and desert.

Key Terms

steppe, taiga

Northern Eurasia sprawls across nearly half of the Northern Hemisphere on two continents. More than three quarters of the region is occupied by Russia, the largest country on earth, more than twice the size of the United States. Within Russia and the surrounding nations are vast and varied landscapes. Endless plains, frozen towns, hot desert dunes, busy seaports, and crowded cities are all part of the enormous spaces of Northern Eurasia.

Physical Regions

Even though Northern Eurasia is so large, its basic landforms follow one overall pattern. The land is flat in the west and becomes increasingly mountainous toward the east. Along the southern border lie great lakes and forbidding mountain chains.

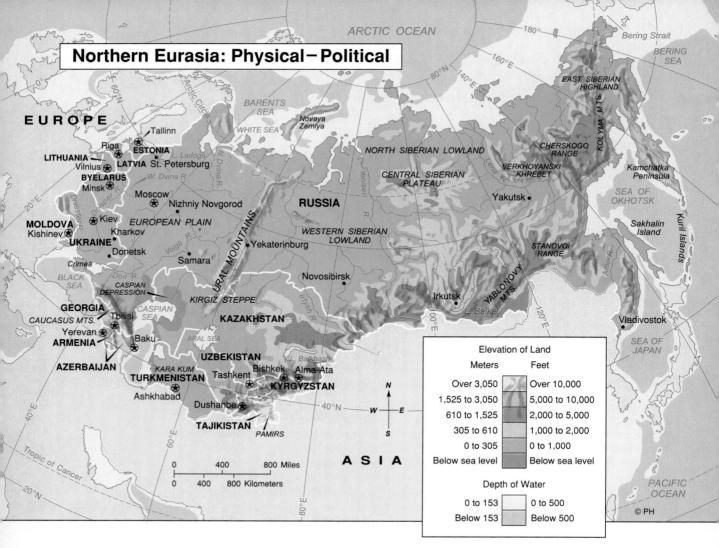

Northern Eurasia: Physical–Political

Elevation of Land

Meters	Feet
Over 3,050	Over 10,000
1,525 to 3,050	5,000 to 10,000
610 to 1,525	2,000 to 5,000
305 to 610	1,000 to 2,000
0 to 305	0 to 1,000
Below sea level	Below sea level

Depth of Water

0 to 153	0 to 500
Below 153	Below 500

© PH

Applying Geographic Themes

Location The Ural Mountains are often described as the border between Europe and Asia. What landlocked waterway lies at the southern end of the Urals? Which sea links the nations of Northern Eurasia with Europe?

Expansive Plains Notice on the map on page 388 that much of Northern Eurasia consists of plains. Its northern plains stretch along the Arctic Ocean. The frozen Arctic Lowlands lie very low at the coast but increase gradually in elevation toward the south. South of the Arctic Lowlands lie two large plains regions that sweep across Russia, separated only by a gentle mountain range, the Ural Mountains.

On the map these mountains may look like a barrier running from north to south, but they actually form no more than a small ripple across the vast Russian plain. The Ural ranges average only 3,000 to 4,000 feet (915 to 1,220 m) in elevation. The highest peak rises only 6,214 feet (1,894 m). One travel writer once wrote: "It is easy to cross the Urals by car or train without ever realizing that you are in the mountains."

Another writer noted, "I have twice flown directly over the Urals and missed seeing them entirely." Some geographers designate the Urals as the dividing line between Europe and Asia, while others suggest that the entire landmass should be seen as the single continent of Eurasia.

The western plain is the continuation of the North European Plain, which stretches across Poland, Germany, and other northern European nations. The land is flat, gently rolling, and very fertile. Just to the east of the Urals lies another lowland region—the Siberian Lowland. It covers more than 1 million square miles (2.6 million sq km) and is the largest area of unbroken plain in the world. The eastern boundary of the West Siberian Lowland is the Yenisei (yen uh SAY) River.

Many rivers flow across these plains. The Dnieper (NEE puhr), the Don, and the Volga flow from the north toward the south. The Volga, the longest river in Europe, flows from the Moscow area into the Caspian Sea—a salt lake, which is the world's largest inland body of water. East of the Urals, the Irtyish (ir TISH) and Ob flow north into the Arctic Ocean.

Siberia   The West Siberian Lowland and the Central Siberian Plateau make up the region of Russia known as Siberia. Siberia is the Asian part of Russia. It includes all the land lying east of the Urals as far as the Pacific Ocean.

The Yenisei and the Lena rivers form the boundaries of the Central Siberian Plateau. Although this plateau is not as level and monotonous as the European and Siberian plains, it also can be considered one of the flat lands of Northern Eurasia. The waters of Lake Baikal (by KAHL) in southern Siberia measure an astounding 1 mile (1.6 km) in depth, making it the deepest lake in the world.

Eastern Mountains   Far beyond the Central Siberian Plateau, rugged mountains erupt from the landscape. A jumbled mass of ranges and ridges, steep valleys, and volcanic mountains, these Eastern Highlands stretch all the way to the Pacific Ocean. Look on the map on page 388 to locate the Kamchatka Peninsula. Many of the mountains on this large peninsula, which juts into the Pacific Ocean, are volcanic. Some peaks tower higher than 15,000 feet (4,572 m).

The Southern Border   A variety of landscapes are present in the nations that rim the southern edge of Northern Eurasia. The Black Sea serves as an important trade route to the oceans of the world. Not far from it lies the Caspian Sea. Between these two seas stand the rugged peaks of the Caucasus Mountains.

East of the Caspian Sea is a lowland of deserts and grasslands. Some of these middle-latitude grasslands are called steppes (STEPZ), a term used to describe the temperate grasslands

---

The Kamchatka Peninsula

This reindeer-herding family lives amid the volcanic mountains of the Kamchatka Peninsula. Why are there volcanoes in this region?

in Northern Eurasia. Range after range of forbidding mountains fold up east of the steppes. The highest peaks in the region rise where the borders of Tajikistan, China, and Afghanistan meet in the lofty Pamir mountains, where the tallest peak stands almost 25,000 feet (7,620 m) above sea level.

Climate

As you might expect in a region as large as Northern Eurasia, climates are varied. In general, climates are influenced less by landforms and more by the region's location in the high latitudes of the Northern Hemisphere. Its huge size also affects climate variations.

Linking Climate and Location Because of the region's huge size, most locations in Russia and surrounding nations are far from oceans. The Arctic Ocean to the north is frozen for most of the year. Located so far from the moderating effects of ocean water, much of Northern Eurasia's interior has a continental climate. Temperatures in the region have ranged from a record high of 122°F (50°C) in central Asia to a record low of −90°F (−68°C) in Verkhoyansk (VYER kuh YAHNSK), Siberia.

WORD ORIGIN

Arctic
The word *arctic* comes from ancient Greek. The Greeks knew that as one traveled northward, the Little Bear constellation rose higher in the sky. The Greek word for "bear" is *arkto*, so they called this northern region *arktikos*.

Regional Variations The far northern reaches of Eurasia, open to the chilling winds from the Arctic Ocean, have a polar climate. A subarctic climate extends from northeastern Russia across most of Siberia, while the southern part of the European Plain has a humid continental climate with warm summers. Use the map on page 391 to compare the climate of Moscow with the climate of Irkutsk (IR kutsk), in southern Siberia. Except for Antarctica, Siberia has the coldest winters of any land area in the world.

Farther south is a climatic zone extending from southern Ukraine through southern Russia into Kazakhstan, Kyrgyzstan, and Tajikistan, where the semiarid climate is characterized by warm winters and hot summers.

Linking Climate and Precipitation As is true in the interiors of other continents, much of Northern Eurasia is relatively dry. Some of the moisture carried by winds blowing from the Atlantic Ocean reaches European parts of the region, however. The Baltic nations of Estonia, Latvia, and Lithuania receive an average of 20 to 25 inches (51 to 64 cm) of rain a year.

One season that is commonly rainy on the European Plain is autumn. A geographer described autumn in the region in this way:

> At the end of September it is already cold, and then the miserable period of autumn rain begins. . . . The frost comes as a relief, and the people are as delighted with the first crisp snow as though spring had come.

The greater precipitation and colder temperatures of northwestern Russia and nearby nations means that snowfalls are frequent and heavy. Snow usually lies on the ground in the city of Moscow for six months of the year.

The influence of the Atlantic decreases toward the east and south. Most of the precipitation in the southern European and Asian parts of the region falls as summer rains, but it amounts to less than 10 inches (25 cm) a year. The dusty deserts of the nations east of the Caspian Sea result from arid climates, with some locations in the Kara Kum of Turkmenistan benefiting from less than 3 inches (8 cm) of rain each year.

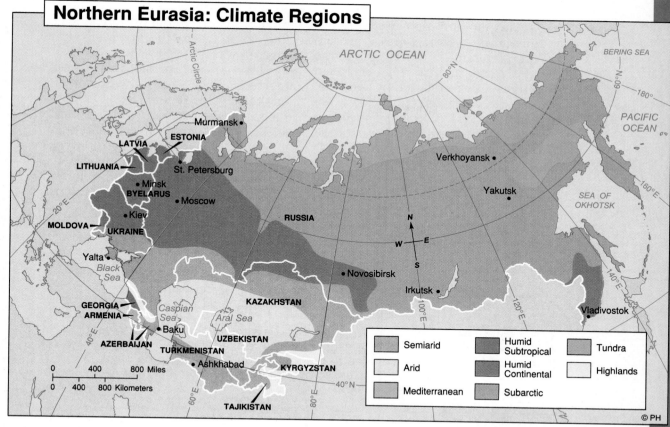

Northern Eurasia: Climate Regions

Legend:
- Semiarid
- Arid
- Mediterranean
- Humid Subtropical
- Humid Continental
- Subarctic
- Tundra
- Highlands

© PH

Applying Geographic Themes

Place Warm climates are rare in Northern Eurasia because most regions are far from the equator and from moderating oceans. Around which body of water are temperatures warmest?

Natural Vegetation

In general, the natural vegetation regions form four broad bands that stretch across Northern Eurasia from east to west. Beginning at the Arctic Ocean, these are tundra, forest, grassland, and desert.

Tundra The polar conditions near the shoreline of the Arctic Ocean support only tundra vegetation—grasses, mosses, and lichens that in turn support the region's reindeer and small groups of reindeer herders. Tundra plants have shallow root systems and thrive during the short summer, when the surface of the permafrost—perma-nently frozen soil—thaws. In some parts of Siberia, the permafrost extends to 1,300 feet (396 m) below the ground. A recent visitor to Yakutsk, talked to a local geologist who told him about finding permafrost there:

An early 19th century merchant, . . . digging a well, could not reach the bottom of the frozen layer of soil. He made a fire in the hole to thaw the ground for easier digging. In ten years, [he] dug down 116 meters—381 feet—and still the earth was frozen. . . . Today the well is a national monument.

Today, modern high-rise buildings in Siberia stand six feet off the ground on special pilings, or posts, so that the frigid air can circulate beneath them and diffuse the heat that they generate. These precautions are now taken because the first tall buildings erected on the frozen earth collapsed after their heat thawed the soil around their foundations.

Forest South of the tundra, running from west to east, sprawls the largest forest region in the world. It covers nearly half of Northern Eurasia—more than 4 million square miles (10

million sq km). About one out of every five trees on earth grows in these Russian forests.

The northern part of this forested region is thinly scattered with coniferous trees. This type of forest is called the **taiga**, the Russian word for "little sticks." Few people live in the taiga, where forest fires can rage for weeks unnoticed. Farther south, where the climate is warmer, both coniferous and deciduous trees grow in broad, dense stands.

Grassland South of the forests stretch the steppes, the rich, rolling

Applying Geographic Themes

1. Regions The vegetation of Northern Eurasia stretches from east to west in several bands. Which band of vegetation lies to the far north?

2. Interaction Compare this map to the population density map on page 396. How do the tundra and desert scrub regions shown on this map influence settlement patterns?

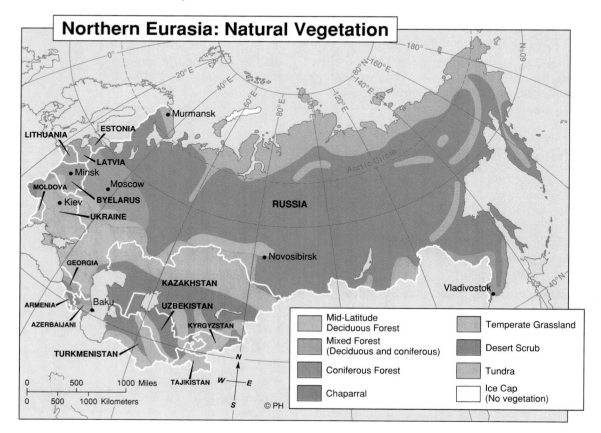

Northern Eurasia: Natural Vegetation

Legend:
- Mid-Latitude Deciduous Forest
- Mixed Forest (Deciduous and coniferous)
- Coniferous Forest
- Chaparral
- Temperate Grassland
- Desert Scrub
- Tundra
- Ice Cap (No vegetation)

Where Winter Rules

When winter comes to Siberia, shown at right, keeping warm requires more than an extra sweater. One visitor to Yakutsk describes the effects of Siberian cold in this way:

The air burns. Sounds are brittle. . . . Every few minutes people on the streets put gloved hands over their mouths and noses to keep the flesh from freezing. The smallest children are wrapped in layer after layer so that little more than their eyes are exposed. Buildings have triple windows and triple doors. . . . In this climate, . . . rubber soles on boots can break cleanly in half.

1. Describe Siberia's winters.
2. Why might people choose to live in Siberia?

temperate grasslands of Northern Eurasia. Soils beneath the grasses are very fertile and productive. This soil is called chernozem (CHER nuh zuhm), or "black earth."

Desert The arid areas east of the Caspian Sea contain two large sandy deserts, as well as areas of scattered shrubs and tufts of grass. One writer described the Kara Kum landscape south of the Aral Sea as "varying from scrub growth to desolation." When water is brought to this soil through irrigation ditches and canals, however, the desolate land can be made very productive. Unfortunately, poor management of irrigated farmlands in these areas has created many environmental problems.

SECTION 1 REVIEW

Developing Vocabulary
1. Define: **a.** steppe **b.** taiga

Place Location
2. Which sea lies to the south of the Ural Mountains?

Reviewing Main Ideas
3. What is the dominant landscape in the European part of this region?
4. What are the four major vegetation zones of Northern Eurasia?

Critical Thinking
5. **Drawing Conclusions** Where would you expect most people in Northern Eurasia to live?

✔ Skills Check

☐ Social Studies
☐ Map and Globe
☑ Reading and Writing
☑ Critical Thinking

Recognizing Bias

When you study history or any other social science, you should be on guard for signs of bias in what you read. Biased descriptions of an event, issue, or person often present only one point of view. A bias may be favorable or unfavorable.

The following questions will help you determine if something you read is biased. Follow these steps to see if any of the five statements in the box below is biased.

1. **Does the statement include positive or negative images that may be unjustified?** Biased statements sometimes use exaggerated images that are not backed up by evidence in order to persuade the audience to some prejudice. What phrase in statement A is an exaggerated image?

2. **Does the statement present only one side of an issue, while suggesting it is a full view?** Which one of statements B and C presents only one side of an issue?

3. **Does the statement consist mostly of opinions?** Which statements do you think contain mostly opinions, with little supporting evidence?

4. **Does the statement contain hidden assumptions that are not justified?** What hidden assumptions are made by the writer of statement E?

5. **Determine if the statement is free of bias.** Which two statements below are mostly free of bias?

Statements

A. In 1237, invading Mongols destroyed Kiev. The Mongols had a history of acting like wild animals once they invaded an area.

B. *Glasnost* provided an outlet for people in the Soviet republics to voice their complaints. But because the Soviet government permitted too much unrest, the Soviet Union was destroyed.

C. Before the Revolution, the Russians had everything to gain and nothing to lose from the violent overthrow of the czars. The Russian people were about as low as they could be. Any change in their condition was bound to be an improvement.

D. Since the end of World War II, the countries of Western Europe have enjoyed the longest period without war in history. While Western European nations look with hope toward the fuller economic union of the European Community in 1992, old rivalries are surfacing once again in Eastern Europe.

E. The smartest thing that political dissidents in the old Soviet Union could do was to go along with whatever the government said. Then later, once they were out of the country, they could criticize the Soviet system freely.

2 Human Geography

Section Preview

Key Ideas

- The population density of Northern Eurasia is greatest in the west.
- Bands of higher population density exist near natural resources and along rivers and railways.
- The Russians are the most numerous of the many national groups in Northern Eurasia.

Key Terms

soviet, *glasnost*

As the twenty-first century approaches, the population of Northern Eurasia nears 300 million people. More than ninety ethnic groups speaking more than two hundred different languages are scattered across this land.

The Russian Capital of Moscow
Industry and fertile farmland nearby support a large population in the city of Moscow, where more than nine million people live.

Linking Population Density to the Environment

Although Northern Eurasia has a large population, it has a low population density. Areas of high population density do exist, however.

Areas of Dense Population With some knowledge about the physical regions and climates of Northern Eurasia, one might guess where the most densely populated land is located. Regions of denser settlement and fertile chernozem soils overlap. The rich soil of the steppes is so well suited for growing crops that the area has been populated continuously for about four thousand years.

Areas of Sparse Population Compare the population map on page 396 with the climate map on page 391.

Notice that few people live in the desert areas near the Aral Sea. People have settled the deserts only in places where a canal, a river, or a moisture-catching mountain range makes farming possible.

The intensely cold climate of Siberia also discourages settlement. A recent traveler to the Siberian town of Mirny described his visit:

We flew into . . . town during a winter heat wave, only 42°C below zero, and were met by . . . a young reporter on the newspaper. [He said,] "We always say that southern towns grow, middle towns are built, and northern towns are

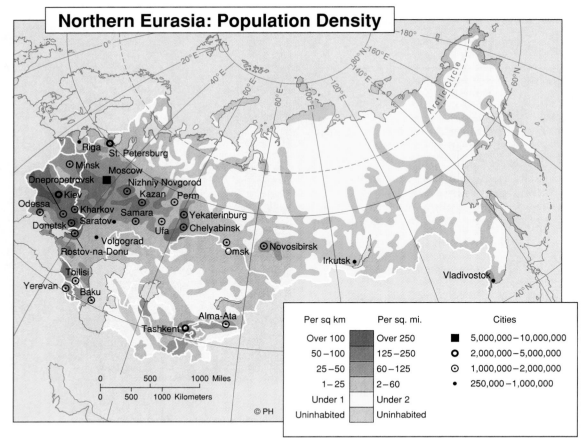

Northern Eurasia: Population Density

Per sq km	Per sq. mi.	Cities	
Over 100	Over 250	■	5,000,000–10,000,000
50–100	125–250	○	2,000,000–5,000,000
25–50	60–125	⊙	1,000,000–2,000,000
1–25	2–60	•	250,000–1,000,000
Under 1	Under 2		
Uninhabited	Uninhabited		

© PH

Applying Geographic Themes

Interaction Compare this map to the map on page 409. What resources might draw settlers to the less populated regions shown on this map?

---

brought in. It was so [here in 1956]: A forty-truck convoy came from the Trans-Siberian Railroad, building a winter road as they came. Six weeks to get here."

Today, Mirny has a population of more than 75,000 people.

Ribbons of Population Density

More than 150,000 rivers meander through Northern Eurasia. The Ob, Yenisei, and Lena rivers flow northward across Siberia to the Arctic Ocean. These rivers transport ships carrying lumber cut from the Siberian forests. The population map on this page shows ribbons of higher population following the paths of these rivers.

Overland transportation has also attracted people to Siberia. For example, the Trans-Siberian Railway, completed in 1905, spurred significant increases in population in the lands along its length.

People and Republics

Over hundreds of years a series of Russian emperors, or czars, built an empire by conquering the homelands of more than one hundred different groups. By the beginning of the twentieth century, Russia controlled nearly all of Northern Eurasia.

After a revolution overthrowing the czar in 1917, a new system of government was set up based on the communist beliefs of Karl Marx. Under the new government the homelands of the fifteen largest ethnic groups were designated as republics, with all republics participating as members of the Union of Soviet Socialist Republics. According to the constitution of the Soviet Union, each republic had a soviet, or governing council. All of these soviets were overseen by the Supreme Soviet in Moscow, which made national laws.

Following the selection of Mikhail Gorbachev (GOR buh chawf) as head of the Communist party in 1985, however, a wave of reforms began. While Gorbachev hoped that these reforms would restore confidence in the Soviet Union, they encouraged other changes. The policy of *glasnost* (GLAHZ nost), or "openness," permitted Soviet citizens to say what they wished. Given such freedom, many people called for an end to communism and the domination of the central government. By the end of 1991, each of the former Soviet republics chose to dissolve the Soviet Union and declared themselves independent countries. Representatives from most of these newly created countries then met to discuss the formation of a new and much looser association known as the Commonwealth of Independent States. The nature and purpose of the new commonwealth remained unresolved in early 1992.

Russian Domination Even with the division of the Soviet Union into more than a dozen separate nations, one group dominates Northern Eurasia— the Russians. The Russians comprise slightly more than one half of the region's population. Russia includes about three fourths of Northern Eurasia's total land area, including much of the European Plain and *all* of Siberia in the east.

Because Russians are the region's dominant group, Russian is the most commonly spoken language throughout Northern Eurasia. The communist government of the Soviet Union tried to eradicate the languages of other nationalities, and people from other ethnic groups were forced to learn Russian in order to obtain prestigious jobs. The preference given to Russian language and culture caused much unrest among other ethnic groups and fueled their demands for independence.

A Variety of Cultures Cultures and ways of life differ widely from one nation to another in Northern Eurasia. In Latvia, Lithuania, Estonia, Ukraine, and Byelarus, for example, city dwellers are akin to Europeans in culture and live in much the same way as their neighbors in Poland or Czechoslovakia. Azerbaijanis share the religion of Islam with their neighbors across the Iranian border. Even within Russia, dozens of native groups continue to live, especially in Siberia.

■■■ SECTION **2** REVIEW ■■■

Developing Vocabulary
1. Define: **a.** soviet **b.** *glasnost*

Place Location
2. Describe the location of Russia.

Reviewing Main Ideas
3. What factors influence population density in Northern Eurasia?
4. To which ethnic group did most government officials in the Soviet Union belong?

Critical Thinking
5. **Predicting Consequences** What problems did the Soviet Union encounter in trying to balance the demands of many ethnic groups with the need to preserve a strong, unified country?

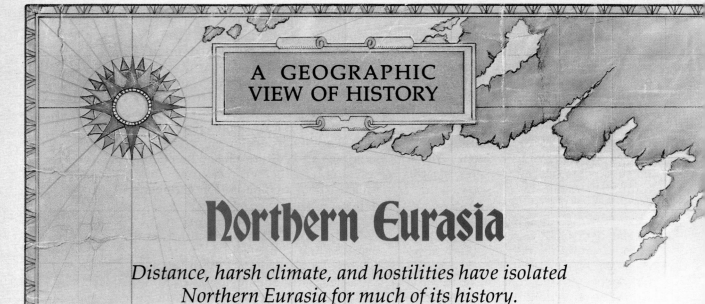

Northern Eurasia

*Distance, harsh climate, and hostilities have isolated
Northern Eurasia for much of its history.*

Winston Churchill, a prime minister of Great Britain, once described the Soviet Union as "a riddle wrapped in a mystery inside an enigma." Northern Eurasia has remained relatively isolated from the rest of the world despite its enormous size and the fact that its nations share borders with many nations in other regions. At the same time, contacts with outsiders through invasion, expansion, and trade have left a lasting mark on the variety of cultures in modern Northern Eurasia.

The Early Slavs

A small band of Slavs came out of the forest onto the bank of the Dnieper River. They looked fearfully at the men in the ship, some of whom wore battle axes and swords. The sailors were Varangians, a fierce Viking people who controlled this river as part of a vital north-south trade route.

The Slavs had come to trade with these Varangians. They carried with them lumps of beeswax, a pouch full of amber gathered from a lake in the endless forest, wooden buckets of honey, and the furs of animals they had trapped during a brutal winter.

The Slavs were forest dwellers. They could not imagine the places these travelers had seen already—the Baltic Sea, the Western Dvina River, and the Dnieper. Nor did the Slavs know where the Varangians were bound: through the Black Sea to Constantinople, the rich capital of the Byzantine Empire. Yet over time the culture of this Mediterranean empire—especially its Eastern Orthodox Christianity—was carried north by traders, where its influence can still be seen.

A Flat Land Without Barriers

This scene, or something similar to it, probably occurred often in the 800s. These Slavs were to become the central players in the drama of Northern Eurasian history. During this early period, the Slavs held the forested areas pierced by the great river system running on a north-south line through modern-day Byelarus and Ukraine. In earlier centuries they occupied the broad, fertile steppes to the south, but wave after wave of invaders drove them northward into the forests.

The flatness of the Slavic heartland left it always exposed and open to invasion. Even if the Slavs could have controlled their part of the North European Plain, they would still have faced danger. The steppes had few natural barriers to protect the area from the rest of Europe and Asia.

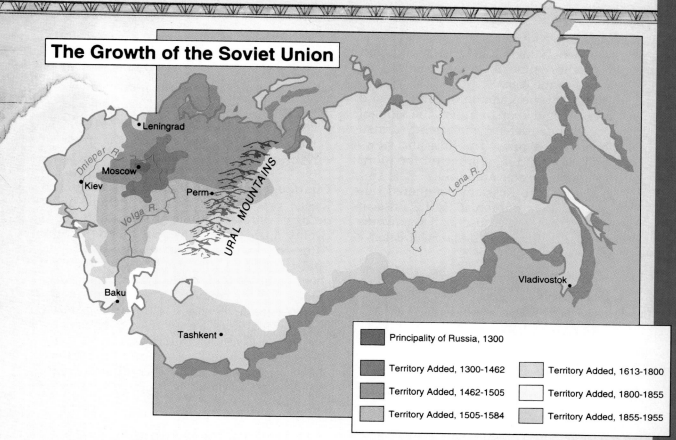

The Growth of the Soviet Union

Legend:
- Principality of Russia, 1300
- Territory Added, 1300-1462
- Territory Added, 1462-1505
- Territory Added, 1505-1584
- Territory Added, 1613-1800
- Territory Added, 1800-1855
- Territory Added, 1855-1955

Map labels: Leningrad, Moscow, Kiev, Dnieper R., Volga R., Perm, URAL MOUNTAINS, Lena R., Baku, Tashkent, Vladivostok

Applying Geographic Themes

Movement This map shows the territories that were annexed by Russia and the Soviet Union between 1300 and 1955. Which city formed the historical core of the Russian and Soviet empires? During which period did Russia first acquire a port on the Pacific coast?

To the Russian down through history the steppe has spelled danger, invasion, the enemy. The forest has spelled safety, security, a formidable stronghold. He has retreated into its depths protected by swamps, marshes, and morasses, there to live out his days until the sway of the alien conqueror has passed.

Flatness Unifies In 879 a Varangian prince named Oleg marched southward out of the northern city of Novgorod and captured Kiev, one of the most important centers of trade. Kiev was built on a cliff overlooking the great waterway of the Dnieper River. In doing so he unified the Slavs and created the beginnings of a Russian state.

Four hundred years later, however, the Mongols, warrior horsemen, rode into Russia from the east. They forced the Russians to pay heavy tribute, or taxes, for 250 years. But they could not break the Russians' unity.

Expansion Brings Wealth After the Russians finally shook off Mongol rule, they pushed outward from their capital, Moscow. They drove the borders of Russia as far north as the Arctic Ocean and as far south as the Caucasus Mountains. The Russians were determined to control all the land within the natural barriers of water and mountains. Within a short time they crossed the Ural Mountains to the east.

Beyond lay the khanate of Sibir, a land held by a Mongol khan, or ruler. The Mongol army rushed to defend their land, armed with bows and arrows. The Russians, however, had guns, and they drove off the Mongols with their greater firepower.

As soon as the Russians conquered this eastern land they offered it to the Russian ruler, Czar Ivan IV. Because Ivan was working to expand the Russian state, he eagerly accepted.

Throughout Ivan's reign . . . the Russian people continued to extend their country's boundaries. . . . It was the result of a spontaneous movement of people, who,

like their American counterparts, were seeking either wealth or freedom in an unexploited land. But, unlike the Americans, who had a straight frontier along the Atlantic seaboard and could proceed only toward the setting sun, the Russians had a circular frontier radiating from Moscow.

Challenges of Diversity and Distance   As Russia expanded, it grew to include hundreds of different ethnic groups. More than twenty such groups lived in Siberia alone. After the Russians created and ruled the Soviet Union in the twentieth century, the sheer number and diversity of people, together with the country's vast size and harsh climates, made the empire difficult to govern. The Soviet Union covered eleven time zones from east to west. Communication was very difficult over such distances and time differences. As a result people were often isolated, or cut off, from others in their own country.

Connection, Isolation, and Protection

The same challenges of climate and distance that made it difficult for the Russians to govern their own nation isolated it from most of the world. At times the Russians and Soviets worked tirelessly to end such isolation. At other times such barriers protected them from perilous invasions.

Movement Toward the Outside World
The earliest means of movement toward the outside world were the rivers that flowed to the Baltic and Black seas. As the Russian Empire grew, its leaders also sought the benefits of trade with a technologically advanced Europe. Although the Russians carried on limited trade with Western Europe, their only ports were on the Arctic Ocean, which was frozen most of the year. The Russians fought numerous wars against their western neighbors in an effort to win a "window on the west"—a warm-water port that would open Russia to better ties with

The Winter Palace, St. Petersburg

At the Hermitage, a wing of the Winter Palace, Catherine the Great kept informed of all the latest political and cultural events in Western Europe by corresponding with Voltaire and other noted philosophers.

Europe. When such a port was finally captured in 1709, the Russian emperor, Peter the Great, built the city of St. Petersburg. He also created a navy to defend the port. Russian trade with the rest of the world subsequently increased 700 percent.

Not long after Peter's rule, Catherine the Great captured land on the Black Sea, giving Russia control of another route to the outside world. Both Peter and Catherine worked to make Russia part of European trade and politics.

Unfortunately, as Russia focused more on Europe, it was also drawn into wars that were continually shaking that continent. Fortunately, the distance and climate that had for so long isolated the country now protected it against invaders.

Napoleon Is Defeated by Climate In 1812, the French emperor, Napoleon Bonaparte, attacked Russia with 450,000 soldiers. Because no landforms served as obstacles, the French army seemed to have a clear path to Moscow. The Russians therefore had to rely on the size of their land as a defense. They retreated as Napoleon advanced, luring him farther and farther into Russia. When winter came, Napoleon found himself and his army trapped deep in enemy territory and cut off from their European supply lines.

Winter struck the French army with all its Russian ferocity. As early as November the snow was six feet deep. The temperature fell to below −20°F (−30°C). Napoleon struggled back over the border into Poland with fewer than 50,000 soldiers. "We are victims of the climate," Napoleon declared.

Hitler Is Defeated More than a century later, in the spring of 1941, the German dictator Adolf Hitler struck at the Soviet Union. Once again a vast army was assembled and poured into the country along an exposed front that stretched from the Arctic Ocean to the Black Sea.

Once more, however, the harsh climate of the Soviet Union proved too much for the invader. Delayed in accomplishing its objectives, Hitler's army was caught by the coldest

Frozen Artillery
The bitter cold Russian winters proved a good defense against Hitler and the German army during World War II.

and earliest winter within memory. By December 5, the temperature was −35°F (−37°C). The engines of tanks, trucks, and locomotives froze and burst. Hitler could offer his troops no relief. In a matter of months, one million German soldiers were dead. Once again, the region's physical geography had provided a shield that served both to protect and isolate.

TAKING ANOTHER LOOK

1. What effect did the flatness of the Russian land have on the young country?
2. What are some changes that expansion brought to Russia?
3. How did the rulers of Russia work to end their country's isolation?

Critical Thinking

4. **Distinguishing False from Accurate Images** Why did people often refer to all citizens of the Soviet Union as Russians?

20
REVIEW

Section Summaries

SECTION 1   Land, Climate, and Vegetation
Northern Eurasia's landscape is dominated by plains in the west, becoming mountainous in the east. The Ural Mountains divide Europe and Asia. Rugged mountains, deserts, and grasslands mark the southern border. Interior locations are dominated by a continental climate, while the climate in the west is moderated by the Atlantic Ocean. Four bands of vegetation stretch from east to west: tundra, a treeless, frozen plain; a band of forest, including the sparser taiga and denser mixed forests; the steppes, a fertile grassland; and sandy and rocky deserts.

SECTION 2   Human Geography   The most densely settled part of Northern Eurasia is the west—regions of fertile soil and much industry. Other areas of high population are near rivers, canals, railways, and mines. The Russians are the largest of the more than ninety national groups found within Northern Eurasia. Following the breakup of the Soviet Union, fifteen nations became independent. Russia is the largest of the nations. Many ethnic groups and nationalities live in the nations of Northern Eurasia. Cultures and lifestyles vary widely from one nation to another.

Vocabulary Development

Match the definitions with the terms below.
1. a region of dense coniferous forest
2. governing council
3. a policy of "openness"
4. temperate grasslands
   a. steppe   c. taiga
   b. soviet   d. *glasnost*

Main Ideas

1. What landform dominates the landscape of Northern Eurasia?
2. What does the location of Northern Eurasia tell you about its climate?
3. The forest region is sometimes found mixed with the vegetation of the region to the north or the vegetation of the region to the south. What are those two vegetation regions?
4. Why does the region of chernozem soils overlap with the most populated region?
5. How was the government of the Soviet Union structured?
6. What was the dominant language in the Soviet Union?
7. What effect do buildings have on permafrost?
8. What is the relationship between natural resources and population density?
9. What is the importance of the Black Sea?
10. What effect did the Trans-Siberian railway have on population density?

Critical Thinking

1. **Expressing Problems Clearly**   The Ural Mountains form a boundary, but the Caucasus Mountains form a barrier. Explain this statement.
2. **Making Comparisons**   How were the multicultural characteristics of the Soviet Union different from the multicultural characteristics of the United States?
3. **Expressing Problems Clearly**   Name two advantages and two disadvantages of being a Russian in Northern Eurasia.
4. **Demonstrating Reasoned Judgment**   What are the advantages and disadvantages of locating industrial areas in Siberia?

Practicing Skills

1. **Recognizing Bias** Look again at statement C on page 394. Rewrite the passage to make it free of bias. After reading your revision of the passage to a classmate, explain how you removed the bias from the original statement.

2. **Using Time Lines** Following are eight events from the history of the Soviet Union. Use library research to find the year in which each event occurred. Then draw a time line and place each event in order.

a. Stalin introduces forced collective agriculture.
b. Russia loses the Russo-Japanese War.
c. The Bolsheviks take control of Russia.
d. The Soviet Union launches Sputnik 1, the first spacecraft to circle the earth.
e. Karl Marx first publishes the *Communist Manifesto*.
f. Alexander Solzhenitsyn is imprisoned for his criticism of Stalin.
g. The Soviet Union is dissolved as its fifteen republics declare independence.
h. The cold war between the United States and the Soviet Union begins.

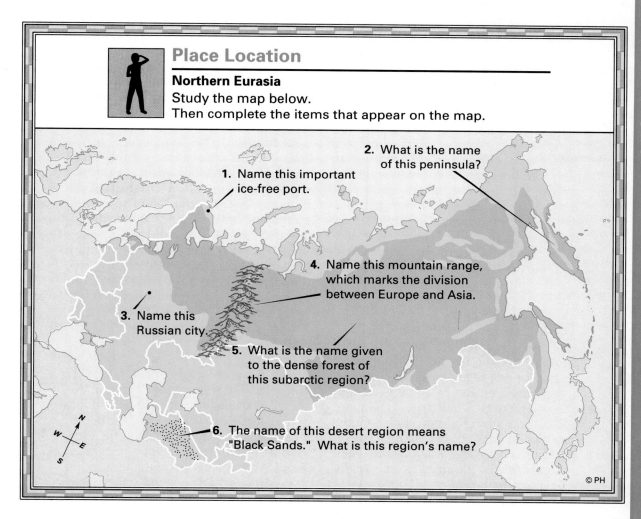

Place Location

Northern Eurasia
Study the map below.
Then complete the items that appear on the map.

1. Name this important ice-free port.
2. What is the name of this peninsula?
3. Name this Russian city.
4. Name this mountain range, which marks the division between Europe and Asia.
5. What is the name given to the dense forest of this subarctic region?
6. The name of this desert region means "Black Sands." What is this region's name?

© PH

Place Location: WHERE ON EARTH?

Where To Eat in Northern Eurasia

You're standing outside the airport terminal in Moscow on a cold winter night. Snow streams past the windows of the taxi as you ride to your hotel. Leaning back in the seat, you review the purpose of your trip. As director of operations for a major American restaurant chain, you must choose the company's next location in Northern Eurasia. The first location, right here in Moscow, has been a huge success.

In your briefcase you have a list of twelve Northern Eurasian cities, all candidates for the site of your company's next restaurant. Key officials of the nations in which those cities are located have dreams of duplicating that success—of bringing American investment and new jobs to their nation.

In your hotel room, you pull the typewritten list out of your briefcase and scan it quickly. The names are familiar to you by now: Moscow (a second outlet here might make sense, given how well the first one is doing), St. Petersburg, Vladivostock, Kazan, and Omsk, all in Russia; Tbilisi, Georgia; Tashkent, Uzbekistan; Minsk, Byelarus; Alma-Ata, Kazakhstan; Kiev and Odessa, Ukraine; and finally Baku, Azerbaijan.

Officials from each of these Northern Eurasian nations will be competing for your business. Knowing this, you must carefully scrutinize the arguments they present. After all the work your company has put into its expansion plan, a poor choice could be financially disastrous.

The next day you have arranged to meet one-by-one with officials from each of the nations. Your first appointment with the Russian official begins and your translator Peter relays his opening words: "Your letters inform us that you have twelve cities under consideration. We can save you some time and effort by telling you we believe that you should not give your golden opportunity to a city located on a coast. Your restaurant would profit better if it were located in an interior city where population densities are higher." "Thank you," you reply and after some small talk the appointment ends.

After checking your population density map of the region you agree with the Russian official and cross four coastal cities off your list: now eight cities remain.

Translating for the Georgian official, Peter begins your next appointment by saying, "We sincerely hope you will give full consideration to Georgia in making your decision. After all, your restaurant would probably be most successful if it were located in a capital city like ours." Then he adds, "We also believe that your restaurant should be located somewhere other than Moscow. The economy there may not be able to support a second franchise." After some thought you agree that a capital city would probably be

the best setting for the new restaurant and that you probably should not risk a second store in Moscow at this time. Mentally, you cross Moscow off the list as well as two cities that are not national capitals. Now five cities remain from which to choose.

Translating for the official from Byelarus, Peter says, "After reviewing your menu we think you should note that some of your meals would be prohibited in nations where Islam is the dominant religion."

You know from the map in your briefcase that two of the five remaining choices have just been eliminated.

Following this appointment you comment to Peter that you have eliminated all but three cities from your list. Peter offers you his own advice saying, "Would it not make sense to locate your restaurant in the remaining city with the largest population?" Given that one of the remaining cities does have a larger population than the other two, you tend to agree. After some careful double-checking you make your call to company headquarters—to announce where in Northern Eurasia your next restaurant will be located.

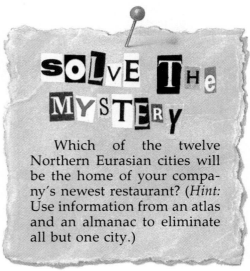

SOLVE THE MYSTERY

Which of the twelve Northern Eurasian cities will be the home of your company's newest restaurant? (*Hint:* Use information from an atlas and an almanac to eliminate all but one city.)

405

21

Russia and Other Independent States

Chapter Preview

Locate the countries covered in these sections by matching the colors on the right with the colors on the map below.

Sections	Did You Know?
1 **RUSSIA**	Slightly more than 80 percent of Russia's 150 million residents are ethnic Russians.
2 **OTHER INDEPENDENT STATES**	Ten of the fifteen nations in Northern Eurasia have total populations less than that of the city of Moscow.

The Trans-Siberian Railroad
This railway links the resources of Siberia with more populated areas in Europe. The Russian government hopes that expansion of this railway will bring higher standards of living to towns along its tracks.

1 Russia

Section Preview

Key Ideas

- Russia must overcome environmental challenges in order to develop its resources.
- Governmental control hampered Russia's ability to improve the standards of living of its people.

Key Terms

ideology, command system, state farm, collective farm, heavy industry, *perestroika*, demand system

The mammoth size of Russia is matched by the nation's potential. It has more natural resources—and potentially more wealth—than any country on earth. Much of the history of Russia in recent centuries has been the struggle to transform the nation's natural wealth into higher standards of living for its people.

Geographical Challenges

Almost every known natural resource can be found somewhere in Russia. It often is difficult, however, to extract these resources or convert them to usable goods.

Distant Locations Maps cannot give you a true sense of the enormous size of Russia. Use a globe and compare Russia's area with those of the world's other large countries—Australia, Brazil, Canada, China, and the United States. Russia is nearly twice the size of any of these other nations.

Another way to view Russia's size is to think about the great distances that must be traveled. Russia is so immense that it includes ten time zones. Before people in St. Petersburg have gone to bed on a Tuesday evening, people are waking up on the Kamchatka Peninsula.

As the economic map on page 409 shows, many of Russia's resources are found in remote locations. Huge deposits of coal, oil, natural gas, and even gold and diamonds can be found beneath the permafrost in the barren reaches of Siberia. For Russia to benefit from them, however, these precious natural resources must be transported to industrial centers, or industrial factories must be built near the resources. Russians frequently have tried both methods.

The city of Magnitogorsk (mag NEET uh gorsk) in the Ural Mountains is an example of an industrial center that was built to take advantage of resources nearby. A rich iron ore deposit was found, so a city was constructed in 1929 to house the workers who mined and processed the ore. More than 400,000 people now live in Magnitogorsk, the site of one of the world's largest metal refineries.

Interaction: Balancing Industry and the Environment In addition to distance and location, the fragile natural environment also presents obstacles to the development of resources. Russian misuse of natural resources often has damaged the environment. In their drive to increase economic production, leaders of the Soviet Union often placed production ahead of environmental concerns. Such practices were criticized loudly after Gorbachev instituted *glasnost*, however. In 1989 he commented:

> *You'd be mistaken if you think people are not troubled by the environment, by the conflict between industry and nature. Their concerns have caused one thousand factories to be shut down.*

One place where the factories had to change or cease their operations was along the shores of Lake Baikal in eastern Russia. Locate Lake Baikal on the map on page 388. Paper mills were built to use both the abundant forest resources of the Baikal district and the crystal-clear waters of the lake. But when the mills were found to be discharging wastes into this great natural resource, public outcry forced the government to put an end to the destructive dumping.

Movement: Gaining Access to Resources

Because of its size and harsh conditions, Russia has had limited success in connecting its distant places with efficient means of transportation.

Challenges of Climate Roads in Russia are of limited use in bringing resources to industrial centers. More than three quarters of all raw materials come from Siberia, where winter frosts buckle concrete and summer thaws turn roadways into mushy swamps. Siberian roads often are covered only with gravel. "No speed limit," said a driver in Mirny, "only limit is the road itself."

River traffic for ships and barges is also discouraged by the climate, because most rivers are covered by thick ice for months at a time. Frozen rivers and other bodies of water do serve as highways, however. One journalist describes how he discovered this feature of Russian life:

> *In a sturdy old Volga taxi we headed for the Lena River. . . . As we approached the river bank and were preparing to cross, I suddenly realized: no bridge. But the driver plunged onward and soon we were bounding along on the river in an icy, rutted but well-prepared track.*

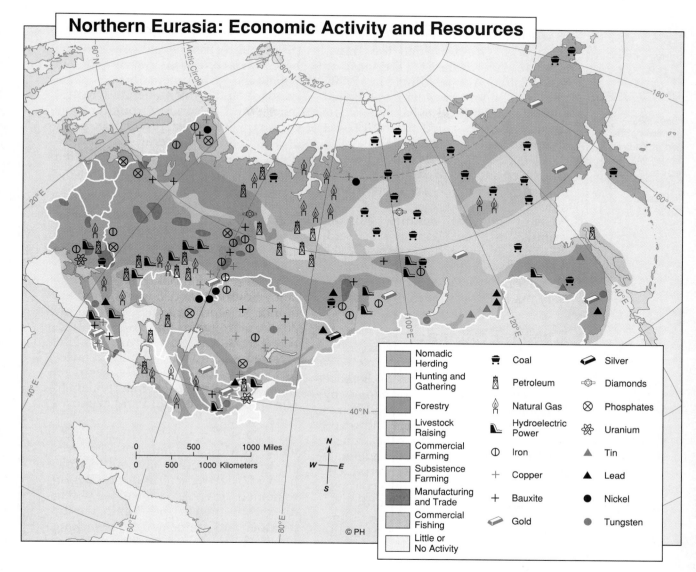

Northern Eurasia: Economic Activity and Resources

Legend:

- Nomadic Herding
- Hunting and Gathering
- Forestry
- Livestock Raising
- Commercial Farming
- Subsistence Farming
- Manufacturing and Trade
- Commercial Fishing
- Little or No Activity

- Coal
- Petroleum
- Natural Gas
- Hydroelectric Power
- Iron
- Copper
- Bauxite
- Gold
- Silver
- Diamonds
- Phosphates
- Uranium
- Tin
- Lead
- Nickel
- Tungsten

0   500   1000 Miles
0   500   1000 Kilometers

© PH

Applying Geographic Themes

Interaction Compare this map to the climate map on page 391. How do you think climate influences the development of resources and economic activity in the eastern parts of Northern Eurasia? Where are manufacturing activities in the region concentrated? Where are most farming activities concentrated?

Another solution is to fly over the vast lands. Air travel is expensive, however, and it is not a suitable alternative for every material. Oil and natural gas, for example, cannot be flown out. Bringing these resources to the western areas where they are needed requires still another solution—the maze of pipelines that crisscross the frozen wilderness.

Access by Rail Despite the value of these other methods of transportation, railroads are the greatest movers of goods in Russia. Because rail transport is inexpensive, Russian rail lines carry nearly half of all the railroad freight in the world. One rail line, the Trans-Siberian, runs from St. Petersburg to Vladivostok. Another line runs 2,000 miles (3,220 km) between Lake Baikal and the Amur River, near the Pacific coast. This line, called the BAM, was completed in the early 1980s. The route crosses 3,700 bodies of water and seven mountain ranges. Some of the railway's tunnels are more than 9 miles (14 km) long.

An eighteenth-century Russian scientist, Mikhail Lomonosov, said, "Siberia will make Russia strong." The region's tremendous wealth of natural resources, however, has not automatically made Russia a wealthy nation.

Freeing Economics from Politics

Many of the challenges of distance and climate have been overcome by technology. Despite major advances during the last half century, however, the Russian people still have trouble obtaining nutritious food, adequate housing, and consumer products. Their dissatisfaction with economic shortages in a land of plentiful resources was a major factor that doomed the communist government of the Soviet Union.

Communist Controls on Production The Soviet government that was established after the revolution in 1917 was based on an ideology—the ideas or principles on which a political system is based. That ideology was modeled on the ideas of Karl Marx, a German economist who died in 1883.

Marx and his followers believed that all land and businesses should be owned by all the people in common (from which the word *communism* comes). By following Marx's philosophy, the communist government leaders hoped to establish a society in which there would be no inequality.

From the late 1920s until the 1950s, Soviet dictator Josef Stalin (STAH lihn) established an economy based on a command system—one in which the government dictates what goods should be manufactured. Farms, mines, and factories throughout the country were given production goals and told how to operate by government officials in Moscow who often knew little about local problems.

Changes in Agriculture Farming in Russia and other parts of the Soviet Union was also transformed. In the early 1900s, peasants in Russia owned about 25 million small farms. The Soviet government took control of the land, however, and farmers were required to work together on roughly 20,000 state farms and 26,000 collective farms. On state farms, workers received wages as they would in factories, while on collectives, workers shared any surpluses that remained after products were sold and expenses were paid.

Because many decisions were made in Moscow, and few incentives existed to encourage farmers to work hard, Soviet agricultural production remained low. As a result, Russian diets continued to be dominated by potatoes and grains and to be lacking in meats, fruits, and fresh vegetables.

Despite strict central control, many Soviet farmers were permitted to cultivate gardens on small plots. These gardens became major sources of fruits and vegetables, and with the end of communist control, they likely will become even more important sources of food in the future.

Industrial Development Soviet changes in farming were matched by the rapid expansion of industry. Stalin's policies emphasized the development of heavy industry—the production of goods such as steel and machines that are used by other industries. Many workers were forced to labor on the assembly lines of heavy industries, often in distant places far from their homes.

These policies seemed to yield impressive results. By 1940, the Soviet Union was the second-largest producer of iron and steel in Europe. The country industrialized at great cost, however. Heavy industry was emphasized at the expense of the production of consumer goods. Many citizens had trouble obtaining consumer goods such as clothing or food. Over time the differences between the quality of life in the Soviet Union and in such industrialized nations as the United States and Japan became glaring.

During the late 1980s, Soviet leader Mikhail Gorbachev instituted new policies calling for economic restructuring. These policies were called perestroika (per uh STROY kuh), a Russian word for "a turning about." The new policies called for a gradual change from a command system to a demand system—one in which the production of goods is determined by people's demand for goods in the marketplace. The proposed changes were very similar in character to the market economies of Western nations.

Although the Soviet Union broke apart before Gorbachev's policies could be implemented fully, the newly elected Russian president, Boris Yeltsin, was also committed to changing the Russian economy. He too favored a shift from a command economic system to a demand economic system. Whether the elimination of government imposed price controls and activation of other free-market measures would solve the economic problems of the Russian people was uncertain, however. While the new policies might offer greater prospects for long-term growth, they erased many of the forms of security on which many people had become dependent. As a result, the economic future of Russia and other former Soviet republics remained clouded as they took their first independent steps.

Radical Changes in 1991
Russians in Moscow dismantle a statue of Feliks Dzerzhinsky, a founder of the Soviet secret police.

SECTION 1 REVIEW

Developing Vocabulary
1. Define: **a.** ideology **b.** command system **c.** state farm **d.** collective farm **e.** heavy industry **f.** *perestroika* **g.** demand system

Place Location
2. Near what body of water is the city of Irkutsk located?

Reviewing Main Ideas
3. Why are trains more useful than trucks for transporting resources out of Siberia?
4. Why were consumer goods scarce in the former Soviet Union?

Critical Thinking
5. **Making Inferences** Why did the former Soviet Union encounter so many problems with central control of economic production?

✔ Skills Check

☐ Social Studies
☐ Map and Globe
☐ Reading and Writing
☑ Critical Thinking

Distinguishing False from Accurate Images

We all believe some things that are not true. Often a false image or stereotype can gain force and extend its influence widely. It is critical, therefore, to make a habit of testing our images of the truth against new information that we judge to be reliable.

Use the following steps to read and to analyze the passage below.

1. **Identify the main point of the passage.** New information allows you to examine a widely held belief about a person, place, or thing, and determine whether or not this belief is based in fact. In order to judge the significance of any new information, you must first identify its central message. Answer the following questions: (a) What is the main point of the passage? (b) How does this main point differ from popular American beliefs about Mikhail Gorbachev's leadership of the Soviet Union?

2. **Evaluate the reliability of the source.** Popular beliefs or images of a person or place arise from many sources, and not all of them are reliable. It is important to determine if a source is reliable or biased. Answer the following questions: (a) What is the source of this passage? (b) How would you describe the reliability of the source?

3. **Look for evidence.** A point of view presented without evidence to support it may be just opinion. Many false images of people or places are based on opinion, not fact. Answer the following questions: (a) What evidence does the author cite to support his statements? (b) What other sources could you consult to check the accuracy of his references?

For all the effusiveness that the West showered on Gorbachev's policies of *glasnost* and *perestroika,* most of his programs didn't work. By January 1988, he had started to pepper his speeches with the word democracy. When the Central Committee voted to approve a plan bringing greater democracy to the Soviet Union, it didn't know what it was getting into. The idea was to supply fresh air, new blood, but not to kick over Marxism-Leninism. Gorbachev wanted to build "regulated democracy" or "democracy within the socialist choice." He couldn't feel the society under him stirring. . . .

But from Gorbachev down, Soviet leaders misunderstood their own system. Reform doesn't mix with totalitarianism. It is impossible to have a *little less* totalitarianism, or *a new improved* totalitarianism. When the people take to the streets—and it is no longer possible to con enough of them or shoot enough of them—the system collapses.

—Tom Mathews, *Newsweek*,
December 30, 1991

2 Other Independent States

Section Preview

Key Ideas

- Ukraine, Byelarus, Moldova, and the Baltic nations long have been influenced by Europe and Russia.
- About 50 million Muslims live in Northern Eurasia, mostly in Azerbaijan and in five nations in central Asia.

Key Term

nationalism

After the end of World War II, the Soviet Union seemed indivisible. A strong central government obscured the diversity that ultimately caused the union to break apart in 1991. While Russia remains the largest and most powerful of the nations to emerge from the former Soviet Union, an equal number of people reside in the other independent nations of Northern Eurasia.

Location: Three Nations on the Baltic

The nations of Lithuania, Latvia, and Estonia are tucked along the eastern edge of the Baltic Sea. The three

Applying Geographic Themes

Place How do the boundaries of the countries shown on this map compare to where the ethnic groups live?

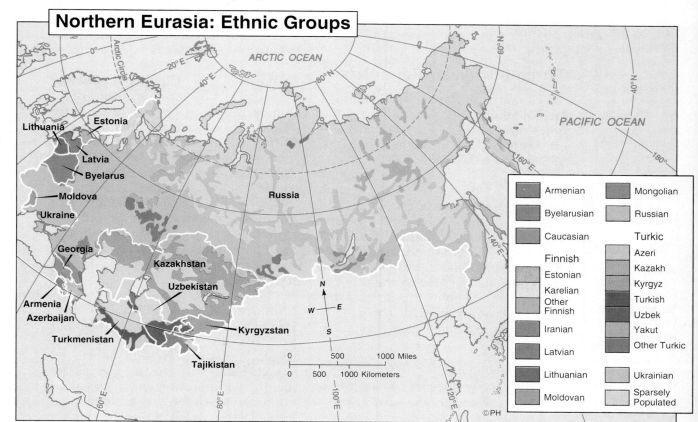

Northern Eurasia: Ethnic Groups

nations are small—together they would fit within the boundaries of the state of Missouri—and their combined populations are less than 8 million. Like the nations of Eastern Europe, their location astride major trade routes favored them at times, but they also were subject to frequent conquests by other powers.

Over the centuries, their cultures evolved in different yet related ways. Lithuanians and Latvians speak similar languages, but Lithuania is largely Catholic while Latvia is more Lutheran. Estonians also tend to be Lutherans, but their language is a distinctive tongue closely related to Finnish.

All three nations existed as independent nations during the period between the two World Wars, but the Soviet Union annexed them in 1945. They were the first Soviet republics to declare their independence in the late 1980s, and their more advanced economies gave them promising hopes for the future.

Three European Republics

Ukraine, Byelarus, and Moldova are the other nations that lie along the border of Northern Eurasia and Eastern Europe. Look at the map on page 388 to locate these three nations.

Ukraine   Despite a long and rich history, Ukrainians rarely have been masters of their own land. Caught between the Orthodox Christian heartland of Russia and the Catholic stronghold of Poland, Ukraine usually was controlled by one or the other of these powers. Submission to the Russians often was particularly galling. One writer explained their feelings in this way:

The Ukrainians consider themselves more cultivated [advanced] than the Russians, and never cease reminding them that the Ukrainian

city of Kiev was the country's capital when Moscow was "a wheel track in the forest."

Although Ukraine is slightly smaller than the state of Texas, its 50 million residents help produce about one fifth of all of the food and industrial products of Northern Eurasia. Ukraine's agricultural productivity results from the fact that it includes those parts of the chernozem soil belt with the greatest precipitation.

Ukraine's farms were harmed severely by the 1986 Chernobyl (CHAYR noh bul) disaster, when a nuclear reactor run by the Soviet government was destroyed by fire. At least twenty-six people were killed and hundreds more were injured in the accident. Some 12.3 million acres (5.9 million hectares) of land were polluted by radiation from the damaged plant. The slow and poorly explained response of the Soviet government to this disaster led many Ukrainians to conclude that the nation must become independent if it was to serve the needs of its people.

Byelarus   The Chernobyl disaster also affected Byelarus, a nation of 10 million people that is roughly the size of the state of Kansas. More than one fifth of Byelarus's farmland was contaminated by radioactivity. The Orthodox Slavs who comprise the bulk of the population, therefore, will have to rely more heavily on development of industries that process imported raw materials. Byelarussians also will seek to develop their oil reserves and potash deposits, which are used to make fertilizer.

Moldova   Moldova once was a province of the Eastern European nation of Romania. Its four million residents speak Romanian. Despite efforts by Soviet leaders to foster a separate identity within the region, the breakup of the Soviet Union was accompanied by calls for Moldova's incorporation

Saying *Nyet* to Alcohol

Thirty men and women are seated in a large room at a community center in Moscow. The anti-drinking group these people belong to, which meets daily, was the first such group ever to be permitted to meet in the Soviet Union.

Alcoholism was a problem in many of the former Soviet republics for years. But it was never faced directly until 1985, when Mikhail Gorbachev launched a campaign against alcoholism. The campaign was typical of the openness in Soviet society under *glasnost*, where citizens were suddenly free to discuss societal and personal problems.

One member of the meeting explained the change in attitude this way: "We are rediscovering how to help ourselves, and how to help each other. In this country, we had forgotten how to do that."

1. Why do you think alcoholism in the Soviet Union had been ignored for so long?
2. When did the change in attitude toward alcoholism begin?

back into Romania. Whatever its future status, Moldova will rely heavily on production of wine, sugar beets, and seed oils.

Three Nations in the Caucasus

Locate Georgia, Armenia, and Azerbaijan (ah zur by ZHAHN) on the map on page 388. These nations lie in the Caucasus Mountains between the Black and Caspian seas. Residents of these nations share many characteristics, but significant differences have led to conflicts among them.

Georgia Nestled along the southern slopes of the Caucasus, about five million Georgians live in a nation that is slightly larger than the state of West Virginia. The warmest winters and heaviest rainfall in Northern Eurasia are along the Black Sea coastline, so Georgia is a center for the production of wine, tobacco, silk, and some citrus crops. Georgians generally adhere to Orthodox Christianity, like the Russians, with whom they have traditionally maintained close ties. Despite these ties, however, Georgia was among the first of the Soviet republics to declare its independence in 1991.

Armenia Armenia is the smallest of the nations in Northern Eurasia. Most of the land is extremely rocky, although farmers grow a variety of crops in southern valleys. Small factories also manufacture goods, especially in the capital of Yerevan. Armenia's people have been Christians since the second century A.D. Caught between the Turkish and Russian empires, they often have suffered, especially so in 1915 when more than one million Armenians were killed in massacres and forced migrations. Armenians therefore are especially concerned about their fate in the new political relationships of Northern Eurasia.

Azerbaijan Pointing like an arrowhead into the western shore of the Caspian Sea, Azerbaijan is blessed with rich petroleum deposits. It also has productive farmlands nourished by a mild climate. Most of Azerbaijan's seven million residents are Muslims. Azerbaijanis were especially aggressive in demanding independence from Russian control in the late 1980s. With that goal achieved, however, they now face two new challenges. First, Armenia has demanded the return of western provinces controlled by Azerbaijan where a majority of the population is Armenian. Second, nationalism—the desire of people to form an independent nation to promote their common culture or interests—has always been a powerful force in Azerbaijan.

Azerbaijanis may want to explore the possibility of reuniting their nation with Azerbaijanis who live in northwestern Iran.

Five Nations in Central Asia

About one fifth of the people of Northern Eurasia are Muslims. Almost all of the region's Muslims live in Azerbaijan and in the five nations of Central Asia—Kazakhstan (kuh zak STAN), Turkmenistan (turk MEN ih STAN), Uzbekistan (ooz BEK ih STAN), Kyrgyzstan (kihr GEEZ stan), and Tajikistan (tah ZHEEK ih STAN).

Nations Based on Forced Ethnicity The nations of central Asia had no formal political identity before they were established as republics within the Soviet Union in the 1930s. Where people once considered themselves to be residents of Turkestan, they were told that they were Kazakhs, Turkmen, Uzbeks, or Kirzigs.

All four groups spoke local dialects of the same Turkic language, but Soviet officials replaced their traditional Arabic script first with Latin and finally with Cyrillic alphabets. As a result, complained an actor from Uzbekistan, "I cannot read the works of my own literature. . . . It is as though an Englishman had been robbed of the ability to read Shakespeare." While the five nations declared their intentions to function independently in 1991, many local leaders called for a return to a unified Turkestan.

The combined area of the five central Asian nations is roughly half the size of the mainland United States. Much of this land is covered by mountains or deserts, however.

Economic Regression and Environmental Damage Prior to their incorporation into the Soviet Union, most people in the central Asian lands lived in rural areas and practiced herding and other self-sufficient activities. Decades of communist rule dramatically altered the way of life of the people, however, largely to the benefit of people elsewhere. Parts of northern Kazakhstan were settled in the 1950s, when the Soviet government mandated the expansion of grain production. These efforts were moderately successful. On average, Kazakhstan grows about 20 percent of Northern Eurasia's grain, although its fields are prone to frequent droughts.

Much more controversial was the expansion of irrigated cotton production south and east of the Aral Sea under Soviet rule. Locate the Aral Sea on the map on page 388. Much of the water within the Amu Darya and Syr Darya rivers was diverted to water cotton fields nearby. In addition, a canal that is 500 miles (800 km) long and 300 feet (90 m) wide was cut across southern Turkmenistan to irrigate patches of the Kara Kum desert.

Cotton yields in the region did increase substantially. The Soviet Union became the world's second greatest cotton producer, but residents of central Asia objected to these developments. They noted that most raw cotton was shipped out of the region and woven into cloth elsewhere, depriving local residents of the opportunity to benefit and profit from their initial production of cotton.

The cultivation of irrigated cotton has also resulted in health and environmental problems in the region. Repeated planting of cotton has reduced soil fertility and heavy doses of chemical fertilizers and pesticides continue to create serious health risks for workers who perform many operations by hand. Furthermore, rivers that once flowed into the Aral Sea now are dry before they reach it. As a result, water levels in that large inland sea are much lower, and the exposed salts, sands, and chemicals in the sea bed frequently are blown about in harmful dust storms.

Maintaining Cultural Traditions
A Muslim family from Tajikistan works together to paint a carpet by hand. Carpet designs and weaving secrets have been handed down from generation to generation.

SECTION 2 REVIEW

Developing Vocabulary
1. Define: nationalism

Place Location
2. What body of water borders Azerbaijan?

Reviewing Main Ideas
3. How has the location of the Baltic nations made life there different from life in other parts of Northern Eurasia?
4. Why do some people suggest that the nations in central Asia should be united?

Critical Thinking
5. **Synthesizing Information** In what ways may the different nations of Northern Eurasia still have to work together?

Case Study on Current Issues

Lithuania's Dream of Independence

Some observers said it was foolish, others that it was brave; some said that it was high time, and others said that it was too soon. Everyone, however, was amazed when little Lithuania declared that it was no longer a part of the mighty Soviet Union.

The Clash

The struggle began in earnest in 1989, when Lithuania declared that it would no longer take orders from Moscow about what to produce and sell. Lithuania then changed the laws of the republic so that the Communists were not the only party allowed to hold power.

Mikhail Gorbachev, the President of the Soviet Union, protested that the Lithuanians had "stabbed *perestroika* in the heart." He meant that Lithuania had not given Gorbachev's new policies a chance to succeed. In January 1990, sensing that Lithuania was about to secede or withdraw from the Soviet Union, Gorbachev made a dramatic three-day trip to the Baltic republic. He scolded government leaders, argued with crowds in the street, and challenged workers in factories. Everywhere, he delivered the same message: Be patient. Don't cause trouble. *Perestroika, glasnost,* and democratic reforms will bring change soon enough.

On March 11, however, Lithuania voted to secede from the Soviet Union. Gorbachev demanded that the republic reverse its decision. A war of nerves began. Gorbachev had propaganda leaflets dropped over Vilnius, the Lithuanian capital. Soviet troops captured the Communist party headquarters.

Lithuanian leaders held fast. Finally, Gorbachev ordered all oil supplies to the republic shut off. All but one gas pipeline were closed. Vital raw materials were withheld by the Soviets. The message to the Lithuanians was clear: Back down or else.

The Soviet View

Gorbachev claimed that he did not object to Lithuanian independence. He insisted, however, that secession occur according to Soviet law. Important questions of property had to be settled. For example, the Soviet Union owned ninety-five factories in the republic. What was to become of them? Another question was the status of the Russians who made up almost 10 percent of the population of Lithuania.

Soviet leaders also noted that the Union supplied 90 percent of the raw materials and energy used by Lithuania at a fraction of the cost Lithuania would have had to pay for them on the world market. Oil, for example, was selling at about $24 a barrel around the world; Lithuanians had been buying oil from the Union for roughly $6 a barrel. It was clear that Lithuania would never be able to pay either for facilities built by the Soviets or for the raw materials it used.

The Lithuanian View

The Lithuanians countered that the Soviet Union had stolen their entire country by forcing it to join the Union in 1940. Why should they have to pay the Soviets for what was their own property? "They have been lining their pockets with profits made on Lithuanian soil for fifty years," grumbled one Lithuanian leader. As for the Russians in Lithuania, the Lithuanian people had little sympathy for them.

Lithuania Falters but Later Prevails

In spite of their courage, the outlook was not good for the Lithuanians. More than 40,000 people were laid off from their jobs because of insufficient fuel and raw materials. The republic simply had no goods it could exchange for oil and gas from other countries. The goods they produced were of good quality by Soviet standards, but they were poorly made in comparison with products from Germany, Japan, or the United States. In short, their products could be sold only to other republics in the Soviet Union.

By the end of June 1990, the Lithuanians agreed to suspend their declaration of secession for one hundred days while talks with the Soviet government took place.

Suspending the declaration did not mean that Lithuanians gave up their dream of independence, however. When an attempt to remove Gorbachev from power failed in August 1991, the Communist party was totally discredited. Lithuania's declaration of independence was recognized by countries throughout the world and within three months, the Soviet Union ceased to exist.

TAKING ANOTHER LOOK

1. What questions did Gorbachev feel had to be settled before the Lithuanians seceded?
2. Why did the Lithuanians feel that they did not owe the Soviet Union anything for Soviet-built facilities?
3. What events finally led to recognition of Lithuanian independence?

Critical Thinking

4. **Recognizing Ideologies** Besides the reasons Gorbachev gave, what underlying reasons might he have had for opposing Lithuania's secession? Do you think his reasons were valid?

An Anti-Soviet Rally in Vilnius *Lithuanians demanded independence in 1990.*

21
REVIEW

Section Summaries

SECTION 1 Russia Russia has more natural resources than any nation on earth. Problems of location, distance, climate, and political ideology have prevented full development of these resources, however. Under communism, agriculture was practiced on large, relatively inefficient state and collective farms. Consumer goods were scarce because of an emphasis on heavy industry. Upon becoming an independent nation, Russia entered a rugged period of transition to a market-driven or demand economic system.

SECTION 2 Other Independent States
The small Baltic nations are oriented toward Europe and light manufacturing. Ukraine and Byelarus are important agriculture and industrial producers tied more closely to Russia because of their closely related ethnic heritage. The nations of the Caucasus are homes to diverse groups of people who have fought each other in the past. In central Asia, common ethnic and religious ties are drawing together people in different nations.

Vocabulary Development

Match the definitions with the terms below.

1. ''a turning about''; economic restructuring
2. the production of goods such as steel and machines that are used by other industries
3. farms owned by the government whose workers are paid wages for their work
4. the ideas or principles on which a political system is based
5. farms owned jointly by farmers, who share the surpluses produced on the farm
6. an economic system in which people's demands for goods in the marketplace determine which goods will be produced
7. the desire of people to form an independent nation to promote their common culture or interests
8. an economic system in which the government makes decisions about the use of resources and the production of goods

a. collective farm   e. ideology
b. heavy industry   f. command system
c. *perestroika*   g. demand system
d. state farm   h. nationalism

Main Ideas

1. How does winter affect transportation in Northern Eurasia?
2. How did the Soviet Union try to establish separate republics in central Asia?
3. Describe the forces that caused the Soviet Union to break apart.
4. How did President Gorbachev try to stop the Lithuanians from declaring their secession from the Soviet Union?
5. What happened at Chernobyl in 1986?
6. How will different nations in Northern Eurasia try to change their economic systems?
7. Why will Russians continue to be a powerful influence in the new political geography of Northern Eurasia?

Critical Thinking

1. **Distinguishing Fact from Opinion** Explain why you agree or disagree with the statement, ''Life under a communist system is little better than slavery.''

2. **Making Comparisons**   How is Azerbaijan similar to the five nations of central Asia? How is it different?
3. **Formulating Questions**   Make up three questions the Soviet government might have printed on the propaganda leaflets scattered over Vilnius.
4. **Perceiving Cause-Effect Relationships**   How did *glasnost, perestroika,* and the democratic policies of the Soviet Union influence events in Eastern Europe?
5. **Expressing Problems Clearly**   What are the major economic problems in Northern Eurasia?

Practicing Skills

1. **Distinguishing False from Accurate images**   Many people's image of Northern Eurasia consists of a frozen landscape and Russian people bundled in fur hats and coats trying to keep warm. Explain why this image is not accurate for all of Northern Eurasia.
2. **Reading a Time Zone Map**   Use the time zone map on page 115 to answer the following questions: What is the time difference between Moscow and Tokyo? Between Moscow and New York?

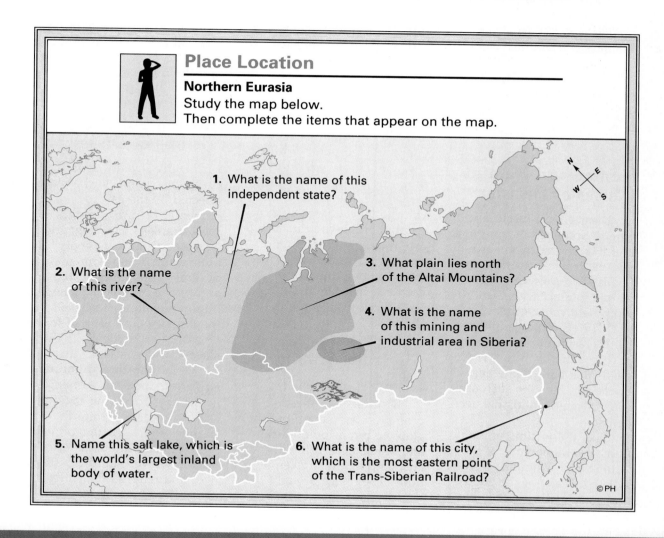

Place Location

Northern Eurasia
Study the map below.
Then complete the items that appear on the map.

1. What is the name of this independent state?
2. What is the name of this river?
3. What plain lies north of the Altai Mountains?
4. What is the name of this mining and industrial area in Siberia?
5. Name this salt lake, which is the world's largest inland body of water.
6. What is the name of this city, which is the most eastern point of the Trans-Siberian Railroad?

©PH

MAKING CONNECTIONS

·WHERE REGIONS MEET·

International Relief Efforts

On December 7, 1988, the ground in northern Armenia began to tremble. Within minutes a major earthquake had struck this Caucasian nation—at the time a republic of the Soviet Union. Buildings crumbled and cities were leveled. Nearly 50,000 people were killed and 130,000 injured. A doctor on the scene called the devastation "a vision of horror."

Human and Natural Causes of Disaster

Disasters can occur anywhere. Some result from natural hazards, such as earthquakes, tropical storms, and floods. The ways in which people interact with their environment cause other disasters. For example, famines can be the result of poor farming practices that lead to low crop yields. Humans alone bear the blame for disasters such as wars or major industrial accidents.

Like the victims of many disasters, the victims of the Armenian earthquake were isolated. Many of the goods and services they needed to survive were not immediately available. Massive efforts were required to get food, drinking water, medical supplies, and clothing to victims. Building supplies, hospital equipment, and construction vehicles also were needed at the site of the disaster.

In recent years a number of disasters have rallied the world community. In the mid-1980s a devastating earthquake leveled parts of Mexico City, killing more than four thousand people. About the same time, the Colombian volcano Nevado del Ruiz erupted and melted its snowcap, releasing tons of mud that buried a nearby village. A massive earthquake in northern Iran killed at least forty thousand people in 1989.

Aid from Around the World

Disasters like these prompted people around the globe to put aside political and ideological differences and send aid to stricken nations. They have brought the world's people together in a common effort to save lives and ease human suffering.

Aid sent to nations suffering the effects of disasters can come in the form of official aid from governments or private aid from relief organizations like the United Nations and the International Red Cross.

Moving Supplies to Victims of Disasters

Nearly seventy countries rushed assistance to the places affected by the earthquake in Armenia. Medical supplies, food, blankets, clothing, and heavy earth-moving equipment poured into Armenia. Several nations sent search and rescue groups. This huge international effort marked the largest outpouring of foreign aid to the Soviet Union since World War II.

More than three hundred airplanes landed in Armenia in the days following the disaster. Although getting the supplies to the victims proved to be difficult, the relief effort probably saved hundreds of lives and relieved the suffering of thousands more.

The View from the United States

The United States is a major contributor to international relief efforts. Following the Armenian earthquake the United States government sent eight plane loaded with goods to the disaster site. In addition, private donations from American citizens across the country paid for enough goods to fill a dozen more airplanes.

The United States usually is well prepared to handle disasters that occur within the country, partly because of advance planning. Such planning was rewarded when a serious earthquake occurred in the San Francisco Bay area of California in 1989. Strict construction codes, backup water supplies and communication systems, and well-practiced rescue responses saved lives and limited damage to buildings in the area.

TAKING ANOTHER LOOK

1. What types of goods and services did the victims of the Armenian earthquake need?
2. How does international cooperation help victims of major disasters?

Critical Thinking

3. **Identifying Central Issues** Why do you think many nations are poorly prepared for disasters?

Rescue Efforts in Armenia

Although the 1988 earthquake in Armenia was centered near the city of Leninaken—now called Kumayri—the map below shows that its shock waves toppled buildings hundreds of miles away.

The Middle East and North Africa

CHAPTERS

A wise old man in the desert was handed this riddle: How can any people survive in a place where rainfall is slight, thirst is terrible, and almost nothing blooms? The old man rocked back on his heels and laughed before giving his response: They can thrive by making their desert into a garden.

Since ancient times, the people of the Middle East and North Africa have followed that old man's advice; they have thrived by making the desert into a garden. In this unit, you will discover the ways in which people have adapted to the climate of this region.

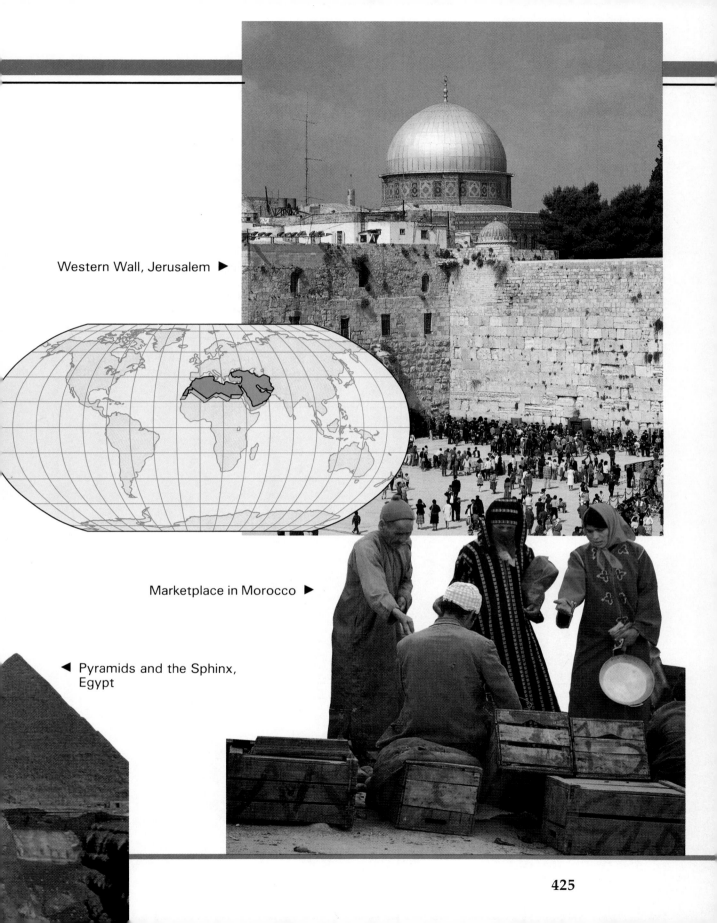

Western Wall, Jerusalem ▶

Marketplace in Morocco ▶

◀ Pyramids and the Sphinx, Egypt

425

Regional Atlas: The Middle East and North Africa

Chapter Preview

Both sections in this chapter provide an overview of the countries of the Middle East and North Africa, shown in red on the map below.

Sections	Did You Know?
1 **LAND, CLIMATE, AND VEGETATION**	The Nile is the longest river in the world. It is 4,187 miles (6,737 km) in length.
2 **HUMAN GEOGRAPHY**	About 99 percent of Egypt's people live on only about 3.5 percent of the country's total land.

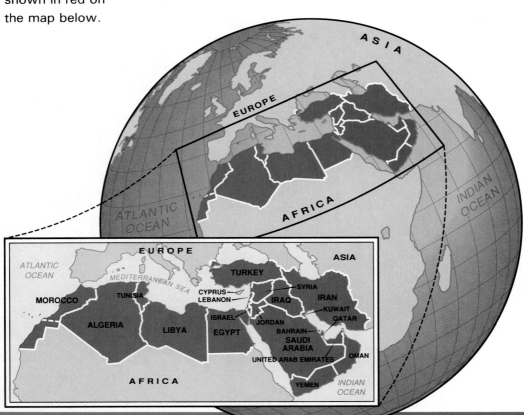

A Desert Community
Wells and springs provide life-giving fresh water to North African oases villages such as this one in Algeria. The scarce water sustains plants, animals, and people.

1 Land, Climate, and Vegetation

Section Preview

Key Ideas

- Nearly two thirds of the land in the Middle East and North Africa is dry, hot desert.
- Water is one of the most precious resources in the Middle East and North Africa.
- Mountains and plateaus in the Middle East and North Africa have an important effect on the climates of the region.
- Mountains in the region play a part in creating the deserts.

Key Terms

erg, wadi, oasis

Picture a hot and endless desert where water is as precious as gold. Winds sweep across the dry ground, blowing pebbles and bending small desert shrubs. Next, imagine a blue river with narrow strips of green along its banks winding its way through the desert. Finally, picture rocky mountains and plateaus where rain is common and grasses and forests grow. Together these images characterize much of the land of the Middle East and North Africa.

Look at the globe on the opposite page to see where this region of sand, rock, and water is located. The five countries that are located in North Africa lie south of Europe and the Mediterranean Sea and north of the Sahara. The fifteen countries of the Middle East are located in Southwest Asia. Often Egypt and Libya, while located in North Africa, are also considered part of the Middle East.

A Dry Land

A quick glance at the world climate map on pages 38–39 shows one of the unique characteristics of the Middle East and North Africa. It is the largest dry region on earth. Using the climate map on this page, you can see that much of this huge region has an arid or semiarid climate. Most of the region receives less than 10 inches (25 cm) of rain per year. You can see how little that is by comparing it with other cities. Chicago—an example of a place with ample rainfall—receives an average of 33 inches (84 cm) per year.

Linking Precipitation, Temperature, and Vegetation With so little precipitation, it is no wonder nearly two thirds of the region is desert. The Sahara stretches for 3.5 million square miles (9 million sq km) across northern Africa. It is the largest hot desert in the world, roughly the size of the United States. Use the physical-political map on pages 430–431 to locate the other large deserts of the region: the Syrian Desert and the Rub'al-Khali (rub ahl KAHL ee), which means "the empty quarter."

Because the desert air contains little moisture or humidity, few clouds form over the dry land. Without clouds, the sun's hot rays are able to beat directly onto the land during the day. As a result, temperatures rise dramatically. Temperatures may reach as high as 125°F (52°C) during the day. At night, they may drop to as low as

---

Applying Geographic Themes

Interaction Much of the area shown on this map is arid or semiarid. Compare this map to the population density map on page 436. What conclusions can you draw about the relationship between climate and the places where people live?

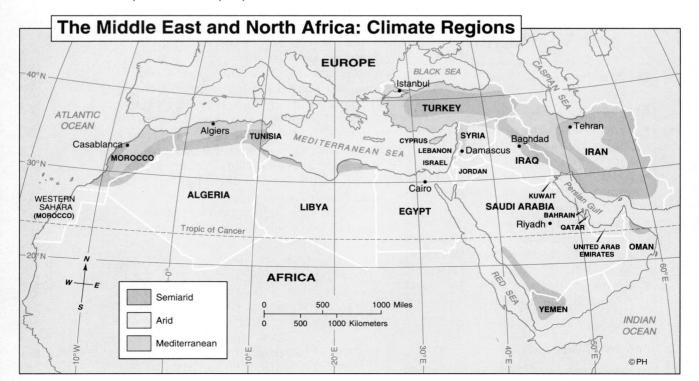

The Middle East and North Africa: Climate Regions

40°F (4°C), because no clouds prevent the daytime heat from rising from the earth.

Extreme temperatures combined with little rainfall make it difficult for humans, plants, and animals to survive in the desert. Yet many life-forms have adapted to severe desert climates. Many varieties of desert plants have shallow, spiderlike root systems that extend out from the plant in every direction to collect water over great distances. Other desert plants survive by growing tap roots—deep root systems that tap sources of underground water. Animals such as the camel have adapted to the dry climate and are able to survive for long periods without much food and water. Even with adaptations, however, life in the desert is spread out over great distances. As a result, deserts are often described as barren or lifeless.

Wind and Sand   With little vegetation to hold the soil in place, wind becomes a powerful sculptor that shapes the landscape of the region. In many of the region's deserts and plateaus, including large areas of the Sahara, the wind carries away all the loose sand particles. Remaining on the ground is what some people call "desert pavement"—a thin surface covering of pebbles, gravel, and boulders. In other areas of the Sahara and in the Rub'al-Khali, large areas of loose sand pile up where the wind carves great expanses of sand dunes, or **ergs**. One British writer described the ergs he saw as he crossed the Rub'al-Khali:

Isolated dunes, two or three hundred feet high, rose in apparent confusion from the desert floor. . . . Each had its own shape that did not vary perceptibly over the years, but all had certain features in common. . . . The surface was marked with [tiny] ripples, of which the ridges were built from

Desert Sands
Sand dunes, or ergs, characterize much of the Sahara. The spreading sands of the Sahara cover more land area than the United States.

the heavier and darker sand, while the hollows were of smaller paler-coloured grains. It was the blending of these colours that gave such depth and richness to the sand: gold with silver, orange with cream, brick-red with white, burnt-brown with pink, yellow with grey. There was an infinite variety of colours and shades.

Notice the Grand Erg Occidental, the Grand Erg Oriental, and the Erg Chech (shesh) shown in the Sahara on the map on the next page. Some ergs, like these and the ones described above, are stationary and can be identified on a map. The location, shapes,

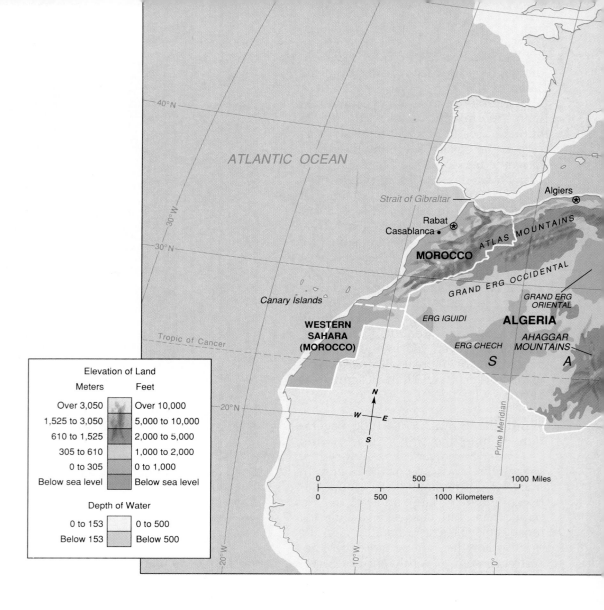

Elevation of Land

Meters		Feet
Over 3,050		Over 10,000
1,525 to 3,050		5,000 to 10,000
610 to 1,525		2,000 to 5,000
305 to 610		1,000 to 2,000
0 to 305		0 to 1,000
Below sea level		Below sea level

Depth of Water

0 to 153		0 to 500
Below 153		Below 500

and sizes of most dunes, however, are constantly altered by the wind. Most types of dunes advance only a few feet a year. In the case of a violent wind storm, however, a dune may move 60 feet (18 meters) or more in a single day. Like lava flowing from a volcano, the sand dune covers everything in its path—plants, fences, and villages.

Water in a Dry Land

Most people in the United States take it for granted that water will stream out of the faucet whenever they turn it on. In the Middle East and North Africa, water is not used so casually. Water is one of the most precious resources in the Middle East and North Africa. In most places the land is parched and brown. But add water in the form of rain or rivers, and life survives, creating curving, green ribbons on a rough, brown canvas.

Dry Riverbeds Water, like wind, plays a part in shaping the dry land, even though very little rain falls. What little rain does fall often comes down intensely over a short period of time.

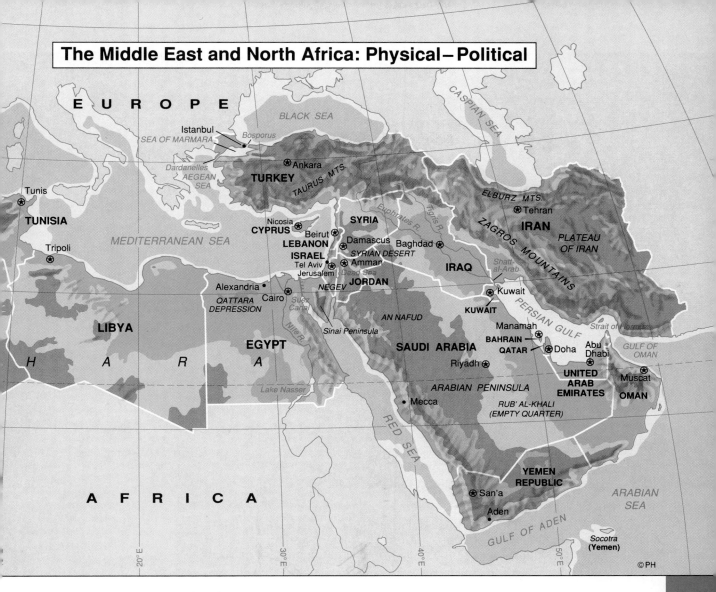

The Middle East and North Africa: Physical–Political

EUROPE

BLACK SEA

CASPIAN SEA

Istanbul
SEA OF MARMARA
Bosporus
Dardanelles
AEGEAN SEA
Ankara ⊛
TURKEY
TAURUS MTS.

ELBURZ MTS.
Tehran ⊛
ZAGROS MOUNTAINS
IRAN
PLATEAU OF IRAN

Tunis ⊛
TUNISIA

Tripoli ⊛

MEDITERRANEAN SEA

Nicosia ⊛
CYPRUS
Beirut ⊛
LEBANON
ISRAEL
Tel Aviv •
Jerusalem ⊛
SYRIA
Damascus ⊛
SYRIAN DESERT
Amman ⊛
Dead Sea
NEGEV
JORDAN

Euphrates R.
Tigris R.
Baghdad ⊛
IRAQ
Shatt-al-Arab

Kuwait ⊛
KUWAIT

PERSIAN GULF
Strait of Hormuz

Alexandria •
Cairo ⊛
QATTARA DEPRESSION
Suez Canal
Sinai Peninsula
Nile R.

AN NAFUD

Manamah ⊛
BAHRAIN
QATAR
Doha ⊛
Abu Dhabi
UNITED ARAB EMIRATES

GULF OF OMAN
Muscat ⊛
OMAN

LIBYA

EGYPT

Lake Nasser

S   A   H   A   R   A

SAUDI ARABIA

Riyadh ⊛

ARABIAN PENINSULA

Mecca •

RUB' AL-KHALI
(EMPTY QUARTER)

RED SEA

AFRICA

YEMEN REPUBLIC
San'a ⊛
Aden •
GULF OF ADEN

ARABIAN SEA

Socotra
(Yemen)

©PH

20°E   30°E   40°E   50°E

Applying Geographic Themes

Regions   Most of the land of the Middle East and North Africa is desert. In which physical and political regions can Egypt be placed?

The sun bakes and hardens the surface of the ground. When sudden downpours take place, water runs over the surface—the same way as a spilled glass of water spreads out on the floor —rather than soaking into the ground. As the water flows across the land to lower elevations, it cuts sharp gullies, or riverbeds, in the ground. In the Middle East and North Africa, these gullies are known as **wadis** (WAH deez). Wadis are filled with water temporarily. Within a short time after the rain stops, the water empties, and the wadis become dry riverbeds again until the next brief downpour.

Desert Springs   Rainfall is not the only source of water in the region. In a few places, deep underground springs

Sources of Water

This parched wadi in Israel (right) sustains little life, whereas areas fed by the Tigris and Euphrates rivers (left) are fertile. Ancient trading centers that began on the riverbanks of the region grew into great centers of civilization.

force their way up to the surface. Each of these springs forms an **oasis** (o AY sis), or a place where the supply of fresh water makes it possible to support life in a dry region. In other places people have created their own oases by digging deep wells that tap the waters of an underground spring.

At an oasis, lush green plants sprout up, providing food and shade for people and animals seeking relief from the harsh, dry climate. Some oases supply enough water to support only a small cluster of trees or a small village or town. Others have a water

supply large enough to maintain an entire city. The Sahara has approximately ninety oases, some of which have provided relief to travelers crossing the desert for centuries.

Rivers of Life Although rainfall and underground springs are valuable sources of water, the region's rivers are by far the most important source. The largest of these rivers are the Nile in northern Africa and the Tigris and Euphrates (u FRAYT eez) rivers of Southwest Asia. Find these three rivers on the map on pages 430–431.

In a region where water is so important, even small rivers have great worth. The Jordan River, which flows into the Dead Sea, is one example of a small but life-giving river. As you can see when you look at these rivers on the vegetation map, the water and rich soil that they deposit bring fertility to the nearby lands.

Linking Elevation, Climate, and Vegetation

Although the Middle East and North Africa are commonly known for their dry deserts, rocky plateaus, and lush river valleys, several mountain ranges also mark the landscape of the region. A British writer described one such mountainous area on the border of Turkey and Iraq.

I felt an uplift of the spirit as we climbed over the pass, into Iraq. . . . Never have I seen such country. . . .Everywhere one range was superimposed upon another. The knees of the mountains, and the valley sides up to six thousand feet, were wooded. . . . In the valley bottoms the Zab, the Little Zab and other smaller rivers and torrents flowed down to join the far-off Tigris, foaming ice-cold through narrow gorges of polished rock, and swirling among great boulders tumbled from the cliffs above; or calm in deep green pools under grassy banks and overhanging willows.

Applying Geographic Themes

Regions This geographic region is dominated by desert. Most of the land sustains only sparse life forms. Yet some areas are richer in vegetation and denser in population than others. How can such differences be explained?

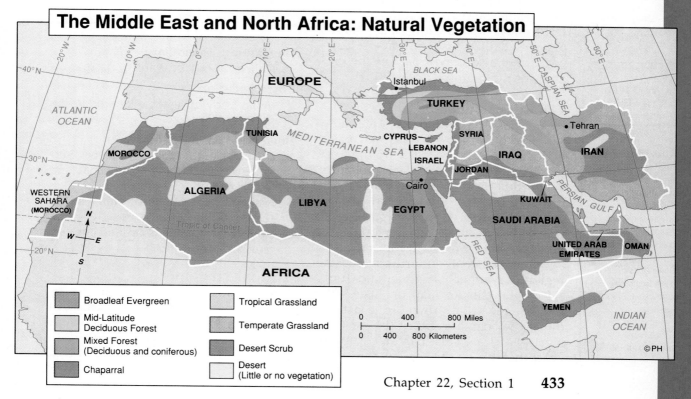

The Middle East and North Africa: Natural Vegetation

Broadleaf Evergreen

Mid-Latitude Deciduous Forest

Mixed Forest (Deciduous and coniferous)

Chaparral

Tropical Grassland

Temperate Grassland

Desert Scrub

Desert (Little or no vegetation)

0   400   800 Miles
0 400 800 Kilometers

©PH

The Taurus Mountains of Turkey
The population of the Middle East is concentrated in areas near water as well as in highland areas where the climate is cooler and precipitation is greater.

the Atlas Mountains receives the rain that falls as the air rises up the mountains. With this rainfall, plants bloom and crops flourish. On the far, or leeward, side of the mountains, the sands of the deserts resume their harsh command of the environment.

Other Highland Areas Several parts of the region have high elevations. These areas include the mountains of Lebanon, the Taurus Mountains, and the Zagros Mountains. Locate these highlands on the map on pages 430–431 and then locate the same areas on the climate map. How does the climate in these places differ from that of most other parts of the region? This difference in climate also affects vegetation. By looking at the same area on the vegetation map on page 433, you can see that forests and grasslands are able to grow in highland areas that have milder climates and greater amounts of precipitation.

Mountains like those described above have an important effect on the region's climate. Like oases and rivers, mountains help to create pockets of fertile land in a region noted for its dryness. Unlike oases and rivers, however, they also play a part in the creation of the region's deserts.

WORD ORIGIN

Atlas Mountains
Atlas is a mythical Greek character who was thought to support the heavens on his shoulders. According to myth, Atlas was turned to stone by one of the Greek gods. The early Greeks believed that Atlas had stood on the mountains of North Africa, so they called them the Atlas Mountains.

The Atlas Mountains Find the Atlas Mountains of Morocco and Algeria on the map on pages 430–431. Air passing over the Atlantic Ocean to the west of the mountains picks up moisture. When this moisture-laden air reaches the mountains it is forced to rise up over the slopes. As it does so, the air cools. Because cool air can hold less moisture than warm air, rain falls.

Now locate the northern coastal region of Morocco, Algeria, and Tunisia on the climate map on page 428. Notice how that fringe of coastline differs in climate from the rest of North Africa. This narrow strip of coast that lies on the windward side of

SECTION **1** REVIEW

Developing Vocabulary
1. Define: **a.** erg **b.** wadi **c.** oasis

Place Location
2. What body of water is located to the north of Egypt?

Reviewing Main Ideas
3. Describe two types of desert surfaces found in the Middle East and North Africa.
4. Name three sources of water in the region.
5. How do mountains affect the region's climate?

Critical Thinking
6. **Distinguishing False from Accurate Images** Many people think of deserts as lifeless. Why do you think people have this image of deserts?

✔ Skills Check

☐ Social Studies
☐ Map and Globe
☑ Reading and Writing
☐ Critical Thinking

Making an Outline

An outline is a plan for organizing the main points and details of an essay in a clear and logical order. The text of this book has been organized under headings and subheadings that outline the information in each section.

Outlining can also be a useful tool for studying information that you have read. A partial outline of the section you have just read is presented in the next column. Copy it onto a separate piece of paper. Then follow the steps below to complete the outline.

1. **Give the outline a title.** Outlines generally follow a format similar to the one shown on the right. The title of the outline describes the main subject of the essay. What is the main subject of the outline to the right?

2. **Use roman numerals to list the main points of the essay.** The section you have just read is divided into three subsections, each of which begins with a main heading. In the outline to the right, main headings I and II have been filled in for you. Review the section to find the main heading for roman numeral III.

3. **Use capital letters to list examples or evidence supporting the main points.** Three of the main points in the section you have just read are further broken down into subheads. The text following each subhead describes examples of the main point. Some of the subheads are shown on the outline to the right, others are not. Read the text and fill in the missing subheads under roman numerals II and III.

4. **Use arabic numerals to introduce detailed points.** The outline below shows two details under the subhead "Linking Precipitation, Temperature, and Vegetation." Use the text to fill in the missing details under the remaining subheads.

Land, Climate, and Vegetation of the Middle East and North Africa

I. A Dry Land
  A. Linking Precipitation, Temperature, and Vegetation
    1. With little precipitation, two thirds of the region is desert.
    2. Extreme temperatures and little rainfall make it difficult for humans, plants, and animals to survive in the desert.
  B. Wind and Sand
    1. In some places wind carries away all the loose sand and only "desert pavement" remains.
    2. _____

II. Water in a Dry Land
  A. Dry Riverbeds
    1. During downpours, water cuts sharp gullies or wadis in the ground.
    2. _____
  B. _____
    1. Deep underground springs, known as oases, force their way up to the surface.
    2. _____
  C. Rivers of Life
    1. Rivers are the most important source of water in the region.
    2. _____

III. _____
  A. The Atlas Mountains
    1. Land on the windward side of the Atlas Mountains receives rainfall.
    2. _____
  B. _____
    1. These areas have milder climates than most other parts of the region.
    2. These areas have more abundant vegetation because of the differences in climates.

2 Human Geography

Section Preview

Key Ideas

- The population of the Middle East and North Africa is not evenly distributed. People live in places where water is available.
- The predominant culture in the region is that of Islam.

Key Terms

arable, fellaheen, monotheism, Muslim, mosque, muezzin, minaret, prophet, Hajj

Many people have an image of sandy deserts, colorfully striped tents, and camel caravans in mind when they think about life in the Middle East and North Africa. As you will read in this section, however, these images do not accurately depict the region today.

Linking Population Density to Climate and Vegetation

Notice on the map below that the population of the region is not evenly distributed. Most of the three hundred million people in the region live near water. They live beside rivers, along coasts, in the mountains and plateau regions, and at a few oases.

Applying Geographic Themes

Regions Which areas of the Middle East and North Africa are the most densely populated? What are the physical characteristics of these areas?

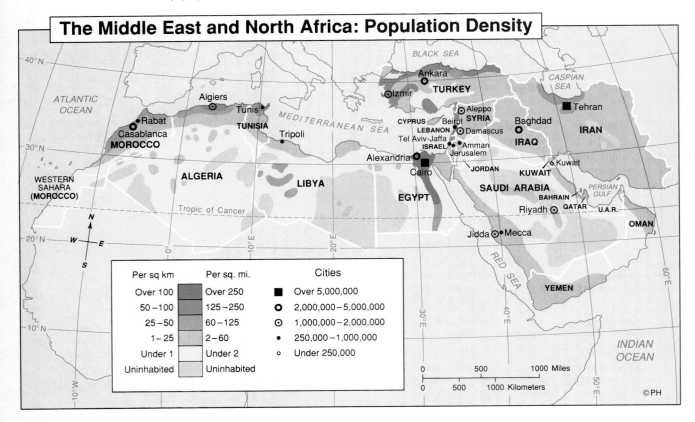

The Middle East and North Africa: Population Density

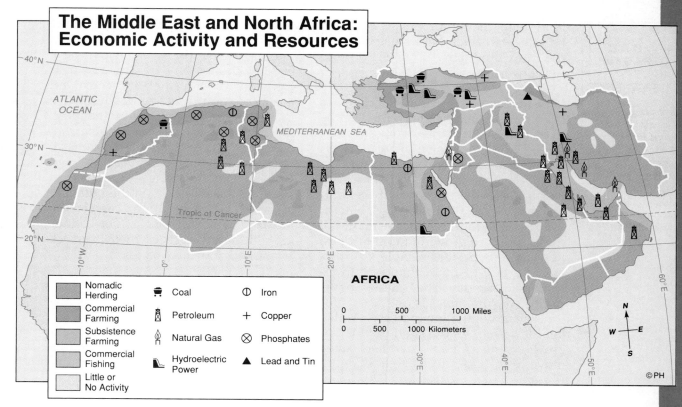

The Middle East and North Africa: Economic Activity and Resources

ATLANTIC OCEAN

MEDITERRANEAN SEA

AFRICA

Tropic of Cancer

Legend:
- Nomadic Herding
- Commercial Farming
- Subsistence Farming
- Commercial Fishing
- Little or No Activity
- Coal
- Petroleum
- Natural Gas
- Hydroelectric Power
- Iron
- Copper
- Phosphates
- Lead and Tin

0   500   1000 Miles
0   500   1000 Kilometers

©PH

Applying Geographic Themes

Place Vast supplies of oil brought great economic and political changes to parts of the Middle East and North Africa. Around which body of water are most of the oil deposits found?

By comparing the population density map with the climate and vegetation maps in Section 1, you will see that, in addition to living near water, people live in areas where the climate is cool, rainfall is greater, and vegetation is abundant. For example, look at northwest Africa and the mountain regions of Iran. If you could place the climate and vegetation maps over the population map, you would see that the areas of greatest population closely resemble the areas that are the coolest and most fertile.

Three Ways of Life

Three ways of life predominate in the Middle East and North Africa: farming, city living, and nomadic herding. Compare the map on this page with the climate map on page 428. Notice that there are distinct patterns of economic activity. In areas where there are reliable sources of water, people farm or live in cities. In areas with little water, people live as nomadic herders, moving frequently from place to place.

Farming Only a small percentage of the region is made up of **arable** (AR uh buhl) land, that is, land that can be farmed. But farmers who live in small towns and villages make up more than half of the working population. Many of these farmers rely on simple tools, irrigation systems, and hours of hard work to meet their needs. By building dams and irrigation systems in the

bedouin
The word *bedouin*
comes from the
Arabic word *badawi*
meaning "desert
dweller." Today
bedouins make up
about one tenth of
the population of
the Middle East.

Nile Delta, the *fellaheen* , as Egyptian farmers are called, have created one of the most productive stretches of land in the world.

City Living   Some of the oldest cities in the world are located in the Middle East and North Africa. For centuries, these cities have drawn people together and served as centers for trade.

Look again at the population map on page 436. Locate some of the largest cities in the region—Casablanca, Cairo, and Tehran. Today many of these cities are growing rapidly. Cairo (KY roh), Egypt, for example, is growing at a rate of more than one thousand people a day. The limited opportunities in rural areas is one reason for this urban population increase. Arable land is scarce. As the rural population grows, some family members must move to the cities to make a living.

Nomadic Herding   Few people in the region today live as nomadic herders. The bedouins (BED uh wins) have lived in the region's deserts and near oases herding camels, sheep, and goats for centuries. The bedouins live in small family groups. In the past, these groups moved on camel from place to place—using their extensive knowledge of the location of water and seasonal climate changes. Today, most bedouins have settled on the outskirts of small villages. Often, camels are replaced by pickup trucks. The bedouins sell the products of their livestock at markets in nearby cities.

The Birthplace of Three Major Religions

Three of the world's major religions trace their origins to the Middle East — Judaism, Christianity, and

Applying Geographic Themes

Regions   Sunni Muslims are the largest religious group in this region. What other Muslim sect is found in many parts of the region?

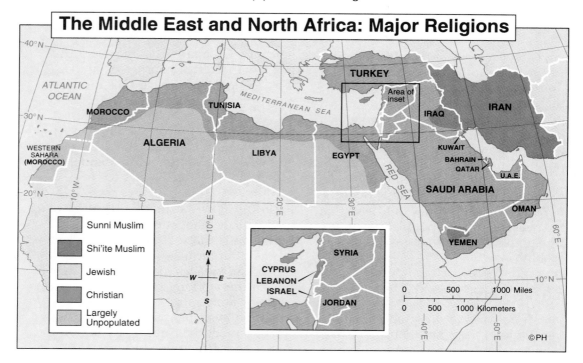

The Middle East and North Africa: Major Religions

DAILY LIFE

A Traditional Marketplace

In cities throughout the Middle East and North Africa, the old contrasts with the new. Passing under an archway, you leave behind the hustle and bustle of city life and wander into another world. Here, lining the maze of narrow passages is the *souk*, or marketplace.

Every day, people in business attire mingle with those dressed in traditional long robes and veils. Striped canvas awnings protect shoppers from the hot sun. Here you see bolts of bright yellow cloth, and there brown, roasted coffee beans. Across the way are squawking chickens. Sounds blend with smells in a mall-like atmosphere that has been a part of Arab life for thousands of years.

1. What is the Arab name for the marketplace?

2. In what ways are *souks* similar to shopping malls in the United States?

Islam. As the map on page 438 shows, followers of all three religions live in the region today. Judaism developed as the religion of a people known as the Hebrews. Five holy books, together known as the Torah, record the early history of the Hebrews and their religion. Unlike other religions at that time, Judaism was monotheistic. **Monotheism** is a belief in one God. The Hebrews lived along the eastern shore of the Mediterranean, where they established the kingdoms of Judah and Israel at around 1000 B.C. A large temple, dedicated to the Hebrews' God and located in the city of Jerusalem, served as the religious center for the Jews, followers of Judaism.

Christianity first developed around A.D. 30 in the same region as Judaism, along the eastern shore of the Mediterranean. At that time, the region was known as Palestine. Christianity is based on the teachings of Jesus, a Jew who traveled throughout Palestine spreading his beliefs. Because Jesus' teachings were viewed as threatening to many people, Jesus was tried and crucified in Jerusalem. Christians consider Jerusalem to be a holy city.

Chapter 22, Section 2 **439**

The third major religion to develop in the Middle East was Islam. Islam first developed in Arabia around A.D. 600. Like Judaism and Christianity, it is a monotheistic religion. Today Islam has two main branches—Shiite (SHEE eyt) and Sunni (SOO nee).

The Spread of Islamic Culture In the period from about A.D. 650 to 900 the leaders of Islam conquered and ruled a vast empire that stretched from the Indus River in what today is Pakistan all the way across North Africa and into Spain. Gradually, people of many different ethnic backgrounds became Muslims, followers of Islam, the religion of their conquerors. They also adopted the Arabic language, which is the language of the Koran, the holy book of Islam. Except in Turkey and Iran, where the people kept their own languages, the use of Arabic spread along with the Islamic religion. Hundreds of years later, visitors can see signs everywhere of the impact that Islamic culture continues to have on everyday life in the region.

The Influence of Islam Some of the most visible signs of Islamic culture are the mosques—Islamic places of worship. Five times a day, a muezzin (MYU ez in) climbs the minaret, the tall, thin tower attached to the mosque, to call the people on the streets below to prayer. Muslims who hear the cry kneel, bow, and touch their foreheads to the ground as a sign of their submission to their God, Allah.

Performing daily prayers is one of the Five Pillars of Islam by which all Muslims are expected to live. According to the Pillars, Muslims are also required to give to charity and to state their belief that there is but one God. The Five Pillars are described in the Koran. The Koran is believed to be a record of the word of God as it was revealed to his prophet Muhammed. A prophet is a person whose teachings are believed to be inspired by God.

Muhammed was born in Mecca, on the southwest coast of Arabia, around A.D. 570. Today, millions of Muslims from around the world travel to Mecca to worship at the Great Mosque there. This journey to Mecca is known as the Hajj (HAJ). The Hajj is also one of the Five Pillars of Islam. All Muslims are required to fulfill the Hajj once in their lifetime, if they can do so.

Another of the Pillars of Islam requires that Muslims fast from sunrise until sunset during the holy month of Ramadan (RAM uh dahn), which is the ninth month of the Islamic year. A visitor to the Middle East and North Africa during Ramadan will notice

Dome of the Rock, Jerusalem

Muslim women visit the most important mosque in the holy city. It was built around A.D. 690 and straddles a site sacred to both Muslims and Jews.

The Wealth of the Desert

Income from oil refineries, like this one in Saudi Arabia, propelled the once nomadic culture of the Persian Gulf into the modern world. In what ways has oil changed the region?

that many shops are closed during the day and that many festivities take place at night after sunset when each day's fast is broken.

Oil in the Middle East and North Africa

The countries of the Middle East and North Africa have 40 to 60 percent of all the planet's known oil reserves. With only 5 percent of the world's population, the region has far more oil than it needs for its own people. But not all countries in the region have oil reserves. Those countries that have developed their priceless oil resources, however, have achieved great wealth. Saudi Arabia and Iran especially have become two of the world's leading oil-producing countries. Some neighboring countries, however, have no oil and suffer extreme poverty, making the region one of great contrast.

Today, many people throughout the world depend on oil from the Middle East and North Africa for use as fuel for factory machinery, cars, and jets and for heating buildings. This dependence has provided the oil-rich countries of the region with wealth as

well as political power. As you will read in the next chapter, oil frequently has placed the Middle East in the spotlight of international affairs.

SECTION **2** REVIEW

Developing Vocabulary
1. Define: **a.** arable **b.** fellaheen **c.** monotheism **d.** Muslim **e.** mosque **f.** muezzin **g.** minaret **h.** prophet **i.** Hajj

Place Location
2. What is the largest country that lies to the east of the Persian Gulf?

Reviewing Main Ideas
3. Where do most people in the Middle East and North Africa live?
4. What is the predominant culture in the region?

Critical Thinking
5. **Expressing Problems** Why do you think that most people in the Middle East and North Africa live in places where water is readily available?

The Middle East and North Africa

*The Middle East was the setting for some of the most
important developments in human history.*

Imagine a world almost entirely unchanged by people: a world of thick
forests, blue-green oceans, rushing rivers, and wind-blown grasslands. A
world with no roads, no cities, and no farms.
Until quite recently in geological time, that
was how the earth looked. People lacked the
skills, knowledge, and tools necessary to
make many changes in the land.

Perhaps most surprising of all, people did
not even know how to plant and grow their
own food until around 8000 B.C. Instead,
people everywhere on the earth lived as
hunters and gatherers. Most were nomadic,
moving from place to place according to the
seasons, following herds of migrating animals, and seeking new supplies of wild
plants.

Changing the Landscape

One of the first places where this way of
life began to change was in the Middle East.
It was in the mountainous area of the Middle
East that a few groups of people first developed simple farming skills. These people
actively began to change the land to meet
their needs in a way that no one before them
had done. They plowed the earth to loosen
the soil, they planted seeds, and they pulled
out weeds. They began to tame and raise
their own animals.

In time, this extraordinary change in the
relationship between people and the environment led to even greater changes in how
people lived. Most important, the development of farming and a steady supply of food
allowed people to settle permanently in one
place.

Early Farming

Once people were able to settle down in
one place, they began to form small villages.
Many of the earliest farming villages were
located in the northern mountains, where
the amount of rainfall was sufficient to grow
wheat and barley. In some of these villages,
people developed simple irrigation skills to
supplement the rainfall and to produce more
bountiful crops. Early farmers irrigated their
fields by digging ditches through the banks
of a river and thus diverting some of the
water into their fields.

Once people had developed the technology needed to irrigate crops using water
from a river, they no longer needed to depend on rainfall for farming. As a result, they
could move south onto the plain of the lower
Tigris and Euphrates rivers, where rainfall
was scarce but the waters of the rivers were
abundant. In ancient times, the Greeks called
this flat, treeless plain between the two rivers
Mesopotamia.

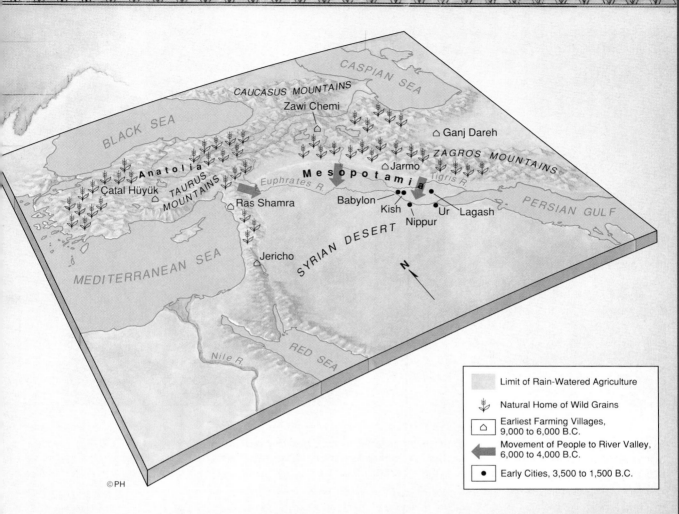

©PH

Map labels: BLACK SEA, CASPIAN SEA, CAUCASUS MOUNTAINS, Zawi Chemi, Ganj Dareh, ZAGROS MOUNTAINS, Anatolia, TAURUS MOUNTAINS, Çatal Hüyük, Euphrates R., Mesopotamia, Jarmo, Tigris R., Babylon, Kish, Nippur, Ur, Lagash, PERSIAN GULF, Ras Shamra, SYRIAN DESERT, MEDITERRANEAN SEA, Jericho, RED SEA, Nile R., N

Legend:
- Limit of Rain-Watered Agriculture
- Natural Home of Wild Grains
- Earliest Farming Villages, 9,000 to 6,000 B.C.
- Movement of People to River Valley, 6,000 to 4,000 B.C.
- Early Cities, 3,500 to 1,500 B.C.

Early Farming and Cities in the Middle East

Applying Geographic Themes

Interaction When early farmers moved south, they built some of the world's first cities on the banks of the Tigris and Euphrates rivers. Why were these sites especially well suited to the development of complex cultures?

The move south to Mesopotamia was especially appealing to early farmers, because the land along these rivers was some of the most fertile in the world. In most years the rivers flooded, depositing new layers of nutrient-rich soil along their banks. This meant that farmers could plant crops in the same fields year after year without using fertilizer.

The Growth of Cities

In time, the combination of irrigation and rich soil allowed farmers in the river valley to produce surpluses—more than they and their families needed— of their crops. This surplus of food meant that some people in the area could make a living as artisans, or

Linking Irrigation to the Growth of Cities

Simple irrigation techniques (right) allowed people in Mesopotamia to build some of the world's earliest cities. In many of these cities ziggurats (left), or step-like pyramids, served as political and religious centers.

craftspeople, instead of as farmers. Artisans could exchange the goods they produced for food from farmers.

As irrigation technology improved, more and more food could be produced on smaller and smaller pieces of land. With animals to carry loads and pull carts, it also became easier to transport the food from the fields to the people. With each of these advances in technology, the population grew. In time, it was possible to support entire cities of people with food from the surrounding lands. Some of the earliest cities in Mesopotamia were Ur, Lagash, and Babylon.

Centers for Trade People living in cities depended on farmers for their food supply. They also depended on traders from nearby

lands to supply them with building materials such as stone and wood, which were lacking on the treeless plain. As the early cities of Mesopotamia grew, they became bustling centers of trade.

A Division of Labor Within the cities, people began to specialize in different types of work. One group of specialized workers kept track of the flow of products. In doing so, they developed a new technology—a system of writing. With the development of more and more complex systems of writing, people were able to store information for long periods of time without having to rely on memory. In this way, knowledge of all kinds was passed on and built upon from one generation to the next.

Government and Written Laws

Other specialized workers were needed to plan and maintain the complex system of dams for irrigation and flood control. Another group was needed to plan and maintain a system of defense for the city. The people who administered each of these tasks were part of the earliest systems of organized city governments.

To aid each of these groups in their tasks, government leaders developed laws to help settle disputes among traders or farmers whose needs for water conflicted. For example, if a farmer upstream diverted a great deal of water, less was available for farmers living further downstream. One of the oldest records of a law code, an organized system of written laws, was compiled by the scribes of Hammurabi, a Mesopotamian leader in the city of Babylon around 1750 B.C. The 282 laws in Hammurabi's Code are organized under headings such as trade, labor, and personal property. These laws reveal a great deal of information about urban life in early Mesopotamia. The following extracts from Hammurabi's Code show how law and order was imposed around 1750 B.C.

If a person has broken into a house, that person shall be killed and buried there.

If a farmer owes interest on a debt, and the Weather God floods his field and destroys his crop or does not send enough water for the crop to grow, the farmer shall not pay interest for that year.

If a son has struck his father, the son's hand shall be cut off.

If a builder has constructed a house and the walls fall down, the builder shall repair the walls. If the house collapses, killing the owner, the builder shall be put to death.

In short, many aspects of life in the world today have their origins in the river valley of the Tigris and Euphrates rivers. Because of the geography of the Tigris-Euphrates river valley—rich water and soil resources—the

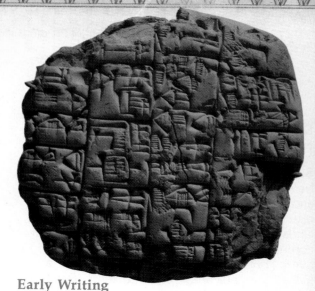

Early Writing
The wedge-shaped marks, or cuneiform, on this clay tablet from ancient Sumer are an example of one of the earliest forms of writing.

people living in the region were able to develop an entirely new way of life. This way of life was based on agriculture, permanent settlement and city life, trade, the division of labor, writing, government, and laws.

TAKING ANOTHER LOOK

1. What two important resources were present in the Tigris-Euphrates River Valley in ancient times?
2. What developments enabled people to specialize in their work?
3. Why did Mesopotamian cities become centers of trade?

Critical Thinking

4. **Making Comparisons** In what ways is life today similar to and different from life in an ancient Mesopotamian city?
5. **Perceiving Cause-Effect Relationships** How did the rich soil and water resources of the Tigris-Euphrates river valley indirectly contribute to the development of an urban way of life?

Section Summaries

SECTION 1 Land, Climate, and Vegetation
Nearly two thirds of the land in the Middle East and North Africa is desert, making it the largest dry region on earth. The wind shapes the landscape by forming ergs out of sand. Water comes from rainfall, underground springs, and rivers, but it is very scarce. Mountains and plateaus create areas of more temperate climate and more abundant vegetation than is found in the desert.

SECTION 2 Human Geography
Population is greater in areas near water, where the climate is cooler and rainfall more abundant. Three ways of life—farming, city living, and herding—dominate the region. People of many different cultures and religions live in the region, but Islamic culture predominates. Recent wealth from oil has brought many changes to some countries.

Key Terms

Match the definition with the terms below.

1. able to be cultivated
2. a dry riverbed
3. an Islamic place of worship
4. a region of sand dunes
5. a fertile place in a dry area
6. a follower of Islam
7. a person whose teachings are believed to be inspired by God
8. Egyptian farmers
9. a person who calls Muslims to prayer
10. a religious journey to the Great Mosque at Mecca
11. a tall tower attached to a Mosque
12. a belief in one God

a. oasis
b. wadi
c. prophet
d. erg
e. arable
f. fellaheen
g. mosque
h. minaret
i. Muslim
j. Hajj
k. muezzin
l. monotheism

Main Ideas

1. Why are rivers of great importance in the Middle East and North Africa?
2. What is the relationship between population density and water?
3. Explain how the region's mountains affect the climate and vegetation.
4. What are some of the visible signs of Islam in the Middle East and North Africa?
5. What three ways of life are most common in the region?
6. How did the development of writing increase knowledge?
7. How do ancient cities compare to cities today?

Critical Thinking

1. **Perceiving Cause-Effect Relationships** Living in an extremely hot, dry region requires skill and the ability to adapt to a demanding environment. How have plants, animals, and humans adapted to living in the Middle East and North Africa?
2. **Synthesizing Information** How do the oil resources of the Middle East and North Africa translate into political power for the oil-rich nations of the region?
3. **Drawing Inferences** Based on what you have read in this chapter, why do you think geographers place North Africa in the same

region as the Middle East? Supply evidence from both the text and the maps in this chapter.

4. **Analyzing Information** Surplus food allowed people in early cities to specialize in their work. What do you think were some of the advantages and disadvantages of specialized work both for individuals and society as a whole?

Practicing Skills

Making an Outline On the skill page in this chapter you completed an outline of Section 1. Now complete the outline of Section 2 that follows. Use the text to fill in the missing details under the remaining subheads. Refer to the outline on page 435 as a guide.

Human Geography
I. _____
II. Three Ways of Life
  A. Farming
    1. _____
    2. _____
  B. _____
    1. _____
    2. _____
  C. _____
    1. _____
    2. _____
III. _____
  A. The Spread of Islamic Culture
    1. _____
    2. _____
  B. _____
    1. _____
    2. _____
IV. _____

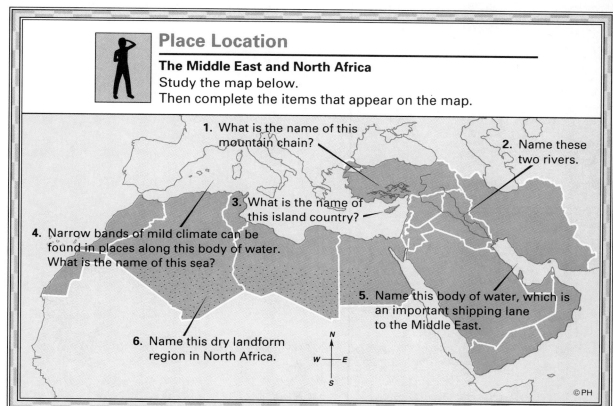

Place Location

The Middle East and North Africa
Study the map below.
Then complete the items that appear on the map.

1. What is the name of this mountain chain?

2. Name these two rivers.

3. What is the name of this island country?

4. Narrow bands of mild climate can be found in places along this body of water. What is the name of this sea?

5. Name this body of water, which is an important shipping lane to the Middle East.

6. Name this dry landform region in North Africa.

N
W — E
S

©PH

23

The Middle East

Chapter Preview

Locate the countries covered in these sections by matching the colors at right with those on the map below.

Sections		Did You Know?
1	**CREATING THE MODERN MIDDLE EAST**	The Middle East is the birthplace of three of the world's major religions—Judaism, Christianity, and Islam.
2	**ISRAEL: A DETERMINED COUNTRY**	Hebrew and Arabic, the languages of the Jewish majority and Arab minority, are Israel's official languages.
3	**JORDAN, LEBANON, SYRIA, AND IRAQ**	One of the earliest civilizations developed in Mesopotamia, now Iraq. Mesopotamia means "land between the rivers."
4	**THE ARABIAN PENINSULA**	The countries of the Persian Gulf own about 60 percent of the world's known oil reserves.
5	**TURKEY, IRAN, AND CYPRUS**	Istanbul, Turkey's largest city, is built on both banks of the Bosporus Strait, straddling Europe and Asia.

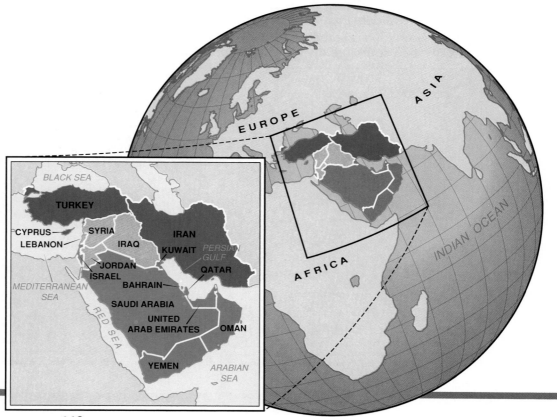

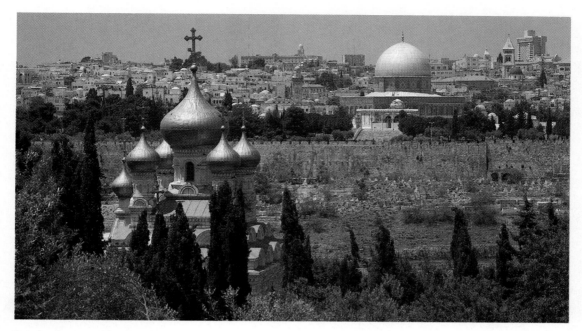

Jerusalem—The Holy City

Although sacred to Muslims, Jews, and Christians, Jerusalem has known little peace in its long history. It has often been caught in bitter religious and territorial disputes.

1 Creating the Modern Middle East

Section Preview

Key Ideas

- With the fall of the Ottoman Empire, ethnic and religious groups began demanding political independence.
- World War I greatly influenced the modern history of the Middle East.
- The 1948 war between Israel and the Arab countries left the Palestinians without a homeland.

Key Terms

mandate, Zionist

The Middle East has a long and turbulent history. More than three thousand years ago, the area's great wealth and location at the center of trading routes between Europe, Africa, and Asia made it an important source of power. The area was repeatedly conquered by groups from within and outside it. Great empires emerged in the Middle East, some expanding to rule lands as far away as India and North Africa.

The movement of conquering peoples across the Middle East gave the region a unique character. It became a tangle of diverse ethnic groups and religious beliefs. Eventually with the fall of the Ottoman Empire after World War I, many groups in the Middle East began to demand their own homelands. As a result, today the Middle East is divided into several independent countries.

Regions: Uniting Peoples

When the followers of Muhammad swept out of the Arabian Peninsula into the ancient lands of Mesopotamia,

Muslim Astronomers at Work in Istanbul
The map at right shows the Islamic Empire at its height around A.D. 750. The astrolabe, shown at the top right, was a Muslim invention.

Palestine, and Persia in the mid-600s, many different ethnic and religious groups were living in these areas. Although most of the people adopted the Islamic religion and the Arabic language, the area still remained a rich mosaic of cultures. Groups of Christians and Jews continued to practice their religions. The Persians, Kurds, and Armenians maintained their strong cultural identities.

For over 150 years Islam was successful in governing these different people as one political region. But beginning in the tenth century, the Arabs could no longer control their huge empire in the Middle East. Within a short time, large numbers of Turks, led by the Seljuks (SEL jooks), conquered almost all of the Middle East. They adopted the Islamic religion and ruled the Middle East for more

than four hundred years before losing control to the region's last great empire builders—the Ottoman Turks.

Under the Ottomans, the people of the region, including the Jews, Christians, and Muslims, became part of the Ottoman *rayah*, or "flock." The flock was divided into groups according to religion. The Ottomans did not impose Islamic law on non-Muslims. Christians and Jews governed aspects of their lives, such as marriage and death, according to their religious beliefs.

Beginning in the late 1700s, discontent and rivalry developed among the different ethnic and religious groups under Ottoman control. Many of these groups were anxious to establish independent homelands. The Ottoman leadership was no longer powerful enough to hold its empire together.

At the same time, European nations were anxious to exert political influence in the Middle East and gain new markets for their products. By the mid-1800s, the Ottoman Empire was being called "the sick man of Europe." And Great Britain, France, and Russia were impatiently waiting for it to die.

The Impact of World War I

In 1914, World War I broke out. Great Britain, France, and Russia, known as the Allies, were on one side. On the other side were Germany and Austria-Hungary, with whom the Ottomans joined. Although World War I was mainly fought in Europe, it greatly affected the course of modern Middle Eastern history.

Soon after the war started, the Allies began secret negotiations to decide how to divide the Ottoman Empire when it was defeated. They agreed that, except for the Arabian Peninsula, each of them would control different parts of the empire. The Arabs in the Arabian Peninsula would be given their independence when the war ended. Great Britain, eager to exert its power in the area, entered into other, separate agreements as well.

In 1915, Sir Henry McMahon, a representative of the British government, began to correspond with Husayn ibn 'Ali. Husayn was the Arab ruler of the sacred cities of Mecca and Medina on the Arabian Peninsula. He was an important leader among the Muslim Arabs who wanted to break away from the Ottoman Empire and establish an independent Arab homeland. In his letters, McMahon hoped to convince the Arabs to support Great Britain in its fight against the Ottomans. Letters went back and forth between the two men for almost a year.

Finally, Husayn agreed to revolt against the Ottomans in exchange for British support of a homeland for all Arabs, including Christians. From the letters that had passed between him and McMahon, Husayn believed that almost all of the area from southern Turkey to southern Arabia, and from the Mediterranean Sea east to the borders of Iran would be one vast Arab country.

Unknown to Husayn, however, Great Britain and France were secretly working out another agreement for dividing the Ottoman Empire. This agreement, known as the Sykes-Picot Agreement, limited the independent Arab state to the area that is now Saudi Arabia and Yemen. It gave the French control of Syria and allotted Palestine and Iraq to Great Britain. When the Arabs discovered this they felt Great Britain had broken its promise to them.

The Struggle Between Arabs and Jews

At the peace conference following World War I, the once great empire of the Ottomans was reduced to a single independent country—Turkey. As arranged in the Sykes-Picot Agreement, France and Great Britain divided the rest of the Ottoman Empire between them. France took Syria—including the area that would become the country of Lebanon—as a mandate. A mandate referred to land to be governed on behalf of the League of Nations until it was ready for independence. Great Britain was given Palestine—the land that is now Israel, Jordan, part of Egypt—and Iraq as mandates. By the mid-1940s, Iraq, Jordan, Syria, and Lebanon had been established as independent countries. The political future of what remained of Palestine after the creation of Jordan was still to be decided, however.

The issue of independence for Palestine created a dilemma for Great Britain. Two groups claimed Palestine as their homeland—the Arabs and the Jews. The Arabs had lived for centuries in Palestine. Many of them traced their ancestry back to the area's earliest settlers. But the Jews also had ancient historical ties to Palestine.

Jewish Movement to Palestine Beginning around 1900 B.C. to 1700 B.C., migrating nomadic people in the Middle East—the ancestors of today's Jews—began to settle in Canaan, the land that was later called Palestine. By 1000 B.C. these people were known as the Hebrews. The Hebrews established a kingdom in Canaan, which later split into two kingdoms and then was defeated in a succession of military conquests. Beginning with their defeat and exile by the Babylonians in 586 B.C., the Jews began to move to other lands. Over the centuries most Jews settled in other places, although some remained in Palestine.

By the late 1800s, there were about ten million Jews scattered throughout the world. In many of the places they lived, they were discriminated against and cruelly persecuted. In eastern Europe and Russia, where more than half of the world's Jews lived, they faced increasing oppression. Afraid of what lay ahead, Jews began to emigrate. Some called themselves **Zionists,** after the mountain in Jerusalem to which Jews had always prayed to return.

WORD ORIGIN

Zionists
The word *Zionist* comes from *Zion,* in Hebrew *Tsiyon,* the name of a hill in ancient Jerusalem where one of the first Jewish temples was located.

Applying Geographic Themes

Regions Which modern-day countries were part of the British mandate following World War I? Which modern-day countries were part of the French mandate?

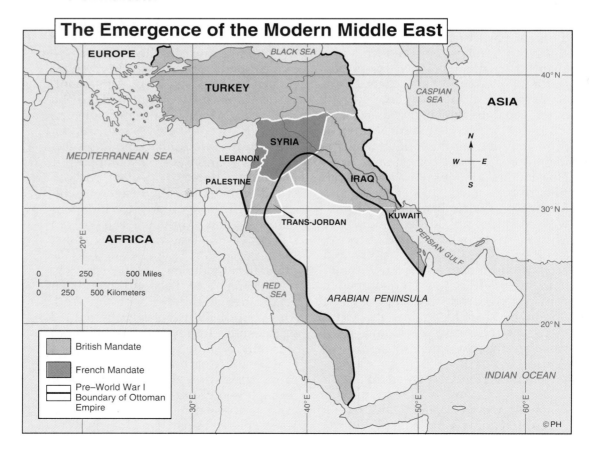

The Emergence of the Modern Middle East

They believed that the only way to solve the problem of oppression was by returning to the place they considered their homeland—Palestine—and creating their own country.

In 1882, the first group of Zionists immigrated to Palestine. Their numbers had reached almost 85,000 by 1914. As Jewish immigration increased, the Arabs who were living in Palestine under Ottoman rule grew more and more fearful of losing their land.

Two Peoples, One Homeland The Zionists put increasing pressure on Great Britain and other European nations to support their plan for an independent homeland. In 1917, in the midst of World War I, the British government issued the Balfour Declaration. It stated Britain's support for the creation of a Jewish national home in Palestine without violating the rights of Arabs living there:

His Majesty's Government views with favor the establishment in Palestine of a national home for the Jewish people, and will use their best endeavors to facilitate the achievement of that object, it being clearly understood that nothing shall be done which may prejudice the civil and religious rights of non-Jewish communities in Palestine or the rights and political status enjoyed by Jews in any other country.

The Arabs were shocked and dismayed by the content of the declaration. They had been led by the British to believe that all Arabs would be granted the right of self-determination, and that Palestine would become part of a larger, independent Arab country. The British sent representatives to Arab leaders to assure them that Great Britain's goal was still self-determination for the Arabs. The Brit-

ish formally promised the Arabs that their political rights would not be affected by the Balfour Declaration.

When Great Britain took up the League of Nations mandate for Palestine in 1920, both Arabs and Jews asked Great Britain to fulfill its promises to them. Jews believed the Balfour Declaration meant that Great Britain would encourage Jewish immigration to Palestine in order to establish a Jewish national home. However, Great Britain had also promised the Arabs the right to self-government. The Arabs felt that their political future as an independent Arab country was threatened by increased Jewish immigration. It became clear that the goals of Jews and Arabs were at odds.

While Britain searched for a way to solve the problem, the struggle between Jews and Arabs in Palestine grew increasingly violent. As Jewish settlements increased, the Arabs grew more frustrated and afraid. They revolted by boycotting Jewish businesses and burning bridges and crops. The Jews retaliated. People on both sides were killed.

Meanwhile, Hitler came to power in Germany in 1933. As Nazi Germany began to persecute Jews, thousands fled to Palestine. By 1939, the number of Jews living in Palestine had increased from 85,000 to 445,000.

Tensions between Great Britain, the Palestinians, and the Jews accelerated. Great Britain decided to limit Jewish immigration to the area, leaving Jews stranded in Germany and other parts of Europe. The Jews in Palestine began a campaign of guerrilla warfare against the British.

The Creation of Israel More than six million Jews had perished in Nazi concentration camps by the time World War II ended in 1945. Thousands of survivors had no place to go. When the world learned of the Holocaust, there was an outpouring of support for a Jewish homeland in Palestine.

Jewish Refugees in Haifa
These Jewish refugees are being deported to Cyprus by the British navy after trying to land illegally on the coast of Palestine in 1947.

However, the Arabs made up 70 percent of Palestine's population. They were bitterly opposed to the creation of a Jewish state in Palestine. Why, they wondered, should they give up their land because of what the Nazis had done?

In 1947, realizing it had no hope of finding an acceptable solution, the British government announced that it was withdrawing from Palestine and turning the problem over to the United Nations. Immediately the United Nations formed a special committee to find a solution to the problem. After months of debate, the committee recommended that Palestine be partitioned into two states—one Arab and one Jewish. The city of Jerusalem, sacred to Jews, Christians, and Muslims would be designated an international city.

The Jews accepted the United Nations plan. However, the Arabs were furious. According to the plan, the Jewish state would include more than half the total land of Palestine, though less than one third of the population was Jewish.

Arab leaders warned that any action taken to divide Palestine would result in war. One Arab leader stated, "We Arabs shall not be losers. We shall be fighting on our ground and shall be supported . . . by 70 million Arabs around us . . ."

Nevertheless, the United Nations voted to approve the partition of Palestine. In May 1948, David Ben-Gurion, leader of the Palestinian Jews, announced the independent, new state of Israel. In a matter of hours, neighboring Arab countries attacked Israel. The Jews were ready. By the end of 1948, Israel controlled almost three fourths of Palestine, including land in the Negev Desert and half of Jerusalem. Instead of there being an independent Arab state in what was left of Palestine as the United Nations had proposed, Jordan and Egypt divided the rest of Palestine between them. So when the war ended, the Palestinians were left not with half a country, but with no country at all. You will read more about the Palestinians in the Case Study on pages 460–462.

■ SECTION **1** REVIEW ■

Developing Vocabulary
1. Define: **a.** mandate **b.** Zionist

Place Location
2. Where is Jerusalem located?

Reviewing Main Ideas
3. What effect did World War I have on the Middle East?
4. What was the Balfour Declaration?

Critical Thinking
5. **Demonstrating Reasoned Judgment** Why do you think the Arab nations were opposed to the creation of the nation of Israel?

2 Israel: A Determined Country

Section Preview

Key Ideas

- Israel has turned swamps and desert into productive land.
- Israel is a leader in high technology.
- Many different cultural, ethnic, and religious groups give Israel a diverse character.

Key Terms

drip irrigation, Knesset

There was nothing but desert and swamp. [Our parents and grandparents] had to clear it and build it—propulsed forward by the sheer force of their dreams and their faith. They never stopped to say, "Should we clear a potato patch here, or raise sheep over there?" . . . No, our founders said with breathtaking simplicity: "Let there be a potato patch. . . . Anywhere, everywhere, and right away."

We [today] do not say, "Let there be a potato patch, and scratch it into the nearest soil." We must say, "Should there be a potato patch? And, if so, where is the best place to put it? . . . How much irrigation will it need? . . . Or do we need more cotton, more tools, or is there a more nutrient, efficient food than potatoes?"

Gideon Samet, a young Israeli journalist, used these words to describe the changing character of Israel. Today Israel is a very different place than when its founders first dug potatoes from the land. Its landscape is different. Its economy is different. Even the character of its people has changed. In less than half a century, Israel has raced along a path of urgent development to become one of the most technologically advanced countries in the world.

Interaction: Relationship with the Land

When the first Zionist settlers arrived in Palestine, people were already living along the fertile coastal plains and in the rich valleys of the highland

---

Irrigation in Israel

In their efforts to transform desert into fertile farmland, the Israelis developed many methods of irrigation including drip irrigation, a process in which precise amounts of water drip onto plants from pipes.

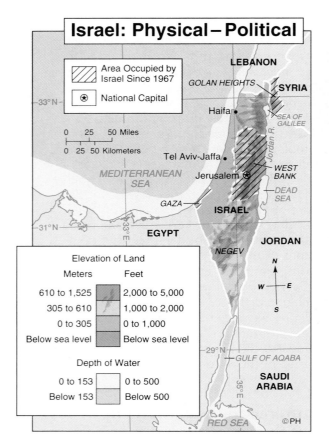

Israel: Physical – Political

Area Occupied by Israel Since 1967

⊛ National Capital

LEBANON
GOLAN HEIGHTS
SYRIA
Haifa
SEA OF GALILEE
Jordan R.
Tel Aviv-Jaffa
MEDITERRANEAN SEA
Jerusalem ⊛
WEST BANK
GAZA
DEAD SEA
ISRAEL
EGYPT
NEGEV
JORDAN
GULF OF AQABA
SAUDI ARABIA
RED SEA

33°N
31°N
29°N
33°E
35°E

0 25 50 Miles
0 25 50 Kilometers

N
W — E
S

Elevation of Land

Meters		Feet
610 to 1,525		2,000 to 5,000
305 to 610		1,000 to 2,000
0 to 305		0 to 1,000
Below sea level		Below sea level

Depth of Water

0 to 153		0 to 500
Below 153		Below 500

©PH

Applying Geographic Themes

Interaction Tons of salt are extracted from the Dead Sea (shown at right) each year. Which two countries border the mineral-rich waters of the Dead Sea? Name the desert in southern Israel that has been transformed by an extensive irrigation system.

regions. Much of the land that was available to the immigrants was either mosquito-infested swamp or barren stretches of desert. In the 1880s settlers began the long, slow process of reclaiming the land. Acre by acre they drained the swamps. Patiently they coaxed water into the desert.

Since 1948, when Israel became independent, the Israeli government has viewed the desert as a challenge to its existence. David Ben-Gurion, Israel's first prime minister, said, "If the state does not put an end to the desert, the desert is likely to put an end to the state." Then and now, an important part of Israel's national policy has

been directed at turning the desert into land that can be used for agriculture, industry, and settlement.

Technology Transforms the Desert
The Negev Desert is Israel's driest region. It covers almost half of the country. Here the Israelis have built a system of pipelines, canals, and tunnels almost 100 miles (167 km) long called the National Water Carrier. Water from the Sea of Galilee is pumped southward through the system to irrigate parts of the Negev. A region that was once barren stretches of sand is now striped with huge tracts of fertile green land.

The Israelis have also invented other scientific methods for growing crops in desert soil. A process called **drip irrigation** preserves precious water resources by letting precise amounts of water drip onto plants from pipes. Agricultural production in Israel has increased greatly in the last forty years. Today Israel produces about three fourths of its own food.

Mining the Dead Sea Between Israel and Jordan lies the Dead Sea. The Dead Sea is actually a huge saltwater lake. Because of the quantity of minerals in the sea, fish or other animals cannot live in it. Even the surrounding land is a dry, lifeless wilderness. The Dead Sea is the lowest point on earth —12,580 feet (786 m) below sea level.

However, the Dead Sea has vast amounts of mineral resources. Since it is fed by mineral springs, its waters include many dissolved minerals. The Israelis have built processing plants to extract these minerals. First the water is pumped into huge pools. Then heat from the sun evaporates the water, leaving behind the raw materials that contain the minerals. These raw materials are processed to yield potash— which is used in explosives and fertilizer—table salt, bromine, and other minerals. Israel exports millions of tons of these minerals annually all over the world.

Encouraging Movement to the Desert Despite the Israeli government's drive to develop the Negev and other desert areas, it has been difficult to attract people to these places to work. Few people wanted to live and raise families away from the conveniences of modern life and in such an isolated area. New towns, such as Arad, had to be built. Workers had to be offered high pay and extra time off.

Still, feelings about living in the desert are mixed. One Israeli couple, Zvi and Rebecca Rubin, had differing views about their life in Arad.

I came to Arad because I was offered a high salary, a good flat, and low taxes. . . . This is a good place to live, work, and put money aside. However, Rebecca [Zvi's wife] replied, "For him it is a good place to work and live. For me it is the desert. . . . I wish he could find a job back in Haifa."

Rich Human Resources Israel has successfully developed its few natural resources. However, a decade ago, the tiny country knew that its agricultural and chemical industries could not produce enough to support its growing population of over four million people. It had to develop new industries.

Haifa, Israel

The city of Haifa is located on the Bay of Haifa in northwest Israel. The city is a major industrial center and Israel's major seaport.

Israel turned to its most important available resource: its people. Israelis are well-educated. Many of them are scientists or engineers. With the help of grants and loans from other countries, Israel looked in part to high technology as the answer to its struggling economy.

Today Israel is a world leader in medical laser technology. And, because of its concerns about security, Israel develops some of the world's most sophisticated weaponry, including aircraft and aerospace equipment. From 1984 to 1989, Israel exported over $1.5 billion worth of military equipment.

Soviet Jews Arrive in Israel, 1990
Under Mikhail Gorbachev, the former Soviet Union relaxed its emigration policies, and thousands of Soviet Jews moved to Israel.

Place: Citizens of Different Backgrounds

Israel's citizens come from a great variety of backgrounds. Perhaps the most obvious distinction is between Jew and Arab. Eighty percent of Israel's population is Jewish. But at any public gathering you will recognize that great differences exist even among the Jews of Israel. If you listen, you will hear Hebrew spoken with a variety of accents—Russian, American, Turkish, German. And if the meeting happens to be a political one, you will probably hear widely differing political views.

Israel's Jews Until recently, two groups of Jews—European Jews and Oriental, or *Sephardic* Jews—formed a sharp division in Israeli society. Almost all the Jews who immigrated to Israel before 1948 came from Europe. As a result, when Israel was established it had a modern, Westernized character. However, after 1948, more than half of the Jews immigrating to Israel were *Sephardic* Jews, or Jews from countries in the Middle East, North Africa, and Asia.

These *Sephardic* Jews had a hard time fitting into Israeli society. Because they came from less advanced countries, they were generally poorer and less educated than the rest of Israel's citizens. Most of them worked as unskilled laborers. They earned less money and had a lower standard of living than the European Jews. But in recent years, especially as *Sephardic* Jews have gained political power, the gap between the two groups has begun to close.

Political divisions in Israeli society seem to be widening, however. Representatives in the **Knesset,** Israel's democratically elected parliament, range from ultra-Orthodox Jews to the nonreligious. Ultra-Orthodox Jews adhere strictly to Jewish religious tradition and believe that Israel should be governed accordingly. The nonreligious believe that religion should not dictate the running of the state and interfere with people's daily lives. In between these two groups are a number of other groups. They have serious political conflicts with one another in complicated, coalition governments. It is always difficult, and often impossible, to reach any kind of agreement on important issues.

Israel's Arabs Seventeen percent of Israel's population is Arab. It is a diverse population that includes Christians, Muslims, and Druze (a faith that combines elements of Christianity, Islam, and Judaism). As a minority they hold a very different place from that of the Jews in Israel.

Although Israeli Arabs are citizens of Israel and have the right to vote, they cannot participate fully in Israeli society. Jews and Arabs live in separate communities and usually do not attend the same schools. And, because Arab countries and Israel are in conflict, the only Arab Israelis allowed to serve in the army are the Druze. In the last few years, Arabs living in Israel have begun to demand a greater voice in Israeli society.

Despite these differences, most Jews and Arabs manage to coexist peacefully. Shmuel Toledano, an ex-adviser on Arab affairs to three of Israel's prime ministers, had this to say about relations between its Jewish and Arab citizens:

> *Sometimes I am amazed . . . at how 600,000 Arabs—with intellectuals, with people suffering, with people thinking that they're second-class—have such quiet behavior. People are living—not loving each other but living together. . . . On the other hand, I don't think that we [Israel] succeeded in real integration. We have coexistence, but not integration.*

Praying at the Western Wall
This wall in East Jerusalem is the only remaining part of an ancient Jewish temple that was destroyed by the Romans in A.D. 70.

SECTION **2** REVIEW

Developing Vocabulary
1. Define: **a.** drip irrigation **b.** Knesset

Place Location
2. Where is the Negev Desert located?

Reviewing Main Ideas
3. How has Israel made use of technology in developing its land?
4. What are the main cultural, ethnic, and religious groups in Israel today?

Critical Thinking
5. **Making Comparisons** In what ways are Israel and the United States alike? In what ways are they different?

The Palestinians

Palestine is my home and the path of my triumph. My homeland will remain the passion of my heart and the yearning melody on my lips. Strange faces are in my stolen land. They are selling my crops and occupying my home. I know my path and my people will return to my grandfather's home, to my warm cradle. Palestine is my home and the path of my triumph.

That passage was taken from a textbook that is used to teach schoolchildren in Jordan and, secretly, in the Israeli-occupied West Bank. It explains the feelings of the Palestinians—the people who lost their land when Israel was created.

A Refugee Camp in Syria *Many Palestinians live in refugee camps like this one.*

An Uprooted People

In 1947, immediately after the United Nations voted to partition Palestine into an Arab state and a Jewish state, civil war broke out between Palestinians and Jews. The fighting was bitter on both sides. When Israel declared itself an independent country in 1948, other Arab countries entered the war. The violence and chaos grew worse. By the end of the war in 1949, most of the Palestinians had lost their homes and property. Hundreds of thousands had fled to other Arab countries. Palestinian society was shattered.

The Palestinians took refuge in Egypt (Gaza), and the countries that border Israel—Jordan, Syria, and Lebanon. However, when Arab countries attacked Israel in 1967, Israel took control of the West Bank in Jordan and the Gaza Strip in Egypt. Many Palestinians who had escaped to these areas in the late 1940s fled again. Permanent residents fled, and some were forced out of their villages during the war. After the war the Palestinians were not allowed to return to their homes.

Approximately 2.5 million Palestinians now live outside of the area that was once Palestine. About 1.7 million Palestinians reside in Jordan. Most of the rest of the refugee population is divided among Lebanon, Syria, Saudi Arabia, and Kuwait.

Many Palestinians lead fairly normal lives in other Arab countries. But large numbers live permanently in the refugee camps spread throughout the West Bank, Gaza Strip, and the neighboring Arab countries. Conditions in the camps are crowded and sanitation is poor. In the Deheisha refugee camp in

Hebron on the West Bank, for example, over eight thousand Palestinians live packed together in houses made of concrete blocks. The houses are so close together that their tin roofs touch. Rainwater and sewage flow down the paths of the camp. The houses have no running water and, often, no electricity.

Living in temporary shelters in refugee camps has become a symbol to the Palestinians. The camps are daily reminders that they are exiles waiting to return home. And while they wait, they dream of the places they left.

A Government-in-Exile

In the mid-1960s, many of the refugee camps in Lebanon, Syria, and Jordan became bases for the new Palestinian Liberation Organization (PLO), which eventually became the Palestinians' government-in-exile. The PLO—which is best known in the West for its guerrilla fighters—runs many social services for Palestinians in camps in the Arab countries. The PLO operates hospitals, schools, and a national welfare system. Its members include associations of doctors, lawyers, teachers, students, and laborers.

For many years the PLO refused to recognize Israel as a country and demanded an end to its existence. In its place it proposed a democratic Palestinian country in which, it claimed, Jews, Christians, and Muslims would live together in peace. PLO guerrillas gained worldwide attention for their cause by hijacking planes, kidnapping and killing Israelis, and conducting raids on Israeli settlements.

However, in the mid-1980s, the PLO appeared to take a more moderate position toward Israel. Yasser Arafat, the longtime leader of the PLO, stated that the Palestinians would agree to recognize Israel's right to exist if Israel agreed to accept an independent state in the occupied territories of the West Bank and Gaza Strip. The PLO stated that they would not use violence as a means to achieving their goal of having

an independent Palestinian state in 1989. As Ziad Abu Zayyad, a Palestinian lawyer said:

There is a deep change in the area and inside our people. Until 1977 no Arab was ready to speak with Israel. Now, we are talking with them and arguing with them. Israel is a fact. I can't tell the Jews, "Take your suitcases and go home" . . . therefore I say, let's divide it. I am after a Palestinian state beside Israel, not instead of Israel.

The Struggle for a Solution

Various solutions to the challenge of creating a homeland for the Palestinians have been proposed. However, for more than a decade, the creation of an independent country in the West Bank and Gaza Strip has been viewed as the most likely alternative. In 1978, Egypt and Israel took the first step toward the creation of some kind of Palestinian homeland when they signed the Camp David Accords.

Included in the terms of the agreement was the provision that Israel would allow the Palestinians in the West Bank and Gaza Strip some kind of autonomy. At the end of five years, a final decision about the status of the Palestinians in these areas would be reached. More than a decade has passed since the Accords were signed. However, the Israelis have taken no initiative to try to implement self-government for the Palestinians.

Part of the problem lies in Israel's unwillingness to accept the PLO as the legitimate representative of the Palestinian people. Poll after poll taken in the West Bank and Gaza Strip has shown that the Palestinians overwhelmingly support the PLO as their political representative. However, the present Israeli government refuses to talk with the PLO, which it continues to regard as nothing more than a terrorist organization. Although the PLO has acknowledged Israel's right to exist, Israel insists that the PLO cannot be trusted and that it still threatens Israel.

Many Israelis are determined to hold onto the occupied territories. Some of these believe that Israel has a historic right to the West Bank and the Gaza Strip because it is the land inhabited by the Jews during biblical times. Other Israelis have more practical reasons. They do not want an independent Palestinian state that might grow militarily strong and attack Israel one day. Also, since 1967, when the Israelis captured the West Bank and the Gaza Strip, over half of the Palestinian land has been claimed by the Israeli government. Jewish settlements have been constructed on much of this land. As of 1989, there were 118 settlements in the West Bank and Gaza. People living in these settlements want the West Bank and the Gaza Strip to become part of Israel.

The Intifadeh *Palestinian frustration boiled over in 1987 when growing resistance to Israeli occupation took a violent turn.*

However, almost as many Israelis think that the occupied territories should be given their independence. They argue that the Palestinians should be granted their independence because it is their political right. Others, however, maintain that annexing the occupied territories would mean the end of Israel as a Jewish nation because Israel would have to absorb the 1.5 million Palestinians who live in the West Bank and the Gaza Strip. Together with the 800,000 Palestinian Arabs who are already Israeli citizens, Palestinians would eventually become the majority in Israel.

With no solutions in sight, the Palestinians have become more and more frustrated. In December 1987, the patient waiting that had for so long characterized the Palestinians in the occupied territories ended. Youths in Gaza picked up rocks and began hurling them at Israeli soldiers. Soon the uprising, or *intifadeh* (in tee FAH dah) as it is called in Arabic, spread to the West Bank. The Israeli government responded harshly, using military force to try to end the rebellion.

In 1988, spokespersons for the Palestinians declared the West Bank and Gaza an independent country. Israel still occupies the area, however, and the Palestinians wait.

TAKING ANOTHER LOOK

1. Why did the Palestinian refugees leave Palestine?
2. How has the PLO's stated position toward Israel changed over time?
3. What are the Israeli arguments in favor of making the occupied territories part of Israel? What are some of the arguments against annexing the occupied territories?

Critical Thinking

4. **Drawing Conclusions** Do you think the West Bank and Gaza Strip should become an independent Palestinian state? Why or why not?

3 Jordan, Lebanon, Syria, and Iraq

Section Preview

Key Ideas

- Jordan and Lebanon have been greatly affected by wider conflicts in the Middle East.
- Syria and Iraq are working to increase agricultural production.
- Oil is important to Iraq's economy.

Key Terms

militia, anarchy

Along with Israel, the land that is now the modern countries of Jordan, Lebanon, Syria, and Iraq made up the center of the ancient Middle East. An arc of rich land known as the Fertile Crescent ran through this area, where farming and the first civilizations developed. These countries remain at the center of the Middle East today, and are often the focus of political, economic, and social challenges that affect the entire Middle East region.

Jordan: A Fragile Kingdom

Notice on the map below that Jordan is bordered by Israel, Syria, Iraq, and Saudi Arabia. This positions it in

Applying Geographic Themes

Place The borders of these four countries were established following World War I. Where is the waterway known as the Shatt-al-Arab located? Between which two countries does it form part of the border?

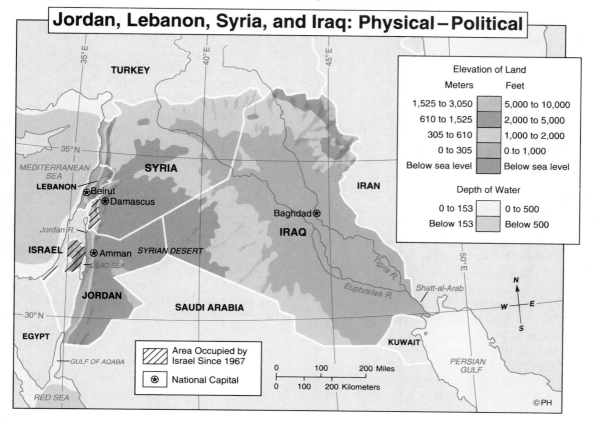

Jordan, Lebanon, Syria, and Iraq: Physical–Political

the middle of the struggle between Israel, the Palestinians, and neighboring Arab countries. Since 1948, Jordan has been greatly affected by the conflicts in the region.

Changing Boundaries Affect Jordan's Economy   When Jordan was given its independence in 1946, almost all of its land was dry, rocky desert. However, after the 1948 war between the Arab countries and Israel, Jordan annexed the West Bank. The addition of the West Bank to Jordan's territory supplied it with fertile farmland for growing crops and much needed mineral resources. Jordan developed the West Bank's resources in particular by increasing agricultural production. Workers built irrigation canals and farmers used modern methods to grow vegetables, fruit, and wheat. Herders raised large flocks of sheep and goats. Jordan also opened mines and other industries in the area. By the mid-1960s, 45 percent of Jordan's gross national product came from the West Bank. Then in 1967 Jordan, Egypt, and Syria attacked Israel. Israel gained control of the West Bank. Jordan also lost its second-largest city, East Jerusalem. The impact on Jordan's economy was devastating. Jordan lost a huge part of its agricultural production, its banking business, and its industry.

Movement of Palestinians to Jordan   The Arab-Israeli wars of 1948 and 1967 also had a significant effect on Jordan's population. After each of these wars, many Palestinian refugees fled to Jordan. Today almost half of Jordan's population of four million people are Palestinian Arabs. Unlike other Arab countries, Jordan encouraged Palestinians to become part of its society. Most of them became Jordanian citizens. The Palestinians are a strong political force in Jordan. In the past, they have challenged Jordan's government, which is a constitutional monarchy. Palestinian political groups

threatened to overthrow the king if he did not support them in their struggle for a homeland. Although Jordan's king is sympathetic to their need for a homeland, he is concerned about his own country's political stability. Jordan does not want the *intifadeh* in the occupied territories to spread to Palestinians in Jordan.

An Enduring Country   Despite the economic and political challenges of the last decades, Jordan has established itself as a modern Middle Eastern country. Since the late 1960s its economy has begun to improve. Many of Jordan's workers are employed in its growing service industries, such as banking and tourism. And, although its relations with the Palestinian population are at times difficult, Jordan has maintained its national unity.

Lebanon: A Place in Conflict

The tiny country of Lebanon was looked upon for many years with a mixture of awe and envy by people from other Middle Eastern countries. It was unlike any other place in the region. Lebanon had a mild climate, beautiful beaches, and an open social and political atmosphere. It also had a thriving, Western-style economy. Lebanon's capital, Beirut, was a center of international tourism, banking, and trade. A glamorous and free-spirited city, Beirut was often referred to as the "Paris of the Middle East." However, in recent years, struggles between a multitude of different groups have almost destroyed the country. The gentle, sophisticated city of Beirut has been turned into a battlefield. Many of Lebanon's people have left to live in other countries.

The Beginning of the War   The chaos in Lebanon grew out of a breakdown in the political system. Since Lebanon became independent of France in 1943, its many religious

Jordan
The region of Jordan known as the East Bank was formerly called Transjordan, which means "beyond the Jordan," because it was located across the Jordan River from Palestine.

Beirut, Lebanon: Before and After War

Before war began in 1975, Beirut was a thriving center of tourism and education (left). Years of war have left much of Beirut in ruins (right).

groups had shared responsibility for governing the country. Power was divided based on a census taken in 1932. According to that census, the Maronite Christians formed the largest single religious group. Therefore, the president was to be a Maronite. The prime minister and the speaker of the legislature were to be chosen from among the next two largest religious groups, the Sunni Muslims and the Shi'ite Muslims. The Greek Orthodox Christians and the Druze were also allotted government offices. For many years this system of government worked well. But as the Muslim population grew, Muslims began to demand a greater role in the government. At the same time, growing economic inequality between groups in different parts of the country created social tensions. In southern Lebanon the Shi'ite Muslims felt that government policies particularly discriminated against them. A civil war broke out in 1958. A compromise was reached, and the political system remained unchanged. Then, in 1975, civil war broke out again.

A Kaleidoscope of Terror The situation had grown far more complicated by 1975. Thousands of Palestinian refugees had made their homes in Lebanon. The PLO set up military bases in Lebanon from which they conducted raids across the border to Israel. The Israelis, in turn, struck back at PLO forces in Lebanon. Many Lebanese, Christians and Muslims alike, opposed the presence of the PLO in their country. However, some Muslim and Lebanese radical groups formed an alliance with the PLO.

When the civil war erupted, other countries quickly became involved. The United States and Israel came to the aid of the Maronite Christians. Syria also came to the defense of the Maronites. However, as the war progressed, Syria switched its support to the Muslim radicals.

In 1982, Israel invaded Lebanon to drive out the PLO. First Israeli troops destroyed much of southern Lebanon where the PLO had established bases. Then they advanced to Beirut. The Israelis bombed the city heavily for weeks. Thousands of Lebanese and Palestinians were killed. Large sections of the city were turned to rubble.

An international force was sent in to establish peace in Beirut, but the country slid further and further into chaos. Thousands of Shi'ite Muslims, who had lost their homes and whose farmland had been destroyed by the Israeli invasion of southern Lebanon, filled Beirut. Muslim and Christian groups split into different factions. Each faction had its own militia, or army. Not only did Muslim militias fight against Christian militias, Christians warred against other Christians and Muslims warred against Muslims. Alliances between groups changed overnight.

By the mid-1980s, Lebanon was in a state of anarchy, or lawlessness. No government, army, or police force could maintain order. Bands of militia roamed the streets kidnapping members of other groups and foreign citizens whom they held hostage, sometimes for years. In the middle of the day, fights broke out between militia on crowded streets. Families installed steel doors on their houses and apartments and bought machine guns to protect themselves. Today the Lebanese people are still struggling to free themselves from the cycle of violence and regain their national identity.

Syria: A Promising Relationship with the Land

Since the time of its earliest settlers, Syria has been a prosperous land. Its location on the eastern edge of the Mediterranean between Europe, Africa, and Asia has made cities like Damascus, the capital, and Aleppo busy centers of trade. For thousands of years, Syria's people have taken advantage of its rich farmlands and its thriving cities to make a living.

Today, about half of Syria's population of 12.1 million people live in large cities or towns in western Syria.

Many of them own or work in shops or exporting businesses. Others work for the government. And still others work in growing industries such as food processing, textiles, and oil.

The other half of Syria's people are rural and make their living by farming. Many of these farmers live in western Syria. Here, a mountain range runs north to south, parallel to the coast. Winds coming off the Mediterranean Sea drop their rainfall on the mountains' western slopes. Streams running off the mountains water the plains along their eastern edge. A large number of farmers also live in northeast Syria, through which the Euphrates River and its tributaries run.

However, in recent decades more and more Syrians have left their farms to work in the cities. Although Syria is fortunate to have fertile farmland, many farming methods are out of date. Few farmers have modern machinery, and only about one third of the farms are irrigated. Most of Syria's farmers depend on rainfall to water their crops of cotton, wheat, fruit, and vegetables. However, rainfall is unreliable. When droughts occur, farmers are unable to make a living.

The Syrian government is trying to improve farming methods in the hope of encouraging farmers to stay on their land. It has given money to farmers to help them buy modern machinery. In the last decade, Syria has also been focusing more attention on research to improve crop output.

Most important, Syria is developing strategies to make better use of its water resources and increase the amount of land under irrigation. Syria has built dams in the northeast and northwest, where farmers depend on water from the Euphrates River and its tributaries. The water in the lakes formed by these dams irrigates thousands of acres of land. In an effort to increase agricultural production, Syria plans to bring even more land under irrigation by the year 2000.

Iraq's Rich Resources— Land and Oil

Iraq is located in a fortunate place. A large part of the country lies on the well-watered plain between the Tigris and the Euphrates rivers. Grains, fruits, and vegetables grow easily. For hundreds of years farming was the most important economic activity in this land. Then, in the 1930s, large quantities of oil were discovered.

The Iraqi government has spent billions of dollars of money from oil to develop the country. It has built roads, airports, and hospitals, and improved communication. It has opened new schools and universities. It even rebuilt its capital, the ancient city of Baghdad. Today freeways crisscross the city.

Iraq has also spent a great deal of money on irrigation and dams to improve agricultural output. As in Syria, outdated farming methods have affected Iraq's agricultural output. Today Iraq, once the granary of the Middle East, must import much of its food.

However, since 1980, when war broke out between Iran and Iraq, Iraq's financial resources have been severely strained. Iraq was forced to close its ports on the Persian Gulf because of the war, so it could not ship its oil to foreign buyers. Also, the war cost billions of dollars. Iraq purchased huge quantities of weapons from countries everywhere, including the United States, France, and the former Soviet Union. When the war with Iran ended in 1988, Iraq was heavily in debt.

Iraq was also heavily militarized. It had great stockpiles of weapons and a huge army. In 1990, Iraq attacked the neighboring country of Kuwait. Amidst worldwide protest, it declared Kuwait a part of Iraq, thus putting itself in control of a large percentage of the world's oil. War broke out in the region in early 1991 when armed forces led by the United States and supported by the United Nations attacked Iraq and liberated Kuwait.

Modern Baghdad
A poster of Iraqi President Saddam Hussein towers over a Baghdad street.

SECTION 3 REVIEW

Developing Vocabulary
1. Define **a.** militia **b.** anarchy

Place Location
2. Identify the countries that border Jordan.

Reviewing Main Ideas
3. How have conflicts in the Middle East affected Jordan?
4. What were the causes of Lebanon's civil war?
5. Describe the role that oil plays in Iraq's economy.

Critical Thinking
6. **Perceiving Cause-Effect Relationships** How have outdated farming methods contributed to an increase in the number of people moving from the countryside to cities in Syria and Iraq?

4 The Arabian Peninsula

Section Preview

Key Ideas

- Oil has brought great wealth to many countries on the Arabian Peninsula.
- The oil-rich countries of the region have used their oil wealth to pay for modernization.
- Oman and Yemen are the least developed countries in the region.

Key Terms

desalinization, infrastructure, *falaj* system

The Arabian Peninsula is a land of superlatives—of *largests* and *leasts*. Among the features that fall under the *largest* category is its desert, an enormous stretch of sand called the Rub' al-Khali, or the Empty Quarter. At 250,000 square miles (647,500 sq km), the desert is about the size of Texas. It is the world's largest sand desert. Among the *leasts* in the Arabian Peninsula is water. Without one single body of fresh water, the peninsula has the least amount of water of any large landmass. Instead of water, it has the world's largest known petroleum reserves. Since oil is an important resource, the peninsula has seen the most change in the least amount of time of any place in the world.

Changing Human-Environment Interactions

Before the discovery of oil, people in Saudi Arabia, Kuwait, Bahrain, Qatar, and the United Arab Emirates existed in much the same way as they had for centuries. Along the coasts, they fished and traded using *dhows*, Arab sailing ships. In the fertile oases of the desert, they lived in small towns and villages in houses made of sun-dried bricks. There they grew wheat, vegetables, and dates. They also tended small herds of camels, goats, and sheep. Groups of Bedouin herders roamed the deserts surrounding the oasis settlements. They made daring raids on other Bedouin groups and on towns to capture livestock and other bounty.

However, the discovery of oil in the Arabian Peninsula in the 1930s greatly changed traditional ways of life. It brought enormous wealth to the people there. Today, most of the people have moved to cities such as Riyadh, Saudi Arabia's capital, and Abu Dhabi, the capital of the United Arab Emirates. There many of them live in modern, air-conditioned houses and apartments. Some work in gleaming chrome and glass buildings as engineers, computer programmers, and executives of international corporations.

The Bedouin Arena

In spite of its oil wealth, life seems unchanged for many of the Bedouin inhabitants of the Arabian Peninsula.

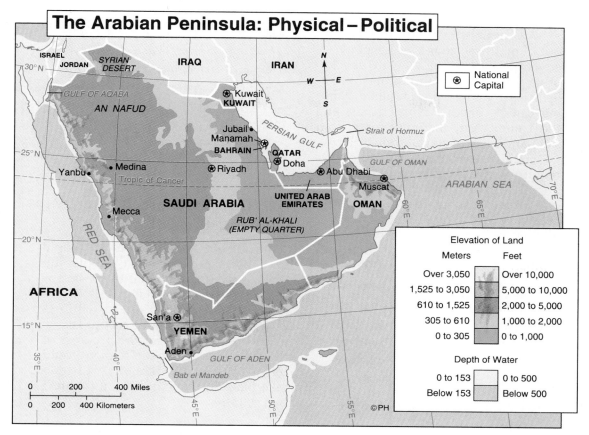

The Arabian Peninsula: Physical–Political

National Capital

Elevation of Land

Meters		Feet
Over 3,050		Over 10,000
1,525 to 3,050		5,000 to 10,000
610 to 1,525		2,000 to 5,000
305 to 610		1,000 to 2,000
0 to 305		0 to 1,000

Depth of Water

0 to 153		0 to 500
Below 153		Below 500

©PH

Applying Geographic Themes

Movement   Many of the oil-producing countries of the Arabian Peninsula rely on the open passage of ships and oil tankers through the Persian Gulf to export their oil to countries around the world. Where is the Strait of Hormuz located?

Oil Pays for Modernization   The countries of the Arabian Peninsula have used the wealth from their oil to pay for modernization. Locate these countries in the table on pages 751–761. Notice that they have small populations. With relatively few people, these countries have been able to spend large amounts of money to improve the lives of their citizens. Hospitals, schools, roads, and apartment buildings have been built. Such services are free or heavily subsidized by the oil-rich governments.

These countries have also spent billions of dollars to create more of their scarcest resource—water. Industrial plants have been constructed to remove the salt from seawater so that it can be used for drinking and irrigation. This necessary process is called **desalinization.** By 1990, Saudi Arabia had built twenty-three desalinization plants and had become the world's largest producer of desalinated water.

Looking to the Future   The countries of the Arabian Peninsula will not always be able to depend on oil to support their economies. Some experts believe that Saudi Arabian and Kuwaiti oil will last another fifty to sixty years. Qatar and Bahrain's oil may last for only another twenty to thirty years.

Aware that they will one day run out of oil, these countries have invested large sums of money to develop other industries. Bahrain has established itself as an international banking center. Saudi Arabia, Qatar, and the United Arab Emirates have built steel and petrochemical industries.

Such massive development efforts require workers. But because their own populations are so small and often lack the necessary skills, the oil-rich countries have had to hire huge numbers of foreign workers. In some countries on the Arabian Peninsula, foreigners outnumber citizens. As one author wrote:

My hotel in Jiddah [a city on the western coast of Saudi Arabia] was typical. The receptionist was Lebanese . . . Pakistani construction workers were building an ex-

tension to the hotel under a Palestinian foreman. When I came to leave, a Jordanian made up the bill. But it was a Saudi Arabian who drove me to the airport because taxi driving, like the army and police, is reserved for nationals.

Saudi Arabia: Straddling Past and Present

Until the mid-1960s most people in Saudi Arabia lived a simple life either in rural settlements or as nomads. In 1966, their king, Faisal, told them, "We are going ahead with extensive planning, guided by our Islamic laws and beliefs, for the progress of the nation." Few people then could have imagined the changes that would take place in their country in the next two decades.

Decades of Change Beginning in the late 1960s, the Saudi Arabian government spent billions of dollars of oil revenue to build the country's infrastructure. An **infrastructure** is a country's basic support facilities, including its roads, schools, and communication systems. The Saudis strengthened their infrastructure by building modern highways, airports, seaports, and a telephone system. Villages that had always depended on oil lamps for light were hooked up to electricity.

Saudi Arabia opened schools throughout the country and provided children with free education. Universities began educating Saudi Arabians in engineering, science, and medicine so that one day they could run their own country. The government also increased opportunities for women to pursue higher education.

On the west and east coasts of Saudi Arabia, construction was started on two giant industrial cities—Yanbu and Jubail. There oil and gas would be

Jidda, Saudi Arabia
Traditional Islamic designs are often incorporated into Saudi Arabia's modern-day architecture.

Modern Saudi Arabia

The family is still the most important social unit in Saudi Arabia. There are no public places of entertainment such as movie theaters or concert halls. Most people spend their free time at home with their families or visiting relatives.

Women, as wives and mothers, have an honored position in Saudi society. But they are limited members of society in other ways. As more and more women attend universities, they are working in jobs outside the home. However, Islamic custom prohibits them from associating with men outside of their immediate family. As a result, they must find professions where they are in contact only with other women, such as teaching in girls' schools, as shown at right, or treating women patients.

1. How do most people in Saudi Arabia spend their free time?
2. What types of jobs do Saudi women hold outside the home?

gathered, processed, and shipped. Petrochemical and other industries would also be developed. With the creation of Yanbu and Jubail, Saudi Arabia planned to diversify its economy so that it would not always be dependent on oil.

Saudi Arabia also spent further billions of dollars on irrigation and desalinization to increase agricultural production so that it would not have to rely on other countries for food. By the early 1980s, Saudi farmers were supplying much of the country's vegetables and poultry and almost all of its wheat.

Islam and Modernization In less than two decades, Saudi Arabia transformed itself from an ancient desert kingdom into a modern country. However, it did so cautiously. The government was careful not to let modernization upset the Islamic and other traditions upon which life in Saudi Arabia is rooted.

Hajj
Hajj, also spelled
Hadj, is the Arabic
word used to de-
scribe the holy
pilgrimage to Mecca
that every Muslim
is required to make
at least once, if
possible. The Hajj
includes several cer-
emonies that last
for many days.

Saudi Arabia has tried to create a harmonious balance between change and tradition. This can be seen in its role as guardian of Islam's most sacred cities, Mecca and Medina. Each year two million Muslims from all over the world visit Saudi Arabia for the *Hajj,* or pilgrimage to Mecca. Pilgrims are greeted in modern airline terminals, then taken by bus to huge tent cities that have been set up especially for the *Hajj.* Here, hundreds of thousands of people are provided with sanitation and medical facilities. Closed-circuit televisions monitor the crowds for emergencies. Saudi Arabia has used modern technology to support what Muslims believe to be the single most moving and meaningful religious ritual in Islam.

Oman and Yemen: On the Edge of Change

Unlike the other countries on the Arabian Peninsula, life for most people in Yemen and Oman has changed little since ancient times. Although Yemen has some oil deposits, they have not been developed yet. Oman is just be-ginning to use money it has earned from oil to improve life for its people. Neither Oman nor Yemen has under-gone large-scale modernization like neighboring countries on the Arabian Peninsula.

Yemen is the poorest country on the Arabian Peninsula. It is also the newest. Until the spring of 1990, the present country of Yemen was divided into two countries—North Yemen and South Yemen. The new Yemen's lead-ers hope the merger of the two coun-tries will improve the economy and eventually help Yemenis to live a bet-ter life.

San'a, the former capital of North Yemen, is the new political capital. However, it is the port city of Aden in what was formerly South Yemen that is the economic capital. It currently provides most of Yemen's income. As the map on page 469 shows, Aden is strategically located at the entrance to the Red Sea. Ships on their way through the Suez Canal in Egypt use the port for refueling, repairs, and transferring cargo.

Most people in both Yemen and Oman make their living by farming and herding. However, it is a difficult existence. Most of Oman is desert. Farmers depend on an ancient system of underground and surface canals called the *falaj* system for water. These canals carry water from the mountains to villages miles away.

Since the 1970s, the government of Oman has used money from oil ex-ports to improve life for its people. It has begun to update irrigation systems and build roads, hospitals, and schools. Slowly, change is coming to the people of Oman.

■ SECTION 4 REVIEW ■

Developing Vocabulary
1. Define **a.** desalinization **b.** infrastructure **c.** *falaj* system

Place Location
2. Which countries lie south and south-east of Saudi Arabia?

Reviewing Main Ideas
3. In what ways have the oil-rich countries of the Arabian Peninsula used their oil profits?
4. How are Yemen and Oman different from the other countries on the Arabian Peninsula?

Critical Thinking
5. **Recognizing Ideologies** In 1966, King Faisal told the people of Saudi Arabia, "We are going ahead with extensive planning, guided by our Islamic laws and beliefs, for the prog-ress of the nation." What did the king mean when he said this?

✓ Skills Check

☑ Social Studies
☐ Map and Globe
☐ Reading and Writing
☐ Critical Thinking

Reading Circle Graphs

Circle graphs are useful for comparing the parts of a whole. Relationships among many categories can quickly be seen. Use the following steps to study the circle graph below.

1. **Identify the kind of information on the graph.** Answer the following questions: (a) What is the subject of the circle graph? (b) How many segments are shown on the graph? (c) Which segment of the graph is the smallest?

2. **Practice reading the information shown on the graph.** Answer the following questions: (a) What percentage of world oil reserves does Saudi Arabia claim? (b) What percentage is held by the United States? (c) What regions are included in "Others"?

3. **Look for relationships among the data.** Answer the following questions: (a) What percentage of world oil reserves do the combined Persian Gulf reserves of Saudi Arabia, Kuwait, and Iraq equal? (b) What percentage of world oil reserves is held by nations outside the Persian Gulf? (c) How many times greater than United States oil reserves is the Saudi share?

4. **Use the graph to draw conclusions.** Answer the following questions: (a) What is the most powerful global region in terms of the world oil reserves it controls? (b) How likely is it that other nations of the world could do without oil from this region? (c) Why does the United States depend so much on imported oil?

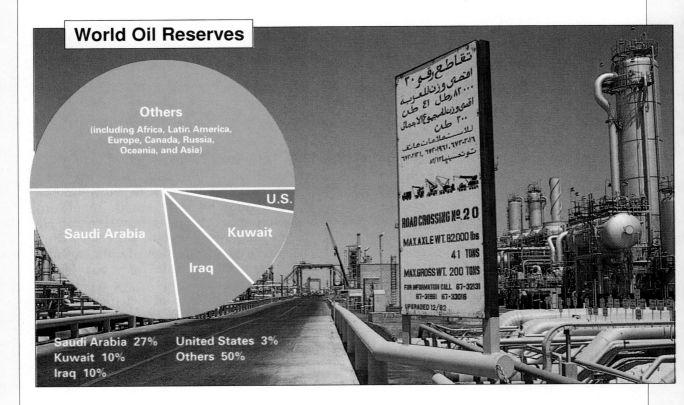

World Oil Reserves

Others
(including Africa, Latin America, Europe, Canada, Russia, Oceania, and Asia)

U.S.

Saudi Arabia

Kuwait

Iraq

Saudi Arabia 27%   United States 3%
Kuwait 10%   Others 50%
Iraq 10%

5 Turkey, Iran, and Cyprus

Section Preview

Key Ideas

- Turkey is a modern nation whose economy depends on industry and agriculture.
- Iran's Islamic revolution seriously affected the country's development.
- Ethnic unrest has divided Cyprus.

Key Term

ayatollah

Turkey, Iran, and Cyprus are different from the other countries in the Middle East. Although the majority of people in Turkey and Iran and a large number in Cyprus are Muslims, they are not Arabs. They speak different languages, and they trace their ancestors back to different roots.

Turkey: Breaking with Islamic Tradition

The Persians, the Greeks, and the Romans—at various stages in history—all controlled what today is Turkey. However, it is the Turks, the last empire builders in the Middle East, from whom most of the people in Turkey today trace their ancestry.

The Turks originally came from central Asia. Scholars and merchants who traveled from the Middle East to central Asia introduced them to Islam. When the Turks began conquering the Middle East, they came as Muslim warriors. However, although they were Muslims, their language and culture were Turkish, not Arab.

The "Father of the Turks" When World War I ended in 1918, the victors broke up the Ottoman Empire. Turkey kept only Asia Minor and the area around Istanbul. The Dardanelles were kept under international control. Greece was given some islands in the Aegean, but was determined to regain Constantinople—the capital of the Byzantine empire renamed Istanbul by the Ottomans. Many Turkish nationalists were furious that a weak sultan had given up so much of their territory. They also realized they stood in danger of losing all of Turkey to Greece and its supporters. Mustafa Kemal, a fiery young army officer, began a movement to establish Turkey as an independent republic. The revolutionaries finally succeeded. In 1923, they overthrew the sultan, declared Turkey a republic, and elected Kemal as president.

Immediately Kemal set about making Turkey a modern country. Along with other Turkish leaders, he believed that Turkey would not survive without sweeping social and political changes. In a passionate speech he pleaded with the Turkish people:

Remain yourselves, but learn how to take from the West what is indispensable to an evolved people. Admit science and new ideas into your lives. If you do not, they will devour you.

One of Kemal's first steps toward modernization was to break the bond between Islam and the government. Religious leaders were no longer involved in running the government. He replaced Islamic law with laws from European countries. Public secular schools replaced the traditional Islamic schools.

Many of Kemal's changes unraveled the fabric of traditional life. He outlawed the *fez,* a brimless, flat-topped, red hat worn by men, and he ridiculed the custom that required women to wear veils in public. After centuries of subservience, women were given many political and legal

WORD ORIGIN

Dardanelles
The strait known as the Dardenelles, which is located between Europe and Asia, takes its name from the legendary Greek king Dardanus who, according to myth, founded a city on its shores.

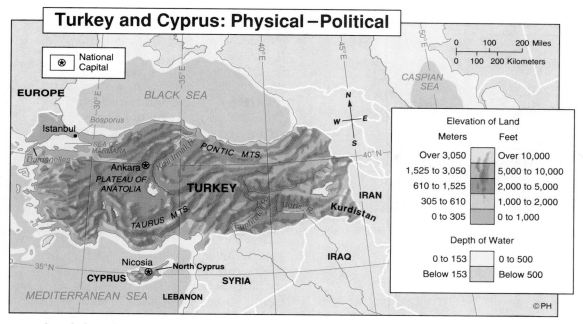

Turkey and Cyprus: Physical—Political

Applying Geographic Themes

1. Place   What are the names of the two straits leading to the Black Sea?

2. Movement   Access to the Black Sea is guaranteed by international treaty. Which of Turkey's giant neighbors relies particularly on this access?

rights. They could vote and hold office. And everyone in Turkey, including women, was encouraged to go to school. "It is schoolmasters, and they alone, who can save the people," Kemal proclaimed.

By the time Kemal died in 1938, Turkey was well on its way to becoming a modern country. Kemal was such a strong force in establishing Turkey's identity after World War I that the Turks gave him the surname Atatürk, meaning "father of the Turks."

Turkey Today   When Atatürk came to power in the 1920s, Turkey was almost entirely an agricultural country. Today Turkey is the most industrialized country in the Middle East. It has more than thirty thousand factories. Still, agriculture remains important. More than half of Turkey's population of 55.4 million make their living from the land.

Islam and Change in Iran

Nearly 3,500 years ago, the Persians arrived in the area that is now Iran. Today their descendants are the dominant cultural group in Iran. The Persians once ruled a vast empire that stretched west into what is now Libya and east to what is now the country of Pakistan. In 331 B.C., Alexander the Great conquered the empire. Later, the Persians won back much of their territory. By the time conquering Arab armies reached them in the mid-600s, the Persian culture was well-established in Iran.

For about six hundred years, the region of Persia was part of the Islamic empire. Even though most of Persia's people converted to Islam, they were not Arabs. They maintained their links to their Persian past and continued to speak Farsi, the language of their Persian ancestors.

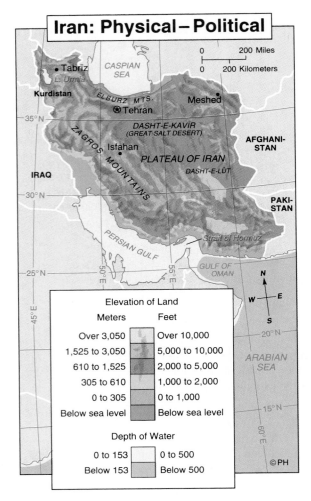

Iran: Physical–Political

0   200 Miles

0   200 Kilometers

Elevation of Land

Meters		Feet
Over 3,050		Over 10,000
1,525 to 3,050		5,000 to 10,000
610 to 1,525		2,000 to 5,000
305 to 610		1,000 to 2,000
0 to 305		0 to 1,000
Below sea level		Below sea level

Depth of Water

0 to 153		0 to 500
Below 153		Below 500

©PH

Applying Geographic Themes

1. Location Which countries border Iran to the north and east?

2. Place Iran is a region of both mountains and plateaus. Which mountains run along the Caspian Sea?

Iran Becomes a Modern Country In 1925, an army officer, Reza Khan, seized power and declared himself Iran's ruler, or shah. When he came to power most of Iran's people were either nomadic herders or farmers, barely able to make a living from Iran's dry land. In the years of the shah's rule, Iran began to change. Like Atatürk, the shah opened schools, built roads and railroads, encouraged industry, and gave women more rights.

When his son, Mohammad Reza Pahlavi, took over during World War II, he was even more determined than his father to make Iran into a modern, Westernized nation. In a very short time, great changes took place in Iran. Profits from its huge oil industry were channeled into industrial and agricultural development. Land was distributed to peasants. Teachers and medical workers traveled into the villages to improve literacy and health care. Women began to vote, hold jobs outside the home, and dress in Western-style clothing.

However, resentment against the shah's rule developed. Although many Iranians benefited from the shah's reforms, many more still lived in great poverty. Some Iranians believed that Iran should be run as a democracy. Others, especially conservative religious leaders, known as **ayatollahs** thought Iran should be governed in strict obedience to Islamic law. But the shah ran the country as a dictatorship. No one dared oppose the government for fear of being put in prison or exiled.

An Islamic Revolution In 1979, the people of Iran revolted. The shah and his supporters fled. The Ayatollah Khomeini, an exiled religious leader, returned to Iran and set up a new government. He declared Iran an Islamic republic.

Immediately, Khomeini's government set out to rid the country of all Western influences, which they saw as a threat to Islam. Westerners were forced to leave the country. Alcohol was outlawed. Women were discouraged from wearing Western-style clothing and once again donned their long, black cloaks, called *chadors.*

Iran's ayatollahs belong to the Shi- 'ite branch of Islam, as do the majority of Iranians. However, in most other Middle Eastern countries the Shi'ites are a minority, often oppressed by the Sunni majority. The new rulers of Iran called on Shi'ites everywhere to overthrow their governments and establish Islamic republics. In 1980, angered by Khomeini's attempts to provoke Shi'ites in Iraq to take such action, Iraqi leaders launched a war against Iran. Hundreds of thousands of Iranians were killed in the fighting, which lasted eight years.

Iran Today   Since Ayatollah Khomeini died in 1989, Iran has begun to take a less extreme position. The Iranian government realizes that some change is necessary if it is to solve the country's serious economic problems.

The revolution and the war with Iraq severely affected Iran's economy. Thousands of technicians and managers who opposed the Islamic government fled Iran along with the shah. Industries that depended on highly skilled workers were forced to close. Also oil exports, which supply most of Iran's income, dropped sharply because of damage to storage terminals and refineries during the war. Without money to pay for basic necessities, such as food, Iran's rapidly growing population of almost 54 million faced serious shortages.

Cyprus: A Divided Land

Cyprus is an island country in the eastern part of the Mediterranean Sea. Greek colonists came to this island as early as 1200 B.C. Today about four fifths of the Cypriot people speak Greek as their first language and are Greek Orthodox Christians. However, the Turks are also a part of the history of Cyprus. Cyprus was a part of the Ottoman Empire from the 1570s until the British occupied it in 1878. One fifth of the island's people trace their roots back to Turkey and follow Islam.

In the 1970s, civil war split the island in two. Some Greek Cypriots wanted Cyprus to unite with Greece. In 1974, Turkey sent troops to Cyprus to prevent this. The Turks took over a large part of northeastern Cyprus.

Rugs on Display in Iran
These rugs on display at the Darbe-Imam Mosque in Iran are a part of the Iranians' ancient Persian culture.

SECTION 5 REVIEW

Developing Vocabulary
1. Define: ayatollah

Place Location
2. What country lies east of Iraq?

Reviewing Main Ideas
3. How did Atatürk change Turkey?
4. Why did Iranians revolt in 1979?
5. What two ethnic groups live on the island of Cyprus?

Critical Thinking
6. **Making Comparisons**   How is the role of Islam in present-day Turkey different from its role in Iran?

23

REVIEW

Section Summaries

SECTION 1 Creating the Modern Middle East For more than four hundred years, most of the Middle East was part of the Ottoman Empire. However, with the Ottomans' defeat in World War I, the region was divided into the countries that make up the Middle East today.

SECTION 2 Israel: A Determined Country Since Israel became an independent nation in 1948, it has created one of the most technologically advanced countries in the Middle East. Its diverse people have proven to be a rich resource. Challenges facing Israel include deep political and social divisions among its Jewish citizens, the need for greater participation of its Arab population, and the hostility of the neighboring Arab countries.

SECTION 3 Jordan, Lebanon, Syria, and Iraq These countries are at the political center of the region. Jordan, Syria, and Iraq are struggling to become modern countries.

SECTION 4 The Arabian Peninsula Oil has brought great wealth to the countries of the Arabian Peninsula. It has also brought great changes in a very short time. Countries such as Saudi Arabia have not let modernization upset their traditional way of life.

SECTION 5 Turkey, Iran, and Cyprus After World War I, both Turkey and Iran set out to become modern, industrial countries. Today, Turkey is the most industrialized country in the Middle East. However, in 1979, Islamic leaders in Iran led a revolt against the shah and increasing Western influences in their country. Ethnic unrest between Greek and Turkish Cypriots divides the island country of Cyprus.

Vocabulary Development

Choose the italicized term in parentheses that best completes each sentence.

1. An agreement under which a foreign power er governs a country until it is ready for independence is called a (*mandate/treaty*).
2. Jews who believe that the return of Palestine to the Jewish people is their historic right are (*Zionists/refugees*).
3. A method of allowing precise amounts of water to drip from pipes onto plants is called (*fertilization/drip irrigation*).
4. The Israeli parliament is called the (*Knesset/Sephardic*).
5. A state of lawlessness, or political disorder, is called (*anarchy/stability*).
6. A name for an army is (*militia/civilian*).
7. Purifying seawater of salt is called (*desalination/irrigation*).
8. A country's basic support services are called its (*infrastructure/communications*).
9. The ancient system of underground and surface canals in Oman is called the (*falaj system/desertification*).
10. Conservative Islamic religious leaders are sometimes called (*ayatollahs/chadors*).

Main Ideas

1. What role did World War I play in the creation of the modern Middle East?
2. How has technology been important to Israel's development?
3. Why is the diversity of Israel's population both a challenge and a strength?
4. What impact has constant warfare had on the economy of Iraq?
5. What impact has Lebanon's civil war had on the Lebanese people?

Critical Thinking

1. **Predicting Consequences** Israeli Jews and Palestinian Arabs consider Israel to be their homeland. Design a peace treaty that addresses the demands of both groups, and explain why your plan could be accepted by both sides.
2. **Analyzing Information** Water and oil are the two most important resources in the Middle East. Do you agree or disagree? Support your argument with examples from different countries in the region.

Practicing Skills

Reading Circle Graphs Study the circle graph on page 473 again. Convert the data on the circle graph to a bar graph. Label the vertical axis "Percent of World Oil Reserves," and label the horizontal axis "Countries." Then answer the following questions: (a) In general, which form, circle graph or bar graph, makes it easier to grasp the relationships among many parts? Explain your answer. (b) If you could add more data to your graph, what would it be? Explain your choices.

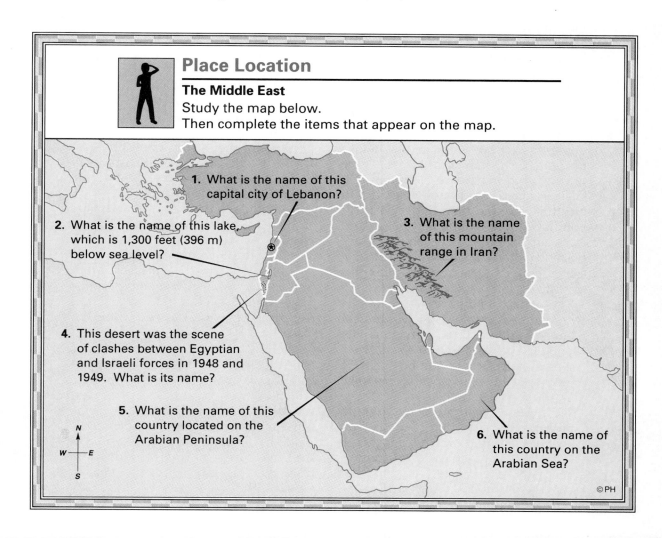

Place Location

The Middle East
Study the map below.
Then complete the items that appear on the map.

1. What is the name of this capital city of Lebanon?
2. What is the name of this lake, which is 1,300 feet (396 m) below sea level?
3. What is the name of this mountain range in Iran?
4. This desert was the scene of clashes between Egyptian and Israeli forces in 1948 and 1949. What is its name?
5. What is the name of this country located on the Arabian Peninsula?
6. What is the name of this country on the Arabian Sea?

©PH

Place Location: WHERE ON EARTH?

Destination Unknown

You're baffled, confused, dismayed, and mystified. You're also trying not to panic. Here it is, 9 P.M. and tomorrow you're supposed to turn in the minutes of tonight's Travel Club meeting—a meeting at which the guest speaker described a trip to seven countries in the Middle East and North Africa without once naming any of them!

Your friend Amy had asked you to fill in as Travel Club secretary for her while she was out of town. "It couldn't be easier," she had assured you. "One of the club members, Matthew St. Clair, is speaking about his last trip to North Africa and the Middle East. All you have to do is tape his speech and write up minutes. The minutes are placed in the club book as sort of a travel guide in case other members want to take the same trip."

As your friend gave you details of where and when the club would next meet, she added, "I can't imagine why Mr. St. Clair was asked to speak. He's traveled quite a lot, but he has one annoying habit—he never seems to be able to finish a sentence."

You hadn't paid much attention to Amy's last observation, but as you listened again, tape recorder in hand, to Mr. St. Clair's speech, you understood what she meant. Mr. St. Clair rarely completed a sentence and spoke in a rambling manner that gave an overall effect of vagueness. And either his memory wasn't very good or he was absentminded, because he never mentioned one place he visited on his trip. Here's what you heard:

"Good evening, ladies and gentlemen, and let me say that I'm so pleased to be here. . . . Tonight I'll be telling you about my last trip to . . . oh, yes, we had a wonderful time . . . the thing was, we began in that country, uh, just west of Egypt right on the Mediterranean Sea. My, was that water warm!

"We next arrived in . . . oh, yes, that was where we sailed on the Nile; what an experience that was . . !

"From there we traveled east across the Red Sea into—I can't tell you how *hot* it was there—and the desert! Was that a sight!

"Quite a contrast to the landscape of the next country we visited . . . that country with the two ancient rivers that flow into the Persian Gulf.

"Just south of the Turkish border we visited some fascinating ruins, right on the Mediterranean Sea . . . we spent two whole days exploring the place.

"One of the best parts of the whole trip was in—boy, that's a small country; not as small as Lebanon, though, which is just north of it . . . the Western Wall, the Dome of the Rock, the sense of history there is just incredible.

"The last country we visited—well, we saw it north to south, right from the Caspian Sea to the Persian Gulf."

Head in hands, you realize you will have to spend time poring over an atlas, geography textbook, or even an

SPAIN
ITALY
GREECE
ROMANIA
UKRAINE
MEDITERRANEAN SEA
CRETE
CYPRUS
LEBANON
ISRAEL
Suez Canal
JORDAN
SYRIA
BLACK SEA
TURKEY
GEORGIA
ARMENIA
AZERBAIJAN
RUSSIA
CASPIAN SEA
TURKMENISTAN
IRAN
IRAQ
Euphrates
Tigris River
River
SAUDI ARABIA
RED SEA
KUWAIT
PERSIAN GULF

encyclopedia to identify the countries in Mr. St. Clair's vague descriptions. Sighing, you open the atlas, rewind the tape, and listen for Mr. St. Clair's first description. In just a short while, you'll have the names of all the countries he visited—and you'll be able to write up your minutes.

Travel Club Speech MIDDLEAST - N. AFRICA

September

Travel Club Meeting

Guest Speaker

Matthew St. Clair

SOLVE THE MYSTERY

What seven countries did Mr. St. Clair visit on his trip? *Hint:* Your textbook has maps that will help you match Mr. St. Clair's descriptions with the countries of North Africa and the Middle East.

24

North Africa

Chapter Preview

Locate the countries covered in these sections by matching the colors on the right with the colors on the map below.

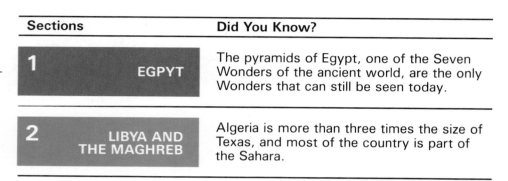

Sections	Did You Know?
1 EGPYT	The pyramids of Egypt, one of the Seven Wonders of the ancient world, are the only Wonders that can still be seen today.
2 LIBYA AND THE MAGHREB	Algeria is more than three times the size of Texas, and most of the country is part of the Sahara.

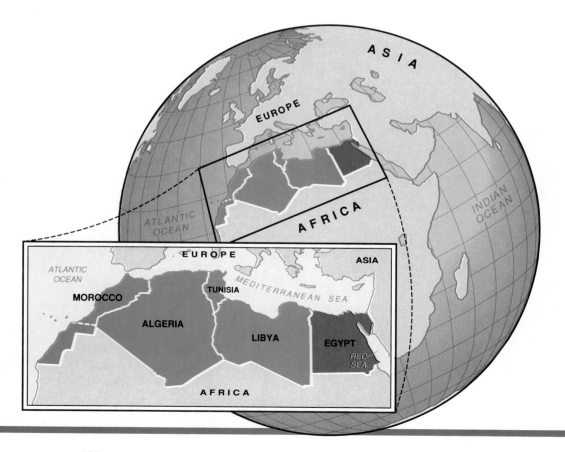

Egypt's Lifeline
These sailing vessels, known as *feluccas*, are a familiar sight on the Nile River. Egyptians have used *feluccas* since ancient times.

1 Egypt

Section Preview

Key Ideas

- Most Egyptians live in the Nile River valley, the Nile Delta, or the Suez Canal zone.
- Urbanization and rapid population growth are major challenges in modern Egypt.
- Although Egypt is still dependent on the export of raw materials, it is developing its industries.

Key Terms

bazaar, pharaoh, basin irrigation, perennial irrigation, reservoir, capital

As the globe on the facing page shows, the Arab nation of Egypt is located in the northeast corner of Africa, where Africa and Asia meet. It is about one and a half times the size of Texas, and it is the most populous country in the Arab world. Its large population and strategic location enable Egypt to play an important role in world affairs.

Egypt's Location and Regions

Egypt is sometimes referred to as the "Gift of the Nile." Without the Nile River, all of Egypt would be a desert. The Nile River valley, which runs the length of Egypt, is flanked on both sides by large and forbidding deserts.

A Ribbon of Green The Nile, which flows through Egypt from south to north, is the world's longest river. Beginning in central Africa, it flows northward for 4,160 miles (6,695 km), emptying into the Mediterranean Sea. Near

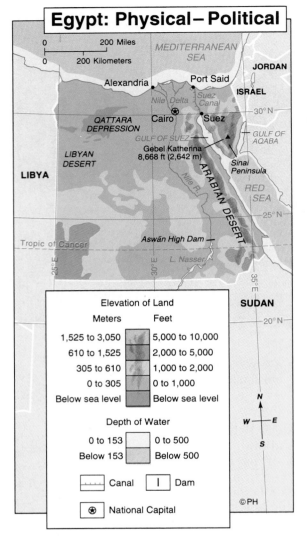

Egypt: Physical–Political

0   200 Miles
0   200 Kilometers

MEDITERRANEAN SEA

JORDAN

Alexandria   Port Said

ISRAEL

Nile Delta   Suez Canal

30°N

QATTARA DEPRESSION   Cairo   Suez

GULF OF SUEZ

GULF OF AQABA

Gebel Katherina
8,668 ft (2,642 m)

LIBYAN DESERT

Sinai Peninsula

LIBYA

ARABIAN DESERT

Nile R.

RED SEA

25°N

35°E

Tropic of Cancer

Aswān High Dam

L. Nasser

25°E

30°E

Elevation of Land

Meters	Feet	
1,525 to 3,050		5,000 to 10,000
610 to 1,525		2,000 to 5,000
305 to 610		1,000 to 2,000
0 to 305		0 to 1,000
Below sea level		Below sea level

Depth of Water

0 to 153		0 to 500
Below 153		Below 500

SUDAN

20°N

N
W E
S

Canal   I Dam

National Capital

©PH

Applying Geographic Themes

1. Movement Egypt has an important location. Which bodies of water are joined by the Suez Canal?

2. Place What type of landform lies both east and west of the Nile River?

---

the end of its course, the river forks into two major branches. The land between these two branches is known as the Nile Delta. A delta is formed by the sediment that is dropped as the river slows and enters the sea.

The delta, which has been fertilized by the Nile for centuries, is astoundingly fertile. Impressive crops are grown without the aid of modern machinery. Even plows are rare. Most cultivation is done by hand in Egypt, where human labor is plentiful.

The majority of Egypt's fifty-four million people live either along the Nile or just to the east in the Suez Canal zone. Along the Nile's cultivated banks, population density averages close to 5,000 people per square mile (8,045 per sq km).

Egypt has two major cities: Cairo, the capital city, straddles the Nile; Alexandria, Egypt's second largest city, is a major seaport and resort on the Mediterranean Sea.

The Desert Regions To the east and west of the Nile Valley are harsh deserts. The Western Desert region is part of the Libyan Desert. The Eastern Desert region is a continuation of the Arabian Desert. The Sinai (SY ny) Peninsula, which lies in Asia to the east of the Suez Canal, is part of Egypt's Eastern Desert region.

Strong winds blow constantly across the Sahara. In the early summer a special wind, known in Egypt as the *khamsin* (kam SEEN), creates sandstorms. It blows hot air, dust, and grit into the Nile Valley, including the city of Cairo. In bad years the *khamsin* blows so hard that the Egyptian sky turns orange with flying sand.

Oases dot the desert regions. The largest oasis in Egypt is at Fayyum (FAY yoom), where water buffaloes lie in the swamps and canals, and brilliantly colored birds flit across the water. Fayyum has been an important agricultural settlement since ancient times. Today thousands of Egyptians farm the irrigated land at Fayyum. Local factories then process the cotton, fruits, and cereals that are grown there.

Oases are the only arable land in the desert. But the desert does offer other natural resources. Phosphates, for example, which are used to make fertilizer, are extracted from the desert, as is some oil.

A Summer Night in Cairo

On summer nights, people of Cairo, shown at right, love to walk by the cool of the river, enjoying the sunset, the scenery, and the pageant of their fellow citizens. A traveler to Cairo in 1990 described the scene:

Under a clump of trees, a vendor of drinks and candies has placed his television out on a rickety chair so he and an impromptu [unplanned] audience of adults and children can watch the evening's programs. . . . A family that could not afford a wedding in one of the expensive hotel ballrooms has invited friends and relatives down to the river's banks. . . . The orchestra music from the parked showboats and faded casinos along the river wafts up into the night air. . . . The long barges, small fishing boats, and . . . sailboats that crowd the river during the day have been docked for some time. Only a few small boats are traveling on the dark, slow-moving waters.

1. How would you describe an evening along the Nile in Cairo?
2. Cite two phrases from the passage that provide evidence of wealth in Cairo.

Egyptian Ways of Life

It has been estimated that 40 to 50 percent of Egypt's population live in rural areas. At the same time, Egypt's urban areas have grown tremendously in recent decades.

Village Life In some respects, village life in Egypt has remained unchanged for hundreds of years. The *fellahin*, or farmers, live in small, low houses made of sun-dried mud bricks. These houses have one to three rooms. They are often built around a central courtyard in which household chores are done. If rural families can afford to, they keep domestic animals, such as chickens, goats, and donkeys, which provide food and transportation.

Life in the Cities Cairo and Alexandria provide a wide variety of jobs and opportunities for education, culture, and entertainment. For these reasons, they attract more people than they can comfortably hold.

New arrivals from the countryside are often unable to find jobs or housing. Unwilling to return to the villages, these people live in tents and other makeshift housing.

The constant arrival of people from rural communities makes Cairo a fascinating blend of old and new. Only blocks away from modern department stores that display the latest Paris fashions are the traditional Arab open-air markets, or **bazaars**. There, merchants and artisans sell brass, leatherwork, tapestries, perfume, and jewelry. On the Nile, ancient sailboats called *feluccas* share the water with modern barges and motorboats.

The Influence of the Past

Egypt was one of the birthplaces of civilization. Well over 5,000 years ago an advanced civilization developed along the banks of the Nile River. Both this ancient Egyptian civilization and the cultures of the invaders that ruled after its downfall have shaped the distinctive culture of modern Egypt.

Ancient Egypt   The deserts that flank the Nile Valley protected the people of ancient Egypt from invasions. Undisturbed for centuries, the ancient Egyptians developed a unique and lasting civilization. They were among the first people in the world to set up an organized government and religion and to invent a written language.

Southwest of Cairo, on the edge of the desert, are the world-famous pyramids. Dozens of these enormous structures dominate the landscape. The pyramids symbolize ancient Egypt in the same way that the Eiffel Tower symbolizes France or the Statue of Liberty does the United States.

The pyramids were built as tombs for the **pharaohs**, the rulers of ancient Egypt, who were worshiped as gods.

Egyptians believed that a person's spirit might need to return to its body after death. Therefore, they preserved the bodies of the pharaohs in a process known as mummification. Egyptians also believed that a person's spirit might need nourishment in the afterlife. As a result the pharaohs were placed in their pyramids together with many valuable objects, including food, furniture, jewelry, and gold.

The civilization of ancient Egypt lasted more than three thousand years. It was not until about 1700 B.C. that the first invaders descended upon Egypt. The Hyksos, using horse-drawn chariots, crossed the desert from Asia and conquered the kingdom of the pharaohs. The Egyptians recovered from the invasion of the Hyksos and went on to build their own empire along the Mediterranean Sea. However, Egypt's location at the crossroads of Asia, Africa, and Europe made it a tempting target for the waves of invaders that followed. In the centuries that followed, Egypt was ruled by Alexander the Great as well as the Romans.

Muslim Rule   The Arabs conquered Egypt in A.D. 641. As a result of this invasion, Arabic became Egypt's official language and Islam its official religion. Today more than ninety percent of Egyptians are Muslims. Most of the remaining minority are Copts, a very early Christian sect.

For more than one thousand years Egypt was ruled as part of various Muslim empires. The last of these was that of the Ottoman Turks, who conquered Egypt in the early 1500s.

European Interventions   By the late 1700s, the Ottoman Turks' power was in decline, and European nations began to intervene in Egyptian affairs. In 1798, French troops invaded Egypt in an attempt to disrupt Great Britain's trade routes in the Mediterranean. The French withdrew a few years later, however.

During the 1800s Britain's interest in Egypt increased. The Suez Canal, which linked the Mediterranean and Red seas, opened in 1869 and made Egypt a vital link between Britain and its eastern colonies in Asia. In 1875 when Egypt's ruler faced heavy debts, Great Britain gladly purchased Egypt's share of ownership in the Suez Canal. In 1879 Egyptian nationalists revolted, determined to regain control of the Canal. Nationalism is the desire of people to form an independent nation to protect their common culture and interests. Britain responded by invading and defeating the new government in 1882. British troops remained in Egypt for decades.

Following World War I, Egyptian nationalists pushed for independence. In 1922 Britain agreed. The British effectively continued to control Egypt, and Egyptian rulers had little real power.

Independent Egypt   In 1952 a group of nationalist army officers overthrew the government of Egypt. Colonel Gamal Abdel Nasser emerged as the new ruler. Nasser was determined to end Western domination of Egypt, modernize the country, and make it a major influence in world politics.

In 1956, Nasser seized control of the Suez Canal, creating an international crisis. Israel, Britain, and France jointly invaded Egypt in an attempt to retake the waterway. Both the United States and the Soviet Union supported a United Nations resolution demanding a cease-fire and the withdrawal of outside forces from Egyptian territory. This action forced the Western nations to call off the attack. Nasser held the Canal, and the British left Egypt in 1957. For the first time in more than two thousand years, Egypt was ruled solely by Egyptians.

Nasser formed close ties with the Soviet Union, using Soviet money and experts to help with his many modernization projects. He encouraged the development of industry, and worked to reduce Egypt's dependence on cotton, its main export crop.

When Nasser died in 1970, Anwar Sadat became president of Egypt. Sadat ended Egypt's alliance with the Soviet Union and forged new ties with the West.

Egypt and Israel   After World War II, Egypt developed closer ties with the Arab Middle East. The major cause of this trend was the establishment of the state of Israel in 1948. The Arabs were united in their opposition to the existence of Israel. As a large nation that shared an important border with Israel, Nasser's Egypt took a major role in the 1948, 1967, and 1973 wars with Israel. Egypt suffered defeat in all three wars. In 1967 its air force was destroyed, and it lost control of the oil fields in the Sinai Peninsula.

When Egypt was defeated for the third time in 1973, Sadat decided to seek a permanent peace with Israel. In 1979 Egypt became the first Arab

Business as Usual on the Suez Canal
When completed in 1869, the 100-mile (160-km) Suez Canal cut in half the length of the journey between Europe and Asia.

nation to recognize Israel's right to exist. In return, Israel agreed to return the Sinai Peninsula to Egypt by 1982.

Sadat's peace treaty with Israel was harshly criticized by other Arab nations, who considered that he had betrayed the Arab cause. In 1981 Sadat was assassinated. His successor, Hosni Mubarak, continues to honor the Egypt–Israeli peace treaty.

Interaction: Controlling the Nile

Faced with a harsh environment, the Egyptians have had to adapt to survive. In the past, larger-than-normal Nile floods often brought disaster to the areas along the river. But every year the floods also brought water and silt which formed rich, fertile soil.

Irrigation   Farmers built walls around their fields to trap the water and silt as the river overflowed its banks. This form of irrigation, basin irrigation, enabled the farmers to raise crops.

Beginning in the 1800s, dams were built to control the Nile's flooding and to set up a new system of irrigation. The dams allowed the building of a perennial irrigation system—one that provided water year round.

The Aswan Dam   In 1960 President Nasser undertook an enormous new water project. He built a dam that would store Nile floodwaters in an enormous reservoir, or artificial lake. The waters of Lake Nasser, as the reservoir was called, could be used year-round to irrigate the deserts. They could also be tapped to provide extra water for Cairo and to generate electricity for the modern industries Nasser hoped to develop in Egypt.

The Aswan High Dam was completed in 1970. As a result, flooding on the Nile ended and irrigation water from Lake Nasser has allowed more and more desert to be reclaimed for farming. Egyptian farmers are now able to plant two or three crops every year. But, the dam caused some problems. Floodwaters no longer carry silt to fertilize the land on the banks of the Nile. Farmers now are forced to use chemical fertilizers to make up for the lack of natural fertilization.

Another problem caused by the dam is that salt, which once washed away during the floods, now accumulates in the soil. Some 35 percent of Egyptian farmlands now suffer from too high a salt content. To solve the problem, huge, costly drainage systems are being installed.

A Changing Population and Economy

An important trend in Egypt today is urbanization—the movement of rural people to the nation's cities. Only twenty years ago, 60 percent of Egyptians lived in rural areas, 40 percent in cities. Today the majority has shifted in the opposite direction: 53 percent live in cities, 47 percent in rural areas.

Population Growth   A second important trend is rapid population expansion. Egypt's population is growing at an annual rate of 2.5 percent, which means a doubling of the population every thirty years. The Egyptian population increases by about a million people each year. Feeding, housing, educating, and providing other services for this fast-growing population strain the economy. But stemming population growth is not easy. Many Egyptians need the labor of every member of their large families. They resist government efforts to limit family size.

One of the major problems of Egypt's population growth is that it is outstripping the country's food supply. In 1970, Egypt produced nearly all its own food. Today, it imports about 60 percent of the food its people eat. The fertile land along the Nile is already intensively farmed. The Egyptian government has plans to increase the amount of arable land by irrigating the desert, but this land will not be as naturally fertile as the Nile Valley. In order to make it productive, large amounts of money will have to be spent on chemical fertilizers.

Egypt's Exports In the past, Egypt's economy depended on a single export: cotton. When international cotton prices were high, Egypt prospered; when they fell, it faced economic disaster. Now oil and petroleum products have taken first place among Egyptian exports.

This change has not solved Egypt's economic problems. The country is still dependent upon the export of raw materials rather than manufactured goods. Most experts agree that to prosper, Egypt needs an industrial base that will provide much-needed jobs and produce goods for sale abroad.

Obstacles to Development Efforts to promote industrialization began in the late 1950s and increased in the 1960s. But several factors have limited the growth of industries. One factor is the country's limited number of skilled workers. Although Egypt has the largest pool of educated people in the Arab world, it frequently loses these professionals to wealthier countries, where salaries are much higher.

A second major challenge facing Egypt is lack of capital—money that is invested in building and supporting new industries. Average annual per capita income in Egypt is about $710, compared with about $7,000 in Saudi Arabia or $18,430 in the United States. With such relatively low incomes, few Egyptians have money left over after paying for their basic needs of food and housing to invest in new factories or industries. Lacking the oil reserves of some of its more fortunate neighbors, Egypt depends heavily on aid from Western and other Arab nations.

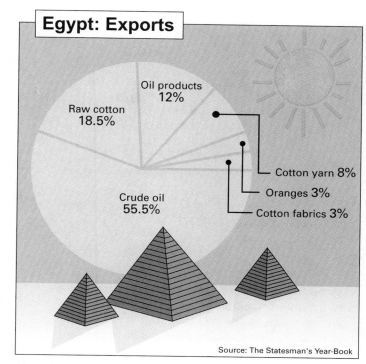

Egypt: Exports

Oil products 12%
Raw cotton 18.5%
Crude oil 55.5%
Cotton yarn 8%
Oranges 3%
Cotton fabrics 3%

Source: The Statesman's Year-Book

Graph Skill
Egypt is still dependent on the export of raw materials rather than manufactured goods. What is Egypt's largest export?

SECTION 1 REVIEW

Developing Vocabulary
1. Define: **a.** bazaar **b.** pharaoh **c.** basin irrigation **d.** perennial irrigation **e.** reservoir **f.** capital

Place Location
2. Which major city in Egypt is located on the Mediterranean Sea?

Reviewing Main Ideas
3. What challenge does Egypt's rapidly growing population present?
4. What was President Sadat's contribution to Egypt's relationship with Israel today?

Critical Thinking
5. **Perceiving Cause and Effect** How do you think life in Egypt might be different without the Nile River valley?

✔ Skills Check

☐ Social Studies
☐ Map and Globe
☑ Reading and Writing
☐ Critical Thinking

Paraphrasing and Summarizing

There are many ways to say essentially the same thing. Paraphrasing and summarizing are two ways of restating a piece of writing. To paraphrase a passage is to restate it in different words. To summarize a passage is to present its major points in brief. Study the following passage and its sample paraphrase and summary. Then follow the steps below to practice the skills.

Original: Egypt's geography helped to protect it from enemy invasions and allowed its civilization to develop without interruption. For example, deserts on either side of the Nile River and the Sinai Desert in the northeast discouraged enemy invasions. The Mediterranean Sea offered further protection in the north.

Paraphrase: Egypt had many geographical advantages that helped to protect its civilization. The deserts served as a barrier to invasion from east or west. The Mediterranean Sea was an obstacle to invasions from the north.

Summary: Both land and water barriers helped protect Egypt's civilization from enemy invasion.

1. **Paraphrase a passage by including all its ideas and details.** A paraphrase should not include any quotes from the original and should be approximately the same length. You may present information in a different order in your paraphrase, but the new version should be as detailed as the original. Write a paraphrase of paragraph 1, below.

2. **Summarize a passage by selecting its most important ideas and details.** The goal in summarizing is to restate only the main ideas. Summaries should be brief—usually no more than about one third of original length. Details and supporting evidence should be left out of a summary. Write a summary of paragraph 2, below.

Paragraph 1:
The Nile River, the world's longest river, begins in Central Africa and flows northward for 4,160 miles (6,695 km). Most of the river is navigable, except between the cities of Khartoum in the Sudan and Aswan in Egypt. Between Khartoum and Aswan, the river drops in a series of six cataracts, or waterfalls.

Paragraph 2:
When the Nile overflowed its banks, it left behind fertile, muddy soil in which the farmers grew wheat, corn, and barley. Egypt became the breadbasket of the ancient world because it sold much of its grain to other countries year after year without needing to fertilize. Rome, for example, imported huge quantities of grain from Egypt. Roman emperors gave away as much as 450,000 tons of grain each year in order to prevent violence and maintain their positions. Egyptian grain and control of Mediterranean trade were essential to the Roman Empire.

2 Libya and the Maghreb

Section Preview

Key Ideas

- North Africa's culture has been influenced by African, Arab, and European cultures.
- The major cultural divisions in Libya and the Maghreb are between urban and rural areas.

Key Term

medina

The North African countries west of Egypt are Libya and the Maghreb nations—Tunisia, Algeria, and Morocco. The word *Maghreb* comes from an Arabic term meaning "land farthest west." For a thousand years, these countries were the westernmost outposts of an Islamic empire that stretched across Asia, the Middle East, Africa, and into Europe. Today they retain close ties to other Islamic countries, especially those of the Middle East.

The North African nations are similar in many respects. The majority of their people are Arabic-speaking Muslims who live along the Mediterranean coast. Away from this narrow coast,

WORD ORIGIN

Maghreb
Maghreb is an Arabic word that means "the land farthest west." The term was first used by Arabs to refer to the lands west of the center of Arabic culture.

Applying Geographic Themes

1. Location Name the four North African capital cities found on the map. What geographic factors explain their locations?

2. Place Which major mountain range forms a chain to the west?

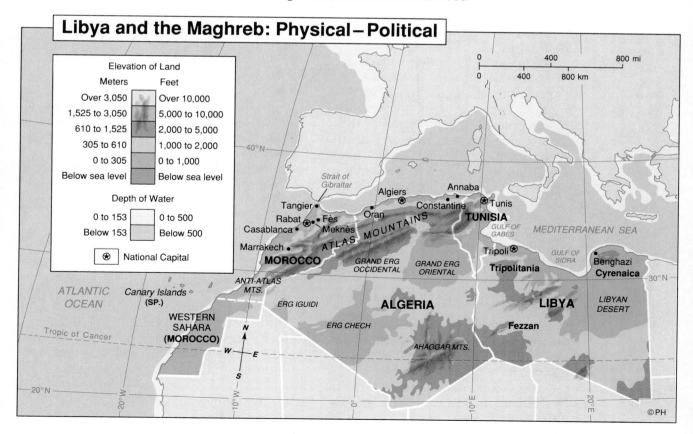

Libya and the Maghreb: Physical–Political

their lands are more arid, forming the northern margins of the Sahara. The shared presence of the desert and their similar history give the cultures of these four countries many things in common.

There are, however, also important differences among these nations. For example, Libya is a large country with rich oil reserves and very little arable land. Tunisia is small and much more agricultural, but it lacks oil.

The Landscape of North Africa

In the coastal areas of North Africa the climate is Mediterranean, with hot, sunny summers and cool, rainy winters. Away from the seacoast, the extremely dry climate of the Sahara prevails. But the landscape of the desert varies from area to area—sandy dunes flow into gravel and bare rock deserts. Low basins gradually rise to meet high, windswept plateaus and then mountains.

People who lived along the coast of North Africa found it easier to have contact with other countries than with interior regions of their own country. The people of the interior had little contact with one another or with the outside world. No navigable rivers connected these places. The mountains and the desert were formidable barriers to travel and communication. For these reasons, the people of those interior regions have tended to maintain traditional ways.

A Blend of Cultures

Located on the southern coast of the much-traveled Mediterranean Sea, North Africa has long been influenced by contact with other peoples. Today the region has a distinctive culture that is a blend of African, European, and Asian influences.

The Berbers Sometime after 5000 B.C. knowledge of agriculture spread from southwest Asia through Egypt to the Berbers—the original inhabitants of North Africa. The Berbers, who had been nomads, became farmers and herders. They settled in villages along the Mediterranean coast and on the northern mountain slopes. Only a small portion of the population lived in oases or continued to live as nomads.

Carthaginian and Roman Rule About three thousand years ago, the Phoenicians—seafarers from the eastern Mediterranean—established trading posts on the North African coast. In 814 B.C. they founded the city of Carthage, on the coast of what is now Tunisia. Carthage became a wealthy and powerful city, trading with other nations throughout the Mediterranean world.

By 400 B.C., Carthage had gained control of the entire North African coast, as far east as Egypt. The Carthaginians' wealth and power brought them into conflict with the Romans, the other great Mediterranean power. In the long and bitter Punic Wars the Romans finally defeated the Carthaginians. Victorious Roman soldiers burned the city of Carthage to the ground and, according to legend, sowed its ground with salt so that nothing would ever again grow there.

During the period that they ruled North Africa, the Romans extended irrigation systems, built dams and wells, and developed large farms. They also enlarged coastal cities and built new inland cities.

Desert Caravans During this period of Roman rule, camels imported from Central Asia were introduced to North Africa. Camels have been called "ships of the desert." They are well adapted to desert conditions. Even in very hot weather, camels can travel for several days without water. Their large, flat feet allow them to walk over sand

An Oasis in Algeria

This camel caravan has stopped at an oasis to collect fresh water and supplies. Until recently, camels were the lifeline of desert communities. How has this situation changed?

dunes much as snowshoes allow people to walk over snow. A strong camel can travel as much as 30 miles (48 km) a day carrying a burden of 500 pounds (227 kg).

Camels changed the geography of North Africa. For the first time, North Africans established regular trade with the people living south of the Sahara. Southbound desert caravans carried salt, which was very valuable to people in tropical climates. Northbound caravans carried slaves and exotic cargoes like ivory, gold, and feathers. They also transported wild animals such as hippopotamuses and elephants for the contests staged in Roman amphitheaters.

The Spread of Islam A dramatic change occurred in Libya and the Maghreb during the mid-600s A.D. Arab armies conquered Egypt in 641. Next they invaded the rest of North Africa. The Arab's impact upon the region was tremendous. They brought with them a new religion, Islam, and a new language, Arabic. Most North Africans accepted both the new religion and the new language.

Today, Arabs form the majority of the population in North Africa, with Berbers a substantial minority. Berbers in Algeria, for example, make up only 17 percent of the population. The Arab conquest was the start of a long golden age for North Africa. The new governments repaired roads and irrigation systems that had deteriorated after the fall of Rome. North Africa became a vital center of trade between Europe, Africa, and Asia and an important center of learning and scholarship.

European Influence In the nineteenth century some European powers sought to control North Africa. In 1830 France invaded Algeria. Armed Algerian resistance to French rule lasted more than fifty years, but the rebels were eventually defeated.

During the late 1800s, France also extended its empire to Tunisia. European conquest of the area was completed in 1912, when France gained control of Morocco and Italy conquered Libya.

Following Italy's defeat in World War II, in 1951 the United Nations declared Libya an independent nation.

A Mosque in Morocco
Muslims form a majority of the population in North Africa today.

However, Algeria, Tunisia, and Morocco had to fight for their independence. In 1956 Morocco and Tunisia gained their freedom, followed by Algeria in 1962.

North African Ways of Life

The most important cultural divisions in North Africa today are not those between countries but rather the differences between urban and rural ways of life.

Rural Life North Africans traditionally have made their living as farmers and settled herders. In rural areas throughout North Africa, people live in much the same way. The center of social and economic life is the *souk,* or market. The market is usually located along a trade route near a major crossroad, where many people can reach it. It runs for about a week every month. Early in the morning on each market day stalls and tents are set up either in the open air or in an enclosed area.

Country people come to the *souk* to buy needed supplies, to sell animals, crops, or craftwork, and to meet their friends and neighbors. Visitors to a North African souk may sit on the ground and listen to ancient tales told by a professional storyteller. Musicians, traditional dancers, and snake charmers may also be found attracting crowds in the *souk.* Children watch the antics of trained monkeys, who perform in exchange for coins.

Farmers living in Libya and the Maghreb still live in small rural villages in mud or stone houses that may have only one room. For the sake of privacy, these houses usually do not have windows that face on the street. Instead, windows face the family's open courtyard. Water often comes not from a tap but from a goatskin bag that hangs on the wall of the house. The family's supply of water must be carried from the village well each day.

People rise at dawn to begin their work. In the middle of the day, when temperatures are hottest, North Africans rest for several hours. Even in the cities, a three-hour midday break is the custom. When the sun's glare begins to lessen, people return to work until dusk. Some farmers own or rent small plots of land, raising wheat, barley, and livestock. The tools they use often are the same kind their ancestors used centuries ago. Wooden plows drawn by camels are not uncommon sights. Other villagers hire themselves out to work on larger, more modern farms for someone else.

Desert Nomads Some North Africans have always followed a nomadic way of life. One of the most distinctive nomadic groups is the Tuareg (TWAR ehg), who live in small groups throughout the southwest Sahara. The Tuareg speak their own language—the only Berber language that can be written. And they practice a unique form of Islam that preserves many elements of their previous religion.

The Tuareg's name for themselves means "free men." They have resisted giving up their nomadic ways to come under the control of any government. Recently, however, severe droughts in the Sahara have forced many of the Tuareg to settle in villages and work on farms in order to survive. It is possible that their ancient way of life—and that of the other remaining North African nomads—will soon disappear.

Urban Life Like Egypt, the rest of North Africa is undergoing rapid urbanization. Recent estimates show that more than half of the populations of Algeria, Libya, and Morocco live in urban areas. Tunisia is close behind, with nearly 49 percent of its people living in urban centers.

The older Arab sections of North African cities, called medinas, usually are centered around a great mosque. Urban *souks*—narrow streets and alleyways lined with many shops and workrooms—wind out from the mosque. One visitor recently described the streets of a medina:

You walk past endless walls shiny from having been polished by generations of human beings wedged into narrow alleys. . . . Exquisite and often sumptuous houses are hiding behind these walls amid scented gardens filled with the murmur of fountains. . . .

Like Cairo, the major cities of Libya and the Maghreb attract more rural people than they can absorb. Housing and jobs for unskilled laborers are scarce.

Since the 1950s, when European control of North Africa ended and oil wealth began, modern parts of cities have grown rapidly. Modern sections of North African cities look much like cities in Europe or the United States, with broad avenues, modern skyscrapers, internationally known stores, and corporate offices.

Studying the Koran
The children of Ghat in southwestern Libya gather for a lesson in Arabic so they can read the Koran.

North Africa Today

Since independence, the four nations of North Africa have taken different paths politically and economically.

Libya After years of Italian control, in 1951 newly independent Libya had virtually no technicians, doctors, engineers, or skilled workers. In the whole country, there were only a handful of college graduates, and 90 percent of the population were illiterate. The Italians did not give Libyans opportunities for education. As a result, Libya was one of the poorest nations in Africa. Its revenues came almost entirely from foreign aid and rent from British and American military bases. That changed abruptly with the discovery of oil. By 1961 Libya's first oil wells were in production.

More than any other North African nation, Libya benefited from its oil reserves. Average annual per capita income zoomed from about $40 a year in 1951 to more than $6,000 in 1976, and $7,000 by 1989. Today, oil brings in about 80 percent of the Libyan government's revenues and makes up 95 percent of the country's exports. Oil revenue has transformed life in Libya. It paid for roads, schools, modern housing, hospitals, and airports. It has brought electricity and new water wells to rural villages. It also provided many farmers with modern machinery. The impact of oil money is also reflected in higher wages. One writer noted that "Libyan oil-field laborers [now earn] more in a month than they formerly earned in a year as farmers or city workers."

It is hard to say which has changed Libya more, oil wealth or the government of Colonel Muammar Qaddafi. In 1969 Qaddafi led a military coup that overthrew the pro-Western king and abolished the monarchy. Qaddafi established a unique form of socialism that combined strict adherence to Islamic traditions with modern economic and political reforms. One of Qaddafi's goals was a more equal distribution of wealth in Libya. For example, he ordered that no Libyan could have more than one house or more than 1,000 dinar (about $3,400) in savings. The government seized the property of anyone who had more than it allowed.

Another of Qaddafi's goals was to root out Western influences, which he thought were unhealthy. His government closed bars and nightclubs. It banned blue jeans for men and any kind of pants or short skirts for women. All signs in English or Italian were ordered to be removed from public streets and buildings. To bring the country back to its Islamic traditions, Qaddafi established Islamic law as the law of the land.

Algeria   When Algeria won its war of independence from France, almost all of the million French colonists who had lived there left the country. This flight was disastrous for Algeria. Algerians had no opportunities to educate themselves while under European rule, and they depended on educated French settlers for teachers, doctors, lawyers, engineers, and government administrators. The new government began massive training programs so that Algerians could take over the positions abandoned by the French.

Oil and natural gas, which were first discovered in the 1950s, make up about 90 percent of the value of all Algerian exports. Though Algeria's average annual per capita income ($2,650) is less than half of Libya's, oil revenues have raised the general standard of living.

Although the oil industry produces most of Algeria's revenues, it employs only about 5 percent of the country's workers. With the population growing

New Construction in Libya

Libya's government has used money from the sale of oil to pay for the construction of modern housing, hospitals, and roads.

rapidly, too few jobs are available. Unemployment is a widespread problem. Many Algerians have emigrated to Europe to work, especially to France.

The Algerian government is trying to encourage rural Algerians to continue farming instead of flocking to the cities. If it is successful in its efforts, it will accomplish three goals. First, fewer Algerians will be unemployed because agricultural workers are in great demand. Second, Algeria will be able to reduce its expensive dependence on food imports. Today the country has to import 60 percent of its food. And third, the severe problems of overcrowding in Algeria's coastal cities will be reduced. At present, two-room apartments house on average nine occupants.

Tunisia and Morocco Unlike Libya and Algeria, Tunisia and Morocco do not have large oil reserves. Some inhabitants view this as a blessing. As one Tunisian business leader stated, "We are lucky we didn't find much oil. Otherwise we wouldn't have worked so hard to develop our people." Tunisia spends about one third of its money on education, and education is free from the primary grades through the universities.

One recent visitor to the Tunisian desert observed:

> *It is very touching to see groups of tiny children . . . trudging sturdily to classes across a wide, dusty landscape in which, as far as the eye can see, there is no obvious sign of home or school.*

Another important source of wealth for Tunisia and Morocco is minerals. In both countries, phosphates are the most valuable mineral export. Since independence, both countries have increased their profits by building up their chemical industries so that they can process the phosphates before exporting them.

A Country Market in Tunisia
All North African *souks* follow a similar pattern. Consumers can find everything to meet their needs squeezed into tiny spaces.

SECTION 2 REVIEW

Developing Vocabulary
1. Define: medina

Place Location
2. Which body of water do all the countries of North Africa border?

Reviewing Main Ideas
3. What outside groups have influenced the North Africans during their long history?
4. What is the most important natural resource of Libya and Algeria? Of Morocco and Tunisia?

Critical Thinking
5. **Drawing Conclusions** How might North African governments go about encouraging people to remain in rural areas instead of crowding into the cities?

24

REVIEW

Section Summaries

SECTION 1 Egypt Egypt is a desert region except for the Nile River and a few oases. The Nile Delta is formed by sediment that is dropped as the river enters the Mediterranean. In rural areas, the *fellahin* live much as their ancestors did. In the cities, some have adopted Western ways, others live more traditionally. Cairo, Egypt's capital, is among the largest and most populated cities in the world. Egypt was the site of one of the world's earliest civilizations. The country was ruled by other Muslims and Europeans before gaining full independence during the 1950s. As a result of Muslim rule centuries ago, the majority of Egypt's citizens are Muslims. Arabic is Egypt's official language. Today, it faces the challenges of rapid urbanization, population growth, and food shortages. Egypt's border with Israel made it a major player in the Middle East conflict until 1979, when it became the first Arab nation to make peace with Israel.

SECTION 2 Libya and the Maghreb Libya, Tunisia, Algeria, and Morocco are located in North Africa to the west of Egypt. The majority of the people live along the coast. The interior forms part of the Sahara. North Africa has been ruled by many outsiders. Today the culture is a blend of African, Asian, and European influences. The most important influence has been that of the Arabs. Today, most of the people of the region are Muslims, whose official language is Arabic. In rural areas traditional ways are still dominant. Both modern and traditional ways of life coexist in urban areas. After gaining independence from their European rulers, the new governments stressed education and training programs to modernize their countries. Oil wealth has significantly changed ways of life in Libya and Algeria.

Vocabulary Development

Match the following definitions with the terms listed below.

1. a system of irrigation in which walls are built around a field to trap the water and silt from a river overflowing its banks
2. money that is invested in building and supporting new industries
3. a system of irrigation which provides water throughout the year
4. the older, Arab section of North African cities that is usually centered around a great mosque
5. a traditional, Arab open-air market
6. ruler of ancient Egypt
7. an artificial lake

a. bazaar
b. pharaoh
c. basin irrigation
d. perennial irrigation
e. reservoir
f. capital
g. medina

Main Ideas

1. What natural forces influenced settlement in ancient Egypt?
2. Why is Egypt's rapidly increasing population a challenge?
3. What are Egypt's major exports? What parts of its economy is the government trying to develop?
4. Explain why, with the exception of Egypt, most people in North Africa live along the coast.
5. Name three groups that have contributed to creating the culture of contemporary North Africa.

6. How has oil wealth changed everyday life in Libya and Algeria?
7. What are some of the challenges the people of all the North African countries face today?
8. Explain why Tunisia spends about one third of its money providing free education for its citizens? How does Tunisia differ from Libya economically?

Critical Thinking

1. **Expressing Problems Clearly** The Aswan High Dam has brought both benefits and disadvantages to Egypt. Explain why you think the dam should or should not have been built.

2. **Drawing Conclusions** Why are there strong ties today between North Africa and the Middle East?
3. **Perceiving Cause-Effect Relationships** When North African nations gained independence, they had few educated people. Why do you think that their European rulers did not allow North Africans to become educated?

Practicing Skills

Paraphrasing and Summarizing Find the paragraph on page 495 with the boldfaced heading ''Libya.'' Using the skills you learned on page 490, first paraphrase, then summarize, the content of the paragraph.

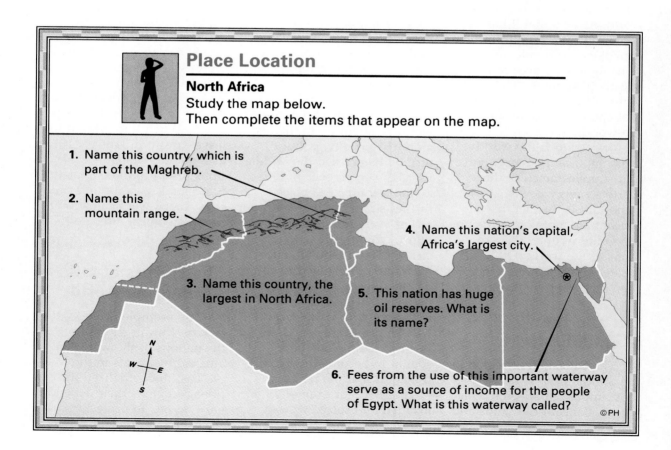

Place Location

North Africa
Study the map below.
Then complete the items that appear on the map.

1. Name this country, which is part of the Maghreb.
2. Name this mountain range.
3. Name this country, the largest in North Africa.
4. Name this nation's capital, Africa's largest city.
5. This nation has huge oil reserves. What is its name?
6. Fees from the use of this important waterway serve as a source of income for the people of Egypt. What is this waterway called?

© PH

MAKING
CONNECTIONS
·WHERE REGIONS MEET·

Energy, Oil, and the Middle East

Energy—it is the lifeblood of the industrialized nations of the world. The manufacturing and transportation that keep nations going require huge and steady amounts of energy. The question of who controls energy supplies is so important to the stability of world markets that it has become as much a military concern as an economic one. Nations fight wars over control of energy supplies. The availability of energy is an issue that connects regions around the world.

Petroleum is the world's most important energy source and supplies much of the world's energy needs. More important is the fact that significant amounts of petroleum are found in only a few regions of the world. Oil has a greater strategic importance than any other energy source.

Oil and the Middle East

The Persian Gulf region in the Middle East is the world's most important oil-producing region. Beneath the sands and waters of the Persian Gulf na-

tions lie 45 percent of the world's known oil reserves. The Persian Gulf countries produce about 30 percent of the world's total oil needs. The map on the next page shows the region's role as a world supplier of oil. This role gives the Persian Gulf region enormous international importance. It has made many Middle Eastern nations very wealthy and has given them great power in the world.

However, the Middle East's oil has also brought unwelcome attention to the region. Historically, political turmoil has racked the Middle East. The emergence of oil as a vital resource has placed further strain on the region. In recent years wars, border disputes, ethnic strife, and longstanding international feuds have often threatened to cut off the flow of Middle Eastern oil. When Middle East stability—and, with it, oil security—have been threatened, industrialized nations have sought to become involved in the politics of the region.

In 1990, for example, Iraq invaded and seized Kuwait, thus gaining control of about

20 percent of the world's oil. Iraq insisted that its action concerned a regional dispute and that the region should be left alone to resolve it. Yet Iraq's act pushed up world oil prices. Many people feared that if one country controlled too much of the world's oil supply, it could hold the countries of the world as economic hostages. The United States was in the forefront of those nations that immediately sent troops to defend Saudi Arabia from a possible Iraqi attack. When Iraq refused to leave Kuwait in January 1991 as required by a United Nations resolution, the United States and several other countries waged war against Iraq in an attempt to liberate Kuwait.

The Politics of Oil Around the World

The example of Iraq's attack on Kuwait shows how the nations of the world are affected by events in the Middle East. Industrialized nations recognize their dependence on oil. To protect themselves against possible interruption of their Middle Eastern oil supply, many countries have tried to build up oil reserves. The United States has stockpiled about 590 million barrels of oil. Japan has about 150 million barrels in storage.

The View from the U.S.A.

 The United States consumes a huge amount of the world's energy. It relies on foreign oil for about 52 percent of its needs. Saudi Arabia is its single largest supplier. United States leaders have long been uncomfortable with this dependence on imported oil. Crises like the Iraqi invasion of Kuwait have caused many people to call for a national energy policy aimed at both conserving existing oil supplies and developing alternative energy supplies. Such alternatives include the development of domestic oil supplies and of other sources of energy. The goal of these efforts is to reduce United States dependence on other regions.

TAKING ANOTHER LOOK

1. Explain why oil is such an important energy supply.
2. How do events in the Middle East affect the rest of the world?

Critical Thinking

3. **Checking Consistency** Oil has given the Middle East enormous power and wealth, and it has made the Middle East subject to interference from other countries. Explain how these two statements are consistent.

Applying Geographic Themes

Movement A great percentage of the world's oil comes from the Persian Gulf. Which two areas of the world import large amounts of oil from the Middle East?

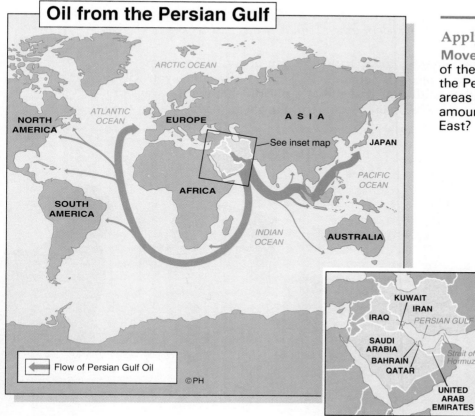

Oil from the Persian Gulf

Flow of Persian Gulf Oil

©PH

UNIT
8
Africa South of the Sahara

CHAPTERS

Dry, flat plateaus, basins, and valleys, a snowy mountain peak near the equator, swift-running rivers, lush and tropical vegetation . . . near-lifeless deserts, vast savannas, rushing waterfalls . . . These images of Africa South of the Sahara stand in sharp contrast to the modern urban centers of this vast region.

Africa South of the Sahara is today truly a region of contrast where tradition comes face to face with change. As you read this unit, you will discover the role that landforms and climates play in shaping the ways of life of the many people—African, Asian, and European—who live in the region.

Carrying grain in Senegal ▶

◀ Harvest in
Zimbabwe

◀ City skyline, Nairobi, Kenya

503

Regional Atlas: Africa South of the Sahara

Chapter Preview

Both of these sections provide an overview of the countries of Africa south of the Sahara, shown in red on the map below.

Sections	Did You Know?
1 **LAND, CLIMATE, AND VEGETATION**	Tropical rain forests grow twenty times slower than forests in the middle latitudes.
2 **HUMAN GEOGRAPHY**	In Botswana, people greet each other with the word *pula,* which means ''rain.'' The unit of currency is also called the *pula.*

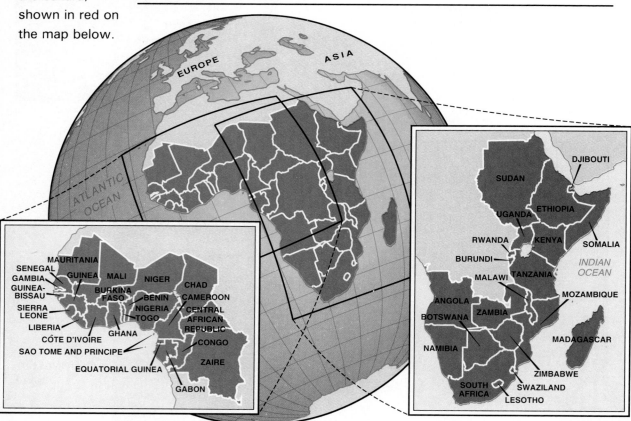

Africa's Great Rift Valley

One of the most spectacular features of Africa's physical geography is the Great Rift Valley, a wide, deep trench edged in many places by steep mountain walls. Africa's most fertile soils are located within the valley.

1 Land, Climate, and Vegetation

Section Preview

Key Ideas

- Most of Africa south of the Sahara is an elevated plateau that drops sharply to the coast.
- Similar bands of climate and vegetation stretch across Africa north and south of the Equator.

Key Term

cataract

Africa is the world's second-largest continent. The United States could fit inside its boundaries nearly four times. Most of this landmass is an enormous plateau, fringed by a narrow strip of coastal plain.

Africa's Tilted Plateau

Africa's low coastal plains are very narrow. Just a short distance inland, the land rises sharply to a plateau, which is higher in the east than in the west. This sharp rise forms an escarpment, or steep cliff, where the plateau meets the coastal plain. The escarpment was a major factor in discouraging European exploration of Africa before the 1800s. European ships attempting to navigate African rivers encountered many cataracts, or waterfalls, at the escarpment. African rivers roar down the enormous cataracts on their way to the Atlantic or Indian ocean.

The Great Rift Valley Most of the terrain of Africa's great plateau is flat or gently rolling. A spectacular exception is the Great Rift Valley, which slices through Jordan in Southwest Asia and then through eastern Africa

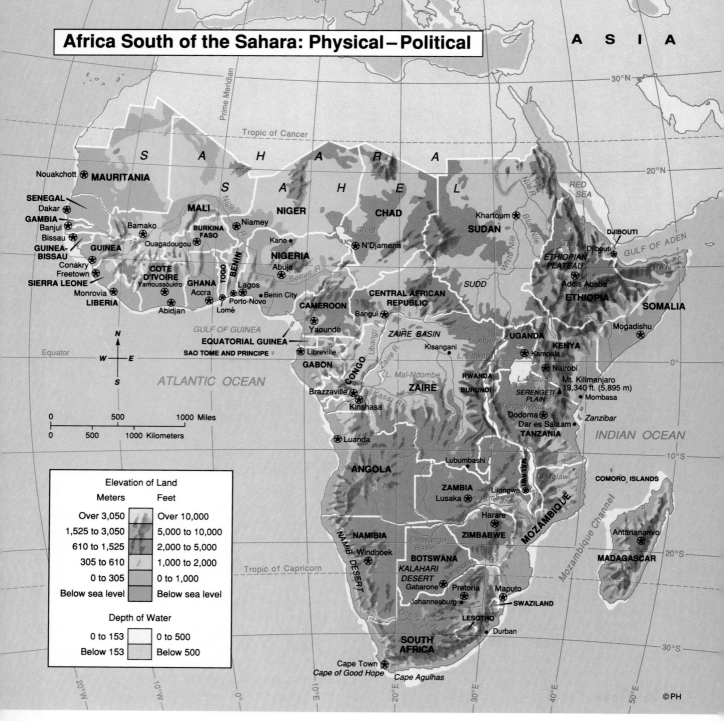

Africa South of the Sahara: Physical–Political

ASIA

MAURITANIA — Nouakchott

SENEGAL — Dakar
GAMBIA — Banjul
Bissau
GUINEA-BISSAU — Conakry
GUINEA
Freetown
SIERRA LEONE — Monrovia
LIBERIA

MALI — Bamako
BURKINA FASO — Ouagadougou
NIGER — Niamey
Kano
NIGERIA — Abuja
COTE D'IVOIRE — Yamoussoukro
GHANA — Accra
TOGO
BENIN
Lagos
Lomé
Porto-Novo
Benin City
Abidjan

CHAD — N'Djamena
L. Chad

SUDAN — Khartoum
SUDD

CAMEROON — Yaoundé
Bangui
CENTRAL AFRICAN REPUBLIC

GULF OF GUINEA
EQUATORIAL GUINEA
SAO TOME AND PRINCIPE
GABON — Libreville
CONGO — Brazzaville
ZAIRE BASIN
ZAIRE
Kisangani
Kinshasa
Kasai R.
Ubangi R.
Zaire R.
L. Mai-Ndombe

RWANDA
BURUNDI

DJIBOUTI — Djibouti
GULF OF ADEN
RED SEA

ETHIOPIAN PLATEAU
Addis Ababa
ETHIOPIA
SOMALIA — Mogadishu

Nile R.
Blue Nile
White Nile

UGANDA — Kampala
L. Albert
L. Edward
KENYA — Nairobi
L. Victoria
Mt. Kilimanjaro 19,340 ft. (5,895 m)
Mombasa
SERENGETI PLAIN
Dodoma
Dar es Salaam
TANZANIA
L. Tanganyika
Zanzibar

Luanda
ANGOLA
Lubumbashi
ZAMBIA — Lusaka
L. Malawi
MALAWI — Lilongwe
Zambezi R.
Harare
ZIMBABWE
MOZAMBIQUE
COMORO ISLANDS

INDIAN OCEAN
Mozambique Channel

NAMIBIA — Windhoek
NAMIB DESERT
Okavango Basin
BOTSWANA — Gabarone
KALAHARI DESERT
Pretoria
Johannesburg
Maputo
SWAZILAND
Orange R.
LESOTHO
Durban
SOUTH AFRICA
Cape Town
Cape of Good Hope
Cape Agulhas

MADAGASCAR — Antananarivo

ATLANTIC OCEAN
Equator

Prime Meridian
Tropic of Cancer
Tropic of Capricorn

N
W — E
S

0   500   1000 Miles
0   500   1000 Kilometers

Elevation of Land

Meters		Feet
Over 3,050		Over 10,000
1,525 to 3,050		5,000 to 10,000
610 to 1,525		2,000 to 5,000
305 to 610		1,000 to 2,000
0 to 305		0 to 1,000
Below sea level		Below sea level

Depth of Water

0 to 153		0 to 500
Below 153		Below 500

©PH

Applying Geographic Themes

1. Location Africa south of the Sahara is divided into nearly fifty countries. Nearly one third of all the political units in the world are located here. Name the landlocked countries of Africa south of the Sahara.

2. Place Nearly all of the African continent is a plateau that is higher than 1,000 feet above sea level. Is the plateau higher in the eastern part of the continent or in the western part?

from the Gulf of Suez in the north to the country of Malawi (muh LAH wee) in the south. Scientists believe that the valley is a rift zone which forms when two tectonic plates are moving apart far beneath the earth's surface, much as they do in the undersea rift valleys. Along the rift valley, the land drops suddenly—sometimes far below sea level. The valley's sides may reach as high as 1 mile (1.6 km), and in some places the valley floor is more than 20 miles (32 km) wide. Eventually, in a few million years, ocean water will fill the entire length of the Great Rift Valley.

Because the Great Rift Valley is so deep, the climate and vegetation of the valley floor is often quite different from the plateau that surrounds it. A string of large lakes lies along the rift zone in this region. Look at the map on page 506 to locate these lakes.

Linking Elevation to Climate Eastern and southern Africa are more elevated than western Africa. Highlands in the east rise thousands of feet above the plateau. Although most of the continent lies within the tropics— between the Tropic of Cancer and the Tropic of Capricorn—temperatures at high elevations are comfortable. Look, as an example, at the climate map on page 510 for the city of Nairobi (ny RO bee), which is in the Kenyan highlands. Although Nairobi is located close to the Equator, its temperatures are cooler in summer than most cities in the southern United States.

The highlands also contain a few high, isolated volcanic peaks. The tallest of these is Mt. Kilimanjaro (KIL uh mahn JAHR o), which towers 19,340 feet (5,895 m) above the surrounding plain. Temperatures on these peaks— even at the Equator—are so low that the mountains are permanently snow-capped. As one writer has noted, " . . . one might stand on the equator and get frostbite, or for that matter freeze to death."

Africa's Steamy Rain Forest
Dense tangles of vegetation grow where sunlight is able to reach the rain forest's floor.

Linking Climate and Vegetation Regions

Many people mistakenly imagine all of Africa to be the steamy rain forest or the hot dry desert they have seen in movies. In truth, both these images are accurate. Africa's climate regions do not differ much in their average temperatures, which are all hot except at higher elevations. But the regions do differ in their rainfall amounts.

The map on page 509 shows that Africa straddles the Equator. If two travelers in Africa parted company at the Equator, one going north while the other traveled south, both would pass through the same types of climate and vegetation regions. They would each first travel through lush rain forests, next grasslands, and then through nearly lifeless deserts.

Rain Forests Much of equatorial Africa, from Guinea (GI nee) on the west coast to the Great Rift Valley in the

Africa's Grasslands

Africa's savannas are home to some of the continent's communities of wildlife. Many parks and wildlife reserves, such as the Serengeti Plains in Tanzania and Kenya's Nairobi National Park, are visited by many tourists.

east, is tropical rain forest. The rain forest has no seasons. It is always hot and very rainy, with as much as 100 inches (254 cm) of rain each year. Thunderstorms crash through the region almost every afternoon. One observer described them in this way:

> To be out in a really first-class rainstorm here is rather like walking around at the bottom of a lake; visibility is only a few feet, and it seems as if anyone who carelessly took a deep breath would drown.

One of the most striking features of the rain forest is its vegetation. Plants thrive in the climate of the rain forest. More than three thousand species of plants may be found in a single square-mile area. Towering trees allow little light to reach the forest floor, where thick, fast-growing vines, palms, ferns, and shrubs vie for space. And, as Winston Churchill observed, "Birds are as bright as butterflies; butterflies are as big as birds."

The African rain forest, much like the Brazilian rain forest, is shrinking every decade as land is cleared for farms, towns, and factories. Environmentalists around the world are working together to save the rain forests, as well as the thousands of species of plants and animals that dwell in them.

Grasslands Much of Africa north and south of the rain forest is savanna, which has tall grasses with scattered trees. Like the equatorial rain forest, the weather in the savanna is hot all year. But unlike the rain forest, the savanna has a dry season. Rain forest vegetation cannot survive dry periods, even when the total rainfall in a region is high. Moving farther away from the rain forest, at the northern and southern edges of Africa's grasslands, average annual rainfall drops and the dry season gets longer. These areas usually contain shorter grasses and fewer trees.

One journalist had this to note about rainfall in Africa:

savanna
The name for these grasslands is from the Spanish *sabana*. Interestingly, the Spanish word for bedsheet, an item that is large and flat, is also *sabana*.

There are three main rules of thumb to keep in mind about African rain. One is that it is seasonal. . . . The second rule is that the amount of rain varies . . . with the distance from the equator: the closer the wetter, and the farther the dryer. . . . The third rule is not to put much faith in the other two. They are riddled with so many exceptions, . . . that the only safe generalization is, "It all depends."

Applying Geographic Themes

Regions Wide bands of vegetation stretch across Africa. The colors on the diagram below correspond to the vegetation key above. How do prevailing winds influence Africa's precipitation and vegetation?

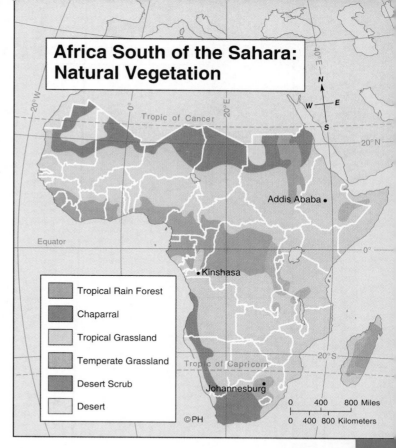

Africa South of the Sahara: Natural Vegetation

- Tropical Rain Forest
- Chaparral
- Tropical Grassland
- Temperate Grassland
- Desert Scrub
- Desert

0   400   800 Miles
0   400 800 Kilometers

©PH

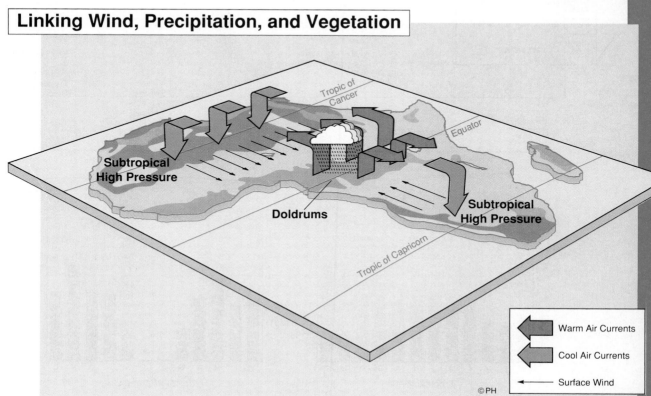

Linking Wind, Precipitation, and Vegetation

Subtropical High Pressure

Doldrums

Subtropical High Pressure

Warm Air Currents

Cool Air Currents

Surface Wind

©PH

Africa South of the Sahara: Climate Regions

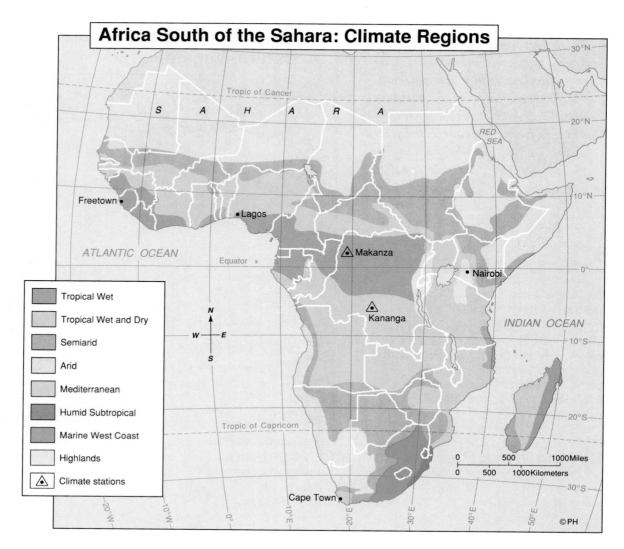

- Tropic of Cancer
- 30°N
- 20°N
- S A H A R A
- RED SEA
- 10°N
- Freetown
- Lagos
- ATLANTIC OCEAN
- Equator
- Makanza
- Nairobi
- 0°
- Kananga
- INDIAN OCEAN
- 10°S
- Tropic of Capricorn
- 20°S
- Cape Town
- 30°S

Legend:
- Tropical Wet
- Tropical Wet and Dry
- Semiarid
- Arid
- Mediterranean
- Humid Subtropical
- Marine West Coast
- Highlands
- △ Climate stations

Scale: 0 500 1000 Miles / 0 500 1000 Kilometers

©PH

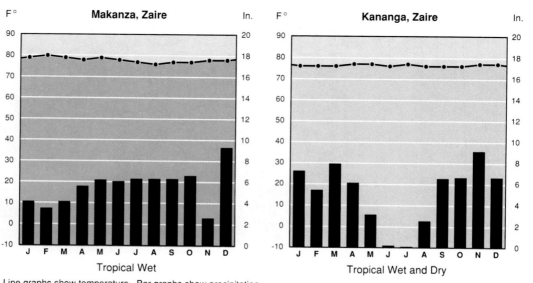

Makanza, Zaire — Tropical Wet

Kananga, Zaire — Tropical Wet and Dry

Line graphs show temperature. Bar graphs show precipitation.

The savanna is home to some of Africa's most precious natural resources. Lions, giraffes, rhinoceroses, elephants, and dozens of other fascinating animals live on the African savanna.

Arid Regions Chapter 22 describes the huge Sahara, which covers much of North Africa. South of the Sahara is a broad band of semiarid land known as the Sahel (suh HEL). *Sahel* is the Arab word for shore or border—in this case, the shore that edges Africa's Sahara. Conditions in the Sahel are very dry; the area receives an average of about 20 inches (51 cm) of rain per year. The dryness of the Sahel is made worse because of the high temperatures.

Far south of the Sahel, in southwestern Africa, two deserts spread across the land, the Namib (NAHM eeb) and the Kalahari (kal uh HAHR ee). Although smaller than the Sahara and not as hot, these two deserts are equally dry, barren, and forbidding.

Moderate Climate Regions On the southern tip of Africa, as on the northern coast, a small area enjoys mild climates. Locate this area on the climate map on page 510. Most of this mild climate region falls within the nation of South Africa. It also includes Swaziland (SWAH zee LAND) and the southern tip of Mozambique (MO zuhm BEEK). Partly because its mild climate is good for farming, this area is one of the most heavily populated in Africa. The next section discusses the people who live south of the Sahara and the challenges they face.

Applying Geographic Themes

Place In general, Africa is the world's driest continent. What type of climate does Nairobi, Kenya have? Compare the graphs below the map. Does Makanza or Kananga receive more rain each year?

People of the Kalahari
The people of the Kalahari Desert traditionally have been nomadic hunters and gatherers. Today, most migrate frequently to herd cattle.

■ SECTION **1** REVIEW ■

Developing Vocabulary
1. Define: cataract

Place Location
2. Name three lakes that are located in the Great Rift Valley.

Reviewing Main Ideas
3. What are the major landforms of Africa south of the Sahara?
4. What is the pattern of rainfall on the continent of Africa?

Critical Thinking
5. **Distinguishing Fact from Opinion** Is the following statement one of fact or opinion? Explain your answer. "Even if it delays modernization in equatorial Africa by fifty years, the preservation of the rain forests is well worth the cost."

Chapter 25, Section 1   **511**

✓ Skills Check

Composing an Essay

The model for composing an essay is essentially the same as that used in writing a paragraph. An essay needs to be clear, unified, and coherent. It should contain effective transitions so that it reads smoothly, and each point made should follow logically from what has gone before. Use the following steps to practice composing an essay.

1. **Determine the essay topic.** For an essay to be unified and coherent, it must address a single topic. Answer the following questions: (a) What is the one single topic that unifies the eight sentences below? (b) Which sentence would you select to be the first one in the essay?

2. **Arrange the supporting evidence in a logical order.** An essay should always follow a logical outline that makes the point of the essay easy to grasp. Answer the following questions: (a) How would you outline the three main divisions into which the sentences below fall? (b) In what order should the sentences be arranged to create an essay that has clarity, unity, and coherence?

3. **Use effective transitions for clarity.** Transitions knit together the various threads of an essay. Common transition words such as *but, however, although,* and *nevertheless* are used to signal a relationship among the points being made. Answer the following questions: (a) What transition words or phrases can you identify in the sentences below? (b) What do these transitions have in common?

Sentences

A. Sometimes scholars are able to determine the approximate date on which these natural disturbances occurred.

B. One method has been to visit the modern peoples and listen to the stories the tribal elders tell.

C. Almost none of the early peoples of sub-Saharan Africa developed a written language.

D. These stories have been passed down from generation to generation for thousands of years.

E. In this way, scholars can determine relative times and even pinpoint important dates.

F. Often, the stories describe earthquakes, floods, or eclipses of the sun or moon.

G. Thus, historians, archaeologists, and anthropologists have had to use other methods to discover the histories of the ancient kingdoms, empires, and people of Africa.

H. They then compare the legends with events found in the records of ancient Egypt, Greece, or Rome or in the writings of European or Arabic traders and adventurers.

2 Human Geography

Section Preview

Key Ideas

- Africa's rapidly growing population is becoming more urban.
- Agriculture is often unsuccessful because of insufficient rain and poor soil.
- Modernization in Africa is necessary but difficult to achieve.

Key Terms

leaching, diversify

Africa may be rich in natural resources, but its people are not wealthy. African governments believe that modernization is the key to wealth and to improving life for their people. But modernization, which took centuries to develop in Europe, is a difficult and expensive process. Governments are learning what works and what does not through a painful process of trial and error.

Population Patterns

Africa south of the Sahara is the world's third most populous region and the fastest-growing region in the world. Today, more than 460 million people live in the region; some countries are much more crowded than others. Nigeria has by far the largest population—more than 108 million people. A seasoned traveler wrote this about Africa:

There are . . . places [like Nigeria], with its hillsides . . . looking in the rainy season as if they had been freshly upholstered in bright

Fishing in Zaire

Zaire lies within the basin of the Zaire River and is laced with lakes and streams. The people of Zaire catch more than 100,000 tons of fish per year, almost entirely from inland waters.

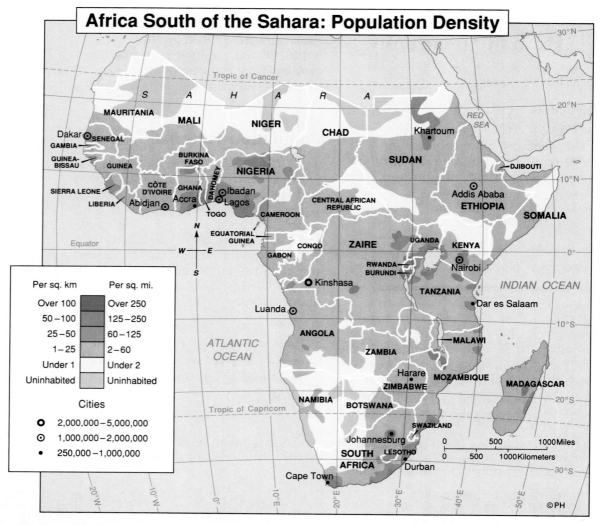

Africa South of the Sahara: Population Density

Per sq. km	Per sq. mi.
Over 100	Over 250
50 – 100	125 – 250
25 – 50	60 – 125
1 – 25	2 – 60
Under 1	Under 2
Uninhabited	Uninhabited

Cities

- 2,000,000 – 5,000,000
- 1,000,000 – 2,000,000
- 250,000 – 1,000,000

Applying Geographic Themes

Interaction Today, one of every eight people on earth lives in Africa. The region south of the Sahara is the fastest growing region in the world, yet food production per person has been falling in the last fifteen years. Where are the areas of highest population density?

WORD ORIGIN

Sahara
Sahara is from the Arab word *sahra,* which means "desert." Saying "Sahara Desert," therefore, is redundant.

green plush. But elsewhere, evidences of human life sometimes are hard to find . . . here and there, an occasional village. . . . It is all the more startling, therefore, to enter Africa's cities.

Nearly three quarters of the people in Africa south of the Sahara live in rural villages. As more farmers are

discouraged by droughts and famines, however, the urban population is increasing rapidly.

Most of the large cities south of the Sahara are located along the coast or near major rivers. Several cities have also developed around natural resources. Johannesburg, South Africa, for example, was founded after gold was discovered in the area in the late 1800s. Today, Johannesburg is an important industrial and commercial city. Look at the map on page 514 to locate other major cities and densely populated regions.

Overburdened Farmers

The majority of Africans are still, as they have been for centuries, subsistence farmers. Many problems make it difficult for Africa's farmers to grow enough food. As a result, African countries experience famine, malnutrition, and even starvation.

Drought In recent years, droughts—prolonged periods with a severe lack of rain—in the Sahel have caused crop failures and untold suffering. Lack of rainfall is turning more and more land into desert, where no crops can be grown. As land is lost to the advancing Sahara, starving families abandon their farms and flee to the cities. Insufficient rain is only one of the factors causing the desert to grow.

Poor Soil Many people are surprised to learn that most of Africa has poor soil, which makes farming difficult. Even soil that has supported dense rain forests is not productive when trees are cut down to create farmland. Tropical soil is more fragile than the soil of temperate regions. Constant heavy rains in a tropical wet climate cause leaching, the dissolving and washing away of nutrients contained in the soil. Without the constant fertilization provided by leaves decomposing on the forest floor, the soil is quickly worn out. Erosion is another serious problem. The leaves of the rain forest canopy normally protect the layer of topsoil. But the topsoil is quickly washed away when the trees are cleared.

Away from the rain forest, the soil is equally poor and the climate is dry. One visitor summed up the problem of farming in Africa as follows:

Taking it all together—the sultry swamps and backwaters and quagmires of the rainy parts, the cruelly unrewarding lands of permanent drought, along with the gullies and rocky protrusions and the leached out, burnt-out, worn-out soils—probably a third or more of [sub-Saharan] Africa has offered its inhabitants little more than a marginal existence.

Linking Land Use to Economic Problems

Like many developing countries all over the world, African countries are trying to modernize rapidly. To do so, they must break out of old trading patterns.

Forestry in the Congo
Tropical rain forest covers 60 percent of the Republic of the Congo where large quantities of lumber are harvested and exported.

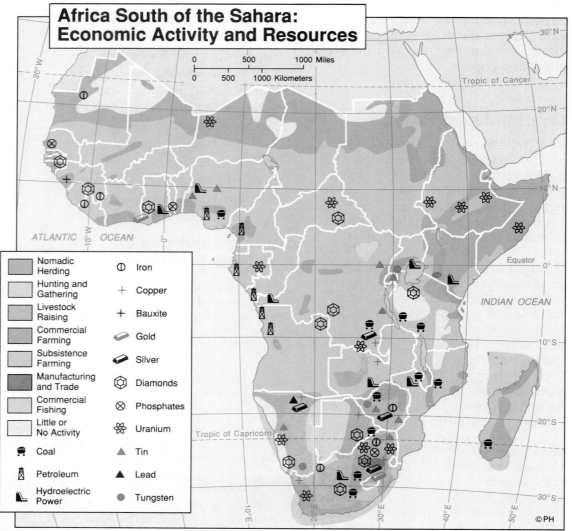

Africa South of the Sahara: Economic Activity and Resources

0   500   1000 Miles
0   500   1000 Kilometers

Legend:

- Nomadic Herding
- Hunting and Gathering
- Livestock Raising
- Commercial Farming
- Subsistence Farming
- Manufacturing and Trade
- Commercial Fishing
- Little or No Activity

- Coal
- Petroleum
- Hydroelectric Power

- Iron
- Copper
- Bauxite
- Gold
- Silver
- Diamonds
- Phosphates
- Uranium
- Tin
- Lead
- Tungsten

ATLANTIC OCEAN
INDIAN OCEAN
Tropic of Cancer
Tropic of Capricorn
Equator

©PH

Applying Geographic Themes

Regions Subsistence farming and herding require large areas of land to support small numbers of people. Improvements in farming methods are necessary to support Africa's increasing population. In which region of Africa south of the Sahara is economic activity most developed?

Limited Exports In colonial times, Africa was used as a source of raw materials for European industry. The products of African mines and forests were exported, as were cash crops such as rubber, silk, and cotton.

Today, African nations are still dependent on the export of raw materials. This dependence creates two problems. First, entire African economies are often based on just one or two major exports. So if the world price for these items drops, those economies crumble. Second, profits from selling raw materials are not as high as those from the sale of manufactured goods. For these reasons, African governments are trying to **diversify,** or increase the variety of, exports and to industrialize their economies further.

The Burden of Debt After winning their independence, many African governments borrowed large sums of money. They expected that the modernization projects on which they spent the money would greatly increase their countries' wealth. They planned to use some of this wealth to pay for more modernization and have plenty left over to pay back the original loans.

Many cases have not turned out that way. Modernization projects have not always brought the expected results in new revenues. In addition, Africans do not have control over many aspects of their economies. For example, when world oil prices skyrocket, as they did in 1974 and 1990, governments have to spend money for energy that they had planned to use for development.

Loan repayments are often a major part of a country's budget. Governments that find it hard to meet basic human needs must still spend the little money they do have to repay loans. They know that they must repay old loans to be eligible for the further aid they desperately need.

The Challenge of Keeping Africa's Population Healthy

Life expectancy in Africa is lower than in any other continent. In some countries it is only thirty-five years; and many children die before the age of five. One of the reasons for the low life expectancy is Africa's high rate of disease.

African governments recognize that disease takes a tremendous toll in human suffering. It also saps a country's economic strength by making its people less productive. One of the primary goals in Africa, therefore, is to improve public health. For example, health-care workers teach villagers to eat more balanced diets. The constant problems of drought and famine, however, often cause crops to fail before

they have a chance to grow. Without an adequate diet, a person's ability to fight disease is weak.

Ethnic Diversity

Many African nations are challenged by conflicts among ethnic groups within their borders. Africa south of the Sahara is a region of tremendous diversity. More than two thousand ethnic groups, speaking a total of more than eight hundred languages, live in the region.

Ethiopian Medical Clinic
Families in this village visit the medical clinic when they are ill and to learn how to stay healthy. These children are being weighed to check that they are nourished and growing.

DAILY LIFE

Banking on Diamonds

When Botswana (bahts WAH nuh) gained its independence in 1966, paved roads were few. Public education did not exist. The United Nations ranked Botswana as one of the world's least-developed countries.

Diamonds, discovered in the Kalahari Desert in 1974, changed all that almost overnight. The precious gems now provide more than three quarters of the country's revenue. Between 1980 and 1990, diamonds helped raise the average income from $290 to $1,690 a year.

Profits from the sale of diamonds have been used to provide health care, roads, schools like the one shown at right, and irrigation. In addition, enough money for thirty months' food is held in savings. This policy kept people from starving when Botswana suffered six consecutive years of drought beginning in 1981.

1. When did Botswana achieve its independence?
2. How have diamond profits been used?

The people of Africa belong to hundreds of small ethnic groups with different histories and traditions. During the period of European colonization of Africa, borders were set without regard to the interests or customs of the diverse ethnic groups. As African countries won their independence years later, however, the arbitrary borders imposed by the Europeans remained unchanged.

Most new independent countries contained many different ethnic groups. As a result, citizens of one country did not necessarily speak a common language or share a common religion. In many cases some groups had long histories of warfare and mistrust. These circumstances made it difficult for new governments to instill a sense of national pride and unity in their citizens.

Agricultural Development

Zimbabwe's agricultural program is a model for the rest of Africa. Today, the country's farms produce 1,000 times more than they did in 1980, resulting in a higher standard of living and longer life expectancies for Zimbabwe's people.

The Challenge of Political Instability

Economic problems and ethnic conflict contribute to yet another challenge faced by many African countries. In hard times dictators have seized power by force, and in some countries entire governments are frequently overthrown.

Political instability causes many serious and difficult problems for citizens. People suffer because of harsh laws, economic disruptions, and civil war. In addition, political instability interferes with a country's ability to modernize.

The countries of Africa present a varied picture. The next two chapters provide more information about the individual countries that make up the continent's regions.

SECTION **2** REVIEW

Developing Vocabulary
1. Define: **a.** leaching **b.** diversify

Place Location
2. What is the name of the large island country that lies east of the African mainland in the Indian Ocean?

Reviewing Main Ideas
3. Name two problems faced by African farmers.
4. Where are most of Africa's large cities located?

Critical Thinking
5. **Making Comparisons** Compare the ethnic conflicts in the United States with those experienced by Africans.

Africa South of the Sahara

*Large–scale migrations of people and European
imperialism have had a great impact on Africa.*

An important part of the history of Africa south of the Sahara has been the history of large-scale movements of people within Africa, out of Africa, and to Africa. The three movements of people described below were each very different from one another. All of them had an important impact on the characteristics of the continent today.

The Bantu Migrations

The Bantu people came originally from western Africa, near the present-day countries of Nigeria and Cameroon. Over a period of about two thousand years, they migrated south and east, eventually populating much of the continent.

Pioneers of the Rain Forest Around 500 B.C. knowledge of ironworking reached Africa from southwest Asia. At about the same time, the population of western Africa expanded rapidly, forcing the people of the region to seek new lands to live on and to farm. To the west was the Atlantic Ocean. The plains to the north and east were already occupied. The only direction in which expansion was possible was southeast, toward the rain forest.

The people who ventured into the rain forest left the world that was familiar to them and set off to make a life in an unknown and almost uninhabited wilderness. These pioneers used their knowledge of ironworking to make tools with which to clear the rain forest, hunt, and fight off enemies and predators.

Later generations of Bantus fanned out from the rain forest in all directions. By A.D. 500 they had populated a vast belt of central African lands between the Atlantic and Indian oceans. After another thousand years, they had settled as far south as the Cape of Good Hope.

Cultural Exchange In the course of their migrations, the Bantu came into contact with many different people who helped them adapt to new environments. For example, from people in eastern Africa they learned to grow root crops, such as yams and taro. Unlike the grains the Bantu had previously grown, root crops flourished in areas with heavy rainfall.

At the same time, the Bantu introduced other people to their knowledge of ironworking, farming, and fishing. They also taught other groups their language and religion. Many of the people with whom they came into contact adopted Bantu culture.

Bantu migrations continued for two thousand years. Many groups of migrants had little or no contact with other Bantu groups after they left their homeland. Over time, and as a result of adaptation to new environments, these groups developed differently from one another. For example, although their ancestors spoke a common language, Bantu people today speak as many as four hundred different languages. Although these languages are all related, speakers of one language cannot understand the others.

Bantu people differ from group to group in many other respects, as well. They grow different crops, wear different clothing, and build different houses. The adaptability that made the Bantu such hardy pioneers also served, over time, to make them strangers to one another.

It was not until European philologists—students of language—began to study the languages of Africa that the connections among the different Bantu people were understood. Even the term Bantu, referring to the whole family of languages, was coined by philologists and not by the Bantu themselves. During their study the language experts noticed that almost all of the different groups in southern Africa used the same word, *abantu,* to mean "the people." Since then, the forgotten history of the Bantu migrations has been reconstructed.

Today Bantu-speaking people dominate a large section of Africa south of the Sahara, reaching north from the nation of South Africa to Cameroon in the west and to Kenya in the east.

Slavery: A Forced Migration

A very different kind of large-scale movement of people was the slave trade. From the 1500s to the 1800s, millions of Africans were brought against their will to work as slaves in the Americas. The slave trade affected African ways of life in many profound ways.

The Slave Trade The Europeans who settled the Americas in the 1500s faced a severe labor shortage. They needed large numbers of workers to work on their plantations, where they grew cash crops such as sugar and cotton. At first, the settlers forced Native Americans to work the plantations. But so

Enslaved Africans
Thousands of Africans died during their voyage to the Americas in the crowded and disease-ridden conditions of the slave ships.

many Native Americans had been killed or had died from disease that the settlers were later forced to look for other sources of labor. In Africa they found the slaves they wanted.

Slavery had existed in much of the world, including Africa, since ancient times. Prisoners of war and conquered people were commonly enslaved by their enemies. In addition, East African leaders had for centuries sold their people as slaves to Arab traders who carried them to the Middle East.

Some Africans, therefore, were bribed into selling other Africans as slaves by promises of such prized goods as cloth, guns, jewelry, and whiskey. Historians estimate that between 1500 and 1870 almost ten million Africans were forced to journey to the Americas. Many of them, however, did not survive the crowded, disease-ridden trip across the Atlantic.

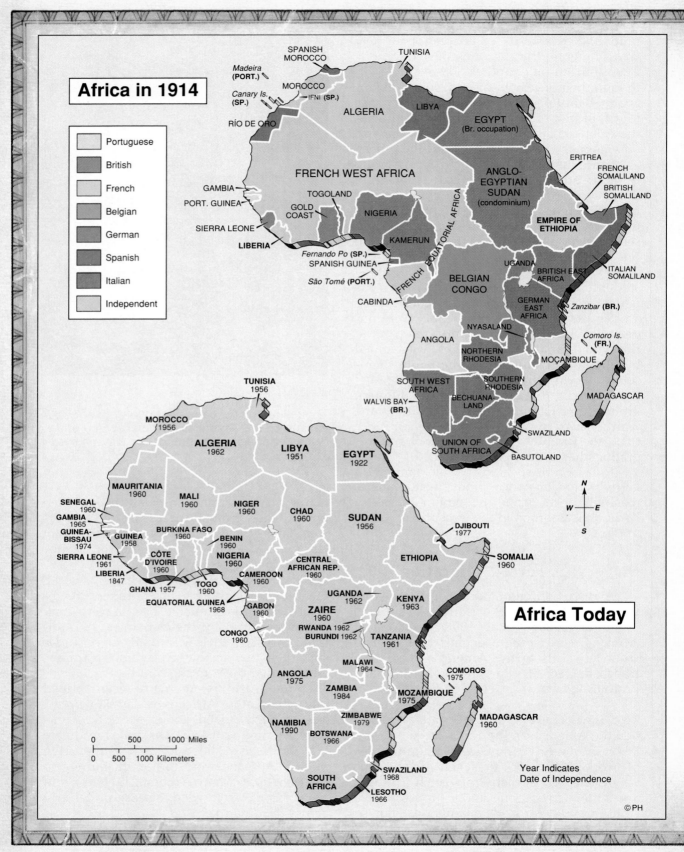

Africa in 1914

Legend:
- Portuguese
- British
- French
- Belgian
- German
- Spanish
- Italian
- Independent

Madeira (PORT.)
SPANISH MOROCCO
TUNISIA
Canary Is. (SP.)
MOROCCO
IFNI (SP.)
ALGERIA
LIBYA
EGYPT (Br. occupation)
RÍO DE ORO
ERITREA
FRENCH SOMALILAND
FRENCH WEST AFRICA
ANGLO-EGYPTIAN SUDAN (condominium)
BRITISH SOMALILAND
GAMBIA
TOGOLAND
PORT. GUINEA
GOLD COAST
NIGERIA
EMPIRE OF ETHIOPIA
SIERRA LEONE
LIBERIA
KAMERUN
Fernando Po (SP.)
SPANISH GUINEA
São Tomé (PORT.)
CABINDA
FRENCH EQUATORIAL AFRICA
BELGIAN CONGO
UGANDA
BRITISH EAST AFRICA
ITALIAN SOMALILAND
GERMAN EAST AFRICA
Zanzibar (BR.)
NYASALAND
Comoro Is. (FR.)
ANGOLA
NORTHERN RHODESIA
MOÇAMBIQUE
SOUTH WEST AFRICA
SOUTHERN RHODESIA
MADAGASCAR
WALVIS BAY (BR.)
BECHUANALAND
SWAZILAND
UNION OF SOUTH AFRICA
BASUTOLAND

Africa Today

TUNISIA 1956
MOROCCO 1956
ALGERIA 1962
LIBYA 1951
EGYPT 1922
MAURITANIA 1960
MALI 1960
NIGER 1960
CHAD 1960
SUDAN 1956
DJIBOUTI 1977
SENEGAL 1960
GAMBIA 1965
GUINEA-BISSAU 1974
GUINEA 1958
BURKINA FASO 1960
BENIN 1960
NIGERIA 1960
CENTRAL AFRICAN REP. 1960
ETHIOPIA
SOMALIA 1960
SIERRA LEONE 1961
CÔTE D'IVOIRE 1960
LIBERIA 1847
GHANA 1957
TOGO 1960
CAMEROON 1960
UGANDA 1962
KENYA 1963
EQUATORIAL GUINEA 1968
GABON 1960
ZAIRE 1960
RWANDA 1962
BURUNDI 1962
TANZANIA 1961
CONGO 1960
MALAWI 1964
COMOROS 1975
ANGOLA 1975
ZAMBIA 1984
MOZAMBIQUE 1975
MADAGASCAR 1960
NAMIBIA 1990
ZIMBABWE 1979
BOTSWANA 1966
SWAZILAND 1968
SOUTH AFRICA
LESOTHO 1966

Year Indicates Date of Independence

0 500 1000 Miles
0 500 1000 Kilometers

N
W — E
S

©PH

The Slave Wars Because selling slaves was so profitable, African rulers launched wars for the purpose of capturing prisoners to sell. Groups that had lived in peace for hundreds of years raided one another's villages to capture as many prisoners as they could. As many as forty million Africans—four times as many as were sent to the Americas—died in these slave wars.

African rulers tried to stop the terrible slave trade, but it was too late—their countries were too dependent on European trade. European merchants refused to sell their goods unless the Africans agreed to take part in the slave trade. The rulers reluctantly did so. The Africans' growing dependence on European trade was a step toward greater European dominance.

European Rule

Beginning in the late 1800s European countries seized control of African land for their own political and economic purposes. The era of colonialism that followed changed the face of Africa.

In the 1800s, European nations were industrializing at a very rapid pace. They needed raw materials for their factories and new markets in which to sell the goods they manufactured. One of the places where such raw materials and markets existed was Africa. Countries often competed fiercely with one another to take over the richest areas of Africa. In 1884, in order to avert a disastrous war, the German government called for a meeting of the colonial powers—the Berlin Conference.

No Africans were invited to take part in the conference, where the fate of Africa was discussed and determined. Delegates from fourteen European countries and from the United States agreed on the rules that would decide how Africa was to be carved up. They also agreed that they would join forces to put

down any resistance by African people. By 1914 all of Africa, except for Liberia and Ethiopia, was under European control.

The Legacy of Colonial Rule

European colonial powers ruled their colonies in different ways. Most sent their own officials to fill all government posts, while Africans were not given any role in the government. Even in colonies where African officials were allowed to handle day-to-day affairs in traditional ways, they had little real power. As a result of these policies, few Africans had the knowledge or administrative experience to take over the complex business of governing when their countries gained independence after World War II.

The Europeans wanted their colonies to be profitable, but they were not interested in Africa's long-term development. Therefore, neither economic nor educational development was encouraged or even permitted. Today, African governments continue to work to reverse these patterns. Over the past twenty-five years, educational and employment opportunities in the region have expanded greatly.

Although the schools, ports, roads, and railroads that were built during colonial rule are still used, Africans regard them as developments that came at great cost—the cost of freedom. Most parts of Africa gained their independence by the early 1960s, although a few countries remained under European rule longer. In some parts of Africa, the struggle for freedom continues even today.

Applying Geographic Themes

Regions Which two countries held the largest African empires? What African nation became independent most recently?

TAKING ANOTHER LOOK

1. What were the Bantu migrations?
2. How did the slave trade affect life in Africa?
3. Name three ways in which colonial rule affected Africans.

Critical Thinking

4. **Demonstrating Reasoned Judgment** Explain why you think that Africa would or would not be better off today if the continent had never been colonized.

25

REVIEW

Section Summaries

SECTION 1 Land, Climate, and Vegetation
Most of Africa lies on a tilted plateau that rises sharply from a narrow coastal plain. In the northeast, the Great Rift Valley splits the landscape for hundreds of miles. In most of Africa, temperatures at low elevations are hot throughout the year. The major difference between the climate zones is in the amount of rainfall each receives. Generally, the greater the distance a climate zone is from the Equator, to the north or the south, the less rainfall it receives. Areas along the Equator are largely rain forest. The savannas have a rainy and a dry season. The Sahel is semiarid and in danger of becoming desert.

SECTION 2 Human Geography Africans are eager to modernize, but they face many obstacles. The soil is mostly poor and rainfall is scarce. Drought and famine are constant problems for farmers. Urban populations are growing as they look for other ways to survive. Many countries are heavily in debt and unable to repay loans. Disease, political instability, and ethnic conflicts are other obstacles to modernization in African nations.

Vocabulary Development

Match the definitions with the terms below.

1. waterfall
2. increase in type and variety
3. the dissolving and washing away of nutrients from the soil

a. diversify
b. leaching
c. cataract

Main Ideas

1. What are the four major landscapes of Africa south of the Sahara?
2. Why are many of Africa's rivers difficult to navigate?
3. Why is political instability an obstacle to modernization?
4. Where are most of Africa's large lakes?
5. Which regions in Africa south of the Sahara are most heavily populated?
6. Explain why mountains are snowcapped, even at the Equator.
7. Name one economic effect of European colonization.
8. Why did some Africans participate in the slave trade?
9. What patterns of vegetation are found in Africa north and south of the Equator?
10. Where is the Sahel?
11. Why are African countries trying to diversify their economies?
12. How do borders established by Europeans continue to influence African countries?

Critical Thinking

1. **Demonstrating Reasoned Judgment** What problems would a country have to overcome in its first ten years of independence from colonial rule?
2. **Determining Relevance** What do you think is the biggest obstacle to modernization that Africa faces? Explain your reasons.
3. **Distinguishing False from Accurate Images** The rain forest is often perceived as an extremely fertile environment. Explain why this view is or is not accurate.
4. **Drawing Inferences** What information would lead you to believe that the Bantu were technologically advanced people?

Practicing Skills

1. **Composing an Essay** Use library resources to research one of the following African ethnic groups. Then using the steps shown on page 512, write a brief essay on that ethnic group. Be sure to use an outline to arrange your points logically and use effective transitions.

    Asante   Kikuyu
    Fulani   Swazi

2. **Reading Circle Graphs** Draw a circle graph showing the makeup of Rwanda's exports using the following data. Be sure to label each piece of the circle graph and to give the graph a title. (Refer back to the Skills Check, "Reading Circle Graphs," on page 473. Use the circle graph on that page as a guide for preparing your graph.)

    Coffee   55%
    Tea   18%
    Tin   8%
    Other   19%

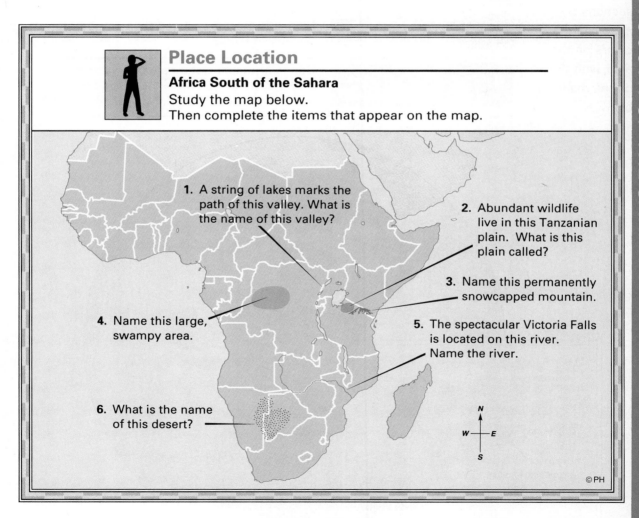

Place Location

Africa South of the Sahara
Study the map below.
Then complete the items that appear on the map.

1. A string of lakes marks the path of this valley. What is the name of this valley?

2. Abundant wildlife live in this Tanzanian plain. What is this plain called?

3. Name this permanently snowcapped mountain.

4. Name this large, swampy area.

5. The spectacular Victoria Falls is located on this river. Name the river.

6. What is the name of this desert?

© PH

West and Central Africa

Chapter Preview

Locate the countries covered in each of these sections by matching the colors on the right with those on the map below.

Sections		Did You Know?
1	**THE SAHEL: THE REGION, THE ENVIRONMENT**	*Sahel* is the Arab word for ''shore'' or ''border''—in this case, the shore that edges Africa's Sahara.
2	**THE COASTAL COUNTRIES**	Liberia was founded in 1822 by freed American slaves. The country's motto is ''The Love of Liberty Brought Us Here.''
3	**NIGERIA: AFRICA'S HOPE**	Nigeria is more populated than any other country in Africa.
4	**CENTRAL AFRICA**	Congo's transportation system, which runs 320 miles from Brazzaville to the coastal city of Pointe-Noire, is one of the longest in Africa.

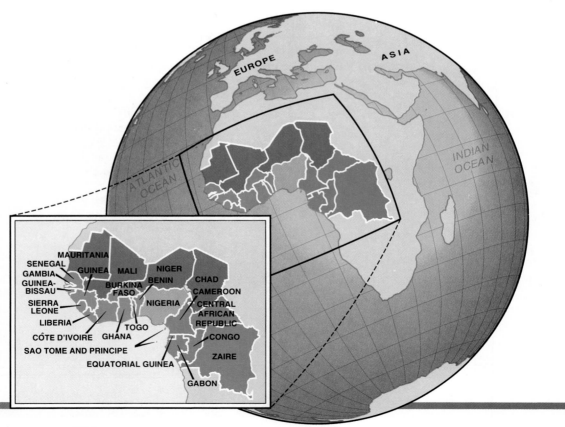

The Challenge of Life in the Sahel
The threat of drought and famine is ever present in the sparsely populated Sahel. Intermittent droughts have plagued the region since 1986.

1 The Sahel: The Region, the Environment

Section Preview

Key Ideas

- Many empires have flourished in the Sahel, a region defined by location, climate, and vegetation.
- The interaction between people and the environment in the Sahel has had negative consequences.
- The people of the Sahel are determined to withstand the harsh environment, develop their natural resources, and preserve their culture.

Key Terms

shifting agriculture, forage, deforestation, desertification, refugee, landlocked, inland delta

The Sahel is the name of the region in Africa stretching in a broad band from the Atlantic coast in the west to the Red Sea in the east. The Sahel is the borderland between the Sahara in the north and the tropical rain forests of coastal West Africa and Central Africa.

Because the Sahel is a transitional zone between desert and forest, it is in some ways similar to both these regions. For the most part it is savanna. Since there are as many as thirty different kinds of savanna in Africa, however, simply using this term to describe the Sahel does not do justice to the tremendous variety of landscapes within the region.

Here is one journalist's description of the many landscapes that exist across the region:

The forest thins out until it turns into . . . savanna—undulating grasslands dotted with individual

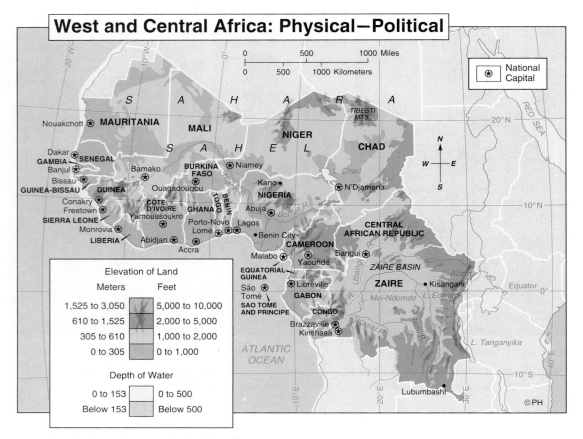

West and Central Africa: Physical—Political

Elevation of Land

Meters	Feet
1,525 to 3,050	5,000 to 10,000
610 to 1,525	2,000 to 5,000
305 to 610	1,000 to 2,000
0 to 305	0 to 1,000

Depth of Water

0 to 153	0 to 500
Below 153	Below 500

National Capital

Applying Geographic Themes

1. Location Which important river basin is centered on the Equator? Which countries border Lake Chad?

2. Movement What river begins in the Guinea Highlands, flows through several west African countries, and enters the Atlantic Ocean in Nigeria?

trees and occasional groves. . . . But this in turn shades off into sparser country, with scrubby trees and bushes and mottled patches of bare earth, and then into desert lands speckled only with thorn bushes and other tough growths and scarred by gullies and dry sand rivers; and at the final extreme, rocky, sandy, barren desert.

Many people in the West think of the Sahel as an arid region. They believe, as well, that its history and culture are as barren as its climate. In fact, the Sahel was for centuries a busy crossroads and a meeting point for different cultures. Today the area contains more than a dozen independent countries, each with its own vision of past, present, and future.

The Past: Power and Learning

One of the many surprising facts about West Africa is that the Sahara was not always a desert. Rock paintings found there show that as recently as ten thousand years ago people hunted hippopotamus in the region's rivers and chased buffalo on its wide, grassy plains. Over time, however, the climate grew drier. Some people of the Sahara moved north toward the Mediterranean Sea; others moved south toward the Sahel. Eventually, vast stretches of desert developed and became a barrier between them.

Trade Links and Empires   The two groups never entirely lost touch with each other, however. Over the sea of sand from the north came merchants bringing salt to trade. They sought ivory, slaves, and—most important of all—the gold that patient miners panned from the two great rivers of the Sahel region, the Senegal and the Niger. Because of its central location, the trade routes across the Sahel became a bridge between the Mediterranean coast and the rest of Africa.

The chiefs of the people of the Sahel found that they could grow wealthy by taxing the traders passing through their kingdoms. By A.D. 400 a great kingdom had emerged in the Sahel, known as Ghana, the land of gold. By the year 800 its capital, Koumbi-Saleh, was a city of twenty thousand people. An Arab traveler described the ruler of Ghana in these words:

When he gives audience to his people . . . he sits in a pavilion around which stand his horses . . . in cloth of gold; behind him stand ten pages holding shields and gold-mounted swords; and on his right hand are the sons of the princes of his empire, splendidly clad and with gold plaited into their hair.

A Center of Learning   Ghana was defeated by conquerors from the desert in 1076. But new empires soon took its place in the Sahel.

Mali (MAH lee) was just one of these. At its height, in the early 1300s, Mali was the second-largest empire in the world. Its most famous emperor, Mansa Musa, journeyed to Mecca and brought back Arab doctors and teachers. Under his rule the capital of the empire, Tombouctou (TOM book TOO), became an important university city, rich in the knowledge and arts of Islam.

When Mali fell, still another empire, the Songhai Empire, became dominant in the region. Its rulers revived the learning of Tombouctou. One report of its universities at the time stated, "In Tombouctou there are numerous judges, doctors, and clerics, all receiving good salaries from the king. He pays great respect to men of learning." Confident in its armies and in its wealth of knowledge, Songhai remained a great power in the Sahel until four hundred years ago.

The Present: War with the Desert

Today the Sahel is divided not into empires but independent countries. Mauritania (mawr i TAYN ee uh), Mali —named after the ancient kingdom— Niger (NY jer), Burkina Faso (boor KEE nuh FAH so), and Chad are the five northernmost countries of the Sahel. To their south lie eleven countries that fit like jigsaw pieces around the shore of the Atlantic Ocean. Although these nations are discussed mainly in the following section on coastal countries, most of them have at least some savanna in the interior. They are, therefore, linked with the Sahel region. Indeed, one of them is named Ghana, after the ancient kingdom that lay deep in the Sahel.

Making a Living   Many people of the Sahel support themselves by farming. Two environmental factors determine how they farm the region: the dry climate and the poor soil.

Farmers cope with the dry climate by growing crops during the short rainy season. They meet the challenge

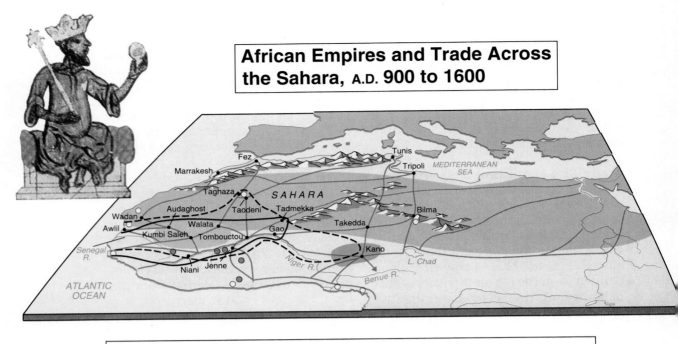

African Empires and Trade Across the Sahara, A.D. 900 to 1600

Empire Boundaries:

- Ghana
- Mali
- Songhai
- Hausa States

Resources:

- Gold
- Salt
- Trade Routes
- Desert

©PH

Applying Geographic Themes

Interaction   Mansa Musa (inset), the Muslim king of the ancient kingdom of Mali, made Tombouctou a great center of Islamic learning. According to the map, which regions were joined by trade routes across the Sahara?

of poor soil by using **shifting agriculture.** Under this system a site is prepared and used to grow crops for a year or two. Having thus exhausted the soil, the farmer moves on to a new field. The first field is then abandoned. The farmer moves on to clear a new area of forest and does not return to the previous farmed land.

Two grains, millet and sorghum (SAWR guhm), are the vital crops that keep Sahel farmers alive. For cash they grow peanuts, which some sell to the distant cities of the coast for use there or for export.

Instead of farming, many people in the Sahel herd camels, cattle, and sheep. At first glance the savannas of the Sahel might seem ideally suited to herding. They have low grasses and

other edible plants, as well as trees such as the baobab (BAY o BAHB) and acacia (uh KAY shuh) with leaves that provide **forage,** or food for grazing animals. Unfortunately, however, such animals can destroy plants and trees when they are crowded too closely into the same range. This overgrazing has had a grim impact on the environment.

The Encroaching Desert   Overgrazing harms the Sahel by destroying the plants that hold the sandy soil in place. The endless search for firewood to use for cooking and the tremendous demand for charcoal by a growing urban population has further damaged the environment by stripping the land of its trees. This is a process called

deforestation. When there is a drought, vast areas of the Sahel may suffer a loss of all vegetation, called **desertification**. In effect, the savanna turns to desert. Desertification is very difficult to reverse.

A severe drought struck the Sahel in 1968 and lasted into the early 1970s. Farming in much of the region simply ended, and more than half the cattle were wiped out. A quarter of a million people died of starvation. The desert conditions of the Sahara spread south, claiming up to 60 miles (96 km) of savanna in some places.

Then, in the early 1980s, the worst drought in 150 years struck the region. Affecting almost half of Africa, this drought caused the desert to spread over farms and pastures at the rate of 27,000 square miles (70,000 sq km) per year. It was as if the United States were to lose an area of farmland larger than the state of West Virginia year after year. Most of the desertification took place in the Sahel. To make matters worse, brush fires raged through the parched savannas.

Worldwide Connections In addition to adding to desertification, the drought exposed 150 million Africans to the risk of starvation. The Sahel nations of Mauritania and Mali and the coastal nations of Senegal (SEHN uh GAHL) and Ghana were particularly hard hit by the drought. Half the land area of Mauritania and Mali already lay within the Sahara even before the drought. Some experts wondered if these would be the first modern nations to be completely desertified. As the desert spread, people throughout the Sahel fled to the cities, turning what had been modest urban clusters into huge refugee camps. A **refugee** is a person who flees his or her home to escape danger or unfair treatment.

During the famine the world rallied to the aid of Mauritania, Mali, and the other nations of the Sahel. Most of the Sahel countries are **landlocked**, that is, cut off from the sea. Their transportation links with the coast are so poor that it was hard for relief supplies to get through. In Mauritania the main road through the country could be kept open only by shoveling the sand off it every day.

When the drought finally began to end in 1985, the people of Mali and other nations throughout the Sahel and the rest of Africa celebrated. By 1989, Mali was once again producing good crops. In other areas of Africa, however, the drought soon returned.

According to some experts, it may be thousands of years before the desertified areas can be reclaimed. Some believe that once the land lost its cover of vegetation, reflected solar energy kept rain from falling.

Protecting the Future

The nations of the Sahel are directing their energies toward three goals: withstanding the harsh environment; developing natural resources; and making the most of their current human resources and culture.

Holding On The Sahel countries need continuing foreign aid. Food, medicine, and technical help are always in demand. Some sixty agencies around the world are working against desertification in Niger alone. In that country's Majia Valley, foreign aid has helped to plant hundreds of miles of trees as a windbreak for thousands of acres. Now the soil does not blow away during the dry season, so plants have a better chance of growing when rain does fall.

Developing Resources One water resource has helped the people dwelling in the Sahel for thousands of years —the region's rivers. The Senegal and Niger rivers and their tributaries provide both transportation and water for irrigation.

WORD ORIGIN

refugee
The word *refugee* comes from two Latin words, *fugere*, meaning "to flee," and *re*, which means "back." A refugee is one who flees from danger to seek safety elsewhere.

On the Banks of the Niger, Mali
As is their custom, Fulani families assemble on the banks of the Niger before moving on in search of food for their animal herds.

have reserves of iron ore. Bauxite, the ore from which aluminum is made, is an important resource in Mali, just as Mauritania's huge reserves of copper are in that country. Niger has one of the world's most valuable deposits of uranium.

Human Resources The descendants of the people of the Sahel's ancient empires still dwell in the region today. One of the larger groups is the Mossi of Burkina Faso. In Niger and in other countries of the Sahel, many of the Fulani (FOO lah nee) live as herders, and the Hausa (HOW suh) are famous as traders. In Mali, the Malinke (MAH lin kay) and the Songhai peoples are known for their music and dance. The Mandingo (man DIN go) craft magnificent jewelry and the Bambara carve graceful objects of wood. The Moors of Mauritania have preserved their traditional Islamic culture for centuries.

Although the Niger's source is located in the mountains of the nation of Guinea only 150 miles (240 km) from the Atlantic Ocean, the river flows inland for 2,600 miles (4,180 km) before reaching the ocean. On its journey it brings water to countless villages within the Sahel. In Mali the Niger expands into an inland delta, an area of lakes, creeks, and swamps away from the ocean. Here people grow rice, wheat, corn, and vegetables.

The Sahel countries also possess mineral resources that can be sold to buy food. Mauritania and Mali both

SECTION 1 REVIEW

Developing Vocabulary
1. Define: **a.** shifting agriculture **b.** forage **c.** deforestation **d.** desertification **e.** refugee **f.** landlocked **g.** inland delta

Place Location
2. Name the capitals of Mali and Niger.

Reviewing Main Ideas
3. Describe two ways in which people have contributed to the desertification of the Sahel.
4. Why is the Niger River important to the nations of the Sahel?

Critical Thinking
5. **Perceiving Cause-Effect Relationships** Describe some of the ways that the people of the Sahel are directing their energies toward protecting the future of the region.

2 The Coastal Countries

Section Preview

Key Ideas

- The coastal nations of West Africa have long taken advantage of their location to trade with foreign nations.
- West Africans are acting to improve their economies on the local level.
- Women in West Africa contribute to the economy by running the local markets.

Key Terms

coup, ancestor worship, animism

Besides the four Sahel nations, West Africa contains twelve other countries. One, Cape Verde, is a small island nation. The others ring the coastline of West Africa, beginning in the west with Senegal and continuing along the Atlantic coast to Nigeria.

Location Leads to Trade

Because of their location, the coastal countries of West Africa have two advantages over those of the Sahel. First, they have a wetter climate. Adequate rainfall allows successful farming and the growth of valuable trees. Second, they have access to the sea. Freetown, in Sierra Leone, has the third-largest harbor in the world but it does not rank as a leading port.

Resources for the World The coast of West Africa attracted European traders from the 1500s. They came for gold, slaves, ivory, and palm oil. This trade made trade across the Sahara less important. The coastal kingdoms fought each other for slaves and for control of the new foreign trade.

Today the nations of the West African coast export only a few products and raw materials. Senegal, Gambia (GAM bee uh), and Guinea-Bissau (GI nee bee SOW) export peanuts. Côte d'Ivoire (KOT dee VWAHR)—also called Ivory Coast—Ghana, Sierra Leone, and other nations largely depend upon the export of cocoa beans. Liberia exports iron ore.

Unequal Trade The economies of the West African countries suffer in part because their exports total less than their imports in value. Also, the African countries are heavily in debt. Africa as a whole needs 9 billion dollars every year just to pay the interest on its debts.

Governments: Too Much and Not Enough

European colonial powers ruled most of Africa until the 1960s. When the African countries gained their independence, their economies were often in very weak condition. Few new governments in Africa have been able to overcome or recover from these economic burdens.

Too Many Governments When governments are weak, the army often steps in and takes over. Sometimes, different army groups fight for power.

In Benin five coups—sudden political takeovers—took place from 1963 to 1972. Lieutenant Colonel Ahmed Mathieu Kerekou (AKH muhd mat YUH ker uh KOO), stayed in control from 1972 until 1990. In that year, with Benin's economy failing, Kerekou was faced with strikes and unrest. He then called for a new constitution that allowed others to share power.

The year 1990 brought some other signs that one-man rule in West Africa was changing. When world cocoa prices fell, the government in Côte d'Ivoire introduced harsh taxes to provide an alternative source of income to help

WORD ORIGIN

Sierra Leone
Sierra Leone was named "Mountains of the Lions" by Portuguese sailors who thought they saw towering mountains and heard lions roaring. They probably saw clouds and heard the roar of the surf.

The Tonight Show

Every night they walk miles along dirt paths to Zak's Villa at the edge of Tamale, Ghana. For an admission price of about twelve cents, hundreds of villagers will be able to watch television outside.

Television is very popular here. The government of Ghana is taking advantage of this by insisting that popular shows be mixed with educational programs. A typical broadcast may include advice on how to keep grasshoppers from infesting one's crops, followed by a short film entitled *The Man Who Never Felt Like Going to the Capital.* This film tries to discourage villagers, like those shown at right, from abandoning their farms and moving to nearby cities—a problem plaguing many African nations. Explains a local television reporter: "For a country like Ghana to develop, television is as important as the school."

1. What is television like in Ghana?
2. What do you think the reporter means by the phrase "a country like Ghana"?

pay off its huge debts. Citizens protested. The president was forced to repeal the taxes and give up some power to other political parties after ruling single-handedly for thirty years.

Too Little Governing West Africans have learned that their governments can do little to improve economic conditions. One writer described the consequences of this realization:

There are signs that some Africans already are taking matters into their own hands. As rural people have become disillusioned [disappointed] with outsiders and with their own governments, millions of them have begun grass-roots efforts . . . to organize local resources.

534   Chapter 26, Section 2

The key to this new economic approach is its grass-roots beginnings. Grass roots means that the effort begins with people, not governments. And in the region of West Africa, increasingly, it is women who make grass-roots efforts work.

Women's Work

Many of the women of West African countries, just as in the rest of the continent, are front-line troops in a hard-fought battle: they grow crops in the war against hunger. They also run an important part of the economy—the markets where food is bought and sold. As Africa modernizes, women are constantly expanding their traditional roles and are becoming owners of small businesses.

Children are also valuable workers in West African countries, helping to grow and harvest crops. Children are important for another reason, as can be seen in the case of the Asante (ah SAHN tay), a group of people who live in southern Ghana.

The Asante, like many African peoples, believe that if their children continue to respect and honor them after death, they will live on in the spirit world. An African chief once described his people as "a vast family, of which many are dead, few are living, and countless members are unborn." This belief in the spirits of the dead is called ancestor worship.

Ancestor worship is one aspect of the Asante religion; another is animism. According to this belief, ordinary things—the sky, rivers, trees—all contain gods or spirits.

In Africa, as in other places, social custom, religious beliefs, and economic conditions sometimes translate into large families and a fast-growing population. The birth rate in Ghana is about 50 percent greater than the average for the rest of the world. The population of Africa is growing faster than anywhere else on earth.

An Asante Chief, Ghana
The splendor of the West African past is shown at the court of an Asante chief. The Asante are known for their gold designs and skillful weavings.

SECTION 2 REVIEW

Developing Vocabulary
1. Define: **a.** coup **b.** ancestor worship **c.** animism

Place Location
2. How does Gambia's location affect jobs in that country?

Reviewing Main Ideas
3. How did the fall of cocoa bean prices affect the government of Côte d'Ivoire in 1990?
4. What new economic approach is finding success in West Africa?

Critical Thinking
5. **Making Comparisons** Compare women's roles in Africa with women's roles in the United States.

✔ Skills Check

☐ Social Studies
☐ Map and Globe
☐ Reading and Writing
☑ Critical Thinking

Perceiving Cause-Effect Relationships

Understanding the relationship between cause and effect is basic to an understanding of geography. Use the following steps to practice identifying statements that tell about cause and effect.

1. **Identify the two parts of a cause-effect relationship.** A cause is an event or an action that brings about an effect. Authors usually indicate a cause-effect relationship by using words such as *so, thus, because,* and *as a result.* Read statements A through D below and answer the following questions: (a) Which statements are cause-effect statements? (b) Identify the cause and the effect in each cause-effect statement. (c) Which word or words signal the cause-effect relationship?

2. **Remember that an event can have more than one cause and more than one effect.** Several causes can combine to create one event, just as one cause can bring about several effects. Read statement E below and answer the following question: What are the causes and the effects presented in the statement?

3. **An event can be both a cause and an effect.** Causes and effects can form a chain of events that extend over a period of time. You can diagram such a chain as follows: People relied on livestock as a food source. → Overgrazing of livestock destroyed vegetation. → Lack of vegetation led to soil erosion. Read statement F below and draw a diagram of its causes and effects as shown in the example.

Statements

A. Because of heavy rainfall, a band of tropical rain forest covers the continent from the ''bulge'' of western Africa to the interior.

B. Africa is a huge continent of many countries and many different kinds of people.

C. Rival chiefs were eager to make themselves monarchs of a great land. As a result, tribal warfare interfered with rebuilding the kingdom, and Ghana began to decline.

D. There are as many forms of African music as there are African languages.

E. Because European trade was expanding and African trade routes shifted from the Sahara to the coast, the Saharan grasslands declined in importance, coastal communities became more powerful, and contact with Europeans increased.

F. In several African kingdoms, gold and the control of trade produced great wealth. With this wealth, the rulers built up their military might. Thus, each kingdom was able to conquer neighboring areas and demand payment from these subject areas. The demands for payment caused unrest among the conquered areas, and eventually the unrest led to the conquered areas breaking free.

3 Nigeria: Africa's Hope

Section Preview

Key Ideas

- Nigeria suffers from conflict between regions and a lack of national unity.
- Dependence on oil affects the economy and political stability of the country.
- Nigeria may someday help to improve the economies of the entire West African region.

Key Term

structural adjustment program

A Foot Bridge, Lagos, Nigeria
This settlement is located on the outskirts of Lagos, Nigeria's capital, chief port, and center of industry and trade.

Nigeria is, in many ways, all of Africa in one country. Its ancient past was a rich one, and its recent history has been stormy but hopeful. Nigeria is the hope for Africa's future. It promises to become a great economic engine that will fuel a higher standard of living for people throughout the West African region.

Varying Regions of Climate and Vegetation

Of the coastal nations of West Africa, Nigeria, which is twice the size of California, has the most varied climate and vegetation regions. From south to north, first coastal swamps give way to tropical rain forest; then a large area of savanna gradually changes to desert scrub. Rainfall varies widely over the country too. Southern regions may receive up to 120 inches (305 cm) of rain a year, while the parched north gets only 20 inches (50 cm).

A variety of crops is also grown throughout Nigeria. In the south, cocoa trees, oil palms, and rubber trees thrive. In the drier north, peanuts are

cultivated. The middle belt of the country supports few crops because of poor soil. These variations affect where people live.

Settlement Patterns Historically, the more powerful groups took control of the valuable land. For example, the Yoruba (yaw ROO buh) settled in the southeast, the Ibo (EE bo) lived in the southwest, and the Hausa traders and Fulani herders controlled the most fertile areas of the north. Smaller, weaker groups were left with the least fertile lands in the middle belt of the country. Although English is the official language of Nigeria, more than 180 different languages are spoken in the middle belt, an area about the size of New Mexico.

Population Movement The many groups living in Nigeria total more than 115 million people, giving Nigeria about a quarter of the population of Africa south of the Sahara. Many of Nigeria's people migrate from one region to another, seeking work in the industries of the cities or on cocoa or rubber plantations.

In addition, people from other countries often move to Nigeria seeking a better life. Foreign workers hold so many of the jobs in Nigeria that the government periodically expels a number of foreigners from the country. Between 1984 and 1986 Nigeria closed its borders to keep illegal immigrants out.

Regional Conflicts One of the difficulties Africa faced after its return to independence was the arbitrary bor-

ders of its countries. African countries have had to work hard to unify the many groups within these borders.

A few years after Nigeria became independent in 1960, members of the Ibo group in the eastern part of the country declared that they were forming a new country. They named it Biafra (bee AH fruh). The other regions chose to go to war to stop the Ibo. Before the Ibo were defeated, thousands of people had been killed and famine had caused many more to starve to death.

Dependence on One Resource

Nigeria drilled its first oil well in 1956, and by 1980 the country was the largest oil producer in Africa south of the Sahara. The country earned twenty-two billion dollars a year in oil revenue. Unfortunately, the export of almost all other goods ended as Nigeria focused on oil profits. This dependence on a single export soon led to serious problems.

The Consequences of Dependence Countries that depend on selling only one crop, product, or resource often suffer economic disaster when prices fall on the world market. Such a disaster befell Nigeria when oil prices tumbled from 1981 to 1983.

As the economy broke down, the military staged a coup. Tens of thousands of government workers were fired. Leaders who had stolen public money were brought to trial, among them one state governor who had stolen forty-five million dollars.

Nigerians soon saw that their new military rulers were not improving the economy. But when the people protested, harsh laws were passed.

A New Chance Finally, in 1985, another general, Ibrahim Babangida (bahb ahn GEED uh), took over the government. At that time, a **structural**

Nigerian Gold

The oil industry brought prosperity to Nigeria. A decrease in demand for oil led the government to diversify the economy in the 1980s.

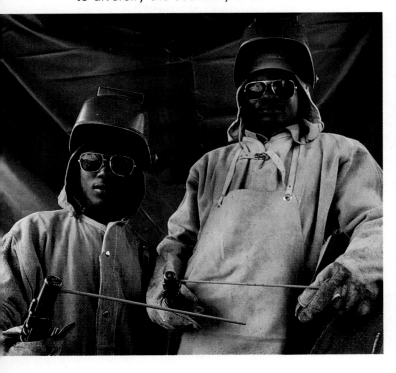

adjustment program, a program imposed on a country by the World Bank to make its economy work better, was begun. The government sold state-run businesses to private companies, fired some government workers, and did not allow wages and prices to rise.

During Nigeria's period of structural adjustment, students and workers often protested the hard times the program caused for them. Babangida was forced to declare the program over in 1988, although he continued much of the plan anyway. Rising oil prices at the beginning of the 1990s gave the Nigerian economy an unexpected but welcome boost.

Babangida promised Nigerians that he would give up power to an elected government in 1992. He was determined, however, that each of Nigeria's ethnic groups should not simply form a political party of its own which could once again lead to civil war.

A Region of Promise

Others also see in Nigeria the shape of Africa's future. In spite of its dependence on oil, the Nigerian economy is strong. The total value of the goods and services it produces every year is nearly forty billion dollars. In the whole continent, only South Africa has a stronger economy. A team of journalists summed up Nigeria's importance in this way:

> *The best hope for Africa is that the continent's two giant economies— South Africa and Nigeria—can be harnessed. . . . Together, they could become a giant market to absorb the rest of Africa's products.*

An African proverb says, "If you get your bundle ready, you will be helped to carry it." African countries are getting their economies ready. Perhaps Nigeria will be the country to help carry them.

Nigeria's Future

Nigerians can take advantage of one of the best school systems in Africa. Part of the country's oil revenues are committed to education.

SECTION 3 REVIEW

Developing Vocabulary
1. Define: structural adjustment program

Place Location
2. Which countries border Nigeria?

Reviewing Main Ideas
3. Why did Babangida insist that each of Nigeria's ethnic groups not form its own political party?
4. In what way is Nigeria's progress important to the future of all of West Africa?

Critical Thinking
5. **Recognizing Ideologies** Why did Babangida declare an end to the structural adjustment program while intending to continue many of its policies?

Hydroelectricity in Nigeria: Kainji Dam

The great interior plateau from which rivers plunge to the coast gives Africa the greatest hydroelectric potential of any continent on earth. Hydropower is Africa's most important natural resource—it depends on a cheap, renewable source of energy, it doesn't pollute the air, and it eliminates the need to import expensive oil for the production of electricity. But beginning in the 1980s, conflict has been growing between developers, who view hydroelectric projects as a way of boosting their country's economy, and conservationists, who fear the negative results of disrupting or destroying Africa's natural environment.

One of the earliest efforts to harness the power of Africa's rivers took place in Nigeria. The results of the interaction between humans and the river environment have been both satisfying and sobering.

The Advantage of Location

The plan to tame the Niger River took shape in the late 1950s, when Nigeria decided to build a dam at Kainji. The location had certain advantages. For one thing, a dam at this spot would back water up into a broad valley between two plateaus. The new lake, the largest artificial body of water in Africa at the time, would extend 90 miles (144 km) up the river and cover 500 square miles (1,300 sq km). This huge reservoir would ensure a constant supply of water for generating electricity at the dam.

Electricity would not be the only benefit of the dam. Planners estimated that four years after the dam was completed a new lake fishing industry would be catching 10,000 tons (9, 070 m tons) of fish annually, adding a rich source of protein to the diet of many Nigerians. Furthermore, the dam would provide a new road across the river for herders bringing cattle to the cities of the south from the dry grazing lands in northern Nigeria.

The High Costs of Technology

The location chosen for the dam presented a number of problems as well. The dam would have to be enormous, stretching 1,800 feet (548 m) across the river valley at a height of 215 feet (66 m). Its cost would eventually reach 190 million dollars; most of the money would have to be borrowed.

Many towns lay in the area that would be flooded. At least fifty thousand people would have to move and resettle in new towns. Crews would have to cut timber from a heavily forested island that covered 107,000 acres (43,320 ha) of the bed of the future lake. Dynamite, dredges, and human sweat would have to clear out five sets of rapids downstream from the dam site so that building materials could be transported upriver.

Taking on the Challenge

Engineers and workers took on the problems posed by the project one by one and solved them. Construction began in 1964. Two years later, the new dam was tested when the heaviest rains in fifty years swelled the Niger to record levels. By 1969 Kainji Lake was full, and the hydroelectric plant at the dam was supplying new industries in Nigeria with cheap and ample power.

Kainji's Lessons

In addition to its rewards, however, the dam brought surprising losses. Predictably, the floods that had refreshed riverside farms with new soil every year came to an end. But nobody had foreseen that without the fresh soil, harvests would shrink by 50 percent. The fish population also fell by 50 percent.

From Kainji, experts learned that dams should be used not to end downstream flooding but to control it. The natural cycle of flooding and dry spells had kept the river environment at its most productive. New discussions between developers and conservationists resulted in a set of planning guidelines. The Manantali Dam on the Senegal River was built in Mali in 1988 with controlled flooding in mind. Lessons learned from Kainji have brought African developers closer to the goal of working with the environment instead of against it.

TAKING ANOTHER LOOK

1. What advantages did the site at Kainji offer?
2. What did Kainji teach planners about human interaction with the environment?

Critical Thinking

3. **Predicting Consequences** The Manantali Dam uses controlled releases of water to mimic natural flooding. How would this benefit people downstream?

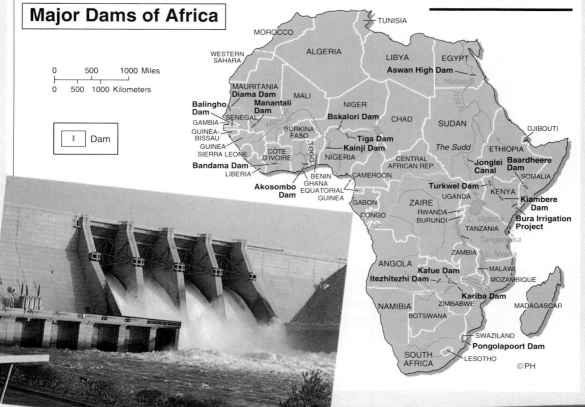

Africa's Hydropower
Where is the Kainji Dam (shown below) located?

Major Dams of Africa

CASE STUDY ON CURRENT ISSUES

541

4 Central Africa

Section Preview

Key Ideas

- Movement in Central Africa is affected by the region's rivers, forests, and grasslands.
- Environmental damage has resulted from the misuse of hydroelectric power and forest resources.
- Mineral wealth is important to Central African countries, especially Zaire.

Key Term

mercenary

East of Nigeria, the coast of Africa turns sharply to the south. Along this southward stretch lie the seacoasts of Cameroon, Equatorial Guinea, Gabon, Congo, and Zaire (zy EER). Offshore is the island nation of the Republic of Sao Tome (SOW tuh MAY) and Principe (preen SEEP). Beyond the coastal nations, deep within the continent, is the Central African Republic. Together these seven countries make up Central Africa. They range in size from the Republic of Sao Tome and Principe, a little larger than New York City, to Zaire, which is one quarter the size of the United States.

A Region Built by Movement

Movement of people has affected this region possibly more than any other in Africa. And the region's physical characteristics have, in turn, affected the ways in which that movement has taken place.

The Big River The largest river of the region is the Zaire, formerly called the Congo River. Like the Niger River, even though its source is only a short distance from the ocean, it flows inland 2,900 miles (4,640 km) through a huge basin before finding its outlet in the Atlantic Ocean. The Zaire and its many tributaries total about 8,000 miles (12,800 km) of waterway. The entire Zaire River system is a great living highway that provides food, water, and transportation for much of the region.

Most of the Zaire River is located in the country of Zaire. Boats can travel from Boyoma Falls in the northeast of the country to Zaire's capital, Kinshasa (kin SHAHS uh), located in the west. Below Kinshasa the course of the river is blocked by cataracts. Because boats cannot pass this stretch of the river, goods are carried overland by the railroad that links Kinshasa with the huge port of Matadi on the lower reaches of the Zaire.

Movement Through Rain Forest and Savanna The basin that feeds the Zaire River system is over 1 million square miles (3 million sq km) in area. In the center of the basin is a dense rain forest. It is easy to see why people migrating in early times shied away from entering this dark and forbidding maze of trees.

Although the forest presented a frightening barrier to movement, on the savannas travel was relatively easy. These grasslands stretch around the rain forests in a broken ring to the north, east, and south. From ancient times people of the savanna were able to trade, communicate, or conquer others without obstacles.

Today the forest is still a barrier to travel. Its many valuable kinds of wood, such as mahogany, ebony, walnut, and iroko, can be harvested only along the rivers or where a railroad has been carved through the forest.

In spite of the rich vegetation it supports, the forest soil is actually of little use for farming. Soil in the savanna lands, too, is often poor. People have migrated away from these areas either to plantations located on fertile soil or to great cities like Kinshasa or Brazzaville, the capital of Congo.

Movement to an Urban Area   Migration has turned Kinshasa into a major world city. Its estimated population of three million is similar in size to that of Rome or Berlin.

The city grew explosively in the second half of the twentieth century. Some of the people who flocked to Kinshasa during those decades found wealth by working in the city's businesses or in the national government and built expensive homes on Kinshasa's tree-lined avenues. Others continued the subsistence life they had known in the countryside, scraping together a living in Kinshasa's vast slums.

Rich and poor alike, however, take part in the culture that has grown up from ancient African roots and mixed with the modern world. For example, Kinshasa has gained an international reputation for its popular music, which is a lively blend of African, rock, and pop rhythms.

Movement Fosters Interdependence   Just across the Zaire River from Kinshasa lies Brazzaville. Although their two countries frequently disagree politically, the two cities share the river that forms the border between them.

Brazzaville, like Kinshasa, has a rail connection with the coast, and not surprisingly the route is dotted with industrial towns. The railroad also serves the inland nations of Chad and the Central African Republic, which ship mineral resources down the Ubangi (yoo BANG ee) and Zaire rivers to Brazzaville and from there to the Atlantic Ocean.

Many countries of West and Central Africa belong to an African financial community known as the CFA. The CFA countries use a currency, or form of money, called the CFA franc, which has solid value on international markets because it can be exchanged for the French franc. Use of this common currency promotes trade, travel, and general interdependence among countries in the region.

Rich Environmental Resources

The rivers and forests of Central Africa affect resource use as well as movement. In some African countries, the physical landscape is one of the most valuable resources. Unfortunately, the natural resources of the region have not always been wisely used.

Hydroelectric Power   The continent of Africa consists of a group of basins set in a vast plateau. Where the rivers that drain the basins cut through the

The Zaire River

Zaire's economy is heavily dependent on the Zaire River. In spite of its many waterfalls, much of this water route is navigable.

edge of the plateau to the coastal plain, they drop sharply. At this escarpment, the rivers have great hydroelectric potential, or power that can be used to create electricity. The region's hydroelectric power is discussed more fully in the case study on pages 540–541.

Saving the Forests Another way in which Central Africans use and change their environment is by cutting down trees. Rain forests are cut down faster than they can replace themselves. Deforestation has been less of a problem in Central Africa than in the coastal nations of West Africa where loggers can get to forest areas more easily. In Côte d'Ivoire, for example, the rain forest is only half the size it was in 1970. At this rate, it may be completely wiped out early in the twenty-first century.

The rain forests of Africa are valuable for many reasons. In addition to supplying lumber, they provide habitats for thousands of animal species and shelter thousands of plants. They also absorb carbon dioxide. Increasing amounts of carbon dioxide left unabsorbed in the air may lead to what scientists call the "greenhouse effect," a gradual rise in global temperatures.

The nations of Central Africa are still in a position to control logging within their borders. But planting new forests costs more than most African countries can afford. In all of Africa about one acre is replanted with trees for every twenty-nine that are cut.

The World's Landfill A nation such as Congo faces a powerful temptation in an age where disposal of waste is a growing problem. Because most of Congo's population live in the cities, industrial nations in Europe view the country's thinly inhabited rural regions as an ideal place to dispose of toxic waste. In the late 1980s, foreign industries contacted Congo's government, offering 300 million dollars to use the country as a waste dump.

Because the Republic of the Congo needed the money, the temptation to accept the offer was great. Like both Nigeria and its neighbor Gabon, Congo depended heavily on oil for its income. Lower oil prices and declining production caused Congo's oil income to fall from 810 million dollars in 1985 to only 124 million in 1988. The country owed foreign banks 4.5 billion dollars and in addition, was importing four times as much as it exported.

Congo signed the contracts. But when the news became public people were outraged by the government's lack of concern for their health and safety. As a result of the scandal, officials were fired, including the minister in charge of environmental affairs. In the end, Congo said it was not willing to become the world's landfill. Other nations in Central Africa, however, remained targets for countries with millions of dollars to spend and the need to dispose of thousands of tons of toxic waste.

A Giant in the Region

The smallest nations in Central Africa are Equatorial Guinea and the Republic of Sao Tome and Principe. Sao Tome and Principe are islands, while Equatorial Guinea is made up of five islands in addition to a small area on the continent. Both these nations are very poor in resources.

The larger Central African nations, however, have large deposits of minerals such as bauxite and iron ore. Zaire has huge copper reserves in the Shaba region in the south. Zaire produces more cobalt than any other country in the world, and 50 percent of the world's industrial diamonds lie hidden within its borders. It also produces about seven million barrels of oil more than it uses every year.

Because of these resources, Zaire plays the same role in Central Africa that Nigeria plays in West Africa—that of a potential market for the region.

WORD ORIGIN

Côte d'Ivoire
Cote d'Ivoire got its name from French traders who went there for ivory in the 1400s. *Côte d'Ivoire* means "Ivory Coast" in French.

Difficult Times The history of Zaire is marked by periods of civil war and coups. Within a week of gaining independence from Belgium in 1960, it faced a revolt by its armed forces. At the same time the province of Shaba chose to leave the country.

The country was torn apart for four years as Belgian troops, United Nations forces, rebel armies, and **mercenaries**—hired soldiers—battled for power. Eventually a general named Mobutu Sese Seko established himself as dictator.

Under Mobutu, Zaire won back the province of Shaba and improved its mining and industries. By the 1980s, however, the nation fell deeply into debt. Mobutu was forced to begin a structural adjustment program, but it did little good. By 1990 Zaire owed foreign banks 8.5 billion dollars. In the same year Mobutu's personal wealth was estimated at 5 billion dollars. The citizens of Zaire had little difficulty guessing where their wealth had gone.

Changes Come The wind of democratic change that swept over the world in the late 1980s seemed to blow through Zaire as well. In April 1990, its dictator of twenty-five years announced that he was permitting the formation of other political parties and allowing a prime minister to rule in his place. "It is time to let go little by little," said Mobutu.

Some observers felt that these words were just one more lie. One expert said this of Mobutu's government in 1988:

The system has worked well for Mobutu over the past twenty-two years and he is unlikely to abandon it, even under duress.

As the final decade of the century began, it remained unclear whether Zaire would throw off the corrupt leadership of Mobutu and win the place it deserved in Africa.

Copper Mines in Zaire
The discovery of immense copper reserves brought heavy industrial development to the Shaba region in southern Zaire.

SECTION 4 REVIEW

Developing Vocabulary
1. Define: mercenary

Place Location
2. Which eight countries border Zaire?

Reviewing Main Ideas
3. How has the forest of Central Africa affected human movement in the region?
4. What type of natural resource contributes the most to Zaire's wealth?

Critical Thinking
5. **Identifying Central Issues** What course do you think the countries of Central Africa should follow with regard to toxic waste?

CHAPTER

26

REVIEW

Section Summaries

SECTION 1 The Sahel: The Region, the Environment The Sahel lies between the Sahara of northern Africa and the tropical rain forests of coastal West Africa and Central Africa. It is mostly savanna with a semiarid climate. In ancient times important trading empires flourished in the region. People have changed the environment through overgrazing and deforestation, and the region is in danger of desertification. The Sahel countries need foreign aid to fight drought and famine and to develop their resources.

SECTION 2 The Coastal Countries The location of these West African countries enabled them to trade with Europeans from the 1500s. Today the West African nations have huge debts. Women are a key element in a new kind of economy. Social custom, religious beliefs, and economic conditions contribute to a high population growth rate.

SECTION 3 Nigeria: Africa's Hope Nigeria was settled by various ethnic groups, leading to a lack of unity in the country. The country experimented with a structural adjustment program to help it recover from problems caused by dependence on oil exports. Nigeria is an economic powerhouse that may be able to help lift all of West Africa out of its economic difficulties.

SECTION 4 Central Africa In recent times many people have moved away from areas with poor soil and into the region's cities. Movement has also encouraged interdependence among Central African nations. Much of the wealth of the region is built on the environment and on mineral resources. Great mineral wealth has made Zaire the leader of the region.

Vocabulary Development

Match the following definitions with the terms they define below.

1. a belief in the spirits of the dead
2. an area of lakes, creeks, and swamps away from an ocean
3. a hired soldier
4. a program to change the structure of an economy to make it work better
5. destruction of forest vegetation
6. a sudden takeover of a government
7. the practice of preparing and growing crops on a site for only a year or two
8. a person who leaves his or her home to escape danger or persecution
9. without access to the ocean
10. a loss of all vegetation
11. food for grazing animals
12. the belief that natural objects contain spirits or gods

a. inland delta
b. ancestor worship
c. landlocked
d. coup
e. refugee
f. deforestation
g. shifting agriculture
h. desertification
i. structural adjustment program
j. mercenary
k. animism
l. forage

Main Ideas

1. What factor most affects the life of people in the Sahel?
2. Name one problem concerning export trade from West African countries today.
3. How has ethnic diversity affected Nigeria?
4. What steps did Nigeria take in its structural adjustment program?

5. How does the Zaire River affect life in Central Africa?

Critical Thinking

1. **Making Comparisons** Describe two ways in which the modern nation of Ghana differs from the ancient kingdom of Ghana.
2. **Testing Conclusions** Do you think Africa is beginning a slide into isolation? Explain your answer.

3. Perceiving Cause-Effect Relationships How would Africa's hydroelectric potential be different if most of the continent were at the same altitude as the coastal plain?

Practicing Skills

Perceiving Cause-Effect Relationships Now look again at statement F on page 536. Identify the words or phrases in this passage that signal cause-effect relationships.

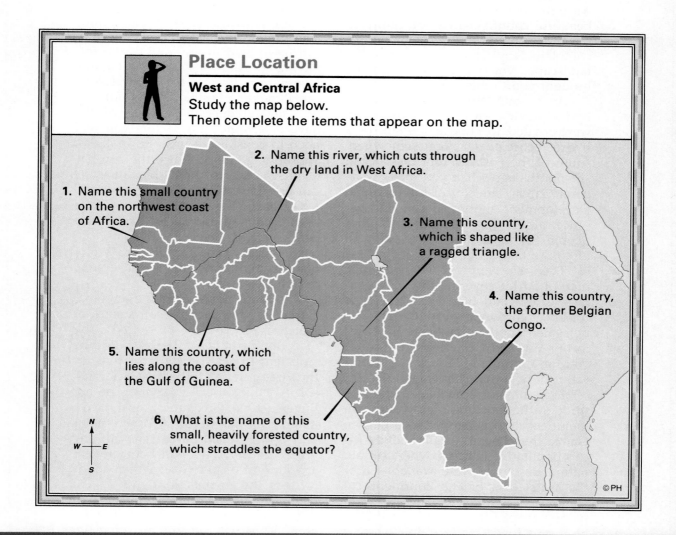

Place Location

West and Central Africa
Study the map below.
Then complete the items that appear on the map.

2. Name this river, which cuts through the dry land in West Africa.

1. Name this small country on the northwest coast of Africa.

3. Name this country, which is shaped like a ragged triangle.

4. Name this country, the former Belgian Congo.

5. Name this country, which lies along the coast of the Gulf of Guinea.

6. What is the name of this small, heavily forested country, which straddles the equator?

N
W—E
S

©PH

Place Location:
WHERE ON EARTH?

Assignment in Africa

Down to the riverbank you bounce in the dented, rusted pickup truck. The wild grasses around you ripple and wave in the breeze. You will miss this beautiful country when you return to the United States. But your job here is complete. Now, as you gaze across the landscape, you think back to what brought you to this place.

As a staff geologist for a major hydroelectric power company, you were invited to choose the site of a hydroelectric power plant somewhere south of the Sahara. Your task began six months ago when your boss, Margaret Garcia, Director of International Development, motioned you into her office, along with half a dozen other people. She closed the door, asking all of you to take seats.

Then she approached the huge map of Africa covering the wall. "Ladies and gentlemen," Garcia said briskly, "corporate headquarters has delegated another challenge in Africa to this division." She looked at all of you for a moment. "You're the team I want to meet that challenge."

Excitement immediately crackled through the room. You and your team members almost snapped to attention. Garcia then began to talk about Africa's population clusters, where certain natural resources were available, how the waterways of the continent cut through rain forest to the sea. Her pointer tapped the map left and right, high and low. Garcia didn't waste a

word or a movement. At the end of her talk, everyone but you had an assignment. You waited, expectantly.

Then she turned to you and said, "I want you to select the country with the best possible site for the power plant. The plant is the cornerstone on which our development and expansion plans rest, so the site must be well chosen. It can be located in any one of the following ten countries: Senegal, Ghana, Nigeria, the Congo, the Central African Republic, Zambia, Zaire, Botswana, Mozambique, and Zimbabwe. We have determined the factors you will need to consider in order to make an appropriate choice. Once the site has been selected, corporate headquarters can present the plan to that country."

Over the top of her glasses, Garcia eyed each team member. "Ladies and gentlemen, I need hardly say that headquarters expects a superb set of plans from this division. We have three months. Get busy."

As team members left the room, Garcia handed each person a stack of materials. To you she presented a roll of curled-up maps. "These should be sufficient to get you started. Here are the three criteria for the site selection," she explained, ticking them off on her fingers. "One, the country must have a major river that does *not* form a major part of a political boundary with any other country. This will help prevent any kind of dispute over which country has the ownership rights to the dam. Two, the river must flow through a densely populated sub-Saharan region. This will ensure an adequate

548

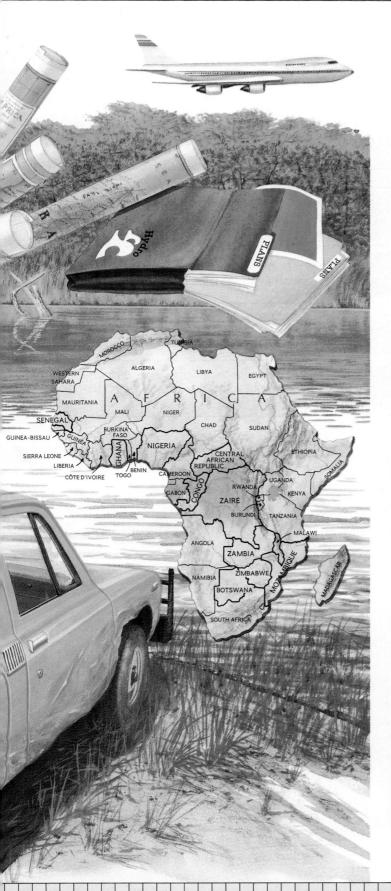

demand for power in coming years. Three, to make communication with workers as easy as possible, the official language of the country should be English. Any questions?"

You walked slowly back to your desk. How could you narrow the possibilities? You started with a political map showing the major rivers of sub-Saharan Africa. First you noted which rivers formed a major part of a political boundary. Next you studied a map showing population distribution. Finally, you consulted an almanac to find countries in Africa where English is the official language. Then you made your choice.

After you and other team members prepared and submitted plans to Garcia, the team flew to Africa to inspect your suggested site. The setting appeared ideal. You and the team had no problem convincing corporate headquarters of the site's suitability and the project's practicality.

Your contacts in Africa are excited about the project you and your team have brought to life. The groundbreaking for the plant is scheduled for early next month. As you sniff the moist air by the river, you can already see the plant rising over the river and hear the water pouring over the dam.

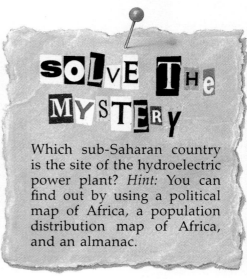

SOLVE THE MYSTERY

Which sub-Saharan country is the site of the hydroelectric power plant? *Hint:* You can find out by using a political map of Africa, a population distribution map of Africa, and an almanac.

27

East and Southern Africa

Chapter Preview

Locate the countries covered in each of these sections by matching the colors on the right with those on the map below.

Sections	Did You Know?
1 KENYA	As part of a conservation movement, villagers in Kenya plant tree seedlings and are paid for each tree that survives.
2 OTHER COUNTRIES OF EAST AFRICA	Lakes cover more than one sixth of the area of the country of Uganda.
3 SOUTH AFRICA	By the year 2000, black South Africans will make up 80 percent of the total population of South Africa.
4 OTHER COUNTRIES OF SOUTHERN AFRICA	*Zimbabwe,* which means "house of stone," was a city whose ruins now stand near the city of Masvingo.

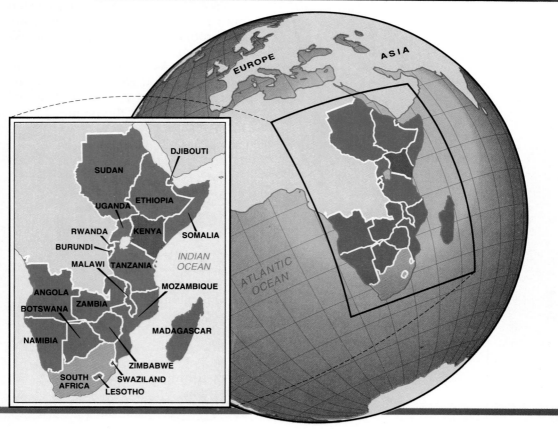

The Masai of East Africa
More than 110,000 nomadic Masai live in the Great Rift Valley of Kenya and Tanzania. Today the Masai are contesting the expansion of wildlife reserves into their grazing lands.

1 Kenya

Section Preview

Key Ideas

- Kenya's most fertile land has been the focus of movement through the area for centuries.
- Kenya today has a solid economy built on cash crops but faces malnutrition, rapid population growth, and unemployment.

Key Term

malnutrition

Many people think of Kenya when they think of Africa, although they may not even realize they are doing so. That is because Kenya has many features that have become symbols of Africa—rolling savanna lands, highland coffee plantations, the Masai (mah SY) people, and spectacular national parks where elephants, lions, and other wildlife roam freely.

Kenya is, of course, more than just these symbols. It is a vibrant country with a population of more than twenty-three million and a varied and beautiful landscape.

Where the Rift Crosses the Equator

Kenya is located on the east coast of Africa and extends deep into the interior of the continent. As the map on page 552 shows, the Equator runs right through the center of the country, so that parts of Kenya are bathed in steamy heat. In addition, the Great Rift Valley slices through Kenya's highlands, where elevation makes the climate cooler.

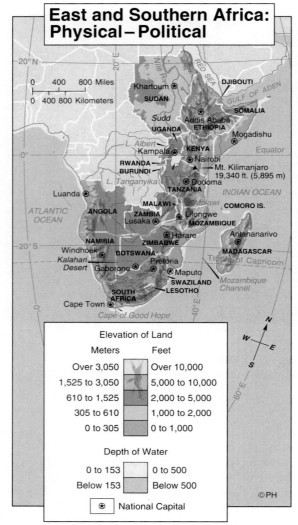

East and Southern Africa: Physical–Political

Elevation of Land

Meters	Feet
Over 3,050	Over 10,000
1,525 to 3,050	5,000 to 10,000
610 to 1,525	2,000 to 5,000
305 to 610	1,000 to 2,000
0 to 305	0 to 1,000

Depth of Water

0 to 153	0 to 500
Below 153	Below 500

⊛ National Capital

©PH

Applying Geographic Themes
Location Which countries border Zimbabwe? In which country is Addis Ababa located?

WORD ORIGIN

plateau
From the French *plat*, meaning "flat," this word has come to mean "an elevated land area with a flat top."

live in this dry land. Groups of nomadic people roam the region in search of grazing land for their livestock, but the soil is too dry for farming.

Great Lakes and Fertile Farmlands
Most of Kenya's people live in the highlands in the country's southwest region. Forests and grasslands cover much of the highland region. In the westernmost corner of the country, Kenya's border is the magnificent Lake Victoria, the largest lake in Africa and the third largest in the world. Two other nations, Uganda and Tanzania, also border Lake Victoria, which forms a watery link among these three East African nations.

Movement: Seeking Fertile Land

The most fertile land in Kenya is found in the central highlands on either side of the Great Rift Valley. These fertile farmlands have been the focus of movement through Kenya for centuries. In the 1700s, a group of herders called the Masai moved into the area around the central highlands, raiding farming villages in the process. One large group, the Kikuyu (ki KOO yoo), was successful in holding on to its farmland in the highlands.

Interaction: The Railroad Comes
Kenya came under British rule in 1895. By the late 1890s, disease and famine reduced the populations of both the Masai and the Kikuyu. Both groups were forced by Europeans to give up and leave much of the land they held. Under British rule, Africans lost their most fertile farmland and all political power. In an effort to encourage economic development and to gain access to the rich farmland in the central highlands, in 1901 the British decided to build a railroad from the coast to Lake Victoria. The building of the railroad was no simple task, as one writer described it.

Dry Lowlands As well as giving the country access to the sea, Kenya's coastal lands have beautiful sandy beaches. Patches of rain forest also line the coast. But for the most part these lands are not especially productive. The plateau that leads toward the center of the country, gradually rising as it goes west, is the driest part of Kenya. These lands, like those in the north and the northeast where rainfall is uncertain, are prone to drought. Few people

It entailed laying more than six hundred miles of track over difficult and hazardous terrain: across a sterile, waterless desert; over unmapped savanna and scrubland teeming with lions and buzzing with tsetse flies; through a volcanic region dissected by the yawning chasm of the Great Rift Valley; and through one hundred miles of quagmire [swamp].

About thirty thousand workers were brought from India, also a British colony at the time, to build the railroad. More than two thousand Indians lost their lives to the hazards of the task, including twenty-eight who were killed by lions. The railroad was completed by 1903, and many of the Indian workers stayed in Kenya to become shopkeepers and traders.

The new transportation link across Kenya brought immense changes. A new town built on the line, Nairobi (ny RO bee), grew rapidly and has since become Kenya's capital. With a population of three million, Nairobi today is a bustling, modern city that hums with industry and commerce.

Critics complained that the growth along the railroad was not enough to justify building the line, which had cost thousands of lives and millions of dollars. Hoping to silence lingering criticism of their project, the British encouraged their citizens and other Europeans to settle in Kenya and develop the central highlands through which the railroad ran. White settlers, some from South Africa, were eager to move into the cool climate of the highland area.

Kenyans Challenge the British The white settlers took much of the land the Kikuyu still considered their own. Other groups were forced to work on farms run by whites or to move from their homelands. If they resisted, they were brutally punished by the British.

In the 1950s, the Kikuyu went to war against the British settlers in Kenya in a fierce confrontation called the Mau Mau Rebellion. The rebellion was crushed, but one of the leaders of the Kikuyu, Jomo Kenyatta, became president when Kenya emerged from British rule in 1963. Under Kenyatta, the Kikuyu regained some of their farms in the central highlands.

Characteristics of Place

Kenya's flag has stripes of three colors: black, for the people of Kenya; red, representing their struggle for independence; and green, symbolizing the country's agriculture. The seal of the Republic of Kenya today shows two lions leaning on a shield. Beneath them is the single Swahili word *harambee,* which means, "Let's pull together." The ideal of *harambee* explains in part why Kenya is one of the most successful countries in Africa.

Harambee **at Work** After independence in 1963, Kenyatta encouraged all parts of the economy—the government, privately held companies, and individuals—to work together to make Kenya's economy strong. *Harambee* grew as a grass-roots movement of people pulling together to help themselves and each other. Foreign investors were pleased with Kenyatta's attitude and willingly pulled together with the Kenyans.

The result was solid economic growth. Because Kenya has little mineral wealth, the growth was based mostly on agriculture. Kenyatta encouraged farmers to raise cash crops—coffee and tea—which grow well on the central highland farms of the Kikuyu. Many government officials were Kikuyu, and they soon grew wealthy from their own farming. "The government of Kenya," noted one expert, "has, from the moment of independence, been a government of farmers, by farmers and for farmers."

Land Use in East Africa
The protection of wildlife has attracted intense international interest. But fast-growing African populations often compete for the same land.

Starvation in a Healthy Economy Today many Kenyans do not get enough food. The government has concentrated on growing cash crops—luxuries like coffee and flowers—rather than food. People who cannot afford food crops suffer from malnutrition, or the disease caused by not having a healthy diet. The United Nations estimates that 30 percent of Kenya's population suffer from malnutrition. Kenya produces about 40 percent less food per person today than it did in the 1950s.

One commentator had this to say about Kenya's dependence on cash crops:

As more land is devoted to cash crops—coffee, tea, flowers—less land is available for subsistence agriculture. Some of the worst malnutrition in Kenya borders rich coastal lands used . . . to grow pineapples for export.

Regions and Politics During the 1980s, the population of Kenya grew at an astounding rate, perhaps as much as 4 percent a year. The government was not able to supply this growing population with jobs. In 1989, for example, 300,000 Kenyan young people left school to work, but only 67,000 jobs were available. High unemployment often results in social unrest—and social unrest can mean trouble for any government.

At the beginning of the 1990s, the president of Kenya refused to allow movement toward democracy in Kenya. He punished independent judges, threw critics in jail, and closed newspapers that protested his policies. Many Kenyans believed that making the country a one-party system was taking *harambee* too far. Pulling together was fine, but it was important that Kenyans do it of their own free will.

SECTION 1 REVIEW

Developing Vocabulary
1. Define: malnutrition

Place Location
2. Why are the central highlands of Kenya cool even though they are located on the equator?

Reviewing Main Ideas
3. What changes did the British railroad bring to the region?
4. What is *harambee* and how did it benefit Kenya?

Critical Thinking
5. **Predicting Consequences** How might the government of Kenya meet the challenge of the country's growing population?

2 Other Countries of East Africa

Section Preview

Key Ideas

- The seacoast countries of East Africa are working to improve their economies by building networks of trade.
- Regional issues have caused lasting and bitter conflicts in several East African nations.
- Tanzania has changed its ideology to promote economic growth.

Key Terms

strategic value, ethnocracy, villagization

Kenya shares the land of East Africa with many other countries. Several of them border the Indian Ocean, the Red Sea, or the Gulf of Aden. Some are island nations, and others are landlocked.

Coastal Locations

As in West Africa, it was centuries ago in the coastal region that powerful empires based on trade developed in East Africa. Today the seacoast nations of East Africa are working to rebuild their economies by again building networks of trade.

The island country of Mauritius (maw RISH ee uhs) is leading the way in creating an economy based on trade. More than half of the people in Mauritius are of Indian descent. Smaller groups of Chinese, French, and British descent are also counted among the island's population. Both cultural makeup and relative location give Mauritius strong ties to Asia. Investors from Hong Kong have helped Mauritius build light industry on the island.

Economists believe that the nearby island nation of Madagascar could also raise its people's standard of living. The country's average per capita income is only 200 dollars per year, compared with 1,810 dollars in Mauritius. At the same time, many people are concerned about the quality of life on these islands. As they develop a more industrial economy and way of life, the people of these islands have also begun to experience many of the stress-related illnesses common in industrialized countries.

Location: Countries with Strategic Value

Besides giving East Africa the opportunity for trade, the region's coastal location also gives some of its countries strategic value—the value for nations planning large-scale military actions. The countries of Ethiopia, Djibouti (ji BOO tee), and Somalia, which are located on a landform known as the Horn of Africa, have particularly strategic locations. They lie near both the oil supplies of the Middle East and the shipping lanes of the Red Sea and the Gulf of Aden. These countries also lie at the midpoint on the journey from Europe to Southeast Asia.

Djibouti Djibouti is also a vital link between neighboring Ethiopia's capital city of Addis Ababa and the sea. Djibouti earns most of its income from its ports. Also, France pays large fees to Djibouti for the right to maintain a military base in the country.

Ethiopia Of all the nations in the Horn of Africa, Ethiopia is the oldest. Ruins and ancient Egyptian writings record the history of the Kushite civilization in Ethiopia about 3,500 years ago. The region's high, fertile plateaus, which enjoy temperate climates, rise like massive walls above the deserts of the Sudan to the west and Somalia to the east. In recent years, regional conflict and drought have brought this ancient nation almost to the edge of collapse.

A drought in 1984 caused famine and starvation in Ethiopia. In addition, war with neighboring Somalia, as well as civil war in the coastal province of Eritrea, caused grave crises. World nations sent aid to Ethiopia, but the civil war prevented food from reaching those who needed it. During these troubles huge numbers of people died. Others fled wherever they could to escape violence or famine.

The Sudan

Ethiopia's neighbor, the Sudan, also has an ancient past and a troubled recent history. The Sudan is the largest country in area in all of Africa.

Regional Differences   The Sudan is much like the Sahel nations discussed in Chapter 26. To the north the country is largely a desert of bare rock or shifting sand dunes called ergs; in the south are clay plains and an extensive swamp area called the Sudd, which means ''The Barrier.''

Most northerners are Muslim Arabs; most southerners are Africans, many of whom practice animism or Christianity. From 1899 to 1955, when the British ruled the country, the north was cut off from the south as a matter of government policy. With independence in 1956, the Muslim government in the north began trying to impose Islamic beliefs on the southerners. Except for a period of relative peace from 1972 to 1983, the country has been at civil war ever since.

Farming in the Sudan
Sparse vegetation and a parched landscape attest to the fact that desertification is overtaking farming settlements in the Sudan. Many villages are abandoned as people are forced to travel in search of food.

War and Famine Civil war in the Sudan has led to destruction of the economy and great human misery. In late 1990 the Sudan was faced with famine on an almost unimaginable scale. Some people thought that one reason for the food shortage might be that the country's president, General al-Bashir, traded 300,000 tons of food to Libya and Iraq for weapons to use in his war against the southern region.

The Landlocked Countries

Uganda, Rwanda, and Burundi are the landlocked countries of East Africa. All three are heavily populated, agricultural countries. Coffee is the most important crop grown for export, but both Rwanda and Burundi lack the means to consistently move their goods to foreign buyers.

Uganda: A Thriving Agricultural Economy Located to the west of Kenya, Uganda is for the most part a plateau with fertile soils. After the British claimed the area in 1894, the land was reserved for agriculture. Within twenty years Uganda was producing cotton so successfully that the colony no longer required financial aid from Britain. In time, coffee came to replace cotton as Uganda's number one export.

As in Mauritius, Asians entered the country and invested their money in new businesses and industries. This economic growth was aided by Uganda's transportation system. The country had well-packed dirt roads that eased travel over the mostly flat lands. The Rift Valley lakes, the Albert Nile, and the railroad through Kenya gave the landlocked colony several routes to export its products.

Disruption of Civil War When Uganda gained independence from Britain, civil war broke out and disrupted the country's success. Northern people, who had won most of the military power, struggled against southern groups, which had most of the economic might. Under a ruthless dictator, Idi Amin (EE dee ah MEEN), as many as three hundred thousand Ugandans died or "disappeared" in these struggles during the 1970s. Roads and communication links that were destroyed by war were not rebuilt until the late 1980s. Today Uganda's government promises a return to democracy sometime in the mid-1990s.

Rwanda and Burundi

An ethnocracy is a government in which one ethnic group rules over others. Rwanda (roo AHN duh) and Burundi (boo ROON dee), two of the smallest African nations, are both ethnocracies.

In Rwanda, 90 percent of the population belongs to the Hutu (HOO too) group. Most of the remainder are Tutsi (TOOT see), sometimes called Watusi (wah TOO see). The Hutu have been firmly in power since they overthrew the Tutsi-controlled government in 1959, killing some one hundred thousand of the minority.

In Burundi, on the other hand, the Tutsi are still in power, although they are clearly a minority—only 13 percent of the population. In Burundi, the Tutsi give about 80 percent of the university places to their own people. The Tutsi also make up the army in Burundi, and they use their power with fatal consequences. In 1972, and again in 1988, the army massacred thousands of Hutu families.

Tanzania

To the east of Rwanda and Burundi lies Tanzania (tan zuh NEE uh). Like many African nations, it is a land of great potential wealth. Its soils are fertile in many areas. The range of its lands includes the hot, humid coastal lands, the cool highlands, the varied others. Industry and some land were nationalized, that is, taken over by the government. Tanzania's rural people were subjected to villagization—forced to move into towns and to work on collective farms.

***Ujamaa* in Trouble** In the 1980s Tanzania saw that its political ideology was failing. Tanzanians did not want to grow food because the government-controlled price was too low. The production of sisal, a natural fiber used to make rope, fell from 200,000 metric tons in the 1960s to 30,000 in 1987. The plantations where sisal had been grown were neglected and were overtaken by tangled trees. The country could not afford the fuel to run tractors and trucks, and harvests rotted in the fields.

Nyerere stepped down, *ujamaa* ended, and the economy turned around. The key to recovery was paying farmers a fair price for their crops. When they saw that they could profit by growing corn and cotton, they once again farmed land that had been idle for years.

The Sisal Industry in Tanzania
Coffee, tea, cotton, diamonds, and sisal, a natural fiber used to make rope, are some of Tanzania's major export items.

terrain around Lake Victoria, and the dry central plains where Masai herders still live among antelopes, wild buffalo, and lions. Beneath its surface lie quantities of iron ore, coal, diamonds, and other minerals.

An Ideology for Development In 1961 Tanzania became independent of Britain. Its leader, Julius Nyerere, like many African leaders wished his people to work together, rather than individually, toward equal wealth for all. He felt that socialism was the best means to achieve such progress. He called his socialist program *ujamaa*, a Swahili word that means something like "neighborliness" or "familyness."

In practice, *ujamaa* meant that the government took over the thriving plantations built by the British and

SECTION 2 REVIEW

Developing Vocabulary
1. Define: **a.** strategic value
   b. ethnocracy **c.** villagization

Place Location
2. Why does the Horn of Africa have strategic value?

Reviewing Main Ideas
3. How does the Sudan's geography contribute to its regional differences?
4. Why did Tanzania change its political ideology?

Critical Thinking
5. **Making Comparisons** Compare north-south differences in the Sudan and in Uganda.

3 South Africa

Section Preview

Key Ideas

- A white minority controls the South African government, as well as about two thirds of the land.
- The apartheid system was created in an attempt to preserve the power of whites in South Africa.
- Under pressure by blacks and international sanctions, the government is moving toward change.

Key Terms

apartheid, segregation, sanction

The Republic of South Africa is one of the powerhouse economies of the continent. In fact, it is the wealthiest, most highly developed nation in Africa south of the Sahara. Yet it is not a stable nation—rather it is a nation where colonial rule by people of European descent has endured despite years of protest and violence.

A Country Divided by Race

Much can be told about the Republic of South Africa with a few numbers. Some 68 percent of South Africa's population is black; about 3 percent is Asian; 18 percent is white; and about 10 percent is of mixed race. The five million white people who live in South Africa make up a small minority when compared with the thirty-two million residents who belong to the three other officially designated races. In South Africa, however, it is the white minority that holds power.

Minority Rule The white minority controls the South African government, as well as about two thirds of the land in South Africa and most highly

paid jobs. The whites own the gold mines, the diamond mines, and the mines where some seventy other minerals are dug from beneath the soil.

White South Africans own the best farmland as well. The Republic of South Africa is mostly a high plateau. Around the edges of the plateau is an escarpment that drops to a narrow coastal plain. The plateau itself is very dry, but in places where there is good rain corn, wheat, and a wide variety of fruits grow abundantly.

Whites also own the thriving industries of South Africa, where the metals of the mines are manufactured into machines and other goods. The wealthiest whites—about 5 percent of the population—possess 88 percent of all the personal wealth in the nation. How did this country's minority come to possess so much wealth and power?

Cape Town, South Africa
One of the consequences of South Africa's policy of apartheid is the harsh living conditions for the majority of black South Africans.

Movement into Black Lands The inequality of ownership in South Africa came about, first of all, through movement. Europeans came to South Africa beginning in the 1600s—first the Dutch, then some Germans and a few French. Over time these groups together came to be known as Afrikaners (AF ri KAHN erz), or Boers, speaking their own distinctive language, Afrikaans (AF ri KAHNZ). The Afrikaners pushed the Africans inland, gradually claiming the Africans' land by treaty and by force.

In the late 1800s the discovery of diamonds and gold in the area around what is now the city of Johannesburg brought many changes. Great numbers of people, mostly British, arrived in South Africa. Friction between the Afrikaners and the newcomers led to a series of confrontations. The Afrikaners continually moved into African lands to escape from British control, and the British moved after them repeatedly, reasserting British rule.

After three years of fighting the Boer War, the Afrikaners accepted British rule. The final effect of this movement was a combined colony of Afrikaners and English-speaking settlers. The majority African population was driven into separate lands called reserves or put to work on plantations or in factories owned primarily by whites and Asians.

Movement into White Lands By the time South Africa left the British Commonwealth to become an independent republic in 1961, a new pattern of movement had appeared. Blacks were moving out of the reserves into the cities. The reserves promised nothing but subsistence farming on arid land, while jobs were available in the cities.

From about 1940 until 1980 the economy of South Africa grew faster than that of any country on the continent, and faster than that of most nations in the world. There were four reasons for this growth.

First, South Africa had an inexpensive energy source based on its abundant coal reserves. Second, the country also had capital, or money, to invest. Foreigners saw that they could make money from investing in South Africa's resources, and South Africans themselves were willing to reinvest profits they had made. Third, South Africa's excellent connections with Britain and the rest of Europe provided the technology, or knowledge and skills, that South Africans needed to build factories and mills.

The fourth and most important element in South Africa's great expansion was the Africans themselves. They formed a vast pool of labor, and worked for very little because they had no choice.

Artificial Regions

The white government was frightened by the movement of Africans toward the cities. Whites were afraid that the Africans who were crowding into the townships, or settlements near the cities, might claim a right to live there permanently.

In order to control the Africans, the South African government created arbitrary regions called "homelands." Under the homelands plan, the blacks —nearly 70 percent of South Africa's population—were forced to live on only 13 percent of the country's land. Every black in the nation was assigned to a homeland and was supposed to stay in it unless a pass had been issued allowing them to live somewhere else.

An Unworkable Solution The homelands were not based on the traditional homelands of the Africans. In most cases these homelands were the least desirable lands, where the Africans had been forced to move in previous centuries.

Hand in hand with the homelands plan was a system of laws known as apartheid (uh PAR tyd), which means

Multiracial Schools

Traditionally, blacks and whites in South Africa have been taught in separate schools. But a handful of schools are trying to break the pattern of racial isolation under apartheid.

One example of this approach is St. Barnabas, a multiracial school near Johannesburg, like the school at right. "I want to use this school as a base for helping to create graduates who will contribute to healing the wounds of this country," says headmaster Michael Corke.

A large percentage of the school's graduates go on to college. Many, like one recent black graduate named Kevin Flusk, plan to work for the betterment of South Africa once they have their degrees. "I would like to help change people's attitudes," Kevin confirms.

1. What effect has apartheid had on public education in South Africa?
2. How have certain private schools tried to promote racial healing?

"apartness." Under apartheid, blacks are **segregated,** or forced to live apart from whites. By law blacks are required to use separate public facilities of all types, including schools and colleges. The facilities for blacks were never as good as those that were available for whites. For example, in 1990 the black school system was so crowded that the average class had fifty-four students. That same year the white schools were so underused that nearly eighty schools across the nation stood closed and empty.

International Backlash Apartheid and the homelands plan was an unjust system, and much of the world refused to let it continue without protest. In the 1980s South Africa's major trading partners, Europe and the United States, placed economic sanctions against South Africa. **Sanctions** are measures that punish a country for actions that the international community does not approve of.

The United States sanctions prohibited Americans from investing in South Africa and banned the import of

Negotiations for Peace

In May 1990 South African President F.W. de Klerk met with African National Congress Deputy President Nelson Mandela.

reformer, and in spite of angry opposition from some whites, he started making changes.

Moving Toward Negotiation   One of de Klerk's most important actions was the release of prominent African activist Nelson Mandela from prison, where he had been held for twenty-seven years for his anti-apartheid activities. As a leader of the African National Congress (ANC), Mandela entered into negotiations with the white government on behalf of black South Africans. Although there was no instant settlement of the great inequalities present in South Africa as a result of Mandela's release, the1990s hold great promise for change. In early 1991, de Klerk announced that his government would begin to dismantle the laws comprising apartheid.

certain South African products. In the first few years of the sanctions, half the American states, 80 cities, and some 150 colleges sold off their stocks in companies that did business in South Africa. Imports from South Africa fell 40 percent in the first nine months. One expert estimated that sanctions were costing South Africa two billion dollars a year.

The South African economy began to fall. Africans in the country's black townships kept up pressure of their own with unrest and protests that not even open police violence could stop. Finally, some South African whites became aware that political and economic changes would have to take place. Officials in the government began to realize that apartheid was suffocating their economy.

Winds of Change

In 1989 a new prime minister named F. W. de Klerk came to power in South Africa. He proved to be a

Developing Vocabulary
1. Define: **a.** apartheid
   b. segregation **c.** sanction

Place Location
2. Where is Pretoria located in relation to Cape Town?

Reviewing Main Ideas
3. What movement caused problems for white South Africans during the twentieth century?
4. Give two reasons for the recent change in the South African government's apartheid policy.

Critical Thinking
5. **Perceiving Cause-Effect Relationships**   The international community sometimes uses economic sanctions to force a government to change its policies or actions. Do you think economic sanctions are successful? Why do you think they are sometimes controversial?

✔ Skills Check

- [ ] Social Studies
- [ ] Map and Globe
- [ ] Reading and Writing
- [✔] Critical Thinking

Drawing Conclusions

Drawing conclusions means figuring out information that is suggested but not stated directly. The ability to do this enables you to go beyond what is presented in the text and form new insights. Use the following steps to practice drawing conclusions.

1. **Study the facts and ideas in the passage.** Before you can draw conclusions you must clearly understand the basic facts and ideas. Read the two passages below and answer the following questions: (a) How do most rural Kenyans support themselves? (b) What are some actions that have been taken against South Africa by nations critical of that country?

2. **Make a summary statement from the contents of the passage.** After reading for facts and ideas, try to summarize the basic information in the passage. Answer the following questions: (a) From passage A, what can you conclude about the nature of Kenyan agriculture? Are most Kenyan farmers able to support themselves through farming? (b) From passage B, what can you conclude about the effects of international criticism on white South Africans?

3. **Consider whether or not you can draw a conclusion based on what is stated.** Depending on the information in the passage, you may or may not be able to draw valid conclusions. Answer the following questions: (a) Given what you read in passage A, can you conclude that the number of small farmers in Kenya is more likely to grow or to shrink in the years ahead? (b) Given what you read in passage B, what conclusion can you draw about the future of apartheid in South Africa?

Passage A

In the rolling countryside of Kenya, most people live in small villages. Here, families grow crops and raise livestock. Few people in this country can survive through farming alone. It is common for rural residents to hold part-time jobs to supplement their small, unreliable farm incomes. They may work as village carpenters or blacksmiths, or on large coffee and tea plantations.

The cities are also a magnet for many. Each year, thousands of rural Kenyans move to nearby cities in search of better-paying jobs in factories, stores, and offices.

Passage B

South Africa's racial policies have long been criticized by the world community. International criticism is more than just talk; it carries an economic 'punch.' Many nations have greatly restricted their trade with the embattled country.

White South Africans generally support segregation and resent the criticism that has placed their nation in such a negative light. They say that black people live better in South Africa than in neighboring countries. A growing number of white South Africans, however, are now calling for an end to apartheid.

4 Other Countries of Southern Africa

Section Preview

Key Ideas

- The countries of Southern Africa are affected by the wealth and policies of the Republic of South Africa.
- Angola and Mozambique share a background of conflict and a potential for future prosperity.
- The current condition of Zambia and of Zimbabwe can be explained in part by their different attitudes toward farming.

Key Term

land redistribution

A School Day in Botswana

As in many countries in Africa, Botswana's government believes that a good system of education is a key to the country's success.

The Republic of South Africa casts its shadow over the entire region of Southern Africa. Most of the region has been affected by its apartheid policies or its economic and military might. The presence of the Republic of South Africa is one of the characteristics that defines Southern Africa as a region.

A Region in Shadow

In some cases South Africa's superiority has overwhelmed its neighbors. Lesotho (luh SO to), for example, is completely surrounded by and dependent upon the Republic of South Africa. Swaziland, although richer in resources than Lesotho, is in much the same position.

Namibia, on the west coast, was almost a colony of South Africa until recently. It even had its own version of apartheid, including black homelands.

Landlocked Malawi (muh LAH wee), a crowded nation on the western shore of Lake Nyasa in the Great Rift Valley, has worked to remain on good terms with South Africa because many of its migrant workers have labor contracts in South Africa. Malawi is a member of the Southern African Development Coordination Conference (SADCC), a group of countries that work together to end the region's economic dependence on the Republic of South Africa.

Like Malawi, Botswana relies on economic ties to South Africa and must keep relations friendly with the apartheid-based government there. Botswana, however, is less dependent on South Africa than Malawi is because it is wealthier.

A comparison of Malawi and Botswana reveals the impact of physical geography on their economies. Malawi is fertile and has an excellent water supply, so that over time it has attracted a large population. Its resources must be stretched to meet the needs of more people. Botswana, on the other hand, is an arid country, but it is sparsely populated. Its yearly profits from the sale of diamonds, copper, coal, and the millions of beef cattle it raises every year benefit a large part of its population.

Botswana, too, is a member of SADCC. It also allowed the anti-apartheid African National Congress to maintain bases within its borders during the time the congress was banned in South Africa. Botswana's leaders see their country as a "bridge nation" between white-controlled South Africa and the black-controlled nations around it.

Angola and Mozambique

Angola and Mozambique, although separated from one another by the other countries of Southern Africa, share similar characteristics. Both are coastal states that were once Portuguese colonies. Both countries won their independence very suddenly in 1975 when Portuguese settlers fled, taking their wealth with them. This "white flight" made the task of the new governments doubly difficult.

A Tangled Web Reacting to the problems that colonialism and capitalism had created in their countries, both governments committed themselves to a communist economic system. This angered South Africa and Western countries such as the United States. In Angola a rebel group known as Unita waged war against the new government; in Mozambique a group known as Renamo played a similar role. South Africa backed both rebel groups with weapons, money, and, in the case of Angola, troops. Countries on four continents became involved in both conflicts.

Paying the Costs The human cost of these wars was horrifying. In Mozambique about nine hundred thousand

Civil War in Angola
Countries on four continents became involved in a rebel group's fight against a communist government in Angola.

people died. Under the stress of civil war, the communist economies of both nations fell apart. Angola, once able to feed its people, soon had to import most of its food. This is how one historian described the condition of Angola at the end of the 1980s:

A generation of young Angolans has grown up in a country at war. There are one million refugees, many packed into urban slums. Disease and malnutrition are widespread, and one child in four dies before the age of five. Once prosperous southern settlements are now ghosttowns, reduced to rubble.

Peace and Potential By the beginning of the 1990s hope for peace was emerging. South Africa, Cuba, the

WORD ORIGIN

malnutrition
The condition resulting from not having enough to eat comes from two words: the French *mal,* meaning "bad" or "sick," and the Latin *nutrire,* "to nourish, or feed."

United States, and the Soviet Union had ended their military involvement with Angola. In Mozambique, the government expressed a willingness to include the rebels in the political process. Today the outlook for both countries is excellent.

Angola may again produce enough food for its population. The country has valuable oil and diamonds from which it might profit. Mozambique has a huge labor force, an excellent port, and good transportation connections with South Africa and other African nations. It also has energy for industry in the form of tremendous reserves of coal and its huge hydroelectric dam on the Zambezi River.

In Pursuit of Independence

Two countries in the region, Zambia and Zimbabwe (zim BAHB way), have tried with some success to keep themselves out of South Africa's shadow. Each of these countries knows that having an independent economy that is not dependent on South Africa gives it more freedom within the region. Although they have pursued the same goal, they have fared very differently.

A Missed Opportunity Half a billion years ago seas that washed over present-day Zambia laid down over 800 million tons of copper in what is now called the Katanga deposit. Early in this century, European and American companies began to exploit this wealth. Agriculture in the country developed substantially, too, until Zambia achieved independence in 1964. Then the commercial farmers, most of them white, fled the country in uncertainty about their future.

The president of Zambia, Kenneth Kaunda, sure that Zambia's copper would always provide the nation with

Victoria Falls, Zambezi River, Southern Africa
The Zambezi River divides Zimbabwe and Zambia. Waterfalls like this one, formed along the edge of the continent's great interior plateau, make many of Africa's rivers difficult to navigate.

money to buy food, took few steps to rebuild Zambia's agricultural economy. At first it seemed he was right; Zambia became the fourth-largest producer of copper in the world.

During the 1970s and 1980s, however, the price of copper on the world market plunged. Zambia became poor as fast as it had become wealthy. Today the average household in Zambia earns only about 60 percent as much as it did in 1964. Kaunda tried a structural adjustment program to save the economy, but backed out of it when his people rioted.

Making Agriculture Work Zimbabwe's experience was different. After the nation won full independence in 1980, its leader, Robert Mugabe, was very cautious about making changes. The highly successful commercial farmers—most of whom were white—did not abandon the country as they had in Zambia.

These sixty thousand white farmers owned over three times as much land as the five million black farmers. For this reason, Mugabe pursued a policy of land redistribution. Under such a policy, land is taken from those who have plenty and given to those who have little or none. According to Zimbabwe's constitution, white farmers had to be paid for their land, and they had to be willing to sell it before it could be taken from them.

Land redistribution went slowly in Zimbabwe. This gave the government more time to develop the necessary infrastructure for the new farmers. An infrastructure is a country's basic support systems—transportation, education, water and electricity supply, and other necessities that keep an economy operating.

As a result of Zimbabwe's caution, the nation's farmers have continued to be among the most productive in the world. After one recent harvest Zimbabwe had enough reserves of corn to feed its people for two years.

Zambia's Copper Industry
Coal, zinc, and copper, shown above, are among Zambia's most valuable mineral resources.

SECTION 4 REVIEW

Developing Vocabulary
1. Define: land redistribution

Place Location
2. Why does Lesotho's location make it dependent upon the Republic of South Africa?

Reviewing Main Ideas
3. List three ways in which Angola and Mozambique are similar.
4. Why did agriculture fail to develop in Zambia but develop successfully in Zimbabwe?

Critical Thinking
5. **Predicting Consequences** Some observers in Zimbabwe fear that Mugabe will accelerate the pace of land redistribution in the 1990s. What consequences might this have on the country's agriculture?

27

REVIEW

Section Summaries

SECTION 1   Kenya   Kenya's most fertile land is located in the central highlands. It once belonged to the Kikuyu people, but was taken over by white settlers brought in by the British to justify the cost of the railroad across Kenya. The Kikuyu won back some of their land after Kenya achieved independence in 1963. Since that time *harambee,* pulling together, has been the guiding principle of Kenya. The result has been solid economic growth. Too much emphasis on cash crops for export, however, has left some 30 percent of Kenyans suffering from malnutrition. Kenya's future looks uncertain because the population is growing faster than the economy can create jobs for it.

SECTION 2   Other Countries of East Africa   East Africa's location is vital as a site for potential trade and because it has great strategic value. Among its many different nations is Mauritius, an island nation forging new trade links with other countries. In Ethiopia, both civil war in the seacoast provinces and a border war have caused the movement of hundreds of thousands of refugees. The Sudan and Uganda have both fought prolonged civil wars caused by rivalries between north and south. In Rwanda and Burundi the Hutu and the Tutsi groups respectively have established ethnocracies. Tanzania has had to adapt its ideology to make up for the failure of its socialist program, *ujamaa.*

SECTION 3   South Africa   Long periods of dispute within South Africa resulted in a predominantly black African nation ruled by a white minority. Under the apartheid system, blacks were assigned to live in arbitrarily created "homelands." They were not allowed opportunity for a fair share in South Africa's economic wealth, which was built in part on the cheap labor they supplied. Under the pressure of international sanctions and black unrest within South Africa, the white government has been moving toward a settlement that may allow its black majority true political and economic power.

SECTION 4   Other Countries of Southern Africa   The presence of the Republic of South Africa is one of the defining characteristics of the region of Southern Africa. Some nations are highly dependent on the Republic of South Africa. The Republic of South Africa has intervened in conflicts in the region, such as those in Angola and Mozambique. Zambia and Zimbabwe have each tried different routes to achieve prosperity and independence from South African influence. Zambia chose to depend on its copper resources to fund its development, but this policy failed when copper prices fell. Zimbabwe has been able to make its agriculture among the most productive in the world.

Vocabulary Development

Use each of the following terms in a sentence that shows its meaning.

1. malnutrition
2. strategic value
3. ethnocracy
4. villagization
5. apartheid
6. sanction
7. land redistribution
8. segregation

Main Ideas

1. How did Kenya's course change socially and politically during the 1980s?
2. Why does Mauritius have strong ties to Asia?

3. Why are Rwanda and Burundi considered ethnocracies?
4. List four reasons why the economy of the Republic of South Africa boomed in the period from 1940 to 1980.
5. What is the Southern African Development Coordination Conference?

Critical Thinking

1. **Determining Relevance** How did British rule over the Sudan contribute to later regional conflict in that country?

2. **Formulating Questions** Write two questions that a white South African might have about any proposed settlement of South Africa's inequalities, and two questions a black South African might have.

Practicing Skills

Drawing Conclusions Look again at Passage B on page 563. Using the steps you have learned, what conclusions can you draw about the effect on South Africa of international actions against apartheid?

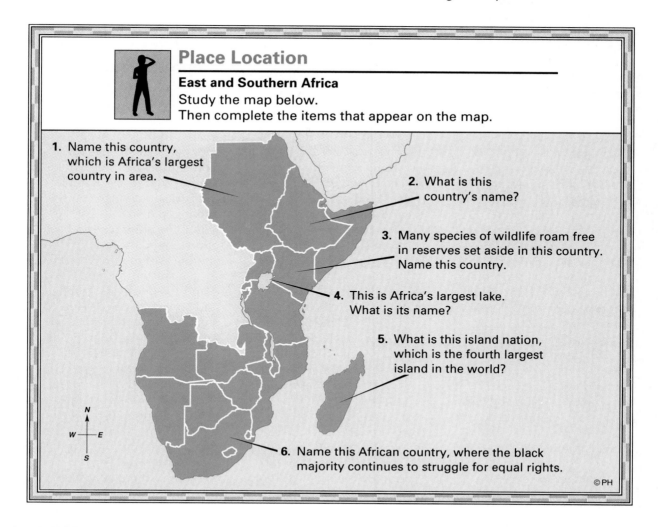

Place Location

East and Southern Africa
Study the map below.
Then complete the items that appear on the map.

1. Name this country, which is Africa's largest country in area.

2. What is this country's name?

3. Many species of wildlife roam free in reserves set aside in this country. Name this country.

4. This is Africa's largest lake. What is its name?

5. What is this island nation, which is the fourth largest island in the world?

6. Name this African country, where the black majority continues to struggle for equal rights.

©PH

MAKING CONNECTIONS

•WHERE REGIONS MEET•

Saving Endangered Animals

Reading the list of names— orangutan, panda bear, blue whale—it is hard to imagine a world without them. Yet the reality is that these and many other of the world's magnificent creatures are threatened with extinction. Concern over the growing list of endangered species connects every region of the world.

Why are so many animal species threatened? There are two major causes. First, humans have moved into the habitats of many animal species. The growth of cities has led to many changes in the natural landscape. The damming of rivers, the draining of swamps, the clearing of forests, and the pollution that often comes with human habitation have destroyed animal habitats and food sources.

Second, hunters have devastated many species. Historically, humans mainly hunted for food and killed relatively few animals. Today, however, many people hunt for fun or profit. Even in areas where animals are legally protected, poachers (illegal hunters) prey on endangered species to make money.

Africa's Endangered Elephants

During the 1980s, the plight of endangered species in Africa began to cause great concern among scientists and animal lovers, and became a symbol of endangered species everywhere. These animals include cheetahs, gorillas, and white and black rhinos.

By the late 1980s, even the once vast herds of African elephants were endangered. In fact, in the ten years between 1980 and 1990, poachers and hunters reduced the African elephant population by more than 50 percent. Even within protected parks and game reservations, efforts to stop the slaughter were only partially effective.

Like most endangered African species, elephants were hunted mainly for economic reasons. Consumers around the world have long prized products made from the ivory of elephant tusks. As a result, ivory brought high prices on the world market, and poachers could earn huge profits from killing elephants. In addition, many African farmers, who consider the elephant a dangerous nuisance that can trample a season's crop in minutes, turned a blind eye to the poaching.

The World Reacts to Elephant Slaughter

People around the world have become aware of the destruction of the African elephant, and they have responded. One result: the Convention on International Trade voted in 1989 to place the African elephant on the endangered species list. As a result, the shipment and sale of any elephant products, including ivory, is now banned. As a result of the ban, the demand for ivory has dropped, and so has its price. This change means that poachers no longer make huge profits.

Environmentalists and international wildlife protection groups hope that the recent success in preserving the African elephant is just one chapter in an ongoing story. These groups are continuing to focus attention on the African elephant and thereby awaken people to the plight of other endangered species around the world. The map below shows many of these endangered animals. The map also illustrates that the problem is worldwide.

A number of international groups have responded to the plight of the world's endangered species. The World Wildlife Fund raises money for conservation programs, particularly in Latin American, Asian, and African rain forests. In addition, an organization called Wildlife Conservation International operates a number of animal preservation projects in at least thirty-four nations.

The View from the U.S.A.

In addition to supporting worldwide efforts to protect elephants and other species, the United States has its own share of endangered animals to protect. These include grizzly bears, timber wolves, bald eagles, spotted owls, and California condors. Conservation groups and zoological societies are working to breed some of these animals in captivity and eventually return them to the wild. Their efforts have met with some success but the future is still unknown.

TAKING ANOTHER LOOK

1. Give two reasons why animals become endangered.
2. Why is the fate of the African elephant especially important to some international wildlife groups?

Critical Thinking
3. **Determining Relevance** How are the attitudes of poachers who hunt elephants and farmers who ignore poaching related?

Applying Geographic Themes

Interaction African elephants are an endangered species. Name three endangered species in North America.

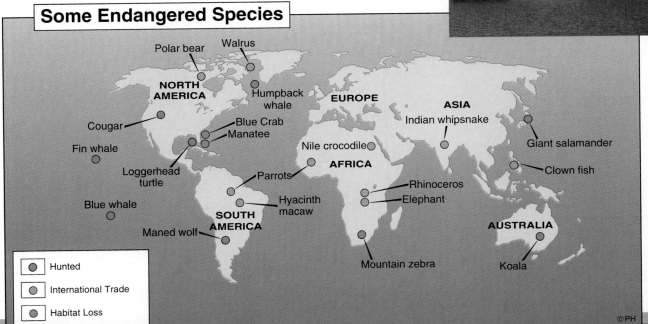

Some Endangered Species

Polar bear
Walrus
NORTH AMERICA
Humpback whale
EUROPE
ASIA
Indian whipsnake
Cougar
Blue Crab
Manatee
Giant salamander
Fin whale
Nile crocodile
AFRICA
Clown fish
Loggerhead turtle
Parrots
Rhinoceros
Elephant
Blue whale
Hyacinth macaw
SOUTH AMERICA
AUSTRALIA
Maned wolf
Mountain zebra
Koala

Hunted
International Trade
Habitat Loss

©PH

UNIT
9
South Asia

CHAPTERS
28 Regional Atlas: South Asia
29 The Countries of South Asia

Magnificent mountains six miles above sea level . . . broad plateaus stretching to the horizon . . . a rich alluvial plain drained by three great rivers . . . torrential, seasonal rains that hold the key to life . . . masses of people hurrying through crowded city streets . . . solitary monks praying in the silence of a temple . . .

The contrasts of South Asia's physical geography are as striking as its culture—a mix of traditional and modern. In this unit, you will discover how landscapes and climate affect the everyday lives of the region's millions of people.

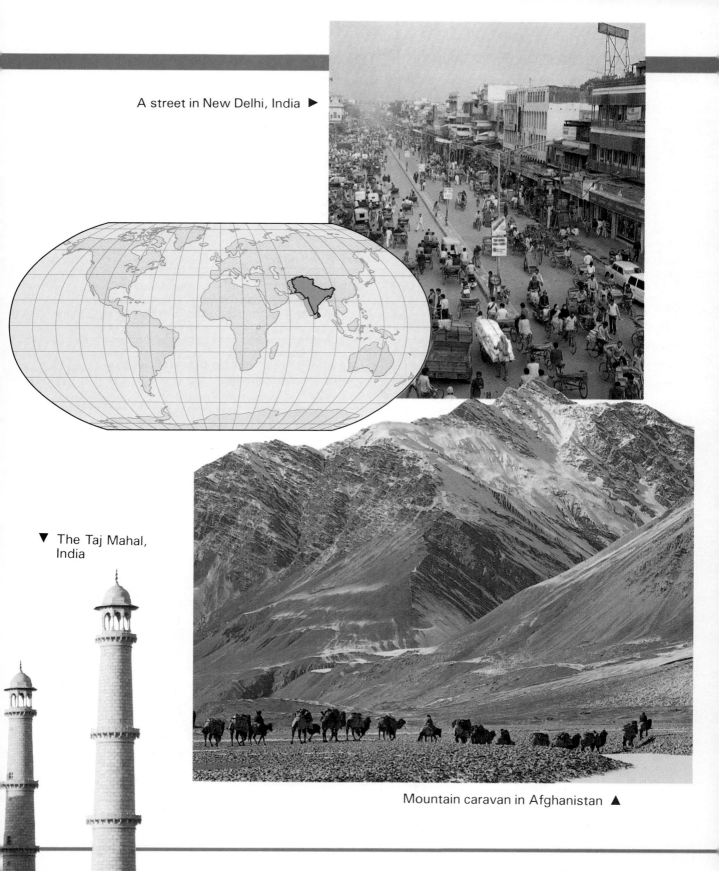

A street in New Delhi, India ▶

▼ The Taj Mahal, India

Mountain caravan in Afghanistan ▲

28

Regional Atlas: South Asia

Chapter Preview

Both sections in this chapter provide an overview of South Asia, shown in red on the map below.

Sections	Did You Know?
1 LAND, CLIMATE, AND VEGETATION	The Himalayan peak of Mount Everest is the highest mountain in the world. It is nearly twenty times higher than the Empire State Building.
2 HUMAN GEOGRAPHY	One of every five people on earth lives on the Indian subcontinent.

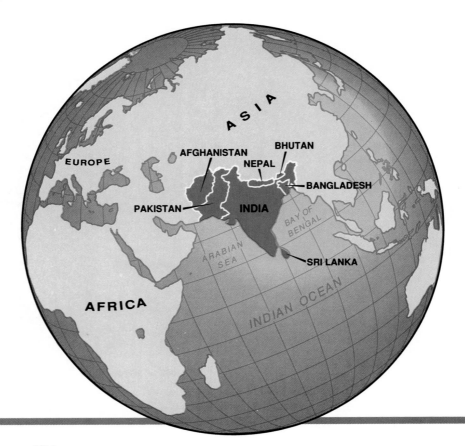

The Himalaya

The Himalayan range includes thirty of the world's highest mountains and are often called the ''rooftop of the world.'' Scientists believe that the Himalaya started pushing upward millions of years ago when the tectonic plate carrying what is now India collided with the Eurasian plate.

1 Land, Climate, and Vegetation

Section Preview

Key Ideas

- Monsoons bring vital rains as well as dangerous flooding to South Asia.
- The mountain ranges of South Asia determine where the rains fall.

Key Terms

subcontinent, alluvial plain, monsoon

Like an enormous arrowhead, the Indian subcontinent points southward from the sprawling landmass of Asia. A **subcontinent** is a large landmass, a major subdivision of a continent. The colossal Himalayas, the highest mountains in the world, form a barrier between the Indian subcontinent and the rest of Asia. Look at the map on page 576 to locate the Himalayas. Then locate the seven countries of the Indian subcontinent—India, Pakistan, Bangladesh, Nepal (nuh PAWL), Bhutan (BOO tahn), Afghanistan, and the island country of Sri Lanka (sree LAHNG kuh). This area of the world is called South Asia.

Although smaller than the United States, the Indian subcontinent is much more varied in its landforms, vegetation, and climate. Lush rain forests spread across the slopes of India's west coast. Deserts stretch throughout Pakistan. Glacier-covered mountains overlook the isolated villages of Nepal. Low, steamy marshlands cover much of Bangladesh. Geographically, South Asia is a region of dramatic contrasts.

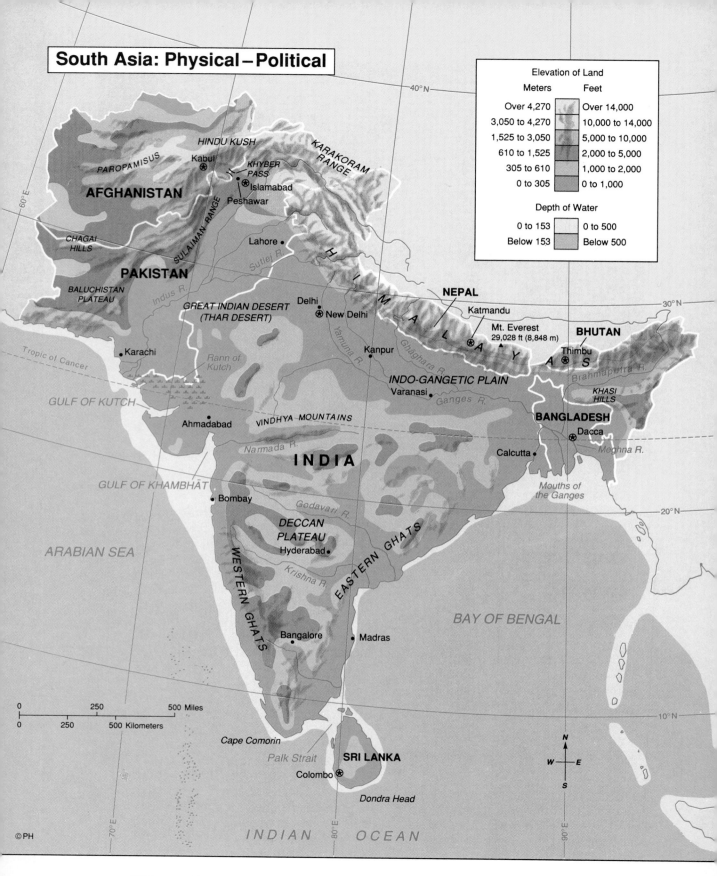

South Asia: Physical–Political

Elevation of Land

Meters	Feet
Over 4,270	Over 14,000
3,050 to 4,270	10,000 to 14,000
1,525 to 3,050	5,000 to 10,000
610 to 1,525	2,000 to 5,000
305 to 610	1,000 to 2,000
0 to 305	0 to 1,000

Depth of Water

0 to 153	0 to 500
Below 153	Below 500

40°N

HINDU KUSH
PAROPAMISUS
KARAKORAM RANGE
KHYBER PASS
Kabul
Islamabad
Peshawar
AFGHANISTAN
CHAGAI HILLS
PAKISTAN
BALUCHISTAN PLATEAU
Lahore
Sutlej R.
SULAIMAN RANGE
Indus R.
H I M A L A Y A S
NEPAL
Katmandu
Mt. Everest 29,028 ft (8,848 m)
BHUTAN
Thimbu
30°N
GREAT INDIAN DESERT (THAR DESERT)
Delhi
New Delhi
Yamuna R.
Kanpur
Ghaghara R.
Brahmaputra R.
Karachi
Rann of Kutch
Tropic of Cancer
INDO-GANGETIC PLAIN
Varanasi
Ganges R.
KHASI HILLS
GULF OF KUTCH
VINDHYA MOUNTAINS
BANGLADESH
Dacca
Ahmadabad
Narmada R.
INDIA
Calcutta
Meghna R.
GULF OF KHAMBHĀT
Mouths of the Ganges
20°N
Bombay
Godavari R.
DECCAN PLATEAU
Hyderabad
EASTERN GHATS
ARABIAN SEA
WESTERN GHATS
Krishna R.
BAY OF BENGAL
Bangalore
Madras
0   250   500 Miles
0   250   500 Kilometers
Cape Comorin
Palk Strait
SRI LANKA
10°N
Colombo
Dondra Head
N
W   E
S
60°E   70°E   80°E   90°E
INDIAN OCEAN

©PH

A Land Shaped by Plate Movement

Until fairly recently in the history of the earth's landforms the Himalayas did not exist at all. In their place was the ocean. The Indian subcontinent, riding atop the Indo-Australian plate, was once an island separated from Asia by the sea. About fifty million years ago, the Indo-Australian and the Eurasian plates began to collide. In the resulting crush, the earth's crust folded up like an accordion. The land that was pushed skyward then became the ridges of the Himalayas and nearby ranges.

Towering about 5.5 miles (8.8 km) above sea level, the Himalayas are among the highest mountains in the world. As Ashvin Mehta, an Indian photographer, wrote on first seeing the Himalayas, "Apart from clouds or stars, I had never seen anything on earth that high before."

Snow-fed Rivers   Look at the map on page 576; notice that three great rivers cross the northern region of the Indian subcontinent: the Indus, the Ganges (GAN jeez), and the Brahmaputra (BRAHM uh POO truh). These rivers begin their journeys to the sea as trickles down the icy crags of the Himalayan slopes. When they reach the lower, flatter land of the plains, they slow and unload the silt that they picked up along their journey. After millions of years of depositing their cargoes, the rivers have formed alluvial plains. **Alluvial plains** are broad expanses of land along riverbanks. They are made of the rich, fertile soil left by the flooding of the rivers. Consequently, parts of the Indus, Ganges, and Brahmaputra valleys are excellent areas for farming.

The Deccan Plateau   South of the fertile Indo-Gangetic Plain in the central part of the Indian subcontinent lies the Deccan Plateau. The Deccan, which means "south" in Sanskrit, stretches southward and westward through the heart of India. Find this area on the map on page 576. Two sets of mountains, the Western and Eastern Ghats, frame the Deccan, outlining the enormous triangle that forms the tip of the Indian subcontinent and gives it such a distinctive shape.

Interaction: South Asia's Monsoon Climate

Two key characteristics describe the climate of much of South Asia: the scorching heat—which can reach a temperature of 110°F (43°C) or more —and the monsoons. Monsoons are seasonal shifts in the prevailing winds.

In winter the winds blow from the northeast and bring dry continental air from Asia's mainland to much of South Asia. But in the summer, the winds reverse direction and pick up moisture from the warm Indian Ocean. When the moisture-laden monsoon winds move over the land, they release heavy rains.

One traveler to South Asia wrote that the monsoons bring rains that

. . . drop like great sheets, like lakes raised and then let loose upon the land below. They come endlessly, one feels, . . .

Applying Geographic Themes

Regions   Before gaining independence in 1947, Bangladesh and Pakistan were also a part of India. The subcontinent's northern boundary lies in the highest mountain ranges in the world. What are the names of these two northern ranges?

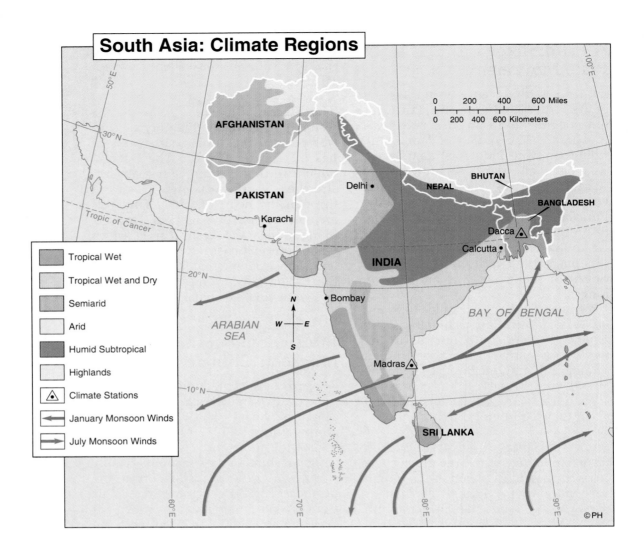

South Asia: Climate Regions

Legend:
- Tropical Wet
- Tropical Wet and Dry
- Semiarid
- Arid
- Humid Subtropical
- Highlands
- Climate Stations
- January Monsoon Winds
- July Monsoon Winds

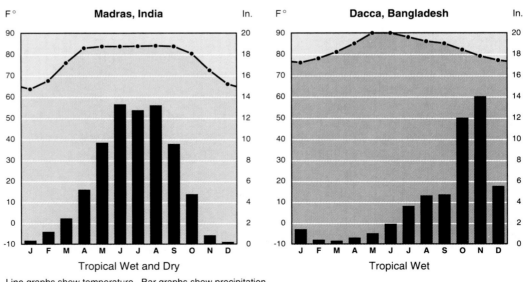

Madras, India

Tropical Wet and Dry

Dacca, Bangladesh

Tropical Wet

Line graphs show temperature. Bar graphs show precipitation.

DAILY LIFE

Waiting for the Monsoon

Just before the monsoon, the rice fields of northern India, shown at right, are brown and dry. It is an anxious time of waiting and watching the skies.

Suddenly, the first big rains lash across the land. Whole families rush into the fields. They must plant their rice seedlings while the ground is soggy. Women bend down in the rain to do the planting while men guide the water buffalo along the rows of rice, loosening the clumps of soil with their plows. Men, women, and children work to the point of exhaustion. By summer's end the fields are green.

1. Why do farmers stockpile food before the arrival of the monsoon?
2. What must happen in the rice fields once the rains begin?

The monsoons affect nearly every region of South Asia. In the north, the Himalayas form an unyielding barrier when the monsoon winds blow from the southwest. Blasting against the high wall of mountains, the winds drop their moisture with fury, sometimes causing dangerous landslides. Terraced fields, homes, and entire villages can be swept away by landslides during monsoon season.

Applying Geographic Themes

Regions The monsoons bring practically all the rain that falls on South Asia. Use the two climate graphs to compare seasonal patterns of rainfall in Madras and Dacca.

The lowland population also awaits the southwest monsoon with fear. If the rains hit hard, these people face the danger of storm surges and floods. A storm surge is a sudden rise in the water level of the ocean during a storm. Coastal areas, such as those in Bangladesh, can be particularly hard hit by storm surges.

The monsoons are not only life-threatening but also life-giving. Much of South Asia is hot and dry for half the year. For the many farmers who cannot afford to irrigate, the annual coming of the monsoon rains means survival. When they arrive on time, farmers can begin planting. But when the rains are late, people face drought and hunger.

Chapter 28, Section 1   **579**

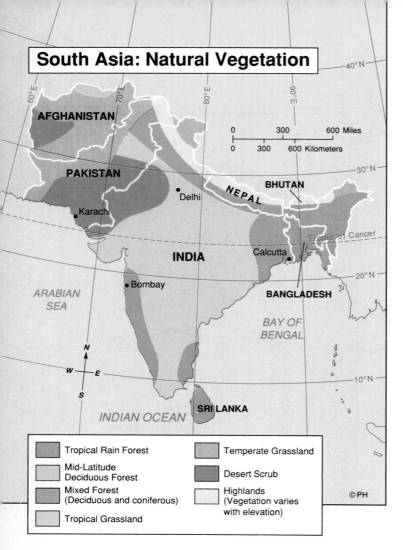

South Asia: Natural Vegetation

AFGHANISTAN

PAKISTAN

• Delhi

NEPAL

BHUTAN

• Karachi

INDIA

Calcutta

Tropic of Cancer

ARABIAN
SEA

• Bombay

BANGLADESH

BAY OF
BENGAL

N
W—E
S

INDIAN OCEAN

SRI LANKA

0   300   600 Miles
0   300   600 Kilometers

40°N
30°N
20°N
10°N
60°E 70°E 80°E 90°E

©PH

Tropical Rain Forest

Mid-Latitude
Deciduous Forest

Mixed Forest
(Deciduous and coniferous)

Tropical Grassland

Temperate Grassland

Desert Scrub

Highlands
(Vegetation varies
with elevation)

Applying Geographic Themes

Place   When scant rain falls on the subcontinent's interior, the grasses temporarily turn from brown to green. What type of natural vegetation is dominant on India's southwest coast? What type is dominant in the area around Karachi?

WORD ORIGIN

monsoon
The word *monsoon* is derived from an Arabic word meaning "time" or "season."

Linking Elevation, Climate, and Vegetation

The physical geography of South Asia determines where the monsoon winds drop their rain and where the forests grow. Compare the physical-political map on page 576 with the natural vegetation map on this page. You will see from the vegetation map

that a thick rain forest spreads across the Malabar Coast of southwest India. This rain forest thrives because of the abundant rainfall brought by the monsoons. When the southwest monsoon winds rising from the Arabian Sea meet the barrier of the Western Ghats, they drop heavy rains, supporting the tropical vegetation of the coast.

Similarly, whenever the monsoon winds blow across the Bay of Bengal to the east of India and hit the Himalayas, enormous amounts of rain fall in Bangladesh and the eastern states of India. Average rainfall in this part of the world reaches an astonishing 450 inches (1,143 cm) per year. That's almost 40 feet (12.2 m) of water—four times the height of a basketball net!

Because the Western Ghats block the monsoon rains, the interior of the subcontinent is hot and dry. The desert scrub vegetation is typical of a semiarid climate region—dry, coarse grasses growing in tufts and short tree growth.

SECTION 1 REVIEW

Developing Vocabulary
1. Define: **a.** subcontinent **b.** alluvial plain **c.** monsoon

Place Location
2. Identify the mountain range that acts as a barrier between the Indian subcontinent and the rest of Asia.

Reviewing Main Ideas
3. Why is the Indo-Gangetic Plain well suited for farming?
4. How do the mountains of India affect where the monsoon rains fall?

Critical Thinking
5. **Determining Relevance** Imagine that you lived in a monsoon climate. How might the rains affect your daily life?

2 Human Geography

Section Preview

Key Ideas

- South Asia is one of the most densely populated regions of the world.
- Hinduism and Islam are the predominant religions in South Asia.
- Conflict among religious groups has caused great turmoil in South Asia.

Key Terms

reincarnation, caste system

The human geography of South Asia is a blend of many cultures. Throughout its long history, many different peoples have settled in South Asia, bringing with them their own distinct cultures. Today the culture, religion, and architecture of the region reflect South Asia's immense variety.

Linking Population to Climate

As the population density map on page 582 shows, South Asia has one of the most densely settled populations on earth. By 1990, South Asia counted more than 650 people per square mile (1,684 per sq km), compared with 67 people per square mile (174 per sq km) in the United States. That means the region is almost ten times as crowded as the United States. Remember too, that the population density figure is an average and doesn't reflect the conditions in cities, where the overcrowding can be much greater.

Compare the population density map with the climate map on page 578; notice that the greatest population is concentrated in areas of abundant rainfall. These include coastal regions, as well as northeastern India and Bangladesh. The fertile land of the Indo-Gangetic Plain is one of the most densely populated areas of the subcontinent.

Some areas of South Asia are sparsely populated. Populations are lower in areas where it is more difficult for people to survive. For example, the mountainous and heavily forested areas of the north and northeast, including Nepal and Bhutan, are sparsely populated. The deserts and semiarid regions of the far west are also much less crowded than coastal cities with more moderate climates.

Much of South Asia is rural, a land of villages. But the region is becoming increasingly urban. Around 1900, only about 10 percent of the population lived in cities. Today the figure is slightly more than 25 percent.

Despite this trend, agriculture still dominates South Asia's economy. The vast majority of people—about three fourths of the working population—depend directly on the land for their livelihood and survival.

Urban Growth in Bombay

Many newcomers to Bombay live in packed settlements on the outskirts of this coastal city.

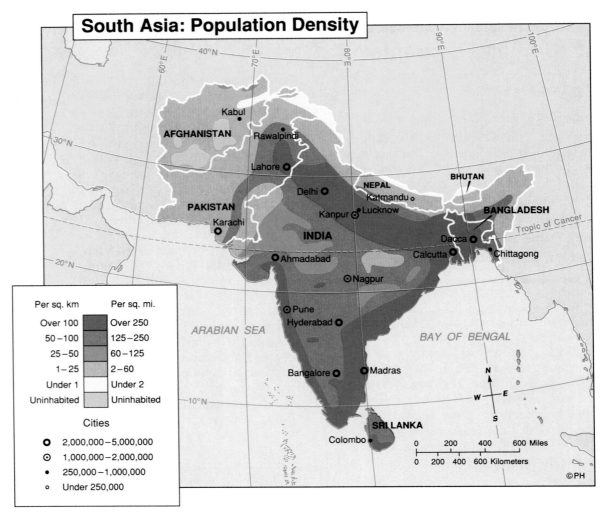

South Asia: Population Density

Per sq. km		Per sq. mi.
Over 100		Over 250
50–100		125–250
25–50		60–125
1–25		2–60
Under 1		Under 2
Uninhabited		Uninhabited

Cities

- **O** 2,000,000–5,000,000
- **⊙** 1,000,000–2,000,000
- **•** 250,000–1,000,000
- **o** Under 250,000

Applying Geographic Themes

Regions Population predictions estimate that the Indian subcontinent will be inhabited by more than 1 billion people by the year 2000. On the whole, which is the most densely populated country in South Asia?

A Land with Many Languages

About half the people in India understand a language called Hindi (HIN dee). Many different languages are spoken in South Asia, however. The languages of South Asia generally belong to two families—the Dravidian (druh VID ee uhn) languages, spoken in southern India, and several Indo-European languages, one of which was brought to South Asia by a group called the Aryans. Within these families are a great variety of languages.

The map on page 583 shows the major native languages spoken in India today. Many people also speak English. In addition, about 850 dialects are spoken. A dialect is a local form of a language. People often cannot understand someone who is speaking the same language but using a dialect different from their own.

The Indian government has been making efforts to increase communication between different regions in India by encouraging the use of Hindi as well as English.

Cultures and Conflict

The majority of people in India practice Hinduism, which is an ancient polytheistic religion. That is, its followers believe in many gods. Hinduism teaches the unity of all life. Hindus believe that every living thing has a spirit, or soul, which comes from the Creator, Brahma. Because every creature possesses a soul, Hindus treat animals with great respect. Many Hindus are vegetarians because they do not believe that animals should be killed to supply people with food. Cows are especially sacred to Hindus and are allowed to wander freely through city streets.

Hindus also consider the Ganges River to be holy. The Ganges is believed to purify the souls of the people who bathe in or drink its water. As a result, the banks of the Ganges often are lined with Hindus who have come for healing.

According to Hinduism, the final goal of every living thing is unity with Brahma, a state of bliss without change or pain. In order to achieve this goal, the soul passes through cycles of reincarnation. Reincarnation is the belief that the souls of human beings and animals go through a series of births, deaths, and rebirths. Hindus believe that the soul does not die but passes from body to body until it becomes pure enough to be united with Brahma.

The Caste System For hundreds of years, Hindu society has been organized according to the caste system. This system is a social hierarchy in which people are born into a particular group that has been given a distinct rank in society. The caste system most

likely grew out of the social patterns of the Aryans who invaded India about 3,500 years ago.

At the top of the caste system are Brahmins—the priests, teachers, and judges. Beneath the Brahmins are the Kshatriyas (kuh SHAHT ree yuhz), or warriors. Below these two groups are the Vaisyas (VEEZ yuhz), farmers and merchants. The fourth group are the Sudra, craftworkers and laborers.

A group called "untouchables," or outcastes, holds the lowest rank in the caste system. Traditionally, untouchables do the work that is considered "unclean," such as street sweeping and tanning hides. Untouchables have traditionally been denied common privileges and lived bleak lives outside the social system.

Applying Geographic Themes

Place Newspapers in India are printed in about fifty languages. What is the dominant language in the region around the city of Delhi?

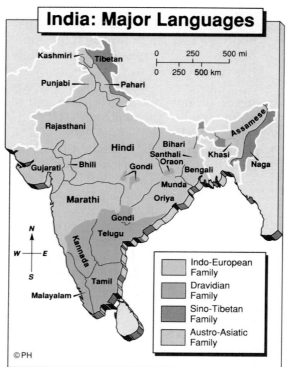

India: Major Languages

Sikhism's Most Sacred Shrine
The Golden Temple in the northern Indian city of Amritsar is surrounded by a lagoon of holy water in which religious Sikh's bathe.

Over time the caste system became very rigid. Traditionally, people could marry only within their caste. There were complicated rules for eating and sharing food. Today the system has become less rigid. Although some people take up professions that follow the traditions of their caste, many do not. However, social relationships are often, though not always, confined to people within the same caste.

In 1952, all Indians won official equality under the country's constitution. But the caste system continues to shape people's lives. Children of untouchables continue to have fewer opportunities in education and employment than their contemporaries who belong to higher castes. Efforts are being made by the Indian government, however, to offer greater employment and educational opportunities to that group.

Islam and Other Religions Islam is the major religion in Pakistan, Afghanistan, and Bangladesh. As explained in Chapter 22, Islam is based on the beliefs of Mohammed, who lived in the seventh century. Followers of Mohammed, called Muslims, are monotheistic. That is, they believe in one God. In contrast to Hindus, Muslims believe in the equality of all people.

Other religions practiced in South Asia include Christianity, Sikhism (SEEK iz uhm), and Jainism (JY nihz uhm). Sikhism began as a movement to combine Hinduism and Islam. Sikhs are not divided into castes. Today many are farmers in their native province of Punjab in northwest India. Jainism, which also developed from Hinduism, teaches that violence of any kind is wrong.

Conflict among religious groups has caused great turmoil in South Asia. At times the distrust—between Hindus and Muslims, Muslims and Sikhs, and most recently, Hindus and Sikhs—has boiled over into rioting and violence. Chapter 29 describes how the conflict between Hindus and Muslims resulted in dividing the subcontinent.

WORD ORIGIN

caste
The word *caste* comes from a Portuguese word meaning "race" or "lineage." Its early roots go back to a Greek word meaning "to split" and a Sanskrit word meaning "cuts to pieces."

SECTION 2 REVIEW

Developing Vocabulary
1. Define: **a.** reincarnation **b.** caste system

Place Location
2. Which body of water lies south of Pakistan?

Reviewing Main Ideas
3. How does climate affect population density in South Asia?
4. How do most people in South Asia make a living?

Critical Thinking
5. **Drawing Conclusions** What conflicts do you think might develop in a society with a rigid social structure like the caste system?

✓ Skills Check

- ☐ Social Studies
- ☑ Map and Globe
- ☐ Reading and Writing
- ☐ Critical Thinking

Analyzing Cartograms

Countries on a cartogram are not drawn in proportion to their land area. Instead, some other feature—such as population growth, or GNP—determines the size of each nation. On the cartogram below, the size of a country's population determines its size on the cartogram.

Use the following steps to study and analyze the cartogram below.

1. **Identify the kind of information represented on the map.** Look first for the title and the key that may help you interpret what you are looking at. You will need to compare a cartogram with a conventional land-area map—as shown on pages 762–763—to assess the degree of distortion involved. Answer the following questions: (a) What kind of information does this cartogram show? (b) When compared to the land-area map, what countries appear to be very different on this map?

2. **Practice reading the information shown on the map.** Answer the following questions: (a) Which countries on the map show a very large population? (b) Which show very small populations? (c) Which two have the largest populations?

3. **Look for relationships among the data.** How would you explain the relatively large sizes of Japan, Taiwan, the Philippines, and Indonesia on the cartogram in contrast to the tiny size of Australia?

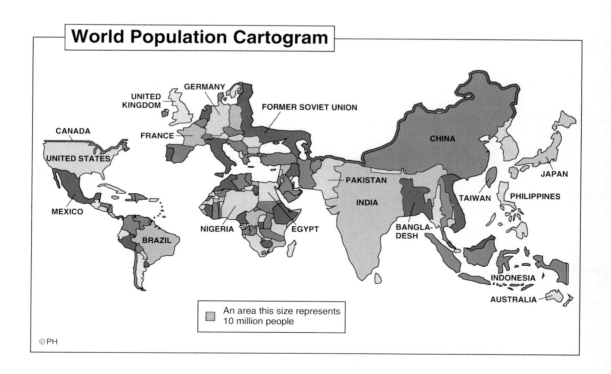

World Population Cartogram

An area this size represents 10 million people

©PH

South Asia

*The movement of people into South Asia has had a
lasting influence on culture in the region.*

South Asia is one of the world's most ancient societies. Throughout its five-thousand-year history, many groups have invaded, ruled, and lived on the subcontinent. The interaction between these groups and the geography of the land they settled has shaped much of modern South Asian history.

Laying the Cultural Foundation

Two ancient groups—the Dravidians and the Aryans—influenced the development of South Asian culture. Some scientists speculate that the Dravidians lived in the south of India while the Aryans lived to the north. It is further believed that these groups remained separate for hundreds of years, isolated by distance and by the hills of the Vindhya Range on the northern rim of the Deccan Plateau. The physical features and languages of people in India today reflect the different backgrounds of the Dravidians and Aryans.

Flourishing Culture in the Indus Valley
One of the world's earliest civilizations evolved along the Indus River in what is now Pakistan. Like the Nile and the Tigris and Euphrates, the Indus forms a ribbon of life as it passes through the surrounding desert. Beginning in about 2500 B.C. and continuing for 700 to 900 years, Indus culture was prosperous and sophisticated. Irrigated agriculture supported thousands of people. Cities were laid out in well-planned streets that formed city blocks, and people lived in brick houses several stories high. Houses even had sewage disposal systems.

No one knows for sure what happened to the Indus civilization. It may have been destroyed by flooding of the Indus River. Or the Indus dwellers might have been pushed southward by invaders from the north.

North-South Influences Some scientists believe that the Indus dwellers were Dravidians who migrated to South Asia from East Africa in prehistoric times. Scientists further speculate that the Dravidians at first settled throughout India and Pakistan and were later confined to the Deccan Plateau and the southern tip of India.

The languages of the people of southern India today—Tamil and Telugu, for example—reflect their ties to the ancient Dravidians. These languages are quite distinct from the languages of northern India.

Many people in southern India trace their ancestry to the Dravidians, while most people in northern areas of the subcontinent are descendants of the Aryans and other Asian people. The Aryans are related to the ancient Persians and Europeans. Many Indian people have both Aryan and Dravidian roots.

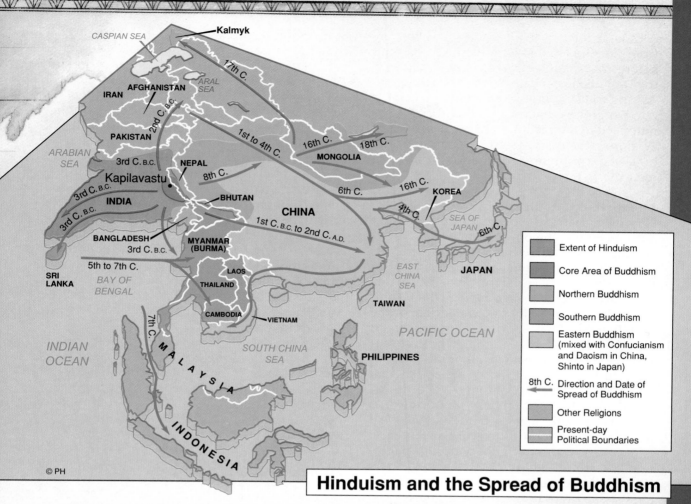

Hinduism and the Spread of Buddhism

Applying Geographic Themes

Movement   Both Hinduism and Buddhism originated in South Asia. In what century did Buddhism spread to Japan?

Beginning around 2000 B.C., the Aryans entered the subcontinent from the northwest, probably pushing the Dravidians southward.

The language of the ancient Aryans, called Sanskrit, is related to Greek and Latin. Sanskrit became the basis for most of the languages of the subcontinent, except those that are Dravidian in origin.

Religious Roots

The patchwork of religious beliefs found on the Indian subcontinent is rooted in the region's history.

Hinduism   The Aryans worshiped nature gods such as the Sun, Moon, Wind, Water, and Fire. The Aryan belief in these gods eventually developed into the Hindu religion and worship of many gods. Aryan worship also involved many religious rituals. The priests performing these rituals held great authority in Aryan society. This priesthood developed into the powerful Brahmin caste of Hinduism.

Tensions caused by clashing ways of life may have created the caste system. The Dravidians were primarily settled farmers, while the Aryans were nomadic cattle herders.

Although the Aryans learned many lessons from the Dravidians, who were settled long before the Aryans arrived, they considered the Dravidians beneath them. As a result, the Aryans tried to limit social contact between the two groups. Eventually, this social separation led to the caste system. As described in Section 2, the caste system, though officially outlawed, continues to influence life in India today.

The Birth of Buddhism   By about 600 B.C. Hinduism spanned most of the subcontinent and was the religion of Aryans and non-Aryans alike. Many Indians, however, opposed the caste system and were critical of the power held by Hindu priests. Among this group was a noble named Siddhartha Gautama. Siddhartha—later called Buddha, or the Enlightened One—turned his back on wealth and prestige. Although he was opposed to castes, he believed in reincarnation. He also believed that life was evil, but that one could escape evil through good deeds, self-knowledge, and giving up worldly desires. He taught that each individual must make a conscious effort to avoid all violence.

Buddhism did not gain wide acceptance until the third century B.C., when it flourished and coexisted alongside Hinduism for almost one thousand years. Missionaries carried Buddhism around the Himalayas, across the Plateau of Tibet, south to Sri Lanka, and east to China and Japan. Through Buddhism, India's influence spread throughout Asia. In India itself, however, Buddhism was gradually absorbed back into Hinduism.

Movement: The Spread of Islam   Throughout India's history, many invaders in addition to the Aryans entered the subcontinent—Persians from the west, Alexander the Great's armies from Greece, and Huns from Central Asia. The last of these invaders were the Muslims from the Middle East.

During the eighth century, Muslims began to settle near the mouth of the Indus River in what is now Pakistan. Over the next two centuries, they conquered territory in the Indus valley but did not have much effect on India. Then, in 998, a Muslim army began to raid Indian territory from Afghanistan. The Indians were no match for their Muslim conquerors, who used swift war horses against the slow Indian elephants. In addition, the Indians were divided among themselves and weakened by the caste system, which allowed only one section of the population—the warrior caste—to fight. Hindus of lower caste were attracted to Islam because it is based on the belief that all people are equal. Thus, Islam conquered some through the power of persuasion, rather than by the sword.

The Muslims continued to raid India for many years. Gradually most of the country came under Muslim rule. In 1526, Muslim rulers founded the Mogul Dynasty, which became the most powerful in India's history. Muslim rule in India lasted until the early eighteenth century.

Under the Moguls, Islamic culture, art, and architecture became a vital part of India's culture. Today there are mosques throughout India. One of the most famous Indian buildings, the Taj Mahal (TAHJ muh HAHL), is an example of Islamic architecture during the Mogul Dynasty.

Invasions and Empires

Notice from the map of South Asia on page 576 that the Himalayas seem to protect the subcontinent from invasion from the northeast. To the northwest tower other ranges, such as the Hindu Kush. But deep-cut passes—such as the Khyber Pass in Afghanistan—penetrate the mountains. The majority of conquerors, including the Muslims, invaded the subcontinent through these passes.

All of the subcontinent's invaders ruled over roughly the same territory: the Indo-Gangetic Plain. To the north, they were confined by the Himalayas. To the south, the Vindhya Range formed enough of a barrier to discourage large migrations south. Most newcomers to India stopped north of the Deccan Plateau. Consequently, the Dravidian culture in southern India remained to a large extent culturally and ethnically distinct from the rest of the subcontinent.

India and Britain

As early as the 1500s, Europeans—Portuguese, Dutch, French, and British—began to arrive on the shores of South Asia. The raw materials and natural resources that attracted the Europeans to South Asia were available only in a tropical climate: indigo (a blue dye), jute (a tropical plant used for making rope and burlap), sugar, tea, cotton, ginger, pepper, and other spices.

For many years, Europeans held only a few ports. However, as the last Muslim empire disintegrated in the mid-1700s, the subcontinent was left with no central ruler. Great Britain and France then vied for control of it. The French were soundly defeated in 1757. This defeat marked a new period in Indian history—the period of British rule.

After 1757, the British East India Company virtually ruled India. Under the company's rule, the subcontinent was divided into areas directly subject to British rule and others ruled by local princes. In the latter areas, the British controlled foreign policy.

At first the British had no wish to change the Indian way of life. They merely wished to acquire raw materials and sell British manufactured goods in India. However, the British soon began introducing reforms. For example, the British outlawed slavery and the killing of female babies. Suttee—the practice in which widows burned themselves on their husbands' funeral pyres—was also declared illegal. The British changed India's school system, introducing secondary schools and universities. English became the language used in schools. The Indian people became second-class citizens in their own country.

Many Indians were angered by British policies and resented British interference. Finally, in 1857, certain groups of Indian troops rebelled. As a result of this rebellion, the rule of the British East India Company ended. The British government itself took over. India officially became a British colony and, in 1877, Queen Victoria of England was proclaimed Empress of India.

India remained a British colony until 1947. Between 1857 and 1947 India's economy, land, and society changed dramatically.

British Imperialism
This painting, completed in 1846, shows British officials in India. What do you think might have been the purpose of this meeting?

The British built the largest railway system in Asia, constructed roads, and irrigated large areas of land. Health care improved and famines were prevented. Yet, the improvements brought by the British did not reach the poor and unskilled workers. Discontent with British rule festered throughout the subcontinent, among Hindus and Muslims alike. Chapter 29 explains how this discontent finally turned into a struggle for India's independence.

TAKING ANOTHER LOOK

1. What were some of the accomplishments of the Indus civilization?
2. Why did Dravidian culture remain separate from the rest of India?
3. Which geographic factors drew the British to South Asia?

Critical Thinking

4. **Identifying Central Issues** What factors caused Indians' anger toward the British?

CHAPTER

28

REVIEW

Section Summaries

SECTION 1 Land, Climate, and Vegetation

South Asia is a region of dramatic contrasts in geography and climate. Much of South Asia, however, has a monsoon climate. During part of the year, the climate is dry and very hot. During the other part of the year, the monsoon brings almost continuous rain. The people of South Asia are dependent on the monsoon for almost all their rainfall. Drought and famine result if the monsoon is late or does not come. But the monsoon can also bring dangerous flooding and landslides. When the monsoon winds hit the Himalayas or the Western Ghats, they drop their moisture, creating a tropical climate and abundant vegetation. Areas where the monsoon does not fall are hot and arid.

SECTION 2 Human Geography   Most of

South Asia is very densely populated, and most people live in rural areas and work in agriculture. The small country of Bangladesh is one of the most densely populated countries in the world. Despite improvements in food production, hunger is still a major concern throughout the region. Religion—primarily Hinduism and Islam—is an important cultural force in South Asia. The Hindu belief in reincarnation and the Hindu organization of society into the caste system have been particularly important in shaping Indian society and culture as well as the Indian economy.

Vocabulary Development

Match the definitions with the terms below.

1. a vast subdivision of a continent
2. a seasonal wind that brings heavy rains
3. a broad expanse of land made up of fertile soil that has been left by floods

4. the Hindu belief that the souls of all living creatures go through a series of births, deaths, and rebirths
5. a social system in which groups of people are separated from each other according to ancestry and occupation

a. alluvial plain
b. reincarnation
c. monsoon
d. subcontinent
e. caste system

Main Ideas

1. What are the main geographical regions of South Asia?
2. How did the Himalayas form?
3. Name three of South Asia's major rivers.
4. Why are parts of the river valleys of South Asia excellent areas for farming?
5. How do the monsoons affect the lives of the people of South Asia?
6. Why is the interior of the Indian subcontinent hot and dry?
7. What is the relationship between population density and the Indo-Gangetic Plain?
8. Where are the least populated areas of South Asia?
9. What are some of the basic beliefs of Hindus?
10. How has Hinduism affected the way Indian society is organized?
11. In what basic way does Hinduism differ from Islam?
12. Why do many of the people of northern and southern India have different languages and cultural traditions?
13. Why did most invaders of India concentrate their empires in the Indo-Gangetic Plain?
14. Why did the British try to rule India?
15. How did British rule affect India?

Critical Thinking

1. **Drawing Conclusions** Many of the most common words for *year* in India mean "rain" or "rainy season". How would you explain this and what does this characteristic of the language indicate about the impact of climate on culture in India?
2. **Demonstrating Reasoned Judgment** Although a majority of South Asians work the land, hunger is a major problem. How would you account for this contradiction?

3. **Identifying Assumptions** As you have read, the British used India as a market for British goods. What did this action suggest about the way some British leaders viewed India?

Practicing Skills

Analyzing Cartograms Look again at the cartogram on page 585. Compare India on the world map on pages 762-763 with India on the cartogram. What does this comparison tell you about population density in that country?

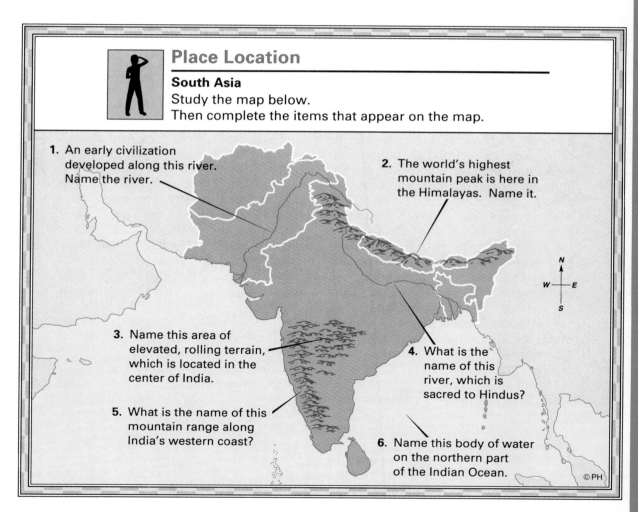

Place Location

South Asia
Study the map below.
Then complete the items that appear on the map.

1. An early civilization developed along this river. Name the river.

2. The world's highest mountain peak is here in the Himalayas. Name it.

3. Name this area of elevated, rolling terrain, which is located in the center of India.

4. What is the name of this river, which is sacred to Hindus?

5. What is the name of this mountain range along India's western coast?

6. Name this body of water on the northern part of the Indian Ocean.

© PH

Place Location: WHERE ON EARTH?

The Peace Corps Assignment

Only one person in the world knows where you will be spending the next two years of your life. Her name is Agnes Cremaldi, and she is an assistant director of the Peace Corps training center where you're preparing for your first Peace Corps assignment. Officially she's not saying anything. But she might be willing to scatter a few hints in your direction. If she is, in addition to finding out whether to bring sandals or hiking boots, you could start preparing yourself psychologically for the challenge that lies ahead.

Thinking back, you remember how you made up your mind to join the Corps. For the past six years you had worked as a master carpenter. Your handcrafted furniture had been in demand all over the state. During the last year, however, you made two important decisions. You wanted to travel abroad. At the same time, you wanted to continue working as a master carpenter.

After doing some research in the local library, you realized that volunteering for the Peace Corps would allow you to do both. As a Peace Corps volunteer, you would live in a developing country and use your carpentry expertise to help improve living conditions. You lost no time filling out an application form, and, shortly afterward, you were accepted.

At the Peace Corps training center, you learned how to conduct yourself as a representative of the United States on your own in a foreign country. Now you're ready for more specialized training in the language and culture of your host country—whatever country that is.

That brings you back to Agnes Cremaldi and the question hanging over your head: Where in the world will you be spending the next two years of your life? Volunteers won't be notified officially until next week. But you have heard through the grapevine that Ms. Cremaldi knows. And though she can't tell, she's been dropping hints.

Assistant Director Agnes Cremaldi is always reading spy novels and loves to be mysterious. You heard that yesterday, for example, a volunteer named Mark pressed her for more information. She told him that he might want to start practicing his Amharic. Mark looked up the unfamiliar word in a dictionary ("Amharic—a southern Semitic language, the official language in Ethiopia"), and knew where he was headed.

You're in luck—here comes Ms. Cremaldi now. The possibilities are driving you crazy. You have to know. "Excuse me, Ms. Cremaldi," you say. "I was wondering if—"

"You were wondering if I would tell you the name of your assigned country. You know I can't do that," she says sharply. Yet she pauses momentarily and looks at the ceiling, as if she's thinking.

AFGHAN

PAKIST

ARABIAN

SEA

Just as quickly, she snaps back into focus. "I will, however, tell you this much," she says. "The nation where you are going is in South Asia. One of South Asia's three major rivers is in this country. The country has two climate regions, arid and semiarid. And the dominant religion of the country is Islam."

South Asia! But where in South Asia? What did she say? The country has one of the region's three major rivers, arid and semiarid climates, and Islam is the dominant religion. You hurriedly jot these clues down. Then you head for the training center's resource room.

You decide to look at a physical map of South Asia to locate the major rivers. You will also need a climate map to find a country with both arid and semiarid climates. Finally, you will need to find information on the predominant religions in each of the countries of South Asia.

SOLVE THe MYSTeRY

In which country is the Peace Corps assignment? *Hint:* Check the maps in Chapter 28 of your textbook. Looking up each of the countries of South Asia in an almanac will also help you.

The Countries of South Asia

Chapter Preview

Match the colors on the right with the colors on the map below to locate the countries covered in this chapter.

Sections	Did You Know?
1 NEW NATIONS IN THE SUBCONTINENT	Although separated by 1,100 miles (1,770 km), Bangladesh and Pakistan formed one country from 1947 until 1972.
2 INDIA'S PEOPLE AND ECONOMY	The Himalayan peak Mount Kailas is holy to Hindus and Buddhists because it is said to contain the thrones of their gods.
3 OTHER COUNTRIES OF SOUTH ASIA	Cherrapunji, Bangladesh receives an average of 432 inches of rain each year— more than almost any other place on earth.

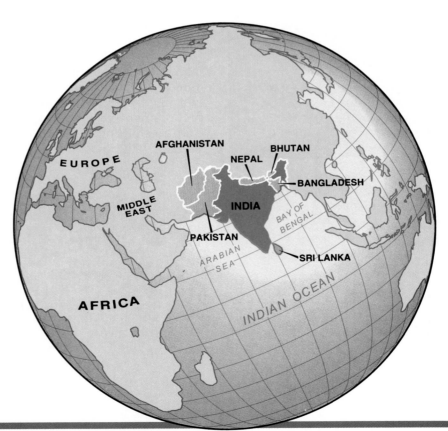

The Spirit of Independent India
Rajiv Gandhi (far right), prime minister in the 1980s, is shown with India's three previous leaders—Mohandas Gandhi, Jawaharlal Nehru, and Rajiv's mother, Indira Gandhi, who was assassinated in 1984.

1 New Nations in the Subcontinent

Section Preview

Key Ideas

- Mohandas Gandhi used nonviolent methods to help lead India to independence from Great Britain.
- India was divided into two countries when it became independent in 1947: India and Pakistan.
- After a civil war in 1971, the eastern part of Pakistan became the nation of Bangladesh.

Key Terms

nonviolent resistance, boycott, partition

The date was August 14, 1947. A drenching rain was falling in New Delhi. But the thousands of Indians crowded outside the Assembly building ignored the weather. They were listening to a dignified man speak these words:

At the stroke of the midnight hour, while the world sleeps, India will awake to life and freedom. A moment comes, which comes but rarely in history, when we step out from the old to the new, when an age ends, and when the soul of a nation, long suppressed, finds utterance.

The speaker was Jawaharlal Nehru (juh WAH huhr lahl NAY roo), the first Prime Minister of India. After ninety years of struggle, his country was just hours away from independence.

Indian Independence

Great Britain had controlled and brought many changes to India since the mid-1700s. Some of these changes, such as the abolition of slavery and the construction of a large railroad network, benefited India. Other changes, however, did not.

Before the Europeans arrived, India had a flourishing textile industry. The Indians were the first people in the world to grow cotton. Indian artisans produced new fabrics such as calico, cashmere, chintz, and muslin. The British, however, wanted to use India as a market for their own cheaper, machine-made textiles. They imported raw cotton from India, made it into cloth, and shipped the cloth back to India for sale. As a result, India's textile industry was almost completely wiped out, and millions of people lost their livelihoods.

In addition, the British like other colonizers, did not treat their subjects as equals. For example, both the government and the army were organized with British officials in positions of power and Indians at the lower levels.

WORD ORIGIN

boycott
Boycott comes from the name of Irish land agent Charles C. Boycott. In 1880, people *boycotted*, or refused to do business with him, because he wouldn't reduce his rents.

This situation understandably caused anger and resentment among some Indians.

Mohandas Gandhi During the late 1800s, Indians became more familiar with their own history and developed a strong sense of nationalism. In addition, Western ideas of individual rights and self-government began to spread among the country's English-speaking middle class—its lawyers, doctors, and teachers. Many middle-class Indians traveled to England to study during this time. One of these men was a young law student named Mohandas Gandhi (moh HAHN dahs GAHN dee). It was Gandhi—later called *Mahatma*, meaning "the Great Soul"—who led India through its great struggle toward independence.

Gandhi's belief in using nonviolent resistance against injustice was his most powerful weapon against the British. **Nonviolent resistance** means opposing an enemy or oppressor by any means other than violence. Gandhi believed that peace and love were more powerful forces than violence. Everywhere he went, he won the hearts of the Indian people.

One way that Gandhi peacefully resisted British rule was to **boycott**— to refuse to purchase or use—British cloth. Gandhi stopped wearing Western clothes. Instead, he wore clothes made from cotton cloth he had spun himself. He devoted two hours each day to spinning his own yarn and urged other Indians to follow his example. The spinning wheel became a symbol of national pride. As a result of Gandhi's leadership and the boycott by the Indian people, the sale of British cloth in India fell sharply.

Gandhi's program of nonviolent resistance developed into a mass movement involving millions of Indians. In spite of Gandhi's pleas to avoid violence, however, some protests against British rule led to riots. Hundreds of people were killed or hurt.

Mohandas Gandhi
During British rule of India, Gandhi advocated self-sufficiency by spinning his own thread.

Gandhi and his followers attracted sympathy in many parts of the world. In 1935 the British gave in to mounting Indian and international pressures and agreed to establish provinces that were governed entirely by Indians.

Religious Conflict

In the early 1940s the long-standing conflict between India's Hindus and Muslims deepened. For hundreds of years, the relationship between the two groups had often been hostile. Most recently, economic differences divided the two groups. The Muslims were generally the poorer peasants or landless workers, while the Hindus were often landowners.

For a time, Hindus and Muslims worked together for independence. But as they drew nearer to their goal, both groups began to fear being ruled by the other. In 1946 Great Britain offered independence to India on condition that Indian leaders could agree on a form of government. But Hindus and Muslims were unable to reach an agreement. Riots broke out in which thousands of people died.

Gandhi yearned for a united India, but the violence persisted. Finally, in 1947 British and Indian leaders agreed that the only solution to the conflict was to **partition**—divide into parts—the subcontinent into separate Hindu and Muslim states. Part of the subcontinent became the mostly Hindu Republic of India. The northwestern and northeastern parts of the subcontinent, where most Muslims lived, formed the nation of Pakistan.

Partition Brings Further Violence
India and Pakistan finally became independent countries on August 15, 1947. The event brought joyous scenes of celebration. But independence also brought confusion and suffering. In one of the greatest migrations of refugees in history, twelve million people

The Bitter Reality of Partition
Partition of India and Pakistan in 1948 created millions of refugees. Hindus fled to India, and Muslims to Pakistan.

moved—Hindus to India, Muslims to Pakistan. They moved to avoid being ruled by a majority religion to which they did not belong. For many, the journey was long and torturous. An eyewitness to the migration gave this account:

> They passed in eerie silence. They did not look at each other. . . the creak of wooden wheels, the weary shuffling of thousands of feet, were the only sounds rising from the columns.

Most of the refugees were forced to leave their possessions along the road or to give them away in exchange for lifesaving water. Many people, weakened by hunger, thirst, or exhaustion, died. In addition, an estimated one million more were killed in fighting between Hindus and Muslims.

Since independence, India and Pakistan have fought two wars. In 1965 India was forced to defend her northern border against Pakistan. The second war, in 1971, led to the creation of the new nation of Bangladesh.

The Perils of Nature
By enriching the soil, rivers sustain the people of Bangladesh. But floods and tidal waves took about 300,000 lives in the early 1970s.

The Creation of Bangladesh When Pakistan became independent, it consisted of two regions—West Pakistan and East Pakistan—separated by 1,100 miles (1,770 km) of Indian territory. The boundaries of East and West Pakistan were not based on any physical landforms but rather on the predominance of Islam in these two regions. In fact, Islam was the only thread that connected the two regions. In all other cultural respects, the two regions were very different. The residents of West Pakistan belonged to several ethnic groups. But the residents of East Pakistan were mostly Bengalis (ben GAHL eez). People in West Pakistan spoke mostly Urdu (OOR doo), which became the official language of the new country. This upset the East Pakistanis, who were very proud of their Bengali language and literary tradition.

Economics and politics further complicated the situation. West Pakistan contained some factories, while East Pakistan was largely agricultural. Despite being economically less developed, East Pakistan paid more in taxes than West Pakistan. At the same time, more than half the national budget was spent in West Pakistan, where the government was located. Most positions in the government and the army were held by West Pakistanis.

As many people in East Pakistan began to feel that their region was merely a colony of West Pakistan, unrest grew. Then, in 1971, about three hundred thousand East Pakistanis died in a devastating flood caused by a cyclone and tidal wave. Many people in East Pakistan accused the government of delaying food and relief supplies to the victims.

The disaster touched off fighting between the two regions. India joined the conflict on the side of the East Pakistanis. In the face of such opposition, the West Pakistani forces surrendered and on December 16, 1971, East Pakistan became the independent country of Bangladesh, meaning "Bengali Nation." Bangladesh is shown on the map on page 594.

SECTION 1 REVIEW

Developing Vocabulary
1. Define: **a.** nonviolent resistance **b.** boycott **c.** partition

Reviewing Main Ideas
2. How did Gandhi use nonviolent resistance to oppose British rule?
3. Why was the Indian subcontinent partitioned in 1947?
4. Why did East Pakistan want independence from West Pakistan?

Critical Thinking
5. **Perceiving Cause-Effect Relationships** How did economic differences contribute to the conflict between West Pakistan and East Pakistan?

✔ Skills Check

☐ Social Studies
☐ Map and Globe
☐ Reading and Writing
☑ Critical Thinking

Predicting Consequences

Social scientists sometimes attempt to predict the consequences of events and trends. Practice your skill at predicting consequences by following the steps below.

1. **Have a clear and accurate understanding of the event or trend.** Before you can predict the consequences of an event or trend you need to understand what has taken place. Read Passage A and answer the following questions: (a) In what year was the British East India Company founded? (b) What was the nature of the company's dealings with the people of India?

2. **Predict and evaluate all possible consequences.** Many events have more than one consequence and some consequences are often more likely to occur than others. Answer the following questions: (a) What are three possible consequences of the trend described in Passage B? (b) How likely is it that each of them will happen?

3. **Use the knowledge that you already have when predicting consequences.** Often your evaluation of a consequence is based on your prior knowledge or experience. Review Passage A and answer the following question: What development in early American history resembles the events of this passage?

4. **Predict consequences for different groups involved.** The consequences of an event or trend often vary depending on whom or what they affect. Read Passage C and answer the following questions: (a) What consequences would you predict for the untouchables? (b) For other Indians competing for jobs and education? (c) For government officials who introduced the quota system?

Passage A

The British East India Company was founded in 1600 by London merchants as a trading exchange for Indian goods. Making profits was not easy. For one thing, India was a long way from England. For another, since India was broken up into small, independent states, the company had to do business with many Indian princes. Without government control, the East India Company took it upon itself to do what it thought best. If the company wanted to increase its land holdings, it took over new land. If it decided to increase taxes or make war, it did these things.

Passage B

A rich variety of people whose ancestors migrated from China, India, or Tibet long ago live in the countries of Southeast Asia. Historians think that over hundreds of years various groups of people crossed the mountains and traveled down the rivers. Gradually, they settled throughout Southeast Asia.

Passage C

Competition for jobs and education among India's 835 million people is intense. In recent years, the Indian government has sought to advance the lower classes within the nation—in particular, the more than one hundred million untouchables—by means of a quota system that gives preference to low-caste.

2 India's People and Economy

Section Preview

Key Ideas

- The majority of Indians live in rural villages where traditional ways of life prevail.
- Urbanization and the rise of a middle class are major trends in modern India.
- India's government is working hard to raise the country's standard of living.

Key Terms

purdah, joint family system, cottage industry

"The city air makes a man free," runs a medieval European saying. And, adds one modern journalist, "It is in the cities that twentieth-century India is casting off . . . the past." But most Indians still live in small, rural villages and carry on traditional ways of life. Many of these villages, especially those in the Indo-Gangetic Plain, have much in common.

Tradition Prevails in India's Villages

About seven out of ten Indians live in rural villages and depend on agriculture to make a living. Most Indian villages consist of a group of houses surrounded by fields. Dirt paths may lead to the village school, the pool that is used for washing clothes, and the small vegetable gardens owned by individual families. A larger path may lead to the neighboring village. Sometimes a bus goes by on its way to a larger town. Many people own bicycles; almost no one has a car. Each section of the village shares a well.

Rural Housing Houses belonging to the more prosperous families in a village are often built of brick and have tiled roofs and cement floors. Houses owned by poorer villagers may be made of mud and thatched with dried grass. The floor is usually made of packed earth. Mud houses have no windows, which would only let in wind and rain. And, in the half of India's villages without electricity, the inside is dark. Usually, the only furniture is a *charpoy* (CHAR poy), a wooden bed frame with knotted string in place of a mattress. Most families move the *charpoy* outside to the courtyard when the weather turns especially hot. The cooking is often carried out in the courtyard, as well.

Food The majority of Indians follow a primarily vegetarian diet, for both religious as well as economic reasons. Because the cow has religious significance, Hindus generally do not eat beef, and Muslims are forbidden to eat pork. As a result, most meat dishes in India tend to be made with goat meat or chicken. Indians who live near rivers or the sea also eat fish.

Most Indians eat rice every day. With it, they may eat a lentil soup called *dal* (DAHL) and some form of pancake or bread. In northern India, the people make *rotis* (RO teez), or flat cakes of wheat or sorghum that are baked on an iron griddle. In southern India, the people eat *idli* (ID lee), or steamed pancakes of rice.

Clothing Because the climate in most of India is so hot and humid, clothing styles tend to be light and loose. Many Indian women wear traditional dress—some form of the *sari* (SAH ree). A *sari* is a length of brightly colored cloth that is draped over one shoulder like a long dress and can be easily adjusted to the wearer's size. Some women cover their face completely with a veil when they are outside their home. This custom is called *purdah* (PUR duh) and is followed by Hindus as well as Muslims, among whom it originated.

Family Life Families in India are generally large. When a man marries, he usually brings his new wife to live in his parents' house. Often the household includes uncles, widowed sisters, and other relatives. This arrangement is known as the **joint family system**.

Everyone in the family has a role. Even the youngest children can take care of the family's chickens, goats, or sheep. Older children carry water and help their parents in the fields. People too old to help in the fields do light jobs around the house, such as shelling peas or washing rice.

Life can be particularly demanding for village women. Here is how one writer describes a typical day for women in a village in southwest India:

The women work an 18-hour day, which begins at four in the morning with millet grinding—two women to a stone for more than two hours. After breakfast, the dung must be cleared and carried to the fields. Then there is water to fetch and firewood to chop, the children to dress, and always a pile of clothes to wash and mend, not to mention the toil in the fields—planting, weeding, clearing stones, harvesting, gathering fodder and fuel.

Some modern technologies, however, have made their way into many Indian villages. Radios, movies, and television, for example, have brought new ideas to the attention of villagers. Television in particular has been so effective that the government has installed public television sets in thousands of villages. India's leaders hope that as villagers come to know more about better farming techniques, they will be able to produce more food. The short-

age of food in the face of an increasing population is a concern for the Indian government.

India's Towns and Crowded Cities

India's urban areas are growing rapidly because of widespread immigration from rural villages. India's urban areas range from towns with twenty thousand inhabitants to enormously crowded cities that swell with populations of more than ten million.

Life in the Towns Many of India's people live in small or medium-sized towns. India's towns are far more populated and lively than its rural villages.

WORD ORIGIN

purdah
The word *purdah* comes from the Persian word for veil or screen, *pardah*. Purdah is a custom followed by women who cover their faces with a veil when outside their homes.

City Life in India
Because of Hindu beliefs, India's cattle are considered sacred. Cows are as accepted in Indian cities as they are in the countryside. As a result, human labor, rather than animal labor, is often used to move goods.

South Asia: Economic Activity and Resources

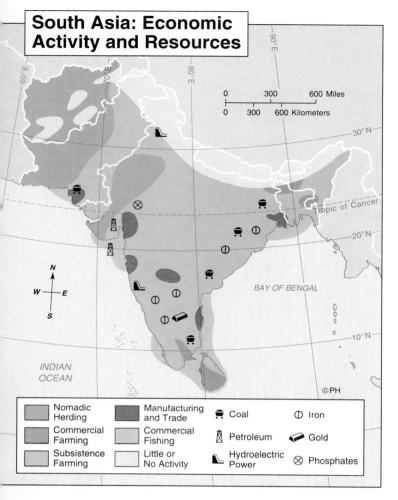

0 | 300 | 600 Miles
0 | 300 | 600 Kilometers

BAY OF BENGAL

INDIAN OCEAN

©PH

Nomadic Herding	Manufacturing and Trade	Coal	Iron
Commercial Farming	Commercial Fishing	Petroleum	Gold
Subsistence Farming	Little or No Activity	Hydroelectric Power	Phosphates

Applying Geographic Themes

1. Interaction Where are India's major hydroelectric power plants located?

2. Location Where are India's major manufacturing centers located?

A writer who taught in India described a typical town:

> . . . cows wander through the streets, washermen bang clothing against rocks in nearby streams, homes built of mud and tar paper and corrugated tin and planks and cardboard lean against one another, ready to be toppled by the first big storm. But the pace of the Indian town nearly terrifies the

villager. The streets are often a wild free-for-all, with buses bearing down on pedestrians, dogs and goats scurrying out of the way of three-wheeled taxis and cars, bicycles weaving past the carters who wearily push their loads of flour sacks uphill.

Life in the Cities One reason for the pace of life in India's cities is that an incredible number of people live in a relatively small area. New York, one of the most crowded cities in the United States, is home to 10,000 people per square kilometer. But Bombay, India's second-largest city, has 44,000 inhabitants per square kilometer! The writer V. S. Naipaul described Bombay's bulging population:

> In Bombay there isn't room for the newcomers. There is hardly room for the people already there. The older apartment blocks are full; the new skyscrapers are full; the small, low huts of the squatters' settlements on the airport road are packed tightly together.

Despite the extreme crowding and poverty that exist, most families consider themselves better off in a city than in a village. The cities offer far more opportunities for work and education than rural areas. India's rural population therefore has been drawn to many of India's large cities. Bombay, on India's west coast, is the country's busiest port and its financial center. Madras (ma DRAHS) and Calcutta (kal KUH tuh), both on the east coast, are also centers of commerce and shipping. New Delhi, India's capital and center of government, is centrally located in the country's interior.

Because the city of Varanasi (vah RAH na see) is built on the banks of the Ganges, Hindus regard it as the holiest city in the world. Anyone lucky enough to die in Varanasi, Hindus believe, is released from the cycle of birth, death, and rebirth. Devout Hindus hope to visit the city at least once within their lifetime to wash away their sins in the sacred Ganges River.

Improving Economic Conditions

Since independence, the government of India has worked to provide a higher standard of living for its people, whether they live in remote villages or teeming cities. It has been successful in many of its efforts, but enormous challenges remain.

Agriculture Advances One of India's main goals after independence was to produce food to feed its growing population. More land was brought under cultivation. Better farming methods, increased irrigation, and higher quality seeds helped to yield more and better crops. More information is given on pages 616–617.

Perhaps the major obstacle to improved prosperity for farmers in India is the small size of most of the farms. Only a few families own enough land to support themselves. Almost half of the farmers in India do not own any land at all.

One answer to the problem is the development of cottage industries. People employed in these industries produce goods in their homes using their own tools and machines. They may spin yarn and weave cloth, or they may produce such items as brassware, jewelry, leather goods, and pottery. These goods can then be sold in the cities and towns.

Expanded Industry Although 70 percent of India's people are farmers,

Rush Hour Crowds in Bombay, India
Named during the period of British rule, busy Victoria Station is a legacy of India's colonial era.

One City, Two Worlds

Fourteen-year-old Shushma and seventeen-year-old Amit live just a few miles apart from each other in New Delhi, India. But they move in different worlds.

Amit, a Brahmin, lives in a comfortable neighborhood of the city, spins around town on a motorcycle, like the person shown at right, and listens to rock music in his spare time. Amit plans to go to college and study art.

Shushma, who works as a maid, comes from a family of untouchables. She is forbidden to play with Brahmin children. Shushma and her family of seven live in a shack beside a polluted pond. With little education and few contacts outside her caste, her future is bleak.

Discrimination based on caste has been outlawed since 1952. Although the government has struggled to advance the low castes by giving them preference in hiring and college admissions in recent years, Hindu tradition continues to exert a powerful hold on Indian daily life.

1. How are the lives of these two Indian teenagers different?
2. Why does the caste system endure?

the country is one of the ten largest industrial nations in the world. India has made great advances in the computer industry and in space research. The country has recently placed its first communications satellite in orbit. It plans to use the satellite to improve the nation's telephone system and to beam educational programs to rural areas.

A major growth industry in recent years has been the production of consumer goods—color televisions, videocassette recorders, wristwatches, and automobiles. The growth of the consumer goods industry is largely due to the growth of the urban middle class.

Traditionally, Indian society has been sharply divided between a wealthy minority and a poor majority. Over the past decade, however, many people, employed as teachers, writers, doctors, and government workers, have become part of a growing middle class. Others have moved into the middle class after building successful businesses of their own.

Education In 1950 only about 16 percent of India's population could read and write. In 1980 the figure had reached 37 percent and was still rising. This improvement was the result of intensive government efforts to expand education. Almost every village now has a primary school. Despite the government's efforts, however, many children either do not attend these schools or drop out very quickly. Often their families need them to work in the fields. Girls stay home to care for younger brothers and sisters.

Health Care India's government has also given high priority to improving people's health. In 1950, the average Indian's life expectancy was only 32 years. By the 1980s it had risen to 55 years, and the government is hoping to increase the figure to 64 years by the year 2000. Unfortunately, many people in India are too poor to afford food, proper medical care, or even a home. Homeless people live on the streets of India's cities, where some beg for food. In cities like Bombay, several thousand sleep on the sidewalks.

One of the government's efforts to improve health has been directed at improving the water supplies in rural areas. In the past, most Indians drank from open wells, which were breeding grounds for bacteria. In the early 1980s, the government began drilling deep, machine-made wells, covered to reduce the risk of contamination. Most

Industrial India

In New Delhi, a technician assembling computer pieces is part of India's growing industrial base.

villages now have a safe water supply and as a result, diseases such as malaria and cholera have become much less common.

SECTION **2** REVIEW

Developing Vocabulary
1. Define **a.** *purdah* **b.** joint family system **c.** cottage industry

Reviewing Main Ideas
2. Describe some differences between rural and urban life in India.
3. What effect has growth of the middle class had on India's economy?
4. Why do many Hindus visit Varanasi?

Critical Thinking
5. **Drawing Inferences** In what ways is education in India related to the country's economic improvement?

India's Growing Population

Indira (IN duhr uh) Gandhi, a former prime minister of India, often said that India was adding "an Australia" to its population each year. She was not exaggerating. In 1989 Australia's population was a little under seventeen million people. During the 1980s India's population grew at the rate of almost sixteen million people a year. One Indian baby was born every two seconds. If the population continues to grow at such a rate, India will have more than one billion people by the year 2000.

Rapid Population Growth

For much of its history India has had a high birthrate. But until the mid-1900s, the death rate was also very high. As a result, the overall population remained fairly stable.

During the 1950s improvements in health care and increased food production lowered the number of deaths. But this drop in the death rate was not matched by a lower birthrate. More children were staying alive longer and eventually having large families of their own.

In rural India, children have traditionally meant security for their parents. Children can work the land or help in a family business. When they grow older, they can work outside the home and contribute income to the family. Most Indians have no provision for care in their old age, such as a pension, social security, or savings. They rely on their children to take care of them when they can no longer provide for themselves.

Problems of Overpopulation

India's vast and growing population affects the quality of life for its people. Many people live in poverty and are unemployed. The Indian government thought that industrialization would reduce unemployment, but the work force continues to exceed the number of available jobs.

Overpopulation also puts pressure on the country's resources. Farms in many areas are being overplanted,

---

India's Population   *What does the shape of this graph tell about India's population?*

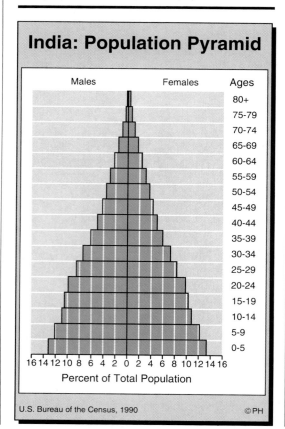

India: Population Pyramid

Males	Females	Ages
		80+
		75-79
		70-74
		65-69
		60-64
		55-59
		50-54
		45-49
		40-44
		35-39
		30-34
		25-29
		20-24
		15-19
		10-14
		5-9
		0-5

16 14 12 10 8 6 4 2 0 2 4 6 8 10 12 14 16
Percent of Total Population

U.S. Bureau of the Census, 1990   ©PH

Where People Live

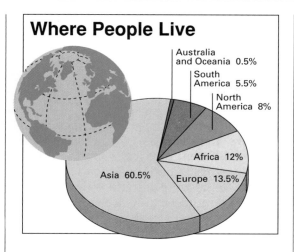

Australia and Oceania 0.5%
South America 5.5%
North America 8%
Africa 12%
Europe 13.5%
Asia 60.5%

The World's Population *Where do the majority of the world's people live?*

so that the land is losing its fertility. India's forests are rapidly being cut down to create more farmland and to provide wood for fuel. After a forest is cleared, there are no trees to stop rivers of mud from pouring down the slopes when the monsoon rains come. Tremendous landslides result. As India's forests disappear, its deserts expand and more land becomes unusable.

Limiting Population Growth

The Indian government is encouraging people to have only two children instead of five or six. Billboards on city streets show a father, a mother, and two children, together with the words, "A small family is a happy family." In late 1989, the Indian government proposed two ways to limit population growth.

The first was to imitate a strategy that is used in China. The government would offer jobs, education, housing, and medical care to couples who had no more than two children.

The second approach was for the government to offer retirement payments to the elderly. The government hoped that people would then rely on government payments, and not their children, for security in their old age.

As long as India's population continues to grow, the country's leaders face many difficulties in their attempts to fight poverty and hunger. India has successfully met challenges in the past. Slowing population growth is now one of the most urgent.

TAKING ANOTHER LOOK

1. Why have Indians traditionally wanted many children?
2. What are some of the challenges facing India because of its rapid population growth?
3. Describe two ways in which the Indian government is working to limit population growth.

Critical Thinking

4. **Demonstrating Reasoned Judgment** India hopes that family size can be reduced by increasing retirement payments to the elderly. What changes in people's beliefs and attitudes would be necessary for this plan to succeed?

दो या तीन बच्चे... बस
क्टर की सलाह मानिए

Indian Family Poster *The government uses murals of the "ideal" family to persuade Indian couples to have only two children.*

607

3 Other Countries of South Asia

Section Preview

Key Ideas

- Water—too little or too much—has a major influence on the lives of people in Pakistan and Bangladesh.
- Physical geography has been a major factor in shaping the histories of Afghanistan, Nepal, Bhutan, and Sri Lanka.

Key Term

buffer state

India's neighbors face many of the same challenges as India—poverty, overpopulation, and conflict between ethnic and religious groups. But from the crowded cities of Bangladesh to the remote mountaintop villages of Afghanistan, each country is physically and culturally distinct.

Pakistan: A Dry and Rugged Land

The physical map on page 576 shows that Pakistan is made up of three different physical regions. Along its northern and western borders, some of the world's highest mountains —the Hindu Kush—rise majestically toward the sky. Several passes cut through the mountains, making transportation possible. The Khyber (KY ber) Pass allows movement between Peshawar (puh SHAH wuhr) in northwest Pakistan and Kabul (KAH bool), the capital of Afghanistan. Just as in India, these towering mountains keep out the cold air from central Asia during the winter. As a result, except at high elevations, temperatures in Pakistan are generally warm or hot. The high temperatures in the city of Islamabad average 62°F (17°C) in January and 98°F (37°C) in July.

Much of western Pakistan is covered by the rugged Baluchistan (bah LOO chi STAN) Plateau. To the east lie barren stretches of the Thar Desert and brown, dusty plains. Sandwiched between these two forbidding regions is the fertile valley created by the Indus River as it flows south out of the Himalayas to the Arabian Sea.

The Struggle for Water Drought has always been a major problem in Pakistan. The Indus River is the lifeline of an otherwise dry and rugged country. Most of the population live in the Indus River basin. The basin also contains most of the country's agricultural areas, as well as its major hydroelectric power stations. From the Indus, a vast irrigation system laces the countryside.

Interaction: Creating a Water Supply One of the keystones of the irrigation system is the Tarbala Dam. Completed

Linking Two Countries

The Khyber Pass is an ancient trade route between Pakistan and Afghanistan.

in 1976, it is three times the size of the Aswan High Dam in Egypt. And like the Aswan Dam, the building of the Tarbala Dam has had both positive and negative results.

Most parts of Pakistan receive less than 10 inches (25 cm) of rain each year, making agriculture dependent on irrigation. The regular supply of water that is ensured by the Tarbala Dam has turned millions of acres of arid desert into lush crop land. Wheat is Pakistan's major crop, and most of it is used for food within the country. The country's major cash crop is cotton, which supplies a growing textile industry.

The dam, although successful, has also created problems. The Indus River picks up silt as it flows through northern Pakistan's loose, eroded soil. The silt is slowly piling up behind the dam. Engineers estimate that within twenty years it will be useless as a source of irrigation water. Unless a solution is found, Pakistan will once again face its traditional water shortage.

Traditional Ways of Life As in India, most of the people in Pakistan live in farming villages. Almost all Pakistanis are Muslims. Prayers are an important part of daily life.

Tradition also plays an important role. For example, women generally have far fewer freedoms and opportunities than men. Many women avoid contact with men outside the home and cover their face with a veil in the presence of strangers.

National Challenges Islam is the tie that holds the people of Pakistan together. Yet differences among the Muslims sometimes threaten to split the country apart. For example, only about 7 percent of Pakistan's people speak Urdu, the national language. Almost 64 percent speak Punjabi, the language of the Punjab province. In addition, disputes among the country's various ethnic groups often flare up into violence.

Farming in the Punjab Region, Pakistan
Lack of modern farm machinery indicates that little has changed since ancient times for Pakistan's subsistence farmers.

Pakistan's literacy rate is lower than India's, while its rate of population growth is higher. Pakistan's leaders have a hard road ahead as they work to improve their people's standard of living and hold their nation together.

Bangladesh

The map on page 576 shows that most of Bangladesh is an enormous delta. The delta has been formed over hundreds of years by three powerful rivers—the Ganges, the Brahmaputra, and the Meghna (MAYG nuh). As a result, the soil is very fertile. However, because the elevation of the country is so close to sea level, flooding occurs very often.

WORD ORIGIN

Pakistan
Pakistan was coined in 1930 to reflect these regions: *P* for Punjab, *A* for Afghania, *K* for Kashmir, and *S* for Sind. *Pak* is Urdu for "pure" and *stan* is "land," meaning "land of the pure."

Dacca, Bangladesh
Rickshaws, small covered carts pulled by a person, serve as showpieces of local folk art as well as a means of transportation in Indian cities.

The Challenge of Climate The climate of Bangladesh is humid subtropical. Temperatures do not usually drop below 80°F (27°C). Because of the monsoon winds, large amounts of rain fall within a three- to four-month period.

The climate and geography of Bangladesh create a delicate balance between survival and disaster. In good times, the warm temperatures, abundant water supply, and fertile soil enable farmers to plant and harvest three crops a year on the same land. In bad times, however, the rivers overflow, and fierce storms sweep in from the Bay of Bengal, submerging the land in salt water.

In 1989 torrential monsoon rains poured down on Bangladesh, leaving 75 percent of the country under water and an estimated thirty million people homeless. Millions of people became ill from lack of food and clean drinking water. In addition, power lines were knocked down and roads, bridges, and railway lines were washed away. Transporting food and medicine from one part of the country to another proved almost impossible.

Overpopulation Like India, Bangladesh is struggling with overpopulation. It is the eighth most populous country in the world and the second most densely populated. With 120 million people, it has about half the population of the United States squeezed into an area the size of Wisconsin.

Overpopulation and natural disasters have created another problem for Bangladesh: hunger. Famine occurs after periods of flooding. But malnutrition—a lack of adequate food or an unbalanced diet—is an almost constant problem. As Bangladesh looks ahead, it faces some of the greatest challenges of any country in the world.

Movement Through Afghanistan

In many ways, Afghanistan's history has been influenced by its location and terrain. The map on page 576 shows that the towering Hindu Kush forms the central backbone of the country. It also marks the boundaries of three regions. The first region consists of the mountains themselves. At their feet lie several fertile valleys where most of the people live. North of the Hindu Kush is a region of semiarid plains. The land south of the Hindu Kush is mostly desert.

For centuries, merchants and soldiers crossed through Afghanistan on their way between China and the Middle East or into India. Although the Hindu Kush formed a barrier, it had many passes, the most well-known being the Khyber Pass. The Hindu Kush received its name, which means "Indian Killer," in the thirteenth century. At that time, soldiers from the northwest were coming through the Khyber Pass and raiding Indian villages on the Indo-Gangetic Plain. When the soldiers returned home after their raids, they carried Indians with them to sell as slaves. But many Indians were unable to withstand the cold of the Hindu Kush, and thousands of them died.

A Diverse Population As a result of successive invasions and migrations, the population of Afghanistan includes people of many ethnic groups. The country has two official languages, but the people speak many other languages as well. Over the centuries local groups isolated themselves in pockets of land as protection against invaders. Each group developed its own language and customs. As a result, some groups are unable to communicate with one another.

Despite their ethnic differences, however, the people are united by their Islamic faith. Religious holidays are national holidays. Mosques are educational and social centers as well as religious ones.

A Buffer State Over the last two centuries, Afghanistan has been invaded several times. During the 1800s, when most of India was under British rule, both Britain and Russia competed for influence in central Asia. Anxious to protect the northern approaches to India against the Russians, the British moved into Afghanistan. This led to the first of three British-Afghan wars. At the same time, Russian troops made several attempts to move south.

However, neither Britain nor Russia was able to conquer the people or the terrain of Afghanistan. Finally, both powers agreed to honor certain boundaries and recognize Afghanistan as an independent nation. In this way, Afghanistan became a buffer state—a country that separates two political enemies.

Afghanistan's Islamic Shrine
White birds at the Blue Mosque, an important Afghan landmark, provide a moment of peace and harmony in this war-scarred country.

In 1979 Soviet troops marched into Afghanistan to help put down a revolt at the request of the country's government. The rebels included traditional Muslims as well as strong nationalists who resented their government's pro-Soviet attitude, as well as its attempts at modernization.

The Soviet invasion created a wave of refugees from Afghanistan. About 4.5 million people fled. Many of them crossed the border through the Khyber Pass and settled in or near Peshawar, Pakistan.

In 1989, after ten years of fighting, the Soviets announced that they were pulling out of Afghanistan. Various Afghan groups immediately began to fight for control of the country. In the face of this violence, most refugees remained in Pakistan.

A Wool Market in Katmandu, Nepal
Colorful yarns are displayed at a market in Nepal's largest city. The livelihood of most of the people of the Himalayas depends on raising sheep, goats, and yaks.

Nepal and Bhutan: Location at the Top of the World

Both Nepal (nuh PAHL) and Bhutan (boo TAHN) are located high in the Himalayas. But both countries span a great range in altitude, from a low elevation of about 600 feet (183 m) to the highest mountains in the world. Mt. Everest, in Nepal, towers 29,029 feet (8848 m) high. And there are more than one hundred other peaks that rise over 20,000 feet (6096 m).

The southern lowlands of Nepal and Bhutan are extremely hot and humid. Monsoon rains pour down every summer. Tropical crops flourish here, including citrus fruits, sugarcane, and rice. Water buffalo pull plows through the rice paddies. In high elevations where the temperatures are cooler, people grow wheat, millet, and potatoes and use long-haired yaks as beasts of burden. The yaks also provide wool, milk, and meat. Most crops are grown in terraced fields built into the hillsides. Strong winds and temperatures below freezing are common on the snow-covered peaks of the highest elevations.

Religion reflects both countries' diverse physical and human geography. Nepal is about 85 percent Hindu, while Bhutan is about 75 percent Buddhist. Hinduism tends to be practiced in the lowlands, while Buddhism is the religion of the mountainous areas. In general however, throughout both countries each religion has influenced the other. For example, people celebrate festivals honored by both religions. Buddhists in Nepal, unlike Buddhists elsewhere in the world, are divided into castes.

Both high mountains and politics kept Nepal and Bhutan somewhat separated from the rest of the world until the middle of the twentieth century.

Today, Bhutan continues to discourage contact with tourists and other foreigners in an effort to save its traditional culture. Nepal, on the other hand, is happy to have a thriving tourist trade.

Sri Lanka: Ancient Roots of a Modern Conflict

The pear-shaped island of Sri Lanka, meaning "Magnificent Island," is located in the Indian Ocean 33 miles (53 km) off the southern tip of India. Sri Lanka is often referred to as "a tear dropped off the subcontinent of India." As the climate map on page 578 shows, Sri Lanka's climate is tropical, but is made cooler by ocean breezes.

The heaviest rains fall in the southwest part of the island, which contains the estates where plantation crops are grown for export. The most important plantation crops are coconuts, which are grown in the lowlands, and rubber and tea, which come mostly from the higher slopes of the island's mountains. Sri Lanka produces about one eighth of the world's tea.

At one time, Sri Lanka was covered with a thick rain forest. Today, about 45 percent of the island is covered by forest. The rest has been cut down to make room for farming and other development. Scientists think that this deforestation may have contributed to changes in the island's weather patterns and caused severe droughts. One of the challenges facing the government of Sri Lanka is to restore the island's forest. Another is to maintain peace among the island's people.

About four fifths of Sri Lankans are Sinhalese (sin hah LEEZ), descendants of Aryans who migrated from northern India about 500 B.C. Later, the Tamils (TAHM uhlz), a people of Dravidian origin, came to Sri Lanka from southern India in several waves. At present, the Tamils are concentrated in the northern and eastern parts of the island.

Over the centuries, the Sinhalese and the Tamils often fought each other. Religion and language further split the two groups. The Sinhalese are Buddhists, while the Tamils are Hindus. The two groups speak different languages and have different alphabets.

Since Sri Lanka gained its independence from the British in 1948, the Sinhalese have controlled the government. The Tamils, feeling that their interests have not been considered, have demanded the establishment of a separate Tamil state. The conflict has led to bloody fighting between government forces and Tamil guerrillas. Many civilians have been caught in the crossfire. The economy has also suffered as money has been diverted from agriculture to the military. About 45 percent of the population's labor force makes a living from agriculture. The fighting has disrupted farming and fishing and a political solution does not seem likely in the near future.

■■■ SECTION 3 REVIEW ■■■

Developing Vocabulary
1. Define: buffer state

Place Location
2. Locate and name the capital city of Bangladesh.

Reviewing Main Ideas
3. What challenges do climate and landforms in Bangladesh create?
4. How has Afghanistan's location affected its history?
5. What effect do the Hindu Kush have on the climate of Pakistan?

Critical Thinking
5. **Distinguishing False from Accurate Images** Explain whether or not you think the following statement is accurate. "The monsoons bring life to the countries of South Asia."

CHAPTER

29

REVIEW

Section Summaries

SECTION 1   New Nations in the Subcontinent   The British established an empire in India in order to take advantage of resources and to find a market for their goods. During the late 1800s and early 1900s, Indians became increasingly dissatisfied with British colonial rule. Led by Mohandas Gandhi, they practiced nonviolent resistance against their overlords. Finally, Great Britain granted independence to India in 1947. At the same time, India was divided into two separate countries—a mostly Hindu India and a mostly Muslim Pakistan. In 1971 East Pakistan split away to become the country of Bangladesh.

SECTION 2   India's People and Economy   People living in rural areas of India live in traditional ways. Because of the hardships of rural life, many villagers have migrated to the increasingly crowded cities, where a middle class is rapidly growing. The Indian government is working hard to improve the country's standard of living and its agricultural and industrial production.

SECTION 3   Other Countries of South Asia   Because of its elevation, Bangladesh is vulnerable to devastating flooding. It is one of the poorest and most densely populated countries on earth. Afghanistan has served as a crossroads for merchants and soldiers moving between China and the Middle East or from central Asia into the Indian subcontinent. Afghanistan also served as a buffer between British India and Russia. Until recently, the Himalayas have kept Nepal and Bhutan isolated from much of the rest of the world. Sri Lanka was invaded by the Sinhalese and Tamils from nearby India, and the differences between the two groups have resulted in continuous fighting on the island.

Vocabulary Development

Match the definitions with the terms below.

1. to divide a country into separate parts
2. a country that separates political enemies
3. a Muslim custom according to which women keep their faces covered outside the home
4. to oppose an enemy or oppressor by any means other than violence
5. a small business that uses a person's home as a place to produce goods
6. the refusal to buy or use certain products
7. a large family made up of several related families living together

a. nonviolent resistance
b. partition
c. *purdah*
d. boycott
e. joint family system
f. cottage industry
g. buffer state

Main Ideas

1. Why was Gandhi called *Mahatma*?
2. How did the creation of Pakistan reflect the hostility between Hindus and Muslims?
3. What was the reason for the great migration of people after Indian independence?
4. What challenges face people living in India's villages?
5. Why do many people think they are better off in an overcrowded city than in a village?
6. How do the landforms and climate of Pakistan differ from those of Bangladesh?
7. How do ethnic conflicts continue to shape the history of Sri Lanka?
8. What are some of Sri Lanka's most important plantation crops?

Critical Thinking

1. **Drawing Conclusions** Why do you think Gandhi's use of nonviolent resistance was successful in helping India to achieve independence?
2. **Determining Relevance** If you lived in a village in India, do you think you would remain there or move to the city? Give reasons for your answer.
3. **Identifying Central Issues** What do you think are the main challenges facing Nepal?

Practicing Skills

Predicting Consequences Overpopulation continues to be a serious challenge for the people in India and for the country's leaders. Two steps have been proposed to attempt to slow the rate of population growth. One is to reward couples with no more than two children. The other suggestion is to increase retirement payments to all Indians. What consequences do you predict each suggested step will have for India in the next ten years?

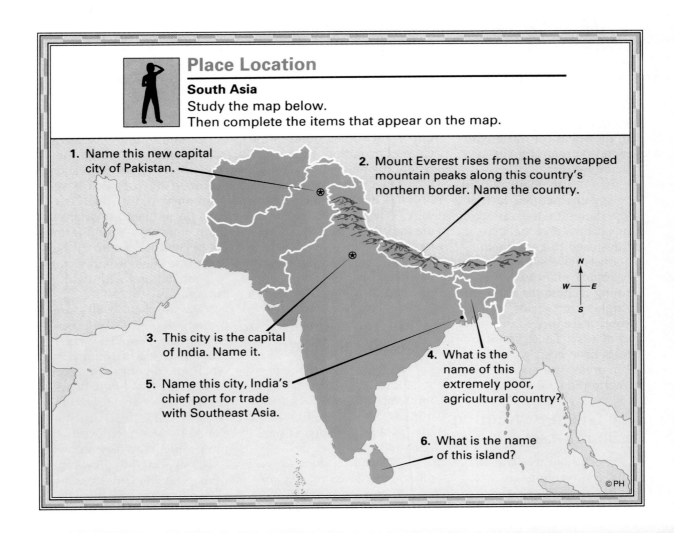

Place Location

South Asia
Study the map below.
Then complete the items that appear on the map.

1. Name this new capital city of Pakistan.
2. Mount Everest rises from the snowcapped mountain peaks along this country's northern border. Name the country.
3. This city is the capital of India. Name it.
4. What is the name of this extremely poor, agricultural country?
5. Name this city, India's chief port for trade with Southeast Asia.
6. What is the name of this island?

©PH

MAKING CONNECTIONS

•WHERE REGIONS MEET•

The Green Revolution

The ability of a nation to grow crops successfully depends on many factors. These include the availability of good land, the prevailing climate, the presence of diseases and pests that can harm plants, and the weather. The lack of any one of these factors creates the potential for failure in a nation's food-producing capacity. And if crops fail, the result for those who depend on them for survival may be hunger.

Two factors that can help to overcome nature's lack of cooperation are the variety of crops grown and the techniques used to grow them. Scientists have succeeded in developing plant varieties and farming techniques that result in higher crop yields and greater resistance to drought and disease. These scientists' efforts have been very successful in areas of the world that had chronic food shortages.

The Green Revolution in South Asia

The golden wheat fields of India, for example, are feeding more people today than they were several decades ago. This important development began with the work of a few plant experts who developed new and better varieties of grain. The introduction of these strains of rice and wheat into the agriculture of developing nations, begun in the 1950s, is known as the Green Revolution.

South Asia has been an important battleground in the Green Revolution. In areas where the new plant varieties have been used, food production has often risen dramatically. Wheat production in India and Pakistan, for example, has tripled since 1967. Bangladesh is another nation that has increased its grain yields through the Green Revolution.

The Green Revolution Around the World

Many nations of South Asia have been successful in implementing the Green Revolution. Elsewhere, however, the revolution's success has been mixed.

China and Indonesia have greatly increased their grain supply by planting new seed varieties and using sophisticated techniques of crop cultivation. However, the program's focus on irrigation systems, costly fertilizers and pesticides, and transportation of crops to markets has put the benefits of the Green Revolution beyond the financial reach of many of the developing nations.

Most of the countries of Africa South of the Sahara have profited little from the Green Revolution. Rice and wheat—two crops that have been improved by scientists— have never been important in the region. As a result, farmers in countries south of the Sahara struggle to grow crops in poor soil and with light rainfall.

Despite the overall gains brought about by the Green Revolution, people are still hungry—even in countries where the revolution has been successful. This limited success is attributable to several factors: lack of food distribution networks; the poverty of the people, who are too poor to buy the food that their nations produce; and huge population increases. Because there is little new land available for farming, more food will have to be produced on existing farmlands. As many people agree, the Green Revolution must be improved and expanded to more nations.

The View from the U.S.A.

 United States scientists have been in the forefront of agricultural research and the development of superior crop varieties. Their efforts have reinforced this nation's status as the world's leading agricultural producer. Experts here agree that such research must continue if the developed nations of the world are to help prevent famine.

In addition to research at home, the United States continues to share its expertise with developing nations. The United States Agency for International Development and the Peace Corps are two government agencies that promote agricultural development worldwide. The United States also supports efforts of the United Nations Food and Agriculture Organization. In addition, private organizations such as Global 2000 and Oxfam America continue to work to advance agriculture around the world.

TAKING ANOTHER LOOK

1. Describe the Green Revolution and explain how it has succeeded in nations such as India and Pakistan.
2. Why has the Green Revolution failed in some nations in Africa south of the Sahara?

Critical Thinking

3. **Demonstrating Reasoned Judgment** Much of the population increases projected for coming decades are expected to occur in the areas hardest hit by hunger. Why does this fact suggest a more urgent need for Green Revolution-style reforms?

Research and Results
A scientist in India (left) works to develop plants that will lead to higher crop yields in many parts of the world, including Indonesia (below).

A nation more populous than any other . . . urban centers more crowded than those in any other region . . . rivers that flood their banks with rich, yellow soil . . . islands close to the mainland, some rocky and mountainous, others thick with tropical vegetation . . .

These images describe much of the region of East Asia, yet a region this large and varied defies simple description. In this unit, you will discover East Asia—its landscapes, its people, and their cultures.

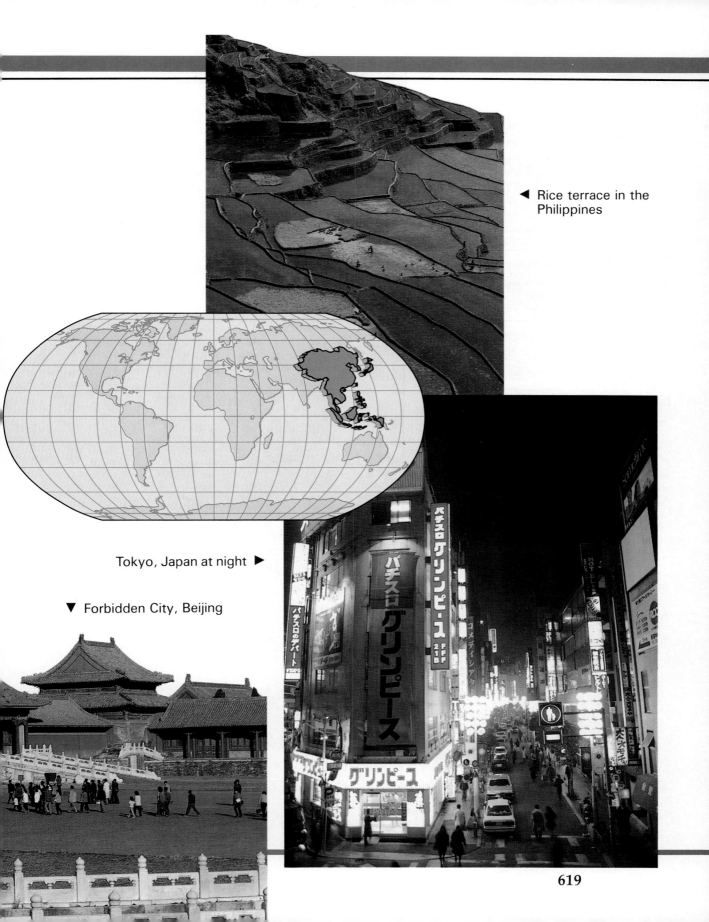

◄ Rice terrace in the Philippines

Tokyo, Japan at night ►

▼ Forbidden City, Beijing

619

Regional Atlas: East Asia

Chapter Preview

Both sections in this chapter provide an overview of East Asia, shown in red on the map below.

Sections	Did You Know?
1 LAND, CLIMATE, AND VEGETATION	The average Chinese farm is three hundred times smaller than the average American farm.
2 HUMAN GEOGRAPHY	The earliest known version of the Cinderella story dates to ninth century China. The heroine's name was Yeh-hsien.

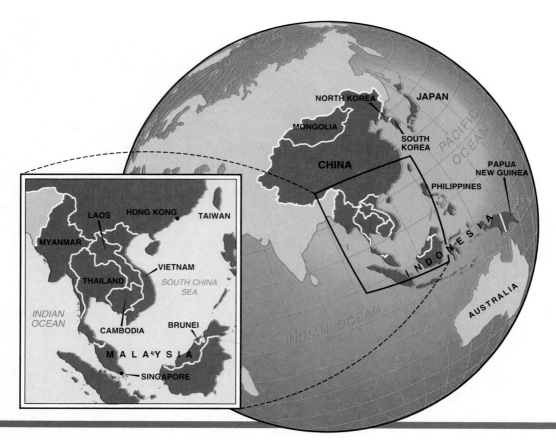

Houseboating on the Li River, China
Water buffalo graze on the banks of the Li River in China's Guangxi province as a houseboat makes its way downstream. In the background, China's steep and jagged limestone hills rise as high as 600 feet (183 km).

1 Land, Climate, and Vegetation

Section Preview

Key Ideas

- Tectonic activity created much of East Asia's mountainous landscape.
- Although lowlands occupy little of East Asia, most of the region's human activity takes place there.
- Landforms, ocean currents, and monsoon winds influence the climate of East Asia.

Key Term

seismic

China, like a huge earthen giant, anchors the eastern third of the great Eurasian landmass. Locate China on the physical-political map of East Asia on page 622. The other countries of the region appear tiny in comparison with this immense country.

Connecting Plate Boundaries and Landforms

Tectonic forces created much of East Asia's physical landscape. As described in Chapters 2 and 28, the collision of the Eurasian and Indo-Australian plates caused the earth's surface to fold and buckle, creating the majestic Himalayas. The vast Plateau of Tibet and China's western mountain ranges—the Kunlun Shan, the Altun Shan, and the Tian Shan—rose skyward in the same crush of plates. Carving the valleys in this massive, mountainous area, China's major rivers flowed eastward to the sea.

The same folding of the earth's crust also squeezed up parallel mountain ridges, such as the Arakan Yoma

Chapter 30, Section 1   **621**

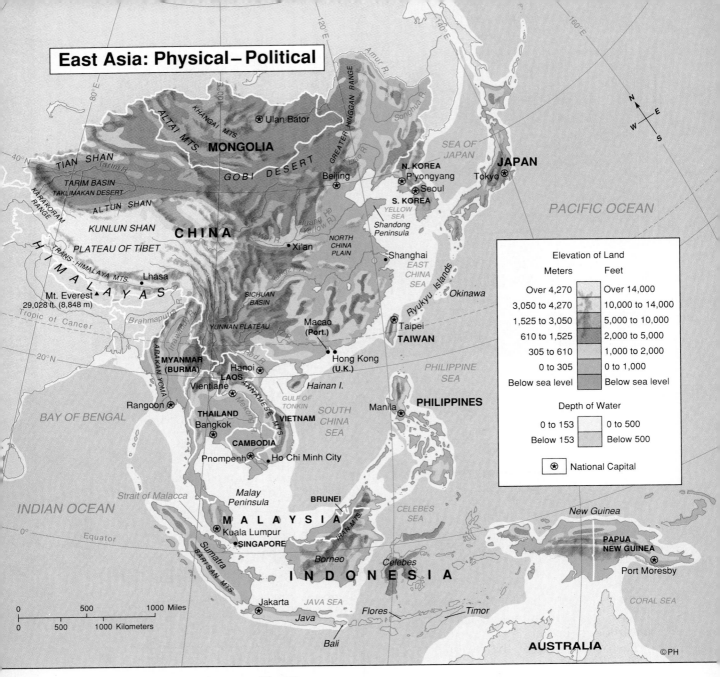

East Asia: Physical–Political

Elevation of Land	
Meters	Feet
Over 4,270	Over 14,000
3,050 to 4,270	10,000 to 14,000
1,525 to 3,050	5,000 to 10,000
610 to 1,525	2,000 to 5,000
305 to 610	1,000 to 2,000
0 to 305	0 to 1,000
Below sea level	Below sea level

Depth of Water	
0 to 153	0 to 500
Below 153	Below 500

⊛ National Capital

Applying Geographic Themes

1. **Place** Which two major rivers border the North China Plain?
2. **Location** Where is Taiwan located in relation to China? Which small island country is located at the southern end of the Malay Peninsula?

and the Annamese Mountains in Southeast Asia. Like long, bony fingers, these mountain ridges point seaward, directing the great Irrawaddy, Chao Phraya (chow PRIY uh), and

Mekong rivers southward to the Indian Ocean and the South China Sea.

Tectonic activity also helped to create the Malay and Japan archipelagoes —groups of islands—that edge the mainland. These islands form part of the Ring of Fire—a line of volcanic and **seismic** (earthquake-related) activity that encircles the Pacific Ocean. As explained in Chapter 2, volcanoes and earthquakes occur as plate boundaries meet. The tectonic plate map on page 22 shows how the islands of Indonesia, the Philippines, and Japan align along plate boundaries. The islands that make up these countries are the cones of volcanoes rising high above the ocean's surface.

Among the highest of these cones is Mount Fuji, an inactive volcano on the Japanese island of Honshu that stands 12,388 feet (3,776 m) above sea level. An English traveler said of Mount Fuji:

It is seen at its best from the sea at dawn. At such times it seems to tower over everything, its perfect snowcapped cone a purplish green in the light of early morning, seeming to be suspended in the sky. The beauty of Fuji is due to the simplicity of its outlines and the fact that it stands alone.

A great deal of movement occurs on the western boundary of the Pacific plate, making it a region of endless tectonic "fireworks." Java, an island in Indonesia not much more than 100 miles (161 km) wide, has more than fifty active volcanoes. Almost 20 percent of the world's earthquakes occur in Southeast Asia.

Interaction: The Importance of Plains and Rivers

Highlands dominate the landscape of East Asia, but most human activity takes place in the region's plains and river valleys. Only China has extensive plains areas. Note the far northeast of China on the physical-political map of East Asia on page 622. Locate the extensive lowland area between the

Mount Fuji, Japan
The majestic, snow-crowned volcanic cone of Mount Fuji is Japan's highest peak but it is only one of Japan's many volcanoes.

Songhua and Liao rivers. This area, known as the Manchurian Plain, is an important agricultural region.

Farther to the south lies the North China Plain. The great Huang He, or Yellow River, carries rich alluvial soils to it. Winds blowing across Mongolia deposit rich yellow soil, called loess, along the upper reaches of the Huang He. The river picks up the loess and carries it eastward, depositing it on the North China Plain. These loess deposits make the soil of the North China Plain very fertile.

Climate and Vegetation

Eastern China and Japan have climates comparable to those found in similar latitudes in other parts of the

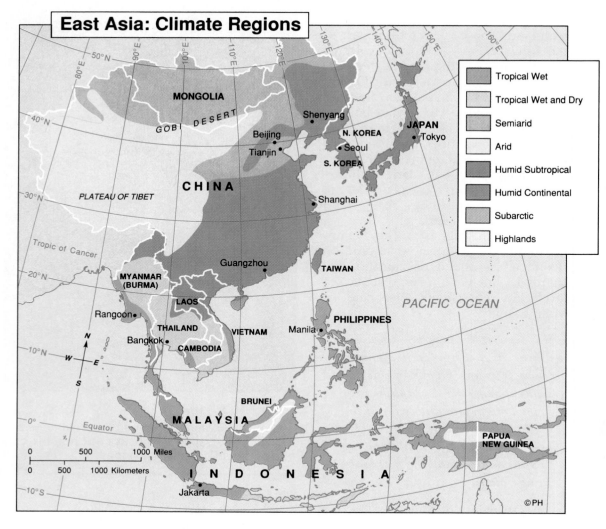

East Asia: Climate Regions

MONGOLIA

GOBI DESERT

Shenyang

Beijing

Tianjin

N. KOREA

Seoul

S. KOREA

JAPAN

Tokyo

CHINA

PLATEAU OF TIBET

Shanghai

Tropic of Cancer

Guangzhou

TAIWAN

MYANMAR (BURMA)

LAOS

Rangoon

THAILAND

Bangkok

VIETNAM

CAMBODIA

PHILIPPINES

Manila

PACIFIC OCEAN

BRUNEI

MALAYSIA

Equator

PAPUA NEW GUINEA

INDONESIA

Jakarta

©PH

	Tropical Wet
	Tropical Wet and Dry
	Semiarid
	Arid
	Humid Subtropical
	Humid Continental
	Subarctic
	Highlands

0   500   1000 Miles
0   500   1000 Kilometers

Applying Geographic Themes

Location Because East Asia is so large, climates reflect the range of latitudes at which the region lies. Climates vary with elevation and proximity to oceans. Climates vary from the cold, dry winds of the Gobi Desert to the tropical conditions of Indonesia. What type of climates are found in Japan?

WORD ORIGIN

Himalayas
The name Himalayas comes from the Sanskrit words *hima,* or "snow," and *alaya,* or "residence." The Himalayas are "abode of the snows" on the India-Tibet border.

world. Consult the map on page 89 and compare eastern China and Japan with the eastern United States. Note that both regions have mostly moderate climates. Factors other than latitude, however, influence the climates found in northern East Asia.

Linking Climate and Elevation in China Turn again to the East Asia physical-political map on page 622.

Locate the mountain region in the west of China. In these highlands, climate varies with elevation. The mountains also influence the type of climates found in northern China.

In the summer, monsoon winds heavy with moisture blow off the Indian Ocean from the southwest. The towering Himalayas and the mountain ranges of China block the winds, forcing them to drop their moisture on the southern slopes. As a result, the areas to the north of the mountain ranges receive little or no rainfall. The East Asia climate map on page 624 shows that those areas have mostly dry climates. The Gobi, a large desert, covers much of the vast, arid area in northwest China.

Linking Ocean Influences and Climate in Japan The waters that completely surround Japan greatly affect the climate of that country. Because water heats up and cools down more slowly than land does, summers tend to be cooler and winters tend to be warmer in Japan than they are in other places located along the same latitudes.

Ocean currents, too, help to moderate Japan's climate. The ocean currents map on page 34 shows the cold current that flows southward into the Sea of Japan. Northwesterly winds blowing across this current keep summers quite cool in much of northwestern Japan. In contrast, winds blowing across the warm Japan Current keep winters mild along Japan's eastern coast as far north as Tokyo.

Tropical Climates The southernmost tip of China and almost all of Southeast Asia lie within the tropics and have tropical climates. Only highland climates in some mountain areas break this pattern.

The Southeast Asian islands on and around the Equator have a tropical, wet climate. The weather here is very easy to predict—hot and humid. Average temperatures exceed 80°F (27°C), and the almost daily convectional rain produces about 80 inches (203 cm) of rain each year.

The far south of China and the Indochina Peninsula—the area occupied by Myanmar (formerly Burma), Thailand (TY land), Cambodia, Laos, and Vietnam—have tropical, wet summers and dry winters. The southwest monsoon winds produce wet and dry seasons.

Linking Climate and Vegetation As discussed earlier, strong links exist between climate and natural vegetation. The moderate climates found in the northeastern sections of the region support mixed forest vegetation.

The dry areas to the north of the mountains have little vegetation. The semiarid zone in the far north—a continuation of the Russian steppes—has short grasses. The arid area of the Gobi Desert, however, has almost no vegetation. Marco Polo, the famed thirteenth-century traveler, wrote the following comment about the Gobi, which covers much of the arid area:

'Tis all composed of hills and valleys of sand, and not a thing to eat is to be found on it. But after riding a day and a night you find fresh water. . . . [I]n some 28 places altogether you will find good water, but in no great quantity.

Mongolia's Gobi Desert
The Gobi Desert is a barren region that covers northern China and nearly all of Mongolia.

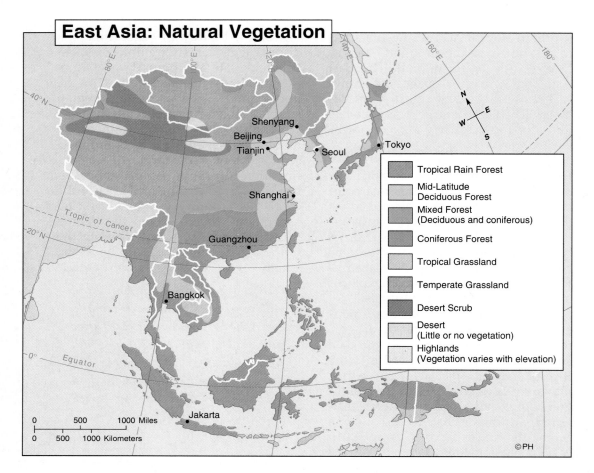

East Asia: Natural Vegetation

▨	Tropical Rain Forest
▨	Mid-Latitude Deciduous Forest
▨	Mixed Forest (Deciduous and coniferous)
▨	Coniferous Forest
▨	Tropical Grassland
▨	Temperate Grassland
▨	Desert Scrub
▨	Desert (Little or no vegetation)
▨	Highlands (Vegetation varies with elevation)

©PH

Applying Geographic Themes

Place   The lush rain forest vegetation of much of East Asia hides the fact that most of the region's soils are infertile. In mountainous areas, agricultural land is also very limited. Describe the natural vegetation of the area around Beijing, China. How does it differ from the area around Bangkok?

Natural vegetation in Southeast Asia is mostly tropical. Around the Equator, for example, the heat and nearly constant precipitation—it rains on about two hundred days of every year—support a thick, tropical rain forest. But the soils of the rain forest are poor. Heavy rains, as they filter down through the soil layers leech, or wash out, nutrients that are important for the growth of vegetation. Yet the rain forest thrives because the plants have adapted to this environment. Their shallow roots absorb nutrients from decaying plant life near the surface of the soil.

SECTION 1 REVIEW

Developing Vocabulary
1. Define: seismic

Place Location
2. Where is the Gobi Desert located?

Reviewing Main Ideas
3. What factors other than latitude influence East Asia's climate?

Critical Thinking
4. **Drawing Inferences** Why might life in parts of East Asia be hazardous?

✔ Skills Check

☐ Social Studies
☐ Map and Globe
☑ Reading and Writing
☐ Critical Thinking

Answering an Essay Question

Clear thinking and clear writing go hand in hand. The same principles of logical organization and clear expression that make other kinds of writing effective also serve to strengthen the writing of essays.

Using the following steps, organize the statements below into a coherent response to the following essay question: What was the significance and nature of the family in ancient China?

1. **Select a topic sentence.** Your answer to any essay question should begin with a general statement of the main point you will be making in the essay. The remaining sentences supply evidence for this main point. Answer the following question: Which of the five sentences below best states the topic?

2. **Identify supporting evidence for the topic sentence.** Once you have selected the topic sentence, consider the remaining sentences for their relative importance to your presentation. The second sentence of your essay should be less sweeping than the topic sentence, yet serve as a bridge to the details to follow. Answer the following questions: (a) Which sentence belongs next, or second, in your essay? (b) What is the common characteristic of the sentences that will follow your second sentence?

3. **Rank details that support your answer.** Details and examples that support your argument should be organized within the essay in order of importance, with the most significant statements first. Answer this question: In what order should you arrange the remaining sentences in your essay?

4. **Check the clarity and consistency of your essay.** Always read through your completed essay to make sure it reads smoothly and to see that you have developed your points logically.

A. Among siblings, a younger child had to obey and respect an older one.
B. Confucius, a great Chinese philosopher of the fifth century B.C., created a series of rules about behavior among family members.
C. People often called each other by their family position, such as "Elder Sister" or "Second Son," rather than by personal names.
D. The family, rather than the individual, was the center of ancient Chinese society.
E. Confucius taught that children had to obey their parents, and that a wife had to obey and honor her husband.

REGIONAL ATLAS: EAST ASIA

2 Human Geography

Section Preview

Key Ideas

- East Asia is one of the most heavily populated regions of the world.
- The region is still predominantly agricultural.
- It is a region containing great cultural diversity.

Key Terms

shantytown, intensive farming, terrace, aquaculture

Like a puzzle with many pieces, East Asia is a region made up of many subregions. It is a region of many different cultures, religions, languages, and styles of living. The region is home to one of the world's most technologically advanced nations. Yet, traditional ways coexist with twentieth-century conveniences.

Place: Examining Population Patterns

Like South Asia, East Asia has a huge population. While covering only 11 percent of the world's land area, it is home to about 34 percent of the world's population. China alone has 20 percent of the world's population.

Linking Physical Geography and Population Densities   Throughout East Asia, the population is unevenly distributed. The population density map of East Asia on page 629 shows that in China most of the population lives in the east. The eastern seaboard has a density of well over 250 people per square mile (100 people per sq km). Western China, however, with a density of less than 25 people per square mile (10 people per sq km), is very sparsely populated.

Compare this map with the physical-political map on page 622 and the climate map on page 624. Notice how the areas of dense population coincide with lowland areas that have mild and wet climates. On the other hand, much of western China consists of mountains and arid regions. Such landscapes cannot support large numbers of people.

Japan, with a population half that of the United States but living in an area no larger than California, has an average population density of 850 people per square mile (328 people per sq km). Most of these people are crowded into the narrow coastal areas of the larger islands. Density can be as high as 5,800 people per square mile (2,239 people per sq km) in these areas.

Average population densities in Southeast Asia tend to be lower than in the rest of the region. The fertile river valleys of Indochina, however, are heavily populated. And the rich volcanic soils of the Indonesian island of Java support a huge population. With more than 1,500 people per square mile (580 people per sq km), Java ranks as the most densely populated island in the world. Even government programs to move people to other, less populated islands have not helped Java's overcrowding problem.

Cities in a Rural World   East Asia is largely a rural region. In fact, 80 percent of China's population is rural. East Asia also claims some of the largest cities in the world. China has twelve cities with populations of two million or more. Located in lowland areas in the east, these cities serve as the country's centers of industry and trade. The port of Shanghai (shang HEYE), with a population well over twelve million, is China's largest city. Unlike most other East Asian people who enjoy freedom of movement, the Chinese cannot move without permission from the government. This policy, which limits the number of people

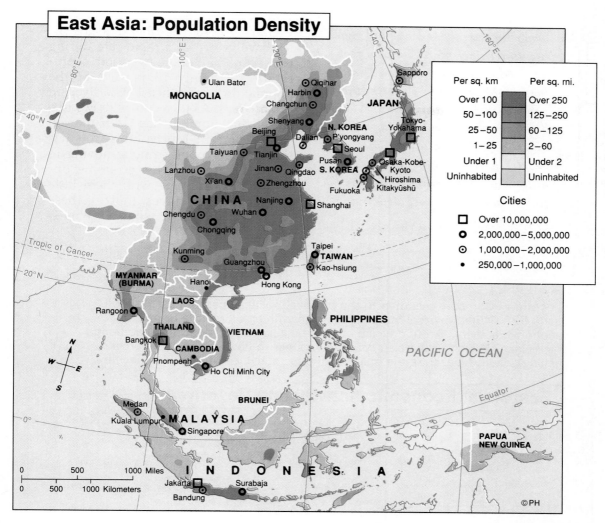

East Asia: Population Density

Per sq. km	Per sq. mi.
Over 100	Over 250
50–100	125–250
25–50	60–125
1–25	2–60
Under 1	Under 2
Uninhabited	Uninhabited

Cities

☐ Over 10,000,000
⊙ 2,000,000–5,000,000
⊙ 1,000,000–2,000,000
• 250,000–1,000,000

©PH

Applying Geographic Themes

1. Place China's size in land area is nearly the same as the United States' but its population is more than four times as large. Where in China is the country's population concentrated?

2. Regions Name three regions shown that have high population densities.

---

who move to the cities, has resulted in controlled urban growth. Despite these laws, China's urban population is as large as the entire population of the United States.

In contrast to China, 80 percent of Japan's population live in cities. Japan has four cities with populations of over two million people, all located on the island of Honshu. More than nineteen million people live in Tokyo's metro-politan area. Such huge numbers of people can make daily activities—going to work, for example—very difficult. American journalist James Fallows wrote the following about a rush-hour commuter train in Tokyo:

It means a kind of crowding that is, well, bestial. [It means] having strange bodies pressed tightly against yours on all sides and

shanty
This word for a
crudely-built wood-
en shelter is from
the Canadian French
chantier, or "trel-
lis," an open,
wooden-frame
structure.

being moved along by crowd momentum rather than your own feet.
. . . Most of the passengers can't
hold onto a strap because they
can't even raise their arms. When
the train slows down suddenly or
takes a curve hard, a mass of
people jolts through the center
aisle. The only reason they don't
fall is that there are too many
bodies in the way.

Urban Growth   Southeast Asia's cities have grown at a rapid rate during
the second half of the twentieth century. In 1950 only 15 percent of the
population of Southeast Asia lived in
urban areas. As the century draws to a
close, that number has risen to well
over 30 percent. Much of this growth is
the result of poor farmers moving to
cities in search of better economic opportunities. Bringing with them little
more than their hopes, most of these
migrants   settle   in   shantytowns—

Applying Geographic Themes

Location   Compare this map to the physical-political map of East Asia on
page 622. What physical features coincide with areas of little or no
economic activity? Around which of China's cities does most manufacturing
take place? Which East Asian countries appear to have major commercial
fishing activity? Describe the major economic activities of people in Taiwan.

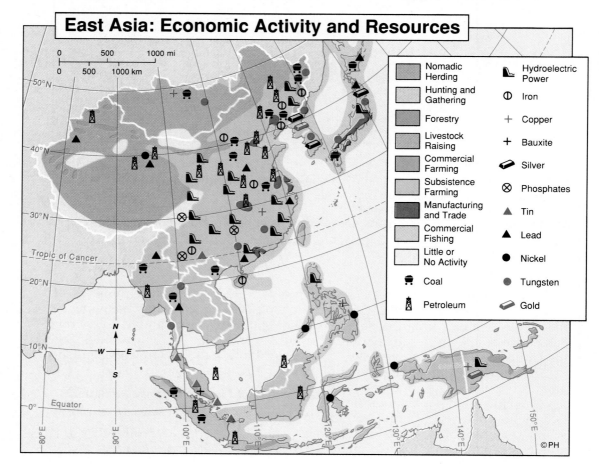

East Asia: Economic Activity and Resources

slums consisting of ramshackle houses made mostly of scrap materials located on the outskirts of cities.

Linking Agriculture and Population

With such a huge population crammed into a relatively small area, the economies of East Asia tend to focus on one problem. That is, how can they produce enough food to feed millions of people living on a limited amount of land?

Intensive Farming Throughout the region, people use intensive farming —farming that requires great amounts of labor—to produce food. To explain their approach the Chinese say, "We use every inch of the land." A look at the economic activity and resources map on page 630 shows that this is true. Vast areas of eastern China have been cleared of their natural vegetation and turned into cropland.

Over millions of years, China's great rivers created alluvial plains along their lower banks. Flat and fertile, these alluvial areas proved ideal for farming. In hilly areas, farmers reshaped the land into terraces, or level, narrow ledges. They built walls of stone and mud parallel to the natural slope. Soil washed down the slopes by rains collected behind the walls, creating level planting surfaces. Steep slopes were thus transformed into land that could produce crops.

In China and throughout Southeast Asia, people, rather than machines, do the agricultural work. In these areas, more than 50 percent of the labor force work in farming. In Japan, however, less than 10 percent of the labor force is in farming. Japanese farmers use modern machinery and the best seeds and fertilizers to increase their crop yields.

East Asian farmers produce a variety of crops, but rice is East Asia's staple food crop. China is the largest

Tea Harvest, Japan
Japanese farmers are able to grow tea on the terraced hillsides of their mountainous islands. People, not machines, harvest the crop.

rice producer in the world. However, when compared to their populations, Thailand and Indonesia actually produce more rice. Climate plays a major role in the success of a rice crop. The monsoons provide the large amounts of rain the crop needs to ripen.

Fishing Fish is another source of food for the people of East Asia, and fishing is a major economic activity in the region. Just as they apply intensive methods to farming, the people of East Asia make every effort to get the most from their fishing resources.

Aquaculture, or fish farming, is perhaps the most productive method used. Fish farmers raise shellfish and other fish in large tanks, reservoirs, ponds, sheltered bays, river estuaries, and even rice fields. China, one of the world's leading fishing nations, nets almost half of its yearly catch from aquaculture.

DAILY LIFE

Korean Affluence

Kim Mun-Cho remembers the hard times after the Korean War. His father, a professor, was paid partly in rice, which his mother then bartered for other kinds of food.

Those days of hardship are a distant memory to many of Mun-Cho's generation. In contrast to their parents, who lived simply and saved up to two thirds of their modest incomes, young South Koreans at the top 10 percent of the economic scale are making great amounts of money and spending it freely. They buy luxury boats, cars, and the latest fashions. They vacation in Europe and the United States and scuba dive in the Pacific islands.

In its rising numbers of young consumers, like those pictured at right, South Korea typifies the region. The number of Asians aged twenty to thirty-nine is expected to reach 600 million—up from about 425 million in the 1970s—by the year 2000.

1. What is different about the way money is handled by Korea's younger and older generations?
2. How is South Korea's population typical of Asia?

Industry The economic activity and resources map on page 630 shows that, except for its cities, East Asia is not highly industrialized. It is a largely rural region that depends heavily on its farming.

Japan, the industrial leader of the region, is sometimes considered a modern miracle. Japan has few important industrial resources and almost no oil. But through cleverness and hard work it has become a leading manufacturing nation. Japan is a world leader in the electronics and automobile industries.

Other East Asian nations, such as Taiwan (teye WAN), South Korea, and Singapore, followed Japan's example.

In recent years, they began manufacturing goods for export. And China, by importing Western technology, hopes to join the Asian giants of industry in the near future.

A Cultural Checkerboard

East Asia is a region of great cultural diversity. Even so, one culture—the Chinese—appears to be the most prominent. In sheer numbers, the Chinese dominate, and in their influence on neighboring cultures over the past 3,500 years, the Chinese are leaders. Yet within its borders, China is a huge mosaic of different ethnic and cultural groups. The Han Chinese are by far the majority—about 93 percent of the population. About sixty different ethnic groups, each with its own language, make up the remaining 7 percent. The distribution of these ethnic groups creates two distinct cultural regions in China. The Han occupy much of eastern China, while most of the minority groups live in the western and northwestern areas.

Cultural diversity is the source of much conflict in Southeast Asia. The religion map above shows that religious boundaries do not align with official national boundaries. During the years of European colonial rule, foreign powers imposed boundaries without any consideration of ethnic differences. Consequently, today's boundaries in Southeast Asia split up some ethnic groups and unite other, unrelated groups that have been traditionally hostile to each other.

The map of religions, however, offers only a general picture. The language map on page 657 shows the five major language families spoken in Southeast Asia. The map does not show that each family includes literally hundreds of separate languages. In Indonesia, for example, 250 languages are spoken, while over 700 different languages can be heard in Papua New Guinea.

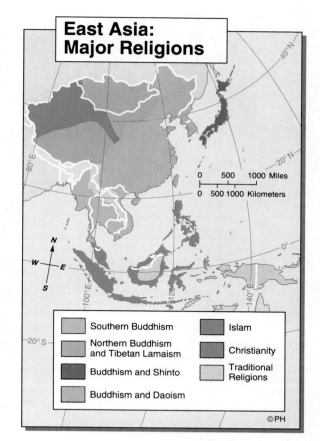

East Asia: Major Religions

0 500 1000 Miles	
0 500 1000 Kilometers	

Legend:
- Southern Buddhism
- Northern Buddhism and Tibetan Lamaism
- Buddhism and Shinto
- Buddhism and Daoism
- Islam
- Christianity
- Traditional Religions

©PH

Applying Geographic Themes

Movement   Buddhism was founded in India in 500 B.C. and spread throughout East Asia. What other religions are practiced in Japan?

SECTION **2** REVIEW

Developing Vocabulary
1. Define **a.** shantytown **b.** intensive farming **c.** terrace **d.** aquaculture

Place Location
2. On which island is Tokyo located?

Reviewing Main Ideas
3. What are the predominant economic activities in most of East Asia?

Critical Thinking
4. **Determining Relevance**   How has physical geography affected the distribution of people in East Asia?

Chapter 30, Section 2 **633**

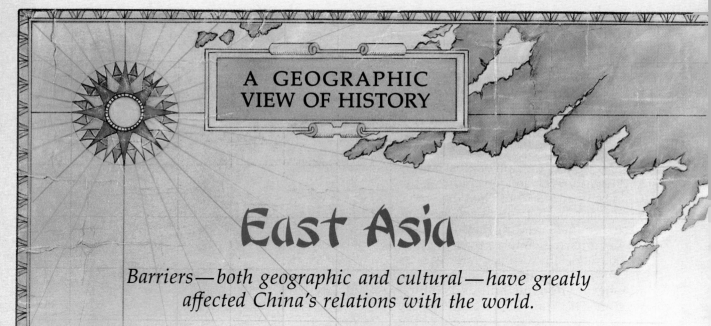

East Asia

Barriers—both geographic and cultural—have greatly affected China's relations with the world.

Located behind natural barriers, China created a culture free from outside influences for many centuries. Even when it made contacts with the outside world, China retained its uniqueness. Unlike much of the Eastern world, China remained almost untouched by Western influences until the nineteenth century.

Barriers to the Outside World

China's physical geography provided the natural barriers that made contact with other centers of civilization difficult. As the map on page 622 shows, the immense, barren lands of the Gobi Desert lie to China's northwest. Another great desert, the Taklamakan, also lies to the northwest. The dry, empty Plateau of Tibet provides over 1,000 miles (1,609 km) of almost unreachable land. Beyond the plateau lies the strongest physical fortress of all—the Himalayas. Completing China's wall of natural barriers are the waters of the Pacific Ocean to the east.

The natural barriers surrounding China did have one weakness. Raiders from the north were able to move south into China's rich agricultural lands. Beginning in about 450 B.C., the Chinese altered the natural terrain in order to patch this hole in their armor. They built a wall across their northern boundary to block invaders. Lengthened and strengthened in later years, this Great Wall—about 23 feet (7 m) high and 15 feet (5.5 m) wide—snakes across 1,500 miles (2,414 km) of China's landscape.

Within this boundary, China created a unique culture—self-reliant, independent, and isolated. The name the Chinese gave their country—the Middle Kingdom—provides a clue to the nature of this culture. The Chinese thought that, in terms of culture and geography, they stood at the center of the world. Therefore, they could gain little through contact with foreigners. In one way, the Chinese were right. Possessing an abundance of natural resources, trade within China was extensive. In fact, China was almost self-sufficient. Consequently, dealings with other countries for the purpose of trading were largely unnecessary.

Internal Developments

Beginning about 1500 B.C., various dynasties—successions of rulers from the same family—reigned over China. These dynasties expanded China from a relatively small empire centered on the Huang He to one that covered great expanses of Asia.

As China grew, internal transportation and communication became a challenge. Because China had first developed along the Huang He, the administrative center of the

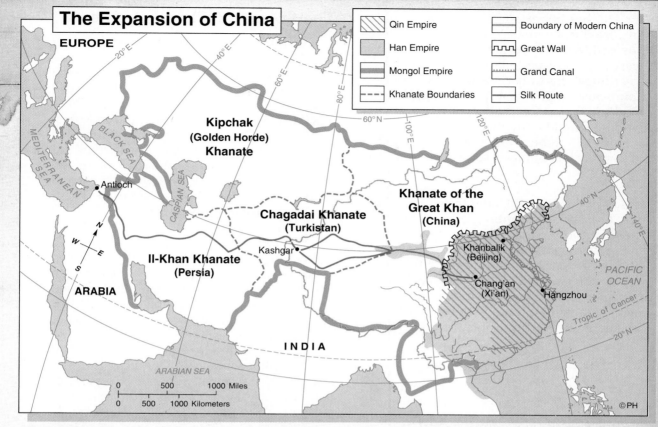

The Expansion of China
Qin Empire
Han Empire
Mongol Empire
Khanate Boundaries
Boundary of Modern China
Great Wall
Grand Canal
Silk Route
EUROPE
MEDITERRANEAN SEA
BLACK SEA
CASPIAN SEA
Antioch
Kipchak (Golden Horde) Khanate
Chagadai Khanate (Turkistan)
Il-Khan Khanate (Persia)
Khanate of the Great Khan (China)
Kashgar
Khanbalik (Beijing)
Chang'an (Xi'an)
Hangzhou
PACIFIC OCEAN
ARABIA
INDIA
ARABIAN SEA
Tropic of Cancer
20°E
40°E
60°E
80°E
100°E
120°E
140°E
60°N
40°N
20°N
N
S
E
W
0 500 1000 Miles
0 500 1000 Kilometers
©PH

Applying Geographic Themes

Interaction Completed around 200 B.C., the Great Wall was an attempt to protect China's farmers from the invaders of central Asia. How far did the borders of the Qin Empire extend?

country remained in this northern area. Farming, however, flourished in the south. It was vital to connect the rice-producing areas of the south with the seat of power in the north. Most of the navigable rivers, however, flowed from west to east.

Once again, the Chinese adapted the natural landscape to their needs. In the fifth century B.C., they undertook a huge project—the construction of the Grand Canal. Eventually completed in the 1300s, the Grand Canal spanned the 1,100 miles (1,760 km) from Beijing in the north to the southern coast. By connecting two of China's major rivers, the Huang and the Yangzi, the Grand Canal served as the empire's major north-south highway.

The Roots of Chinese Culture

Two important philosophies had a great impact on Chinese culture: Daoism (DOW ism) and Confucianism. The ancient Chinese believed that both humanity and the physical world were part of a harmonious order in the universe. Laozi (low DZOO), who lived from 604 to 517 B.C., taught that the path to true happiness lay in following the right way, or Dao (DOW). Laozi's philosophy came to be known as Daoism. This belief that the path to true happiness lies in a harmonious relationship with the natural world is still a powerful part of Chinese culture. Many Daoists try to achieve such harmony by completely withdrawing from the everyday world.

Another important philosopher, Confucius, lived from about 551 to 479 B.C. Confucius believed that society functioned best if every person respected the laws and behaved according to his or her position. For example, Confucius taught that parents should set good examples for their children and that children should obey their parents.

This hierarchical order of life extended to all aspects of Chinese society. Just as the father of a family was absolute ruler in his home, so the Chinese ruler was the absolute master of the people. Under the ruler, society was broken down into groups. The most valued group were scholars, landowners, and government officials. Next in importance were peasants whose agricultural work was highly valued. Below peasants were craftsmen, and at the bottom of society were merchants, whose trade with foreigners was considered unworthy work.

The Silk Road

In time China's physical isolation ended. From the west came reports of civilizations as advanced as China's. In the second century B.C., Chinese merchants crossed China's western physical barriers to find these civilizations, carrying goods to trade with them. Because the journey was so long and dangerous, they took only luxury items—tea, spices, and silks. The route they followed soon became known as the Silk Road.

The map on page 635 shows that the Silk Road stretched some 7,000 miles (11,265 km) into South Asia and the Middle East. Water-rich oases aided the merchants on the vast, dry expanses of the Taklamakan Desert. From China, traders took silks, tea, furs, and spices on the westward journey along the Silk Road. Gold, precious stones, glass, ivory, horses, and wool traveled in the opposite direction.

The Silk Road provided a route for information and ideas as well as goods. Chinese traders carried the philosophies of Confucianism and Daoism to South Asia, where they greatly influenced Buddhism. Later, the Buddhist religion was carried to China from South Asia by way of the Silk Road.

The Spread of China's Influence

As traffic along the Silk Road grew, many goods and ideas were exported by sea as well as by land. The Chinese eventually crossed the Yellow Sea, taking control of the Korean Peninsula in the second century B.C. Very quickly, Chinese culture took hold in Korea.

The Chinese knew of the existence of Japan as early as A.D. 100. However, they did not make solid contact with Japan for another four hundred years, when travelers from Korea sailed across the Sea of Japan. These travelers introduced the Japanese to the Chinese style of writing and to the Buddhist religion. In time, the Japanese adopted other aspects of Chinese culture. They modeled their system of government on the Chinese imperial government and built cities based on the plans of great Chinese cities. They even copied Chinese styles in such areas of everyday life as fashion and cooking.

Chinese culture, especially Buddhism and the philosophies of Confucianism and Daoism, also gradually spread into parts of Southeast Asia, including what today is northern Vietnam.

Contacts with Europe

Despite China's contacts with the outside world, its cultural barriers prevented the flow of influence from other countries into China. As far as the Chinese were concerned, foreign cultures were barbarian and inferior to their own. The few foreign ideas that did influence China were adjusted and modified to fit the Chinese way of life.

China's contact with Europe, however, was to have profound effects on the country. Beginning in the 1600s, the Chinese refused to trade with the European powers. As Emperor Qianlong wrote to King George III of Great Britain in 1792, "We possess all things. I set no value on objects strange or ingenious, and have no use for your country's manufactures." Within a few years, however, China signed trade agreements with a number of European countries. The deadly drug opium, to which many Chinese were addicted, was a major factor in this change.

Opium Arab traders traveling along the Silk Road introduced opium to China around A.D. 600. For one thousand years, it is believed the Chinese used the drug strictly for medical purposes. In the 1700s, however, seeing an opportunity for economic gain, Portuguese, Dutch, French, and British merchants began carrying large quantities of opium from India to China. As a result of European encouragement, some Chinese began smoking opium for pleasure.

The Chinese government limited the opium trade to one port. But opium use spread throughout South China. The small amount of opium that the Chinese authorities allowed into the country could not satisfy the demand. Ever mindful of increasing their profits, some European merchants began smuggling the drug into China.

When the Chinese government took steps to stop the smuggling in the mid-1800s, a series of opium wars broke out. The Chinese were defeated and forced to make a number of concessions to the Western powers. By the end of the 1800s, the Western nations had divided China into separate "spheres of influence"—areas to which a foreign country claimed exclusive trading rights. Not until after World War II was foreign influence completely ended because Chinese communists took over the government in 1949.

China under Communism After the communist takeover, China's leaders tried to turn the country into an ideal communist state. China's communist leaders quickly created an ideological barrier between China and the rest of the world. In the mid-1960s, for example, Chinese leaders launched a Great Cultural Revolution to build a true communist society in China. One goal was to adapt Marxism to the Chinese way of life.

The Great Cultural Revolution proved a disaster, and in the 1970s China again began to establish ties with the West. China adopted a new "open-door" policy to technological and cultural interaction with Europe, the United States, and Japan. China's leaders still remain, however, wary of the ideas and views of the West. For example, a massacre of

Resisting European Power

As this political cartoon shows, the Chinese resisted Europe's desire to divide their country.

political opponents in Beijing in 1989 was condemned by people throughout the world. The Chinese authorities, however, told these critics not to interfere in China's internal affairs.

TAKING ANOTHER LOOK

1. What natural barriers isolated China from the rest of the world?
2. How did the Silk Road act as a channel for information and ideas?
3. How did China's contacts with its Asian neighbors differ from its contacts with European nations?

Critical Thinking

4. **Synthesizing Information** Many people in the West consider China a mysterious country. How do you think Chinese geography and history have contributed to this?

Section Summaries

SECTION 1 Land, Climate, and Vegetation
Tectonic plate activity created much of the landscape of East Asia. Highland areas predominate, but much of the region's human activity takes place in lowland areas. Elevation affects the climate and vegetation of mainland China, while ocean influences modify Japan's climate. Much of Southeast Asia lies in the tropics and therefore has tropical climates and vegetation.

SECTION 2 Human Geography
Population densities are greater in the region's lowland areas, where climates tend to be moderate. Even though East Asia is a largely rural area, it is the location of some of the world's largest cities. Producing enough food to supply the needs of a huge population is the region's major economic activity. Countries in the region work toward this goal through the practice of intensive farming.

Vocabulary Development

Match each definition below with its correct term.

1. related to earthquakes
2. a slum located on the outskirts of a city
3. the use of large amounts of resources in agriculture
4. a level ledge constructed by farmers to increase the area of land under cultivation
5. fish farming

a. aquaculture
b. intensive farming
c. seismic
d. shantytown
e. terrace

Main Ideas

1. What impact did tectonic plate activity have on the landscape of East Asia?
2. In East Asia, what is the relationship between physical geography and population density?
3. What climate conditions prevail in the Gobi Desert?
4. How do elevation and the ocean influence climate in Japan?
5. Describe the climate in southernmost China and Southeast Asia.
6. Why is the soil in the rain forests of Southeast Asia generally poor?
7. How do the countries of East Asia attempt to produce enough food to meet the needs of their large populations?
8. What East Asian nations are rapidly becoming industrialized?

Critical Thinking

1. **Perceiving Cause-Effect Relationships** How might climate changes cause major food shortages in East Asia?
2. **Drawing Conclusions** Why do you think East Asia is not highly industrialized?
3. **Drawing Inferences** Based on what you have read in this chapter, why do you think many geographers consider China the dominant culture in East Asia?
4. **Synthesizing Information** Policy in China does not allow people to move without permission from the government. The result is controlled urban growth. Explain why you think that the advantages gained by limiting urban growth do or do not outweigh the disadvantages of limiting personal choice.

Practicing Skills

1. **Answering an Essay Question** With the paragraph derived from the statements about the ancient Chinese family on page 627 as a model, write five sentences that describe the contemporary American family. Then, using the steps you have learned in this lesson, organize these sentences into an effective answer to the following question: In your opinion, what is the significance and nature of the contemporary American family?

2. **Paraphrasing and Summarizing** Use the steps described on page 490 to summarize the following passage.

The ancient Chinese believed it was vulgar to serve food at the table in the form of an animal carcass and rude to ask dinner guests to cut up their own food. Therefore, they invented chopsticks which were made from ivory, wood, or bone and used to pick up each bite-sized morsel of food. This followed the Confucian philosophy of ''What you do not like when done to yourself, do not do to others.''

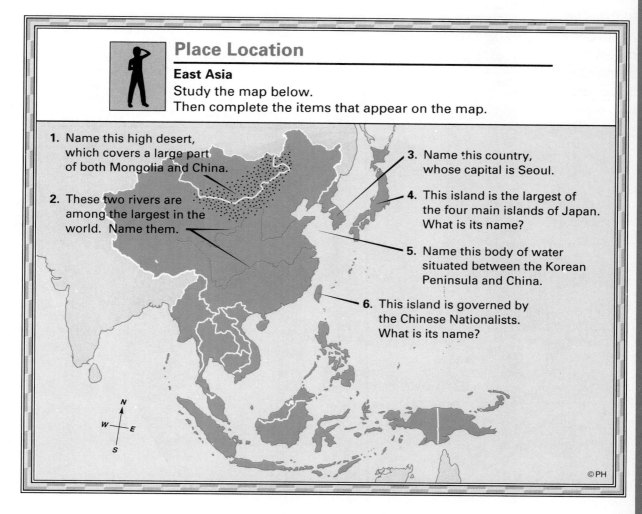

Place Location

East Asia
Study the map below.
Then complete the items that appear on the map.

1. Name this high desert, which covers a large part of both Mongolia and China.

2. These two rivers are among the largest in the world. Name them.

3. Name this country, whose capital is Seoul.

4. This island is the largest of the four main islands of Japan. What is its name?

5. Name this body of water situated between the Korean Peninsula and China.

6. This island is governed by the Chinese Nationalists. What is its name?

© PH

31

China

Chapter Preview

Three sections in this chapter discuss China, shown in red below. The fourth section covers the countries shown in blue on the map below.

Sections	Did You Know?
1 THE EMERGENCE OF MODERN CHINA	The first kite was not a toy but a Chinese invention from around 1200 B.C. that was used as a military signaling device.
2 REGIONS OF CHINA	China is able to support four times as many people as the United States on about the same land area.
3 CHINA'S PEOPLE AND CULTURE	*Pasta,* which means "dough paste" in Italian, was first prepared from rice and bean flour in China three thousand years ago.
4 CHINA'S NEIGHBORS	During the 1200s, Mongolia was the center of the largest land empire ever established, stretching from western Russia to Korea.

Traffic in Beijing, China
Bicycles are a common means of transportation in China's cities. Beijing, northern China's largest city, is the country's national capital and cultural center. This modern city is home to more than 6 million residents.

1 The Emergence of Modern China

Section Preview

Key Ideas

- After a long struggle, Communists gained power in China.
- Mao Zedong (MOW ZHUH DOONG) introduced many programs to construct a Communist state in China.
- The "Four Modernizations" changed the focus of China's economy.
- Calls for democratic reforms in the late 1980s were met with a violent response.

Key Terms

sphere of influence, abdicate, warlord, collective, light industry, martial law

Since its birth along the Huang He (HWAHNG HUH) River in northern China around 3000 B.C., Chinese civilization has been rooted in an agricultural way of life. Guided by the principles of Confucianism, the emperors of China ruled as the fathers of their people. Their main responsibilities were to make sure their peoples' needs were met and to rule by setting an example of fairness.

To make sure that all people were fed, the government asked families to grow rice and other grains. The government stored large amounts of these crops for times of drought and flooding. Flooding was quite common in northern China.

China also set up a large system of trade among the different parts of the country. China's large internal trade network helped the country succeed. Because of this trade, shortages in one region could often be met with surpluses from another region. As a result of both this trade and a large supply of workers, new technology was not needed in China. In fact, Confucian ideals held that moral traits such as cooperation were more important for society than new knowledge.

Lack of military technology proved a serious disadvantage for the Chinese when the industrialized countries of Europe and the United States used their military strength to force their way into China in the early 1800s, as discussed in Chapter 30. The changes forced on China by these Western powers upset the country's internal trade network. These changes combined with a series of natural and other disasters, resulted in widespread famine. A series of rebellions then broke out across the country in the late 1800s. China fell into disorder.

Movement: A Long March to Communism

By 1900, the United States and a number of European powers had carved China into spheres of influence—areas that these countries had control in but did not directly govern. Angered by the treatment they received from the Western powers, many Chinese called for changes. Some favored giving up their traditional ways and accepting the Western culture of their enemies. Others did not agree with this approach. Still a third group wished to take on some parts of Western culture, like technology, as a means to defend and protect their own culture. During this political struggle, a new political party, the Nationalist People's party, emerged. Many Nationalists, although they disliked the foreign powers in China, were greatly influenced by Western ideas.

After a series of revolts in 1911, the Nationalists seized power, forcing the emperor to abdicate, that is, give up his throne. The Nationalists then declared China a republic, choosing Sun Yat-sen (SOON YAHT sen) as the

country's first president. Sun was educated in the United States and wanted China to adopt a government based on Western democratic principles.

A Struggle for Power The Nationalists found governing the country more difficult than seizing control. They first had to deal with threats from the imperial army. To maintain peace and unity, Sun Yat-sen resigned, allowing the commander of the army to become president. But the army then turned on the Nationalists, removing them from the government. The Nationalists retreated to the southern city of Guangzhou (GWAHNG JOH), where they set up a rebel government.

In the meantime, the country rapidly slipped into disorder. With no central governing body in the country, local warlords, regional leaders with their own armies, seized power in their own areas. The Nationalists realized that to gain control of the whole country, they would have to defeat these warlords.

The Nationalists made little progress in this effort until the mid-1920s, when Sun Yat-sen died and Chiang Kai-shek (TS YAHNG KY SHEK) took over the leadership of the party. Chiang, a soldier who had trained with both the Japanese and the Soviet armies, quickly molded the Nationalist troops into a disciplined fighting force. In a two-year campaign, Chiang defeated one warlord after another, taking control of much of the country. And by 1928, he had established himself as president of the Nationalist Republic of China.

The Long March In the 1920s, a split developed in the Nationalist party. Some members of the party had adopted an ideology based on Karl Marx's communism. In their view, Marx's ideas offered an explanation for China's defeat by the Western powers. Marxism also suggested ways in which the government could achieve

The Long March

In 1947 the Communists, under the leadership of Mao Zedong (shown on horseback), made what they called the Long March. As the map above shows, they traveled several hundred miles out of their way to avoid Chiang's armies, while making their way from Jiangxi to the city of Yan'an in Shaanxi province in northern China.

The Long March

(map labels: Beijing, Yan'an, Shanghai, Ruijin)

0 500 1000 Miles
0 500 1000 Kilometers

©PH

prosperity for all—one of the age-old goals and duties of China's government. And finally, Marxism proposed a means of defeating the imperialist powers in China through a revolution led by the working class.

To achieve these goals, Communists in the Nationalist party wanted to give more power to the workers and to give land to the landless peasants. In 1927 Chiang Kai-shek disagreed with this plan and ordered those who sided with him to kill the people in the party who favored communist ideas. Many Communists were killed and those Communists who survived Chiang's death threats fled to the mountainous region of southeastern China. There they built a stronghold in the province of Jiangxi (JYAHNG SHEE).

Over the next six years, the Communists grew in numbers. Fearful that they would soon challenge his hold on the government, Chiang decided to destroy the Communists once and for all. Late in 1933 he sent a huge army against them. After months of fighting, the Nationalists' superior numbers and resources began to tell. The Communists left their positions in Jiangxi and started a year-long northward journey known as the Long March.

The map on this page shows the route of the Long March. In all, the Communists had to cross some eighteen mountain ranges and more than twenty rivers on their 6,000-mile (9,656-km) journey. Hunger, disease, and almost constant attacks by Nationalist troops made the march even more demanding. Of the one hundred thousand Communists who had left Jiangxi, only about ten thousand reached their goal—the northern province of Shaanxi (SHAH AHN SHEE). There, in the mountain town of Yan'an (YAHN AHN), they set up their new headquarters under the leadership of Mao Zedong.

The Communists Take Control

During the early 1930s, the Japanese took advantage of the fighting between the Nationalists and the Communists. They invaded the northern Chinese province of Manchuria, seizing it and renaming it Manchukuo. Then, in 1937, the Japanese moved to take over other areas of China. This foreign intrusion forced the Nationalists and Communists to make an uneasy truce.

The truce lasted only as long as it took to defeat the Japanese. In 1946, the two factions once again fought for control of China. During the war against Japan, the Communists had carried out major social reforms in the areas they controlled. These reforms included lowering the peasants' rents. Many of these peasants now joined the Communist struggle against the Nationalists. By 1949, the Nationalists had been defeated.

Chiang Kai-shek fled the mainland, seeking safety on the island of Taiwan. There he vowed he would one day reconquer China. In Beijing, Mao Zedong made a different statement. On October 1, 1949, he announced the establishment of a new Communist state: the People's Republic of China.

Constructing a Communist State

After thirty years of war, much of China lay in ruins. Even so, Mao indicated he had great plans for the nation.

He wanted to begin changes that would achieve greater agricultural productivity.

Mao believed that improvements in farm production could be achieved only according to the Communist principle of replacing private ownership with common ownership. In 1953, therefore, he called for the establishment of **collectives**—farms on which the land and machinery are pooled and people work together as a group and share the harvest. By 1956, 110 million families—about 90 percent of all peasants—worked on collective farms.

The Great Leap Forward During the mid-1950s, however, China failed to meet Mao's goals. In 1958, therefore, he introduced a new plan: the Great Leap Forward into Communism. Under this plan, the 700,000 collectives were combined into 26,000 People's Communes. These self-sufficient communal settlements—some of which had as many as 25,000 people—contained both farms and industries. Mao hoped that this new economic organization would, in a matter of years, increase China's production greatly.

Life in a People's Commune resembled life in the military. Communist party officials made all the decisions about what goods were made and who received them. The people's task was simply to work in the fields or factories.

The Great Leap Forward resulted, according to one Chinese official, in "a serious leap backward." Production fell rather than increased. The difficult life on the communes offered people no reason to work hard, because they received the same rewards regardless of the amount they produced. In addition, bad weather conditions hindered farm production. The harvests from 1958 to 1960, for example, were among China's worst. The Chinese government abandoned its Great Leap Forward after only two years.

The Cultural Revolution Many political leaders criticized Mao Zedong for the failure of the Great Leap Forward. Even Mao's closest advisers charged him with making mistakes. Deng Xiaoping (DUNG SHEE PING), for example, felt that Mao had tried to do too much too quickly. "A donkey is certainly slow, but at least it rarely has an accident," Deng remarked.

Stung by this criticism, Mao responded that the revolution was failing and that even more drastic measures were needed. In 1966, he called for a Great Cultural Revolution to smash the old order completely and establish a new socialist society. Mao unleashed an army of Red Guards—radical young men and women—to enforce his policies. Their job and command was to destroy the Four Olds—old ideology, old thought, old habits, and old customs.

No part of society was safe from the Red Guards. Communists who favored slower change, teachers, artists, writers—in fact, all those who disagreed with Mao—were publicly humiliated, beaten, and even killed. Those "enemies" who survived the wrath of the Red Guards lost their jobs and were imprisoned or sent to the country to work as peasants.

Farm production fell, factories ground to a halt, and schools closed as the Red Guards moved through the country. Mao approved of their actions by saying, "To rebel is justified." However, as the disorder continued, Mao called for an end to the Cultural Revolution in 1969. He also ordered the army to disband the Red Guards.

The Cultural Revolution was an enormous failure for China. At its end the economy was almost completely ruined. Hundreds of thousands of innocent people were in jail or in remote rural areas. And an entire generation of young people had lost their chance for an education. The loss of all these combined talents made China's economic recovery very hard.

The Red Guards

These young Chinese men and women are members of Mao's Red Guard, a radical group that was established to destroy China's old culture.

The Four Modernizations

A power struggle followed Mao Zedong's death in 1976. On one side stood the Gang of Four, a group of politicians, led by Mao's wife, who wanted to continue the Cultural Revolution. On the other side stood a number of leaders, led by Deng Xiaoping. Most people sided with Deng because they were tired of death and disorder. Deng took a more practical approach to solving China's problems than Mao had taken.

To begin the changes needed to make China productive, Deng set up the Four Modernizations. The goals of the program were to improve agriculture, industry, science and technology, and defense as quickly as possible. To achieve these goals, Deng stated, any ideas would be considered, even if they moved toward the ideas of a free-enterprise economy. As Deng noted, "It doesn't matter if a cat is black or white as long as it catches mice."

Changes in Agriculture Deng first took steps to repair the damage done to farm production during the Great Leap Forward. In place of the communes he established the contract responsibility system. Under this arrangement, the government rented land to individual farm families, who then decided for themselves what to produce. The families contracted with the government simply to provide a certain amount of crops at a set price. Once the contract was fulfilled, the families were free to sell any extra crops at markets for whatever prices they could get.

This chance to make more money by growing more crops greatly increased China's farm production. Since the introduction of the contract responsibility system, Chinese farmers produced about 8 percent more each year than they did in the previous year. And many farmers have benefited greatly from the new plan.

Under the contract responsibility system, families still did not own the land. The long-term leases awarded by the government, however, helped to develop an "owner" attitude among the farmers. As a result, many families have made improvements to the land. One writer who toured northeastern China in the early 1980s found evidence of development in rural China.

> [T]he new brick house . . . belonged to a friendly man named Yin. . . . He had built his house with the help of neighbors, who worked for him on holidays and were paid with "a good meal." . . . Money for lumber and bricks came from his cannery job and from what is officially more and more encouraged as "family sideline production"—known elsewhere as free enterprise. On a patch of land next to his house Yin grows vegetables and a cash crop of pumpkins for seeds.

However, not all of China's farmers enjoy prosperity. Most successful farmers live in eastern China where cities provide a market for their crops.

Industrial Development When the Communists came to power, they used most of China's resources to increase heavy industry. Heavy industries produce goods such as iron, steel, and machines that are used in other industries. At first, heavy-industry production grew rapidly. By the mid-1970s, however, Chinese technology was outdated and inefficient. When Deng took over, he took over an industrial system that was not working well.

Farm Market, Beijing

After selling part of their crop to the state, farmers may sell their surplus at markets.

DAILY LIFE

Popularizing the Law

For much of their history, the people of China have relied on and valued informal means of settling conflicts. They have been less concerned with the enforcement of official written laws. In an effort to change this, the government began a nationwide education campaign. Its goal was to teach 750 million citizens some basic law.

In the countryside, the campaign was designed to present legal knowledge in an entertaining fashion. At rural fairs performers taught legal concepts with songs and stories.

1. What methods did the Chinese use in their legal education campaign?
2. How would you describe the strategy of this campaign?

His program for industry had two goals. First, he wanted people to spend more money on consumer goods. Therefore, he changed the focus from heavy industry to **light industry**, the production of small consumer goods such as clothing, appliances, and bicycles. He also wanted factories to step up production. So he gave more decision-making power to individual factory managers. And he started a system of rewards for managers and workers who found ways to make factories produce more.

The Four Modernizations program has led to much new industrial development in China. But a number of major problems remain. To begin with, China's size and physical geography work against industrial efficiency. Most heavy industries are located in cities along the east coast, while the raw materials they need are located in China's rugged western regions. China does not yet have the transportation systems necessary to solve these problems.

The outdated technology used by the Chinese in their factories has also held up industrial development. To combat this problem, Deng Xiaoping made efforts to get foreign companies to assist in improving old factories and building new ones. Further, he encouraged Chinese students to study science and technology at Western universities.

Deng's economic reforms gave the Chinese a taste of Western lifestyles. Many city dwellers stocked their homes with the kinds of consumer goods commonly found in American

Chapter 31, Section 1   **647**

Goddess of Democracy, Beijing
Students in Beijing erected this statue in a protest for greater political freedom in 1989.

or Western European households—refrigerators, washing machines, and televisions.

Another Political Upheaval

As they became accustomed to economic reform, many Chinese began to demand a "Fifth Modernization"—political freedom. They were eager to enjoy democratic rights, such as the freedom to express their political beliefs openly and without fear. They also called for a greater voice in the running of the government.

Early in 1989, thousands of Chinese, mostly students, began a series of demonstrations in Beijing to demand democratic reforms. As many as one hundred thousand people crowded into Tiananmen Square in the center of the city. There they held political debates, sang songs, or simply sat and talked among themselves. In May, the government decided to end the protests. The country's leaders imposed martial law and ordered demonstrators to leave the square. **Martial law** is the law that is administered during periods of strict military control. Some demonstrators disobeyed the government's orders, and on the night of June 3, the army moved in to clear the few thousand people who remained.

The troops opened fire without warning, killing as many as two thousand people and wounding hundreds more. In the following days, troops rounded up suspected leaders and killed many of them without a trial.

When foreign leaders expressed outrage at such actions, the Chinese authorities accused them of interfering in China's affairs. As a result, relations between China and the West became worse.

After the crackdown, Deng reversed many of his economic reforms and returned to a hard-line Communist stance. With the voices of change silenced and the supporters of old-style communism once again in command, China's future seems uncertain.

■■■ SECTION **1** REVIEW ■■■

Developing Vocabulary
1. Define **a.** sphere of influence **b.** abdicate **c.** warlord **d.** collective **e.** light industry **f.** martial law

Place Location
2. Where is Beijing located in relation to Guangzhou?

Reviewing Main Ideas
3. How did the Communists gain control in China?
4. What was the purpose of the Great Leap Forward?
5. What were the goals of the Four Modernizations?

Critical Thinking
6. **Drawing Conclusions** Why do you think the Chinese authorities responded so violently to the pro-democracy demonstrations in Tiananmen Square?

2 Regions of China

Key Ideas

- The Northeast region serves as China's center of population, industry, and government.
- The Yangzi River is China's major trade route.
- The landscape of the North-Northwest region is stark, rugged, and barren.
- Isolated by huge natural barriers, Tibet developed a distinct, traditional society.

Key Terms

double cropping, theocrat, autonomous region

A journey through China's four major regions provides a vivid picture of the country's geographic diversity. Locate the regions of China on the map on page 650. A densely populated area, the Northeast serves as the country's administrative and industrial center. The Southeast section, an area of fertile plains and river valleys, is China's major agricultural region. China's frontier lands lie to the west. Two sparsely populated regions form this frontier area: the desert North-Northwest and the highland Southwest.

The Northeast: China's Core

The Northeast region includes eastern China from the Amur River in the north to the North China Plain in the south. The region is bounded on the west by the Greater Khingan (SHINJ AHN) Range. China's major lowland areas are in the Northeast.

The Northeast forms China's core. It contains Beijing, the country's capital, and the greatest concentration of China's population. The Northeast was the site of one of the world's earliest culture hearths, centered on the Huang He River. Each dynasty that ruled in China added more and more territory, extending Chinese influence far beyond the country's original boundaries. But no matter how far these empires extended, the capital city remained in the Northeast.

Beijing: A Problem-Plagued Capital

Beijing continues to serve as the seat of power for today's Communist government. Like other cities in the Northeast, Beijing is a major industrial center. In their rush to industrialize in the 1950s and 1960s, the Chinese made few efforts to control pollution from factories. In early spring, Beijing's air quality worsens with the onset of seasonal dust storms. One writer described a visit to Beijing in April:

> [The] annual dust storms were at their peak, and the city was suffocating under tons of fine soil blown in from the Gobi and the arid northwest. The fine yellow powder stung our skin [and] clogged our throats. . . . A golden haze hung over the entire city.

Loess: The Fertile Soil   The fine yellow soil that clouds the skies of Beijing is loess. While it may be an irritant to the inhabitants of the capital, loess also serves a beneficial purpose in the creation of a vast agricultural region in the Northeast.

As described in the previous chapter, strong northwest winds blow this very fine, yellowish soil into the Northwest from Mongolia and the Gobi, depositing it on the Loess Plateau. Rains then wash vast amounts of the soil into the Huang He. In fact, the Huang He is sometimes called "Yellow River" because of the distinctive color the loess gives the water. The loess is

WORD ORIGIN

loess
The term *loess* comes from the Greek word *luein*, "to detach, set loose or free." Loess is unlayered yellow-brown loam spread and deposited by the winds.

Regions of China

Applying Geographic Themes

Regions China's regions are divided between sparsely settled, dry and mountainous lands in the west, and densely settled, well-watered lands in the east. In which region is China's capital?

carried to the Huang He's lower reaches where, during flood periods, it is deposited as silt across the lowland area of the North China Plain.

Loess, which is highly fertile, can be turned into productive agricultural land with the use of irrigation. As a result, the Loess Plateau and the North China Plain are among the most intensely farmed areas in China. Wheat, because of the colder climate, is the dominant crop here.

"China's Sorrow" The Huang He carries rich, fertile soil to the lowland areas. It also serves as an important transportation route. However, the river has also brought death and destruction to the region in the past, earning it the nickname "China's Sorrow."

In years when the spring thaw and rains were very heavy, the river's swollen waters broke their banks, flooding the surrounding areas. Countless numbers of people lost their lives in these destructive floods. Those who survived saw their homes and crops washed away or buried under thick layers of silt. In 1887 flooding along the Huang He River resulted in history's greatest flood disaster, in which close to one million people died.

Consecutive floods over hundreds of years have greatly changed the population density of the Huang He valley. For example, Kaifeng, located about 325 miles (523 km) from the river's mouth, had a population approaching one million at the height of the Song Dynasty, around A.D. 1000. Today, Kaifeng has about one-third that number of people. A visitor to Kaifeng explained how this change came about.

The . . . [r]iver is partly to blame for this melancholy decline. It went on a rampage in 1461. . . . Floodwaters submerged Kaifeng again in 1887. . . . "When my father laid the foundations for this house, he discovered the roof of a former house underneath," said Zhao Pingyu, an oval-faced man of about sixty. . . . "That gives you an idea how much silt was brought by the river."

The Huang He's floodwaters still threaten disaster. However, a system of dikes and dams constructed in the 1950s and 1960s has greatly reduced the damage caused by the river's rampages.

The Southeast

Southeast China stretches from the North China Plain to the country's southern border, and from the eastern coast to the western highland areas. As the maps on pages 622 and 650 show, the Southeast region is more mountainous than the Northeast. In addition, the Southeast has a warmer, wetter climate than the Northeast.

This climate, together with the fertile soil of the region's river valleys, makes the region excellent for farming. Farmers use a number of intensive farming methods to get the greatest yield from the land. In some areas, double cropping—growing more than one crop a year on the same land—takes place. Elsewhere, farmers carve steplike terraces into the slopes of hills to increase the area of arable land. Rice, rather than wheat, dominates agriculture in the Southeast.

Movement Along the Yangzi: China's East-West Road The valley of the Yangzi River is the location of some of China's most productive farmland. With an average population density greater than 5,000 people per square mile (1,930 per sq km), the Yangzi valley ranks among the country's busiest and most crowded areas.

The Yangzi serves as China's east-west highway. Ocean-going ships can navigate some 375 miles (600 km) inland to the city of Wuhan. Small steamers travel even farther upstream, carrying goods to and from towns deep in China's interior. Shanghai, located at the mouth of the Yangzi, illustrates the importance of the river to China's trade. It is China's major port and, with a population of more than twelve million, its largest city. Shanghai handles about 50 percent of the country's overseas cargo and close to 75 percent of its domestic trade goods.

Special Economic Zones Each year, thousands of ships bring foreign cargo to Shanghai. When they depart, these ships take Chinese goods to ports all over the world. In other parts of the Southeast, a different kind of international exchange takes place.

One goal of the Four Modernizations was to spur economic growth by attracting foreign investment and technology to China. To fulfill this goal, in 1980 the Chinese government created four Special Economic Zones in the Southeast, offering special incentives to lure foreign investors and businesses there. The government set

The Yangzi River

The eastern lowlands are China's most productive agricultural region. The Yangzi River serves as China's major east-west highway. Ocean-going ships can travel some 375 miles (600 km) inland to the city of Wuhan.

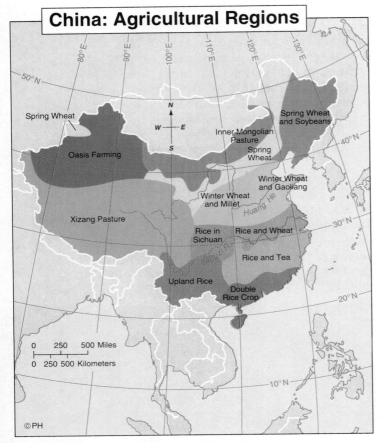

China: Agricultural Regions

Spring Wheat

Oasis Farming

Inner Mongolian Pasture

Spring Wheat and Soybeans

Spring Wheat

Winter Wheat and Gaoliang

Winter Wheat and Millet

Huang He

Xizang Pasture

Rice in Sichuan

Rice and Wheat

Yangtze R.

Rice and Tea

Upland Rice

Double Rice Crop

0 250 500 Miles

0 250 500 Kilometers

©PH

Applying Geographic Themes

Regions Temperatures that vary with latitude affect the crops that farmers are able to grow. What crops are grown in the northern agricultural regions? In the southern latitudes?

The incidents in Tiananmen Square in 1989 left the Special Economic Zones program in doubt. Since that time, Chinese leaders have attempted to limit the country's contacts with the outside world. And, for their part, foreign investors have expressed concern about putting capital into a country that might once again explode in violence.

The Northwestern Regions

The landscape of China's North-Northwest region is stark, rugged, and barren when compared with the landscape of the country's eastern sections. As the map on page 650 shows, the southern rim of the Gobi Desert forms China's northern boundary. Apart from leathery grasses that anchor the thin, sandy soil, very little grows in this rough, rock-strewn desert. Few people inhabit this cold, arid land.

Northwest China has a larger population but is no less rugged. Mountains in this region surround and separate two large basins. The Taklamakan Desert occupies much of the western basin. Steppe grasses cover most of the other basin.

The Silk Road, one of the great trade routes of ancient times, crossed the bare landscape of North-Northwest China. Along the road, way stations developed around oases fed by mountain streams. Over time, some of these way stations grew into large towns. For example, Kashgar, on the western edge of the Taklamakan, has a population of about 175,000. And more than 500,000 people live in Urumqi (OO ROOM CHI), in the foothills of the Tian Shan (TYEN SHAHN).

In these and other oasis towns, many people make a living through farming. For example, in Turpan (TOOR PAHN), about 100 miles (160 km) southeast of Urumqi, a system of underground irrigation canals fed by streams in the Tian Shan has helped to

low tax rates and reduced the number of official forms and licenses required to operate.

Of the four Special Economic Zones, Shenzhen (SHUN JHUN), located on the China-Hong Kong border, initially made the most progress. Prior to its designation, the city had a population of 30,000. By 1986, however, more than 380,000 people lived there. In that short time, Shenzhen's governing committee had signed about 150 foreign-investment agreements worth more than 700 million dollars. In less than a decade, 400 businesses had moved into the city.

make grape growing an important occupation. Even so, nomadic herding is the major economic activity throughout the region. When spring arrives, herders drive their animals to higher elevations in search of fresh pastures. Then, with the onset of cold weather, they return to their lowland meadows.

Xizang: The Southwest

If you look at the map on page 650, you will notice that one landform—the cold, dry Plateau of Tibet—dominates China's Southwest region. With elevations that exceed 12,000 feet (3,658 m) and surrounding mountains that soar above 20,000 feet (6,096 m), the plateau is largely isolated from the rest of the world.

On the plateau behind this huge natural barricade lies Tibet—a distinct, traditional society based on the Buddhist religion. For most of their history the farmers and herders of Tibet lived quiet, simple lives ruled by Buddhist custom and the decrees of their theocratic leader, the Dalai Lama. A theocrat is someone who claims to rule by religious or divine authority.

In 1959 a Chinese invasion ended Tibet's isolation. The Chinese reduced Tibet's Buddhist monasteries to rubble and drove the Dalai Lama into exile in India. They installed a Communist government and designated Tibet an autonomous region—a political unit with limited self-government. The Chinese also gave Tibet a new name—Xizang (SHEE ZAHNG), meaning "hidden land of the west."

In the years after 1959, the Chinese government instituted a policy designed to destroy Tibet's ancient culture. Even so, the Tibetans held on to their way of life. Recent reforms have relaxed restrictions on religion, and the Tibetans once again openly follow the Buddhist faith. However, these changes have not lessened the Tibetans' intense dislike of Chinese domination.

Lhasa, Tibet
Once the home of the Dalai Lama, the Potala Palace is found in Lhasa, Tibet.

SECTION 2 REVIEW

Developing Vocabulary
1. Define **a.** double cropping **b.** theocrat **c.** autonomous region

Place Location
2. Identify and locate the two major river systems of China.

Reviewing Main Ideas
3. Why is the Northeast region considered the core of China?
4. How has Tibet been affected by Communist rule?

Critical Thinking
5. **Making Comparisons** What are the major differences between eastern and western China?

✔ Skills Check

✔ Social Studies
☐ Map and Globe
☐ Reading and Writing
☐ Critical Thinking

Analyzing Primary Sources

Historians often use primary sources in studying the past. A primary source is information produced during or soon after the event, usually by a participant or observer. Examples of primary sources are letters, government documents, and eyewitness accounts.

Primary sources can convey a strong sense of an event or historical period. But the very fact that a writer is personally involved in the event may make the account biased or false. For that reason, you must analyze primary sources critically to determine their reliability.

In the box below are two primary source descriptions of the events that occurred when Chinese authorities used the army to end a mass protest in Beijing's Tiananmen Square in June 1989. Passage A is the statement of a student who was there. Passage B was written by a Chinese officer who helped direct the army. His statement, released in March 1990, was among the few given by Chinese leaders after the killings. Practice analyzing primary sources by following these steps.

1. **Identify the nature of the document.** Primary sources often present one person's point of view which, may be biased. Answer the following question: What is the main point of view of the authors of each of the passages below?

2. **Decide how reliable the source is.** It is important to try to determine the purpose or goal of the author of a primary source and if the author's point of view is biased. Answer the following questions: (a) How convincing does the speaker in each passage seem? (b) Which speaker is more directly responsible for the events in Tiananmen Square? (c) Do you think the two authors or speakers may have any interests that might lead them to conceal or distort the truth? (d) Would you say that Passage A is a reliable source? Is Passage B reliable? Give reasons for your answers.

Passage A

We expected tear gas and rubber bullets. But they used machine guns and drove over people with tanks. . . . It was like a dream. From where I was, the sound of crying was louder than the gunfire, but I kept seeing people falling. One line of students would stand up and then get shot down and then another line of students would stand and the same thing would happen. There was gunfire from all directions. The soldiers were shooting everyone.

Passage B

The People's Liberation Army intervention in Tiananmen was a matter of necessity. . . . At the beginning, we stressed to our forces, ''When beaten don't fight back; when scolded don't reply.'' First the PLA fired into the air as a warning. But a small minority shot at the PLA and snatched weapons. Under these circumstances, the PLA had to fire back in self-defense.

3 China's People and Culture

Section Preview

Key Ideas

- The Communist government has sought to control China's rapid population growth.
- The vast majority of the Chinese people share a common ethnic and cultural heritage.

Key Terms

ideogram, atheism, acupuncture

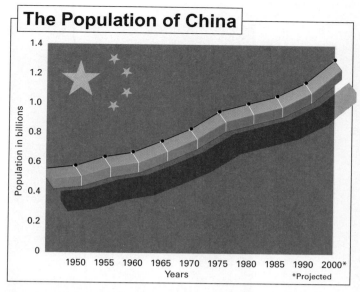

The Population of China

Population in billions (y-axis: 0, 0.2, 0.4, 0.6, 0.8, 1.0, 1.2, 1.4)

Years (x-axis: 1950, 1955, 1960, 1965, 1970, 1975, 1980, 1985, 1990, 2000*)

*Projected

Graph Skill

Describe China's population and growth rate during the period from 1950 to 1990. What is the projected size of China's population in 2000?

Imagine standing on a street corner watching a parade of all the people in your community. If they marched in rows of four, how long do you think it would take for the parade to pass by? Would it take a few minutes, or would the time run into hours, or days? If the whole population of China took part in the parade, you would have to stand on the street corner for more than ten years! With more than one billion people, China ranks as the world's most populous nation. Despite their numbers, the vast majority of China's people share a common language and have the same customs and beliefs.

A Huge Population

A census completed by the Chinese government in 1982 revealed that the country's population stood at 1,008,175,288 people. The population had doubled in the thirty-three years since the Communist takeover. If that rate of growth were to continue, government officials estimated, China would have another two hundred million people by the year 2000.

Mao Zedong, China's first Communist leader, believed that power lay in numbers. A huge number of people, he suggested, could never be overrun by outsiders. He urged the Chinese people to have more children. By 1965 the population of China was growing yearly by 2.85 percent, a rate well above that of most other countries. The growth rate in the United States stood at 0.7 percent.

Mao failed to understand the future results of his policy, though demographers warned him that rapid population growth would mean serious shortages of food and shelter. By the 1960s, their predictions had come true. Overcrowding and hunger were a part of everyday life for many Chinese. The worst overcrowding occurred along China's eastern coast. Population densities in this area rank among the world's highest.

Population Control Policies Realizing his country's predicament, Mao agreed to a new population policy. He called for families to have no more than two children. This slowed China's rate of growth a little, but overpopulation remained a problem.

China's Population Policy
Posters like this one in Shanghai are intended to encourage couples to have only one child.

with posters and billboards that listed the virtues of one-child families. City dwellers living in small apartments generally complied.

People in rural areas, however, proved harder to convince. The contract responsibility system, the farm reform policy described in Section 1, was clearly at odds with government requests to limit population growth. The responsibility system shifted agricultural production away from communes and back to a system of family labor. As a result, couples began to have more children to help raise production by their work in the fields. Children also represented security for parents, who eventually would become too old or sick to work. In rural areas people simply had larger families and accepted the punishments.

By 1985 the annual population growth rate had slowed to 1.1 percent, a significant change. And it remained well below 2 percent for the rest of the decade. However, each 0.1 percent increase in the rate means that China has another ten million mouths to feed. As a result, slowing population growth remains a major challenge for China.

When Deng Xiaoping came to power in the late 1970s, he saw that the only way to improve the standard of living of the Chinese people was to reduce population growth still further. He took strong measures to achieve this goal, setting up a "one couple, one child" policy. Couples who followed this policy received special rewards such as better housing and better jobs or pay increases at work. In contrast, couples who had more than one child faced the threat of fines, wage cuts, or even loss of their jobs.

Outcomes of Population Policies To ensure the success of the new policy, the government started a large publicity campaign, flooding the country

The Chinese Culture

Occupying a vast area and possessing a huge population, China is a land of great ethnic diversity. At the same time, however, the majority of the Chinese people share a common cultural background.

Ethnic Differences About fifty-five ethnic minority groups live in China, mostly in the frontier areas of western China. Each of these groups has its own language; the Chinese government officially recognizes no less than fifty-two separate languages. Different groups also have their own traditions, encompassing everything from the foods they eat to the clothes they wear. And they have many religious faiths.

However, even the largest of these ethnic groups—the Mongols, Uygurs, Tibetans, and Kazakhs—hardly number more than one million people. Together, all the ethnic minorities represent only 6 percent of China's population. The remaining 94 percent, close to one billion people, belong to the Han ethnic group. Taking its name from the Han Dynasty, established about 2,200 years ago, the Han has been the dominant ethnic group in China for centuries.

The Chinese Language The Han people speak Chinese. Written Chinese is unusual in that it is nonphonetic. Most forms of writing use alphabets that give an indication of the sounds of words. The written form of Chinese, however, generally gives no clues to its pronunciation.

Written Chinese involves the use of ideograms, pictures or characters representing a thing or an idea. To perform a simple task like reading a newspaper, a person needs to master as many as 2,000 to 3,000 characters. To achieve a solid grasp of the written Chinese language, however, requires knowledge of at least 20,000 different characters.

People can communicate in writing, regardless of where in China they live. But because the characters give no clues to pronunciation, different regions have developed quite different dialects. In 1955 the government established the northern dialect, known as Mandarin, as the country's official language. When students learn to read and write Chinese characters, they are taught the official pronunciation, even if they speak a different dialect locally.

Religions and Beliefs Officially, China is an atheist state. Atheism denies the existence of God. On coming to power in China, the Communists discouraged all religious practice. According to communism, religion is nothing more than a set of myths and

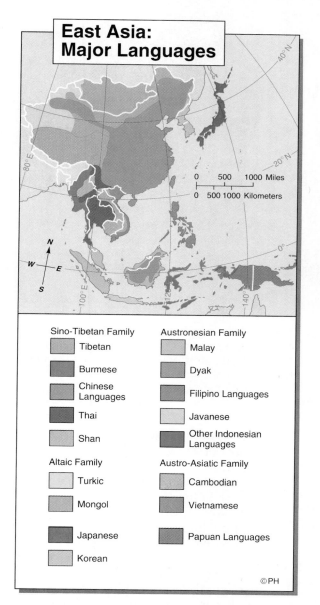

East Asia: Major Languages

0   500   1000 Miles
0 500 1000 Kilometers

Sino-Tibetan Family
- Tibetan
- Burmese
- Chinese Languages
- Thai
- Shan

Altaic Family
- Turkic
- Mongol
- Japanese
- Korean

Austronesian Family
- Malay
- Dyak
- Filipino Languages
- Javanese
- Other Indonesian Languages

Austro-Asiatic Family
- Cambodian
- Vietnamese
- Papuan Languages

©PH

Applying Geographic Themes
Regions Many languages are spoken in East Asia. What languages(s) do people in Taiwan speak?

WORD ORIGIN

superstitions designed to keep workers under the domination of the ruling classes.

The Communists seized the churches, temples, and other places of worship, turning some of them into meeting halls, schools, and museums. During the Cultural Revolution a few

nonphonetic Written Chinese is nonphonetic—the way it looks does not relate to speech sounds. *Phonetic* comes from the Greek *phonetikos*, "to sound with the voice."

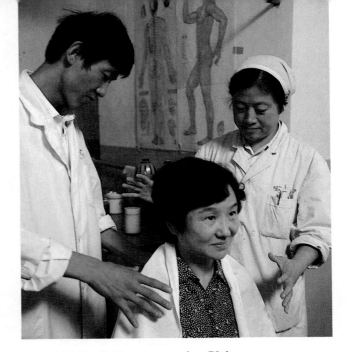

Medical Treatment in China
Modern Chinese medical treatment uses new technologies as it continues to be influenced by traditional Chinese medical practices.

Traditional Medicine Many other old customs live on despite efforts by the Communist government to modernize Chinese society. People in both urban and rural areas still practice traditional Chinese medicine. This discipline, dating back two thousand years, is based on the idea that good health results from harmony between people and the environment.

Traditional practitioners use a variety of treatments for illness. They prescribe special diets and herbal remedies. They also suggest breathing exercises, massage, or acupuncture. Acupuncture is the practice of inserting needles at specific points on the body to cure diseases or to ease pain.

Modern medicine is also practiced in China. However, countless Chinese —especially those who live in rural or remote areas—rely on traditional methods. Modern ideas have so far failed to displace original traditions and customs developed over thousands of years.

years later, the Red Guards destroyed many of these buildings in order to tear down the "Four Olds."

Neither laws nor the violence of the Red Guards, however, could wipe out two thousand years of tradition. Many Chinese people quietly continued to practice their religions. In the 1980s the government eased restrictions. While not encouraging religion, it made little effort to prevent people from following their chosen faith.

The Chinese practice a variety of religions, but the traditional religions —Buddhism, Daoism, and Confucianism—are the most popular. Of these, Confucianism is the most widely practiced faith.

Confucius stressed the importance of honoring and respecting one's ancestors. Today, reverence of ancestors remains an important aspect of Chinese family life. Many homes have altars where candles burn in memory of loved ones, and certain holidays are devoted to ancestors.

SECTION 3 REVIEW

Developing Vocabulary
1. Define **a.** ideogram **b.** atheism **c.** acupuncture

Place Location
2. Where is Shanghai located in relation to Beijing?

Reviewing Main Ideas
3. What steps has Deng's Communist government taken to control China's rapid population growth?
4. Which ethnic group has dominated life in China for hundreds of years?

Critical Thinking
5. **Expressing Problems Clearly** How does the government's contract responsibility system for farmers conflict with its efforts to reduce population growth?

4 China's Neighbors

Section Preview

Key Ideas

- Taiwan ranks as one of Asia's leading economic powers.
- Hong Kong's future is uncertain.
- Mongolia has long been influenced by its two neighbors, China and Russia.

Key Term

provisional

The tiny island states of Taiwan and Hong Kong lie off China's southeastern and southern coasts. Mongolia fringes China's northern border, forming a buffer between China and Russia. China, especially since the Communist takeover in 1949, has tried to influence its three smaller neighbors. And as the twentieth century draws to a close, Taiwan, Hong Kong, and Mongolia have begun to cast a wary eye towards China, trying to guess its intentions.

Taiwan: A World Apart

The small volcanic island of Taiwan lies some 120 miles (193 km) off China's southeast coast. Mountains—the island's major landform—rise in tiers to an elevation of about 12,000 feet (3,658 m). This distinctive landscape gives the island its name. In Chinese, Taiwan means "terraced coastline."

The people of Taiwan and mainland China essentially represent two branches of the same family tree. The Taiwanese, mostly members of the Han ethnic group, share a common history and culture with their mainland neighbors. After the Communist takeover, however, the people of Taiwan and the people of the mainland followed radically different political and economic paths. On the mainland, Marxist collectivism ruled political and economic life. In Taiwan, the government, while not fully democratic, allowed free enterprise to flourish.

The Emergence of Taiwan   Large-scale Chinese immigration to Taiwan began in the mid-1600s, when thousands of people fled the mainland after the Manchus seized control and established the Qing Dynasty. In 1683, the Manchus eventually conquered the island, and it remained under Qing rule for more than two hundred years. In 1895, after its defeat in the Sino-Japanese War, China gave Taiwan to Japan, which controlled the island until the end of World War II.

Another large wave of Chinese migration to Taiwan began in 1949, when the Nationalists, led by Chiang

Rice Paddies, Taiwan

In Taiwan rice is grown on the island's terraced hillsides where the use of intensive farming methods produces large, healthy crops.

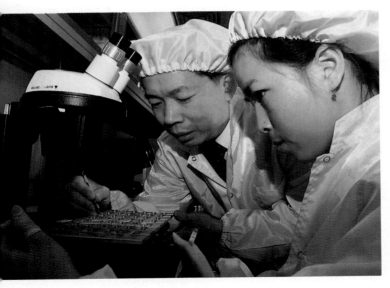

High-Tech in Taiwan
Taiwan today is focusing on high-technology industries to manufacture items such as personal computers and precision electronic instruments.

Kai-shek, fled China and Communist rule. This new group of immigrants, mostly businesspeople and military and government leaders, numbered about two million. On their arrival, Chiang Kai-shek set up a **provisional**, or temporary, government not just for Taiwan, but for all of China.

Since that time, a dispute has raged between the Chinese and Taiwanese governments. The government in Taipei (ty PAY), Taiwan's capital, claims it represents all of China. The Communists claim that the official government of China is located in the mainland capital, Beijing.

For many years, much of the Western world backed the Taiwanese in the hope that Chiang Kai-shek could oust the Communists. During the 1960s, however, many Western powers recognized the reality of communist rule in China and began to seek better relations with the Beijing government. Then, in 1971, the United Nations accepted mainland China as a member and voted to expel Taiwan, which had represented China at the United Nations since 1949. Immediately, most countries recognized Beijing as the legal seat of government for China.

Since the 1970s, Taiwan has lived in a strange international limbo. Much of the world has refused to recognize it as a country. Yet many of the countries that do not recognize Taiwan provide it with money and technical assistance. This aid has helped boost Taiwan to its position as one of the leading economic powers in Asia.

Taiwan's Economy Taiwan's rise to wealth began years ago, while Japan still controlled the province. The Japanese introduced intensive farming methods that changed agriculture from a subsistence to a commercial activity. The Japanese also built factories and began the construction of road and rail systems. After its defeat in World War II, however, Japan lost control of Taiwan. The Nationalists, fleeing Communist rule in China, soon took the place of the Japanese as rulers of Taiwan.

The Nationalist government instituted a sweeping land reform program that placed the land in the hands of tenant farmers. The government also encouraged farmers to increase their use of fertilizers, more productive seeds, double cropping, and other intensive farming methods. As a result of this program, Taiwan's farm production almost doubled.

The Nationalists also set in motion an industrial modernization program. With the help of foreign investment—especially from the United States—Taiwan quickly developed textile, food-processing, plastics, and chemical industries. This industrial growth was truly remarkable, because Taiwan had few natural resources. Nearly all the raw materials for these industries had to be imported.

In recent years, Taiwan has set new industrial goals, concentrating on high-technology industries such as electronics. Selling their products to huge markets in the United States and Europe, Taiwanese companies have greatly contributed to their country's rapid economic growth. This growth has provided most Taiwanese with a high standard of living. In the early 1980s, an American writer noted:

In the most remote villages color television sets blare from the open shops and aerials sprout from farmhouse roofs. In the cities, motorized scooters and cycles swarm like gnats around ever increasing automobile traffic.

Taiwan's Future   Today, Taiwan remains the direct opposite of its giant neighbor to the west in almost every respect. Size alone sets the two countries apart. China ranks as one of the world's largest countries. Taiwan, on the other hand, is about half the size of the state of Maine. In China, the majority of the people—66 percent—make a living through farming, but 83 percent of the Taiwanese work in manufacturing or service industries. In short, China is a developing country, while Taiwan ranks as an industrialized country.

China's Communist government has tried repeatedly to form a partnership with Taiwan, largely because of the advantages such a partnership would offer. In recent years, these efforts have intensified, but Taiwan continues to reject them. The future relationship between these two countries remains uncertain. However, its standing as an economic power in Asia remains strong.

Hong Kong: A Return to China?

The tiny British colony of Hong Kong faces an even more uncertain future. Located on China's southern coast, Hong Kong consists of the Kowloon Peninsula and several islands.

The whole colony covers only about 400 square miles (1,036 sq km), yet it is home to more than 5.7 million people. With average population densities in excess of 14,000 people per square mile (5,500 per sq km), Hong Kong ranks as one of the most crowded places in the world.

The Growth of Hong Kong   Hong Kong did not always bustle with human activity. Before the 1800s, it was largely uninhabited. But during a war between China and Britain from 1839 to 1842, the British used the narrow strait between Hong Kong Island and the Kowloon Peninsula as a harbor for their ships.

City Harbor, Hong Kong
When the Communists took control of China in 1949, Hong Kong used its ports for exporting its own, instead of Chinese, consumer goods.

Realizing the potential of the strait both as a naval base and as a way station for ships sailing to its far-flung Pacific empire, Britain claimed control of Hong Kong Island after winning the war. In 1860, after another war, the British took possession of the Kowloon Peninsula. Then, in 1898, China agreed to lease Hong Kong, Kowloon, and other land in the area to the British for ninety-nine years.

During the twentieth century, Hong Kong became an economic power in Asia. Its deep, natural harbor and its central location on the East Asian sea routes helped the port to become a leader in world trade. Every year more than ten thousand ships discharge and take on cargo at Hong Kong's docks.

Hong Kong also became an important manufacturing center, specializing in textiles, clothing, and electrical appliances. In developing these industries, Hong Kong took advantage of its large pool of human resources. Following World War II, millions of people, fleeing war and political unrest elsewhere in Asia, sought a safe haven in Hong Kong. During the first twelve years of Communist rule in China, Hong Kong took in more than one million Chinese refugees. These refugees provided a vast supply of inexpensive labor for the factories of Hong Kong.

Hong Kong exports about 90 percent of the goods its factories produce. Recent estimates set the value of Hong Kong's trade—both imports and exports—at about 100 billion dollars. In contrast, the value of China's trade was estimated to be only 70 billion dollars.

Relations Between Hong Kong and China Throughout its short history, Hong Kong developed with little interference from China. But, the two countries have developed an interdependence. Hong Kong has long obtained most of its vital resources—fresh water, for example—from the mainland.

Also, China has used Hong Kong as an exchange point for its trade with the West. Since the establishment of the Special Economic Zones, Hong Kong has been a leading investor in the Chinese economy.

The close of the twentieth century will bring the end of Britain's ninety-nine-year lease and a dramatic change in the relationship between Hong Kong and China. On July 1, 1997, Hong Kong will once again become part of China. However, the agreement between Britain and China provides that for the fifty years following that date, Hong Kong is to function as a Special Administration Region, free to carry on as before, both economically and politically.

The people of Hong Kong expressed anger at this arrangement when it was first announced in 1985. The anger turned to fear after the political crackdown in China in 1989. Even more people—especially business owners—began to leave Hong Kong. Experts predict that as many as half a million people may emigrate by 1997, taking with them all their money and investments. Such levels of emigration would deal Hong Kong's economy a crushing blow. As one financial expert noted, ''no economy can withstand that kind of leakage of capital over an extended period.''

Statements by the Chinese government have done little to stop the flow. With hundreds of Hong Kong residents leaving every week, and thousands more seeking visas for other countries, China may inherit little more than an empty shell when 1997 comes around.

Mongolia: A Buffer Between Giants

Mongolia is a vast, dry land about the size of Texas. The Gobi occupies the country's southern areas, while steppe vegetation covers much of the rest of the land. Located between Russia and China, Mongolia acts as a buffer between these two giants. Indeed, the two countries have long used Mongolia as a pawn in their regional power game.

But this was not always the case. In the thirteenth century, Mongolia was one of the world's great powers. The Mongols, under Genghis Khan and his descendants, ruled an empire that extended from China in the east to Hungary and Poland in the west. In later centuries, however, Mongolia came under Chinese rule.

Mongolia remained a province of China until 1911 when, with the backing of the Russians, the Mongols declared their independence. Ten years later, following Russia's example, Mongolia adopted communism. Since then, Mongolia, although a separate country, has been within the Russian sphere of influence.

Traditionally, the Mongols have made a living through nomadic herding. Even today, herding still ranks as the major economic activity on Mongolia's steppe lands. However, with the help of economic aid from the former Soviet Union and Eastern Europe, Mongolia has developed some industries. Coal and copper mining, food processing, leather goods, chemicals, and cement rank among the most important. With industrialization, Mongolia has become more urban. About 45 percent of the population now live in urban areas, mostly in Ulan Bator (oo LAHN BAH tawr), the capital. Gray, concrete apartment buildings—a familiar part of the urban landscape in Communist countries—provide housing for Mongolia's city dwellers.

As the 1990s began, Mongolia joined the other former Soviet satellite countries in calling for freedom and democracy, and Moscow slowly loosened its grip. Whether Mongolia will become a truly sovereign nation or remain under the influence of one of its two giant neighbors is uncertain.

Settlement in Central Mongolia

These circular tent-like structures, called *yurts*, serve as dwellings for the nomadic people of the Mongolian steppe.

SECTION 4 REVIEW

Developing Vocabulary
1. Define: provisional

Place Location
2. Where are Taiwan, Hong Kong, and Mongolia located in relation to China?

Reviewing Main Ideas
3. How did Taiwan rise to economic power in Asia?
4. Why is the future so uncertain for Hong Kong?
5. How have China and Russia influenced Mongolia's development?

Critical Thinking
6. **Recognizing Ideologies** Why do you think that the United States, after supporting Taiwan for so long, decided to recognize the Communist government of China in 1971?
7. **Perceiving Cause-Effect Relationships** How has location contributed to Hong Kong's position as a leader in world trade?

CHAPTER
31
REVIEW

Section Summaries

SECTION 1 The Emergence of Modern China

After the revolution of 1911, a long struggle took place between the Communists and the Nationalists for control of China. The Communists eventually won, and they began to build China into a socialist state. However, such programs as the Great Leap Forward and the Cultural Revolution proved disastrous. In the hope of saving the country's economy, the Communists changed course and started the Four Modernizations. This program had some success. But when the Chinese people called for greater economic and political freedoms, the government cracked down and returned to hard-line Communist policies.

SECTION 2 Regions of China

China can be divided into four basic geographic regions —the Northeast, the Southeast, the North-Northwest, and the Southwest. The two eastern regions are the most densely populated. These regions are also the location of much of China's economic activity. The two western regions are dominated by mountains and have arid climates. These regions are thinly populated, but are home to most of China's ethnic minorities.

SECTION 3 China's People and Culture

With more than one billion people, China ranks as the world's most populous nation. The population has grown rapidly since the Communist takeover, and it continues to do so. Population growth has become such a problem that the government now limits the number of children Chinese couples may have. The Chinese population consists of some fifty-five ethnic groups, yet the vast majority belong to the largest one—the Han. As a result, most Chinese share a common culture and heritage.

SECTION 4 China's Neighbors

Taiwan and Hong Kong, on China's southeastern and southern coasts, are among Asia's leading economic powers. Mongolia, on China's northern border, was for many years a satellite nation of the former Soviet Union. As the twentieth century draws to a close, these three countries are reassessing their relations with their giant neighbor.

Vocabulary Development

Use each term in a sentence that shows its meaning.

1. abdicate
2. collective
3. double cropping
4. ideogram
5. light industry
6. sphere of influence
7. martial law
8. theocrat
9. warlord
10. autonomous region
11. atheism
12. acupuncture
13. provisional

Main Ideas

1. How did the ideas of the Communists and the Nationalists differ?
2. How did Mao Zedong intend to speed up China's progress toward becoming a Communist state?
3. How did the Four Modernizations change China's economic focus?
4. What are the similarities and differences between China's two eastern regions?
5. How are the two western regions of China similar?
6. Why is population growth a pressing problem for China?
7. Why does China have so many dialects? What action did the Chinese government take to overcome this language problem?
8. Why are Taiwan, Hong Kong, and Mongolia wary of China?

Critical Thinking

1. **Identifying Alternatives** What alternative courses could the Chinese government have taken in dealing with the demonstrators in Tiananmen Square? How might a different response have affected China's relations with the West?
2. **Predicting Consequences** Do you think that developments in China and the surrounding countries would have been the same had the United Nations not admitted mainland China and expelled Taiwan?

3. **Making Comparisons** How are Tibet and Mongolia similar? How are they different?

Practicing Skills

Analyzing Primary Sources Use library resources to locate two additional primary-source accounts of the events in Tiananmen Square, Beijing, in June 1989. Then, using the steps on page 654, analyze the sources and present a summary of your analysis to the class.

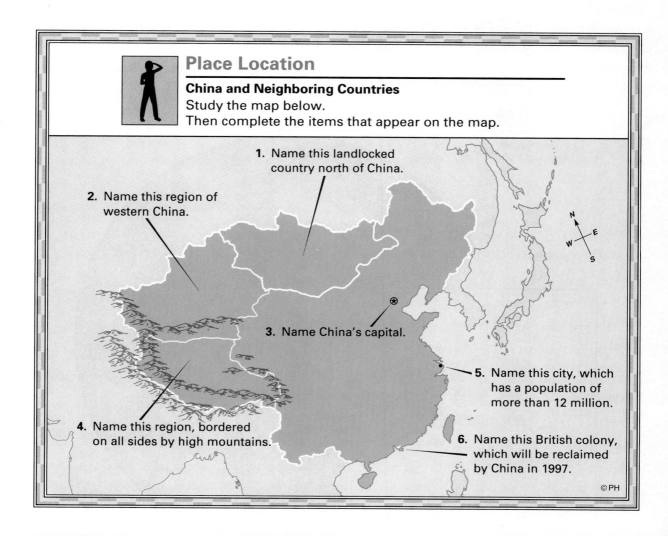

Place Location

China and Neighboring Countries
Study the map below.
Then complete the items that appear on the map.

1. Name this landlocked country north of China.
2. Name this region of western China.
3. Name China's capital.
4. Name this region, bordered on all sides by high mountains.
5. Name this city, which has a population of more than 12 million.
6. Name this British colony, which will be reclaimed by China in 1997.

© PH

CHAPTER

32

Japan and the Koreas

Chapter Preview

Locate the countries covered in this chapter by matching the colors on the right with those on the globe below.

Sections		Did You Know?
1	**JAPAN: THE LAND OF THE RISING SUN**	The McDonald's that has served the most hamburgers in a single day is located in Enoshima, Japan, near Tokyo.
2	**JAPAN'S ECONOMIC DEVELOPMENT**	Japan's successful industries are based on imported foreign minerals, metals, and sources of energy.
3	**THE KOREAS: A DIVIDED PENINSULA**	South Korea has led the world in economic growth for twenty years.

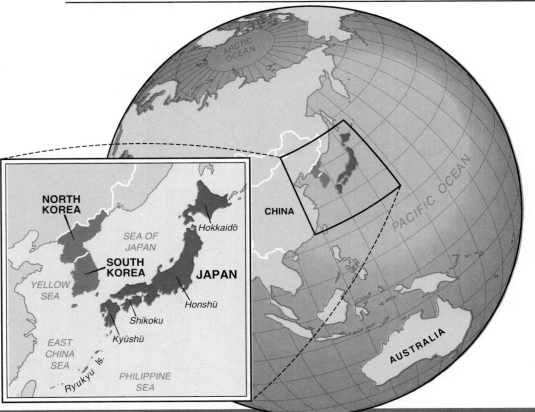

Honoring the Japanese Past

Field trips to the Toshogu Shinto shrine in Nikko, Japan ensure that Japanese schoolchildren understand and learn more about their history. A variety of ancient beliefs and rituals are practiced in Japan today.

1 Japan: The Land of the Rising Sun

Section Preview

Key Ideas

- The islands of Japan are part of the Ring of Fire—a region of tectonic activity surrounding the Pacific.
- Japan is one of the world's most densely populated countries.
- Japan's population is culturally and ethnically uniform.

Key Terms

seismograph, typhoon, homogeneous

In ancient times the Japanese knew of no people who lived to their east. There was only the endless sea. They thought of Japan as the land on which the rising sun first shed its light. According to legend, Amaterasu, the goddess of the sun, was the protector of Japan. The flag of modern Japan, a red circle on a white background, symbolizes this special relationship.

A Country of Islands

Japan consists of an archipelago, or chain of islands, off the coast of East Asia. The archipelago includes thousands of small islands, many of which are little more than large rocks. It also includes four larger islands where almost all of Japan's people live. The largest, Honshu, is home to 79 percent of Japan's population. South of Honshu are the islands of Shikoku and Kyushu. The farthest north of Japan's main islands is Hokkaido.

The islands of Japan are actually the peaks of a great underwater mountain range. Millions of years ago these

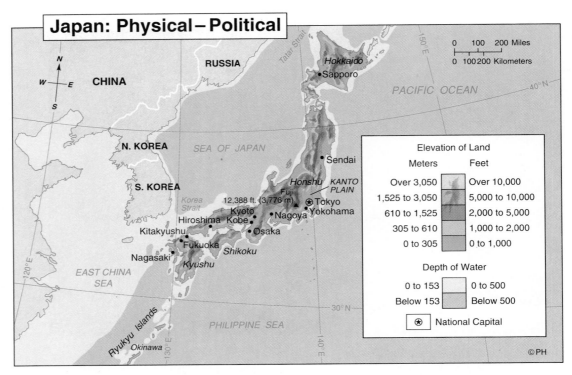

Japan: Physical–Political

RUSSIA

CHINA

N. KOREA

S. KOREA

SEA OF JAPAN

Hokkaido
• Sapporo

PACIFIC OCEAN

• Sendai

Honshu KANTO PLAIN
Fuji
12,388 ft. (3,776 m) ⊛ Tokyo
Kyoto • Yokohama
Hiroshima Kobe • Nagoya
Kitakyushu • Osaka
Fukuoka
Nagasaki Shikoku
Kyushu

Korea Strait

EAST CHINA SEA

Ryukyu Islands
Okinawa

PHILIPPINE SEA

Elevation of Land

Meters		Feet
Over 3,050		Over 10,000
1,525 to 3,050		5,000 to 10,000
610 to 1,525		2,000 to 5,000
305 to 610		1,000 to 2,000
0 to 305		0 to 1,000

Depth of Water

0 to 153		0 to 500
Below 153		Below 500

⊛	National Capital

©PH

Applying Geographic Themes

Place   Japan is a mountainous country frequently threatened by volcanic eruptions and earth tremors. Where is the largest plains area in Japan located?

mountains began pushing up from the ocean floor when two tectonic plates collided in a subduction zone.

Because of its steep, mountainous terrain, only about 15 percent of Japan's land is arable. Small, highly efficient farms are squeezed into small valleys between mountain ridges. Japan's best farmland has been created by the alluvial deposits of its rivers. Over thousands of years, these deposits have filled in parts of its rocky coasts, creating fertile plains. These narrow plains, which make up only about one eighth of Japan's land area, hold most of its population.

The Ring of Fire

Although Japan's islands are millions of years old, they are relatively new parts of the earth's surface. Japan is part of the Ring of Fire—a region of great tectonic activity along the rim of the Pacific Ocean. Earthquakes and volcanoes are common in the region. In fact, Japan experiences more earthquakes than any other country in the world. Sensitive seismographs—machines that register movements in the earth's crust—record about 7,500 earthquakes each year. Only about 1,500 of these are strong enough to be felt by people.

The Japanese have learned to adapt to most of these milder earthquakes. About once in two years, however, Japan has an earthquake that causes serious damage and loss of life. If the epicenter of the earthquake is on land, the earth shifts and buckles, and there are often landslides in mountainous areas. Buildings, farmland, and whole villages may be destroyed.

When an earthquake strikes offshore, the vibration can cause an enormous tidal wave, called a *tsunami,* which can devastate coastal lands.

As part of the Ring of Fire, Japan also has about 170 volcanoes, 75 of which are active. From time to time they erupt, sending showers of hot ash or molten lava down onto the surrounding countryside. Despite the dangers, volcanic activity has benefited Japan. In many places, volcanoes generate heat that warms underground water, creating hot springs. Resorts have been built up around these natural hot tubs.

Variations in Climate

If Japan were set along the east coast of the United States it would stretch all the way from Maine to Florida. Japan's climates, therefore, vary according to latitude. The northern island of Hokkaido has a climate like that of New England, with long winters and cool summers. Northern Honshu's climate is similar to that of the Mid-Atlantic states. On the southern island of Honshu, the climate is similar to that of North Carolina, with hot summers and mild but sometimes snowy winters. Kyushu and Shikoku, on the other hand, have climates more like that of Florida.

The Influence of Monsoons Japan's seasons are affected by monsoons, or prevailing winds. In the summer, the monsoon blows onto land from the east, bringing heavy rains and hot temperatures. From late summer to early fall is the season for typhoons— tropical hurricanes that form over the Pacific Ocean, often causing floods and landslides. In winter the monsoon shifts, blowing in cold, dry air from the Asian mainland.

The Influence of Ocean Currents Ocean currents also affect Japan's climate. The Japan Current, which flows northward from tropical waters along the southern and eastern coasts, warms the air. The cold Oyashio Current, on the other hand, flows southward along the west coast of Hokkaido and northern Honshu, cooling the air above it.

Population Density and Culture

Japan is one of the world's most densely populated countries. Since the late 1800s, its population has quadrupled, climbing from 30 million to more than 120 million. Although Japan's entire area is about the size of California, it has more than four times the population. Adding to the crowding is the fact that three quarters of the population live on the narrow coastal plain between Tokyo and Hiroshima.

Population density has had far-reaching effects in Japan. For example, the shortage of space has driven up the price of land and housing. Many families live in apartments in large high-rise buildings rather than in single-family homes. A family of four may share two or three small rooms.

This trend has had profound effects on the Japanese family. Traditionally, aging parents lived with their eldest son and his family. Now, older people often live by themselves or in special housing for the elderly. But great respect for the elderly is still part of the Japanese way of life.

Cultural and Ethnic Uniformity Japan's population is homogeneous, that is, its people have a very similar heritage. In fact, more than 99 percent of Japan's people have ancestors who lived in Japan thousands of years ago. Koreans are the only ethnic minority in Japan, and they make up only one half of one percent of the population. Both ethnic and cultural similarities have enabled the Japanese to build a strong sense of national identity and unity.

WORD ORIGIN

tsunami
The word *tsunami* is Japanese for "storm wave." A *tsunami* is a destructive wave that sweeps onto the land from the ocean.

Japanese Cherry Blossoms
Cherry blossoms are in full bloom for a very short time, so people gather in parks to capture them on canvas or just to enjoy their beauty.

Similar Religious Traditions Most of Japan's people also share similar religious beliefs and traditions. Japan's earliest people followed a religion known as Shinto. Shintoists worshiped the forces of nature and the spirits of their dead ancestors. Each household had an altar at which family members prayed and offered sacrifices. Today, while most of the holidays the Japanese celebrate are Shinto, the religion is no longer the focus of daily life. Nonetheless, it has had a great influence on Japanese culture, especially the Japanese people's great love of nature. To adapt this love of nature to crowded living conditions, the Japanese create miniature gardens that imitate nature. As one garden architect observed:

> When I am arranging one stone after another, I am always entangling the stone with my dream and pursuing an ideal world of beauty.

The majority of Japanese also practice Buddhist traditions. Buddhism teaches that people should seek spiritual enlightenment or knowledge by overcoming selfishness and living modestly. The Japanese integrated or added Buddhist teachings into their Shinto beliefs.

Japanese culture has also been greatly influenced by Confucianism, a philosophy that began in China. Confucianism stresses respect for the wisdom of older people and obedience to people in positions of authority, such as leaders, employers, parents, and teachers. Japanese society reflects the influence of Confucianism in its belief in the importance of the common good and in the high value it places on loyalty and respect for authority.

A Large Middle Class In most countries, modernization has gone hand in hand with the growth of the middle class. Nowhere is this truer than in Japan. Japan once had an upper class of aristocrats and a lower class of illiterate peasants. Today the vast majority of people belong to the middle class. This fact contributes to the social and economic uniformity, or homogeneity, of the Japanese population.

SECTION 1 REVIEW

Developing Vocabulary
1. Define: **a.** seismograph **b.** typhoon **c.** homogeneous

Place Location
2. On which island is Tokyo located?

Reviewing Main Ideas
3. Name three of Japan's most important physical features.
4. What shared characteristics give Japan's people a strong sense of national unity?

Critical Thinking
5. **Drawing Inferences** Describe some ways in which you think a high population density might influence a region's culture.

2 Japan's Economic Development

Section Preview

Key Ideas

- Japan's limited supply of natural resources and raw materials has influenced the country's history and economy.
- Today Japan is one of the world's leading industrial powers.

Key Terms

tariff, quota

A hundred and fifty years ago, Japan was an agricultural nation that had shut itself off from contact with other cultures. No one then could have foreseen that Japan would become one of the world's most highly industrialized and advanced countries.

First Contacts

At the time of its first contact with the West, Japan had a highly developed civilization. It was a prosperous nation ruled by warrior nobles. Trade between Japan and neighboring Korea, China, and Southeast Asia flourished. Its beautiful textiles were in great demand. From the court of the emperor in the beautiful city of Heian, now Kyoto, came impressive works of art, including the world's first novels, which were written by noblewomen.

In 1547 the first Portuguese trading ships arrived in Japan. Traders were followed by Roman Catholic missionaries, who hoped to bring Christianity to Japan. At first, the Japanese welcomed the Europeans, but soon they began to worry that European nations might try to conquer them. As a result, in 1639 the government closed Japan's doors to the West, ordering all Europeans to leave the country.

A Forced Reopening

Japan's isolation lasted for more than two hundred years. In 1853, however, the United States government sent Commodore Matthew C. Perry to negotiate a trade agreement. Perry's request was backed up by a massive show of force—a fleet of steam-powered warships. The Japanese knew that their weapons were no match for those of the United States navy, so they agreed to Perry's terms. In the next fifteen years, Japan was forced to sign treaties with other Western nations as well. These unequal treaties gave all the economic advantages to foreigners.

Perry through Japanese Eyes

The Japanese artist who drew this picture of Commodore Perry in the 19th century depicted him as menacing and warlike.

An Era of Reforms

In 1868 a new government took control in Japan. Its leaders were determined to modernize and industrialize the country so that it would no longer be at the mercy of foreign powers. The new emperor took the name Meiji (MAY jee), which means "enlightened rule."

During the period of the Meiji reforms, from 1868 to 1912, Japan underwent revolutionary changes. Politically, the country became more democratic. A parliament, called the Diet, was created, and legal reforms made all Japanese men equal before the law. The government also established a new school system, so that all Japanese children could be offered a basic education. To promote rapid industrialization, the government paid for the development of railroads, mines, telegraph systems, and whole new industries. By 1900 Japan had become strong enough to force an end to the unfair treaties and to deal with the West on equal terms.

Although Japan adopted many of the West's political and economic institutions, it did not wish to become a Western society. As they had done throughout their history and continue to do today, the Japanese practiced selective borrowing. They brought in only those ideas and innovations that seemed useful, adapting them to Japanese society. One of the Meiji leaders expressed his attitude in the following poem:

May our country,
Taking what is good,
And rejecting what is bad,
Be not inferior
To any other.

Japanese Imperialism

Lack of natural resources was a serious obstacle to Japan's goal of becoming an industrial power. The two major resources needed for industry, iron ore for steel making and petroleum for energy, are almost nonexistent in Japan. These and other items needed by Japan's developing industries had to be imported. Following the Western example of imperialism, Japanese officials began to try to gain control of weaker countries that were rich in natural resources.

At the turn of the century Japan fought and won wars with China and Russia, thereby gaining new territory and trading rights. In 1910, Korea was forced to become part of Japan. During World War I, Japan joined the Allies. After the war, it was rewarded with control of Germany's former colonies in the Pacific Ocean.

The worldwide economic depression, which began in 1929, took a terrible toll on Japanese industry. Many businesses were ruined, and unemployment soared. Military leaders argued that the way to recovery was through more aggressive expansion in Asia. An overseas empire would provide Japan with markets, raw materials, and new land for its expanding population. As conditions grew worse, militarists were able to gain control of the government. In 1931 Japan invaded Manchuria, and in 1937, China.

World War II

With the outbreak of war in Europe in 1939, Japan's leaders sided with Nazi Germany. And when France and the Netherlands fell under Nazi occupation, Japan seized French and Dutch colonies in Southeast Asia.

On December 7, 1941, Japan attacked the United States naval fleet at Pearl Harbor, Hawaii, and the two countries went to war. Both suffered heavy casualties. In August 1945, Japan surrendered after the United States dropped the world's first atomic bombs on the Japanese cities of Hiroshima and Nagasaki, killing more than 200,000 people.

DAILY LIFE

Japanese Comic Books

The Japanese, like those shown at right, work long hours and often commute a long way to and from work. At the same time, they must find time to digest the constant flow of new information and thus remain competitive. The Japanese have developed an unusual but effective way to assist people in this difficult task. Many companies now produce comic books, called *manga,* which rely on pictures instead of words to convey even complex information. A recent comic on Japanese economics has sold roughly two million copies. As one of Japan's publishers explains, "We must take in so much information in so little time in today's fast-paced world, and *manga* is a quick and easy way of doing it."

1. Why are *manga* an efficient means of relaying information?
2. Why are *manga* popular in Japan?

The American Occupation

From 1945 to 1952, Japan was occupied by the United States army. It was the first time in Japanese history that the country had been ruled by a foreign power.

Americans believed that the best way to ensure that Japan would never again be a threat to the world was to make it democratic. The military leaders of Japan were removed from power. The emperor, who had been worshiped as a god, was stripped of his political powers. Finally, the armed forces were disbanded, and Japan was forbidden ever to rebuild its military.

The United States occupation forces introduced a democratic constitution giving women legal equality with men; large farms and businesses were broken up and sold to poor citizens.

The Economic Boom

In the years after World War II, the Japanese economy grew faster than any other in the world. Instead of seizing raw materials from conquered nations, Japan now obtained them through trade. Japan became known as the world's workshop because it imported raw materials and made them into finished goods for export.

The Japanese Economic Miracle
By improving the quality of their industries, postwar Japan quickly transformed itself from a war-torn economy to one of the richest in the world.

At first, Japanese industries produced poorly made toys and novelties. Then the government encouraged a switch to expensive, high-quality goods such as cameras, electronic equipment, and motorcycles. By studying the methods used in Europe and the United States, the Japanese rapidly increased the efficiency of their factories and the quality of their goods. By the 1960s, Japan had become the most powerful industrial nation in the world after the United States and the Soviet Union.

Sources of Japan's Success

All over the world, people admire Japan's economic success. How did this country, which began with so many seeming disadvantages, succeed so rapidly?

An Educated Workforce Japan's greatest natural resource has turned out to be its people. Education has always been very important in Japan. Today, its people are among the most educated in the world. Almost all of its citizens go to high school, and a third go on to college.

Japanese schools have very high standards. The school day is long, and vacations are short. Students are given a great deal of homework, with additional assignments to complete during summer vacations.

To get into high school, students must take a special examination. To get into college, they must score well on another exam. Because competition for places in the best schools is fierce, many teenage students take extra classes after school to help them do better on these exams.

The Japanese Workplace Another reason for Japan's success is the attitude and cooperative skills of its workers. Japanese employees work hard and put in long hours. They take pride in the success of their company and want to contribute to that success. In return for this dedication, Japanese companies provide many benefits for their employees.

A large Japanese company is often compared to a family. Workers are often hired as soon as they graduate from high school or college. Once hired, Japanese workers are rarely fired or laid off, and very few workers ever quit their jobs.

Companies encourage loyalty and team spirit in a variety of ways. Often, there is an assembly each morning in which workers sing or exercise together. Many companies offer their employees low-cost apartments, so coworkers are also neighbors. Coworkers also often vacation together on trips sponsored by the company. Companies provide other benefits as well, including medical clinics, childcare, and low-interest loans.

Changes in Relative Location Another change brought about by World War II was a shift in patterns of global trade. After the war, Europe, which had been the center of international economic activity, was in decline. The new superpowers were the United States and the Soviet Union, both of whom were Japan's neighbors. At the same time, other Asian nations began to develop economically. As a result, Japan's relative location has changed. Instead of being far from the countries with which it trades, Japan is now at the center of active trade networks.

Government Planning In Japan, the government takes an active role in business. The Ministry of International Trade and Industry (MITI) is made up of leaders from business and government. MITI works to coordinate the efforts of Japan's many companies. For example, it sponsors research to find out what kinds of products are wanted in foreign countries, then shares its findings with companies that might make these products. MITI plans far into the future, deciding what kinds of economic activity will bring the greatest benefit not to individual companies, but to Japan as a whole.

Another way in which the government aids Japanese businesses is by controlling trade. The government has passed laws requiring tariffs. **Tariffs** are taxes on imports and exports that make foreign goods more costly than their Japanese-produced equivalents. The government also sets **quotas,** or fixed total quantities, which limit the number of foreign-made cars or other goods that can be sold in Japan.

In recent years these policies have come under attack. Japanese consumers object because tariffs and other trade barriers keep the prices of many goods high. Foreign governments, whose economies are hurt by these barriers, are also angry. They have begun to insist that Japan open its markets to their products.

An Exercise Break in Japan
At many Japanese companies, workers exercise as part of their workday.

SECTION 2 REVIEW

Developing Vocabulary
1. Define: **a.** tariff **b.** quota

Place Location
2. Where is the old capitol, Kyoto, located in relation to Tokyo?

Reviewing Main Ideas
3. Why did Japan cut off trade with the West in 1639? Why was trade reopened in 1853?
4. What was the goal of the Meiji reforms?
5. Why did Japan seek an empire in Asia?
6. What factors contributed to Japan's rapid economic growth after 1945?

Critical Thinking
7. **Analyzing Information** Describe what might be some positive and negative aspects of working for a large Japanese corporation.
8. **Synthesizing Information** Why do you think foreign countries oppose Japanese trade barriers?

Work and Leisure in Japan

The Japanese have seen many sweeping changes in the twentieth century. They have transformed their country from an agricultural society to one of the world's major industrial powers. Yet change has not disrupted some of the age-old influences of Confucianism in Japanese society. For example, the Japanese consider the well-being of the group, or the common good, to be more important than the needs of the individual. Whereas people in the West often make decisions based on individual preferences, actions taken for reasons of self-interest have no place in Japanese culture. Instead, each member of Japanese society performs a role to ensure the well-being of the group, whether it is the family, the corporation, or the country. Each member of the society feels it is his or her duty to do what is expected, and to do it properly. And in Japan, every role—for women, men, and children—is demanding.

Japanese Schoolchildren   *Even at a very young age, Japanese students face fierce competititon in order to get into the best colleges later.*

The Role of Japanese Women

Japanese women are expected to fill not one, but three important roles in Japanese society—mother, housekeeper, and wife. In addition, many Japanese women now work outside the home. As housekeepers, most Japanese women shop each day for fresh food. They carefully prepare meals, including lunches for their husbands and children to take to work and school. In addition to doing the cleaning, laundry, and cooking, Japanese women manage household finances and make purchasing decisions for the family. They also are expected to care for their elderly parents.

Because a good education is considered the key to many opportunities in Japan, women also play an active role in their children's education. Mothers sometimes attend their children's classes to take notes when their children are sick. They assist their children with homework, and enroll their children in afternoon drill sessions to prepare them for exams that determine who will be admitted to the best schools in the country.

The Role of Japanese Children

With so much importance placed on education, Japanese children have demanding roles to fill. From kindergarten through high school, students are pushed to excel. Competition is fierce and stressful, because examination scores determine which college a student will get into and, therefore, what will be his or her future job. Japanese students spend an average of 240 days a year in school, including Saturdays. In the United States, students go to school for an average of 180 days. Students in both Japan and the United States spend many hours each evening on homework.

The Role of Japanese Men

Japanese men also fill demanding roles. Most men have long commutes on extremely crowded trains and are expected to work evenings as well as days out of loyalty to their companies. Many also do not take their allotted vacation time. If they do vacation, it is often with their co-workers on corporate trips or retreats. A magazine article described the daily life of a corporate executive named Kotaro Nohmura from Osaka as follows:

Six days a week, he gets up at 7:00 A.M. and eats a Western–style breakfast prepared by his . . . wife. Then he is out the door. . . . Each Monday morning at 8:30 A.M. Nohmura and one hundred co-workers assemble for chokai, *a corporate pep rally, where they begin the week by reciting twelve company creeds. [He] calls it a day at about 5:30 P.M.. But Nohmura's work is far from finished.*

The article goes on to describe Nohmura's activities, including evening business entertainment. On many nights, he is required to escort a favored client to an elegant restaurant. He often does not arrive home until nearly midnight and has time only to look in on his four sleeping children before sleeping himself. Sunday is Nohmura's only day off from work. Such a demanding schedule is not unusual for Japanese businessmen.

Signs of Stress and Change

Most women, children, and men perform the duties that Japanese society assigns to them without question or complaint. Yet there are some signs that the stress associated with these duties is beginning to cause people concern. In 1990 one survey found that nearly half of five hundred Japanese employees interviewed feared *karoshi*—early death from overwork. About 70 percent of the people surveyed said they felt stressed.

The Japanese government has begun a campaign to encourage workers to slow down. Banks are now required to close on Saturdays. Government offices now stay open only every other Saturday. Schools are also beginning to lighten the load on students. Some are beginning to give students more choices in the classes they take and to offer leisure-activity classes such as judo, swimming, and flower arranging. In addition, women are gaining a greater voice in government and are working to have laws passed that will benefit Japanese society.

As a result of each of these small changes, Japanese families are finding themselves with more and more leisure time. Leisure industries are experiencing a boom. Sales of campers, sailboats, bicycles, and fishing equipment are up, and recreational centers and parks are expanding. Despite the fact that some of the Japanese are beginning to take it easy, however, one third of the workforce still worked six days a week in the early 1990s.

TAKING ANOTHER LOOK

1. What important roles do Japanese women have in their society?
2. What is a typical workday like for many Japanese workers?

Critical Thinking

3. **Drawing Inferences** How do you think that Japanese work and study habits have influenced the country's standard of living? How have they influenced the country's quality of life?

Work in Japan   A day on the job in Japan often means working until midnight.

3 The Koreas: A Divided Peninsula

Section Preview

Key Ideas

- The Korean Peninsula has been divided into two countries since 1945: North Korea and South Korea.
- Despite differences in political ideology, the people of North and South Korea share a similar language, culture, and heritage.
- Korean culture has been heavily influenced by the Chinese and has in turn transmitted many aspects of Chinese culture to Japan.

Key Term

demilitarized zone

In 1945 the Korean Peninsula became caught up in the Cold War struggle between Communists and non-Communists following World War II. After defeating the Japanese, who ruled Korea from 1910 until 1945, the Soviet Union administered northern Korea and the United States administered southern Korea.

Both powers were expected to remove their troops as soon as Korea was able to govern itself. Instead, the Soviet Union established a Communist government in North Korea. An election was held in South Korea, and American troops pulled out in 1949. Then, in 1950, the North Koreans launched a surprise attack on South Korea. Their objective was to unite the country under the rule of a single Communist government. However, hoping to stop the spread of communism, United Nations forces from fifteen different countries, including the United States, came to the aid of South Korea. For three years the army of the North, assisted by China, and the army of the South fought back and forth across the peninsula. An estimated five million people died in the fighting.

Finally, in 1953, a cease-fire agreement was signed establishing the division between North and South Korea at the thirty-eighth parallel of latitude —the same as it had been before the fighting. The countries were separated by a demilitarized zone—a strip of land on which there are no troops or weapons. The Korean Peninsula remains divided today. North Korea, or the Democratic People's Republic of Korea, is a Communist country; South Korea, or The Republic of Korea, has a non-Communist government. Despite these political differences, the people of the Korean Peninsula share a common history and culture that is thousands of years old.

A Mountainous Land with Varied Climates

The Korean Peninsula extends off the east coast of Asia between China and Japan. To its east is the Sea of Japan. To its west and south is the East China Sea. In area, Korea is about the same size as Minnesota.

The Korean Peninsula is one of the most mountainous areas in the world. There is no part of the country in which a person cannot see mountains in all directions. As a result of the rugged terrain, only about 20 percent of the land is arable.

The monsoons blow from the southeast in summer, bringing hot and wet weather. In July alone, many areas receive as much as 30 inches (76 cm) of rain. In winter, the monsoons blow cold, dry air from the Asian interior. In spring and fall skies are clear and temperatures pleasant throughout North and South Korea. Many festivals, such as Ch'usok, are held in the fall, after the harvest is gathered. At that time of year Koreans say, "The sky is high and the horse is fat."

WORD ORIGIN

Ch'usok
Ch'usok, which means "autumn eve," is a harvest festival that takes place at the end of September in Korea. It is also called "the moon festival."

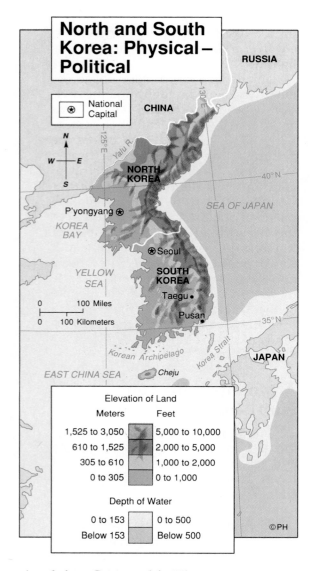

North and South Korea: Physical–Political

⊛ National Capital

RUSSIA
CHINA
NORTH KOREA
P'yongyang ⊛
KOREA BAY
YELLOW SEA
⊛ Seoul
SOUTH KOREA
Taegu •
Pusan •
SEA OF JAPAN
JAPAN
Korean Archipelago
Korea Strait
EAST CHINA SEA
Cheju

0   100 Miles
0   100 Kilometers

Elevation of Land

Meters	Feet
1,525 to 3,050	5,000 to 10,000
610 to 1,525	2,000 to 5,000
305 to 610	1,000 to 2,000
0 to 305	0 to 1,000

Depth of Water

0 to 153	0 to 500
Below 153	Below 500

©PH

Applying Geographic Themes

Regions The Korean Peninsula was divided into two countries in 1953. What physical characteristics are found in both countries?

Although the peninsula is small, there is a great deal of variation in its climates. Because North Korea is located nearer to the Asian mainland, it is influenced by the nearby continental climate regions. In the far north, the climate is similar to that of Siberia, with short cool summers and bitterly cold winters. South Korea, on the other hand, is more influenced by the

moderating effects of the surrounding seas. Parts of southern Korea are subtropical.

Korea's People

South Korea, which is home to more than forty-five million people, is the fourth most densely populated country in the world. Almost a quarter of the population is concentrated in the capital city of Seoul. North Korea, which has only about twenty-two million people, is much less densely populated. P'yongyang, the capital, is the only city with a population of more than one million.

Ethnicity and Language in a Cultural Crossroads Historians believe that the first people who lived in Korea came from Central Asia in ancient times. Because of its location, Korea has played an important role in the history of East Asia. Through the 30,000 years of Korea's history, invading armies have swept through it numerous times. Korea has also served as a link between cultures. Itself heavily influenced by Chinese civilization, it has also acted as an agent, transmitting many Chinese traditions to its eastern neighbor, Japan.

Although their culture was heavily influenced by the Chinese, who ruled Korea for many years, Koreans adapted Chinese cultural ways to their own existing culture. For example, Koreans borrowed extensively from the Chinese writing system and adapted many Chinese words. The Korean language is also related to the Japanese, Mongolian, and Turkish languages.

Religious Traditions As in Japan, many Koreans accepted and integrated more than one religion into their way of life. For example, Taoism came from China. Taoists try to simplify their lives and live in harmony with the natural world. Although few Koreans today would call themselves Taoists,

Taoism has had a deep influence on Korean culture.

Buddhism is the most common religion among Koreans. Like all other people who adopted Buddhism, the Koreans modified its teachings to fit their existing culture. This process continues today with a revival of interest in Buddhism among young Koreans.

Confucianism, which came from China, has also had a major impact on Korean culture. In fact, Koreans were so devoted to the teachings of Confucius that even the Chinese admitted they were his most virtuous followers. And, as in Japan, this philosophy has influenced people to respect and obey those in authority.

In the 1700s the first Christian missionaries were killed by the Korean government. But by the late 1800s Christian missionaries were allowed to build schools and hospitals in Korea.

Today South Koreans have complete freedom of religion. In North Korea, as in all Communist countries, the government discourages people from holding any religious beliefs.

From Agriculture to Industry   Korea was annexed by Japan in 1910. During a thirty-five-year occupation, the Japanese took the best Korean farmland for themselves. This situation forced many Korean farmers to move to the cities to look for work. Japanese rule over Korea also disrupted Korean culture and weakened a rural tradition of strong family ties.

At the same time, industry gained a foothold in pre–World War II Korea, with the northern part of the country developing more rapidly than the south. In spite of industrial development, however, most Koreans still worked as farmers until the division of the Koreas in the mid-1950s. Today less than half the Korean people work in agriculture. North Korean farmers work in huge cooperatives, often with hundreds of other families on one cooperative farm.

At the time of division, South Korea was economically at a disadvantage. The best industries and hydroelectric plants were in communist North Korea, and South Korea was overflowing with battle-weary refugees. Today South Korea has one of the fastest-growing economies in the world.

Natural Resources   Two resources that North and South Korea have in abundance come from the peninsula's waters. The sea around the peninsula is rich in fish, shellfish, and edible seaweeds. These foods make up a large part of the average diet. The peninsula's fast-flowing mountain rivers are another major resource. They have been harnessed to create hydroelectric power. In the more mountainous north, the development of hydroelectric plants is far ahead of similar development in South Korea.

Seoul: South Korea's Capital

Because South Korea's government encouraged private ownership of industry, Seoul has grown much faster than its North Korean counterpart.

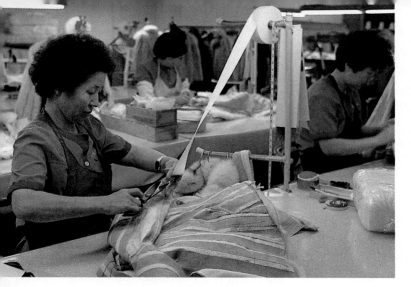

South Korean Industry
Clothing is a major industry in South Korea, where low wages enable the country to compete successfully against Western competitors.

North Korea also has some of the richest natural resources in East Asia. Its mines are owned and operated by the communist government. Coal, copper, iron ore, lead, tungsten, and zinc are among the rich mineral deposits found in North Korea.

North and South Korea Today

South Korea has emerged as one of the new industrial powers of the region surrounding the Pacific Ocean. Over the past forty years, the country has witnessed an impressive rate of economic growth, the development of a new middle class, and an increase in its role in international trade and politics. The textile and clothing industries have made South Korea one of the biggest exporters of the Pacific Rim. Shipbuilding is another major industry of South Korea. The country's corporations have also been especially successful in the export of electronics and automobiles. But rapid industrialization has pulled at the social fabric of Korean culture. In an effort to compete economically with Japan and Western

countries, enormous family-owned businesses in South Korea have often treated workers unfairly. As a result of this conflict, massive labor strikes and political struggles have at times disrupted the country's growth. The United States, which has both military and investment interests there, is South Korea's biggest trading partner.

Under the Communist leadership of Kim Il Sung, North Korea has continued to evolve from an agricultural to an industrial society. Despite the fact that it has greater natural resources than South Korea, however, North Korea lags far behind its neighbor in its standard of living and gross national product.

In late 1991, North and South Korea took significant steps toward reducing tensions between the two countries. They also agreed to meet to discuss steps toward reunification.

SECTION 3 REVIEW

Developing Vocabulary
1. Define: demilitarized zone

Place Location
2. Where is Seoul located in relation to North Korea?

Reviewing Main Ideas
3. Describe the landscape of the Korean Peninsula.
4. How does the climate of the northern part of the peninsula differ from that of the southern part?
5. What caused the division between North and South Korea?

Critical Thinking
6. **Making Comparisons** What cultural and economic factors do you think encourage the people of North and South Korea to unite their countries? What factors do you think discourage the unification of the two countries?

✓ Skills Check

☐ Social Studies
☐ Map and Globe
☐ Reading and Writing
☑ Critical Thinking

Demonstrating Reasoned Judgment

Making connections among the things you know is the basis for reasoned judgment. This skill of critical thinking enables you to move beyond what is stated in a text and to reach your own conclusions. The statement below is followed by seven points of evidence that either support or contradict its message. Use the following steps to analyze these seven points.

1. **Examine the nature of the evidence.** Before determining if the evidence supports or contradicts the statement, determine the nature of the evidence. Answer the following questions: (a) Which of the points concern education? (b) Which points have to do with social attitudes? (c) Which are concerned with national spending? (d) Which point concerns geography?

2. **Look for relationships among the pieces of evidence.** Analyze the evidence and try to infer information that may relate to the statement. Answer the following questions: (a) How would you describe the Japanese attitude toward education, based on points C and G? (b) What attitude toward women is suggested by B and E? (c) What level of employee loyalty is suggested by D?

3. **Identify the relationship of the evidence to the statement.** For each point of evidence ask yourself: Does this seem to prove or disprove the statement? Then decide if the statement has been proven or disproven by the total evidence. Answer the following questions: (a) Which evidence seems to support the statement? (b) Which evidence seems to disprove the statement? (c) Based on the evidence here, do you agree or disagree with the statement?

Statement
Compared with industry in the United States, Japan's industry profits from a number of social and economic characteristics of Japanese life.

Evidence
A. Only about 1 percent of Japan's gross national product is devoted to military spending.
B. Women in Japan are discouraged from seeking jobs.
C. Japanese high school students attend class 220 to 240 days a year; in the United States, students attend about 180 days.
D. Many company managers in Japan wear the company pin even when they're not working.
E. When they do find jobs, few Japanese women rise to positions of management.
F. Japanese industry must import most of its oil across great distances.
G. Teachers in Japan are highly respected and well paid.

32

REVIEW

Section Summaries

SECTION 1 Japan: The Land of the Rising Sun Japan is a chain of islands that actually form the peaks of an underwater mountain range. There are four main islands: Hokkaido, Honshu, Kyushu, and Shikoku. Japan has few natural resources and little arable land. As part of the Ring of Fire, Japan has frequent earthquakes and numerous volcanoes. Its climate, tempered by the surrounding sea, ranges from cold winters and cool summers in the north to hot summers and mild winters in the south. Its high population density affects housing and family patterns. The Japanese people are united by a common ethnic heritage, a common language and shared religious beliefs.

SECTION 2 Japan's Economic Development At the time of its first contact with the West in 1639, Japan had a complex agricultural civilization. Fear of conquest caused the country to close its doors for two hundred years. In 1853 American warships forced Japan to reopen trade. In the late 1800s Japan modernized rapidly, becoming a major industrial power. During the early twentieth century, lack of natural resources caused Japan to adopt imperialist policies. Its aggression in Asia led to war with the United States during World War II. The war ended when the United States dropped atomic bombs on Japan in 1945. After the war, Japan was permanently disarmed. Since 1945 it has enjoyed rapid economic growth.

SECTION 3 The Koreas: A Divided Peninsula The Korean Peninsula is very mountainous. The seas surrounding the peninsula and rivers flowing through the mountain valleys are two of the region's most important resources. The Korean War, which lasted from 1950 until 1953, resulted in the death of nearly five million people and the division of the Korean Peninsula into two countries: Communist North Korea and non-Communist South Korea. Despite differences in political ideology, the people of both North and South Korea share a common history and culture that dates back thousands of years.

Vocabulary Development

Match the following definitions with the terms below.

1. a strip of land that is free of troops and weapons
2. a tropical storm in the Pacific Ocean
3. a fixed number or quantity
4. a schedule of taxes on imports
5. a machine that measures movements of the earth's crust
6. sharing the same characteristics, mostly alike

a. demilitarized zone d. seismograph
b. tariff e. homogeneous
c. quota f. typhoon

Main Ideas

1. What are the four main islands of Japan?
2. Why does Japan experience many earthquakes?
3. In what ways has a high population density affected life in Japan?
4. Why is the Japanese population described as homogeneous?
5. How has a limited supply of natural resources influenced Japan's history?
6. What are some of the factors that have led to Japan's economic success?

7. What impact did the Cold War have on the Korean Peninsula?
8. Why is Korea described as a cultural crossroads?

Critical Thinking

1. **Drawing Conclusions** Why is education especially important to the success of Japan's economy?
2. **Synthesizing Information** Do you think that the end of the Cold War has increased the chances of Korean reunification? The Cold War was in the past responsible for the division of the Korean Peninsula.

Practicing Skills

Demonstrating Reasoned Judgment Look again at sentence A on page 683. Do you believe that the low percentage of gross national product devoted to military spending in Japan is an advantage or a disadvantage to Japanese industry? Explain your answer.

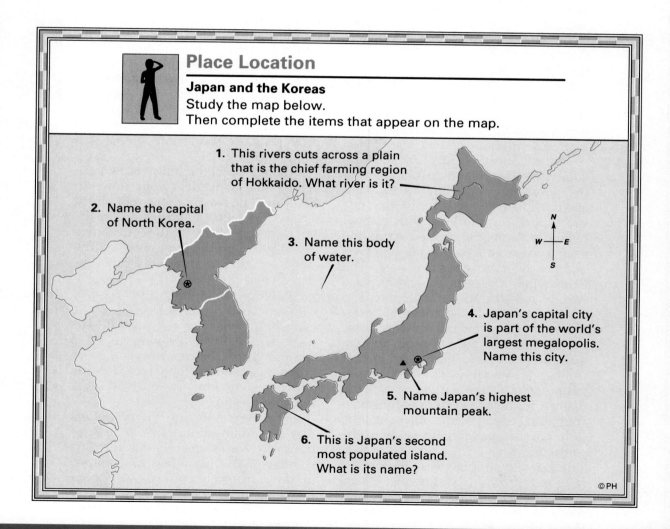

Place Location

Japan and the Koreas
Study the map below.
Then complete the items that appear on the map.

1. This rivers cuts across a plain that is the chief farming region of Hokkaido. What river is it?
2. Name the capital of North Korea.
3. Name this body of water.
4. Japan's capital city is part of the world's largest megalopolis. Name this city.
5. Name Japan's highest mountain peak.
6. This is Japan's second most populated island. What is its name?

©PH

Place Location: WHERE ON EARTH?

Asian Wonder Rice

The reports coming back from the field couldn't be better. Your new strain of rice is exceeding all of your expectations.

You developed it using the latest techniques. It grows faster, is tougher-skinned, and is more nutritious than the standard variety. It also does well without big doses of chemicals—a strong selling point, given today's sensitivity toward the environment. This new rice could be a breakthrough for many Asian nations.

Your colleague Kiso is assisting you with research and development of the promising new grain. As field agent, he is traveling secretly from test site to test site. Working with a small network of scientists dotted across the vastness of East Asia, Kiso is supervising nearly a dozen such sites. The fields where the tests are taking place vary in soil, altitude, and climate.

Both you and Kiso know that careful testing is crucial and must be closely monitored. You must be able to guarantee that you have a grain that can survive under actual conditions in many lands. If the variety proves too delicate, or fails to perform better than competing kinds of rice, there will be little point in pushing ahead with the project.

According to plan, Kiso files weekly reports from his various field sites. To ensure secrecy, these arrive at your office in plain envelopes with no return address. Inside, the sheets of pa-

per are each stamped with a red letter. Each letter stands for the name of a country in East Asia.

In addition to the coding system, Kiso adds a geographic description of each country. Kiso is a geography buff, and this second safeguard was his idea. He had begun his career in geography, then transferred to your field. You're an expert on plant life, fertilizer, and growth cycles. But you agreed when Kiso suggested this method of identifying the countries. A competitor in Texas is also hard at work on new strains of rice. The code system provides basic protection against that competitor.

You slit open the bulky brown envelope that arrived at your office in California just a moment ago. Field reports from five countries are in the same package. Beneath the code letters in each case are lengthy notes on the crop's growth rate, durability, and projected yield per acre.

Rapidly skimming the reports' contents with a professional eye, you can't help whistling to yourself. The new rice really is proving to be a wonder crop. The report says the rice grows fast, shows good root density, and is tough. Plus, it survives in all kinds of terrain.

Then you return to the red letters that are followed by Kiso's descriptions. In Kiso's neat hand, the first country, A, is described as "shaped like an *S*, tropical rain forest, very populous, facing the South China Sea."

Country B earns this note: "a wet land bordered by mountains on west,

Brahmapu

NEPAL

Gange

INDIA

B

INDIAN

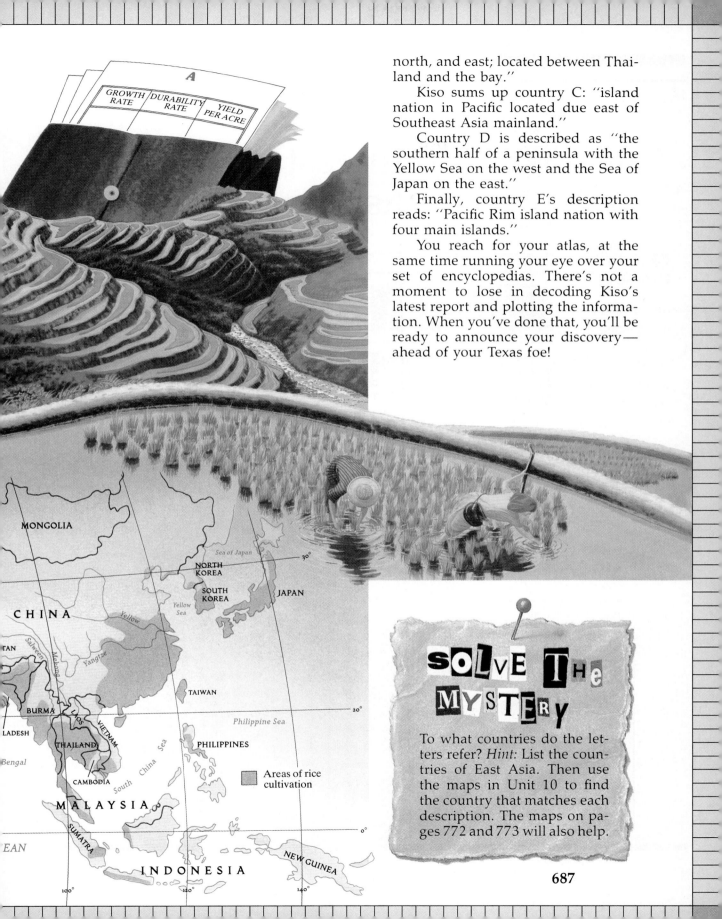

north, and east; located between Thailand and the bay."

Kiso sums up country C: "island nation in Pacific located due east of Southeast Asia mainland."

Country D is described as "the southern half of a peninsula with the Yellow Sea on the west and the Sea of Japan on the east."

Finally, country E's description reads: "Pacific Rim island nation with four main islands."

You reach for your atlas, at the same time running your eye over your set of encyclopedias. There's not a moment to lose in decoding Kiso's latest report and plotting the information. When you've done that, you'll be ready to announce your discovery— ahead of your Texas foe!

SOLVE THE MYSTERY

To what countries do the letters refer? *Hint:* List the countries of East Asia. Then use the maps in Unit 10 to find the country that matches each description. The maps on pages 772 and 773 will also help.

Southeast Asia

Chapter Preview

Both sections in this chapter are about the countries of Southeast Asia, shown in red on the map below.

Sections	Did You Know?
1 **HISTORICAL INFLUENCES ON SOUTHEAST ASIA**	More Muslims live in Indonesia than in any other country in the world.
2 **THE COUNTRIES OF SOUTHEAST ASIA**	Thailand's capital city, Bangkok, is sinking. Experts think that it will be below sea level by the year 2005.

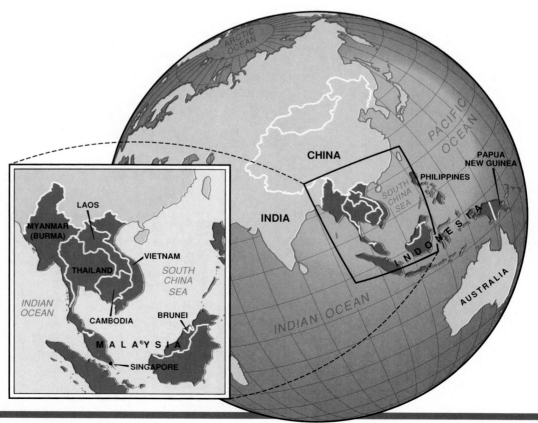

Angkor Wat, Cambodia
Built in the 1100s by the Khmers, the original inhabitants of Cambodia, the Ankhor Wat is a magnificent Buddhist temple. Today, nearly nine tenths of the people in Cambodia are Buddhist.

1 Historical Influences on Southeast Asia

Section Preview

Key Ideas

- Many different groups of people settled in Southeast Asia.
- Hinduism, Buddhism, and Islam influenced the region's culture.
- European control greatly affected the physical and human geography of Southeast Asia.

Key Terms

paddy, indigenous

Southeast Asia's location makes it one of the world's great geographic crossroads. Many groups of people from distant regions have met here, on the seas between Europe and Asia, to trade. The cultures of India, China, the Middle East, and the West all influenced Southeast Asia. This rich variety blended with the cultures of native Southeast Asians to create a diverse and distinct region.

Early Movement Affected the Region's Identity

The earliest inhabitants of Southeast Asia probably migrated to the region from southern China and South Asia. Over hundreds of years, other people slowly moved south from central Asia and southern China into Southeast Asia. Groups such as the Mons, Khmers (KMERS), and Thais made their way down into the peninsula of mainland Southeast Asia. They settled along the rich river valleys and coastal plains.

Indian Influence   By the first century A.D., the Mons, Khmers, and other groups began to establish strongholds in Southeast Asia. No one group ever united the entire region, but rich and powerful kingdoms developed. Attracted by the wealth of these kingdoms, merchants from India sailed along the coasts of Southeast Asia. They brought with them Hindu and Buddhist priests. Through their interaction with the Southeast Asian people, these traders and priests greatly influenced life in the region.

Over the centuries, Indian culture and religion gradually blended with the culture of Southeast Asia. The people of Southeast Asia absorbed Hinduism and Buddhism into their existing religious beliefs. They adopted many Hindu myths and worshiped Hindu gods. Southeast Asian rulers built palaces and temples dedicated to Hindu gods in the Indian architectural style.

Muslim Influence   Sometime between the 1200s and 1400s, a new influence reached Southeast Asia. Traders from Arabia and India brought the Islamic religion to the region. Islam spread quickly along the trade routes. It reached the islands of Indonesia and spread as far east as the southern Philippines. Along with Buddhism and Hinduism, Islam became an important religion in the area.

Chinese Influence   Although many of the people who migrated to Southeast Asia came from China, the Chinese had little impact on the region. One reason was that the Chinese were not interested in exporting their culture. They viewed themselves and their civilization as superior. They were not eager to share their culture with people whom they considered to be barbarians.

There was one exception. Around 100 B.C., China took control of what is today the northern part of Vietnam.

For over one thousand years Vietnam remained under Chinese influence.

During that time, the Vietnamese accepted much of the Chinese culture. Vietnamese language, religious beliefs, art, government, and agriculture all were affected by it to some degree. But the Vietnamese never lost their identity. They kept their own customs, and although many Chinese words entered their vocabulary, they continued to speak Vietnamese.

The Europeans Bring Change to the Region

In the 1500s, Portuguese traders arrived in Southeast Asia and set up trading posts. The Spanish, Dutch, British, and French soon followed them. By the late 1800s, the Europeans had colonized all of Southeast Asia except for Thailand. The Europeans viewed their colonies in Southeast Asia, with their abundance of natural resources, as a means to make their own countries richer.

To take advantage of Southeast Asia's many natural resources, the Europeans drastically changed the region's physical and human geography. They cleared vast areas of forests and established plantations, or large farms, to grow cash crops such as coffee, tea, tobacco, and raw rubber. Europeans also encouraged rich, local landlords to grow rice for export. Rice **paddies,** the wet land on which rice is grown, spanned the deltas of the Irrawaddy, Chao Phraya, and Mekong rivers as far as the eye could see.

Southeast Asian farmers traditionally tended their own small plots of land. However, they could not compete with the large landowners. Many small farmers were forced to leave their land and go to work on foreign-owned plantations and in the paddies of wealthy Southeast Asians.

There were other changes. Europeans financed the construction of inland roads and railroads to carry crops

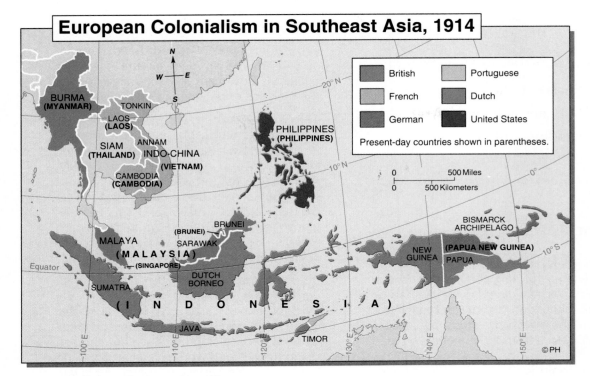

European Colonialism in Southeast Asia, 1914

BURMA (MYANMAR)
TONKIN
LAOS (LAOS)
SIAM (THAILAND)
ANNAM
INDO-CHINA (VIETNAM)
CAMBODIA (CAMBODIA)
PHILIPPINES (PHILIPPINES)
BRUNEI (BRUNEI)
SARAWAK
MALAYA (MALAYSIA)
(SINGAPORE)
DUTCH BORNEO
SUMATRA
(I N D O N E S I A)
JAVA
TIMOR
NEW GUINEA
BISMARCK ARCHIPELAGO
(PAPUA NEW GUINEA)
PAPUA
Equator

British	Portuguese
French	Dutch
German	United States

Present-day countries shown in parentheses.

0 ... 500 Miles
0 ... 500 Kilometers

Applying Geographic Themes

Regions By the late 1800s, the Europeans had colonized all of Southeast Asia except for Thailand. What European power controlled the largest region in Southeast Asia?

and other goods to the port cities for export to Europe. As these once slow, sleepy port cities began to grow rapidly, they attracted large numbers of people from China and India. Tensions sometimes developed between these new immigrants and *indigenous*, or native, Southeast Asians.

Colonization also greatly affected relations among different indigenous groups within Southeast Asia. When Europeans arrived in the region and carved out their own colonies, they paid little attention to existing ethnic boundaries. As a result hostile groups often were united into one colony, while others that had lived together peacefully for centuries were separated. When the colonies finally became independent countries after World War II, many of them thus inherited deep ethnic conflicts.

SECTION 1 REVIEW

Developing Vocabulary
1. Define: **a.** paddy **b.** indigenous

Place Location
2. Where is Southeast Asia located in relation to China and India?

Reviewing Main Ideas
3. From where did Southeast Asia's first immigrants come?
4. What were the major cultural influences on Southeast Asia?

Critical Thinking
5. **Identifying Central Issues** What effect did location have on the development of cultures in Southeast Asia?

✓ Skills Check

☐ Social Studies
☐ Map and Globe
☐ Reading and Writing
☑ Critical Thinking

Drawing Inferences

Drawing inferences means reading between the lines—forming conclusions that are suggested rather than directly stated. Inferences may be either limited or far-reaching in scope. Use the following steps to practice drawing inferences from the material you read.

1. **Find the main idea of a sentence, paragraph, or longer passage.** Before you can infer any information you must clearly understand what is stated. Read the passage below then answer the following question: How can you summarize briefly the main idea contained in the passage?

2. **Think about other facts you know concerning the same subject.** Drawing inferences often requires you to draw upon information you have learned previously. Answer the following questions: (a) From which regions of the world, respectively, did the Buddhist, Muslim, and Christian religions first spread? (b) Historically, which regions have been among the primary sources of trade with Southeast Asia?

3. **Infer information from the text.** Consider the main idea together with other facts that you know. Decide whether all of this information combined suggests additional facts or conclusions about the subject. Using the answers you gave in steps 1 and 2, answer the following question: What can you infer about the effects of trade and invasion on Southeast Asia?

Passage

Southeast Asia includes the island archipelagoes of Indonesia and the Philippines, the Malaysian Peninsula, and the tiny island country of Singapore. It also includes the peninsula of Indochina, which now includes the countries of Vietnam, Laos, and Cambodia. On the mainland of Asia are Myanmar and Thailand. Because landforms have kept most of the countries separated from one another, each country has developed its own traditions and its own ways of life.

A thousand different languages and dialects are spoken in Southeast Asia, and the people there follow many different religions. Indonesia and Malaysia are mostly Muslim; Thailand, Burma, Vietnam, Cambodia, and Laos are mostly Buddhist. In the Philippines 95 percent of the people are Christian; 80 percent of these are Catholic.

Southeast Asia is a vital crossroads for the trade and commerce of many nations as well as a treasury of natural resources including tin, petroleum, bauxite, rubber, tea, spices, and fine kinds of wood. Southeast Asia's location and resources have made the region a target for frequent invasion.

2 The Countries of Southeast Asia

Section Preview

Key Ideas

- Many ethnic groups live within the countries of Southeast Asia.
- Colonialism had a great impact on the modern countries of Southeast Asia.
- The economies of the ASEAN countries are much stronger than the economies of the socialist countries of Myanmar, Vietnam, Laos, and Cambodia.

Key Term

insurgent

Floating Market: Bangkok, Thailand
Bangkok's canals, called *klongs,* are filled with houseboats and floating markets from which Thai merchants sell produce and goods.

Although all of Southeast Asia is now independent, colonialism has left its mark on the region. It was the Europeans who drew the political boundaries of the newly independent countries. In many instances, the groups that were thus united to form one country have been unable to coexist peacefully.

The countries of Southeast Asia can be divided into two groups. Indonesia, the Philippines, Thailand, Malaysia, Singapore, and Brunei all belong to the Association of Southeast Asian Nations (ASEAN). Like members of the European Community, ASEAN countries work together to improve their economic and social conditions. Each of these countries has a free-market economy. Since the 1970s, when the focus of world trade began to shift to the countries bordering the Pacific Ocean, they have benefited greatly. Papua New Guinea also has a free-market economy, but it does not belong to ASEAN. Four other countries—Myanmar, Vietnam, Cambodia, and Laos—all have socialist economies subject to strong government controls. All except Myanmar have ties to China or the Soviet Union.

Myanmar: A Country in Turmoil

Imagine a country only a little larger than Texas where more than one hundred languages are spoken. That is the reality of Myanmar, formerly called Burma. About 68 percent of Myanmar's people are members of the Burman ethnic group and speak Burmese. The rest belong to a variety of different ethnic groups and speak different languages. Sometimes people living in villages only a few miles apart may not speak the same language or share the same customs.

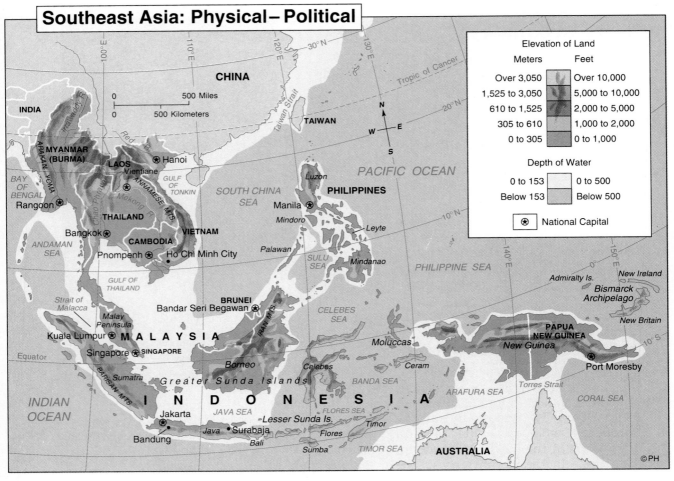

Southeast Asia: Physical–Political

Elevation of Land

Meters		Feet
Over 3,050		Over 10,000
1,525 to 3,050		5,000 to 10,000
610 to 1,525		2,000 to 5,000
305 to 610		1,000 to 2,000
0 to 305		0 to 1,000

Depth of Water

0 to 153		0 to 500
Below 153		Below 500

⊛ National Capital

Applying Geographic Themes

Location Of the countries of Southeast Asia, five are located on the Asian mainland. Others lie entirely on islands. Name the countries that share the Malay Peninsula.

Throughout their history, these groups protected their cultural identities. When the British conquered the region in the late 1800s, they combined the people in Myanmar into a single political unit, but they made little attempt to unify them culturally. They allowed the people a great deal of autonomy. When Myanmar won its independence in 1948, the new country lacked unity.

Since independence, various ethnic groups have fought against the government. Some established their own armies. Some wanted to overthrow the government; others wanted to secede from Myanmar. Although these groups are not numerous, they have disrupted the country. **Insurgents,** people who rebel against their government, attacked factories and damaged government property. They also staged riots and protests that have crippled the government.

As a result of this unrest, Myanmar's economy developed much more slowly than the economies of most other Southeast Asian countries. The country is rich in natural resources such as teak wood and minerals, yet much of this wealth is undeveloped. Instead the Myanmar economy remains based on agriculture, primarily the export of rice.

Rebellious ethnic groups are not the only reasons for Myanmar's weak economy. The frequently changing governments often isolated themselves from the outside world. As a result, foreign aid, including the transfer of new technology from developed countries, was blocked. Much of Myanmar's technology remains outdated. Transportation between cities is slow, and in some rural areas, people still rely on oxcarts for transportation.

Encouraging Movement Helps Thailand Prosper

The population of Thailand is not splintered like that of Myanmar. More than 85 percent of the people speak Thai. Because of their cultural unity, as well as their long history as a free people, Thailand's people have a strong national identity.

Preserving Independence Thailand's people are proud that they have never been dominated by a foreign power. Their country was the one country in Southeast Asia that was not colonized by Europeans. The Thai people say that they are like bamboo that bends in the wind. They have been flexible when dealing with foreign powers in order to keep their independence.

Since World War II, Thailand has had close political ties to the United States. Thailand felt threatened by the Communist revolution in China in 1949. It joined with the United States to stop Communist expansion in Southeast Asia. During the Vietnam War, Thailand allowed the United States to use its country as a base for air attacks against Communist forces in Vietnam, Cambodia, and Laos.

Economic Progress Brings Change
Thailand's ability to "bend with the wind" has helped it to build one of the strongest economies in Southeast Asia. Until recently Thailand's economy was dominated by agriculture and rice was its main export. However, in the 1960s Thailand began to diversify its economy. Today it has industries that produce cement, food products, paper, and textiles. Foreign companies operate plants that assemble automobiles and electronic equipment. Manufactured goods now contribute almost as much to the economy as agricultural products.

Tourism is another major source of income for Thailand. In the last few decades the tourist industry has grown significantly. Thailand welcomes more than two million foreigners a year to enjoy its rich, varied culture.

Thailand's economic development has resulted in great change. Bangkok, Thailand's capital, has become a transportation hub for the entire Southeast Asian region. Nearly forty airlines serve this bustling city of skyscrapers, modern hotels, and noisy expressways. These stand in sharp contrast to the mysterious charm of traditional Bangkok, which one writer described in these words:

I still like Bangkok as a city of secrets. Not far from the railroad station, for example, stands a temple of ordinary exterior, Wat Trimit, containing an image of the Lord Buddha three meters high; it weighs five and a half tons—and is made of gold. Jewel merchants in simple shops may cover a desktop with a fortune in sapphires. . . . And behind the watery moat of the royal palace live the royal white elephants.

Unlike Myanmar, Thailand has opened its doors to the world, reaching out to interact with many other

Tongue-Tied in Singapore

Residents of Singapore shown at right now speak several Chinese dialects. But the government hopes to help the island country's diverse population feel more like a single community by promoting the use of Mandarin, a Chinese dialect. "Start with Mandarin, speak it more often," urges the campaign slogan. The government also hopes thereby to make it easier to do business with mostly Mandarin-speaking China. The campaign seems to be working. According to government figures, when nurses in Singapore's hospitals first talk to a patient, half now use Mandarin. In 1980, the figure was only 18 percent.

1. How will Singapore benefit if more of its population speaks Mandarin?
2. What do you think are the main difficulties of getting people to change their language?

---

countries. As a result of this increased interdependence, Thailand has one of the most successful economies in Southeast Asia.

Vietnam, Laos, Cambodia: Troubled Indochina

Vietnam, Laos, and Cambodia are very different from one another ethnically. The overwhelming majority of the people in Vietnam are Vietnamese. In Cambodia, the vast majority belong to the Khmer ethnic group. Laos is ethnically more diverse. Most people belong to one of four main groups, but there are more than seventy groups in the country.

Despite the differences in their ethnic makeup, Vietnam, Laos, and Cambodia have much in common. All of their cultures were influenced by India, and most of their people are Buddhists. As French colonies, the three countries together formed a region once known as French Indochina.

French influence in the area dates from the 1800s. By the early 1900s, Vietnam, Laos, and Cambodia had become French colonies. During World War II, from about 1940 to 1945, the

> **WORD ORIGIN**
>
> **Vietnam**
> Vietnam means "the far south" in Vietnamese. It lies south of the early Asian cultural centers, on the east coast and southern tip of the Indochina Peninsula.

Japanese took control of Indochina. But when the Japanese surrendered to the Allies in 1945, France was determined to regain its colonies.

Years of War France's attempt to return to power in Indochina marked the beginning of long and bloody wars in the area. In 1945, Ho Chi Minh, a Vietnamese leader, declared Vietnam's independence from France. Ho Chi Minh's forces fought a bitter and fierce war with the French. In 1954, the French were defeated.

After the war, a peace conference was held in Geneva, Switzerland. Instead of ending the conflict, however, the conference laid the foundation for more fighting by dividing Vietnam into two parts. North Vietnam was left to the Communists under Ho Chi Minh. South Vietnam was headed by Ngo Dinh Diem (NGO DIN DEE em), a pro-Western ruler.

But the Communists in North and South Vietnam wanted to reunite the two Vietnams. Another war soon began. The United States entered the war, hoping to keep South Vietnam free of Communist control.

Laos and Cambodia were also drawn into the fighting as Communists in these countries provided a supply line to the Communist insurgents in South Vietnam. The North Vietnamese set up bases in Cambodia. The struggle between Communists and non-Communists in Laos and Cambodia intensified.

The United States finally withdrew from the war in 1973 and South Vietnam fell to the Communists in 1975. Vietnam was reunited one year later. Communists also gained control of the governments of Cambodia and Laos.

New Governments Bring More Bloodshed In all three countries, the new governments attacked their non-Communist enemies. Hundreds of thousands of people were killed. About one million refugees fled their

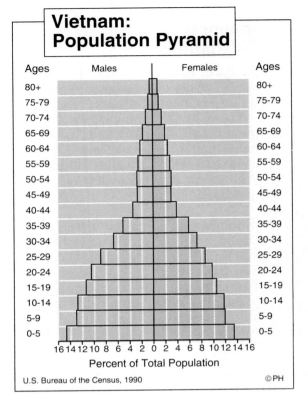

Vietnam: Population Pyramid

Ages — Males — Females — Ages

Percent of Total Population

U.S. Bureau of the Census, 1990   ©PH

Graph Skill
What does the shape of this population pyramid indicate about Vietnam's population?

homelands. Some fled to Thailand. Many others escaped in small, overcrowded boats. Of these, thousands drowned.

Meanwhile, war continued. In 1978 Vietnam attacked Cambodia and installed a new government there. Since then, a civil war has simmered, with various groups fighting for control of the country.

The Economies Suffer Years of fighting seriously affected the economies of Vietnam, Laos, and Cambodia. All three countries rely on agriculture, especially rice farming, to support their economies. But in the upheaval and destruction of war, farming output in these countries fell dramatically.

Indonesia and the Philippines

Both Indonesia and the Philippines consist of thousands of islands. Indonesia has more than 13,660 islands. The Philippines consists of more than 7,000 islands. In Indonesia, more than 250 languages are spoken, compared with 75 in the Philippines. What keeps countries with such variety united?

Indonesia Uniting Indonesia has required great skill. Indonesia has 180 million people living on islands spread over 3,000 miles (4,800 km) of ocean. Throughout its history, no one group of people has ever dominated the area long enough to give it cultural unity. Although the Dutch colonized Indonesia from the 1600s until World War II, they ruled indirectly through local leaders. They never tried to impose their culture on Indonesia.

To help Indonesians form a path to a united, strong country, the government established a national ideology. The *Pancasila,* or Five Principles, are a belief in one God, a concern for human welfare, and commitments to national unity, democracy, and justice.

The Philippines Unlike Indonesia, the Philippines experienced a long period of strong colonial rule. The Spanish ruled the Philippines for 350 years, until their defeat by the United States in the Spanish-American War in 1898. For the next fifty years, the United States controlled the Philippines.

The people of the Philippines were strongly influenced by their colonial rulers. Spanish priests converted the Filipinos to the Roman Catholic religion. More than 80 percent of the current population of the Philippines is Roman Catholic. Many native Filipinos also intermarried with the Spanish. This spread the Spanish culture among the various ethnic groups and helped to unite them.

The Americans also had a great impact on the Philippines. They introduced a new educational system in which English was taught. Along with Filipino, English is one of the country's official languages. The Americans introduced democracy to the country. Combined with their Asian heritage, Western cultural influences help give the Philippines' people a sense of national unity.

Singapore: A Center of World Trade

Represented by just a dot on world maps, Singapore is Southeast Asia's smallest independent country. It consists of a single city located on an island and a few surrounding islands.

Singapore River
The tiny city-state of Singapore is one of the wealthiest countries in Southeast Asia and one of the world's leading ports.

But this tiny country casts a big shadow because of its political and economic power.

Singapore's physical features and relative location played an important role in its success. It has a deep, natural, sheltered harbor. And its location at the southern tip of the Malay Peninsula places it in the center of an important trade route between Europe and East Asia.

Thomas Raffles of the British East India Company first recognized that the location of Singapore could make it important to British trade. The trading post that he established there in 1819 prospered and attracted immigrants from Malaysia, China, and India. Singapore became an independent country in 1965.

Modern Singapore is a thriving center of international trade and an important manufacturing center. It is the world's busiest port. Nearly thirty thousand ships dock at Singapore each year carrying rubber, tin, petroleum products from Singapore's refineries, and many other goods.

Malaysia and Brunei: On the Road to Development

Like Singapore, Malaysia and Brunei have strong economies that are not based on agriculture, unlike those of most of their neighboring countries. Malaysia and Brunei are two of the wealthiest countries in Southeast Asia.

Most of Brunei's wealth comes from its large reserves of oil and natural gas. The income from these natural resources enabled Brunei to modernize. Some remote villages now have electricity and running water. And the government provides free schooling and medical care for its citizens.

Malaysia supports a wide variety of economic activities; rubber, oil, and tin are its leading exports. As in Brunei, Malaysia is using its oil revenues to help develop manufacturing and improve agriculture.

Papua New Guinea: A Region Apart

Papua New Guinea is part of two overlapping regions—Southeast Asia and Oceania. It also seems to straddle two worlds—one modern, the other traditional.

Papua New Guinea is a country of contrasts. Communication is poor because there are no roads to most of the country's villages. Moreover, more than seven hundred languages are spoken across the country. Most of Papua New Guinea's three million people are farmers, and villagers in the remote highlands still plant their crops with traditional tools.

Yet gold and copper ore are mined in great quantities in Papua New Guinea using modern machines. Generators send electricity to one third of the population. And in village stores, young people sell canned beef, rice, and sugar imported from Japan and Australia.

SECTION 2 REVIEW

Developing Vocabulary
1. Define: insurgent

Place Location
2. Locate and name the island country just off the southern tip of Malaysia.

Reviewing Main Ideas
3. What has helped the economy of Thailand to grow?
4. Why are Vietnam, Cambodia, and Laos so poor?
5. What role has location played in Singapore's economic success?

Critical Thinking
6. **Analyzing Information** How does the economy of a small country like Myanmar suffer because it chooses to isolate itself from the rest of the world?

33

REVIEW

Section Summaries

SECTION 1 Historical Influences on Southeast Asia   Southeast Asia is a rich blend of cultures and people. Early in its history, many different ethnic groups settled the region. Later, because of its location at the center of an important trade route and its rich resources, the region was influenced by the cultures of India, China, the Middle East, and Europe. The last group that changed the region's physical and cultural landscape were the Europeans.

SECTION 2   The Countries of Southeast Asia   Today some countries of Southeast Asia are struggling to free themselves from the effects of colonialism and create unified countries. Others have taken advantage of their colonial histories to establish themselves as developing countries. The economies of the countries that belong to ASEAN are much stronger than those of the non-ASEAN countries, which are among the poorest in the world. Besides open trading policies, relative location has favored some countries over others.

Vocabulary Development

Choose the *italicized* term in parentheses that best completes each sentence.

1. The wet land on which rice is grown is known as a (*paddy/swamp*).
2. People who are native to a particular country or culture are (*immigrant/indigenous*).
3. People who act in rebellion against their own government are called (*insurgents/insertions*).

Main Ideas

1. What religions spread to Southeast Asia from India, the Middle East, and Europe?
2. Why is modern Southeast Asia such a diverse mix of people and cultures?
3. What attracted Europeans to Southeast Asia?
4. What changes did the Europeans bring to Southeast Asia?
5. How have past governments partly contributed to Myanmar's poor economy?
6. Why do the Thai people compare themselves to bamboo that bends in the wind?
7. How were Laos and Cambodia drawn into the Vietnam War?
8. How has Indonesia met the challenges of uniting its scattered people?
9. How has Singapore's relative location influenced the economic development in that country?

Critical Thinking

1. **Identifying Central Issues**   Why do you think it is such a challenge to create a feeling of national unity among people in countries like Myanmar and Indonesia?
2. **Perceiving Cause-Effect Relationships**   How has colonialism affected Vietnam's recent history?
3. **Drawing Conclusions**   Why are the economies of the non-ASEAN countries so much poorer than those of the ASEAN countries?
4. **Checking Consistency**   Indonesia and the Philippines experienced very different forms of colonial rule. In what ways do you think this is consistent with the challenges faced by both countries today? Support your answer with clear examples.

Practicing Skills

Drawing Inferences Read the passage below and then answer the questions that follow:

Singapore, which has the world's fourth largest port, is an independent nation. Next to Japan, Singapore now boasts the highest per person income in Asia. While Singapore has become an economic "tiger," it also has developed an authoritarian government. The people do not enjoy many freedoms.

1. Which key factor mentioned in the passage explains Singapore's economic growth in terms of its geographical location?
2. Identify one economic advantage of living in Singapore as opposed to many other countries of the region.
3. What do you think the writer of the passage means by describing Singapore as "an economic 'tiger.'"?
4. Explain why you agree or disagree with the following statement: "The government of Singapore is an excellent model for developing nations."

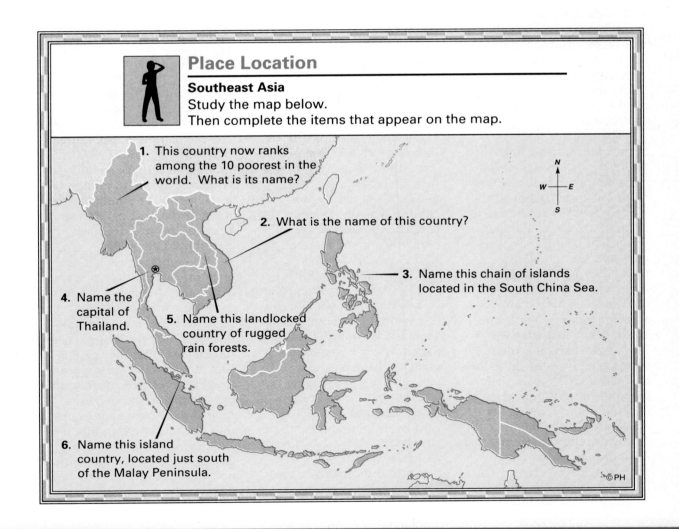

Place Location

Southeast Asia
Study the map below.
Then complete the items that appear on the map.

1. This country now ranks among the 10 poorest in the world. What is its name?

2. What is the name of this country?

3. Name this chain of islands located in the South China Sea.

4. Name the capital of Thailand.

5. Name this landlocked country of rugged rain forests.

6. Name this island country, located just south of the Malay Peninsula.

World Politics and Refugees

Refugees live with constant uncertainty. They are uncertain whether they will ever return to their native country, and often they are uncertain about how or whether they will be received when they reach a new country. Refugees, however, are sure of one thing: they know that they need to find a new place to call their home.

The issue of finding homes for refugees links different regions to one another. For example, refugee settlement has recently linked the United States and Canada to Southeast Asia, as refugees from such countries as Vietnam and Cambodia now make their homes in North America.

Refugees have many reasons for leaving their homelands. Many fear for their lives. They may find themselves caught in the cross fire of a bitter civil war, or helpless and hungry in the midst of a long-term famine. Refugees may also be members of ethnic or religious groups that are suffering persecution.

Most of the refugees who have left their homelands in recent years have done so in search of economic opportunity. Yet these economic refugees apparently inspire less concern than victims of violence, persecution, or famine.

Refugees from Southeast Asia

In the late 1980s, on a small, hilly island off the Hong Kong coast, nearly five thousand Vietnamese people waited for the chance to resettle in another part of the world. These refugees braved choppy seas on dangerously overloaded boats to seek a temporary home in Hong Kong.

More than two thirds of the Vietnamese in Hong Kong said that poor economic conditions had led them to leave Vietnam. Formerly, the United States would have admitted many of these economic refugees. But in 1986, tighter United States immigration laws went into effect. Refugees from Vietnam were now required to show that they had relatives in the United States in order to enter the country. So while officals

allowed some people into the country, others waited in Hong Kong, unsure of just where they were headed.

The Global Refugee Problem

The experience of the Vietnamese boat people is just one example of a growing refugee crisis. The world's refugee population is exploding. Between 1979 and 1989, the number of refugees worldwide rose from 4.6 million to a staggering 14.5 million. The map on the next page shows some of the world's major refugee populations.

The refugee problem is even more troubling in light of the fact that many of the countries that used to welcome refugees have recently changed their positions. These nations fear that they cannot provide the jobs and support that these refugees need without sacrificing the concerns of their own citizens. For example, refugees arriving in a new country often need financial assistance to obtain their new housing, clothing, and food. And, as the pressures of war, persecution, poverty, and hunger shift from one part of the world to another, refugee populations shift with them. For example, 6 million Afghanis fled to refugee camps in Pakistan and Iran because of war in their native land. Bulgarians of Turkish descent are streaming into Turkey at the rate of more than 2,000 a day to escape ethnic persecution. A civil war in Sri Lanka has driven more than 125,000 Tamils from their country since the early 1980s.

The View from the U.S.A.

 Until this century the United States had a long tradition of welcoming refugees. However, this nation has toughened its attitude toward refugees for economic reasons. For example, refugees from Haiti are generally assumed to be economic refugees. Therefore, refugee boats from Haiti are turned back by the Coast Guard.

TAKING ANOTHER LOOK
1. What are the different reasons for which refugees leave their homelands?
2. Why have many nations changed their positions regarding refugees?

Critical Thinking
3. **Drawing Conclusions** What can you conclude from the fact that nations have reconsidered their willingness to accept refugees as the numbers of refugees have increased?

Applying Geographic Themes
Movement This map shows major refugee movements from 1979 to 1989. From which countries did the United States accept many refugees?

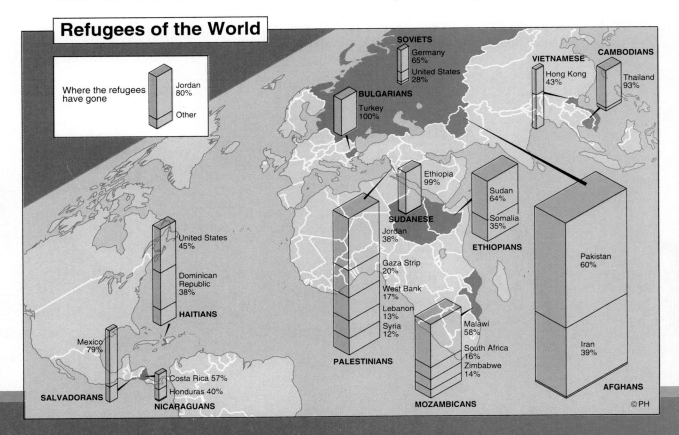

Refugees of the World

Where the refugees have gone — Jordan 80% / Other

SOVIETS: Germany 65%, United States 28%
BULGARIANS: Turkey 100%
VIETNAMESE: Hong Kong 43%
CAMBODIANS: Thailand 93%
ETHIOPIANS: Ethiopia 99%
SUDANESE: Sudan 64%, Somalia 35%
HAITIANS: United States 45%, Dominican Republic 38%
SALVADORANS: Mexico 79%
NICARAGUANS: Costa Rica 57%, Honduras 40%
PALESTINIANS: Jordan 38%, Gaza Strip 20%, West Bank 17%, Lebanon 13%, Syria 12%
MOZAMBICANS: Malawi 58%, South Africa 16%, Zimbabwe 14%
AFGHANS: Pakistan 60%, Iran 39%

©PH

11

The Pacific World and Antarctica

CHAPTERS

A lone, bending palm tree, rustling faintly in the ocean breeze . . . the roar of waves breaking on the fine-grained sand of the shore . . . a golden blaze of sun making the ocean water sparkle with light . . . The rest of the world, with all its bustle and noise, seems far away.

Many people think the Pacific world is unlike any other place on the earth. In some ways it is. Yet life in the region is, in many ways, like life in other parts of the world. As you read this unit, you will discover how people have adapted to the physical geography of this region.

Antarctica's
penguin population ▶

▲ Tropical island landscape in Bora Bora, Polynesia

◀ Opera House, Sydney, Australia

34

Regional Atlas: Australia and the Pacific Islands

Chapter Preview

Both sections in this chapter are about Australia and the Pacific Islands, shown in red on the map below.

Sections	Did You Know?
1 LAND, CLIMATE AND VEGETATION	Australian eucalyptus trees can grow as tall as 500 feet (152 m)—that's taller than the highest known giant sequoia of California.
2 HUMAN GEOGRAPHY	Australia's first human inhabitants—the aborigines—now number less than 1 percent of the country's population.

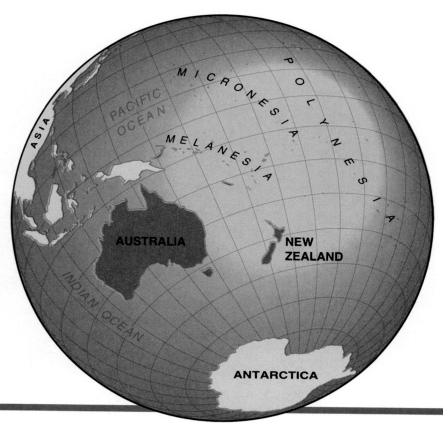

Ayers Rock, Mount Olga National Park, Australia
Carvings and paintings found in small caves at the base of this giant rock help scholars in their study of the culture of central Australia's earliest inhabitants, the aborigines.

1 Land, Climate, and Vegetation

Section Preview

Key Ideas

- Australia is a flat, dry continent.
- New Zealand's physical geography varies widely.
- The Pacific Islands can be divided into two main physical regions: the mountainous high islands and the low coral islands.

Key Terms

outback, geyser, coral reef

Australia and the Pacific Islands are among the world's most exotic places. Imagine thousands of islands with palm-studded shores strewn across the Pacific Ocean. Think of animals as unusual as kangaroos and platypuses, and plants that look like feathers. A century ago an Australian journalist, Marcus Clarke, wrote:

In Australia . . . is to be found the Grotesque, the Weird, the strange scribblings of Nature learning how to write . . . the subtle charm of this fantastic land of monstrosities.

Many aspects of Australia and the Pacific Islands are now familiar to us. However, much about this region still fascinates and mystifies us. In this chapter, you will explore the physical and human characteristics that give the region its unique character.

Australia's Flat, Dry Land

Australia covers an area nearly the size of the continental United States. It is the earth's smallest continent and

Australia and New Zealand: Physical– Political

PACIFIC OCEAN

ARAFURA SEA

TIMOR SEA ★Darwin
Arnhem Land GULF OF
CARPENTERIA
Cape York Peninsula

CORAL SEA

KIMBERLEY **Northern**
PLATEAU **Territory**
KING LEOPOLD
RANGES BARKLY
TABLELAND
TANAMI
DESERT

Great Barrier Reef

GREAT SANDY
DESERT **AUSTRALIA**
MACDONNELL
RANGES GREAT
ARTESIAN
BASIN

GIBSON DESERT MUSGRAVE
RANGES SIMPSON
DESERT

Western Australia GREY
RANGE

GREAT VICTORIA
DESERT ★Brisbane
L. Eyre
South Australia

NULLARBOR
PLAIN FLINDERS
RANGE

Perth ★ Darling R.

Great Australian
Bight Adelaide
★ Lachlan R.

Murray R. ★Sydney
New South Wales
⊕ Canberra

TASMAN SEA Auckland

Victoria
★ Melbourne

Bass Strait North Island

Tasmania Wellington
Hobart★

NEW ZEALAND

South Island

GREAT DIVIDING RANGE

Tropic of Capricorn

20° S

40° S

INDIAN OCEAN

Great Artesian Basin

N
W E
S

0   500   1000 Miles
0   500   1000 Kilometers

120° E   140° E   160° E   180°

©PH

Elevation of Land

Meters		Feet
Over 3,050		Over 10,000
1,525 to 3,050		5,000 to 10,000
610 to 1,525		2,000 to 5,000
305 to 610		1,000 to 2,000
0 to 305		0 to 1,000
Below sea level		Below sea level

Depth of Water

0 to 153		0 to 500
Below 153		Below 500

Applying Geographic Themes

1. Place Australia is sometimes called the "flattest continent," whereas 70 percent of New Zealand is mountainous or hilly. Which mountain range prevents moisture from reaching most of the Australian interior?

2. Location Where are most of the deserts in Australia located? Where is the Great Barrier Reef located?

---

the only one occupied by a single country. It is also the earth's flattest and driest continent.

A Place Without Water Look at the map on this page and find the Great Dividing Range, located about 50 miles (80 km) inland from the eastern coast. The Great Dividing Range is the country's largest highland area. These mountains, which in most places are barely more than hills, extend from the Cape York Peninsula all the way to Tasmania.

Nearly all of Australia west of the Great Dividing Range is arid plain or dry plateau. The Murray River, shown

on the map above, is one of only a few permanent bodies of water in Australia. Most other lakes and rivers become full after heavy rains and then dry up in the burning sun.

Australians often refer to the central and western plains and plateaus as the outback. Find this area on the map on this page.

Australia's vast deserts are mostly uninhabited, although more and more people are now exploring the interior of the continent. Robyn Davidson traveled with four camels and her dog, Diggity, across one of these areas, the Great Sandy Desert in northwest Australia. Her notes describe her trip.

The area was rougher than anything we had crossed before. . . . The setting was lovely, an infinitely extended bowl of pastel blue haze carpeting the desert, with crescent-shaped hills floating in the bowl and fire-colored sand dunes. . . . In the far distance five violet magical mountains soared above the desert. . . .

Linking Climate to Landforms Almost all of Australia is hot and dry. The only areas that get significant rainfall lie along the coast. Compare the climate map below with the physical map on page 708. Notice that Australia's climate changes abruptly at

the Great Dividing Range. These mountains block the flow of moisture carried by the winds blowing westward from the Pacific Ocean. As a result, Australia's interior receives little rain throughout the year.

Linking Plant Life to Climate Compare the climate and vegetation maps on page 709 and page 710. Notice the link between Australia's climate and its vegetation. Where rainfall is heaviest, rain forests grow. In drier areas, the forests thin out into woodlands. These woodlands consist mainly of eucalyptus trees, which Australians call "gums." Eucalyptus have long leaves and they are not bushy, characteristics that help them to survive

Applying Geographic Themes

Interaction Rainfall is heaviest in the northern and eastern coastal regions of Australia. Compare this map to the population density map on page 715. Then explain how climate influences Australian settlement patterns.

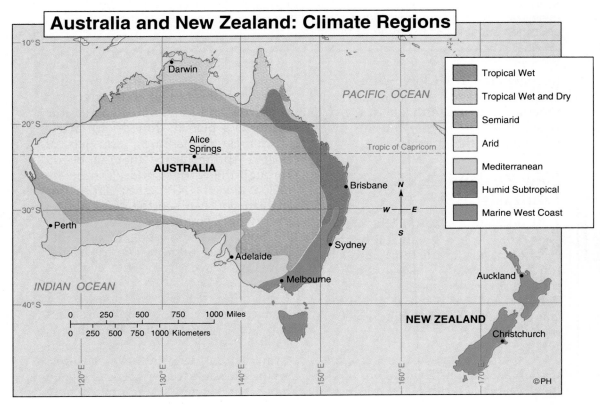

Australia and New Zealand: Climate Regions

Legend:
- Tropical Wet
- Tropical Wet and Dry
- Semiarid
- Arid
- Mediterranean
- Humid Subtropical
- Marine West Coast

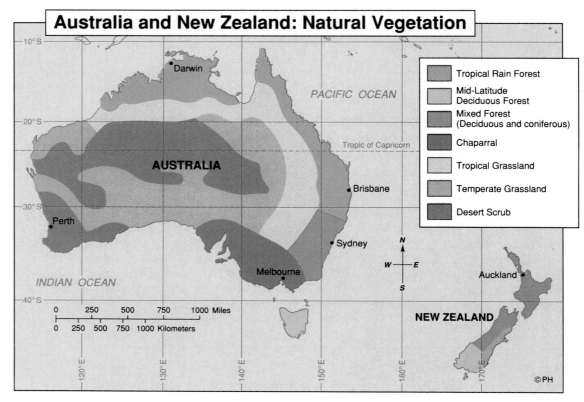

Australia and New Zealand: Natural Vegetation

Tropical Rain Forest

Mid-Latitude Deciduous Forest

Mixed Forest (Deciduous and coniferous)

Chaparral

Tropical Grassland

Temperate Grassland

Desert Scrub

Applying Geographic Themes

Interaction Ranchers in both Australia and New Zealand make a living by raising and selling livestock. What type of natural vegetation makes these countries suitable for this type of farming?

frequent droughts. Closer to the center of the continent, where hardly any rain falls, shrubs and grasses grow. Acacia trees, or "wattles," are also able to survive in the grasslands.

New Zealand: A Place of Contrasts

Find New Zealand on the map on page 708. It lies about 1,200 miles (1,930 km) east of Australia across the rough and windy Tasman Sea. Although New Zealand is part of the Pacific Islands, its physical and human characteristics are very different from the other islands.

Two Islands As you can see from the map, New Zealand is made up of two main islands, North Island and South Island. These two islands are quite different from each other. North Island is narrow and hilly. Spread across the center is a plateau dotted with active volcanoes, hot springs, and jets of heated spring water that shoot up through the earth's surface. These are called geysers (GY zurz). From this plateau you can see the coast almost everywhere you look.

South Island is longer and more mountainous than is North Island. Looking down from the chain of mountains known as the Southern Alps, the scenery is spectacular. Glaciers cover the mountain slopes and drain into sparkling lakes in the valleys below. Far beyond, fjords (fyawrdz) cut deep into the coastline, giving the southwestern coast a ragged look. The rugged geography of both

WORD ORIGIN

geyser
The term *geyser* comes from *geysa,* an Icelandic word in Old Norse meaning "to gush." A hot spring in Iceland is named Geysir, or "gusher."

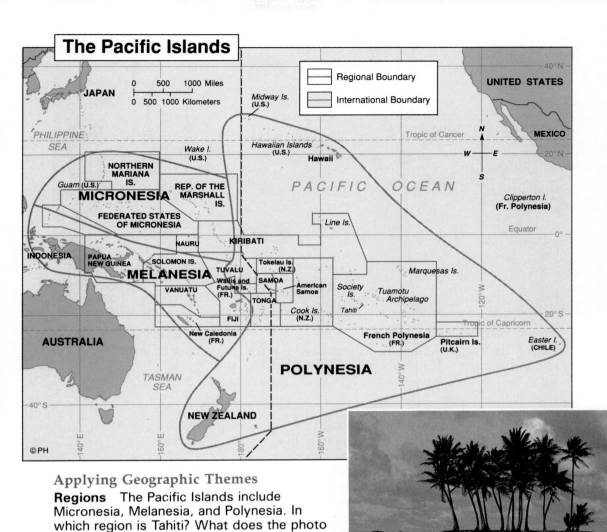

The Pacific Islands

| Regional Boundary |
| International Boundary |

JAPAN
0   500   1000 Miles
0 500 1000 Kilometers

PHILIPPINE SEA

Midway Is. (U.S.)

UNITED STATES

Tropic of Cancer

MEXICO

Wake I. (U.S.)

Hawaiian Islands (U.S.)
Hawaii

PACIFIC OCEAN

NORTHERN MARIANA IS.
Guam (U.S.)
MICRONESIA
REP. OF THE MARSHALL IS.

Clipperton I. (Fr. Polynesia)

FEDERATED STATES OF MICRONESIA

Line Is.

Equator

NAURU
KIRIBATI

INDONESIA
PAPUA NEW GUINEA
SOLOMON IS.
MELANESIA
TUVALU
Tokelau Is. (N.Z.)
Marquesas Is.

Wallis and Futuna Is. (FR.)
SAMOA
American Samoa
Society Is.
Tuamotu Archipelago

VANUATU
TONGA
Tahiti

FIJI
Cook Is. (N.Z.)

New Caledonia (FR.)
French Polynesia (FR.)
Pitcairn Is. (U.K.)
Easter I. (CHILE)

AUSTRALIA

Tropic of Capricorn

TASMAN SEA

POLYNESIA

NEW ZEALAND

©PH

Applying Geographic Themes

Regions The Pacific Islands include Micronesia, Melanesia, and Polynesia. In which region is Tahiti? What does the photo show about the region's vegetation?

South Island and North Island results from their location at a place where two tectonic plates collide.

Settlement Affects Vegetation   In the 1800s, European settlers in New Zealand made great changes in the natural vegetation. Until then, the islands had been covered by forests of pinelike kauri (KOW ree) trees and other kinds of evergreens. The Europeans cut down huge tracts of forest to make room for pastures, farms, and towns. Today only about 25 percent of the land is covered with forests.

The Pacific Islands: Regions in the Ocean

The Pacific Ocean, which covers nearly a third of the earth's surface, is scattered with thousands of islands. Some are barely large enough for a person to stand on. Others cover thousands of square miles. As a group, these islands are called the Pacific Islands, or Oceania. As shown on the map above, the Pacific Islands are divided into three main groups: Micronesia, meaning "small islands"; Melanesia, meaning "black islands"; and Polynesia, meaning "many islands."

Moorea, French Polynesia
Many of the lush green mountains that rise from the ocean floor to form some of the Pacific Islands are active volcanoes.

Rugged Mountains and Coral Reefs
The Pacific Islands can also be divided into high and low island regions based on their physical characteristics. The high islands are mountainous and were created when two large tectonic plates collided. Volcanoes and earthquakes are common on these islands. The low islands are made up of coral reefs. Coral reefs are formed from tiny sea creatures called polyps. These living organisms attach themselves to the skeletons of other polyps. The skeletons accumulate over many thousands of years and eventually form reefs that become exposed when the sea level drops. Below the ocean surface these jewel-like coral reefs are shimmering paradises for tropical fish.

A Tropical Climate Most of the Pacific Islands lie in the tropics and so are warm all year long. Almost all of the islands have a wet season and a dry season.

Large storms called typhoons often race through the Pacific Islands. These cyclonic storms, which are known as hurricanes in the Atlantic, bring violent winds and rain. Small islands rising only a few feet above the water can be severely damaged or even completely washed away.

SECTION **1** REVIEW

Developing Vocabulary
1. Define: **a.** outback **b.** geyser **c.** coral reef

Place Location
2. Identify the landmasses that lie in the southern part of the Pacific Ocean.

Reviewing Main Ideas
3. What are Australia's most important physical characteristics?
4. Compare New Zealand's South Island to its North Island.

Critical Thinking
5. **Making Comparisons** Describe two physical characteristics shared by Australia and New Zealand.

✓ Skills Check

☐ Social Studies
☑ Map and Globe
☐ Reading and Writing
☐ Critical Thinking

Using a Road Map

A road map shows the locations of major highways and secondary roads. It also shows connecting roads between cities and towns. Being able to read a road map makes it possible to reach destinations easily and without delay.

1. **Study the map key or legend.** The road map below shows a section of Australia. According to the map legend, three categories of roads and highways appear on the map. The type size of the name of a city or town indicates how heavily populated the community is. According to the map legend:

(a) What do heavy black lines represent?
(b) What are the route numbers of five major roads?

2. **Determine the scale being used.** Because this map is Australian, distances are expressed in kilometers. One inch equals approximately 50 kilometers. Distances between major intersections—set off with bold red triangular pointers—are shown by numbers in bold red ink. Smaller green pointers and green numbers show the intermediate points within a given distance. For example, the distance from Perth, the state capital city, to The Lakes (on Route 94) is 50 kilometers. An intermediate point is Midland. Perth is 18 kilometers from Midland, and Midland is 32 kilometers from The Lakes. Study the map scale and answer the following questions. (a) What is the distance between Perth and Brookton, a small town in the southeast? (b) What is the distance between Brookton and Dale? (c) between Dale and Kelmscott? (d) between Kelmscott and Perth? (e) What is the total distance when these last three sections of road are added together?

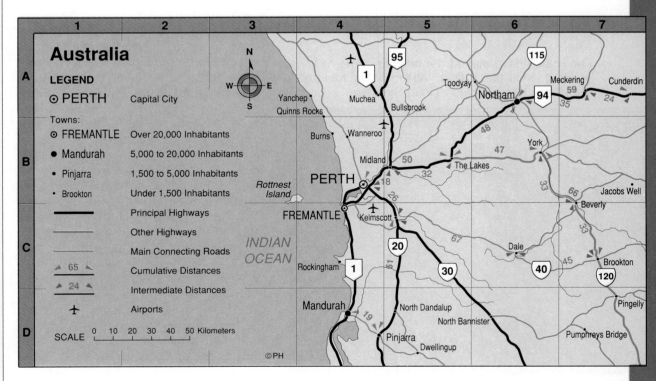

2 Human Geography

Section Preview

Key Ideas

- Although Australia has been linked with Great Britain for much of its history, it is now becoming allied with other nations bordering on the Pacific Ocean.
- New Zealand is a modern country with a culture and economy much like Australia's.
- Most people on the Pacific Islands live at a subsistence level.

Key Terms

aborigine, trust territory

Slightly more than twenty-five million people live in Australia and the Pacific Islands. That's fewer people than live in the state of California, and not many more than live in Mexico City. Yet, despite their small numbers, the different peoples in this area have quite distinct ways of life.

Australia: Yesterday and Today

Australia's culture is dominated by a way of life it inherited from the British. This heritage is seen in its system of government, which, like Great Britain's, has a parliament led by a prime minister and a cabinet. Australia's links with Great Britain can also be seen in the population—most of Australia's people still trace their ancestors to Great Britain. But Australia's ties to Great Britain are growing weaker as it interacts more with other countries that border on the Pacific Ocean.

A History of Movement Scientists think that the first Australians, known as **aborigines** (ab uh RIJ uh neez), crossed a land bridge from Southeast Asia to Australia about fifty thousand years ago. Until Europeans began to settle Australia in the late 1700s, the aborigines, or native people, lived in isolation from the rest of the world. Since then, their history has been similar to that of the Native Americans. Settlers killed them and drove them from their lands. Many died from diseases carried to the island by the Europeans. More than 300,000 aborigines once lived in Australia. Today there are only half that number.

The European settlement of Australia began eighteen years after Captain James Cook landed on the east coast of Australia and claimed it for Great Britain. Britain saw Australia as a solution to the problem of its overcrowded prisons—and a way to rid itself of the poor. In 1787, the first group of prisoners boarded ships for the long journey to the isolated continent. Australia's first European settlers, many of them wearing leg irons, arrived in Sydney Harbour in 1788.

Australia's British Heritage

The British have played the ancient game of lawn bowling since the 1100s. Which ocean can be seen from this bowling club in Sydney?

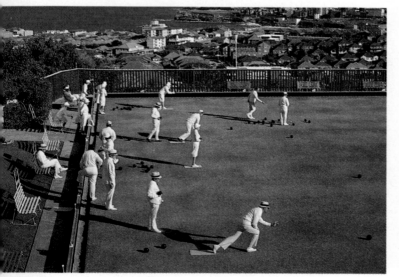

During the next eighty years, more than 160,000 men, women, and children were transported to Australia's distant shores. After their sentences ended, many prisoners stayed. Other settlers from Britain, looking for land on which to raise sheep and grow wheat, joined them.

Beginning in the early 1900s, Australia's population grew steadily. Until the end of World War II, most of Australia's immigrants came from Great Britain. After the war ended, large numbers of immigrants came from Greece, Italy, and other countries in southern and eastern Europe. Today many immigrants come from the near-by countries of Southeast Asia because of Australia's location along the Pacific Ocean, and because of its high standard of living.

Climate Affects Land Use and Population   Australia's hot, dry climate and forbidding interior have greatly affected the country's settlement and land use pattern. Commenting on the pattern of settlement in Australia, one author has noted:

In shape, Australia resembles a ragged square, but the real Australia where people live and work is a ribbon.

Applying Geographic Themes

Regions   Most Australians and New Zealanders live in urban areas along the coasts. Compare this map to the natural vegetation map on page 710. Why do you think the Australian outback is so thinly settled?

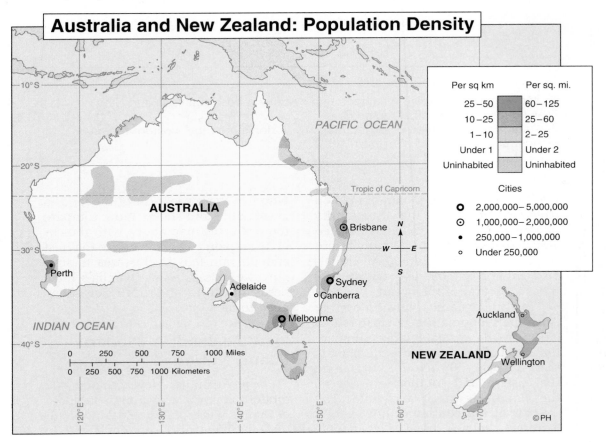

Australia and New Zealand: Population Density

Per sq km	Per sq. mi.
25–50	60–125
10–25	25–60
1–10	2–25
Under 1	Under 2
Uninhabited	Uninhabited

Cities
- ◉ 2,000,000–5,000,000
- ⊙ 1,000,000–2,000,000
- • 250,000–1,000,000
- ○ Under 250,000

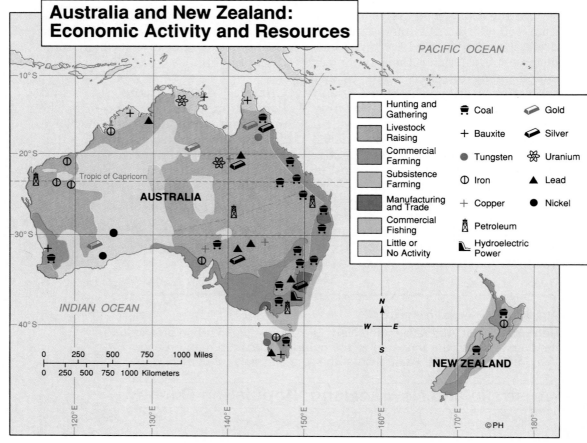

Australia and New Zealand: Economic Activity and Resources

PACIFIC OCEAN

Hunting and Gathering		Coal		Gold
Livestock Raising		+ Bauxite		Silver
Commercial Farming		● Tungsten		※ Uranium
Subsistence Farming		◐ Iron		▲ Lead
Manufacturing and Trade		+ Copper		● Nickel
Commercial Fishing		Petroleum		
Little or No Activity		Hydroelectric Power		

AUSTRALIA

Tropic of Capricorn

INDIAN OCEAN

0 250 500 750 1000 Miles
0 250 500 750 1000 Kilometers

N
W — E
S

NEW ZEALAND

©PH

Applying Geographic Themes

1. Movement   Why are manufacturing centers in both Australia and New Zealand located in coastal regions?

2. Regions   What type of economic activity dominates the western region of Australia?

WORD ORIGIN

New Zealand
The first European to see New Zealand was a Dutch sea captain who arrived at the islands in 1642. He named the islands after a province in the Netherlands.

Look closely at the population map on page 715. Notice that the vast majority of Australians live in cities located along the eastern and southeastern coasts. In fact, 95 percent of Australia's 16.2 million people live within 100 miles (160 km) of the ocean.

To understand why most of Australia's population is clustered in cities on the coastal plains, compare the population map with the climate map on page 709. Notice that along the eastern and southeastern coasts, the climate is moist and mild. The interior of the continent, however, is extremely hot and dry. The few people

who live there use most of the land to graze cattle and sheep. Now compare the resources map above with the climate map on page 709; notice that just enough rain falls in this area to support the grasses that Australia's huge herds of cattle and flocks of sheep need to feed on.

New Zealand: A Modern Island Nation

Like Australia, New Zealand is a modern country with a high standard of living. Much of New Zealand's cultural heritage also can be traced to

Great Britain. Unlike Australia, however, New Zealand has few people who have migrated from areas other than northern Europe.

Polynesian and European Roots   In 1769, Captain Cook landed in New Zealand and established friendly relations with the Maoris (MOW reez), the islands' original inhabitants. His early friendship with the Polynesian Maoris set the stage for future interactions between the British government and the Maoris.

In 1848, the Maoris signed a treaty with the British. In exchange for land rights they agreed to accept British rule. It was the first time Britain, or any world power, had negotiated with native people for political control.

Today New Zealand is an independent country with a government based on the British model. Although 90 percent of New Zealand's population is of European descent, New Zealand has developed a national identity that is rooted in both its Polynesian and British past.

An Urban Population   New Zealand is sparsely populated. Only about 3.3 million people live in the entire country. The population map on page 715 shows that, as in Australia, the great majority of the people live in large cities along the coast. More than 70 percent of New Zealanders live on North Island. Many of these work in industry. Less than 20 percent of the population live in the countryside.

Diversifying an Agricultural Economy

Auckland, New Zealand's largest city, is steadily growing. Sheep raising (inset) may decline as New Zealand and Australia shift their economies toward manufacturing and away from raising livestock. Industrialization will bring growth to urban areas in both countries.

Hinduism in Fiji

Although the landscape and climate are vastly different, daily life on the Pacific island of Fiji is in many ways similar to that of India. This is because, beginning in 1879, workers from India migrated thousands of miles by sea to work as indentured servants in the sugar cane fields of Fiji. With them they brought their religious traditions, as well as their native customs in music, dance, food, and clothing—women's saris, for example.

Today, Indians make up more than half the population of Fiji. Hindu temples, like the one at right, dot the mountainous island terrain and are an important part of the daily lives of many islanders. Inside a temple, where images of the Hindu god Siva are present, men and women kneel on mats to pray. Hibiscus flowers and marigold blossoms lie on the tile altar as offerings to the god.

More than a century after its introduction to the island, Hinduism continues to have a marked impact on daily life in Fiji.

1. How was the Hindu religion first brought to Fiji?
2. What other changes did the Indians bring to these islands?

Notice on the economic activities and resources map on page 716 that nearly half of New Zealand's land is pasture on which sheep and cattle are raised. Wool, beef, and lamb are the country's largest exports.

The Struggling Pacific Islands

The way of life of the Pacific Islanders differs greatly from that of the people of Australia and New Zealand.

On most of the islands, people must struggle to earn a living. There is not enough land for large-scale farming, and few mineral resources exist to encourage industry or trade.

Mysterious Beginnings No one is sure how long people have lived on the Pacific Islands or where they came from. Some scientists think that the region's earliest inhabitants have been in Melanesia for more than fifty thousand years. Some evidence suggests that the ancestors of the peoples of Melanesia, Micronesia, and Polynesia all came from the Malay Peninsula in Southeast Asia.

Most of the world paid little attention to the Pacific Islands until World War II. During the war, Japanese and United States forces fought many bloody battles on the islands. After the war, many of the islands were divided into **trust territories**, or territories supervised by another nation. Guam and American Samoa are both territories of the United States. Many of the islands have now gained their independence.

Living from the Land and Sea The population densities of the Pacific Islands vary greatly. About 772,000 people live on the Islands of Fiji, while some of the smaller islands have only a few hundred people each. Although no large cities exist, many of the islands are overcrowded.

Most Pacific Islanders make their living from farming or fishing. Coconut products, pineapples, bananas, skipjack (a kind of fish), and yellowfin tuna are some of the products exported from the Pacific Islands. Most people, however, live at a subsistence level; that is, they usually grow only enough to feed themselves. Some high islands in Melanesia and Polynesia can support cash crops such as rubber, coffee, and sugar cane. However, in Polynesian cultures especially, people often prefer to maintain their traditional ways of making a living.

Harvesting the Ocean
Plentiful harvests make fish a staple food in the Pacific Islands. These fishers still work in the traditional ways of their ancestors.

Tourism is a growing industry in the Pacific Islands. Vacationers in search of warm, sunny beaches and scenic beauty increasingly head for these islands as travel and communications become faster and easier.

SECTION 2 REVIEW

Developing Vocabulary
1. Define: **a.** aborigine **b.** trust territory

Place Location
2. Identify the area where most Australians live.

Reviewing Main Ideas
3. In what ways are life in Australia and New Zealand similar?
4. How do most people in the Pacific Islands make a living?
5. Who are the aborigines and Maoris?

Critical Thinking
6. **Determining Relevance** Why do most people in Australia and New Zealand live in coastal areas?

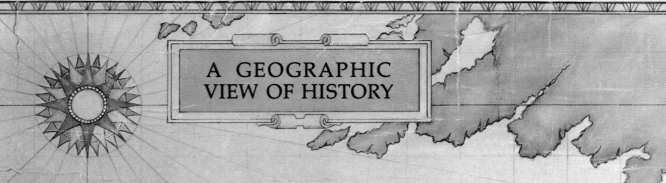

Australia and the Pacific Islands

The relative location of Australia and the Pacific Islands has changed dramatically in recent years.

For millions of years, distance and ocean separated Australia and the Pacific Islands from the rest of the world. The region was the most isolated place on earth—it didn't even appear on some world maps. Unusual plants and animals developed in these places. Early explorers to the area were amazed by the sight of strange animals like the kangaroo. Trees such as the eucalyptus (yoo kuh LIP tuhs) had never been seen in Europe.

European exploration and the eventual settlement of Australia and the surrounding area marked the beginning of the end of the region's isolation. Since that time the barriers of ocean and distance have become less and less important. Today the location of Australia and the Pacific Islands gives it an important place in a dynamic new region: the Pacific Rim.

Australia's Fantastic Animals

Mammals that lay eggs or carry their young in pockets on their abdomens, birds too big to fly, earthworms that reach lengths of 10 feet (3 meters)—these are just some of the fascinating animals native to Australia. For a long time scientists have been curious to learn how these animals developed and why they are so different from animals found in other parts of the world.

Australia's Location Changes Look back at the map in Chapter 2 on page 23. Notice that Australia, Antarctica, Africa, the Arabian Peninsula, India, and South America were once all attached, forming a large continent called Pangaea. As shown on the map, about 130 million years ago this huge landmass began to break apart. Millions of years afterward, Australia started to drift slowly toward its present location in the Pacific Ocean. As Australia moved across the ocean, it carried with it plants and animals from Pangaea. Because other animals and plants could not cross the ocean, these animals and plants developed in isolation.

Marsupials Abound Marsupials (mahr SOO pee uhlz) were the most successful of the new continent's animals. These mammals give birth to babies before they are fully developed. The young are then nurtured in a pouch on the mother's abdomen until they mature. Kangaroos, wallabies, and wombats are just a few of Australia's marsupials.

Marsupials have adapted to every kind of environment on the continent. The ferocious Tasmanian Devil, for example, stalks its prey in the rain forests of Tasmania. In the outback, the kangaroo grazes on grasses and small shrubs. And in the forests along Australia's coasts, the koala bear forages in the treetops for eucalyptus leaves.

Prisoners Arriving in a Remote Land

Australia's first European settlers were prisoners transported from England's overcrowded jails. Why do you think the English selected Australia as a site for releasing their prisoners?

Two Theories: Adaptation or Geographic Isolation Scientists disagree about why marsupials are so common in Australia. Some zoologists, or scientists who study animals, think that marsupials thrive in Australia because they are well suited to the harsh environments of the continent.

Other scientists think that marsupials thrived because their geographic isolation protected them from other species. These scientists say that if land bridges had existed between Australia and other parts of the world, other mammals would have crossed over and destroyed the marsupials.

The Land "Down Under" Moves Up

Two thousand years before Europeans reached the shores of Australia, the Greeks had guessed at its existence. As early as 400 B.C., Greek geographers determined that the

world was spherical. Since they believed that the world was balanced, they thought that a continent existed on the other side of the globe opposite Europe, Africa, and Asia. They called this landmass "Terra Australis Incognita," meaning "unknown southern land."

The Europeans Reach Australia In the early seventeenth century, European explorers finally reached the shores of Australia. They were disappointed to find no gold or spices, or anything else that they imagined might be useful to them. William Dampier, who visited western Australia from England in 1688, described it like this:

The Land is of a dry sandy soil, destitute of water, except you make wells. . . . We saw no Trees that bore Fruit or Berries. We saw no sort of Animal, nor any Track of Beast. . . .

721

It was not until 1770 that Captain Cook sighted Australia and claimed the area for Great Britain. He wrote:

The country itself so far as we know doth not produce any one thing that can become an article in trade to invite Europeans to fix a settlement upon it. However this Eastern Side is not that barren and miserable country that Dampier and others have described the Western side to be. . . . It can never be doubted but what most sorts of Grain, Fruits, and Roots etc. of every kind would flourish here. . . .

Despite Captain Cook's promising description of Australia, the continent and the islands around it were still viewed as the most remote area on earth. Few people were anxious to leave the security of a place they knew for an unknown land on the other side of the globe. Australia's first settlers were convicts, banished from society to a "prison" that covered a continent.

For most people in Europe and North America, Australia and the surrounding Pacific region would remain remote and inaccessible for many years. The area was popularly referred to as either "the antipodes" (ant IP uh deez) or "down under." *Antipodes* actually refers to any two places on opposite sides of the earth, but Europeans and Americans used it only with reference to Australia and New Zealand. The term *down under* refers to the region's location in the Southern Hemisphere. It is based on the assumption that the Northern Hemisphere occupies the "top" half of the globe. As photographs from space show, however, which side of the earth is "up" depends on which way you are looking at the planet.

Australia's New Relative Location   During World War II, the view that Australia and the islands in the Pacific were on the farthest edges of the world changed. Between 1941 and 1945, the area was one of the war's most strategic locations. Battles raged between the Allies and the Japanese, and Japanese forces threatened to invade Australia. People every-

where listened for news from the region. No longer were Australia and the Pacific Islands just exotic locations. They were places where military forces massed, and they were scenes of battles that affected the fates of many other nations thousands of miles away.

A New Region Emerges   Following World War II, the world's nations realigned themselves. The United States and the Soviet Union, formerly allies, established distinct spheres of control. They became the two most powerful nations in the world. And because both superpowers faced the Pacific Ocean, the lands that encircled the Pacific, including Australia and New Zealand, became important strategic locations.

In recent decades, the countries that border on the Pacific have become even more important. Today they are grouped together into a region known as the Pacific Rim. The rapidly growing economies of Japan, South Korea, and other east Asian nations have established this region as an important trading center. The success of many Pacific Rim nations is symbolic of great changes that are taking place in world trade patterns. More than 50 percent of world trade is now conducted among the nations of this region.

Australia, at the southwestern edge of the Pacific Rim, is shifting its focus away from Great Britain and toward other Pacific Rim nations. In the early 1900s, more than 80 percent of Australia's exports went to Great Britain. Less than 4 percent of Australia's exports are now shipped to Great Britain.

Changing Patterns of Movement   Improved communications and transportation have increased the movement of people, goods, and ideas between Australia and the Pacific Islands and the rest of the world. Distance is much less of a consideration for tourists eager to visit the scenic islands of the South Pacific, or for business people wanting to open branches in Sydney. Telecommunications allow business and government leaders to communicate with one another in a matter of minutes. In just hours, planes can transport people and goods thousands of

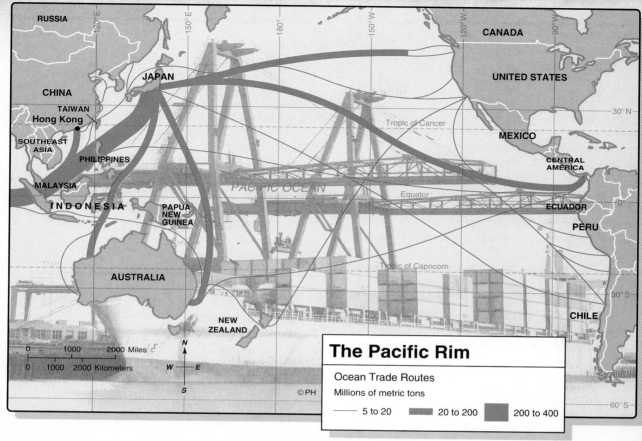

RUSSIA · CHINA · TAIWAN · Hong Kong · SOUTHEAST ASIA · MALAYSIA · INDONESIA · PHILIPPINES · JAPAN · PAPUA NEW GUINEA · AUSTRALIA · NEW ZEALAND · PACIFIC OCEAN · CANADA · UNITED STATES · MEXICO · CENTRAL AMERICA · ECUADOR · PERU · CHILE

Tropic of Cancer · Equator · Tropic of Capricorn · 30°N · 0° · 30°S · 60°S

0   1000   2000 Miles
0   1000   2000 Kilometers

N · W–E · S

©PH

The Pacific Rim

Ocean Trade Routes
Millions of metric tons

—— 5 to 20   20 to 200   200 to 400

Applying Geographic Themes

Regions   This map shows the countries of a new economic region known as the Pacific Rim. How has the relative location of Australia and New Zealand changed as a result of changes in global trading patterns?

miles. Television and radio reports provide people with news from all over the world almost as soon as it happens.

Australia and the Pacific Islands are no longer isolated outposts at the end of the world. The changes that have come to the area are most apparent in Australia. In the 1960s, the majority of its immigrants came from Europe. Now each year almost one third of its newcomers arrive from Asia. Japanese and Korean companies display their names on the entrances of towering skyscrapers. And people can choose everything from noodle dishes from Malaysia to fast-food favorites from the United States for dinner.

TAKING ANOTHER LOOK

1. What are the two most widely held theories concerning the uniqueness of Australia's animals?
2. Why did most Europeans and people living in North America refer to Australia and the Pacific Islands as "down under"?
3. Why are Australia and the Pacific Islands no longer isolated from the rest of the world?

Critical Thinking

4. **Identifying Central Issues**   How has the relative location of Australia and the Pacific Islands changed over the years?

34

REVIEW

Section Summaries

Section 1 Land, Climate, and Vegetation
Almost all of Australia's land is flat and has a dry climate. The only areas with significant rainfall are located along the coast. Unlike Australia, New Zealand's climate is warm and rainy all year long. New Zealand's two main islands are different in their physical characteristics. South Island is more mountainous than North Island. The Pacific Islands are divided into two groups based on physical characteristics. The high islands are mountainous and often volcanic. The low islands are made up of coral reefs. Both groups have tropical climates.

Section 2 Human Geography   Although most Australians still trace their roots back to Great Britain, Australia's population is becoming more varied. Most of Australia's people live in cities along the coasts. Few people live in Australia's hot, dry interior. New Zealand is a modern nation with a national identity that combines both its Maori and British past. Most people in New Zealand live in cities and work in industry.

Vocabulary Development

Match the definitions with the terms below.

1. a jet of heated spring water that shoots up through the earth's surface
2. an area under the supervision of another nation
3. the dry and barren central and western plains and plateaus of Australia
4. a low-lying island or underwater barrier built over time from the skeletons of tiny sea creatures

5. a violent storm with high winds and torrential rains
6. native inhabitant

a. outback	**d.** trust territory
b. coral reef	**e.** aborigine
c. typhoon	**f.** geyser

Main Ideas

1. What are Australia's most distinctive physical characteristics?
2. How did settlement affect vegetation in New Zealand?
3. How has climate affected the settlement patterns of Australia's population?
4. How is Australia's population changing?
5. In what ways is New Zealand's way of life similar to Australia's?
6. Explain why some regions of Australia and New Zealand are less densely populated than others.
7. How does life in the Pacific Islands differ from life in Australia and New Zealand?

Critical Thinking

1. **Drawing Conclusions**   Why do you think the Pacific Islands have not been able to develop their economies?
2. **Making Comparisons**   Compare the populations of Australia and New Zealand. In what ways are they different? In what ways are they alike?
3. **Formulating Questions**   Create a list of questions that you would ask if you were planning to emigrate to Australia.
4. **Predicting Consequences**   Discuss three ways you think new settlers may change Australia in the next twenty years.

Practicing Skills

Using a Road Map Study the Australian road map on page 713 once more. Locate the principal highways of the region. Then answer the following questions.

(a) In what directions do the main highways run: north and south, or east and west?

(b) What geographic features of the continent of Australia might explain this orientation of the highways?

(c) In which direction would you be traveling if you drove from Northam to Fremantle on Route 94?

(d) How many kilometers is it from Mandurah to Pinzarra?

(e) If you were traveling east from Northam to Cunderdin on Route 94, what would the total mileage be? If you made a stop after only 35 kilometers, where would you be?

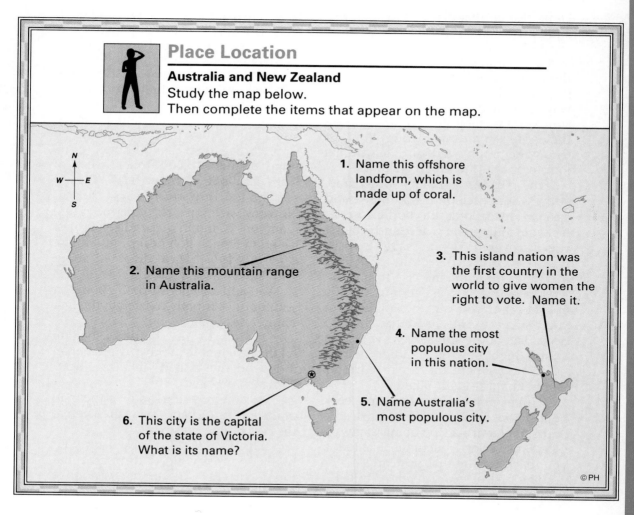

Place Location

Australia and New Zealand
Study the map below.
Then complete the items that appear on the map.

1. Name this offshore landform, which is made up of coral.

2. Name this mountain range in Australia.

3. This island nation was the first country in the world to give women the right to vote. Name it.

4. Name the most populous city in this nation.

5. Name Australia's most populous city.

6. This city is the capital of the state of Victoria. What is its name?

©PH

Place Location: WHERE ON EARTH?

Message in a Bottle

At last, three years of saving money and keeping to a strict budget are paying off. Ocean liner U.S.S. *Holiday* pulls away from the dock, bound for Hong Kong, Bangkok, and Cairo—and you are aboard!

Besides sunning yourself on deck and alternating between the pool and the buffet table, your shipboard plans include doing something you've always dreamed about. Tucked in your suitcase is a small green bottle. Inside the bottle is a message. Your plan is to throw the bottle overboard, leaving its journey up to the ocean currents.

The message contains your name, address, and telephone number and asks whoever finds the bottle to contact you, telling you where and when it was discovered. The message also asks finders to throw the bottle back into the water for someone else to find.

Once the ship is gliding on the open ocean, you dig the bottle from your suitcase and head for the top deck. The bottle spins over the side in a long, high arc. You stand at the rail, wondering where the ocean currents —and chance—will take it.

Three months later, you are back home. You discover several postcards in the mail and a string of messages on your answering machine. Excitedly, you reach for the first postcard.

The first postcard reads: "June 12. Found your bottle on shore of very small island, northernmost of all Polynesian islands and a United States

possession. On a map you would find it at about 30°N and 180°."

"Uh-oh," you think. "I'm going to need the atlas for this one."

The second postcard reads: "June 17. Your bottle floated into the dock where I keep my sailboat. I live in the only United States state capital *not* on the North American mainland."

You open the atlas to a map of the Pacific Ocean. It should be easy enough to trace the bottle's trip so far. You pick up the third postcard, which reads: "Hi. It's June 29. I'm Robert Owen. My wife Sarah and I carried your bottle just past the Tropic of Capricorn—to an island known for its huge statues—2,300 miles west of Chile at about 30°S and 110°W."

The fourth postcard says: "July 7. Now my vacation's *really* perfect. Here I am on Gauguin's paradise, largest island in French Polynesia, and I find a *letter in a bottle*. Have dreamed about both since I was a kid!"

You add another stopping point on your bottle's trip, then listen to the first phone message.

"It's July 23. Your bottle washed up on the beach here in Suva, capital of an independent island in Melanesia located near 20°S and 180°."

Once more you bend over the map of the Pacific Ocean. Then you listen to the last phone message.

"Hi! It's July 27. I picked up your bottle on my way north through equatorial current. I'm on my way to a United States possession in Micronesia near 20°N and 170°E. The island's only three square miles in area and has

no fresh water! I'll toss your bottle back from there."

Before you can turn back to the map, the telephone rings. "Howdy," drawls a voice almost lost in the static. "I'm a member of the United States armed forces. Yesterday I found a little green bottle. Have I got the right number?"

"Yes," you shout. "Where are you?"

"I'm at the largest and southern-most island in the Marianas island group located at about 13°N and 144°E. The name of the place is—" The connection suddenly breaks off, and all you hear is a dial tone.

The caller probably won't call back, but it doesn't really matter. With the help of the atlas, you'll still be able to pinpoint the bottle's latest location.

SOLVE THE MYSTERY

On what Pacific islands has your bottle turned up? *Hint:* An atlas map of the Pacific islands, an almanac, and an encyclopedia will help you match each location with the name of an island.

Australia, Oceania, and Antarctica

Chapter Preview

Locate the regions covered in each of the sections by matching the colors on the right with the colors on the map below.

Sections	Did You Know?
1 **AUSTRALIA**	Australia is the only continent that contains just one country.
2 **NEW ZEALAND AND THE PACIFIC ISLANDS**	On some of the more remote of the Pacific Islands the only mammals that can be found are bats.
3 **ANTARCTICA**	To prevent air pollution caused by exhaust fumes, no aircraft are permitted to fly over wildlife preserves in Antarctica.

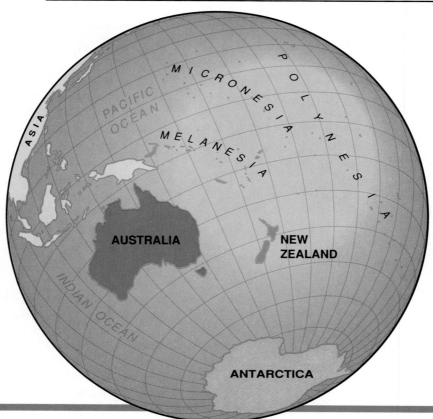

Sydney Harbor and Opera House
After long, harrowing voyages aboard convict ships, Australia's first
European settlers saw a very different Sydney from the one pictured above.
Now Sydney feverishly competes with Melbourne for first place in
Australia's culture, commerce, and industry.

1 Australia

Section Preview

Key Ideas

- Australia's eight major cities, except Canberra, lie along the coast.
- Traditional aboriginal belief holds the outback environment sacred.
- Australians have used the country's interior for raising sheep and cattle and for mining resources.

Key Terms

lagoon, cyclone, artesian well

"You'll come a-waltzing Matilda with me." Written by the folk poet Andrew "Banjo" Paterson in 1895, "Waltzing Matilda" is Australia's un-official national song. The song is not about dancing, however, as speakers of American English might think. In "Strine," as Australians call their particular dialect of English, "to waltz matilda" means "to carry a bedroll or sleeping bag." The song glorifies the way of life of wandering sheep shearers in the rugged Australian outback.

Not all of Australia is untamed wilderness, however. Paterson's outback is only one of Australia's many varied landscapes. A glittering office tower in modern Sydney, a eucalyptus forest on the slopes of the Great Dividing Range, and a ski resort in the Australian Alps provide other real pictures of Australia. The land "down under" is a vast land of breathtaking contrasts, best described by the geographic themes of location and interaction.

Location: A Nation of Coastal Cities

Australia's coastal population clusters in and around eight cities strung along the country's southern and eastern coasts. These eight cities, the nation's largest, include the capitals of Australia's seven states and Canberra, the national capital. Each city gets its own distinct flavor from its location, its landscapes, and its varied people.

Perth and Adelaide The huge state of Western Australia sits astride the Great Sandy, Gibson, and Great Victoria deserts. This region is very sparsely populated with less than ten people per square mile (less than twenty-five persons per sq km). In this vast, empty area, Perth stands out as one of the world's most remote cities. Located on the western coast of Australia, Perth is more than 1,400 miles (2,250 km) from the next major city. That's farther than Boston is from Kansas City. Although the city's more than one million residents live in a remote location, they are by no means cut off. Perth depends on airline routes to connect it to other cities in Australia, Asia, and Africa.

If you flew east from Perth along Australia's southern coast, you would spend hour after hour looking at barren land and nothing more than small, isolated towns—until you reached Adelaide. A city of one million people, Adelaide is the capital and major city of the state of South Australia.

Australia's Urban Rim Three of Australia's most important cities—Sydney, Melbourne, and Canberra—lie within the region known as the Urban Rim. As shown on the map on page 708, this cup-shaped region in southeastern Australia extends from the Great Dividing Range to the eastern coast. Moist winds from the Pacific Ocean and the Tasman Sea rise and cool as they approach the highlands, depositing their moisture in frequent rains. This weather pattern makes the Urban Rim one of Australia's best watered and most fertile regions.

The physical and human characteristics of the Urban Rim combine to present a distinctive Australian "signature" to the world. One such signature is Sydney, the capital of the state of New South Wales and Australia's oldest and largest city. Sydney is best known for its magnificent harbor, laced with small coves and crowned by the sail-like Sydney Opera House.

Melbourne, the capital of Victoria and Australia's second-largest city, has a long-standing rivalry with Sydney. In the late 1800s Melbourne overtook Sydney as the nation's largest city. Although Sydney regained this status in the 1900s, the two cities continue to compete for trade and commerce. Melbourne's south-facing harbor is not as conveniently located for world markets as Sydney's. Still, the factories of the Melbourne area make it a major source of goods for Australia.

Canberra Australia's capital, Canberra, is the country's only major inland city; it lies about 100 miles (160 km) from the coast. Like Washington, D.C., and Ottawa, Ontario, Canberra's location was selected to balance competing political interests in different states. When Australia became a fully independent country in 1901, the government operated from Melbourne while a site for the new capital was debated. By 1927, enough of Canberra had been built to begin transferring the national government there from Melbourne.

Across the Bass Strait The island state of Tasmania hangs off the southeastern coast of Australia like a geographic punctuation mark. Tasmania was not always an island. About twelve thousand years ago, rising

---

WORD ORIGIN

Tasmania
Tasmania was first called "Van Diemen's Land." In 1853 when it was no longer a penal colony the island became Tasmania after a Dutchman, Abel Tasman, who had discovered it in 1642.

DAILY LIFE

A Cattle Station in the Australian Outback

For many people, a job means a desk inside a climate-controlled building and a rush-hour commute. Not so for Cameron and Terrel Weir. As managers of Tempe Downs, a 1.4-million-acre (567,000-hectare) cattle station in the Northern Territory, they face working conditions most people can barely imagine.

The Weirs' station is twice the size of Rhode Island. The land, like that pictured at right, is hot and dry, dotted with scrub. Occupational hazards include heat exhaustion and dehydration, which the Weirs take special care to avoid because any medical care must arrive by plane.

To round up cattle, Cameron first pays a helicopter pilot to locate strays. He then chases them down in his pickup truck. The work is not easy, but the Weirs are making a living in one of the least hospitable places on earth.

1. What are some conditions of life in the Australian outback?
2. Why do you think the Weirs use a helicopter to round up cattle?

ocean levels covered the land that connected it to the mainland and created the Bass Strait.

The island of Tasmania is mountainous and heavily forested, and its capital, Hobart, is cradled in deep blue peaks. With only about 175,000 inhabitants, Hobart is far less cosmopolitan than the mainland cities of Sydney or Melbourne.

The Sunshine Coast Showered with frequent rains from moist trade winds, the east coast of Queensland is Australia's wettest region. Queensland's capital city, Brisbane, is in the heart of Australia's "vacation land." The Sunshine Coast, with its humid subtropical climate and many lovely beaches, attracts millions of tourists to the region each year.

The Great Barrier Reef
In 1979 Australia designated part of the Great Barrier Reef as a government protected park.

North of Brisbane and the Sunshine Coast is the Great Barrier Reef, the largest coral reef in the world. The reef forms a lagoon, a shallow body of water with an outlet to the ocean, between itself and the mainland. The reef extends for 1,250 miles (2,000 km), just about the length of the coast from Maine to North Carolina.

The Tropical North The sparsely populated Northern Territory is mostly too hot and too dry to support human activities. The state's capital, Darwin, however, lies on the northern coast, where the climate is tropical with wet and dry seasons. Darwin is the closest Australian city to Asia, and as flights to other cities become more frequent, it continues to grow. Darwin's location has some disadvantages, however. The city was bombed by the Japanese in World War II, and it has twice been leveled by cyclones, the Australian term for hurricanes. Since the most recent cyclone in 1974, Darwin has been built close to the ground, with few buildings higher than one or two stories.

Interaction: Surviving in "the Bush"

The history of the Australian continent, and especially of the harsh outback region, is a prime example of the geographic theme of interaction between people and the environment. Australians refer to any area away from cities or towns as "the bush." In the bush all Australian people, from aborigines to Europeans, have been put to the test. Strength, bravery, and cleverness are just a few of the traits required to pass this test.

The Dreamtime of the Aborigines
Aborigines, Australia's native people, were the first humans who lived in the outback. They had learned over time how fragile their environment was and felt a sacred obligation to protect it. Aboriginal creation myths teach that in a time long ago, known as the Dreamtime, the ancestors of all living things moved across the formless earth and created the natural world. The ancestors were usually animals, but sometimes they took the form of human beings. Big Bill Neidjie (NAY jee), an aboriginal elder, says that when humans were created, the ancestors gave them responsibility for taking care of the earth: "Now we have done these things, you make sure they remain like this for all time. You must not change anything."

Over countless generations, the aborigines took this responsibility to heart. Like priceless gifts, they handed down ancient knowledge about the sacred sites of each ancestor from father to son. The aborigines learned to take from the land what they needed to survive without destroying their precious earth.

Sheep and Cattle Stations In today's Australia, huge sheep and cattle ranches, called stations, account for most of the economic activity in the outback. Many of these stations are enormous in area. For example, the Anna Creek cattle station in South Australia covers 12,000 square miles (31,000 square km)—which is larger than many of the New England states. Stations must be large, because in some areas it takes 40 acres (16 hectares) of dry, desert scrub vegetation for a single sheep to survive.

Sheep, or "jumbucks," are raised in the cooler plains regions of southeastern and southwestern Australia. Some sheep and lambs are raised primarily for their meat, but the fine, curly wool of merino (muh REE noh) sheep is the most important product. Australia produces 30 percent of the world's merino wool each year.

Cattle are raised in the hotter northern and central regions of Australia, where the native grasses and shrubs provide enough food. Water for these stations comes mainly from **artesian wells**. These wells are bored deep into the earth to tap a layer of porous material filled with groundwater.

The recent growth of the Australian cattle industry reflects changes in both supply and demand. New breeds of cattle that thrive better in hot, dry weather have increased beef yields. And growing demand for meat in Japan, Europe, and other places has provided new markets.

The Search for Resources In 1851, gold was discovered in the outback of New South Wales and Victoria. The gold rush that followed saw Australians swarming out of their cities and immigrants crowding the ports, all eager to join the search. Today gold is only one of many mineral resources that are mined in various locations throughout Australia. Some of these resources are shown on the map on page 716 in Chapter 34.

Keys to Australia's Past
An Aborigine examines an ancient rock painting of his ancestors. This art helps scholars study the geography and history of ancient Australia.

SECTION 1 REVIEW

Developing Vocabulary
1. Define: **a.** lagoon **b.** cyclone **c.** artesian well

Place Location
2. Where is Australia's Urban Rim relative to Perth?

Reviewing Main Ideas
3. What is the distinction of the Great Barrier Reef?
4. How would you describe the aborigines' historical relationship to the land?
5. How have settlers made use of Australia's interior?

Critical Thinking
6. **Demonstrating Reasoned Judgment** Do you think it is fortunate or unfortunate that much of Australia remains unsettled by human beings? Explain your answer.

2 New Zealand and the Pacific Islands

Section Preview

Key Ideas

- New Zealand's economy is based primarily on agriculture.
- Besides participating in New Zealand's affairs, the Maoris are taking steps to preserve their traditions.
- The Pacific Islands were first discovered and settled by expert island navigators.
- Many Pacific Islands rely on tourism for economic survival.

Key Term

atoll

Britain naturalist Charles Darwin wrote these words as he sailed across the Pacific from Tahiti to New Zealand in December 1835:

It is necessary to sail over this great ocean to comprehend its immensity. Moving quickly onwards for weeks together, we meet with nothing but the same blue, profoundly deep, ocean. Even with the archipelagoes, the islands are mere specks, and far distant one from the other. Accustomed to looking at maps drawn on a small scale, where dots, shading, and names are crowded together, we do not rightly judge how infinitely small the proportion of dry land is to the water of this vast expanse.

The Pacific Ocean is indeed immense and deep. On its floor is the deepest point on the earth's surface: the Mariana Trench, about halfway between Japan and New Guinea (see the map on page 777). At 64 million square miles (166 million sq km), the Pacific is more than twice the area of the Atlantic. Set in this watery landscape like an assortment of jewels scattered over blue velvet are the tiny islands of the Pacific. South of the Pacific Islands and east of Australia lie the two comparatively larger islands that make up New Zealand.

New Zealand

The backbone of New Zealand is a string of volcanic mountains formed along the border between the Australian and Pacific tectonic plates. These mountains form two large islands, called simply South Island and North Island. New Zealand's highest mountains tower above South Island. Mystery novelist Ngaio (NY o) Marsh, a native of New Zealand, described the scenery of these mountains—known as the Southern Alps—in this way:

At their highest, they are covered by perpetual snow. Turbulent rivers flow through them, you can see these rivers twisting and glittering like blue snakes through deep gorges, spilling into lakes and pouring across plains to the coast. The westward flanks of the Alps are clothed in dark, heavy forest. It rains a lot over there: everything is lush and green.

An Agricultural Economy Gentle plains slope down from the mountains on both islands. These plains have rich soils, and the marine west coast climate is ideal for farming. Dairy cattle graze on parts of North Island, and sheep roam throughout the country.

Livestock are raised in New Zealand for many of the same reasons as in Australia. They are well suited to the local conditions, and their products can be shipped over thousands of miles to foreign markets. It is not practical to export milk from New

Zealand, because it would spoil or cost too much to transport. But butter and cheese can survive long journeys by boat or yield high enough prices to make the extra cost of air transportation worthwhile. Similarly, wool and frozen lamb and mutton from New Zealand reach buyers in Asia, Europe, and North America.

Kiwifruit are another New Zealand product commonly seen in American grocery stores. The world market for kiwifruit opened in 1952, when some of the fuzzy brown berries were included with a shipment of lemons from New Zealand to England. When the shipment arrived, the lemons had spoiled, but the kiwifruit were in perfect condition. Today New Zealand produces two thirds of the world's kiwifruit.

Life in the Cities About 70 percent of New Zealand's 3.3 million residents live on North Island. Auckland (AWK luhnd), New Zealand's largest city, with 800,000 people, is located there. Auckland's northern latitude places it closer to other nations than any other major New Zealand city, and its airport and ocean port are the country's busiest. Auckland has also developed as a manufacturing center, and nearby farms make it an important agricultural trade center.

Christchurch is New Zealand's second-largest city and the major urban center on South Island. The national capital, Wellington, symbolically unites the nation from its location overlooking Cook Strait which separates North Island from South Island. Buildings in Wellington are designed to survive earthquakes, because the city lies on a fault in the earth's crust.

The First New Zealanders Over one thousand years ago, the Maoris arrived in New Zealand, having navigated their 100-foot (30-meter) canoes across the Pacific Ocean from Polynesia. Starting in the 1840s British set-tlers came to New Zealand in large numbers. Despite their treaty with the Maoris, a series of wars flared between the British and the Maoris through the 1860s. The British finally emerged triumphant, after thousands of Maoris died from fighting or diseases they caught from the British. The Maoris also lost most of their land to the colonists.

Symbol of a European Heritage
Christchurch, on New Zealand's South Island, was named for one of the colleges of Oxford University in England. Europeans in New Zealand built their new homes and cities in the style of their native country.

Today, better education and health care have improved the lives of New Zealand's Maoris, who comprise about 10 percent of the population. Many Maoris today head for New Zealand's cities to go to school, to seek higher-paying jobs, or to exercise their full rights as citizens. Some eventually return to their own territories to rediscover their cultural roots because they realize the importance of nurturing *Maoritanga* (MAW ree tahn gah), or Maori pride, in their young people. With care and patience, the rich cultural heritage of the Maoris will not be lost, but will endure.

Polynesian Immigration Following in the footsteps of the Maoris of long ago, many Polynesians have immigrated to New Zealand in the last century. Many of these Pacific Islanders have married the descendants of European settlers. As a result, more than one third of the total population of New Zealand today has some Polynesian ancestry.

Place: The Pacific Islands

Most people probably think of the Pacific Islands as places where native people stand beneath palm trees and place flowered leis around the necks of tourists as they step from airplanes. In some ways, this image is correct. Thanks to the tropical climate, palms are the most common trees on the coasts of most islands. Many kinds of flowers grow and are often given as gifts. And tourism has become the most important economic activity of many islands.

However, this image is incomplete. Just as color and shape set apart gem-stones, striking differences distinguish the countries and territories of the islands. Take tourism, for example. Fiji actively promotes tourism, and its international airport is a stopping point for airplanes traveling between North America and Australia. Tonga would like to encourage more tourism, but its landscape—relatively flat and level—is not so attractive to foreign visitors. In contrast, Western Samoa is scenic and well situated with respect to air routes, but the government does not want a large tourist industry for fear that visitors might change the native culture.

Differences in Island Formation That the landscape in Tonga differs from that of Western Samoa is an example of another important difference. This is a difference in island formation. As explained in Chapter 34, the Pacific Islands are divided into two main groups: high islands and low islands. Imagine a huge chain of underwater mountains in the Pacific Ocean. Where the tops of these mountains break the surface of the water, they create high islands. Because many of these islands lie along the boundary of the Australian and Pacific plates, they are capped by volcanoes.

On some islands, volcanic cones rise several hundred feet above the ocean's surface. On others, the volcanoes are now beneath the ocean. On these ring-shaped low islands, called **atolls** (AH tawlz), coral reefs surround an inner lagoon. Atolls form most of the Marshall Islands and the islands of Kiribati (KIR uh bas).

An atoll begins as a fringing reef in the warm, shallow waters surrounding a volcanic island. When the volcanic cone falls below the ocean's surface, the coral continues to build up. Then, after millions of years, the volcano disappears, leaving only a ring of coral around a lagoon. Waves crashing over the coral break the top layer into sand. The sand piles up atop the coral and finally forms soil that can support plant and animal life.

Differences Among Island Groups As discussed in the previous chapter, the Pacific Islands are divided into three groups: Micronesia, Melanesia,

An Island of Micronesia
Because they offer only small quantities of fresh water, plant life, and fertile soil, atolls (inset) sustain little or no settlement. Even on inhabited island atolls, Micronesian islanders are mostly fishers or subsistence farmers.

and Polynesia. Each group is shown on the map on page 711. Melanesia was inhabited first—beginning more than 25,000 years ago—probably by people from Southeast Asia. The "small islands" of Micronesia were settled between 3000 and 2000 B.C. by voyagers from the Philippines, Indonesia, and some of the islands north of New Guinea.

The distinct culture and physical characteristics of the Polynesian people developed over a long period of time when they were isolated in the Tonga and Samoa islands. From this base, the Polynesians explored and settled a huge region of the Pacific, mostly to the north and west. Today, the archipelago of Fiji has large numbers of Melanesians and Polynesians,

but about one half of its 800,000 residents are descended from East Indians who were brought to the islands in the late nineteenth century to work on sugar plantations.

Economic Activities The Pacific Ocean continues to influence how Pacific islanders live and work today. With a tropical climate and lush vegetation set against the backdrop of the Pacific, the region's most important economic activity is tourism. But the islands have few natural resources other than their scenic beauty. Many people support themselves by fishing or harvesting coconuts. Some oranges, mangoes, and other fruits are grown for export, but most farm products are used on the islands where they are

The Island of Fiji
Almost half of Fiji's people are Christians. This photograph shows a Catholic church and school on the island.

Cooperation Among Island Nations

Independence has enabled the distinctive cultures of the Pacific Islands to develop in the ways their people feel are best. But many of these nations have also recognized the need for cooperation to achieve common goals. The many joint economic projects have included establishing an airline to promote tourism.

One of the most successful joint ventures has been the University of the South Pacific. Students from nearly all the islands travel thousands of miles each year to study at the main campus in Fiji. But the university also has centers on other islands, using television and postal systems to bridge the distances. In terms of the miles over which its teachers and students move themselves and their ideas, the university is the largest in the world.

produced. Minerals are extracted on a few islands—a large copper mine operates on Fiji, and New Caledonia has a nickel mine.

The Move Toward Independence
Another characteristic many Pacific Islands share is a recent independence. France has kept control of many of its Pacific islands, and the United States still oversees some islands. However, most islands were granted independence in the 1960s and 1970s. Independence helped renew interest in native cultures among the people. Many new national governments were based on traditional forms of leadership. Tonga, for example, chose to remain a kingdom. Western Samoa adopted a parliamentary system, but most representatives are selected by traditional village leaders. Other nations, such as Vanuatu (vahn uh WAH too), operate with fully representative democracies.

SECTION **2** REVIEW

Developing Vocabulary
1. Define: atoll

Place Location
2. Where is New Zealand relative to the Pacific Islands?

Reviewing Main Ideas
3. What are New Zealand's main economic activities?
4. Why must the Maoris make a special effort to preserve their culture from being lost?
5. Why is tourism a vital part of Pacific Island economies?

Critical Thinking
6. **Testing Conclusions** Explain why you agree or disagree with this statement: "Navigation without instruments such as compasses is primitive and unscientific."

✓ Skills Check

☐ Social Studies
☑ Map and Globe
☐ Reading and Writing
☐ Critical Thinking

Interpreting a Contour Map

A hiker poised at the start of an unfamiliar mountain trail needs a special kind of map tucked in his or her backpack: a topographic, or contour, map. This kind of map shows the changes in elevation that lie ahead—and how quickly these changes occur. Does the trail climb steeply for the next mile, or is the grade a slow and steady rise? Will there be serious climbing involved, or can the hiker cover the distance at an easy, arm-swinging pace? And just how far can the hiker expect to go in a single afternoon? These questions are critical to the success and safety of a hike. A good topographic map can help answer them all.

Topographic maps are useful tools with many applications. Backpackers often take them along when they set out on hiking, rock-climbing, and camping trips. Engineers use them when deciding where to build highways and dams. Police and emergency medical personnel often consult topographic maps during search and rescue operations for people who are lost in the woods.

The topographic map below shows the Pacific island of Tahiti. Use the following steps to study and analyze the map.

1. **Understand what isolines measure.** The lines on a topographic map are called *isolines.* An isoline connects all points where elevation is equal. If you were to hike along one of the isolines shown on this topographic map, you would always be at the same height above sea level. Notice that the isolines are labeled with numbers that tell the elevation in feet along that isoline. Now use the map to answer the following questions: (a) What elevation does the isoline that marks the coastline represent? (b) What is the highest point on Tahiti? (c) Are Papeete and Mataiea at about the same or different elevations?

2. **Interpret the relationships among isolines.** When a series of isolines is close together, it means that the elevation of the land is changing rapidly—in other words, the terrain is steep. On the other hand, isolines spread wide apart indicate that the elevation is changing slowly and the land is relatively flat. Answer the following questions: (a) Is the island generally steeper near the top of Orohena or near the coast? (b) Where is the steepest part of the Taiarapu Peninsula, according to the map?

3. **Put the data you have collected to use.** Use the map to answer the following questions: (a) If you and a friend wanted to climb to the top of Orohena, how would you plot the most gradual ascent possible? (b) How would you plot a steeper climb?

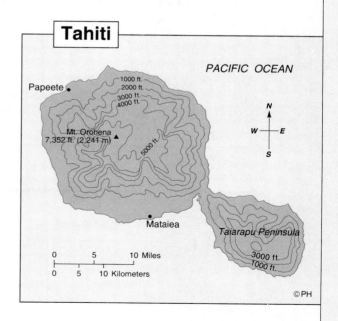

Tahiti

PACIFIC OCEAN

Papeete
1000 ft.
2000 ft.
3000 ft.
4000 ft.
Mt. Orohena
7,352 ft. (2,241 m)
5000 ft.
Mataiea
Taiarapu Peninsula
3000 ft.
1000 ft.

N
W E
S

0 5 10 Miles
0 5 10 Kilometers

©PH

Maori Land Claims in New Zealand

European nations exerted their authority over other lands in a variety of ways during the colonial era. Some nations used military conquest. Others tried to expand their control by using their own system of laws to justify or legitimize their actions. English law allowed ownership and control of property to be transferred through legal contracts and treaties. So as British colonial leaders moved into the southern Pacific, they convinced native leaders to sign treaties that gave the British legal rights to own and use the land.

The Treaty of Waitangi

The legal document by which the British formally established themselves in New Zealand was a treaty signed by more than five hundred Maori tribal chiefs at Waitangi in 1840. The treaty granted the British authority to govern the country. In return, the treaty recognized the authority of the chiefs over their land, people, and other valued articles. Once the treaty was signed, the "pakeha" (as Europeans were called by the Maori) began purchasing land from local chiefs.

Pakeha leaders hailed the Treaty of Waitangi as evidence of British kindness. One colonial leader called it a "Magna Carta" of Maori rights. British Lieutenant Governor William Hobson claimed it was the first time that native people were treated as equals, and he proclaimed, "We are now one people."

Maori Rebellions

Within a decade, however, many Maori tribes violently rebelled against the British, because they felt the British denied them access to their land and resources. The Maori wars finally ended in the 1870s, when British firepower prevailed. Between warfare and disease, the Maori population dwindled to less than sixty thousand. Pakeha acquisition of land reduced Maori possessions to less than 8 percent of New Zealand's total land area. The Land Claims Court established in the 1880s to resolve land disputes seemed just another way for the pakeha to drive the Maori from the islands altogether.

Questioning British Claims

In recent decades, however, New Zealand's government has recognized the legitimacy of many Maori claims. A special court, or tribunal, is currently reviewing land claims dating back as far as 1840. Since the tribunal first met in 1985, Maori tribes have filed more than 150 claims. These claims relate to more than 60 percent of New Zealand's land. They also involve fishing rights for most of the waters surrounding the nation.

The Maori Wars
The Maori surrendered in the 1870s.

Among the land areas in dispute are parts of the wealthy eastern suburbs of Auckland. One tribe is arguing that it should control fishing in most of the waters around the South Island. Early recommendations of the tribunal suggested that it was sympathetic to arguments made by the Maori. Today the tribunal has asked that the New Zealand government return to tribes publicly owned land that was obtained illegally. If the land was privately owned, the tribunal recommended that the Maori receive full financial compensation at current market values. Many New Zealanders fear that too many judgments that favor the Maori will bankrupt the country.

Even more fearful to many pakeha is the fact that some Maori now challenge the legitimacy of the Treaty of Waitangi itself. Key phrases in the Maori translation of the treaty differed in meaning from the English original. Maori attitudes regarding land ownership and sovereignty also contrasted with those of the British. Maori activists therefore argue that their ancestors were tricked into signing a treaty that they didn't understand. As a result, they want the entire treaty to be nullified and all lands returned to the Maori.

Millions of modern New Zealanders are descended from immigrants from other lands. Many others have both Maori and European ancestors. Because of these factors, many observers do not believe that the most extreme Maori claims will be upheld. But the threat remains, and in recent years, many New Zealanders emigrated to Australia and other nations where they feel their future is more certain. Review of the land claims is likely to continue for the rest of this century. To many residents, the debates will be the most important their nation faces, because they will determine who will have the legal right to call themselves New Zealanders.

TAKING ANOTHER LOOK

1. How did the British come to have control over lands in New Zealand?
2. What types of cultural misunderstandings do you think could have taken place during the signing of the Treaty of Waitangi?

Critical Thinking

3. **Expressing Problems Clearly** Do you agree with the Maori tribes' land claims? Why or why not?

The Maori Today
This photograph shows a Maori meeting house near Rotorua, New Zealand.

3 Antarctica

Section Preview

Key Ideas

- Antarctica is a continent covered and surrounded by several different forms of ice.
- Antarctica's climate and terrain have made exploration slow and difficult.
- Scientific knowledge is the greatest antarctic resource.

Key Terms

crevasse, ice shelf, pack ice, convergence zone, krill

Find the continent of Antarctica on a world map, either on the wall of your classroom or in an atlas. It may not be visible at all, if 60°S latitude forms the map's bottom border. Or, if Antarctica is shown, it may appear as a long, ragged strip of white stretching across the southern boundary. In fact, Antarctica is a large, mushroom-shaped continent that accounts for nearly one tenth of the world's land.

To be seen clearly on a map, Antarctica must occupy a central position. Therefore, only the south-polar projection shown in the map on page 743 provides a true picture of Antarctica's shape and size.

The Frozen Continent

Following a visit to Antarctica in the early 1980s, environmental historian Stephen Pyne wrote:

Ice is the beginning of Antarctica and ice is its end. . . . Ice creates more ice, and ice defines ice. Everything else is suppressed. This is a world derived from a single substance, water, in a single crystalline state, snow, transformed into a [crust] composed of a single mineral, ice. This is earthscape transfigured into icescape.

As Pyne's passage conveys, the major feature of the landscape is ice. Ice covers the continent's rocks, and it alters the climate. The ice affects Antarctica's wildlife, because few plants and animals can survive in the frigid conditions. And the ice has greatly limited human activity, leaving Antarctica the only major landmass on the earth without permanent human settlement.

Dense Ice Sheets Ice is found all over Antarctica, but it appears in different forms. In central regions, ice caps, or ice sheets, reach thicknesses of nearly 3 miles (4.8 km). Three miles high is higher than any mountain in North America. The ice is so thick that Antarctica's average elevation is the highest of any continent.

The antarctic ice sheets are incredibly heavy. An acre of ice that is three miles high weighs more than 43 billion pounds (20 billion kg). The weight of the ice is so great that the land beneath it is depressed by hundreds of feet. In fact, the entire globe is distorted by the weight of the ice sheets, becoming slightly pear-shaped rather than remaining a true sphere.

Ice and the Climate The antarctic ice sheets have a significant effect on both the continent's own climate and on weather patterns throughout the Southern Hemisphere. The ice reflects most of the sun's rays back into space rather than absorbing their heat, making temperatures frigid. The *average* annual temperature at one research station is a frigid −70°F (−57°C). The coldest temperature ever recorded on the earth was measured at the same site: −128.6°F (−89.2°C).

While even the glare of six months of summer sun cannot melt them, the antarctic ice caps do not grow rapidly.

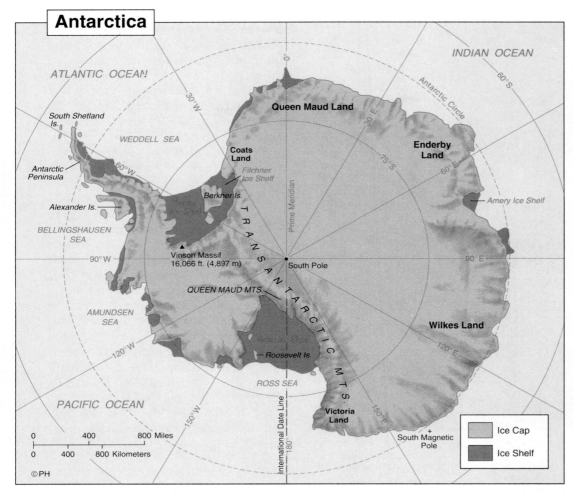

Antarctica

ATLANTIC OCEAN

INDIAN OCEAN

South Shetland Is.

WEDDELL SEA

Queen Maud Land

Coats Land

Enderby Land

Filchner Ice Shelf

Antarctic Peninsula

Berkner Is.

Alexander Is.

Amery Ice Shelf

BELLINGSHAUSEN SEA

▲ Vinson Massif 16,066 ft. (4,897 m)

South Pole

TRANSANTARCTIC MTS.

Prime Meridian

QUEEN MAUD MTS.

AMUNDSEN SEA

Wilkes Land

Roosevelt Is.

ROSS SEA

PACIFIC OCEAN

Victoria Land

International Date Line

South Magnetic Pole

Ice Cap

Ice Shelf

0   400   800 Miles
0   400   800 Kilometers

©PH

Applying Geographic Themes

Place Most of Antarctica lies south of 70°S latitude and is thickly covered with ice. Which mountain range cuts off the southern part of the continent from east to west?

This is because very little snow falls. Like high plateaus on other continents, the region is quite dry, because air loses its moisture as it rises. Air becomes even drier as it gets colder. As a result, the South Pole sees less than 0.2 inches (0.5 cm) of precipitation each year. Although it may seem odd, even the most arid deserts of Africa and Asia usually get more precipitation than Antarctica.

Glaciers Moister and warmer conditions near the coast and in the Transantarctic Mountains permit glaciers to

flow over the land. Antarctic glaciers creep like giant, slow-moving frozen rivers, oozing down from the mountains and the edges of the ice sheet to the coast. Most glaciers do not move more than a few feet each year. Large cracks called **crevasses** often form in glacial ice, but glaciers still provide the most convenient routes to the interior of the continent.

Ice Shelves Antarctica's ice sheets and glaciers are so deeply frozen that the slowly creeping ice extends out over the ocean in several places and

Pack Ice On the fringes of Antarctica, icebergs mix with ice formed in the superchilled waters of the ocean to form pack ice. During the long winter, when the sun shines for only a few hours each day, the pack ice can extend more than 1,000 miles (1,600 km) from the coast. In the summer, the outer reaches of the ice melt, and the pack ice extends only about one tenth as far into the ocean.

The edge of the winter pack ice is close to the convergence zone, where the frigid waters circulating around Antarctica meet the warmer waters of the Atlantic, Pacific, and Indian oceans. This clash of warm and cold waters causes severe storms along Antarctica's coastline. The contrast in temperatures also mixes different layers of water along the edge of the pack ice. Nutrient-rich deep waters rise to the surface, feeding millions of small, shrimp-like creatures called krill. The krill provide ample food for whales and fish; the fish, in turn, become food for seals, penguins, and other animals.

Human Interaction in an Uninhabitable Region

The unique physical geography of Antarctica makes its human geography different from that of any other continent. Because of its remote location and harsh natural conditions, it was the last of the world's continents to be discovered and explored.

Exploration Antarctica was first sighted in the early 1820s by sailors from Russia, Great Britain, and the United States. Explorers reached the Ross Ice Shelf in the early 1840s. But the ice and cold prevented anyone from actually setting foot on the continent until 1895.

Roald Amundsen of Norway and Robert Scott of Great Britain each led major expeditions across Antarctica's

The Last Frontier on Earth
Antarctica is no longer a continent uninhabited by people. Over 2,000 scientists work in Antarctica, and research vessels regularly explore the frozen waters.

WORD ORIGIN

iceberg
The word *iceberg* comes from two Old High German words: *is*, "ice," and *berg*, "mountain." It literally means "mountain of ice."

forms massive ice shelves. The larger shelves cover enormous areas of the Ross and Wendell seas; smaller shelves dot the coastline.

The ice on the shelves is more than 1,000 feet (300 m) thick in many places. The shelves thin out as they extend farther into the ocean, however. Near the edges, the ice often has many crevasses, and large blocks eventually break off and fall into the ocean. In this way, Antarctica produces more than five thousand large icebergs each year. The average berg contains about 1 million short tons (900,000 metric tons) of fresh water.

ice sheets. Both reached the South Pole in the summer of 1911-1912, but Scott and his four companions died on their return trip. Further exploration of the interior began in the late 1920s, when airplanes were built that could withstand Antarctica's high winds and cold temperatures.

Slicing the Antarctic "Pie" Antarctica's unusual conditions also affected the ways in which nations tried to make territorial claims. Many early explorers made claims in the hope of protecting areas rich in whales and seals. But by the late 1880s, the world's most powerful nations agreed that land had to be occupied and actively governed for a national claim to be valid.

Through the first half of the twentieth century, Argentina, Australia, Chile, France, New Zealand, Norway, and the United Kingdom all made claims to parts of Antarctica. Most of these countries' claims were pie-shaped wedges that met at the South Pole. Two nations, the United States and the Soviet Union, refused to make any claims. They also refused to acknowledge the claims of other nations, arguing that settlement had not occurred.

Why did so many nations claim parts of Antarctica? One reason was national pride. Many countries wanted to expand their colonial empires to the Antarctica's frontier or simply to keep other countries from claiming large slices of the continent.

Antarctic Resources Another reason for territorial claims was to claim the ownership of resources. Demand for whales and seals had declined, but it was still possible to find valuable minerals under the ice. Geological discoveries in recent decades suggest that oil, gold, iron, and other minerals may well be present. Coal has already been found, but deposits remain untouched because it would cost too much to

mine and transport them. Other minerals would cost even more to find and to exploit.

People have also looked into the possibilities of tapping some of Antarctica's more available resources. Fishing boats from the Soviet Union and Japan have harvested the krill that live in nearby waters. Krill are high in protein, however they are not very tasty.

Sharing Antarctica's Bounty By far the greatest resource Antarctica has to offer is its wealth of scientific information. The continent is the key to a vast store of knowledge that many countries are now exploring and sharing. More information about how the nations of the world are cooperating to learn more from earth's last frontier is given on pages 748–749.

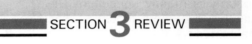
SECTION **3** REVIEW

Developing Vocabulary
1. Define: **a.** crevasse **b.** ice shelf **c.** pack ice **d.** convergence zone **e.** krill

Place Location
2. Why was Antarctica the last landmass to be explored by people?

Reviewing Main Ideas
3. Describe the discovery of Antarctica in the 1800s.
4. What has prevented the United States and the Soviet Union from claiming Antarctica?
5. What are some of Antarctica's resources?

Critical Thinking
6. **Formulating Questions** Imagine that you are planning an expedition to the South Pole. Create a list of questions that you would ask someone who had already completed such a trip successfully.

35

REVIEW

Section Summaries

SECTION 1 Australia Australia's seven state capitals are all located along the coast. The national capital, Canberra, is the only major inland city. Perth, on the west coast, is the most remote city. Sydney and Melbourne, in the Urban Rim, are the most cosmopolitan. The aborigines are experts at finding food and water in the desolate outback. Sheep and cattle stations operate in the more temperate bush regions.

SECTION 2 New Zealand and the Pacific Islands Mountains form the backbone of New Zealand's two main islands. New Zealand's economy is mostly agricultural, although Auckland is a modern industrial city. Many New Zealanders are Maoris, a Polynesian people who are working to preserve their culture. The vast Pacific Island region was settled in waves. Tourism is a vital part of many island economies, since most islands have few mineral resources.

SECTION 3 Antarctica Antarctica is a large continent covered and surrounded by ice. Thick ice sheets blanket the central regions; glaciers and ice shelves spill out into the ocean. Icebergs and other ice form pack ice around the continent, which made early exploration difficult. Nations have agreed not to make claims to Antarctica but to share scientific data instead. Antarctica has coal and other mineral resources, but it is currently impractical to exploit them.

Vocabulary Development

Complete each of the sentences in the next column with the correct term.

1. _____ plays an important role in the antarctic food chain.
2. A large crack in glacial ice is called a(n) _____.
3. As billions of coral polyps attach themselves to a sinking volcano, a(n) _____ is formed.
4. A(n) _____ is formed when glaciers extend out into the ocean.
5. A body of water enclosed by a coral island is called a(n) _____.
6. Australians call the tropical storms that strike their northern coast _____.
7. Warm and cold waters mingle in the _____.
8. _____ is a region of icebergs and floating slabs of ice.
9. Without a(n) _____, a Queensland sheep station could not make use of the dry land.

a. pack ice
b. artesian well
c. cyclones
d. krill
e. atoll
f. crevasse
g. ice shelf
h. lagoon
i. convergence zone

Main Ideas

1. How is life along Australia's coast different from life in the outback?
2. What are some of New Zealand's main economic activities?
3. Who probably first settled the islands of Melanesia and Micronesia?
4. What are some of the different forms ice takes in Antarctica, and where is each found?
5. Why have nations decided not to divide Antarctica into individual claims?

Critical Thinking

1. **Determining Relevance** How did the geography of Antarctica lead to the current situation of international cooperation on the continent?
2. **Predicting Consequences** What do you think might happen in Antarctica if a practical way were found to extract its mineral resources?
3. **Synthesizing Information** Why do you think it is important to the native people of Australia, New Zealand, and the Pacific Islands to preserve their cultural traditions?

Practicing Skills

1. **Interpreting a Contour Map** Refer back to the skill lesson on page 739 to answer the following questions.
   a. When a series of isolines on a contour map is close together, what information is being conveyed about elevation?
   b. What information is conveyed about elevation when isolines are relatively far apart?
2. **Making Comparisons** How are the continents of Antarctica and Australia similar, and how are they different?

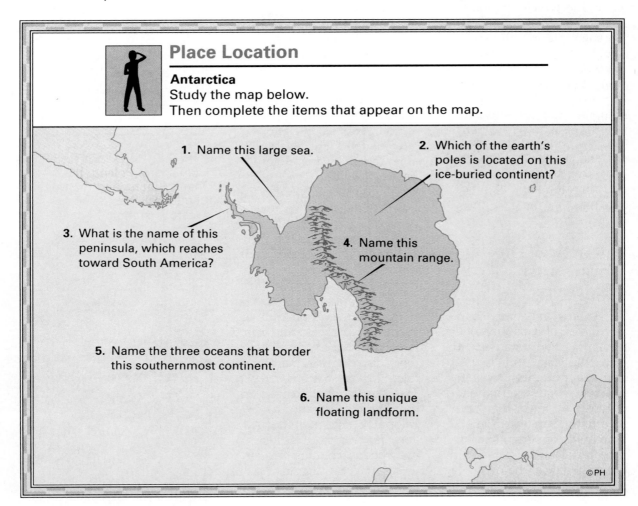

Place Location

Antarctica
Study the map below.
Then complete the items that appear on the map.

1. Name this large sea.
2. Which of the earth's poles is located on this ice-buried continent?
3. What is the name of this peninsula, which reaches toward South America?
4. Name this mountain range.
5. Name the three oceans that border this southernmost continent.
6. Name this unique floating landform.

©PH

Cooperation or Conflict in Antarctica

We live on a planet where people own and occupy nearly every available corner. There are few patches of ground to which one nation or another does not lay claim. Yet some places still exist about which the nations of the world must make a decision: Will they cooperate with one another? Or will they compete among themselves to determine who will benefit most from the land's resources? The answers to these questions are of vital importance, both for the land in question and for the citizens of the world.

Antarctica: The World's Last Wilderness

Antarctica is one of the few places on earth largely unmarked by humans. But because of its scientific and economic importance, Antarctica has become an international outpost. No fewer than sixteen nations maintain research stations there. Scientists find it ideal for the study of earthquakes, the ozone layer, magnetism, solar activity,

and world weather patterns. There they can be certain that their observations have not been influenced by human development.

For the most part, nations have cooperated over Antarctica. In 1961, twelve countries ratified the Antarctic Treaty. This provided for the peaceful use of the continent and the sharing of scientific research. A number of other nations later signed the treaty, and amendments were added to protect wildlife.

Land claims in Antarctica, however, have been a matter of dispute. Seven of the original treaty signers claim ownership of certain land parcels, though they have agreed not to press those claims while the treaty is in force.

Concern over possible oil and mineral development in Antarctica led to the Wellington Convention, signed by twenty treaty nations in 1988. This document set rules governing the exploration and development of Antarctica. Advocates of the convention insist that it contains proper environmental safeguards. But environmentalists fear it

marks the beginning of the exploitation of Antarctica. An alternative plan was recently proposed by France and Australia. This plan suggested that Antarctica be declared a "wilderness reserve." Support for this concept has been growing. Its acceptance, however, will require nations to put the interests of Antarctica

---

Multinational Cooperation
The flags of Poland, the United States, and the United Kingdom fly beside an Antarctic station.

ahead of their own. The extent of their willingness to do this is a key issue for the future of the continent.

The View from the U.S.A.

 The United States McMurdo Station on Ross Island is the largest research station in Antarctica. Recently, American concern has focused on regulating the number of tourists who visit Antarctica. Currently, about 3,500 people travel there every year. Tourists sometimes tramp through wildlife habitats. Other tourists leave trails of litter or disrupt scientific research. New guidelines supported by the United States are aimed at helping to control tourism on the frozen continent.

TAKING ANOTHER LOOK

1. What features distinguish Antarctica from the other continents?
2. What was the purpose of the Antarctic Treaty?

Critical Thinking

3. **Demonstrating Reasoned Judgment** How might Antarctica suffer if nations begin to compete for control of the region?

Applying Geographic Themes

Regions Sixteen countries maintained research stations in Antartica in 1991. Which countries have a station on the South Orkney Islands?

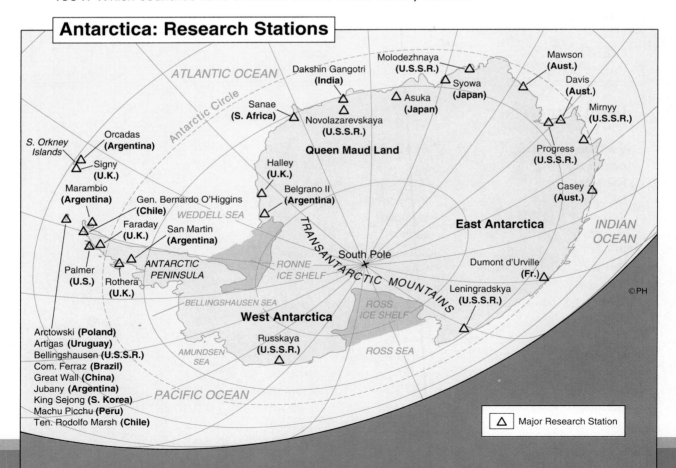

Antarctica: Research Stations

TABLE OF COUNTRIES

The following Table of Countries provides important geographic, economic, and political data for the countries of the world. Countries are listed alphabetically within regions, which are listed in the same order as they appear in the text.

When taken together, the data provide an overview of a country's standard of living, or general quality of life. The table includes the following information for each country:

- **Capital city**
- **Land area**, given in square miles
- **Population** is given in millions. For example, a population of 26.3 means 26.3 million people.
- **Population density** is per square mile. The figure given is based on the total land area of a country divided by the population. For more information on population density, see page 52.

Country	Capital	Land Area (square miles)	Population (millions)	Population Density (per square mile)
The United States and Canada				
Canada	Ottawa	3,851,790	26.6	7
United States	Washington, D.C.	3,615,100	251.4	70
Latin America				
Antigua and Barbuda	Saint John's	170	0.1	377
Argentina	Buenos Aires	1,068,300	32.3	30
Bahamas	Nassau	5,380	0.2	46
Barbados	Bridgetown	170	0.3	1,548
Belize	Belmopan	8,860	0.2	25
Bolivia	La Paz	424,160	7.1	17
Brazil	Brasília	3,286,470	150.4	46
Chile	Santiago	292,260	13.2	45
Colombia	Bogotá	439,730	31.8	72
Costa Rica	San José	19,580	3.0	155
Cuba	Havana	44,220	10.6	240
Dominica	Roseau	290	0.1	294
Dominican Republic	Santo Domingo	18,810	7.2	381
Ecuador	Quito	109,480	10.7	98
El Salvador	San Salvador	8,260	5.3	643
Grenada	Saint George's	130	0.1	2641
Guatemala	Guatemala City	42,040	9.2	219
Guyana	Georgetown	83,000	0.8	9

- **Growth rate,** the speed at which a country's population is increasing, is given as a percentage. To find the actual number of people by which a population is increasing each year, multiply the country's population by its growth rate.
- **Language(s),** the major languages or dialects spoken in the country
- **Per capita GNP**, given in United States dollars, represents an average yearly income for each person. For more information on how this figure is derived, see page 75.
- **Literacy rate,** the percentage of people in a country who can read
- **Infant mortality rate,** the number of infants out of every thousand born who will die before their first birthday
- **Life expectancy**, the average number of years that a person can be expected to live
* **Information not available**

Growth Rate (percent)	Language(s)	Per Capita GNP (U.S. dollars)	Literacy Rate (percent)	Infant Mortality Rate (per 1,000 births)	Life Expectancy (years)
0.7	English, French	16,760	98	7.3	77
0.8	English	18,430	99	9.7	75
1.0	English	2,800	88	24	71
1.3	Spanish	2,640	94	32	71
1.5	English	10,570	89	21.7	71
0.7	English	5,990	99	16.2	75
3.1	English, Spanish	1,460	93	36	69
2.6	Spanish, Quéchua, Aymara	570	75	110	53
1.9	Portuguese	2,280	74	63	65
1.7	Spanish	1,510	96	18.5	71
2.0	Spanish	1,240	88	46	66
2.5	Spanish	1,760	90	17.4	76
1.2	Spanish	1,590	96	11.9	75
2.1	English, French patois	1,650	80	14	75
2.5	Spanish	680	68	65	66
2.5	Spanish, Quéchua	1,080	84	63	65
2.7	Spanish	950	69	54	62
3.0	English	1,370	85	30	71
3.1	Spanish, Indian languages	880	51	59	63
1.9	English, Indian languages	410	86	30	67

Country	Capital	Land Area (square miles)	Population (millions)	Population Density (per square mile)
Haiti	Port-au-Prince	10,710	6.5	607
Honduras	Tegucigalpa	43,280	5.1	119
Jamaica	Kingston	4,240	2.4	575
Mexico	Mexico City	761,600	88.6	116
Nicaragua	Managua	50,190	3.9	77
Panama	Panama City	29,760	2.4	81
Paraguay	Asunción	157,050	4.3	27
Peru	Lima	496,220	21.9	44
St. Christopher-Nevis	Basseterre	140	0.04	288
St. Lucia	Castries	240	0.2	639
St. Vincent and the Grenadines	Kingstown	150	0.1	753
Suriname	Paramaribo	63,040	0.4	6
Trinidad and Tobago	Port of Spain	1,980	1.3	679
Uruguay	Montevideo	68,040	3.0	45
Venezuela	Caracas	352,140	219.6	56

Western Europe

Country	Capital	Land Area (square miles)	Population (millions)	Population Density (per square mile)
Andorra	Andorra la Vella	174	0.05	274
Austria	Vienna	32,370	7.6	235
Belgium	Brussels	11,750	9.9	842
Denmark	Copenhagen	16,630	5.1	309
Finland	Helsinki	130,130	5.0	38
France	Paris	211,210	56.4	267
Germany	Berlin	137,810	79.5	578
Greece	Athens	50,940	10.1	197
Iceland	Reykjavik	39,770	0.3	7
Ireland	Dublin	27,140	3.5	130
Italy	Rome	116,310	57.7	496
Liechtenstein	Vaduz	62	0.03	459
Luxembourg	Luxembourg	990	0.4	374
Malta	Valletta	120	0.4	2,849
Monaco	Monaco-Ville	0.6	0.028	28,072
Netherlands	Amsterdam	14,410	14.9	1,033
Norway	Oslo	125,180	4.2	34
Portugal	Lisbon	35,550	10.4	291

Growth Rate (percent)	Language(s)	Per Capita GNP (U.S. dollars)	Literacy Rate (percent)	Infant Mortality Rate (per 1,000 births)	Life Expectancy (years)
2.2	French, Creole	360	23	122	53
3.1	Spanish, Indian languages	850	56	63	63
1.7	English, Creole	1,080	76	16	76
2.4	Spanish, Indian languages	1,820	81	50	68
3.3	Spanish	830	87	69	62
2.2	Spanish	2,240	90	25	72
2.8	Spanish, Guaraní	1,180	84	42	67
2.4	Spanish, Quéchua	1,440	72	76	65
1.3	English	2,770	80	39.7	68
2.2	English, French patois	1,540	78	21.5	71
1.9	English	1,100	85	24.7	72
2.0	Dutch, Surinamese, English	2,450	80	40	68
2.0	English, Hindi	3,350	95	13.7	70
0.8	Spanish	2,470	94	22.3	71
2.3	Spanish	3,170	88	33	70
2.8	Catalan, French, Spanish	*	100	*	*
0.1	German	15,560	98	8.1	75
0.2	Flemish, French	14,550	98	9.2	74
0.0	Danish	18,470	100	7.8	75
0.3	Finnish, Swedish	18,610	100	5.9	75
0.4	French	16,080	99	7.5	77
0.0	German	*	99	7.8	75
0.2	Greek	4,790	95	11.0	77
1.1	Icelandic	20,160	99	6.2	78
0.6	Irish (Gaelic), English	7,480	99	9.7	74
0.1	Italian	13,320	97	9.5	75
0.89	German	15,000	100	6	70
0.2	Luxembourgish, French, German	22,600	100	8.7	75
0.8	Maltese, English	5,050	83	8.0	75
0.93	French	*	99	*	*
0.4	Dutch	14,530	99	7.6	77
0.3	Norwegian	20,020	100	8.4	76
0.2	Portuguese	3,670	80	14.9	74

Country	Capital	Land Area (square miles)	Population (millions)	Population Density (per square mile)
San Marino	San Marino	23	0.023	975
Spain	Madrid	194,900	39.4	202
Sweden	Stockholm	173,730	8.5	49
Switzerland	Bern	15,940	6.7	420
United Kingdom	London	94,530	57.4	607
Vatican City State	Vatican City	0.17	752 (actual)	5,882

Eastern Europe

Country	Capital	Land Area (square miles)	Population (millions)	Population Density (per square mile)
Albania	Tiranë	11,100	3.3	295
Bulgaria	Sofia	42,820	8.9	209
Czechoslovakia	Prague	49,370	15.7	318
Hungary	Budapest	35,920	10.6	294
Poland	Warsaw	120,730	37.8	313
Romania	Bucharest	91,700	23.3	254
Yugoslavia	Belgrade	98,760	23.7	241

Northern Eurasia

Country	Capital	Land Area (square miles)	Population (millions)	Population Density (per square mile)
Russia	Moscow	6,592,800	147.4	22

■ Other countries are identified on the map on page 406. See an almanac for data.

The Middle East and North Africa

Country	Capital	Land Area (square miles)	Population (millions)	Population Density (per square mile)
Algeria	Algiers	919,590	25.6	28
Bahrain	Manamah	240	0.5	2,172
Cyprus	Nicosia	3,570	0.7	197
Egypt	Cairo	386,660	54.7	141
Iran	Tehran	636,290	55.6	87
Iraq	Baghdad	167,920	18.8	112
Israel	Jerusalem	8,020	4.6	572
Jordan	Amman	37,740	4	109
Kuwait	Kuwait	6,880	2.1	311
Lebanon	Beirut	4,020	3.3	832
Libya	Tripoli	679,360	4.2	6
Morocco	Rabat	172,410	25.6	149

Growth Rate (percent)	Language(s)	Per Capita GNP (U.S. dollars)	Literacy Rate (percent)	Infant Mortality Rate (per 1,000 births)	Life Expectancy (years)
0.1	Italian	8,250	97	10	*
0.3	Spanish, Basque, Catalan, Galician	7,740	97	9.0	77
0.2	Swedish	19,150	99	5.8	77
0.3	German, French, Italian, Romansch	27,260	99	6.8	77
0.2	English, Welsh, Gaelic	12,800	99	9.5	75
0.4	Italian, Latin	*	100	*	
2.0	Albanian	930	75	28	71
0.1	Bulgarian	*	95	13.5	72
0.2	Czech, Slovak, Hungarian	*	100	11.9	71
-0.2	Hungarian	2,460	98	15.8	70
0.6	Polish	1,850	98	16.2	71
0.5	Romanian, Hungarian, German	*	96	25.6	70
0.6	Serbo-Croatian, Slovene, Macedonian	2,680	85	24.5	71
.23	Russian	*	*	17.8	70
3.1	Arabic, French	2,450	52	74	60
2.3	Arabic, English, French	6,610	40	24	67
1.0	Greek, Turkish, English	6,260	99	11	76
2.9	Arabic	650	43	90	60
3.6	Farsi, Kurdish, Arabic	*	48	91	63
3.9	Arabic, Kurdish	*	50	67	67
1.6	Hebrew, Arabic, English	8,650	92	10	75
3.5	Arabic, English	1,500	75	54	69
2.5	Arabic, English	13,680	71	15.6	73
2.1	Arabic, French, English	*	75	49	68
3.1	Arabic	5,410	50	69	66
2.6	Arabic, French, Spanish	750	28	82	61

Country	Capital	Land Area (square miles)	Population (millions)	Population Density (per square mile)
Oman	Muscat	82,030	1.5	18
Qatar	Doha	4,250	0.5	116
Saudi Arabia	Riyadh	830,000	15	18
Syria	Damascus	71,500	12.6	176
Tunisia	Tunis	63,170	8.1	129
Turkey	Ankara	301,380	56.7	188
United Arab Emirates	Abu Dhabi	32,280	1.6	49
Yemen	San'a	203,850	9.8	46

Africa South of the Sahara

Country	Capital	Land Area (square miles)	Population (millions)	Population Density (per square mile)
Angola	Luanda	481,350	8.5	18
Benin	Porto-Novo	43,480	4.7	109
Bophuthatswana	Mmabatho	15,573	1.3	83.5
Botswana	Gaborone	231,800	1.2	5
Burkina Faso	Ouagadougou	105,870	9.1	86
Burundi	Bujumbura	10,750	5.6	525
Cameroon	Yaoundé	183,570	11.1	60
Cape Verde	Cidade de Praia	1,560	0.4	244
Central African Republic	Bangui	240,530	2.9	12
Chad	N'Djamena	495,750	5	10
Comoro Islands	Moroni	690	0.5	663
Congo	Brazzaville	132,050	2.2	17
Côte d'Ivoire	Abidjan	124,500	12.6	101
Djibouti	Djibouti	8,490	0.4	48
Equatorial Guinea	Malabo	10,830	0.4	34
Ethiopia	Addis Ababa	471,780	51.7	110
Gabon	Libreville	103,350	1.2	11
Gambia	Banjul	4,360	0.9	197
Ghana	Accra	92,100	15	163
Guinea	Conakry	94,930	7.3	77
Guinea-Bissau	Bissau	13,950	1	71
Kenya	Nairobi	224,960	24.6	110
Lesotho	Maseru	11,720	1.8	151
Liberia	Monrovia	43,000	2.6	61

Growth Rate (percent)	Language(s)	Per Capita GNP (U.S. dollars)	Literacy Rate (percent)	Infant Mortality Rate (per 1,000 births)	Life Expectancy (years)
3.3	Arabic	5,070	20	100	55
2.3	Arabic	11,610	60	25	69
3.4	Arabic	6,170	52	71	63
3.8	Arabic	1,670	45	48	65
2.0	Arabic, French	1,230	64	59	65
2.1	Turkish	1,280	80	74	64
1.9	Arabic	15,720	56	26	71
3.4	Arabic	540	30	120	50
2.7	Bantu, Portuguese	*	20	137	45
3.2	French, Fon, Yoruba	340	11	110	47
2.8	Setswana, English, Afrikaans	*	*	*	*
2.9	English, Setswana	1,050	54	64	59
3.2	French, Sudanic languages	230	13	126	51
3.2	Kirundi, French	230	23	114	51
2.6	French, English, 24 major African language groups	1,010	55	125	50
2.8	Portuguese	*	37	66	61
2.5	French, Sango	390	33	143	46
2.5	French, Arabic, Sara, Sango	160	20	132	46
3.4	French, Arabic	440	15	94	55
3.0	French, Lingala, Kokongo	930	56	113	53
3.7	French, over 60 African languages	740	35	96	53
3.0	Arabic, French, Afar, Somali	*	9	122	47
2.6	Spanish, Fang, English	350	31	120	50
2.0	Amharic, Tigre, Galla, Arabic	120	15	154	41
2.2	French, Bantu languages	2,970	65	103	52
2.6	English, Mandinka, Wolof, Fula	220	20	143	43
3.1	English, Akan, Ga Moshi-Dagomba, Ewe	400	30	86	55
2.5	French, Malinké, Susu, Fulani	350	38	147	42
2.1	Portuguese	160	9	132	45
3.8	Swahili, English	360	59	62	63
2.8	English, Sesotho	410	55	100	56
3.2	English, Niger-Congo languages	450	35	83	56

Country	Capital	Land Area (square miles)	Population (millions)	Population Density (per square mile)
Madagascar	Antananarivo	226,660	12	53
Malawi	Lilongwe	45,750	9.2	200
Mali	Bamako	478,760	8.1	17
Mauritania	Nouakchott	397,950	2.0	5
Mauritius	Port Louis	790	1.1	1,354
Mozambique	Maputo	309,490	15.7	51
Niger	Niamey	489,190	7.9	16
Nigeria	Lagos (Abuja as of 1990)	356,670	118.8	333
Rwanda	Kigali	10,170	7.3	715
Sao Tome and Principe	São Tomé	370	0.1	337
Senegal	Dakar	75,750	7.4	97
Seychelles	Victoria	110	0.1	629
Sierra Leone	Freetown	27,700	4.2	150
Somalia	Mogadishu	246,200	8.4	34
South Africa	Pretoria	471,440	39.6	84
Sudan	Khartoum	967,490	25.2	26
Swaziland	Mbabane	6,700	0.8	116
Tanzania	Dar es Salaam	364,900	26.0	71
Togo	Lomé	21,930	3.7	168
Transkei	Umtala	15,827	3.6	152
Uganda	Kampala	91,140	18.0	197
Venda	Thohoyandou	2,509	0.4	215
Zaire	Kinshasa	905,560	36.6	40
Zambia	Lusaka	290,580	8.1	28
Zimbabwe	Harare	150,800	9.7	64

South Asia

Country	Capital	Land Area (square miles)	Population (millions)	Population Density (per square mile)
Afghanistan	Kabul	250,000	15.9	63
Bangladesh	Dacca	55,600	114.8	2,064
Bhutan	Thimphu	18,150	1.6	86
India	New Delhi	1,269,340	853.4	672
Maldives	Malé	120	0.2	1,882
Nepal	Katmandu	54,360	19.1	352

Growth Rate (percent)	Language(s)	Per Capita GNP (U.S. dollars)	Literacy Rate (percent)	Infant Mortality Rate (per 1,000 births)	Life Expectancy (years)
3.2	Malagasy, French	180	53	120	54
3.4	English, Chichewa	160	25	130	49
3.0	French, Bambara	230	10	117	45
2.7	Arabic, French	480	17	127	46
1.3	English, French, Creole, Hindi, Urdu	1,810	83	25.2	68
2.7	Portuguese, Bantu languages	100	17	141	47
3.0	French, Hausa, Djerma	310	10	135	45
2.9	English, Hausa, Yoruba, Ibo, Fulani	290	42	121	48
3.4	Kinyarwanda, French	310	49	122	49
2.7	Portuguese	280	54	61.7	65
2.7	French, Wolof, Pulaar, Diola, Mandingo	630	23	128	46
1.7	Creole, English, French	3,800	65	17	70
2.5	English, Mende, Temne, Creole	240	24	154	41
3.1	Somali, Arabic, English, Italian	170	60	132	45
2.7	English, Afrikaans, Bantu languages	2,290	99 (White) 32 (Black)	55	63
2.9	Arabic	340	20	108	50
3.1	English, Swazi	790	68	130	50
3.7	Swahili, Arabic, English	160	85	106	53
3.6	Ewe, Mina, Kabye, Dagomba, French	370	18	114	55
2.2	English, Xhosa, Southern Sotho	86	*	*	*
3.6	English, Swahili, Luganda, Ateso, Luo	280	52	107	49
2.4	Venda, English, Afrikaans	312	*	*	*
3.3	French, English, Kingwana, Tshiluba, Swahili, Lingala, Kikongo	170	40 (male) 15 (female)	108	53
3.8	English, Bantu languages	290	55.5	80	53
3.2	English, Ndebele, Shona	660	77	72	58
2.6	Pashtu, Afghan Persian (Dari), Uzbec, Turkmen	*	12	182	41
2.5	Bengali, English	170	25	120	54
2.1	Dzongkha	150	15	128	48
2.1	Hindi, English, and many other languages	330	36	95	57
3.7	Divehi	410	81	76	61
2.5	Nepali, Newari, Bhutia	170	23	112	52

Country	Capital	Land Area (square miles)	Population (millions)	Population Density (per square mile)
Pakistan	Islamabad	310,400	114.6	369
Sri Lanka	Colombo	25,330	17.2	679

East Asia

Country	Capital	Land Area (square miles)	Population (millions)	Population Density (per square mile)
Brunei	Bandar Seri Begawan	2,230	0.3	115
Cambodia	Pnompenh	69,900	7	100
China, People's Republic of	Beijing	3,705,390	1,119.9	302
Taiwan (Republic of China)	Taipei	12,460	20.2	1,623
Indonesia	Jakarta	735,360	189.4	258
Japan	Tokyo	143,750	123.6	860
Laos	Vientiane	91,430	4	44
Malaysia	Kuala Lumpur	127,320	17.9	140
Mongolia	Ulan Bator	604,250	2.2	4
Myanmar	Rangoon	261,220	41.3	158
North Korea	P'yongyang	46,540	21.3	458
Papua New Guinea	Port Moresby	178,260	4	23
Philippines	Manila	115,830	66.1	571
Singapore	Singapore	220	2.7	12,177
South Korea	Seoul	38,020	42.8	1,125
Thailand	Bangkok	198,460	55.7	281
Vietnam	Hanoi	127,240	70.2	552

The Pacific World

Country	Capital	Land Area (square miles)	Population (millions)	Population Density (per square mile)
Australia	Canberra	2,967,900	17.1	6
Fiji	Suva	7,050	0.8	108
Kiribati	Tarawa	277	0.067	235
Nauru	Yaren	8	0.009	976
New Zealand	Wellington	103,740	3.3	32
Solomon Islands	Honiara	10,980	0.3	30
Tonga	Nuku'alofa	259	0.1	369
Tuvalu	Funafuti	10	0.008	700
Vanuatu	Port-Vila	5,7000	0.2	29
Western Samoa	Apia	1,100	0.2	155

Growth Rate (percent)	Language(s)	Per Capita GNP (U.S. dollars)	Literacy Rate (percent)	Infant Mortality Rate (per 1,000 births)	Life Expectancy (years)
3.0	Urdu, English, Punjabi, Sindhi, Pashtu, Baluchi	350	26	110	56
1.5	Sinhala, Tamil, English	420	87	22.5	70
2.5	Malay, Chinese, English	14,120	45	11	71
2.2	Khmer, French, Vietnamese, Chinese	*	48	128	49
1.4	Chinese	330	76.5	37	68
1.2	Chinese	*	92	17	74
1.8	Indonesian, Dutch, English, Japanese	430	64	89	59
0.4	Japanese	21,040	99	4.8	79
2.5	Lao, French, English	180	28	110	47
2.5	Malay	1,870	75	30	68
2.8	Mongolian	*	90	50	65
2.0	Burmese	*	78	97	55
2.1	Korean	*	95	33	70
2.7	English, Melanesian Pidgin, Motu	770	32	59	54
2.6	Filipino, English	630	88	48	64
1.5	Malay, Chinese (Mandarin), Tamil, English	9,100	86	6.9	73
1.0	Korean	3,530	95	30	68
1.5	Thai, Chinese, English	1,000	85.5	39	66
2.5	Vietnamese	*	78	50	66
0.8	English	12,390	99	8.7	76
2.2	Fijian, Hindustani, English	1,540	86	21	63
1.74	English	480	90	*	*
1.68	Nauruan, English	20,000	99	*	*
0.8	English, Maori	9,620	99.5	10	74
3.5	English, Melanesian dialects	430	60	40	61
0.84	Tongan, English	1,030	95	6	58
7.74	Tuvaluan, English	450	50	42	59
3.2	English, French	820	*	36	69
2.8	Samoan, English	580	90	48	66

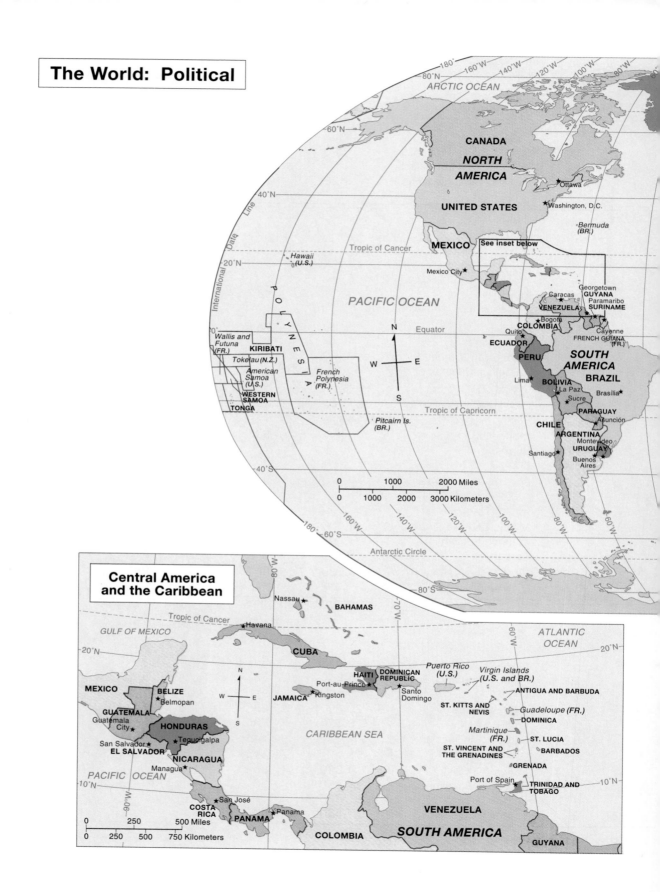

The World: Political

ARCTIC OCEAN

NORTH AMERICA

CANADA

★Ottawa

UNITED STATES

★Washington, D.C.

Bermuda (BR.)

Tropic of Cancer

MEXICO

See inset below

Hawaii (U.S.)

Mexico City★

PACIFIC OCEAN

Caracas★ Georgetown

VENEZUELA★ GUYANA
Paramaribo
SURINAME★

★Bogotá

COLOMBIA

Cayenne
FRENCH GUIANA (FR.)

Equator

Quito★

ECUADOR

PERU

POLYNESIA

Wallis and Futuna (FR.)

KIRIBATI

Tokelau (N.Z.)

American Samoa (U.S.)

French Polynesia (FR.)

WESTERN SAMOA

TONGA

SOUTH AMERICA

BRAZIL

Lima★

BOLIVIA★
La Paz★
Sucre★

Brasília★

PARAGUAY
Asunción★

Tropic of Capricorn

Pitcairn Is. (BR.)

CHILE

ARGENTINA

Montevideo★

URUGUAY

Santiago★

Buenos Aires★

N

W E

S

0	1000	2000 Miles
0	1000	2000	3000 Kilometers

Antarctic Circle

80°S

Central America and the Caribbean

Nassau★

BAHAMAS

Tropic of Cancer

GULF OF MEXICO

★Havana

ATLANTIC OCEAN

CUBA

Puerto Rico (U.S.)

Virgin Islands (U.S. and BR.)

MEXICO

BELIZE
★Belmopan

HAITI DOMINICAN REPUBLIC

Port-au-Prince★ ★Santo Domingo

ANTIGUA AND BARBUDA

GUATEMALA
Guatemala City★

JAMAICA
★Kingston

ST. KITTS AND NEVIS

Guadeloupe (FR.)

DOMINICA

HONDURAS
★Tegucigalpa

Martinique (FR.)

ST. LUCIA

San Salvador★

EL SALVADOR

CARIBBEAN SEA

ST. VINCENT AND THE GRENADINES

BARBADOS

NICARAGUA

GRENADA

PACIFIC OCEAN

Managua★

Port of Spain★

TRINIDAD AND TOBAGO

San José★

COSTA RICA

★Panama

PANAMA

VENEZUELA

N

W E

S

COLOMBIA

SOUTH AMERICA

GUYANA

0	250	500 Miles
0	250	500	750 Kilometers

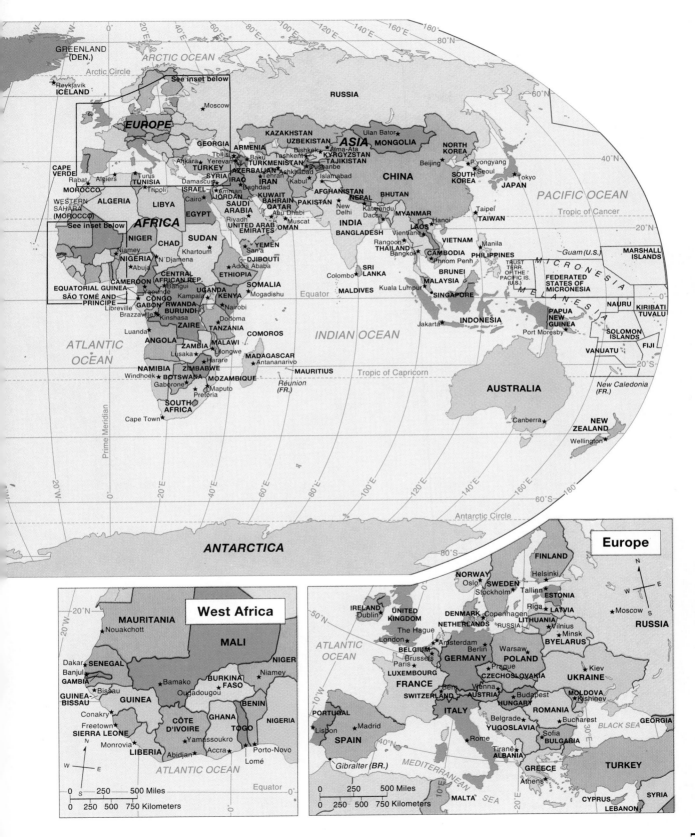

GREENLAND
(DEN.)

ARCTIC OCEAN

Arctic Circle

★Reykjavik
ICELAND

★Moscow

RUSSIA   60°N

EUROPE

See inset below

KAZAKHSTAN

UZBEKISTAN   ASIA   MONGOLIA   Ulan Bator★

GEORGIA   ARMENIA
Tbilisi★ ★Baku Tashkent★ KYRGYZSTAN   NORTH   40°N
★Ankara ★Yerevan   KOREA
TURKEY AZERBAIJAN Ashkhabad★Dushanbe Beijing★ ★P'yongyang
         TURKMENISTAN   ★Seoul
CAPE   Tunis★ Damascus★ SYRIA IRAN Baghdad★ Islamabad★ CHINA SOUTH   Tokyo★
VERDE ★Algiers ISRAEL IRAQ Tehran★ Kabul★ New   KOREA JAPAN
Rabat★ ★Algiers ★Cairo JORDAN Amman★ KUWAIT BAHRAIN Delhi★ AFGHANISTAN PAKISTAN NEPAL BHUTAN PACIFIC OCEAN
MOROCCO Tripoli★   QATAR   Katmandu★   Tropic of Cancer
WESTERN   Riyadh★ Abu Dhabi OMAN INDIA MYANMAR TAIWAN 20°N
SAHARA ALGERIA LIBYA EGYPT SAUDI   UNITED ARAB Muscat★ BANGLADESH Taipei★
(MOROCCO)   ARABIA   EMIRATES YEMEN Rangoon Hanoi★
See inset below AFRICA   San'a★ LAOS Vientiane★
                                    Addis Ababa★ DJIBOUTI THAILAND VIETNAM ★Manila
NIGER CHAD SUDAN   ETHIOPIA Colombo Bangkok★ CAMBODIA PHILIPPINES Guam (U.S.) MARSHALL
NIGERIA Khartoum★   SRI MALDIVES Phnom Penh★ BRUNEI   ISLANDS
Niamey★ ★N'Djamena CENTRAL   SOMALIA LANKA Kuala Lumpur MALAYSIA FEDERATED MICRONESIA
★Abuja AFRICAN REP. Mogadishu★   SINGAPORE STATES OF
EQUATORIAL GUINEA Bangui★ UGANDA KENYA Equator   MICRONESIA 0°
SÃO TOMÉ AND CAMEROON Kampala★   INDONESIA PAPUA NAURU KIRIBATI
PRINCIPE GABON CONGO RWANDA ★Nairobi Jakarta★ NEW TUVALU
Libreville★ Brazzaville★ Kinshasa★ BURUNDI Dodoma★   GUINEA
ATLANTIC Luanda★ ZAIRE TANZANIA COMOROS INDIAN OCEAN Port Moresby★ SOLOMON FIJI
OCEAN ANGOLA ZAMBIA MALAWI   ISLANDS
              Lusaka★ Ilongwe★ MADAGASCAR VANUATU
NAMIBIA ZIMBABWE Harare★ Antananarivo★ MAURITIUS New Caledonia
Windhoek★ BOTSWANA MOZAMBIQUE Réunion Tropic of Capricorn (FR.) 20°S
Gaborone★ ★Maputo (FR.)   AUSTRALIA
SOUTH Pretoria★
AFRICA
Cape Town★   Canberra★
                              NEW
                              ZEALAND
                              Wellington★

ANTARCTICA   60°S

Antarctic Circle   180

80°S

West Africa

20°N   MAURITANIA
★Nouakchott
MALI
Dakar★ SENEGAL NIGER
Banjul★ BURKINA Niamey★
GAMBIA Bamako★ FASO
GUINEA- GUINEA Ougadougou★ BENIN
BISSAU ★Bissau   NIGERIA
Conakry★ GHANA
Freetown★ CÔTE TOGO
SIERRA LEONE D'IVOIRE Accra★ Porto-Novo★
Monrovia★ Yamassoukro★ Lomé★
LIBERIA Abidjan★
ATLANTIC OCEAN Equator

0 250 500 Miles
0 250 500 750 Kilometers

Europe

FINLAND
NORWAY SWEDEN Helsinki★
Oslo★ ★Stockholm Tallinn★
IRELAND UNITED DENMARK ESTONIA
Dublin★ KINGDOM Copenhagen★ Riga★ LATVIA ★Moscow
The Hague★ NETHERLANDS RUSSIA Vilnius★ RUSSIA
London★   Berlin LITHUANIA Minsk★
ATLANTIC BELGIUM Amsterdam★ Warsaw★ BYELARUS
OCEAN Brussels★ GERMANY POLAND Kiev★
LUXEMBOURG Prague★ CZECHOSLOVAKIA UKRAINE
FRANCE Bern★ Vienna★ Budapest★ MOLDOVA
SWITZERLAND AUSTRIA HUNGARY Kishinev★
PORTUGAL ITALY Belgrade★ ROMANIA GEORGIA
Lisbon★ ★Madrid Rome★ YUGOSLAVIA Bucharest★ BLACK SEA
SPAIN   Sofia★ BULGARIA
Gibralter (BR.) Tiranë★ TURKEY
                ALBANIA
MEDITERRANEAN GREECE
MALTA SEA Athens★ CYPRUS SYRIA
                      LEBANON

0 250 500 Miles
0 250 500 750 Kilometers

763

ARCTIC OCEAN
80°N
BEAUFORT
SEA
60°N
BERING
SEA
YUKON R.
HUDSON
BAY
NORTH
AMERICA
CANADIAN SHIELD
Aleutian Islands
ROCKY MOUNTAINS
GREAT PLAINS
Missouri R.
Great
Lakes
St. Lawrence R.
40°N
Mackenzie R.
Colorado R.
APPALACHIAN
MTS.
ATLANTIC
OCEAN
Hawaiian
Islands
Tropic of Cancer
SIERRA MADRE
ORIENTAL
Rio Grande
SIERRA MADRE
OCCIDENTAL
GULF OF
MEXICO
West Indies
20°N
PACIFIC
OCEAN
CARIBBEAN
SEA
GUIANA
HIGHLANDS
AMAZON
Amazon R.
0°
Equator
Polynesia
BASIN
SOUTH
AMERICA
0   1000   2000 Miles
0 1000 2000 3000 Kilometers
BRAZILIAN
HIGHLANDS
20°S
Tropic of Capricorn
ANDES MOUNTAINS
N
W   E
S
PAMPAS
Rio de
la Plata
40°S
PATAGONIA
Cape
Horn
60°S
Drake Passage
ANTARCTIC
PENINSULA
Antarctic Circle
80°S

The World:
Physical

Elevation

Feet		Meters
10,000		3,000
6,600		2,000
3,000		1,000
700		200
0		0
Below Sea Level		Below Sea Level

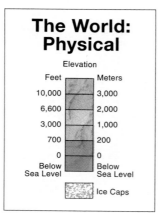

Ice Caps

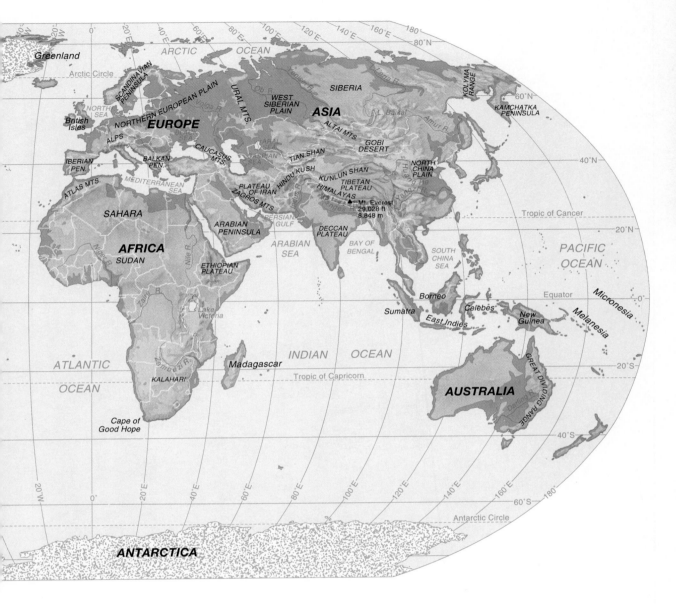

Greenland

ARCTIC OCEAN

Arctic Circle

SCANDINAVIAN PENINSULA

NORTH SEA

British Isles

NORTHERN EUROPEAN PLAIN

EUROPE

URAL MTS.

Volga R.

Ob R.

WEST SIBERIAN PLAIN

SIBERIA

ASIA

KOLYMA RANGE

KAMCHATKA PENINSULA

60°N

40°N

ALPS

IBERIAN PEN.

BALKAN PEN.

CAUCASUS MTS.

BLACK SEA

ALTAI MTS.

GOBI DESERT

TIAN SHAN

NORTH CHINA PLAIN

Amur R.

L. Baikal

ATLAS MTS.

MEDITERRANEAN SEA

HINDU KUSH

PLATEAU OF IRAN

ZAGROS MTS

CASPIAN SEA

KUNLUN SHAN

TIBETAN PLATEAU

HIMALAYAS

Huang He

Mt. Everest 29,028 ft 8,848 m

Tropic of Cancer

20°N

SAHARA

AFRICA

SUDAN

Niger R.

ARABIAN PENINSULA

PERSIAN GULF

ARABIAN SEA

DECCAN PLATEAU

BAY OF BENGAL

SOUTH CHINA SEA

PACIFIC OCEAN

Nile R.

ETHIOPIAN PLATEAU

Micronesia

Equator

0°

Zaire R.

Lake Victoria

Borneo

Sumatra

East Indies

Celebes

New Guinea

Melanesia

ATLANTIC OCEAN

Zambezi R.

Madagascar

INDIAN OCEAN

Tropic of Capricorn

20°S

KALAHARI

AUSTRALIA

GREAT DIVIDING RANGE

Darling R.

Cape of Good Hope

40°S

20°W

0°

20°E

40°E

60°E

80°E

100°E

120°E

140°E

160°E

180°

60°S

Antarctic Circle

ANTARCTICA

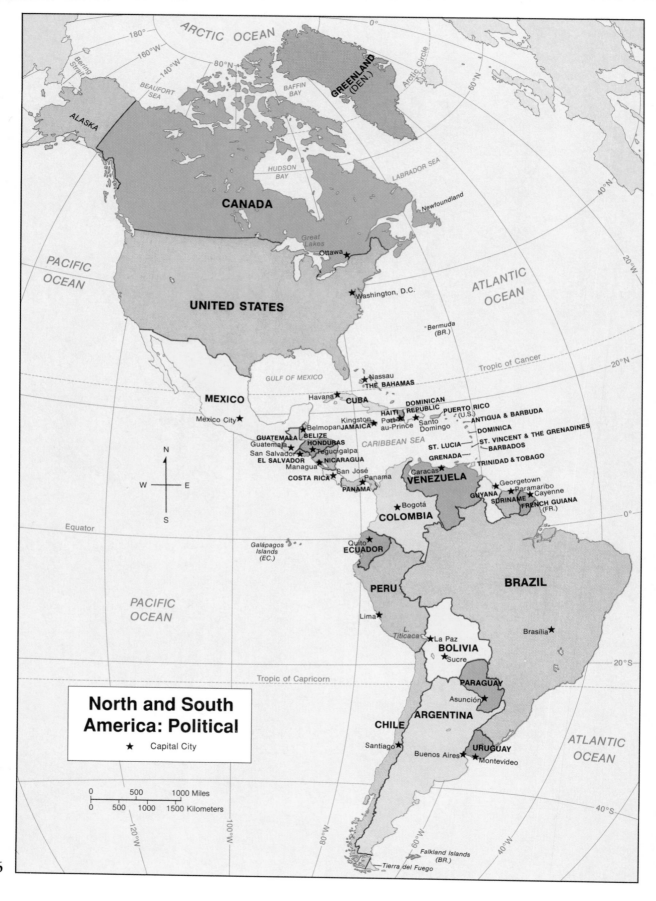

North and South America: Political

★ Capital City

| 0 | 500 | 1000 Miles |
| 0 | 500 | 1000 | 1500 Kilometers |

ARCTIC OCEAN

GREENLAND (DEN.)

ALASKA

BEAUFORT SEA

BAFFIN BAY

CANADA

HUDSON BAY

LABRADOR SEA

Great Lakes

Newfoundland

Ottawa ★

Washington, D.C. ★

PACIFIC OCEAN

ATLANTIC OCEAN

UNITED STATES

Bermuda (BR.)

Tropic of Cancer

GULF OF MEXICO

Nassau ★
THE BAHAMAS

MEXICO

Havana ★ CUBA

DOMINICAN REPUBLIC

Mexico City ★

Kingston
Belmopan JAMAICA
GUATEMALA BELIZE
Guatemala HONDURAS
San Salvador Tegucigalpa
EL SALVADOR NICARAGUA
Managua
San José
COSTA RICA
Panama
PANAMA

HAITI
Port-
au-Prince
Santo
Domingo

PUERTO RICO (U.S.)
ANTIGUA & BARBUDA
DOMINICA
ST. LUCIA
ST. VINCENT & THE GRENADINES
BARBADOS
GRENADA
TRINIDAD & TOBAGO

CARIBBEAN SEA

Caracas ★
VENEZUELA

Georgetown
Paramaribo
GUYANA
SURINAME Cayenne
FRENCH GUIANA (FR.)

Bogotá ★
COLOMBIA

Galápagos Islands (EC.)

Quito ★
ECUADOR

Equator

PERU

BRAZIL

PACIFIC OCEAN

Lima ★

L. Titicaca

La Paz ★
BOLIVIA
Sucre ★

Brasília ★

Tropic of Capricorn

PARAGUAY

Asunción ★

CHILE

ARGENTINA

Santiago ★

Buenos Aires ★

URUGUAY
Montevideo ★

ATLANTIC OCEAN

Falkland Islands (BR.)

Tierra del Fuego

N
W   E
S

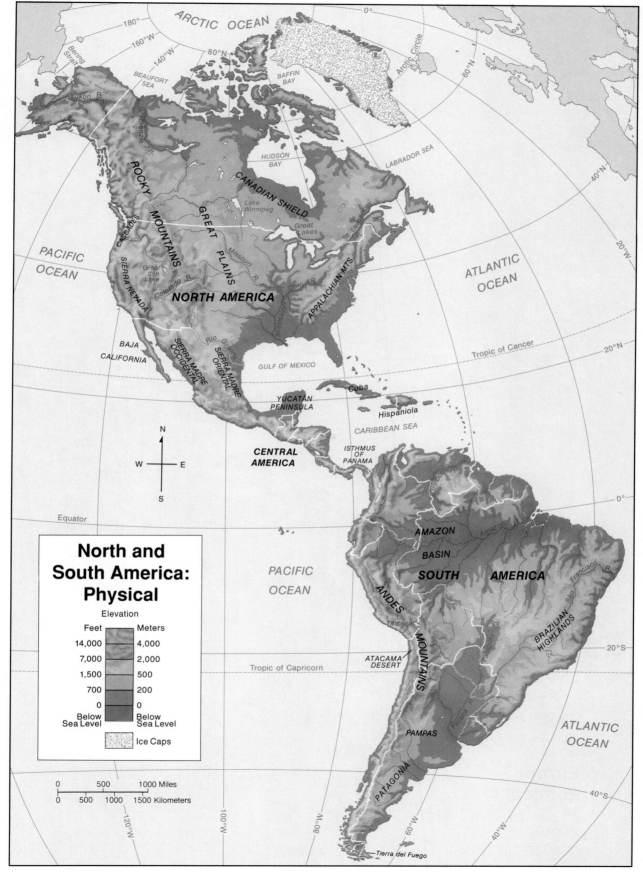

North and South America: Physical

Elevation

Feet		Meters
14,000		4,000
7,000		2,000
1,500		500
700		200
0		0
Below Sea Level		Below Sea Level

Ice Caps

0   500   1000 Miles
0   500   1000   1500 Kilometers

ARCTIC OCEAN

180°
160°W
140°W
80°N
BEAUFORT SEA
BAFFIN BAY
Arctic Circle
60°N
0°
20°W

Yukon R.

BERING STRAIT

Mackenzie R.

HUDSON BAY

LABRADOR SEA

40°N

ROCKY MOUNTAINS

CANADIAN SHIELD

Lake Winnipeg

Great Lakes

St. Lawrence R.

CASCADES

GREAT PLAINS

Missouri R.

Ohio R.

APPALACHIAN MTS.

SIERRA NEVADA

Great Salt Lake

Colorado R.

Mississippi R.

NORTH AMERICA

ATLANTIC OCEAN

PACIFIC OCEAN

BAJA CALIFORNIA

SIERRA MADRE OCCIDENTAL

SIERRA MADRE ORIENTAL

Rio Grande

GULF OF MEXICO

Tropic of Cancer
20°N

YUCATÁN PENINSULA

Cuba

Hispaniola

CARIBBEAN SEA

N
W   E
S

CENTRAL AMERICA

ISTHMUS OF PANAMA

Orinoco R.

0°

Equator

AMAZON BASIN

Amazon R.

SOUTH   AMERICA

São Francisco R.

PACIFIC OCEAN

ANDES

Titicaca

MOUNTAINS

BRAZILIAN HIGHLANDS

20°S

ATACAMA DESERT

Tropic of Capricorn

Paraguay R.

Paraná R.

PAMPAS

ATLANTIC OCEAN

PATAGONIA

40°S

Tierra del Fuego

100°W
120°W
80°W
60°W
40°W

Europe: Political

★ Capital City

● Other City

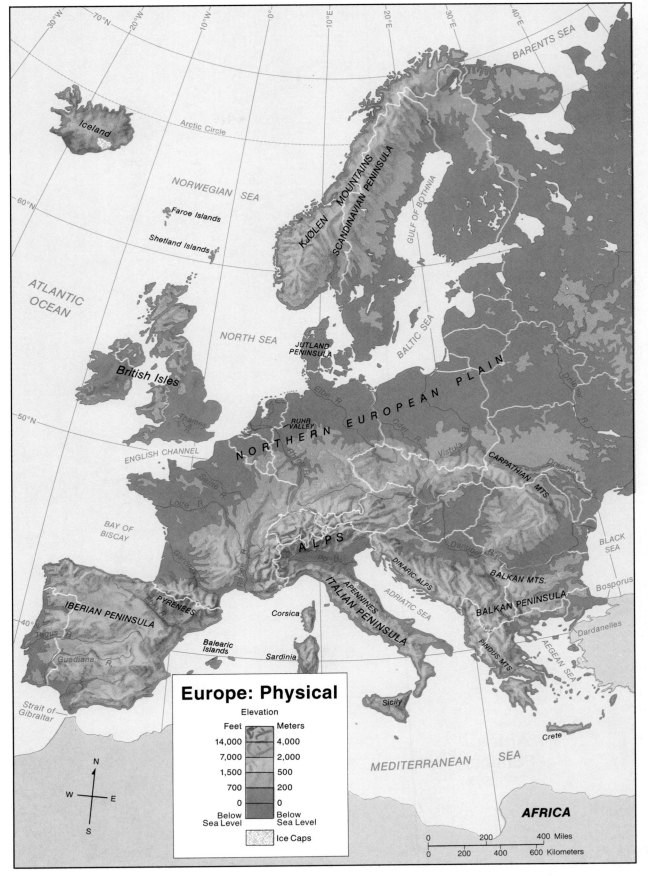

Europe: Physical

Elevation

Feet		Meters
14,000		4,000
7,000		2,000
1,500		500
700		200
0		0
Below Sea Level		Below Sea Level

Ice Caps

BARENTS SEA

Iceland

NORWEGIAN SEA

Faroe Islands

Shetland Islands

ATLANTIC OCEAN

British Isles

NORTH SEA

KJØLEN MOUNTAINS

SCANDINAVIAN PENINSULA

GULF OF BOTHNIA

BALTIC SEA

JUTLAND PENINSULA

Elbe R.

Oder R.

NORTHERN EUROPEAN PLAIN

RUHR VALLEY

Rhine R.

Vistula

CARPATHIAN MTS.

Dniester R.

Dnieper R.

ENGLISH CHANNEL

Thames R.

Seine R.

Loire R.

BAY OF BISCAY

Garonne R.

Rhône R.

ALPS

Po R.

DINARIC ALPS

ADRIATIC SEA

BALKAN MTS.

BLACK SEA

Bosporus

Danube R.

PYRENEES

Ebro R.

IBERIAN PENINSULA

Tagus R.

Guadiana R.

Corsica

Balearic Islands

Sardinia

APENNINES

ITALIAN PENINSULA

BALKAN PENINSULA

PINDUS MTS.

AEGEAN SEA

Dardanelles

Sicily

Crete

Strait of Gibraltar

MEDITERRANEAN SEA

AFRICA

N
W E
S

0	200	400 Miles
0	200	400	600 Kilometers

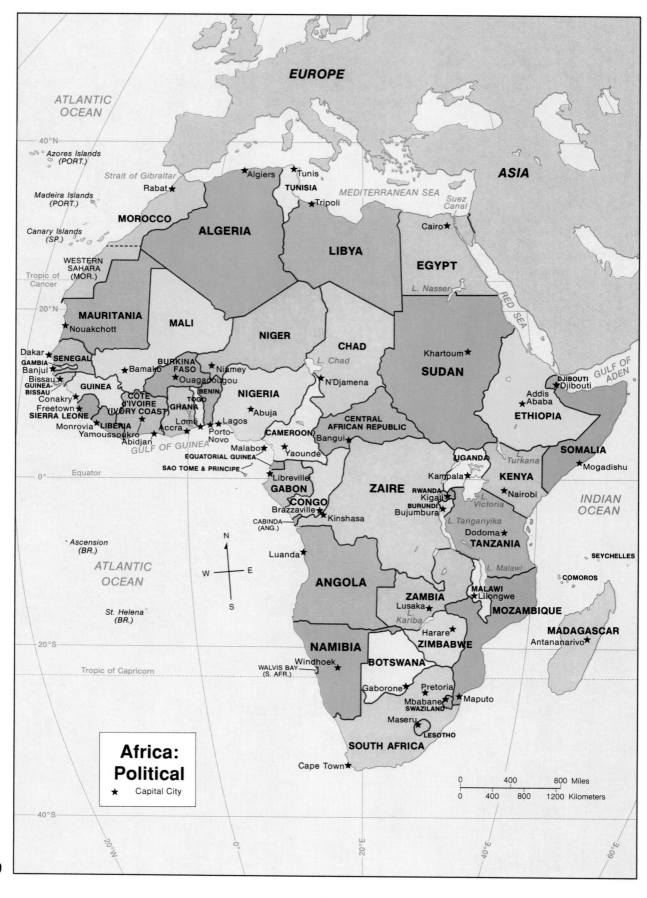

Africa: Political

★ Capital City

770

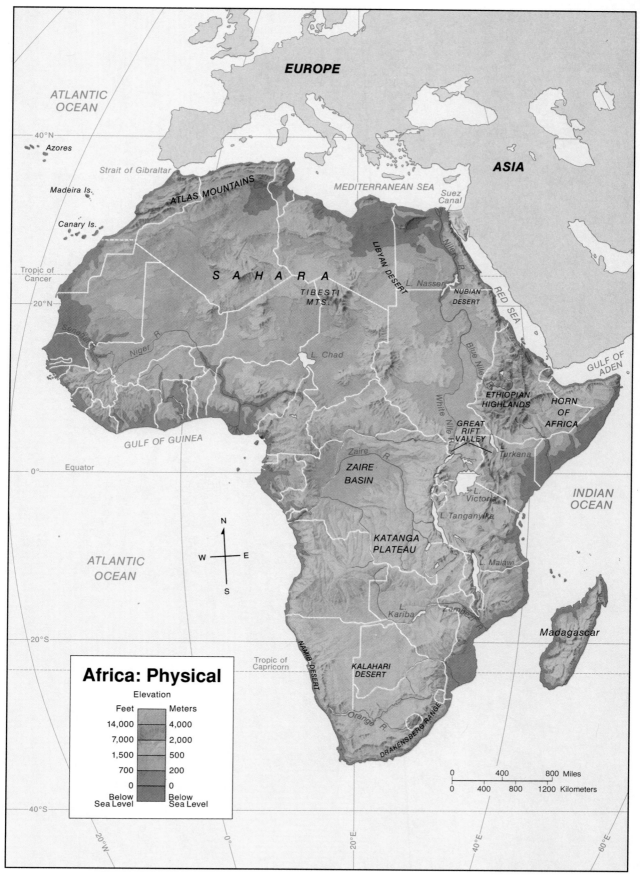

Africa: Physical

Elevation

Feet		Meters
14,000		4,000
7,000		2,000
1,500		500
700		200
0		0
Below Sea Level		Below Sea Level

ATLANTIC OCEAN

EUROPE

ASIA

40°N

Azores

Strait of Gibraltar

MEDITERRANEAN SEA

Suez Canal

Madeira Is.

ATLAS MOUNTAINS

Canary Is.

Tropic of Cancer

S A H A R A

LIBYAN DESERT

L. Nasser

NUBIAN DESERT

RED SEA

20°N

TIBESTI MTS.

Senegal R.

Niger R.

L. Chad

Blue Nile

GULF OF ADEN

GULF OF GUINEA

White Nile

ETHIOPIAN HIGHLANDS

HORN OF AFRICA

Equator 0°

Zaire R.

ZAIRE BASIN

GREAT RIFT VALLEY

L. Turkana

L. Victoria

INDIAN OCEAN

L. Tanganyika

ATLANTIC OCEAN

N
W   E
S

KATANGA PLATEAU

L. Malawi

L. Kariba

Zambezi R.

Madagascar

20°S

Tropic of Capricorn

NAMIB DESERT

KALAHARI DESERT

Orange R.

DRAKENSBERG RANGE

40°S

20°W

0°

20°E

40°E

60°E

0	400	800 Miles
0	400	800	1200 Kilometers

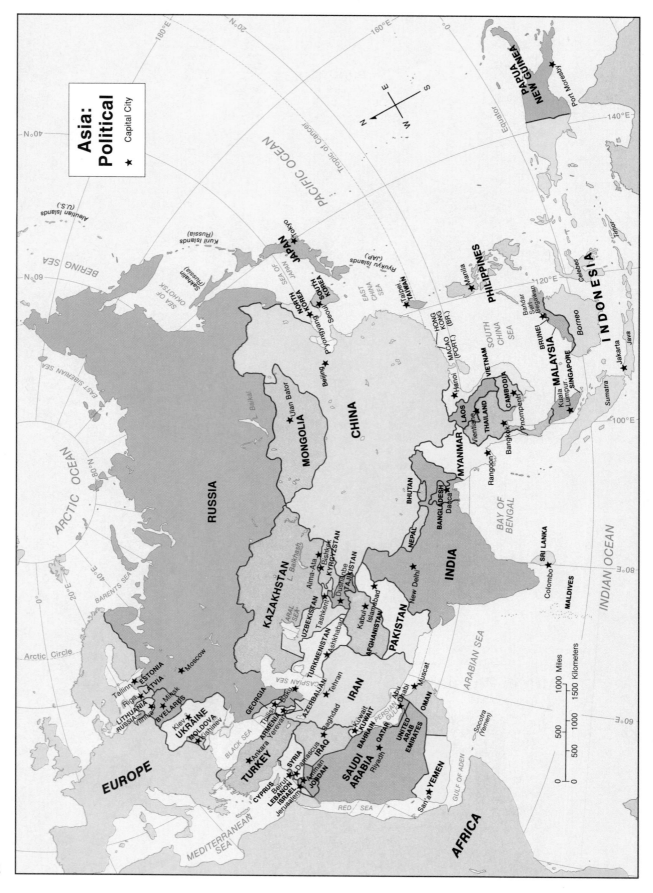

Asia:
Political

★ Capital City

ARCTIC OCEAN

PACIFIC OCEAN

Aleutian Islands (U.S.)

BERING SEA

Kuril Islands (Russia)

Sakhalin (Russia)

SEA OF OKHOTSK

EAST SIBERIAN SEA

RUSSIA

★ Tokyo

JAPAN

SEA OF JAPAN

NORTH KOREA ★ Pyongyang

SOUTH KOREA ★ Seoul

EAST CHINA SEA

Ryukyu Islands (JAP.)

Taipei ★ TAIWAN

PHILIPPINES ★ Manila

PAPUA NEW GUINEA ★ Port Moresby

Equator

★ Beijing

CHINA

★ Ulan Bator

MONGOLIA

L. Baikal

HONG KONG (BR.)

MACAO (PORT.)

SOUTH CHINA SEA

Bandar Seri Begawan ★ BRUNEI

Celebes

INDONESIA

VIETNAM ★ Hanoi

LAOS ★ Vientiane

THAILAND

CAMBODIA ★ Phnompenh

MYANMAR ★ Rangoon

★ Bangkok

MALAYSIA

Kuala Lumpur ★ SINGAPORE

Borneo

Sumatra

Jakarta

Java

Timor

ARCTIC OCEAN

Arctic Circle

BARENTS SEA

BHUTAN

BANGLADESH ★ Dacca

NEPAL

BAY OF BENGAL

INDIA ★ New Delhi

SRI LANKA ★ Colombo

MALDIVES

INDIAN OCEAN

KAZAKHSTAN

ARAL SEA

L. Balkhash

Alma-Ata ★

KYRGYZSTAN ★ Bishkek

UZBEKISTAN ★ Tashkent

TAJIKISTAN ★ Dushanbe

TURKMENISTAN ★ Ashkhabad

Kabul ★ AFGHANISTAN

★ Islamabad PAKISTAN

ESTONIA ★ Tallinn

LATVIA ★ Riga

LITHUANIA ★ Vilnius

RUSSIA

BYELARUS ★ Minsk

★ Moscow

UKRAINE ★ Kiev

MOLDOVA ★ Kishinev

CASPIAN SEA

GEORGIA ★ Tbilisi

ARMENIA ★ Yerevan

AZERBAIJAN ★ Baku

IRAN ★ Tehran

★ Muscat

OMAN

ARABIAN SEA

Socotra (Yemen)

EUROPE

BLACK SEA

TURKEY ★ Ankara

CYPRUS

LEBANON ★ Beirut

SYRIA ★ Damascus

IRAQ ★ Baghdad

JORDAN ★ Amman

ISRAEL ★ Jerusalem

KUWAIT ★ Kuwait

BAHRAIN

QATAR

PERSIAN GULF

★ Abu Dhabi

UNITED ARAB EMIRATES

SAUDI ARABIA ★ Riyadh

YEMEN ★ San'a

GULF OF ADEN

RED SEA

MEDITERRANEAN SEA

AFRICA

1000 Miles

1500 Kilometers

1000

500

500

0

0

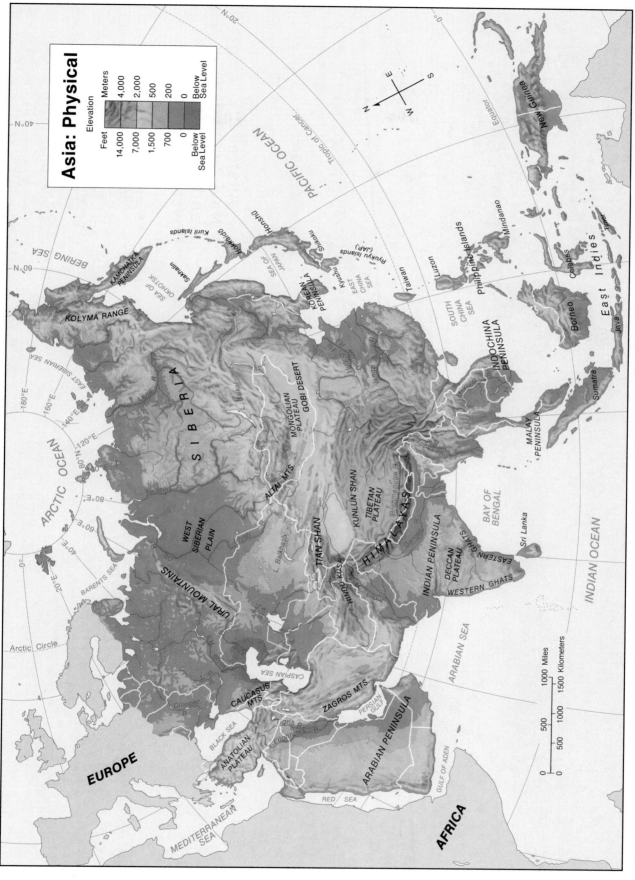

Asia: Physical

Elevation

Meters
4,000
2,000
500
200
0
Below Sea Level

Feet
14,000
7,000
1,500
700
0
Below Sea Level

PACIFIC OCEAN

Tropic of Cancer

Equator

New Guinea

Mindanao

Philippine Islands

Luzon

Taiwan

SOUTH CHINA SEA

EAST CHINA SEA

Ryukyu Islands (JAP.)

Kyushu

Shikoku

Honshu

Hokkaido

Kuril Islands

Sakhalin

SEA OF JAPAN

SEA OF OKHOTSK

KAMCHATKA PENINSULA

BERING SEA

KOLYMA RANGE

EAST SIBERIAN SEA

KOREAN PENINSULA

INDOCHINA PENINSULA

MALAY PENINSULA

Sumatra

Borneo

Celebes

Java

Timor

East Indies

BAY OF BENGAL

Sri Lanka

EASTERN GHATS

WESTERN GHATS

DECCAN PLATEAU

INDIAN PENINSULA

Ganges R.

Brahmaputra R.

HIMALAYAS

Mekong R.

Yangzi

Huang He

Amur R.

Lena R.

GOBI DESERT

MONGOLIAN PLATEAU

L. Baikal

TIBETAN PLATEAU

KUNLUN SHAN

TIAN SHAN

HINDU KUSH

Indus R.

ALTAI MTS.

SIBERIA

Yenisey R.

Ob R.

WEST SIBERIAN PLAIN

L. Balkhash

Irtysh

ARCTIC OCEAN

Arctic Circle

BARENTS SEA

URAL MOUNTAINS

CASPIAN SEA

CAUCASUS MTS.

BLACK SEA

Dnieper R.

Volga R.

EUROPE

MEDITERRANEAN SEA

ANATOLIAN PLATEAU

Tigris R.

Euphrates R.

ZAGROS MTS.

PERSIAN GULF

ARABIAN PENINSULA

RED SEA

GULF OF ADEN

AFRICA

ARABIAN SEA

INDIAN OCEAN

20°N

40°N

60°N

80°N

0°

20°E

40°E

60°E

80°E

100°E

120°E

140°E

160°E

180°E

20°N

0°

0   500   1000 Miles
0   500   1000   1500 Kilometers

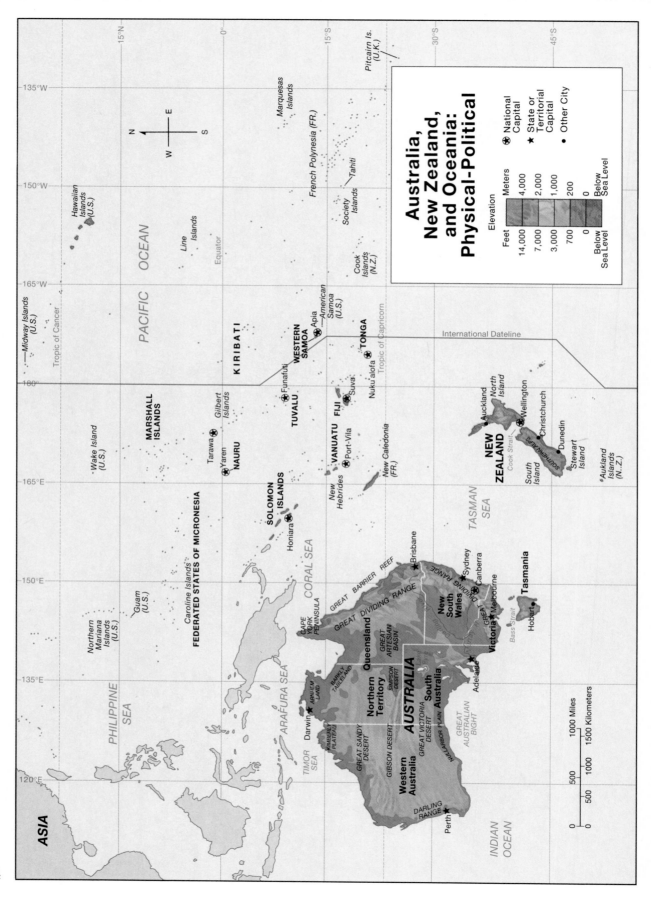

Australia, New Zealand, and Oceania: Physical-Political

Elevation

Feet	Meters
14,000	4,000
7,000	2,000
3,000	1,000
700	200
0	0
Below Sea Level	Below Sea Level

⊛ National Capital
★ State or Territorial Capital
• Other City

N
W — E
S

ASIA

PACIFIC OCEAN

Midway Islands (U.S.)

Hawaiian Islands (U.S.)

Marquesas Islands

Pitcairn Is. (U.K.)

French Polynesia (FR.)

Tahiti
Society Islands

Line Islands

Equator

Cook Islands (N.Z.)

Tropic of Cancer

Wake Island (U.S.)

MARSHALL ISLANDS

Gilbert Islands

KIRIBATI

Funafuti

TUVALU

WESTERN SAMOA
Apia
American Samoa (U.S.)

TONGA
Nuku'alofa

Tropic of Capricorn

International Dateline

Tarawa
Yaren
NAURU

Caroline Islands
FEDERATED STATES OF MICRONESIA

Northern Mariana Islands (U.S.)

Guam (U.S.)

PHILIPPINE SEA

SOLOMON ISLANDS
Honiara

New Hebrides

VANUATU
Port-Vila

FIJI
Suva

New Caledonia (FR.)

CORAL SEA

TASMAN SEA

NEW ZEALAND
Auckland
North Island
Wellington
Christchurch
Dunedin
South Island
Cook Strait
SOUTHERN ALPS
Stewart Island
Aukland Islands (N.Z.)

ARAFURA SEA

CAPE YORK PENINSULA

GREAT BARRIER REEF

GREAT DIVIDING RANGE

Brisbane

Queensland

Sydney
New South Wales
Canberra

TIMOR SEA

Darwin
ARNHEM LAND
Northern Territory

GREAT ARTESIAN BASIN

BARKLY TABLELAND

SIMPSON DESERT

KIMBERLEY PLATEAU

GREAT SANDY DESERT

GIBSON DESERT

GREAT VICTORIA DESERT

Western Australia

South Australia

AUSTRALIA

Adelaide

Victoria
Melbourne

Bass Strait

Tasmania
Hobart

NULLARBOR PLAIN

GREAT AUSTRALIAN BIGHT

DARLING RANGE
Perth

INDIAN OCEAN

0   500   1000 Miles
0   500   1000   1500 Kilometers

774

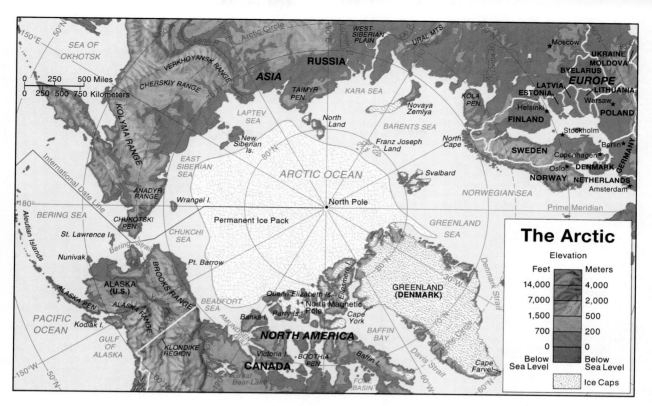

The Arctic

Elevation

Feet	Meters
14,000	4,000
7,000	2,000
1,500	500
700	200
0	0
Below Sea Level	Below Sea Level

Ice Caps

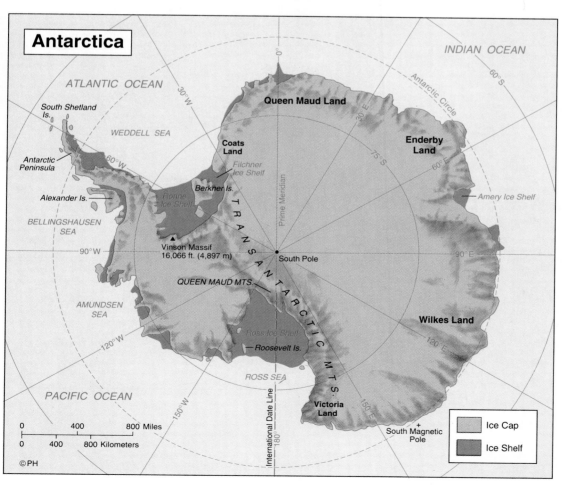

Antarctica

Vinson Massif 16,066 ft. (4,897 m)

Ice Cap

Ice Shelf

©PH

The World: Ocean Floor

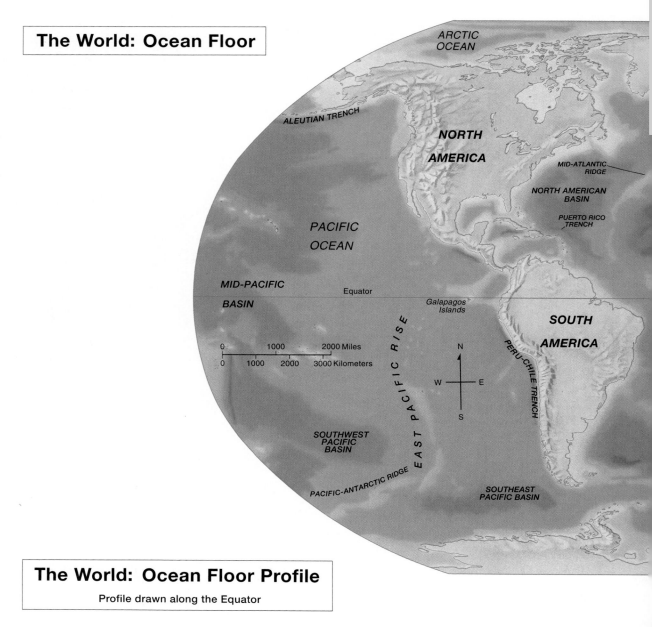

ARCTIC OCEAN

ALEUTIAN TRENCH

NORTH AMERICA

MID-ATLANTIC RIDGE

NORTH AMERICAN BASIN

PUERTO RICO TRENCH

PACIFIC OCEAN

MID-PACIFIC BASIN

Equator

Galapagos Islands

SOUTH AMERICA

EAST PACIFIC RISE

PERU-CHILE TRENCH

SOUTHWEST PACIFIC BASIN

PACIFIC-ANTARCTIC RIDGE

SOUTHEAST PACIFIC BASIN

N
W — E
S

0   1000   2000 Miles
0 1000 2000 3000 Kilometers

The World: Ocean Floor Profile

Profile drawn along the Equator

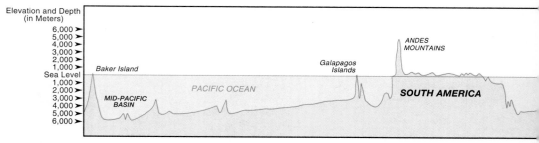

Elevation and Depth (in Meters)
6,000 ➤
5,000 ➤
4,000 ➤
3,000 ➤
2,000 ➤
1,000 ➤
Sea Level
1,000 ➤
2,000 ➤
3,000 ➤
4,000 ➤
5,000 ➤
6,000 ➤

Baker Island

MID-PACIFIC BASIN

PACIFIC OCEAN

Galapagos Islands

ANDES MOUNTAINS

SOUTH AMERICA

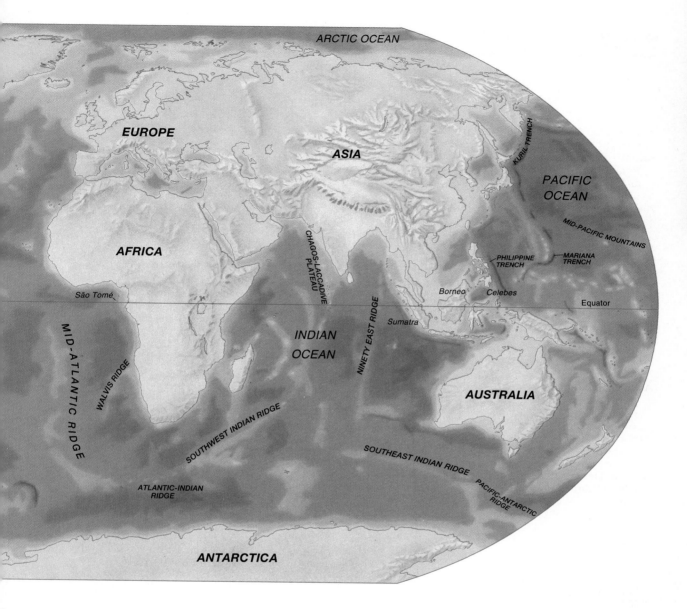

ARCTIC OCEAN

EUROPE

ASIA

PACIFIC OCEAN

KURIL TRENCH

MID-PACIFIC MOUNTAINS

AFRICA

CHAGOS-LACCADIVE PLATEAU

PHILIPPINE TRENCH

MARIANA TRENCH

São Tomé

Borneo

Celebes

Equator

INDIAN OCEAN

Sumatra

NINETY EAST RIDGE

MID-ATLANTIC RIDGE

WALVIS RIDGE

AUSTRALIA

SOUTHWEST INDIAN RIDGE

SOUTHEAST INDIAN RIDGE

PACIFIC-ANTARCTIC RIDGE

ATLANTIC-INDIAN RIDGE

ANTARCTICA

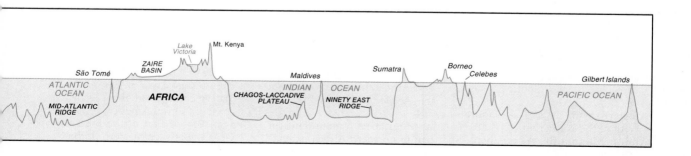

Lake Victoria

Mt. Kenya

ZAIRE BASIN

São Tomé

Maldives

Sumatra

Borneo

Celebes

Gilbert Islands

ATLANTIC OCEAN

INDIAN

OCEAN

PACIFIC OCEAN

AFRICA

CHAGOS-LACCADIVE PLATEAU

NINETY EAST RIDGE

MID-ATLANTIC RIDGE

A

Aden (12°N, 48°E) The chief seaport city of Yemen, p. 431

Adriatic Sea An arm of the Mediterranean Sea between Italy and the Balkan Peninsula, p. 328

Aegean Sea An arm of the Mediterranean Sea between Greece and Turkey, p. 335

Afghanistan, Republic of A country in south Asia, p. 576

Africa The world's second-largest continent, bounded by the Mediterranean Sea, the Atlantic Ocean, the Indian Ocean, and the Red Sea, p. 771

Alaska A state of the United States in northwestern North America, separated from Russia by the Bering Strait, p. 84

Albania, People's Socialist Republic of A European country on the southeast coast of the Adriatic Sea, p. 377

Alexandria (31°N, 30°E) A seaport city in Egypt on the Mediterranean Sea, p. 430

Algeria, Democratic and Popular Republic of A country in northern Africa, p. 430

Algiers (36°N, 3°E) The capital city of Algeria, p. 430

Allegheny Mountains A mountain range of the Appalachian system, located in Pennsylvania, Maryland, West Virginia, and Virginia in the eastern United States, p. 123

Alps A major south central European mountain system, extending through France, Italy, Switzerland, Austria, and Yugoslavia, p. 257

Amazon River A river in northern South America, the longest in the world, flowing from northern Peru to the Atlantic Ocean, p. 173

Amsterdam (52°N, 4°E) A seaport city of the Netherlands, p. 310

Andes Mountains A mountain system extending along almost the entire western coast of South America, p. 173

Andorra A small European country on the border between France and Spain, p. 324

Angola, People's Republic of A country in southern Africa, p. 506

Ankara (40°N, 33°E) The capital city of Turkey, p. 475

Antigua and Barbuda A country consisting of islands in the eastern Caribbean Sea, p. 167

Apennines A mountain system extending the length of Italy and continuing into Sicily, p. 328

Appalachian Mountains A mountain system in eastern North America, extending from southern Quebec, Canada, to Alabama in the southern United States, p. 84

Arabian Desert A desert in eastern Egypt, p. 484

Arabian Peninsula A peninsula in southwest Asia between the Red Sea and the Persian Gulf, p. 430

Aral Sea A landlocked salt-water sea in Kazakhstan and Uzbekistan, p. 388

Argentina, Republic of A country in South America, p. 173

Armenia A country in Northern Eurasia, historically an area of southwest Asia, p. 388

Asia The world's largest continent, bounded by the Arctic Ocean, the Pacific Ocean, the Indian Ocean, and Europe, p. 773

Aswan (24°N, 33°E) A city in Egypt on the Nile River, p. 484

Atacama Desert A desert in Chile, South America; the driest place on earth

Athens (38°N, 24°E) The capital city of Greece, p. 335

Atlantic Ocean A large body of water, separating North and South America from Europe and Africa and extending from the Arctic to the Antarctic, p. 764

Atlas Mountains A mountain system in northern Africa, extending across Morocco, Algeria, and Tunisia, p. 430

Australia, Commonwealth of An island continent in the Southern Hemisphere; a country comprising the continent and Tasmania, p. 708

Austria, Republic of A country in south central Europe, p. 314

Azerbaijan A country in Northern Eurasia, p. 388

B

Baghdad, (33°N, 44°E) The capital city of Iraq, p. 463

Bahamas, Commonwealth of the A group of islands in the Atlantic Ocean off the southeast coast of the United States, p. 167

Bahrain, State of An island monarchy located in the southwest Asia, p. 431

Baja California A peninsula in northwestern Mexico, separating the Gulf of California from the Pacific Ocean, p. 192

Balkan Peninsula A peninsula in southeastern Europe, p. 346

Baltic Sea An arm of the Atlantic Ocean in northern Europe, p. 346

Bangkok (13°N, 100°E) The capital city of Thailand, p. 694

Bangladesh, People's Republic of A country in south Asia, p. 576

Barbados An island country in the eastern Caribbean Sea, p. 167

Beijing (40°N, 116°E) The capital city of the People's Republic of China, p. 622

Beirut (34°N, 35°E) The capital city of Lebanon, p. 463

Belfast (54°N, 6°W) The capital and seaport city of Northern Ireland, p. 278

Belgium, Kingdom of A country in central Europe, p. 310

Belgrade (45°N, 20°E) The capital city of Yugoslavia, p. 346

Belize A country on the eastern coast of Central America, p. 167

Bengal, Bay of Part of the Indian Ocean, between eastern India and southeast Asia, p. 576

Benin, People's Republic of A country in west Africa, p. 506

Bering Strait (65°N, 170°W) A narrow waterway between Russia and Alaska, joining the Pacific and Arctic oceans, p. 84

Berlin (52°N, 13°E) The capital city of Germany, p. 305

Bern (47°N, 7°E) The capital city of Switzerland, p. 314

Bhutan, Kingdom of A country in south Asia, p. 576

Black Sea A landlocked sea between Europe and Asia, connected to the Mediterranean Sea by the Bosporus, p. 388

Bolivia, Republic of A country in South America, p. 475

Bombay (19°N, 73°E) The largest city in India, p. 576

Bosporus (41°N, 33°E) A narrow strait between the Black Sea and the Sea of Marmara, p. 475

Botswana, Republic of A country in southern Africa, p. 506

Brasília (15°S, 48°W) The capital city of Brazil, p. 173

Brazil, Federative Republic of The largest country in South America, p. 173

Brunei Darussalam, State of A country on the northern coast of the island of Borneo in southeast Asia, p. 694

Brussels (51°N, 4°E) The capital city of Belgium, p. 310

Budapest (47°N, 19°E) The capital city of Hungary, p. 372

Buenos Aires (34°S, 58°W) The capital city of Argentina, p. 173

Bulgaria, People's Republic of A country in eastern Europe, p. 377

Burkina Faso A country in west Africa, p. 528

Burundi, Republic of A country in central Africa, p. 552

Byelarus A country in Northern Eurasia, p. 388

C

Cairo (31°N, 31°E) The capital city of Egypt, p. 431

Cambodia, Republic of A country on the Indochina Peninsula in southeast Asia, p. 694

Cameroon, Republic of A country in central Africa, p. 528

Canada A country in North America, consisting of ten provinces and two territories, p. 84

Canberra (35°S, 149°E) The capital city of Australia, p. 708

Cape of Good Hope (34°S, 18°E) Cape in the Republic of South Africa, p. 552

Cape Horn (56°S, 67°W) Cape in Tierra del Fuego, Chile, the southern extremity of South America, p. 173

Cape Verde, Republic of A chain of 15 islands in the Atlantic Ocean, off the west coast of Africa

Caribbean Sea Part of the southern Atlantic Ocean, p. 167

Carpathian Mountains A major mountain system of central and eastern Europe, a continuation of the Alps, p. 346

Cascade Range A mountain range extending from northern California in the United States into southern British Columbia, Canada, p. 136

Caspian Sea A landlocked saltwater sea, which lies in Northern Eurasia, p. 388

Caucasus Mountains A mountain range in Northern Eurasia and Turkey, p. 388

Central African Republic A country in central Africa, p. 528

Central America The part of Latin America that comprises the seven republics of Guatemala, Honduras, El Salvador, Nicaragua, Costa Rica, Panama, and Belize, p. 167

Chad, Republic of A country in central Africa, p. 528

Chicago A major city in Illinois in the midwestern United States, p. 131

Chile, Republic of A country in South America, p. 173

China, People's Republic of A country occupying most of the mainland of east Asia, p. 622

Coast Ranges A series of mountain ranges along the Pacific coast of North America, extending from Baja California to Alaska, p. 136

Colombia, Republic of A country in South America, p. 173

Congo, People's Republic of the A country in central Africa, p. 528

Copenhagen (55°N, 12°E) The capital city of Denmark, p. 291

Corsica A French island in the Mediterranean Sea west of Italy, p. 298

Costa Rica, Republic of A country in Central America, p. 167

Côte d'Ivoire, Republic of A country in west Africa, p. 528

Crete A Greek island in the Mediterranean Sea, p. 335

Cuba, Republic of An island country consisting of the largest of the Caribbean islands, p. 167

Cyprus, Republic of An island country in the eastern Mediterranean Sea, off the coast of Turkey, p. 475

Czechoslovakia, Federative Republic of A country in eastern Europe, p. 372

D

Danube River A river in central and eastern Europe, flowing from Germany east to the Black Sea, p. 346

Dead Sea A salt-water lake on the border of Israel and Jordan, the lowest point on the earth's surface, p. 430

Denmark, Kingdom of A country in northern Europe, p. 291

Djibouti, Republic of A country in east Africa, p. 550

Dominican Republic A country in the Caribbean Sea on the island of Hispaniola, p. 167

Dublin (53°N, 6°W) The capital city of the Republic of Ireland, p. 278

E

Ecuador, Republic of A country in South America, p. 173

Edinburgh (56°N, 3°W) The capital city of Scotland, p. 278

Egypt, Arab Republic of A country in northern Africa, p. 431

El Salvador, Republic of The smallest country in Central America, p. 167

England (*see* **United Kingdom**)

English Channel An arm of the Atlantic Ocean between England and France, connecting the Atlantic Ocean and the North Sea, p. 257

Equatorial Guinea, Republic of A country in central Africa, p. 528

Ethiopia, People's Democratic Republic of A country in east Africa, p. 552

Euphrates River A river flowing from Turkey south through Syria and Iraq, p. 431

Europe The world's second-smallest continent, a peninsula of the Eurasian landmass bounded by the Arctic Ocean, the Atlantic Ocean, the Mediterranean Sea, and Asia, p. 769

F

Fiji, Republic of A country consisting of approximately 332 islands in the southern Pacific Ocean, p. 711

Finland, Republic of A country in northern Europe, p. 291

France, Republic of A country in central Europe, p. 298

French Guiana A country in South America, p. 173

G

Gabon, Republic of A country in west Africa, p. 528

Gambia, Republic of The A country in west Africa, p. 528

Ganges River A river in northern India and Bangladesh, flowing from the Himalayas to the Bay of Bengal, p. 576

Gaza Strip A strip of land at the southeastern end of the Mediterranean Sea, formerly part of Egypt, occupied by Israel since 1967, p. 456

Gdansk (54°N, 18°E) A seaport city in northern Poland, p. 366

Genoa (44°N, 9°E) A major seaport city in northwestern Italy, on the Mediterranean Sea, p. 328

Georgia A country in Northern Eurasia, p. 388

Germany, Federal Republic of A country in western Europe, p. 305

Ghana, Republic of A country in west Africa, p. 528

Gibraltar A British colony at the southern tip of Spain, p. 324

Glasgow (56°N, 4°W) The largest city in Scotland, p. 278

Gobi Desert A desert in Mongolia and northern China, p. 622

Great Lakes A group of five large lakes—Superior, Michigan, Huron, Erie, and Ontario—in central North America, p. 84

Great Sandy Desert A desert in Western Australia, p. 708

Greater Antilles A group of islands in the Caribbean Sea, p. 167

Greece, Hellenic Republic of A country in Mediterranean Europe, p. 335

Greenland A large, self-governing island in the northern Atlantic Ocean, part of Denmark, p. 84

Greenwich The place in England through which the Prime Meridian runs

Guatemala, Republic of A country in Central America, p. 167

Guinea, Republic of A country in west Africa, p. 528

Guinea-Bissau, Republic of A country in west Africa, p. 528

Gulf of Mexico An arm of the Atlantic Ocean, east of Mexico and south of the United States, p. 167

Guyana, Cooperative Republic of A country in South America, p. 173

H

Hague, The (52°N, 4°E) A city in the Netherlands, headquarters of the International Court of Justice, p. 310

Haiti, Republic of A country in the Caribbean Sea on the island of Hispaniola, p. 167

Hamburg (53°N, 10°E) A major seaport city in Germany, p. 305

Havana (23°N, 82°W) The capital city of Cuba, p. 167

Hawaiian Islands A large group of islands located in the northern Pacific Ocean; one of the United States, p. 136

Helsinki (60°N, 25°E) The capital city of Finland, p. 291

Himalayas A mountain system of south central Asia, extending along the border between India and Tibet and through Pakistan, Nepal, and Bhutan, p. 576

Hindu Kush A mountain range in Afghanistan, p. 576

Hiroshima (34°N, 132°E) A seaport city in Japan, on the island of Honshu, p. 668

Hispaniola An island in the Caribbean Sea, divided between Haiti on the west and the Dominican Republic on the east, p. 167

Honduras, Republic of A country in Central America, p. 167

Hong Kong (22°N, 114°E) A British crown colony in East Asia, to become part of the People's Republic of China in 1997, p. 622

Huang He River A river in northern China, p. 622

Hudson Bay An inland sea in the Northwest Territories, Canada, p. 145

Hungary, Republic of A country in eastern Europe, p. 372

I

Iberian Peninsula A peninsula in southwestern Europe, shared by Spain and Portugal, p. 769

Iceland, Republic of An island country in the northern Atlantic Ocean, close to the Arctic Ocean, p. 257

India, Republic of A large country occupying most of the Indian subcontinent in south Asia, p. 576

Indian Ocean The world's third-largest ocean, lying between Africa, Asia, and Australia, p. 764

Indonesia, Republic of A country in southwest Asia consisting of approximately 13,700 islands, including Sumatra, Java, Sulawesi (Celebes), Bali, and the western half of New Guinea, p. 694

Indus River A river in south Asia, rising in Tibet and flowing through India and Pakistan to the Arabian Sea, p. 576

Ionian Sea An arm of the Mediterranean Sea, between Greece and southern Italy, p. 328

Iran, Islamic Republic of A country in southwest Asia, p. 476

Iraq, Republic of A country in southwest Asia, p. 463

Ireland, Republic of A country in northern Europe, occupying part of an island lying west of

Great Britain in the Atlantic Ocean, p. 278

Israel, State of A country in southwest Asia, p. 456

Istanbul (41°N, 29°E) A seaport city in northwestern Turkey on the Bosporus; formerly Constantinople, p. 475

Italy, Republic of A boot-shaped country in southern Europe, including the islands of Sicily and Sardinia, p. 328

J

Jamaica An island country in the Caribbean Sea, p. 167

Japan An island country in the Pacific Ocean off the east coast of Asia, consisting of four main islands—Honshu, Hokkaido, Kyushu, and Shikoku, p. 668

Jerusalem (31°N, 35°E) The capital city of Israel, holy to Jews, Christians, and Muslims, p. 456

Johannesburg (26°S, 28°E) The largest city in the Republic of South Africa, p. 506

Jordan, Hashemite Kingdom of A country in southwest Asia, p. 463

Jordan River A river in southwest Asia, rising in Syria and flowing to the northern end of the Dead Sea, forming the border between Israel and Jordan, p. 463

K

Kalahari Desert A desert plateau in southern Africa, p. 552

Kazakhstan A country in Northern Eurasia, p. 388

Kenya, Republic of A country in east Africa, p. 552

Kuwait, State of A country in southwest Asia, p. 469

Kuwait City (29°N, 47°E) The capital and seaport city of Kuwait, p. 469

Kyrgyzstan A country in Northern Eurasia, p. 388

L

Laos, People's Democratic Republic of A country on the Indochina Peninsula in southeast Asia, p. 694

Latin America The culture region that includes Mexico, Central America, South America, and the Caribbean Islands, p. 767

Lebanon, Republic of A country in southwest Asia on the eastern end of the Mediterranean Sea, p. 463

Lesotho, Kingdom of A country in southern Africa, completely surrounded by the Republic of South Africa, p. 552

Lesser Antilles A group of islands in the Caribbean Sea, p. 167

Liberia, Republic of A country in west Africa, p. 528

Libya, Socialist People's Arab Jamahiriaya A country in northern Africa, p. 491

Liechtenstein, Principality of A country in central Europe, p. 314

Lisbon (38°N, 9°W) The capital and seaport city of Portugal, p. 324

London (51°N, 0°) The capital city of the United Kingdom of Great Britain and Northern Ireland, p. 278

Los Angeles (34°N, 118°W) A seaport city in California in the southwestern United States, p. 136

Luxembourg, Grand Duchy of A country in central Europe, p. 257

M

Madagascar, Democratic Republic of An island country off the southeast coast of Africa in the Indian Ocean, p. 506

Madrid (40°N, 3°W) The capital city of Spain, p. 324

Malay Peninsula A peninsula in southeast Asia, comprising West Malaysia and part of Thailand, p. 694

Malaysia A country in southeast Asia, p. 694

Maldives, Republic of A country consisting of a chain of islands in the Indian Ocean southwest of India, p. 763

Mali, Republic of A country in west Africa, p. 528

Malta, Republic of An island country in the Mediterranean Sea, p. 768

Marseille (43°N, 5°E) A major seaport city in France, p. 298

Mauritania, Islamic Republic of A country in west Africa, p. 528

Mecca (21°N, 39°E) Islam's holiest city, in western Saudi Arabia, p. 469

Medina (23°N, 40°E) Islam's second-holiest city, in western Saudi Arabia, p. 469

Mediterranean Sea A large sea separating Europe and Africa, p. 431

Melbourne (37°S, 149°E) Australia's second-largest city, p. 708

Mexico (United Mexican States) A country in North America, p. 192

Mexico City (19°N, 99°W) Capital and largest city of Mexico; largest urban area in the world, p. 192

Middle East The region that includes southwest Asia and sometimes part or all of northern Africa, p. 430–431

Midway Islands Two islands, Eastern and Sand, in the central Pacific Ocean, administered by the United States, p. 711

Milan (45°N, 9°E) A city in northwestern Italy, p. 328

Mississippi River A river in the central United States, flowing from Minnesota south into the Gulf of Mexico, p. 84

Missouri River A river in the central United States, p. 84

Mojave Desert A desert in southern California in the southwestern United States, p. 136

Moldova A country in Northern Eurasia, p. 388

Monaco, Principality of A country in central Europe, p. 298

Mongolia, People's Republic of A country in east Asia, p. 622

Montreal (45°N, 73°W) A city in the province of Quebec in eastern Canada, p. 145

Morocco, Kingdom of A country in northern Africa, p. 491

Mozambique, People's Republic of A country in southern Africa, p. 552

Munich (48°N, 11°E) A city in southeastern Germany, p. 305

Myanmar, Union of A country in southeast Asia; formerly Burma, p. 694

N

Nagasaki (32°N, 130°E) A seaport city on the west coast of Japan, p. 668

Namib Desert A desert in southern Africa, p. 506

Namibia, Union of A country in southern Africa, p. 552

Nepal, Kingdom of A country in south Asia, p. 576

Netherlands, Kingdom of A country in central Europe, p. 310

Netherlands Antilles Two groups of islands in the Caribbean Sea

New Delhi (28°N, 77°E) The capital city of India, p. 576

New Orleans (30°N, 90°W) A major seaport city in Louisiana in the southern United States, p. 126

New York City (43°N, 75°W) A major seaport city in the state of New York in the northeastern United States, p. 123

New Zealand A country in the southwest Pacific Ocean, consisting of two major islands, p. 708

Nicaragua, Republic of A country in Central America, p. 167

Niger, Republic of A country in west Africa, p. 528

Niger River A river in west Africa, flowing from Guinea into the Gulf of Guinea, p. 528

Nigeria, Federal Republic of A country in west Africa, p. 528

Nile River A river in northeast Africa, rising at Khartoum, Sudan, and flowing north through Egypt into the Mediterranean Sea, p. 430

North America The world's third-largest continent, consisting of Canada, the United States, Mexico, and many islands, p. 767

North Korea, Democratic People's Republic of A country in east Asia, p. 680

North Sea An arm of the Atlantic Ocean between Great Britain and the European mainland, p. 278

Northern Ireland (*see* **United Kingdom**)

Norway, Kingdom of A country in northern Europe, p. 291

O

Okeechobee, Lake A large lake in Florida in the southern United States, p. 126

Oman, Sultanate of A country in southwest Asia on the Arabian Peninsula, p. 469

Oslo (60°N, 10°E) The capital city of Norway, p. 291

Ottawa (45°N, 75°W) The capital city of Canada, located in the province of Ontario, p. 145

P

Pacific Ocean A large body of water, bounded by North and South America on the east and Asia and Oceania on the west, and stretching from the Arctic to the Antarctic, p. 764–765

Pakistan, Islamic Republic of A country in south Asia, p. 576

Palestine (32°N, 35°E) A historical region at the eastern end of the Mediterranean Sea, comprising parts of modern Israel, Jordan, and Egypt, p. 452

Panama, Isthmus of A narrow strip of land linking South and Central America and separating the Atlantic and Pacific oceans; site of the Panama Canal, p. 167

Panama, Republic of A country in Central America, p. 167

Panama Canal An important shipping canal across the Isthmus of Panama, linking the Caribbean Sea (hence the Atlantic Ocean) to the Pacific Ocean

Panama City (9°N, 79°W) The capital and seaport city of the Republic of Panama, p. 167

Papua New Guinea A country in southeast Asia, p. 694

Paraguay, Republic of A country in South America, p. 173

Paris (49°N, 2°E) The capital city of France, p. 298

Persia The historical name for the region around present-day Iran

Persian Gulf An arm of the Arabian Sea, p. 430

Peru, Republic of A country in South America, p. 173

Philadelphia (40°N, 75°W) A city in Pennsylvania in the northeastern United States, p. 123

Philippines, Republic of the A country in southeast Asia, p. 694

Poland, Republic of A country in eastern Europe, p. 366

Portugal, Republic of A country in southwestern Europe, p. 324

Prague (50°N, 15°E) The capital city of Czechoslovakia, p. 372

Puerto Rico, Commonwealth of An island commonwealth of the United States in the Caribbean Sea, p. 167

Pyrenees A mountain range in southwestern Europe forming the border between France and Spain, p. 324

Q

Qatar, State of A country in southwest Asia on the Arabian Peninsula, p. 431

Quebec (46°N, 71°W) The capital city of the province of Quebec in eastern Canada, p. 145

R

Red Sea A narrow sea separating northeastern Africa from the Arabian peninsula, connected to the Mediterranean Sea by the Suez Canal and to the Indian Ocean by the Gulf of Aden, p. 431

Rhine River A river in central Europe, rising in Switzerland and flowing north through Germany to the Netherlands, p. 257

Rhône River A river in central Europe, rising in Switzerland and flowing south through France into the Mediterranean Sea, p. 257

Rio de Janeiro (23°S, 43°W) A major city in Brazil, p. 224

Rocky Mountains A mountain system in North America, p. 84

Romania A country in eastern Europe, p. 346

Rome (42°N, 12°E) The capital city of Italy, p. 328

Rub' al-Khali (Great Sandy Desert) A desert in Saudi Arabia, p. 469

Russia A country in Northern Eurasia, p. 388

Rwanda, Republic of A country in east Africa, p. 552

S

Sahara A desert in northern Africa, p. 430

San Francisco (37°N, 122°W) A seaport city in California in the western United States, p. 136

San Marino, Most Serene Republic of A country in north central Italy, p. 328

Santiago (33°S, 70°W) The capital city of Chile, p. 173

São Paulo (23°S, 46°W) The largest city in Brazil, p. 224

Sardinia An Italian island in the Mediterranean Sea west of Italy, p. 328

Saudi Arabia, Kingdom of A country in southwest Asia occupying most of the Arabian Peninsula, p. 469

Scandinavia A region in northern Europe consisting of Norway, Sweden, Denmark, and sometimes Finland, Iceland, and the Faroe Islands, p. 291

Scotland (*see* **United Kingdom**)

Seine River A river in northern France, flowing through Paris and emptying into the English Channel, p. 298

Senegal, Republic of A country in west Africa

Senegal River A river in west Africa, flowing from western Mali into the Atlantic Ocean.

Seychelles, Republic of An island country in the Indian Ocean, northeast of Madagascar, p. 770

Shanghai (31°N, 121°E) The largest city in China and one of the world's leading seaports, p. 622

Siberia A resource-rich region of Russia, extending east across northern Asia from the Ural mountains to the Pacific coast, p. 773

Sicily An Italian island in the Mediterranean Sea, p. 328

Sierra Leone, Republic of A country in west Africa, p. 528

Sierra Madre A rugged mountain system in Mexico, including the Sierra Madre Oriental (East), the Sierra Madre Occidental (West), and the Sierra Madre del Sur (South), p. 192

Sierra Nevada A mountain range in California in the western United States, p. 136

Singapore, Republic of An island country in southeast Asia, p. 694

Somalia, Democratic Republic of A country in east Africa, p. 552

South Africa, Republic of A country in southern Africa, p. 552

South America The world's fourth-largest continent, bounded by the Caribbean Sea, the Atlantic Ocean, and the Pacific Ocean and linked to North America by the Isthmus of Panama, p. 764

South Korea, Republic of A country in east Asia, p. 680

Spain A country in southwestern Europe, p. 324

Sri Lanka, Democratic Socialist Republic of An island country off the southeast coast of India, p. 576

Stockholm (59°N, 18°E) The capital city of Sweden, p. 291

Sudan, Republic of the A country in east Africa, p. 552

Suez Canal A shipping canal across the Isthmus of Suez, connecting the Mediterranean Sea and the Indian Ocean through the Gulf of Suez and the Red Sea, p. 431

Suriname, Republic of A country in South America, p. 173

Swaziland, Kingdom of A country in southern Africa, p. 552

Sweden, Kingdom of A country in northern Europe, p. 291

Switzerland (Swiss Confederation) A country in central Europe, p. 314

Sydney (34°S, 151°E) The capital of New South Wales, a state of Australia, p. 708

Syria, Arab Republic of A country in southwest Asia, p. 463

Syrian Desert A desert in southwest Asia, p. 469

T

Taiwan (Republic of China) An island country off the southeast coast of the People's Republic of China, p. 622

Tanzania, United Republic of A country in east Africa, p. 552.

Tajikistan A country in Northern Eurasia, p. 388

Tehran (35°N, 51°E) The capital city of Iran, p. 476

Thailand, Kingdom of A country in southeast Asia, p. 694

Tian Shan A mountain system in central Asia, p. 622

Tigris River A river in Turkey and Iraq, p. 431

Togo, Republic of A country in west Africa, p. 528

Tokyo (35°N, 139°E) The capital and largest city of Japan, on the island of Honshu, p. 668

Tunisia, Republic of A country in northern Africa, p. 431

Turkey, Republic of A country in southwest Asia, p. 475

Turkmenistan A country in Northern Eurasia, p. 388

U

Uganda, Republic of A country in east Africa, p. 552

Ukraine A country in Northern Eurasia, formerly a republic of the Soviet Union, p. 388

United Arab Emirates A country in southwest Asia on the eastern coast of the Arabian Peninsula, p. 469

United Kingdom of Great Britain and Northern Ireland An island country of western Europe, consisting of England, Scotland, Wales, and Northern Ireland, p. 278

United States of America A country in North America, consisting of 48 contiguous states, the District of Columbia, and Alaska and Hawaii, p. 107

Urals A mountain system in Russia, forming part of the border between Europe and Asia, p. 388

Uruguay, Oriental Republic of A country in South America, p. 173

Uzbekistan A country in Northern Eurasia, p. 388

 V

Vancouver (49°N, 123°W) A major seaport city in western Canada, p. 145

Vatican City An independent state contained within Rome, Italy; headquarters of the Roman Catholic church

Venezuela, Republic of A country in South America, p. 173

Venice (45°N, 12°E) A seaport city in northern Italy on the Adriatic Sea, p. 328

Vienna (48°N, 16°E) The capital city of Austria, p. 314

Vietnam, Socialist Republic of A country on the Indochina Peninsula in southeast Asia, p. 694

Volga River A river in Russia, rising near Moscow and flowing into the Caspian Sea, p. 388

 W

Wales (*see* **United Kingdom**)

Warsaw (52°N, 21°E) The capital city of Poland, p. 366

Washington, D.C. (39°N, 77°W) The capital city of the United States, p. 126

 Y

Yangzi River A major river in China, p. 622

Yemen, Republic of A country in southwest Asia on the Arabian Peninsula, p. 469

Yucatán Peninsula A low, flat peninsula in southeastern Mexico, p. 192

Yugoslavia, Socialist Federal Republic of A country on the Balkan Peninsula in eastern Europe, p. 377

Z

Zaire, Republic of A country in central Africa, p. 528

Zaire River A river in central Africa, one of the longest in the world, p. 528

Zambia, Republic of A country in southern Africa, p. 552

Zimbabwe A country in southern Africa, p. 552

GLOSSARY

Pronunciation Key

Symbol	Key Words
a	**a**sp, f**a**t, p**a**rrot
ā	**a**pe, d**a**te, pl**a**y, br**ea**k, f**ai**l
ä	**a**h, c**a**r, f**a**ther, c**o**t
e	**e**lf, t**e**n, b**e**rry
ē	**e**ven, m**ee**t, mon**ey**, fl**ea**, gri**e**ve
i	**i**s, h**i**t, m**i**rror
ī	**i**ce, b**i**te, h**i**gh, sk**y**
ō	**o**pen, t**o**ne, g**o**, b**oa**t
ô	**a**ll, h**o**rn, l**a**w, **oa**r
oo	l**oo**k, p**u**ll, m**oo**r, w**o**lf
o͞o	**oo**ze, t**oo**l, cr**e**w, r**u**le
yoo	**u**se, c**u**te, f**ew**
yo͞o	c**u**re, glob**u**le
oi	**oi**l, p**oi**nt, t**oy**
ou	**ou**t, cr**ow**d, pl**ow**
u	**u**p, c**u**t, c**o**lor, fl**oo**d
ʉr	**ur**n, f**ur**, det**er**, **ir**k
ə	**a** as in **a**go
	e as in ag**e**nt
	i as in san**i**ty
	o as in c**o**mply
	u as in foc**u**s
ər	p**er**haps, murd**er**
zh	a**z**ure, lei**s**ure, bei**g**e
nj	ri**ng**, a**n**ger, dri**n**k

A

abdicate To relinquish power or responsibility formally; to surrender one's office, throne, or authority, p. 642

aborigine (ab′ə rij′ə nē′) An original inhabitant; one of the original inhabitants of Australia, p. 714

absolute location The position on the earth in which a place can be found, p. 5

acculturation The process of accepting, borrowing, and exchanging traits between cultures, p. 56

acid rain Rain whose high concentration of chemicals, usually from industrial pollution, that pollutes water, kills plant and animal life, and eats away at the surface of stone and rock; a form of chemical weathering, p. 28

acupuncture The ancient Chinese practice of inserting fine needles at specific body points to cure disease or to ease pain, p. 658

alluvial plain A broad expanse of land along riverbanks, consisting of rich, fertile soil left by floods, p. 577

altiplano (äl′ tē plä′nō) A Spanish word meaning "high plain"; a series of highland valleys and plateaus located in the Andes of Bolivia and Peru, p. 240

anarchy Political disorder and violence; lawlessness, p. 466

ancestor worship The belief that respecting and honoring one's ancestors will cause them to live on in the spirit world after death, p. 535

animism The religious belief that such things as the sky, rivers, and trees contain a spirit, or soul, p. 535

annex To formally incorporate into a country or state the territory of another, p. 288

apartheid (ə pärt′hāt) An Afrikaans word meaning "apartness"; in the Republic of South Africa, the policy of strict racial segregation and discrimination against nonwhites, p. 560

aquaculture Fish farming—the raising of fish for food in enclosed environments such as tanks, reservoirs, ponds, sheltered bays, and river estuaries, p. 631

aqueduct A large pipe or channel designed to transport water from a remote source over a long distance, usually by gravity, p. 137

arable Capable of being farmed, or cultivated, p. 437

archipelago (är′kə pel′ə gō′) A group of islands, p. 213

artesian well A well that is drilled deep enough to tap a layer of porous material filled with groundwater, p. 733

atheism The belief that God does not exist, p. 657

atmosphere A multilayered band of gases, water vapor, and dust above the earth, p. 31

atoll (a′tôl) A ring-shaped coral island surrounding a lagoon, p. 736

authoritarian Descriptive of a system of government in which one person, perhaps a dictator, holds all political power, p. 60

ayatollah A religious leader among Shi'ite Muslims, p. 476

B

basin irrigation A form of irrigation in which land is surrounded by embankments and flooded with water, either by natural or artificial means, p. 488

bayou A marshy inlet or outlet of a lake or a river, p. 125

GLOSSARY

785

bazaar An open-air market; a street lined with shops and stalls, p. 486

bedrock Solid rock underlying all soil, gravel, clay, sand, and loose material on the earth's surface, p. 86

biome (bī′ōm) A region in which environment, plants, and animal life are suited to one another, p. 43

birthrate The number of live births each year per 1,000 people, p. 53

bog An area of wet, spongy ground, p. 282

boycott To refuse to purchase, sell, or use a product or service as an expression of disapproval, p. 596

buffer state A country that serves as a military barrier between two or more hostile countries, p. 611

C

campesino (käm′pe sē′nō) A Spanish word meaning "rural or rustic; a peasant"; in Latin America, a tenant farmer or farm worker, p. 239

canopy The uppermost spreading branchy layer of a forest, p. 177

canton A political division or state; one of the states in Switzerland, p. 316

capital Wealth in the form of money or property, accumulated by an individual, partnership, or corporation, and available for investment in building and supporting property or new industry, p. 489

capitalist Descriptive of an economic system (capitalism) in which the means of production are controlled by private individuals or corporations; also called a market economy in which people, as consumers, help determine what will be produced by buying or not buying certain products, p. 61

cardinal direction One of the four points of the compass: north, south, east, and west, p. 14

cartographer A person who makes maps or charts, p. 12

caste system A social hierarchy in which a person possesses a distinct rank in society that is determined by birth, p. 583

cataract A large waterfall; any strong flood or downpour of water, p. 505

cay (kā, kē) A small, low island or reef of coral, p. 167

census The systematic counting of a population, p. 10

chaparral A type of natural vegetation that . is adapted to Mediterranean climates; small evergreen trees and low bushes, or scrub, p. 46

chemical weathering The process by which the actual chemical structure of rock is changed, usually when water and carbon dioxide cause a breakdown of the rock, pp. 27–28

climate The term used for the weather patterns that an area or region typically experiences over a long period of time, p. 31

collective A state-owned farm or group of farms on which land and machinery are pooled and people work as a group, sharing the harvest under government supervision, p. 644

collective farm A state-owned farm in the former Soviet Union managed by workers who shared the profits from their produce, p. 373

command system An economic system in which the government dictates what goods will be manufactured, p. 411

commercial farming The raising of crops and livestock for sale in outside markets, p. 75

communism A system of government in which the government controls the means of production, determining what goods will be made, how much workers will be paid, and how much items will cost, p. 61

confederation A system of government in which individual political units keep their sovereignty but give limited power to a central government, p. 60

coniferous cone-bearing; a type of tree able to survive long, cold winters, with long, thin needles rather than leaves, p. 45

constitutional monarchy A system of government in which a monarch plays a symbolic role while real political power rests with an elected law-making body, p. 60

continent Any of the seven large landmasses of the earth's surface: Africa, Antarctica, Asia, Australia, Europe, North America, and South America, p. 20

continental climate The type of climate found in the great central areas of continents in the Northern Hemisphere; characterized by extreme temperatures—cold, snowy winters and warm or hot summers, p. 38

continental divide A boundary or area of high ground that separates rivers flowing toward opposite sides of a continent, p. 87

continental drift theory The idea that continents slowly shift their positions due to movement of the tectonic plates on which they ride, p. 23

convection A circular movement caused when a material is heated, expands and rises, then cools and falls, p. 24

convergence zone An area of severe storms where the frigid waters circulating around Antarctica meet the warmer waters of the Atlantic, Pacific, and Indian oceans, p. 744

coral reef A marine ridge in shallow, tropical seas, formed primarily from skeletal fragments of certain marine organisms and the limestone resulting from their compaction, p. 712

cordillera (kôr′ dil yer′ə) A related set of separate mountain ranges, p. 85

core The earth's center, consisting of very hot metal that is dense and solid in the inner core and molten, or liquid, in the outer core, p. 20

Coriolis effect A deflection, or bending, of the winds that move over the earth's surface, caused by the rotation of the earth, p. 34

cottage industry A small-scale manufacturing operation in which people produce goods in their own homes using their own tools, p. 603

coup (kōō) Also called coup d'état (kōō′ dă tä′) The sudden overthrow of a ruler or government, often involving violent force or the threat of force, p. 533

crevasse (kri vas′) A deep crack in glacial ice, p. 743

crust The solid, rocky, surface layer of the earth, p. 20

cultural landscape A landscape that has been altered by human beings and reflects their culture, p. 56

cultural trait A behavioral characteristic of a people, such as a language, custom, or skill, passed on from one generation to another, p. 55

culture The way of life that distinguishes a people, for example, government, language, religion, customs, and beliefs, p. 7

culture hearth A place in which important ideas begin and thereafter spread to surrounding cultures, p. 56

customs Duties, or taxes, charged by one country's government on certain goods imported from another country; a government agency authorized to collect duties on imported goods, p. 155

cyclone The Australian term for hurricane; a violent, rotating windstorm, similar in intensity to a hurricane, p. 732

desalinization The process of removing salt from seawater so that it can be used for drinking and irrigation, p. 469

desertification The loss of all vegetation; the transformation of arable land into desert either naturally or through human intervention, p. 531

developed country A modern industrial society with a well-developed economy, p. 74

developing country A country with relatively low industrial production, often lacking modern technology, p. 74

dialect A variation of a spoken language that has its own distinct grammar, pronunciation, or vocabulary and is unique to a region or community, p. 301

dictatorship A system of government in which absolute power is held by one person, p. 60

diffusion The process of spreading cultural traits from one person or society to another, p. 56

distortion A misrepresentation of the original shape; each map projection used by a cartographer produces some distortion, p. 12

diversify To expand; to give variety to, p. 516

double cropping In farming, growing more than one crop a year, p. 651

drainage basin The entire area of land that is drained by a major river and its tributaries, p. 87

drip irrigation A method of irrigation by which precisely controlled amounts of water drip directly onto plants from pipes, thus preserving precious water resources in dry areas, p. 457

D

death rate The number of deaths each year per 1,000 people, p. 53

deciduous Leaf-shedding; a type of tree that sheds its leaves when winter approaches, p. 44

deforestation The process of stripping the land of its trees, p. 530

delta A flat, low-lying plain formed at the mouth of a river when sediment is deposited by flowing water, p. 28

demand system An economic system in which the demand for goods in the marketplace determines what will be manufactured, p. 411

demilitarized zone A strip of land on which there are no troops or weapons, p. 679

democracy A system of government in which the people are invested with the power to choose their leaders and elected representatives and determine government policy based on the will of the majority, p. 60

demography The study of human populations, including their size, growth, density, distribution, and rates of births, marriages, and deaths, p. 51

dependency A territorial unit ruled by another country, p. 182

E

ejido (e hě′dō) A Spanish word describing farmland owned collectively by members of a rural community, p. 200

environment The physical conditions of the natural surroundings, p. 43

Equator An imaginary line that circles the globe at its widest point (halfway between the North and South poles), dividing the earth into two halves called hemispheres; used as a reference point from which north and south latitudes are measured, p. 5

equinox Either of the two times each year (spring and fall) when day and night are of nearly equal length everywhere on earth, p. 32

erg A great expanse of shifting sands; a sand dune, p. 429

erosion The movement of weathered materials, including gravel, soil, and sand, usually caused by water, wind, and glaciers, p. 28

escarpment A steep cliff that separates two level areas of differing elevations, p. 223

estuary A flooded river valley at the wide mouth of a river; an inlet, or arm of the sea, where freshwater river currents meet saltwater, p. 279

ethanol An alcohol-based fuel, sometimes called gasohol, p. 229

ethnocracy A system of government in which one ethnic group rules others, p. 557

F

falaj **system** In the Arabian peninsula, an ancient system of underground and surface water canals that carry water many miles from mountains to desert villages, p. 472

fault A fracture, or break, in the earth's crust, p. 21

favela A Portuguese word meaning "slum," p. 226

federal Descriptive of a system of government in which political power is shared between a large central authority and smaller political units, such as a national government and its states, p. 60

fellah An Egyptian farmer; a peasant or farm laborer in Arab countries (*pl.* fellaheen), p. 438

fjord (fyôrd) A narrow valley or inlet from the sea between steep cliffs or slopes, originally carved out by an advancing glacier and filled by melting glacial ice, p. 291

floodplain A broad plain or alluvial plain that may form on either side of a river when sediment settles on the banks or riverbeds, p. 28

forage Food for grazing animals, p. 530

fossil The preserved remains or imprint of an animal or plant from a previous geologic period, pp. 23–24

fossil fuel Any one of several nonrenewable mineral resources—coal, oil, natural gas—formed from the remains of ancient plants and animals and used for fuel, p. 68

free enterprise An economic system based on capitalism that allows an individual to own, operate, and profit from his or her own business, p. 109

frost wedging A common type of mechanical weathering in which water in a crack in a rock freezes to ice, expands, and eventually widens the crack or splits the rock, p. 27

G

gaucho (gou′chō) A cowboy who herds cattle in the pampas of Argentina and Uruguay, p. 246

geography The study of the earth's surface and the processes that shape it, the connections between places, and the complex relationships between people and their environments, p. 4

geology The study of the earth's physical structure and history, p. 19

geothermal energy Energy from the earth's intense interior heat, which transforms underground water to steam that can be used to heat homes or to make electricity, p. 69

geyser (gī′zər) A natural hot spring that shoots a column of water and steam into the air, p. 710

ghetto A section of a city in which a particular minority group is forced to live as a result of either economic or social pressures, p. 366

glacier A huge, slow-moving mass of snow and ice, formed over many years from layers of unmelted snow pressing together, thawing slightly, and refreezing, p. 29

glasnost A Russian word meaning "openness"; in the former Soviet Union, a policy allowing citizens to do and say what they wished, p. 411

glen A narrow valley, p. 284

graben (grä′ bən) A long, narrow area that has dropped between two faults, p. 334

grain elevator A tall building equipped with machinery for loading, cleaning, storing, and discharging grain, p. 132

grain exchange A place where grain is bought and sold as a commodity, p. 132

great circle Any imaginary line that circles the earth and divides it into two equal halves, p. 10

gross national product (GNP) The total value of goods and services produced by a country in a year, p. 75

growing season In farming, the average number of days between the last frost of spring and the first frost of fall, p. 132

guerrilla (gə ril′ə) A member of an armed force that is not part of a regular army; relating to a form of warfare carried on by such an independent armed force, p. 211

H

hacienda (hä′sē en′də) A Spanish word used to describe a large Spanish-owned estate in the Americas, often run as a farm or a cattle ranch, p. 198

Hajj (haj) In Islam, a pilgrimage or religious journey to the holy city of Mecca, birthplace of Mohammed, p. 440

heavy industry The production of goods such as steel and machinery used by other industries, p. 410

hemisphere A half of the earth; the Equator divides the Northern and Southern hemispheres. The Prime Meridian divides the Eastern and Western hemispheres, p. 5

hierarchy Rank according to function; a group of persons or things arranged in order of rank, grade, or class, p. 112

hinterland The area served by a metropolis, p. 114

homogeneous (hō′mə jē′nē əs) Having a similar nature; uniform in structure or quality; identical, p. 669

humidity The amount of water vapor in the air, p. 36

hurricane A destructive tropical storm that forms over the Atlantic Ocean, usually in late summer and early fall, with winds of at least 74 miles (119 km) per hour, p. 170

hydroelectricity Electricity that is generated by moving water, p. 97

hypothesis (hī păth′ə sis) An assumption that a scientist makes in order to set about answering a question, p. 9

I

ice shelf A massive extension of glacial ice over the sea, often protruding hundreds of miles, pp. 743–744

ideogram In written language, a character or symbol that represents an idea or thing, as in the Chinese language, p. 657

ideology Ideas or principles on which a political system is based, p. 410

indigenous Native to or living naturally in an area or environment, p. 691

inflation A sharp and continuing rise in prices, p. 304

infrastructure An underlying foundation; the basic support facilities of a country, such as roads and bridges, power plants, schools, and communication systems, p. 470

inland delta An area of lakes, creeks, and swamps away from the ocean, p. 532

insurgent A person who rebels against his or her government, p. 694

intensive farming A farming method that requires a great deal of labor, p. 631

intermediate direction Points on a compass that lie between two of the four cardinal directions: for example, northeast, northwest, southeast, and southwest, p. 14

irrigation The artificial watering of farmland, often by means of canals that draw water from reservoirs or rivers, p. 195

isthmus (is′məs) A narrow strip of land having water on each side and joining two larger bodies of land, p. 207

J

joint family system In India, the custom of housing all members of an extended family together, p. 601

K

karst A soft limestone that is easily dissolved by wind and water, thus producing protruding rock and caverns, p. 348

Knesset The democratically elected parliament of Israel, p. 458

krill Small, shrimplike creatures; food for whales and fish, p. 744

L

lagoon A shallow body of water separated from the sea by coral reefs or sandbars, p. 732

land redistribution A policy by which land is expropriated from those who own large amounts and redistributed to those who have little or none, p. 567

landlocked Almost or entirely surrounded by land; cut off from the sea, p. 531

latifundio (lat′ə fun′dē ō) A Spanish word describing a large commercial farm owned by a private individual or a farming company, p. 200

latitude One of the series of imaginary lines, also called parallels, that circle the earth parallel to the Equator; used to measure the distance north and south of the Equator in degrees, p. 5

lava Magma, or molten rock from the earth's mantle, that breaks through the surface of the earth during volcanic activity, p. 21

leaching The dissolving and washing away of nutrients in the soil, p. 515

leeward Situated on the side facing away from the direction from which the wind is blowing, p. 36

light industry The production of small consumer goods such as clothing and appliances, p. 647

lignite A soft, brownish-black coal having a slightly woodish texture, p. 306

literacy The ability to read, p. 94

lock A section of a canal, equipped with gates, in which a ship may be raised or lowered by raising or lowering the level of the water in that section, p. 146

loess (les, lō′es) Fine-grained, mineral-rich loam, dust, or silt deposited by the wind, p. 29

longitude One of the series of imaginary lines, also called meridians, that run north and south from one pole to the other; used to measure the distance east and west of the Prime Meridian in degrees, p. 5

M

magma Molten, or liquid, rock in the earth's mantle, p. 20

malnutrition The disease caused by lack of proper food; inadequate nutrition resulting from an unbalanced diet or insufficient food, p. 554

mandate After World War I, a commission from the League of Nations authorizing a nation to govern a territory, p. 451

mangrove A tropical tree that grows in swampy ground along coastal areas, p. 125

mantle A thick layer of mostly solid rock beneath the earth's crust that surrounds the earth's core, p. 20

manufacturing The process of turning a raw material into a finished product, p. 73

map projection A way of showing the round earth on a flat surface such as paper; each of the various types of map projections produces some distortion of the earth's surface, p. 12

maritime Bordering on or near the sea; relating to navigation or shipping, p. 144

market economy An economic system in which most goods and services are produced and distributed through the free market, p. 61

martial law The law administered during a period of strict military control, p. 648

mechanical weathering The actual breaking up or physical weakening of rock by forces such as ice and roots, p. 27

medina The old Arab section of a North African city, usually centered around a mosque, p. 495

megalopolis (meg/ə läp/ə ləs) A very large city; a region made up of several large cities and their surrounding areas, considered to be a single urban complex, p. 124

mercenary A person who works as a soldier purely for financial gain; a professional soldier hired by a foreign country, p. 545

mesa (mā/sə) A small, high, flat-topped landform with one or more clifflike sides, p. 166

mestizo (mes tē/zō) A person of mixed Spanish and Native American heritage, p. 178

meteorologist A scientist who studies atmosphere and weather, pp. 170–171

metropolitan area A central city surrounded by smaller communities known as suburbs, p. 54

militia An army; a part of the organized armed forces of a country liable to be called only in an emergency; also, the private army of a particular fighting faction, p. 466

minaret A tall, slender tower attached to a mosque, p. 440

minority A racial, ethnic, religious, or political group whose members remain distinct from the major group in the area, p. 268

monarchy A system of authoritarian government headed by a monarch—a king, queen, shah, or sultan—whose position is usually inherited, p. 60

monotheism The belief in one God, p. 439

monsoon A seasonal shift in the prevailing wind that influences large climate regions, p. 577

moor Broad, treeless, rolling land, often poorly drained and having patches of marsh and peat bog, p. 282

moraine (mə rān/) A ridgelike mass of rock, gravel, sand, and clay carried and deposited by a glacier, p. 30

mosque An Islamic house of religious worship, p. 440

mouth The place where the water of a river enters a lake, a larger river, or the ocean, p. 28

muezzin (myoo ez'in) In Islam, a crier who calls the faithful to prayer five times each day from a minaret, p. 440

mulatto (mə lat/ō) A person of mixed ancestry, p. 236

multiethnic Composed of many ethnic groups, p. 351

multilingual Able to speak several languages, p. 267

Muslim A follower of Islam, p. 440

N

nationalism Devotion to the interests or culture of a nation; the desire for national independence to promote a common culture or interests, p. 416

natural resource A material that humans take from the natural environment to survive and to satisfy their needs, pp. 67–68

natural vegetation The typical plant life that abounds in areas where humans have not significantly altered the landscape, p. 43

navigable Deep or wide enough to allow the passage of ships, p. 324

nonaligned Neutral; not in alliance with either side in a political conflict; during the cold war, tied to neither a communist nor democratic country, p. 378

nonrenewable resource A natural resource that cannot be replaced once it is used—for example, minerals such as fossil fuels, iron, copper, aluminum, uranium, and gold, p. 68

nonviolent resistance The policy of opposing an enemy or oppressor by any means other than violence, p. 596

nuclear energy A type of energy produced by fission—the splitting of uranium atoms in a nuclear reactor, releasing stored energy, p. 69

O

oasis (ō ā/ sis) A place where a supply of fresh water makes it possible to support life in a dry region, p. 432

ore A rocky material containing a valuable mineral, p. 281

outback Remote, sparsely settled, arid rural country, especially the central and western plains and plateaus of Australia, p. 708

P

pack ice Floating sea ice formed by a mix of icebergs with other ice formed in superchilled ocean waters, p. 744

paddy Irrigated or flooded land on which rice is grown, p. 690

paramo (par/ə mō/) A series of highland valleys and plateaus in the Andes of Ecuador, p. 240

partition A division into parts; a separation; to divide into parts, p. 597

peat Spongy material containing waterlogged and decaying mosses and plants, sometimes dried and used as fuel, p. 287

per capita GNP The gross national product of a country divided by the country's total population, p. 75

perennial irrigation An irrigation system that provides necessary water to the land throughout the year, p. 488

perestroika (pə′ ru stroi′ kə) A Russian word meaning "a turning about"; in the former Soviet Union, a policy of economic restructuring, p. 411

permafrost A layer of soil just below the earth's surface that stays permanently frozen, p. 47

pharaoh A ruler of ancient Egypt, worshiped as a god, p. 486

plain A flat or gently rolling area of land with few changes in elevation, p. 20

plant community A mix of interdependent plants that grow naturally in one place, p. 43

plate tectonics The theory that the earth's outer shell is composed of a number of large, unanchored plates, or slabs of rock, whose constant movement explains earthquakes and volcanic activity, p. 22

plateau A raised area of mostly level land with at least one side that rises steeply above the surrounding land, p. 20

polar zone One of two high-latitude zones, stretching from 66½° north and south to the poles, p. 33

polder An area of low-lying land that has been reclaimed from the sea by encircling the area with dikes and pumping the water into canals, p. 310

population density The average number of people living in a given area, p. 52

prairie A temperate grassland characterized by a great variety of grasses, p. 90

precipitation All the forms of water that fall to earth from the atmosphere, including rain and snow, p. 36

prevailing westerlies The constant flow of air from west to east in the temperate zones of the earth, p. 260

Prime Meridian An imaginary line of longitude that runs from the North Pole to the South Pole through Greenwich, England; it is designated 0° longitude and is used as a reference point from which east and west lines of longitude are measured, p. 5

prophet A person whose teachings are believed to be inspired by God, p. 440

province A territory governed as a political division of a country, p. 86

provisional Temporary; pending permanent arrangements, p. 660

purdah (pur′də) The practice among Hindu and Muslim women of covering the face with a veil when outside the home, pp. 600–601

Q

quota A fixed quantity; a proportional share assigned to a group or to each member of a group; the number of immigrants allowed to enter a country in a given period, p. 675

R

rain shadow A region of reduced rainfall causing very dry, even desertlike land on the leeward side of high mountains, p. 36

refugee A person who flees his or her country to escape invasion, oppression, or persecution, p. 531

reincarnation The belief that the soul of a human being or animal goes through a series of births, deaths, and rebirths, p. 583

relative location The position of a place in relation to another place, p. 6

relief The differences in elevation, or height, of the landforms in any particular area, p. 20

remote sensing A method by which airplanes and satellites can produce photographs or computer-generated images of sections of the earth's surface, p. 9

Renaissance The revival of art, literature, and learning that took place in Europe during the 14th, 15th, and 16th centuries, p. 332

renewable resource A natural resource that the environment continues to supply or replace as it is used, p. 68

reservoir A body of water collected in a natural or artificial lake, p. 488

revolution One complete orbit of the earth around the sun. The earth completes one revolution every 365¼ days, or one year, p. 32

rift valley A large split along the crest of an underwater mountain system where small earthquakes and volcanic eruptions frequently occur, p. 24

rotation The spinning motion of the earth, like a top on its axis, as it travels through space, pp. 31–32

rugged individualism The willingness of an individual to stand alone and struggle to survive and prosper, relying on his or her own personal resources and beliefs, p. 109

S

sanction A measure taken by the international community to punish a country for unacceptable actions and to bring pressure on that country to reverse its policy, p. 561

savanna A tropical grassland with scattered trees, located in the warm lands nearest the Equator, p. 46

scientific method A systematic approach to knowledge followed by scientists to collect information that will prove or disprove a hypothesis, p. 9

secede To withdraw formally from membership in a political or religious organization, p. 152

sediment Small particles of soil, sand, and gravel carried and deposited by water, p. 28

segregation The act of imposing the social separation of races, p. 561

seismic Descriptive of earthquakes or earth vibrations, p. 623

seismograph An instrument that measures and records movement in the earth's crust, such as earthquakes and other tremors, p. 668

selva (sel′və) A Spanish word meaning "forest" or "jungle"; in Ecuador, Peru, and Bolivia, a forested region, p. 240

separatism Devotion to the cause of winning political, religious, or racial independence from another group, p. 152

shantytown A town or a section of a city, usually on the outskirts, consisting of ramshackle houses made mostly of scrap materials and inhabited by the very poor, pp. 630–631

shifting agriculture A type of agriculture in which a site is prepared and used to grow crops for a year or two, at which point the farmer moves on to a new site, p. 530

silo A tall, round, airtight building used for the storage of grain, p. 131

socialism A system in which the government owns, manages, or controls the production, distribution, and exchange of goods in such basic industries as transportation, communications, and banking, p. 61

solar energy Energy produced by the sun, p. 69

solstice Either of the two times a year when the sun appears directly overhead at noon to observers at the Tropics of Cancer and Capricorn, p. 32

sovereignty A country's freedom and power to decide on policies and actions, p. 58

soviet In the former Soviet Union, any one of various governing councils that made decisions at various levels, p. 397

sphere of influence An area or country that is politically and economically dominated by, though not directly governed by, another country, p. 642

standard of living A person's or group's level of material well-being, as measured by education, housing, health care, and nutrition, p. 94

state farm A government farm in the former Soviet Union on which workers received wages, p. 410

steppe (step) A temperate grassland, often lightly wooded, found in Europe and Asia, p. 389

storm surge A wave created by a storm rising as much as 25 feet (8 m) above sea level, p. 170

strategic value Importance of a place or thing for nations planning military actions, p. 555

structural adjustment program A program to reform the structure of an economy, pp. 538–539

subcontinent A large landmass; a major subdivision of a continent, p. 575

subduction zone An area where two tectonic plates meet and one plate sinks under the other, p. 24

subsistence farming Farming that provides only enough for the needs of a family or a village, p. 75

suburb A usually residential area or community on the outer edge of a city, p. 123

summit The highest point of a mountain or similar elevation, p. 256

sunbelt The southern and southwestern states of the United States, from the Carolinas to southern California, characterized by a warm climate and, recently, rapid population growth, p. 128

T

taiga A region of Northern Eurasia, thinly scattered with coniferous trees and relatively unpopulated, p. 392

tariff A duty or tax imposed by a government on imported or exported goods, p. 675

temperate zone One of two middle latitude zones that extend from 23½° North to 66½° North and from 23½° South to 66½° South, p. 33

terrace In farming, a flat, narrow ledge of land, supported by walls of stone and mud parallel to the natural slope of the land, usually constructed in hilly areas to increase the amount of arable land, p. 631

theocrat The ruler of a theocracy; someone who claims to rule by religious or divine authority, p. 653

timberline The boundary in high elevations above which continuous forest vegetation cannot grow, p. 174

topography (tə păg′ rə fē) The physical features of a place or region of the earth's surface; the science of graphically representing on a map the exact physical features of a place or region, p. 11

totalitarianism A system of government in which a central authority controls all aspects of society, subordinating individual freedom to state interests, p. 60

tributary A river or stream that flows into a main river, p. 87

tropical depression A storm system built around an organized low atmospheric pressure area, p. 170

tropical storm A storm with winds of at least 39 miles (63 km) per hour, p. 170

tropical zone One of two low-latitude zones that extend up to 23½° North and 23½° South, p. 33

trust territory A dependent colony or territory supervised by another country or countries by commission of the United Nations, p. 719

tsunami (tsoo nä′mē) A huge wave caused primarily by a disturbance beneath the ocean, such as an earthquake or a volcanic eruption, p. 336

tundra A region where temperatures are always cool or cold and only specialized plants can grow—either alpine tundra, in high elevations, or arctic tundra, in high latitudes, p. 47

typhoon A destructive tropical storm occurring in the western Pacific Ocean or the China Sea; similar to a hurricane, p. 669

 # U

unitary A system of government in which one central government holds most of the political power, p. 59

urbanization The growth of city populations; the change from a rural, or countrylike society, to one that is urban, or citylike, in character, p. 54

 # V

villagization A political movement by which rural people are forced to move to towns and work on collective farms, p. 558

volcanism Volcanic activity involving the flow of magma, or molten volcanic rock, p. 21

 # W

wadi (wä′dē) A gully, or usually dry riverbed, cut in the earth by running water after a downpour in arid regions of the Middle East and North Africa, p. 431

warlord A local leader with a military following, p. 642

weather The condition of the bottom layer of the earth's atmosphere in one place over a short period of time, p. 31

weathering The chemical or mechanical process by which rock is gradually broken down, eventually becoming soil, p. 27

windward Situated on the side facing toward the direction from which the wind is blowing, p. 36

 # Z

Zionist A member of a movement known as Zionism founded to promote the establishment of an independent Jewish state, p. 452

United States, 87–89; of Western Europe, 299–300, 328, 335
Cocos Plate, 193
coffee production: in Africa, 553, 557; in Latin America, 200, 210, 225, 228, 239; in Southeast Asia, 690
"cold war," 359
collective, 644
collective farm, 373, 375
collision zone, 25–26, 744
Colombia, 116, 174, 238–239
colonialism, 271
Colorado River, 195
Colosseum (Rome), 330
Columbia, South Carolina, 99
Columbus, Christopher, 98, 271, 327
command system, 411
commercial farming, 75, 105, 200
communication system: of Australia, 722–723; of Canada, 153; of China, 634–635; of Saudi Arabia, 470; of Soviet Union, 382, 400; of United States, 108–109, 383; worldwide, 382–383
communism, 61; in Albania, 376, 378; in Angola, 565; in Bulgaria, 376, 379; in Cambodia, 697; in China, 637, 641–648, 655, 657–658, 660; in Cuba, 217; in Czechoslovakia, 371; in East Germany, 303, 305; in Eastern Europe, 353–355, 359; in Hungary, 371, 374–375; in Laos, 697; in Mongolia, 383, 663; in Mozambique, 565; in North Korea, 679, 682; in Poland, 367–368; in Romania, 376, 379; in Southeast Asia, 695; in Soviet Union, 382, 397, 410; in Tibet, 653; in Vietnam, 697; in Yugoslavia, 376, 378
Communist German Democratic Republic. *See* East Germany
compass rose, 14
computer industry, 230, 604
Confederate States of America, 125
confederation, 60
conformal map, 13
Confucianism: in China, 635–636, 641, 658; in Japan, 670; in Korea, 681; in Southeast Asia, 636
Congo, 515, 542, 544
Congo River. *See* Zaire River
coniferous forest, 44–45; of Mexico, 171; of North America, 90, 135; of Northern Eurasia, 392; of Western Europe, 261
Connecticut River, 122
conquistadors, 198, 241
consequences, prediction of, 599
Constantinople, Turkey. *See* Istanbul, Turkey
constitutional monarchy, 60
consumer goods industry, 604–605
continent, 20
continental climate, 38–41; of Canada, 88; of Germany, 305; of Korea, 680; of Northern Eurasia, 390; of United States, 88, 132
Continental Divide, 87

continental drift theory, 23–24
continental glacier, 30
contour map, 739
contras (Nicaragua), 211
convection, 24, 33
convectional precipitation, 36–37
convergence zone. *See* collision zone
Cook, James, 714, 717
copper resources, 68; of Africa, 532, 544–545, 566–567; of Canada, 150; of Eastern Europe, 366, 372, 378; of Korea, 682; of Latin America, 240
Copts, 486
coral island, 214
cordillera, 85, 135, 166
core, earth's, 20
Coriolis effect, 34–35
corn: domestication of, 184–186; production of, 132, 184, 200
Cortés, Hernán, 191, 198
Costa Rica, 168, 207, 209–210
Côte D'Azur, 300
Côte d'Ivoire, 533, 544
cottage industry, 215, 603
cotton production: in Africa, 489, 557; in Latin America, 210, 215, 225; in South Asia, 589, 595, 609; in central Asia, 416
country, 58; population of, 58; territory of, 58
coups, 533
Crete, 336
crevasse, 743–744
criollos, 198
Croatia, 376–377
Croats, 352, 358, 376–378
crust, earth's, 20–22
Cuba, 129, 211, 213, 215, 217
cultural geography. *See* human geography
cultural landscape, 55–56
cultural region, 8
Cultural Revolution, 637, 645, 657–658
cultural trait, 55
culture, 7; of Australia, 707, 714; of Canada, 94, 152–153, 155–156; changes in, 56; diffusion of, 56; of East Asia, 633–636, 655–668, 680–681, 689–691, 696, 698; of Eastern Europe, 351, 353; Islamic, 440; of Latin America, 182, 186, 197–201, 236; material, 55; of Middle East, 450, 458–459; nature of, 54–56; nonmaterial, 55; of North Africa, 485–486, 492–497; of Northern Eurasia, 397, 400; of Pacific Islands, 737; of South Asia, 583–584, 598, 609; of sub-Saharan Africa, 517–518, 520–521, 543, 555, 559; of United States, 94, 127, 155–156; of Western Europe, 284–285, 287, 292–293, 300–302, 315, 337
culture hearth, 56, 649
cuneiform, 445
Curaçao, 213
customs (tariff), 155–156
cyclone, 732

Cyprus, 477
Cyrillic alphabet, 356, 376–377
Czechoslovakia, 345, 347, 351–353, 356–359, 371–374

D

Dacca, Bangladesh, 610
Dalai Lama, 653
Dallas, Texas, 128–129
Damascus, Syria, 466
Danish language, 268
Danube River, 307, 345, 347, 356, 373, 375–376, 378
Daoism, 635–636, 658
Dardanelles, 474
Darwin, Charles, 734
Darwin, Australia, 732
data: census, 10; collection and display of, 9–11
de Gaulle, Charles, 300
de Klerk, F.W., 562
Dead Sea, 433, 456
death rate, 53, 606
debt, of sub-Saharan Africa, 517, 533, 545
Deccan Plateau, 577, 586
deciduous forest, 44; of North America, 90, 135; of Northern Eurasia, 392; of Western Europe, 261
deforestation, 530–531, 544, 606–607, 613
Delaware, 125
delta, 28–29, 266, 609; inland, 532
demand system, 411
democracy, 60, 109; representative, 60
demography, 51, 53
Deng Xiaoping, 645–647, 656
Denmark, 255, 290–293
Denver, Colorado, 100, 114
dependency (territorial unit), 182
Des Moines, Iowa, 114
desalinization, 469, 471
Descartes, René, 301, 309
desert, 44–46; of Australia, 708; of East Asia, 625, 652; of Latin America, 171, 195, 240, 246; of Middle East, 428–434, 455–457; of North Africa, 427–434, 484, 492; of South Asia, 610; of Northern Eurasia, 389–390, 393, 395; of sub-Saharan Africa, 507, 537, 556; of United States, 90–92; vegetation of, 46, 92, 429
desert caravan, 493–494
desert spring, 431–432
desertification, 531, 556
Detroit, Michigan, 97, 111, 133
developed country, 74–75
developing country, 74–75
DEW Line. *See* Distant Early Warning System
dhows, 468
dialect, 301, 582, 657
diamond resources, 407, 518, 544, 560

dictatorship, 60
Diem, Ngo Dinh, 697
Diet (Japan), 672
diffusion, of cultural traits, 56
dike, 309–310
Dinaric Alps, 334, 348, 376
direction: cardinal, 14; intermediate, 14
Distant Early Warning System (DEW Line), 154
distortion, map, 12
District of Columbia, 125
Djibouti, 555
Dnieper River, 389, 399
doldrums, 34
Dome of the Rock (Jerusalem), 440
domesticated plants, 184–187
Dominican Republic, 213, 215
double cropping, 651
drainage basin, 87
Dravidian language, 582
Dravidian people, 586–588
Dreamtime, 732
Dresden, Germany, 306
drip irrigation, 455, 457
drought, in sub-Saharan Africa, 515, 531
drug war, 116–117, 239
Druze, 459, 465
dry climate, 38–41
dust bowl, 29
Dutch language, 235, 268

E

earth: atmosphere of, 31; core of, 20; crust of, 20–22; geologic history of, 22–26; inner structure of, 19–20; land and water of, 20; mantle of, 20; revolution of, 31–33; rotation of, 31–33; sun and, 31–33; tilt of, 32
earthquake, 22, 24; in Armenia, 422–423; in Japan, 668–669; in Mexico, 193
East Africa, 551–558; coastal region of, 555–556; Tanzania, 557–558; wildlife protection in, 554
East Asia: cities of, 628–631; climate of, 621–626; culture of, 633; elevation of, 624; farming in, 631–633; fishing in, 631; history of, 634–637; human geography of, 628–633; industry of, 632–633; landforms of, 621–626; population distribution in, 628–631; vegetation of, 621–626
East Germany, 267, 303–306, 345
East Pakistan, 598
Easter Island, 24
Eastern Europe, 365–379; climate of, 345–349; communism in, 353–355, 359; culture of, 351; economic activity of, 353–355; farming in, 349, 353; history of, 345, 351–352, 355; human geography of, 351–355; industry of, 353, 356–359; landforms of, 345–349; languages of, 351; religion of, 356–357; vegetation of, 345–349

Eastern Ghats, 577
Eastern Highlands, Russia, 389
Eastern Orthodoxy: in Eastern Europe, 356, 376–377, 379; in Northern Eurasia, 398; in Western Europe, 268
economic activity: of Canada, 96–97, 144, 148, 156; of East Asia, 651–652, 660–663, 671–675, 682, 693–695, 697–698; of Eastern Europe, 353–355, 367–369, 372–373; of Latin America, 200–201, 215–216, 228–231; of Middle East, 437–438, 464; of North Africa, 437–438, 488–489; of Northern Eurasia, 416–419; of Pacific Islands, 734–738; of South Asia, 598, 600–605; of Russia, 407–411; of sub-Saharan Africa, 515–517; types of, 73–74; of United States, 96–97, 105, 110, 156; of Western Europe, 269–270, 289, 293, 306–307, 312, 316–317, 325, 327, 329, 336
economic development, stages in, 74–75
economic system, 60–61
Ecuador, 240–242
Edinburgh, Scotland, 283
Edmonton, Alberta, Canada, 148–149, 155
EEC. See European Economic Community
Egypt, 52–53, 451, 454, 461; cities of, 484–486, 488–489; culture of, 485–486; economic activity of, 488–489; farming in, 438, 484–485, 488–489; Gaza Strip, 460; history of, 486–488; industry of, 489; Israel and, 487–488; language of, 486; natural resources of, 484; population growth in, 488–489; regions of, 483–484; religion of, 486; villages of, 485
Eiffel Tower, 297
ejidos, 200
El Paso, Texas, 125
El Salvador, 209–211
Elbe River, 306
electronics industry, 632, 660–661
elephant, 570–571
elevation, 20, 93; climate and, 38; of East Asia, 624; of Latin America, 193–194, 237; of Middle East, 433–434; of North Africa, 433–434; of South Asia, 580; of Spain, 324–326; of sub-Saharan Africa, 507
Empty Quarter, 428–429, 468
encomienda system, 198
endangered animals, 570–571
energy resources, 68–69; of Brazil, 229–230; of North America, 97, 122–123; worldwide distribution of, 500–501
English language, 94; in Central America, 209; in Guianas, 235; in the Philippines, 698; in South Asia, 589
environment: human-environment interactions, 7; plant, 43

equal-area map, 14
Equator, 5, 32–33
Equatorial Guinea, 542, 544
equidistant map, 14
equinox, 32
erg, 429–430, 556
Erg Chech, 429–430
Ericson, Leif, 144
Erie Canal, 108, 111
erosion, 28–30
escarpment, 223, 505, 544
Estonia, 414
estuary, 279
ethanol (gasohol), 229–230
Ethiopia, 517, 555
ethnocracy, 557
eucalyptus tree, 709–710, 720
Euphrates River, 432, 442, 445, 467
Eurasian Plate, 26, 256, 291, 334, 577, 621
Europe, 266. See also Eastern Europe; Western Europe
European Atomic Energy Commission (Euratom), 312
European Coal and Steel Community (ECSC), 312
European Common Market, 269, 312
European Community (EC), 311–313
European Economic Community (EEC), 157, 312, 327, 329
Europeans: in Africa, 523, 559–560; in East Asia, 642, 671, 690–691, 693; in Latin America, 209–210, 214–215, 241–243; immigration to United States, 123
Everglades, 125

F

Faisal (Saudi Arabian king), 470
falaj system, 472
Falkland Islands, 247
fall-line settlement, 99, 127
family planning. See population policy
famine: in Bangladesh, 610; in Ethiopia, 556; in Kenya, 554; relief efforts, 161; in Sahel countries, 531; in the Sudan, 557; worldwide occurrence of, 160–161
farming: in Canada, 75, 96–97, 145–148, 151; in Caribbean islands, 214–215; in Central America, 208, 210–211; in China, 616, 623, 631–633, 635, 641, 644, 646, 649–653; collective farms, 373, 375; commercial, 75; development of, 442; in East Asia, 75, 631–633, 660, 668, 681, 690, 695, 697; in Eastern Europe, 349, 353, 365, 373, 375, 379; Green Revolution, 616–617; in Guianas, 236; intensive, 631; limits on, 73; in Mexico, 170, 182, 192, 195, 198–200, 202; in Middle America, 182, 184–187; in Middle East, 437–438, 442–443, 455, 457, 467–468, 471–472, 475; in North Africa, 437–

438, 484–485, 488–489, 492, 497; in Northern Eurasia, 387, 395, 416; in Pacific Islands, 699, 719, 734–735, 737; in South America, 182, 184–187, 225, 228–230, 237, 239–243, 246; in South Asia, 575, 579, 581, 589, 601, 603, 606, 608–609, 612–613, 616; in sub-Saharan Africa, 511, 515, 519–520, 529–532, 535, 537, 542, 552–553, 556–560; subsistence, 75, 210; technology and, 106; in United States, 96–97, 100, 105–106, 112–113, 126, 130–132, 200, 617; in Western Europe, 261, 266–267, 279, 287, 289, 293, 298–299, 306–307, 309–311, 316–317, 324–326, 328, 330, 334–335; worldwide distribution of, 160

Farsi language, 475

fault, 21–22, 26; transform, 25

favelas, 226, 228, 230

federal government, 60

Federal Republic of Germany. *See* West Germany

fellaheen *(fellahin),* 438, 485

feluccas, 483, 486

Fertile Crescent, 462

fez, 474

Fiji, 718–719, 736–738

Filipino language, 698

Finland, 256, 258, 290–293

Finnish language, 268

fishing, 73, 75; in Antarctica, 745; in Canada, 98, 144, 148; in East Asia, 631, 681; in Pacific Islands, 719, 737; in South America, 172, 236; in United States, 98, 122, 128, 137; in Western Europe, 283, 293, 300, 336; in Zaire, 513

fjord, 290–291, 710

Flemish language, 267–268, 310–311

floodplain, 28–29

Florence, Italy, 332

Florida, 87, 125, 128–129

food supplies: Green Revolution and, 616–617; worldwide distribution of, 160–161

forage, 530

forest, 43–46; of Canada, 90, 96, 144, 147–148, 341; mid-latitude, 44; of New Zealand, 711; of Northern Eurasia, 392; of Sri Lanka, 613; of sub-Saharan Africa, 515, 544; of United States, 90, 96, 106, 122, 125, 135, 137, 341; of Western Europe, 260–261

fossil fuel, 68–69

fossils, 23–24

France, 256, 258, 263–264; China and, 637; cities of, 302; colonial empire of, 272; culture of, 300–302; economic activity of, 312; farming in, 298–299; fishing in, 300; government of, 301; history of, 300–302, 401, 451; immigrants to Latin America, 180; industry of, 298, 300; invasion of Egypt, 486–487; landforms of, 297; language of, 300–302; mining in, 300; regions of,

297–300; religion of, 268; rule of North Africa, 493; settlements in North America, 99, 129, 151; in Southeast Asia, 690–691, 696–697; tourism in, 299–300; trade by, 302; transportation in, 267

Franco-Prussian War, 304

Frankfurt, Germany, 306

Franks, 300

free enterprise, 109

Free Trade Agreement (United States and Canada), 156

Freetown, Sierra Leone, 533

French Canadians, 151–153

French Guiana, 182, 235–236

French Indochina, 696

French language, 94, 267, 301; in Belgium, 310–311; in Guianas, 235; in Luxembourg, 311; in Switzerland, 315

French Revolution, 301

frontal precipitation, 36–37

frost wedging, 27–28

Fujimori, Alberto, 242

Fulani people, 532, 537

fur trade, in Canada, 151

G

Gabon, 542, 544

Gaelic language, 288

Galveston, Texas, 170

Gambia, The, 533

Gandhi, Indira, 595, 606–607

Gandhi, Mohandas, 595–597

Gandhi, Rajiv, 595

Ganges River, 575, 577, 583, 602–603, 609

Gary, Indiana, 130

gasohol, 229–230

gaucho, 246

Gaul, 300

Gaza Strip, 460, 462

Gdansk, Poland, 369

Geneva, Switzerland, 317

Genghis Khan, 663

Genoa, Italy, 330

geography: definition of, 4; themes in, 4; tools of, 9–15

geologic history, 22–26

geology, 19

Georgia, 415–416

geothermal energy, 69, 291–292

German Empire, 304

German language, 301, 311, 315

Germanic language, 268

Germany, 258, 264; cities of, 305–306; climate of, 305; economic activity of, 306–307, 312; farming in, 266, 306–307; history of, 303–305, 357–358, 401; industry of, 306–307; landforms of, 305–307, 389; manufacturing in, 306; natural resources of, 306; regions of, 305–307; religion of, 268, 303–304

geyser, 710

Ghana, 529, 531, 533–535, 539

Ghat, Libya, 495

ghetto, 366

Gibson Desert, 730

glacier, 29–30, 86, 98, 135, 144, 256, 258, 262, 282–283, 290–291, 743; continental, 30; valley, 30

Glasgow, Scotland, 283–284

glasnost, 411, 415, 418

glen, 284

global warming, 251

globe, 10–11

GNP. *See* gross national product

Gobi Desert, 625, 634, 649, 652, 662

gold resources: of Andean countries, 240; of Antarctica, 745; of Australia, 733; of Canada, 100, 150; of Russia, 407; of sub-Saharan Africa, 514, 529, 560; of United States, 100, 135

Gorbachev, Mikhail, 382, 411, 413, 415, 417–419

Goths, 352, 357

government, 58–60; development of, 445, 486

graben, 334

grain elevator, 132

grain exchange, 132

Grampian Mountains, 282

Gran Chaco, 174, 176, 245

Grand Banks, 98, 122

Grand Canal (China), 635

Grand Erg Occidental, 429–430

Grand Erg Oriental, 429–430

grass roots, 535, 553

grassland, 44–46; of Canada, 90–91, 100; of East Asia, 625; of Latin America, 176, 208, 237, 246; of Northern Eurasia, 389–390, 392–393; of sub-Saharan Africa, 507–511, 542; temperate, 46; tropical, 46; of United States, 85, 90–91, 100, 125, 132

Great Alfold (plain), 375

Great Barrier Reef, 732

Great Britain, 263; China and, 637; in World War I, 451; cities of, 279–280; colonial empire of, 272, 281, 560, 589, 595–597, 661–662, 714, 717, 721–722, 735, 740–741; in Falkland Islands, 247; farming in, 279; government of, 59; immigrants to Latin America, 180; industry of, 279, 281; influence on Southeast Asia, 690–691; invasion of Egypt, 487; landforms of, 277–279; mining in, 281; rural regions of, 277–279; settlements in Kenya, 553; settlements in North America, 99, 151; trade by, 279–281, 699; transportation in, 280; villages of, 277

great circle, 10

Great Dividing Range, 708–709, 730

Great Lakes, 30, 96, 108, 111, 130, 133, 154

Great Lakes–St. Lawrence provinces, of Canada, 145–147

Great Leap Forward, 644, 646

Great Plains, 29, 46, 85, 90–91, 100, 106, 132

Great Rift Valley, 505–507, 551, 557, 564
Great Sandy Desert, 708, 730
Great Victoria Desert, 730
Great Wall (China), 634
Greater Antilles, 167–168, 213–214
Greater Khingan Range, 649
Greece, 58, 256, 348, 376; cities of, 335–336; culture of, 337; Cyprus and, 477; economic activity of, 336; farming in, 266, 334–335; fishing in, 336; history of, 336–337; landforms of, 334–335; religion of, 268; tourism in, 336; trade by, 336
Greek language, 477
Greek Orthodox Church, 465, 477
Green Revolution, 616–617
greenhouse effect, 31, 544
Greenland, 30, 41, 271
Greenwich, England, 5
gross national product (GNP), 75; of Canada, 155; per capita, 75; of Latin America, 183; of United States, 105
growing season, 132
Guadeloupe, 215
Guam, 719
Guangzhou, China, 642
Guarani language, 246
Guatemala, 182, 207–210
Guayaquil, Ecuador, 241
guerrilla, 211
Guiana Highlands, 172, 237
Guianas, 235–236
Guinea-Bissau, 533
Gulf of Corinth, 334
Gulf of Mexico, 192
Gulf Stream, 38, 259
Guyana, 235–236
gypsy, 352, 358

H

hacienda, 198–200
Hague, the Netherlands, 310
Haifa, Israel, 457
Haiti, 213, 215, 217, 703
Hajj, 440, 472
Halifax, Nova Scotia, Canada, 155
Hamburg, Germany, 265, 306
Hamilton, Ontario, Canada, 97, 155
Hammurabi's Code, 445
Han Chinese, 633, 657, 659
Hapsburg dynasty, 357
harambee, 553–554
Hausa people, 532, 537
Havel, Vaclav, 371
Hawaii, 92, 135, 137
hazardous waste, 70
health care, 183; in China, 658; in India, 605; in sub-Saharan Africa, 517
heavy industry, 410
Hebrew language, 458
Hebrews, 439, 452
Heidelberg, Germany, 306
hemisphere, 5
herders, 75, 438, 494

Herzegovina, 376
hierarchy, 112
highland climate, 38–41
Highland Games, 153, 282
Highlands: of South America, 172, 240; of sub-Saharan Africa, 507, 552–553; of Western Europe, 278–279, 282–283
hill, 20
Himalayas, 26, 575, 577, 579, 588, 612, 621, 624, 634
Hindi language, 582–583
Hindu Kush Mountains, 588, 608, 610–611
Hinduism: in Fiji, 718; in Guianas, 236; in South Asia, 583, 587–588, 597, 600–603, 612–613; in Southeast Asia, 690
hinterlands, 114
Hiroshima, Japan, 672
Hispanic people, of United States, 129, 217
Hispaniola, 171, 213
Hitler, Adolf, 304, 358, 371, 401
Ho Chi Minh, 697
Hobart, Australia, 731
Hokkaido, 667, 669
Holland. *See* the Netherlands
Holocaust, 358, 366, 453
Holy Roman Empire, 301
"homelands," of South Africa, 560–561
Homer, 337
homogeneous population, 669
Honduras, 207, 210
Hong Kong, 51, 659, 661–662
Honshu, 623, 629, 667, 669
Horn of Africa, 555
horse latitudes, 35
Hortobaby, 347
Houston, Texas, 128–129
Huang He River, 623, 634–635, 649–651
Hudson Bay, 85–87, 90, 144
Hudson Bay Lowlands, 145
Hudson River, 122
Hudson's Bay Company, 151
human geography, 51–56; of Australia, 714–719; of Canada, 94–97; of East Asia, 628–633, 655–658; of Eastern Europe, 351–355; of Latin America, 178–182, 197–201, 208–210, 236; of Middle East, 436–441; of North Africa, 436–441; of Northern Eurasia, 395–397; of Pacific Islands, 718–719; of South Asia, 581–584, 600–605; of sub-Saharan Africa, 513–519; of United States, 94–97; of Western Europe, 263–269
human-environment interactions, 7
Humboldt Current. *See* Peru Current
humid continental climate, 38–41
humid subtropical climate, 38–41
humidity, 36
Hungarian Basin, 347, 349, 376
Hungarian language, 351
Hungarians, in Yugoslavia, 377
Hungary, 345, 352–353, 356–359, 371, 374–375

hunger. *See* famine
Huns, 352, 357, 588
hunters and gatherers, 75, 184–186, 442, 511
hunting, illegal, 570
hurricane, 170–171
Husayn ibn'Ali, 451
Hussein, Saddam, 467
Hutu people, 557
hydroelectricity, 97, 146, 229, 246, 347, 540–544, 566, 681
Hyksos people, 486
hypothesis, 9

I

Iberian Peninsula, 255, 323
Ibo people, 537–538
Ice Age, 30, 98, 290
ice cap, 38–41
ice sheet, 742–743
ice shelf, 743–744
iceberg, 744
Iceland, 24, 69, 262, 271, 290–293
Icelandic language, 268
ideogram, 657
ideology, 410
Illinois, 132–133
Illinois River, 87
immigration, 8, 56
Inca, 178, 241
incineration, 71
independence: of Albania, 357; of Algeria, 494, 496; of Angola, 565; of Canada, 151–152; of Egypt, 487; of Greece, 337; of Hungary, 358; of India, 589, 595–597; of Iraq, 451; of Jordan, 451, 464; of Latin American countries, 181–182; of Lebanon, 451; of Libya, 493–495; of Mexico, 198; of Morocco, 494; of Mozambique, 565; of Myanmar, 694; of Nigeria, 538; of Pacific Islands, 738; of Pakistan, 597; of Republic of Ireland, 288; of Southeast Asian countries, 691; of Syria, 451; of Tanzania, 558; of Thailand, 695; of Tunisia, 494; of Vietnam, 697; of Zaire, 545
independence movement: in Azerbaijan, 416; in Baltic states, 414; in Eastern Europe, 354–355; in Georgia, 416; in Lithuania, 418–419; in Soviet Union, p. 397
India, 26, 575, 579–582, 586, 589; British rule of, 272; caste system of, 604; cities of, 601–603, 605; economic activity of, 600–605; farming in, 601, 603, 606, 616; government of, 595; health care in, 605; human geography of, 600–605; immigrants to Fiji from, 718; industry of, 595, 603–605; influence in Southeast Asia, 690–691, 696; population of, 58; population policies of, 606–607; religion of, 597, 600–601; schools of, 605; villages of, 600–601

Maracaibo Lowlands, 238
Marco Polo, 3–4, 625
Margrethe (Queen of Denmark), 293
Mariana Trench, 734
marine climate, 37–41; of United States, 88–90; of Western Europe, 259–260, 279, 287, 292
"maritime," 144
Maritime provinces (Canada), 143–145
market economy. *See* capitalism
marketplace: of Central America, 207; of Thailand, 693
Maronite Christians, 465
Marseille, France, 300
Marshall Islands, 736
marsupial, 720–721
martial law, 648
Martinique, 215
Marx, Karl, 397, 410
Maryland, 125
Masai people, 551–552, 558
mass production, 123
Massachusetts, 122–123
Massif Central, 299
Matadi, Zaire, 542
material culture, 55
Mau Mau Rebellion, 553
Mauritania, 529, 531–532
Mauritius, 555
Maya people, 57, 178, 186
Mazatlán, Mexico, 195
Mecca, Saudi Arabia, 440, 451, 472
mechanical weathering, 27
Medina, Saudi Arabia, 451, 472, 495
Mediterranean climate, 38–41, 46; of Andean countries, 240; of Eastern Europe, 349; of North Africa, 492; of United States, 90; of Western Europe, 260, 324, 328, 335
Mediterranean Europe, 322–337
Mediterranean Sea, 255–256, 336, 427, 492
megalopolis, 123–124
Meghna River, 609
Meiji reforms, 672
Mekong River, 622, 690
Melanesia, 711, 719, 736–737
Melbourne, Australia, 730
Mendenhall Glacier, 30
mental map, 13
Mercantile Exchange, 132
Mercator projection, 12–13
mercenary, 545
mercury resources, 353
meridian. *See* longitude
merino sheep, 733
mesa, 166
Meseta (plateau), 324–325
Mesopotamia, 442–445, 449
mestizo, 178, 180, 198, 209–210, 236, 241, 243, 246
meteorologist, 170–171
metropolis, 114
metropolitan area, 52, 54, 74, 110
Mexican Revolution, 198–199
Mexico: cities of, 191–192, 194–195, 200, 202–203; climate of, 168, 193; culture of, 197–201; economic activity of, 200–201; elevation of, 193–194; farming in, 170, 182, 192, 195, 198–200, 202; government of, 198–199; human geography of, 197–201; landforms of, 165–166, 191–195; language of, 199; manufacturing in, 201; mining in, 201; natural resources of, 195; religion of, 199; tourism in, 201; vegetation of, 171
Mexico City, Mexico, 168, 178, 191, 193–194, 197–198, 200–203
micro-climate, 39
Micronesia, 711, 719, 736–737
Mid-Atlantic Ridge, 24
Middle America: cities of, 180, 182, 186; climate of, 165–171; culture of, 182, 186; farming in, 182, 184–187; history of, 178–182, 184–187; human geography of, 178–182; landforms of, 165–171; languages of, 180; natural resources of, 184–187; population distribution in, 178–179; religion of, 180; towns of, 186; vegetation of, 165–171, 184–185
Middle East: cities of, 437–439, 443–445; climate of, 427–434, 436–437; culture of, 450; economic activity of, 437–438; farming in, 437–438, 442–443; history of, 56, 442–445, 449–454; human geography of, 436–441; Jews in, 451–454; landforms of, 427–434; language of, 450; nomads of, 437–438; oil resources of, 441, 500–501; population distribution in, 436–437; religions of, 438–441, 450; vegetation of, 427–434, 436–437; water resources of, 430–433; World War I in, 451, 453
mid-latitude forest, 44
mid-oceanic ridge, 24
Midwest region, of United States, 130–134
migrant workers, Mexican, 200
migration: from Asia to North America, 98; from Europe to North America, 98–99
Milan, Italy, 330
militia, 466
minaret, 440
mineral resources, 68
mining, 73; in Australia, 67, 733; in Canada, 100, 147; in Eastern Europe, 372–373, 375, 378; in Israel, 456; in Korea, 682; in Latin America, 201, 229–230, 236, 243; open pit, 67; in Pacific Islands, 699, 738; in sub-Saharan Africa, 529, 545, 559–560, 567; in tropical rain forests, 250; in United States, 96, 100, 106–107, 135; in Western Europe, 265, 281, 285, 293, 300
Ministry of International Trade and Industry (Japan), 675
Minneapolis, Minnesota, 114, 132–133
Minnesota, 133
minority, 268
Mississippi River, 28–29, 87, 100, 110–111, 133
Missouri, 130
Missouri River, 87, 111
mixed economy, 61, 293
mixed forest, 45; of Mexico, 171; of North America, 90; of Western Europe, 260
Mobutu Sese Seko (of Zaire), 545
moderate climate, 38–41
Mogul Dynasty, 588
Mohammad Reza Pahlavi (Shah of Iran), 476
Mohammed, 584
Mohave Desert, 36
Moldova, 414–415
Mollweide projection, 14
Monaco, 58
monarchy, 60; constitutional, 60
Mongolia, 382–383, 625, 649, 659, 662–663
Mongol people, 399–400, 657, 663
monotheism, 439
Mons people, 689
monsoon: in East Asia, 624–625, 631, 669, 679; in South Asia, 577–580, 609, 612
Mont Blanc, 256, 299
Montenegro, 376
Montevideo, Uruguay, 245
Montreal, Quebec, Canada, 87, 99, 147, 155
Moors, 324, 326, 532
moors (Scotland), 282–283
moraine, 30
Moravia, 372–373
Morocco, 434, 491, 493–495, 497
Moscow, Russia, 390, 395, 399
mosque, 357, 376, 440, 477, 611
Mossi people, 532
Mount Aconcagua, 245
Mount Etna, 328
Mount Everest, 612
Mount Fuji, 21, 623
Mount Kilimanjaro, 38, 507
Mount Olympus, 335
mountains, 20; of East Asia, 621, 624, 679–680; of Latin America, 166, 172, 191–192, 208, 214; leeward side of, 36; of Middle East, 433–434; of North Africa, 433–434; of Northern Eurasia, 389–390; weathering of, 28; of Western Europe, 256, 305, 324, 334–335; windward side of, 36
mouth, of river, 28
movement, 7–8, 10
Mozambique, 511, 565–566
Mubarak, Hosni, 488
muezzin, 440
Mugabe, Robert, 567
Muhammed, 440
mulatto, 236
multiculturalism, 153
multiethnic nation, 351
multilingual region, 267
Munich, Germany, 307
Murray River, 708
Muslims. *See* Islam
Myanmar, 116, 625, 693–694

P

Pacific Islands, 699; climate of, 707–711; culture of, 737; economic activity of, 736–738; farming in, 719, 737; fishing in, 719, 737; formation of, 736; history of, 718–719, 736–738; human geography of, 718–719; island groups within, 736–737; landforms of, 707–711; mining in, 738; population distribution in, 719; relative location of, 720–723; religion of, 718; vegetation of, 707–711
Pacific Plate, 193, 623, 734, 736
Pacific Rim, 24, 722
pack ice, 744
paddy, 690
Pakistan, 116, 575, 584, 586, 588, 597–598, 608–609, 616, 703
Palestine, 439, 450–454; partitioning of, 454, 460
Palestine Liberation Organization (PLO), 461, 465
Palestinians, 453–454; in Jordan, 464; in Lebanon, 465–466
Pamirs Mountains, 390
pampas, 46, 174, 176, 246
pamperos, 176
Panama, 207, 210
Panama Canal, 207, 209–210, 212, 217
Pangaea, 23, 720
Papua New Guinea, 633, 693, 699
Paraguay, 174, 245–246
Paraguay River, 245
Parakana Indians, 231
parallel. See latitude
paramos, 240
Parana River, 245–246
Paris, France, 297–298, 302
Paris Basin, 298
partition, of India and Pakistan, 597
Patagonia, 174, 177, 246
Peace Corps, 161
Pearl Harbor, Hawaii, 672
Pearson, Lester, 157
peat, 287
Peloponnesus, 334
peninsula, of Western Europe, 255–256
peninsulares, 198
Pennine Mountains, 281
Pennsylvania, 122, 124, 133
People's Republic of China, 644
per capita GNP, 75
perennial irrigation, 488
perestroika, 411, 418
permafrost, 47, 90, 155, 391
Perón, Juan Domingo, 247
Perry, Matthew C., 671
Persian Gulf region, 441, 500–501
Persians, 450, 475, 477, 588
Perth, Australia, 730
Peru, 116, 172, 182, 187, 240–242
Peru Current, 172
Peshawar, Pakistan, 608, 611
Peter the Great (of Russia), 401
pharaoh, 486

Philadelphia, Pennsylvania, 110, 122–123
the Philippines, 623, 690, 693, 698
physical region, 8
Pilcomayo River, 245
Pindus Mountains, 256
Piraeus, Greece, 336
place, 6–7
plain, 20; of Australia, 708; of East Asia, 623; of Middle America, 166–167; of Northern Eurasia, 387–389; of South America, 172–176; of Western Europe, 258, 305–306
planned economy. See communism
plant community, 43; desert, 46, 429
plantation: in developing countries, 75; in Latin America, 209, 214, 216–217, 228–229; in Southeast Asia, 690; in Sri Lanka, 613
Plata River system. See Río de la Plata
plate tectonics, 22–26
plateau, 20; of Australia, 708; of East Asia, 621; of Latin America, 166–167, 172, 177, 191–194, 223; of Middle East, 433; of North Africa, 433; of sub-Saharan Africa, 505–507
Plateau of Tibet, 621, 634, 653
plaza, 180, 182
Plaza of the Three Cultures, 197
PLO. See Palestine Liberation Organization
Po River, 258, 329–330, 348
poaching, 570–571
Poland, 347, 353, 356–358; culture of, 353; economic activity of, 367–369; farming in, 365; government of, 368–369; history of, 358, 365–366; landforms of, 365–366, 389; natural resources of, 366; religion of, 367
polar climate, 38–41, 390
polar zone, 33
polder, 310
Polish language, 351
political system, 58–60; of former Soviet Union, 407–413, 418–419; of United States, 109
Polynesia, 711, 719, 736–737
population: of Canada, 94; of country, 58; homogeneous, 669; of United States, 94; world's, 52–53
population density map, 11, 57
population distribution, 51–53; in Australia, 715–716; in Canada, 95–97; in East Asia, 628–631, 655–666, 669–670; in Latin America, 178–179, 183; in Middle East, 436–437; in North Africa, 436–437; in Pacific Islands, 717–719; patterns of, 54; in Russia, 395–396; in South Asia, 581, 610; in sub-Saharan Africa, 513–515, 537–538, 554; in United States, 95–97, 112, 124, 128–129; in Western Europe, 263–269, 310, 329
population growth, 53, 488–489
population policy: of China, 655–656; of India, 606–607
population pyramid, 196

Portugal, 255; China and, 637; climate of, 326; colonial empire of, 271–272, 327, 565; economic activity of, 327; farming in, 326; history of, 326–327; industry of, 327; landforms of, 326; religion of, 268; settlements in Latin America, 178–180, 223–224; trade by, 271, 326–327
Portuguese language, 180, 267
Potala Palace, 653
potato: domestication of, 186; famine in Ireland, 272, 289; production of, 186–187
Potomac River, 129
Prague, Czechoslovakia, 373
prairie, 46, 90, 125, 132
Prairie provinces (Canada), 147–148
precipitation, 36; in Canada, 90; convectional, 36–37; in East Asia, 624–626; frontal, 36–37; in Latin America, 168–169, 174, 193, 214; in Middle East, 428–429; in North Africa, 428–429; in Northern Eurasia, 390; orographic, 36–37; in South Asia, 577–580; in sub-Saharan Africa, 507–509; in United States, 90, 125, 135
Presbyterian Church, 284
prevailing winds, 34–35, 214, 260
Prime Meridian, 5–6
Prince Edward Island, Canada, 143–145
private enterprise, 61
Promontory, Utah, 101
prophet, 440
Protestant Reformation, 288, 303–304
Protestants: in Ireland, 287–288; in Western Europe, 268
Provence, 297
province, of Canada, 86
provisional government, 660
Prudhoe Bay, Alaska, 136
Prussia, 304
Prussian Empire, 357
Puerto Rico, 171, 213, 215, 217
Puerto Vallarta, Mexico, 195
Punjab region, of Pakistan, 609
Punjabi language, 609
purdah, 600
P'yongyang, North Korea, 680
pyramid, 486
Pyrenees Mountains, 256, 323–324

Q

Qaddafi, Muammar, 496
Qatar, 468–470
Qing Dynasty, 659
Quebec, Canada (province), 86, 94, 145–147, 151–153
Quebec, Quebec, Canada (city), 99, 147, 155
Quebecois, 152
Quéchua language, 241
Queensland, Australia, 731–732

questions, formulation of, 262
Quito, Ecuador, 241

R

Radio Free Europe, 382
railroad: in Canada, 101, 148, 153; in Japan, 629–630; in Northern Eurasia, 396; in Russia, 407, 409–410; in sub-Saharan Africa, 542–543, 552–553; in United States, 101, 108, 111, 133, 137; in Western Europe, 267
rain shadow, 36, 90, 324
Raleigh, North Carolina, 99
Ramadan, 440–441
ranchos, 237
Randstat, 310
recycling, 70–71
Red Guards, 645, 658
Red Sea, 472
refugee, 531, 597; global problem of, 702–703
Regina, Saskatchewan, Canada, 148
region, 8
reincarnation, 583, 588
relative location, 6, 720–723
relief, 20
religion, 55; of East Asia, 588, 657–658, 667, 670, 680–681, 689–690, 698; of Eastern Europe, 356–357, 367, 373–374, 376–379; of Latin America, 180, 199, 223, 236; of Middle East, 438–441, 450, 458, 464–466, 471–472, 474, 477, 486; of North Africa, 438–441, 493–494, 496; of Northern Eurasia, 398, 415–416; of Pacific Islands, 718; of South Asia, 583–584, 587–588, 597–601, 609, 611–613; of sub-Saharan Africa, 532, 535, 556; of Western Europe, 268, 284, 287–289, 293, 303–304, 329–332
remote sensing, 9–10
Renamo, 565
renewable resources, 68
Reno, Nevada, 135
representative democracy, 60
Republic of Ireland, 287–289
Republic of Sao Tome and Principe, 542, 544
Republic of South Africa. *See* South Africa
reserves, of Africans in South Africa, 560
reservoir, 488
Reza Khan, 476
Rhine River, 258, 263, 266, 300, 306–307
Rhode Island, 122–123
Rhonddda Valley, 285
Rhône River, 258, 299
rice production: in East Asia, 631, 651, 690, 697; in Italy, 330; in Latin America, 215, 225; in South Asia, 579
rift valley, 24
Ring of Fire, 26, 623, 668–669

Río de Janeiro, Brazil, 223, 225–226, 229
Rio de la Plata, 245–246
river: of East Asia, 621–623, 650; of Latin America, 166–167, 174, 245; of Middle East, 432–433; mouth of, 28; navigable, 324; of North Africa, 432–433; of Northern Eurasia, 389, 396; of South Asia, 577; of sub-Saharan Africa, 505; of Western Europe, 258, 305–306, 324
Rivera, Diego, 178, 199
riverbed, 430–431
Riviera, 299–300
Riyadh, Saudi Arabia, 468
road: in Brazil, 229; in Russia, 408; in United States, 108, 111–112
Robinson projection, 12
Rocky Mountains, 30, 85, 90, 96, 100, 135, 148
Roman Catholicism. *See* Catholicism
Roman Empire, 300, 330, 337, 352, 356, 492
Romance language, 267–268
Romania, 345, 347–348, 356, 376, 378–379, 414
Romansch, 315
Rome, Italy, 330–332
Roosevelt, Theodore, 212
Ross Island, 749
Ross Sea, 744
Rotterdam, the Netherlands, 310
Rub'al-Khali. *See* Empty Quarter
rubber production, 73, 613, 690, 699
rugged individualism, 109
Ruhr River, 306
Ruhrstadt, 306
Russian Empire, 357, 400–401
Russian language, 397
Russia, 407–411, 416–417
Russians, 397–401
Rwanda, 557

S

Sadat, Anwar, 487
SADCC. *See* Southern African Development Coordination Conference
Sahara, 427–432, 484, 492, 527–528, 531
Sahel, 511, 515, 527
Sahel countries: drought in, 531; farming in, 529–532; history of, 528–529; livestock production in, 530; natural resources of, 532; religion of, 532; trade by, 529; transportation in, 531; vegetation of, 527–528
St. Croix, Virgin Islands, 171
St. Lawrence Lowlands, 144–145
St. Lawrence River, 97, 99, 145–147, 151
St. Lawrence Seaway, 145–146, 154
St. Louis, Missouri, 111, 133
St. Moritz, Switzerland, 317
St. Paul, Minnesota, 114
St. Petersburg, 400–401; 409–410

St. Tropez, France, 300
salt production, 456
San Andreas Fault, 26
San Antonio, Texas, 128–129
San Diego, California, 135
San Francisco, California, 88
San Joaquin Valley, 137
San Salvador, 207
San'a, Yemen, 472
sanctions, 561
sand dune, 429–430
Sandinistas, 211
sandstorm, 29
Sanskrit, 587
Santiago, Chile, 60, 243
Santo Domingo, Dominican Republic, 213
São Paulo, Brazil, 42, 226, 229–230, 273
Sarajevo, Yugoslavia, 357–358, 376
Sardinia, 332
Saskatchewan, Canada, 147–148
Saskatoon, Saskatchewan, Canada, 148
satellite image, 9
Saudi Arabia, 451, 460, 468–469, 500–501; modernization of, 470–472; oil reserves of, 441; religion of, 471–472
savanna, 46, 174, 176, 245, 507–511, 527, 537, 542
Scandinavian Peninsula, 255, 258, 260, 290
schools: of India, 605; of Japan, 673; of South Africa, 561
scientific method, 9
Scotland, 277, 280, 282–284
Sea of Galilee, 456
Sea of Japan, 625, 636
sea-floor spreading, 24
seasons, 32
Seattle, Washington, 114, 135
secession, 152; Lithuania from Soviet Union, 418–419
sediment, 28–29, 266, 484
segregation, 560–561
Seine River, 258, 297–298
seismic activity, 623
seismograph, 668
Seljuk people, 450
selva, 240–241
semiarid climate, 38–41; of Latin America, 168, 193; of Middle East, 428; of North Africa, 428; of South Asia, 580; of Northern Eurasia, 390; of Spain, 324; of United States, 125, 135
Senegal, 531, 533
Senegal River, 529, 531, 541
Seoul, South Korea, 680–681
Separatists, French-Canadian, 152
Serbia, 376–377
Serbs, 352, 358, 376–378
Serengeti Plain, 508
sertao, 224–225, 229
service industry, 73–74, 110
Shaanxi province (China), 644
Shanghai, China, 628, 651
shantytown, 630–631

Suez Canal, 484, 487
sugar production: in Latin America, 200, 214–215, 223–225; in South Asia, 589
summit, 256
sun: distribution of heat from, 33–37; earth and, 31–33
Sun Yat-sen, 642
Sunbelt, 128
Sunni Muslims, 440, 465
Sunshine Coast (Australia), 731–732
supercontinent, 23
Supreme Soviet, 397
Suriname, 235–236
suttee, 589
Swaziland, 511, 564
Sweden, 58, 256, 258, 290–293
Swedish language, 268, 293
Swiss Confederation, 316
Switzerland, 256, 261, 267–268, 314–317
Sydney, Australia, 714, 729–730
Sykes-Picot Agreement, 451
Syria, 451, 460–462, 466
Syrian Desert, 428

 T

Tahiti, 739
taiga, 392
Taipei, Taiwan, 660
Taiwan, 632–633, 644, 659–661
Taj Mahal, 588
Tajikistan, 416
Taklamakan Desert, 634, 636, 652
Tamale, Ghana, 534
Tamil language, 586
Tamil people, 613, 703
Tampico, Mexico, 201
Tanzania, 508, 551–552, 557–558
Taoism, 680–681
Tarbala Dam, 608–609
tariff, 155–156, 675
Tasman Sea, 710, 730
Tasmania, 730–731
Tasmanian devil, 720
Taurus Mountains, 434
tea production: in East Asia, 631, 690; in Kenya, 553; in South Asia, 589, 613
technology, 55–56
Tehran, Iran, 438
Telugu language, 586
temperate grassland, 46
temperate zone, 33
temperature, land vs water, 37
Tenochtitlán, 178, 198
Teotihuacán, 57
terrace, 631, 651
Texas, 125, 127–129
textile industry: of India, 595; of United States, 127; of Western Europe, 283, 298, 306, 317
Thai language, 695
Thai people, 689
Thailand, 116, 625, 631, 690, 693, 695

Thames River, 279
Thar Desert, 608
theocrat, 653
Thera, 336–337
Tian Shan Mountains, 621, 652
Tiananmen Square, 648, 652, 654
Tibet, 653
Tibetans, 657
tiempo muerto, 217
tierra caliente, 241
tierra fria, 241
tierra helada, 237
tierra templada, 237
Tigris River, 432, 442, 445, 467
Tijuana, Mexico, 194–195
timberline, 174
Tito, Marshal, 378
tobacco production, 690
Tobago, 213, 215
Tokyo, Japan, 629
Tombouctou, Mali, 529
Tonga, 736, 738
topographical map, 11
topography, 11
Torah, 439
Tornado Alley, 133
Toronto, Ontario, Canada, 97, 146–147, 155
totalitarianism, 60, 353
tourism: in Antarctica, 749; in Latin America, 201, 213, 215–216; in Thailand, 695; in Western Europe, 299–300, 317, 330, 336
trade: by Australia, 722; by Canada, 97; by Czechoslovakia, 373; by East Asia, 634, 636, 641, 651–652, 662, 671, 673, 675, 689–690, 698–699; by European Community, 312–313; by Latin America, 240–241, 287; by Mesopotamia, 444; by North Africa, 492–493; by sub-Saharan Africa, 516, 529, 533; transportation and, 78–79; by United States, 79, 97; vertical, 241; by Western Europe, 271, 279–281, 302, 310, 326–327, 336, 699
trade winds, 35, 169–170
traditional medicine, 658
Trans-antarctic Mountains, 743
transcontinental railroad, 101, 108, 153
Transdanubia, 375
transform fault, 25
transportation, 8; in Australia, 722–723; in Bangladesh, 610; in Brazil, 229; in Canada, 101, 145–146, 148, 153, 155; in East Asia, 629–630, 634–635, 641, 650–651; of food, 160–161; of illegal drugs, 116; in Russia, 407–410; in Saudi Arabia, 470; in sub-Saharan Africa, 531, 542–543, 552–553, 557; trade and, 78–79; in United States, 101, 107–108, 110–112, 123–124, 133, 137; in Western Europe, 266–267, 280
Trans-Siberian Railway, 396, 407, 409–410
Transylvanian Alps, 347

tributary, 87
Trinidad, 213, 215–216
Triple Alliance, 304
Tropic of Cancer, 5–6
Tropic of Capricorn, 5–6
tropical climate, 38–41; of East Asia, 625; of Latin America, 168, 176–177, 193, 208, 224, 235; of United States, 87, 135
tropical depression, 170
tropical grassland, 46
tropical rain forest, 43–44; destruction of, 230–231, 250–251, 508, 544; of East Asia, 626; of Middle America, 171; soil of, 515; of South America, 176–177, 230–231, 237, 239–240, 250; of South Asia, 580; of sub-Saharan Africa, 507–508, 515, 537, 542, 544
tropical storm, 170–171
tropical wet and dry climate, 38–41
tropical wet climate, 38–41
tropical zone, 33
Trudeau, Pierre, 152, 156
trust territory, 719
tsunami, 336, 669
Tuareg people, 494–495
Tucson, Arizona, 114
tundra, 38–41, 44–45, 47; alpine, 47; arctic, 47; of Canada, 147, 155; of Northern Eurasia, 391; of South America, 174; of United States, 135; of Western Europe, 261
Tunisia, 434, 491, 493–495, 497
Turkey, 474–475, 703; Cyprus and, 477; history of, 451; landforms of, 433–434; languages of, 440
Turkmenistan, 416
Turks, 450, 474
Tutsi people, 557
Tweed River, 282–283
typhoon, 669

 U

Ubangi River, 543
Uganda, 552, 557
ujamaa, 558
Ukraine, 398, 414–415
Ulan Bator, Mongolia, 663
Ungava Peninsula, 90
Unita, 565
unitary government, 59
United Arab Emirates, 468, 470
United Kingdom, 264, 277
United Nations Food and Agriculture Organization, 617
United States: acid rain in, 341; in Antarctica, 749; border with Canada, 86, 97, 154–155; cities of, 110–114, 122–124, 129, 132, 137; climate of, 87–92, 125–126, 128, 132, 135; communication system of,

ACKNOWLEDGMENTS

Maps

R.R. Donnelly & Sons Company: Atlas Maps
Richard Sanderson and John Sanderson: Text Maps

Illustrations

Matthew S. Pippin: 13, 35, 59, 74, 86, 113, 127, 149, 180, 194, 216, 225, 242, 267, 280, 315, 331, 354, 374, 393, 415, 439, 471, 485, 518, 534, 561, 579, 604, 632, 647, 673, 696, 718, 731.
Robert Pratt: 21, 25, 37, 64-65, 140-141, 220-221, 320-321, 362-363, 404-405, 480-481, 548-549, 592-593, 686-687, 726-727.
Sanders/Tikkanen: 52, 53, 236, 286, 312, 333, 473, 489, 607, 655

Photos

Front Cover: John Martucci/Martucci Studios

Text: v T, NASA, v M, Marc & Evelyne Bernheim/ Woodfin Camp & Associates, v B, R. Thomas Banff/ FPG, vi, Susan Van Etten, vii L, D. J. Forbert/ Superstock, vii R, Manley Photo/Superstock, viii L, G. Colliva/The Image Bank, viii R, Mike Yamashita/Woodfin Camp & Associates, ix T, Paul Conklin, ix B, Wolfgang Krammisc TSW, x, Travelpix/FPG, xi T, Richard Wood/The Picture Cube, xi B, Jean Kugler/FPG, xii T, Peter Hendrie/The Image Bank, xii B, S. Vidler/Superstock, xiii T, Holton Collection/Superstock, xiii B, Dave Saunders/TSW, xiv, The Granger Collection, xv, Susan Van Etten, xvi, The Granger Collection, xviii, Robert Frerck/Odyssey Productions, xix, Prentice Hall, xx–1, R. Thomas Banff/FPG, 1 T, Alec Pytlowany/Masterfile, 1 B, Lynn Johnson/Black Star, 2, Johnathan S. Selig/Peter Arnold, Inc., 3, The First Satellite Composite View of the World, Tom Van Sant, Inc., Santa Monica, CA, Assistance from NOAA, NASA, STARDENT COMPUTER CORPORATION, 7, P. Pearson/H. Armstrong Roberts, Inc., 8, Cathlyn Melloan/TSW, 9, NASA, 13, Robert Frerck/Odyssey Productions, 18, NASA, 19, J. Novak/Superstock, 26, Kevin Schafer/ Peter Arnold, Inc., 29, Ian Rosenfield/TSW, 30, Alex Langley/DPI, 35, Patrick Ward/Stock Boston, 41, Marc & Evelyne Bernheim/Woodfin Camp & Associates, 47, S. J. Krasemann/Peter Arnold, Inc., 50, Mark Segal/TSW, 51, Rollei/Superstock, 56, Thomas Nebbia/Woodfin Camp & Associates, 59, Rick Smolan/ Stock Boston, 60, Marcelo Montecino/Woodfin Camp & Associates, 66, Steve Vidler/Leo de Wys, Inc., 67, Robert Frerck/Odyssey Productions, 69, Susan Van Etten, 70, Don Smetzer/TSW, 71 L, Paul Conklin, 71 R, W. Hamilton/Superstock, 73, 74, Owen Franken/Stock Boston, 80–81, John Elk, III/Stock Boston, 81 T, Steve Kaufman/Peter Arnold, Inc., 81 B, Susan Van Etten, 83, J. Blank/FPG, 86, Thomas Kitchen/Tom Stack & Associates, 87, J. Messerschmidt/TSW, 87 (inset), George Zimbel/Monkmeyer Press, 92, Stan Osolinski/ TSW, 97, J. Blank/FPG, 99, The Granger Collection, 101, M. McLean/The Nelson-Atkins Museum of Art, 105, Grant Heilman/Grant Heilman Photographers, Inc., 108, The Bettmann Archive, 111, Jim Trotter/ Superstock, 112, Ric Ergenbright Photography, 113, Cathlyn Melloan/TSW, 114, DPI, 117, Thomas Nebbia/Woodfin Camp & Associates, 121, S. H. Johnson/ Superstock124, Gala/Superstock, 127 T, Ray Ellis/ Photo Researchers, Inc., 127 B, Susan Van Etten, 128, Camerique/E. P. Jones, Inc., 130, Terry E. Eiler/Stock Boston, 133, Everett Johnson/Leo de Wys, Inc., 135, Manley Features/Superstock, 143, Ric Ergenbright Photography, 147, M. Julien/Valan Photos, 149, M. Roessler/Superstock, 150, Denis Roy/Valan Photos, 152, Andre P. Therrien/Valan Photos, 154, Bernard Gotfryd/Woodfin Camp & Associates, 157, Pierre Kohler/Valan Photos, 162–163, Manley Photo/Superstock, 163 T, B. G. Silberstein/Superstock, 163 B, D. J. Forbert/Superstock, 165, Chip & Rosa Maria Peterson, 168, Susan Van Etten, 171, Wide World Photos, Inc., 175, Jadwiga Lopez/TSW, 177, Owen Franken/ Stock Boston, 178, Robert Frerck/Woodfin Camp & Associates, 180, Ric Ergenbright Photography, 182, Culver Pictures, Inc., 186 L, Chip & Rosa Maria Peterson, 186 R, Ulrike Welsch/Stock Boston, 187, Martin Rogers/TSW, 191, Robert Frerck/Odyssey Productions, 194, S. Vidler/Superstock, 197, Robert Frerck/ Woodfin Camp & Associates, 199, Laurie Platt Winfrey, Inc., 201, S. Vidler/Superstock, 203, Susan Van Etten, 207, Joe Viesti/Viesti Associates, Inc., 210, Shuster/Superstock, 213, Slim Aarons/Photo Researchers, Inc., 215 L, Katrina Thomas/Photo Researchers, Inc., 215 R, Dave G. Houser/Uniphoto, 216, David Stoecklein/Uniphoto, 223, Will McIntyre/Photo Researchers, Inc., 225 T, Owen Franken/Stock Boston, 225 B, Susan Van Etten, 228, H. Kanus/Superstock, 230, Stephanie Maze/Woodfin Camp & Associates, 231, Randall Hyman/Stock Boston, 235, Ann & Myron Sutton/Superstock, 241, Ric Ergenbright Photography, 242, Robert Frerck/Woodfin Camp & Associates, 243, Marcelo Montecino/Woodfin Camp & Associates, 245, Foto Du Monde/The Picture Cube, 246, Peter Drowne/E. R. Degginger, 251, Claus Meyer/Black Star, 252–253, Mike Yamashita/Woodfin Camp & Associates, 253 T, Alain Choisnet/The Image Bank, 253 B, G. Colliva/The Image Bank, 255, C. H. Rose/Stock Boston, 258, Louis Goldman/Photo Researchers, Inc., 261, Hubertus Kanus/Superstock, 262, Randall Hyman/ Stock Boston, 263, D. & J. Heaton/Stock Boston, 266, Owen Franken/Stock Boston, 267 T, Kurt Scholz/Superstock, 267 B, Superstock, 269, 271, The Granger Collection, 277, Ric Ergenbright Photography, 279, Peter Menzel/Stock Boston, 280, Dr. Gwynne Morgan, 280 (inset), Susan Van Etten, 282, Gerry Cranham/Photo Researchers, Inc., 283, Spencer Grant/Photo Researchers, Inc., 284, Harry Gruyaert/Magnum Photos, 285, Paul Conklin, 286, The Bettmann Archive, 287, Harvey Barad/Monkmeyer Press, 289, Cary Wolinsky/Stock Boston, 290, D. & J. Heaton/Stock Boston, 292, Hubertus Kanus/Superstock, 297, Jean-Marie Truchet/TSW, 299, Richer/TSW, 300, Kemps/TSW, 301, Giraudon/Art Resource, NY, 302, Sepp Seitz/ Woodfin Camp & Associates, 303, Alexandra Avakian/ Woodfin Camp & Associates, 304, UPI/Bettmann, 306, Mike Yamashita/Woodfin Camp & Associates, 307, Sepp Seitz/Woodfin Camp & Associates, 308, The Granger Collection, 309, Farrell Grehan/Photo Researchers, Inc., 311, Robert Davis/Photo Researchers, Inc., 313 L,

Prentice Hall, 313 R, Susan Van Etten, 315, Sam Abell/Woodfin Camp & Associates, 316, John Elk, III/Stock Boston, 323, Robert Frerck/Woodfin Camp & Associates, 325, Paul Conklin, 326, 327, Robert Frerck/Odyssey Productions, 330, Mike Mazzaschi/Stock Boston, 331, Jonathan Blair/Woodfin Camp & Associates, 332, Robert Frerck/Woodfin Camp & Associates, 334, John Baker/Superstock, 337, Nathan Benn/Stock Boston, 341, Gernot Huber/Woodfin Camp & Associates, 342–343, Dallas & John Heaton/TSW, 343 T, Paul Conklin, 343 B, John Eastcott & Yva Momatiuk/Woodfin Camp & Associates, 345, Paul Conklin, 347, Adam Woolfitt/Woodfin Camp & Associates, 348, V. Lefteroff/FPG, 349, Jonathan Blair/Woodfin Camp & Associates, 353 L, Porterfield/Chickering/Photo Researchers, Inc., 353 R, P.A. Enterprises/FPG, 354, 357, Paul Conklin, 358, UPI/Bettmann, 365, Momatiuk/Eastcott/Woodfin Camp & Associates, 367, Wesolowski/Sygma, 368, Tom Haley/Sipa Press, 369, J. P. Laffont/Sygma, 370, UPI/Bettmann Newsphotos, 371, Paul Conklin, 373, Eastcott/Momatiuk/Woodfin Camp & Associates, 374, Paul Conklin, 376, Julius Fekete/FPG, 379, J. Nicholas/Sipa Press, 383 L, Susan Van Etten, 383 R, Travelpix/FPG, 384–385, Soviet Life Magazine, 385 T, T. Peter Turley/Black Star, 385 B, Wolfgang Krammisc/TSW, 387, 389, Tass/Sovfoto, 393, Sovfoto, 395, Mike Peters/FPG, 400, The Telegraph Colour Library/FPG, 401 UPI/Bettmann, 407, 411, Randow/Sipa, 415, Prentice Hall, 417, Novasti/Gamma-Liaison, 419, Manif Vilnius/Sipa Press, 419 (inset), Reuters/Bettmann Newsphotos, 423, Wide World Photos, 424–425, J. Pickerell/FPG, 425 T, Travelpix/FPG, 425 B, Craig Aurness/Woodfin Camp & Associates, 427, Holton Collection/Superstock, 429, Schuster/Superstock, 432 L, G. Glase/Superstock, 432 R, Owen Franken/Stock Boston, 434, H. D. Shourie/Superstock, 439 T, Craig Aurness/Woodfin Camp & Associates, 439 B, D. H. Hessell/Stock Boston, 440, Nathan Benn/Woodfin Camp & Associates, 441, David Austen/Stock Boston, 444 L, B. Norman/Sheridan Photo Library/TSW, 444 R, Leo Touchet/Woodfin Camp & Associates, 445, Ronald Sheridan/Ancient Art & Architecture Collection, 449, World Image/FPG, 450, The Granger Collection, 454, UPI/Bettmann, 455 L, Van Phillips/Leo de Wys, Inc., 455 R, W. Hille/Leo de Wys, Inc., 456, Jill Brown/Superstock, 457, K. Scholz/Superstock, 458, Alon Reininger/Woodfin Camp & Associates, 459, Steve Benbow/Stock Boston, 460, Allen Green/Sygma, 462, Reuters/Bettmann, 465 L, J. G. Ross/Photo Researchers, Inc., 465 R, 467, Reuters/Bettmann Newsphotos, 468, 470, Schuster/Superstock, 471 T, Randa Bishop/DPI, 471 BL, Craig Aurness/Woodfin Camp & Associates, 471 BM, BR, R. & S. Michaud/Woodfin Camp & Associates, 473, Barry Iverson/Woodfin Camp & Associates, 477, Fred J. Maroon/Photo Researchers, Inc., 483, Trevor Wood/TSW, 485, Robert Frerck/Woodfin Camp & Associates, 487, Hubertus Kanus/Superstock, 493, Dieter Blum/Peter Arnold, Inc., 494, Christian Delbert/The Picture Cube, 495, Olivier Martel/Photo Researchers, Inc., 496, Pierre Boulat/Woodfin Camp & Associates, 497, DPI, 502–503, Ron Watts/Westlight, 503 T, Hoa-Qui/Viesti Associates, Inc., 503 B, Alon Reininger/Woodfin Camp & Associates, 505, Susan Van Etten, 507, Peter Arnold/Peter Arnold, Inc., 508, Stephen J. Krasemann/Peter Arnold, Inc., 511, Dieter Blum/Peter Arnold, Inc., 513, William E. Townsend, Jr./Photo Researchers, Inc., 515, Superstock, 517, Bruce Brander/Photo Researchers, Inc., 518 T, Gianni Tortoli/Photo Researchers, Inc., 518 B, G. DeSteinheil/Superstock, 519, Superstock, 521,

The Granger Collection, 527, Steve McCurry/Magnum Photos, 530, The Granger Collection, 532, Richard Wood/The Picture Cube, 534, John Chiasson/Gamma-Liaison, 535, Richard Saunders/Leo de Wys, Inc., 537, Sally Stiles/FPG, 538, Abbas/Magnum Photos, 539, V. Manic/Superstock, 541, E. Reiner/Superstock, 545, Photo Service/Superstock, 551, Stephen J. Krasemann/Peter Arnold, Inc., 554, Ric Ergenbright Photography, 556, Martin Rogers/Uniphoto, 558, Lynn McLaven/The Picture Cube, 559, Byron Augustin/Tom Stack & Associates, 561, G. Mendel/Magnum Photos, 562, Reuters/Bettmann, 564, Paul Conklin, 565, Susan Meiselas/Magnum Photos, 566, Gregory G. Dimijian/Photo Researchers, Inc., 567, Marc & Evelyn Bernheim/Woodfin Camp & Associates, 571, Holton Collection/Superstock, 572–573, DPI, 573 T, Jean Kugler/FPG, 573 B, R. & S. Michaud/Woodfin Camp & Associates, 575, Andy Seltzers/Viesti Associates, Inc., 579, Cary Wolinsky/Stock Boston, 581, S. M. Dudhediya/Superstock, 584, Jehangir Gazdar/Woodfin Camp & Associates, 589, The Granger Collection, 595, Baldev/Sygma, 596, Margaret Bourke-White/Life Magazine © 1946 Time Inc., 597, 598, AP/Wide World Photos, 601, 603, Robert Frerck/Odyssey Productions, 604, Garneau & Prevost/Superstock, 605, Dilip Mehta/Woodfin Camp & Associates, 607, George Holton/Photo Researchers, Inc., 608, 609, Ric Ergenbright Photography, 610, Wanda Warming/Stockphotos, Inc., 611, Ric Ergenbright Photography, 612, Margaret Gowan/TSW, 617 L, 617 R, Robert Frerck/Odyssey Productions, 618– 619, S. Vidler/Superstock, 619 T, Steve Vidler/Leo de Wys, Inc., 619 B, Steve Niedorf/The Image Bank, 621, Anny & Myron Sutton/FPG, 623, A. Freidlander/Superstock, 625, 627, Susan Van Etten, 631, Bruno J. Zehnder/Peter Arnold, Inc., 632, Bob Daemmrich/Uniphoto, 637, The Granger Collection, 641, Ric Ergenbright Photography, 643, 645, Eastfoto, 646, Alon Reininger/Woodfin Camp & Associates, 647, Eastfoto, 648, Alon Reininger/Woodfin Camp & Associates, 651, Lee Day/Black Star, 653, Lindsay Hebberd/Woodfin Camp & Associates, 656, 658, Alon Reininger/Woodfin Camp & Associates, 659, M. Pedone/Superstock, 660, Dilip Mehta/Woodfin Camp & Associates, 661, Chuck Fishman/Woodfin Camp & Associates, 663, George Holton/Photo Researchers, Inc., 667, Alon Reininger/Woodfin Camp & Associates, 670, James Simon/The Picture Cube, 671, The Granger Collection, 673 T, S. Vidler/Superstock, 673 B, © 1978 Fujio-Fujiko, 674, Chuck O'Rear/Woodfin Camp & Associates, 675, 676, Karen Kasmauski/Woodfin Camp & Associates, 678, E. C. Stangler/Leode Wys, Inc., 681, 682, Nathan Benn/Woodfin Camp & Associates, 689, Manley Photo/Superstock, 693, Peter Hendrie/The Image Bank, 696, Suzanne Murphy/TSW, 698, Jack Fields/Photo Researchers, Inc., 703, Christopher Morris/Black Star, 704–705, Dave Saunders/TSW, 705 T, Frank Lane/TSW, 705 B, Fujitsuka/FPG, 707, DPI, 711, T. Nakamura/Superstock, 712, R. Dahlquist/Superstock, 714, Robert Frerck/TSW, 717, Kevin Fleming/Woodfin Camp & Associates, 717 (inset), C. L. B./The Photo Source, 718 T, David Hiser/The Image Bank, 718 B, Susan Van Etten, 719, Mike Holmes/Earth Scenes, 721, The Granger Collection, 723, Cary Wolinsky/Stock Boston, 729, J. Allan Cash Ltd./Superstock, 731, David Austen/Uniphoto, 732, Fred Bavendam/Peter Arnold, Inc., 733, Superstock, 735, S. Vidler/Superstock, 737, Giorgio Ricatto/Superstock, 737 (inset), Jeff Rotman/Peter Arnold, Inc., 738, Fred J. Eckert/FPG, 741, H. Miller/Superstock, 744, Holton Collection/Superstock, 748, Richard Harrington/FPG.